Advances in Experimental Medicine and Biology

Volume 1219

More information about this series at http://www.springer.com/series/5584

Jacinta Serpa
Editor

Tumor Microenvironment

The Main Driver of Metabolic Adaptation

Springer

Editor
Jacinta Serpa
CEDOC, Chronic Diseases Research
Centre, NOVA Medical School |
Faculdade de Ciências Médicas
Universidade NOVA de Lisboa
Lisbon, Portugal

Instituto Português de Oncologia de
Lisboa Francisco Gentil (IPOLFG)
Lisbon, Portugal

ISSN 0065-2598 ISSN 2214-8019 (electronic)
Advances in Experimental Medicine and Biology
ISBN 978-3-030-34024-7 ISBN 978-3-030-34025-4 (eBook)
https://doi.org/10.1007/978-3-030-34025-4

This Springer imprint is published by the registered company Springer Nature Switzerland AG.
The registered company address is: Gewerbestrasse 11, 6330 Cham, Switzerland

Contents

Metabolic Remodeling as a Way of Adapting to Tumor Microenvironment (TME), a Job of Several Holders

1

Jacinta Serpa

Abstract

The microenvironment depends and generates dependence on all the cells and structures that share the same niche, the biotope. The contemporaneous view of the tumor microenvironment (TME) agrees with this idea. The cells that make up the tumor, whether malignant or not, behave similarly to classes of elements within a living community. These elements inhabit, modify and benefit from all the facilities the microenvironment has to offer and that will contribute to the survival and growth of the tumor and the progression of the disease.

The metabolic adaptation to microenvironment is a crucial process conducting to an established tumor able to grow locally, invade and metastasized. The metastatic cancer cells are reasonable more plastic than non-metastatic cancer cells, because the previous ones must survive in the microenvironment where the primary tumor develops and in addition, they must prosper in the microenvironment in the metastasized organ.

The metabolic remodeling requires not only the adjustment of metabolic pathways *per se* but also the readjustment of signaling pathways that will receive and obey to the extracellular instructions, commanding the metabolic adaptation. Many diverse players are pivotal in cancer metabolic fitness from the initial signaling stimuli, going through the activation or repression of genes, until the phenotype display. The new phenotype will permit the import and consumption of organic compounds, useful for energy and biomass production, and the export of metabolic products that are useless or must be secreted for a further recycling or controlled uptake. In the metabolic network, three subsets of players are pivotal: (1) the organic compounds; (2) the transmembrane transporters, and (3) the enzymes.

This chapter will present the "Pharaonic" intent of diagraming the interplay between these three elements in an attempt of simplifying and, at the same time, of showing the complex sight of cancer metabolism, addressing the orchestrating role of microenvironment and highlighting the influence of non-cancerous cells.

J. Serpa (✉)
CEDOC, Chronic Diseases Research Centre, NOVA Medical School | Faculdade de Ciências Médicas, Universidade NOVA de Lisboa, Lisbon, Portugal

Instituto Português de Oncologia de Lisboa Francisco Gentil (IPOLFG), Lisbon, Portugal
e-mail: jacinta.serpa@nms.unl.pt

J. Serpa (ed.), *Tumor Microenvironment*, Advances in Experimental Medicine and Biology 1219,
https://doi.org/10.1007/978-3-030-34025-4_1

Keywords
Cancer cell metabolism · Metabolic network · Metabolic remodeling · Tumor microenvironment (TME) · Glycolysis · Pentose phosphate pathway (PPP) · Glutaminolysis · Fatty acids synthesis · β-oxidation · One-carbon metabolism · Transsulfuration pathway (TSSP)

1.1 Glucose, the Master Metabolic Coin

Since the beginning, teaching metabolism presents the pathways with a central core in glucose catabolism. Hence, we usually think about glycolysis as the main pathway on the course of energy production. However, endogenous metabolism of organic compounds with the objective of producing energy and biomass; is composed by a multipart network of interconnected pathways. Those pathways share constantly organic intermediates.

Because glucose was, since ever, considered the most central organic compound in metabolism, the first documented cancer metabolic alteration is related to glucose catabolism. In 1924, Otto Warburg described that cancer cells were addicted to glucose, and glycolysis was the main metabolic pathway used by cancer cells to sustain energy demands. This metabolic switch was independent of oxygen availability (Warburg 1956).

In the beginning the Warburg effect, as aerobic glycolysis is named, was widely explored in several cancer types, always assuming that cancer cells were incapable of performing the oxidative phosphorylation (OXPHOS). However, the majority of cancer cells fulfil OXPHOS, but not always using substrates originated from glucose (Alam et al. 2016; Guppy et al. 2002; Rodríguez-Enríquez et al. 2000, 2006; Viale et al. 2015; Lopes-Coelho et al. 2017; Silva et al. 2016). Interestingly the hybrid phenotype, glycolysis and OXPHOS simultaneously, contributes for cancer progression, being worth to target both metabolic routes to disturb the metabolic equilibrium in cancer cells (Jia et al. 2019).

1.1.1 Glucose Transport and Glycolysis

The high glycolytic behavior of certain cancer cells, independent of the co-existence of OXPHOS, implies an enhanced uptake of glucose. This is fulfilled by the increased expression of glucose transporters, belonging to two main families: the SGLT (Na^+-coupled glucose transporters) and GLUT (glucose transporter facilitators). Three members of the SGLT family work as sugar transporters (SGLT1 and SGLT2) and sensors (SGLT3). As revised by Madunić et al. (2018), the role of SGLT isotypes is not well explored in cancer, however the up regulation of SGLT1 and SGLT2, in different types of cancer, suggests that they act on cancer metabolic remodeling, being putative therapeutic targets to disrupt the metabolic equilibrium in cancer cells.

From the 14 members of GLUT family, 11 mediate sugar transport and they differ in substrate specificity, kinetics and expression profile. Some GLUT isotypes can function as transporters (e.g., GLUT4 and GLUT8) or they can accumulate a glucose sensor function (e.g., GLUT2) (reviewed Scheepers et al. (2004)). In cancer, GLUT1 is by far the most well studied glucose transporter, and its increased expression is associated to an increased tumor aggressiveness and chemoresistance in different cancer contexts (Chen et al. 2019a; Yang et al. 2019; Sawayama et al. 2019; Wang et al. 2019a; Hamann et al. 2018; Iwasaki et al. 2015; Fujino et al. 2016; Cho et al. 2013; Kim et al. 2013; Nguyen et al. 2008), being its up regulation tightly liked to sirtuins (SIRT) (Chen et al. 2019a; Yu et al. 2019), NFkB pathway (Zhou et al. 2019), Wnt pathway (Wang et al. 2019a), TGF-β1-mTORC1-ATF4 axis (Selvarajah et al. 2019) and HIF-1α stabilization (Chen et al. 2019b).

In addition to GLUT1, some studies give a relevance to GLUT2, GLUT3, GLUT4 and GLUT5 as being part of a panel of metabolic markers in cancer (Hamann et al. 2018; Do et al. 2019; Kim et al. 2017; Mao et al. 2019). A mutated form of GLUT3 (rs7309332C > T) was found to have a prognostic significance in lung cancer patients, being associated with increased overall survival and disease free survival (Do et al. 2019), indicating that a fully functional isoform is a poor prognosis marker.

Glycolysis (Fig. 1.1) occur through two main components (the synthesis of phosphorylated precursors of pyruvate and ATP synthesis) that include three irreversible and rate-limiting reactions: the conversion of glucose into glucose-6-phosphate, catalyzed by hexokinase (HK) and further conversion into fructose-6-phosphate; the conversion of fructose-6-phosphate into fructose-1,6-bi-phosphate by phosphofructokinase (PFK), and the conversion of phosphoenolpyruvate into pyruvate by pyruvate kinase (PK) (Jin and Zhou 2019).

The increased uptake of glucose must be synchronized with glycolysis that is why the overexpression of both GLUT1 and HK2 is coordinated

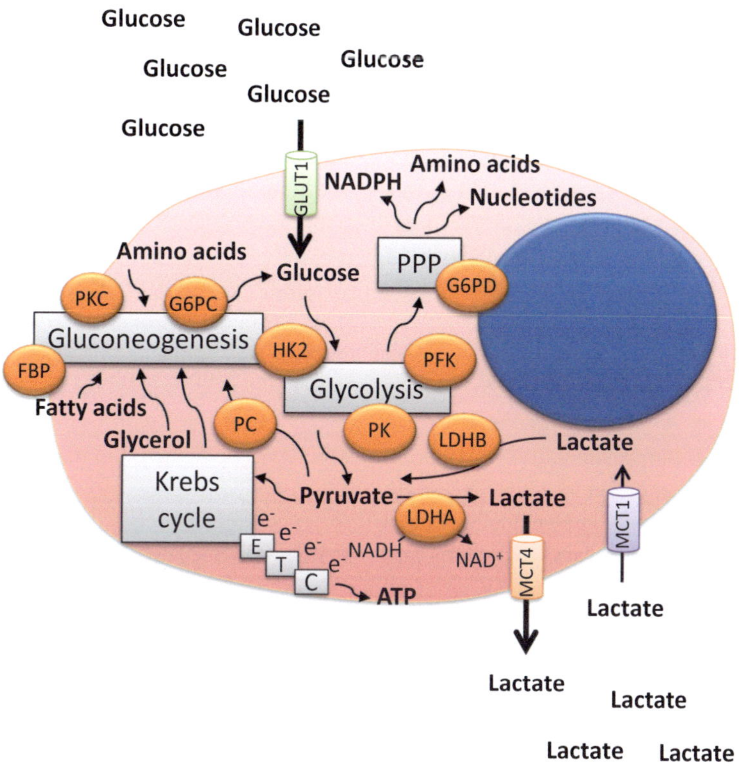

Fig. 1.1 **The scenario of glucose metabolism.** In cancer, GLUT1 is the most studied glucose transporter, being upregulated. Glycolysis have 3 rate limiting irreversible reactions, catalyzed by hexokinase (mainly isoenzyme 2-HK2), phosphofructokinase (PFK) and pyruvate kinase (PK). The increased rate of glycolysis prompts the production and export of lactate, as the easiest way of recovering NAD^+ to sustain glycolysis and to eliminate the pyruvate toxic effects, by converting pyruvate into lactate under the action of lactate dehydrogenase A (LDHA). Lactate export and gradual import is controlled by MCT4 and MCT1, respectively. Lactate is converted into pyruvate by lactate dehydrogenase B (LDHB). Glycolysis intermediate glucose-6-phosphate can be deviated in order to supply phosphate pentose pathway (PPP), being glucose-6-phosphate dehydrogenase (G6PD) the rate controlling enzyme. PPP allows the nucleotides and aminoacids syntheses as well as the production of NADPH to sustain other metabolic pathways as the fatty acids synthesis. Glucose can be synthesized in the cell, through gluconeogenesis, using Krebs cycle intermediates, glycerol, pyruvate, amino acids and fatty acids. Gluconeogenesis is a reversion of glycolysis, though using alternative inverse reactions, catalyzed by pyruvate carboxylase (PC); phosphoenolpyruvate carboxykinase (PKC); fructose 1,6-bisphosphatase (FBP), and glucose 6-phosphatase (G6PC)

(Yang et al. 2019), being regulated by the same signaling pathways (Xintaropoulou et al. 2018). There are 4 isoforms of HK, but in cancer, HK2 is the most expressed (Yang et al. 2019; Mathupala et al. 2009) and has a higher glucose affinity (Wilson 2003), being under the command of HIF-1α (Sato-Tadano et al. 2013) and PKC, TGFβ and PI3K pathways (Xu et al. 2018a; Wang et al. 2018a; Iwamoto et al. 2014). In lung cancer, long intergenic noncoding RNA for kinase activation (LINK-A) is also important in HK2 upregulation (Zhao et al. 2018) as well as the developmental pluripotency-associated 4 (Dppa4) protein, which co-regulates HK2 and GLUT4 expression (Li et al. 2019a). In some tumors, HK appears bound to mitochondria, through the association with the mitochondrial permeability tunnel complex of the voltage dependent anion channel protein (VDAC); this phenomenon increases the efficacy of ATP binding to mitochondria and it is also a mechanism of cell immortalization (Yang et al. 2019; Mathupala et al. 2009; Yan et al. 2013).

PFK (mainly PFK1) activity is increased in different cancer types (Moon et al. 2011; Park et al. 2013), being its activity regulated by the levels of fructose-6-phosphate and ATP (Cabrera et al. 2011), and its expression and stability are increased by PI3K pathway (Lee et al. 2017) and decreased under the action of p53 and cyclin D3/cyclin dependent kinase 6 (CDK6) (Wang et al. 2017a, 2018a).

There are two isoforms of PK, M1 and M2, being M2 the most expressed in cancer cells (Israelsen et al. 2013). Moreover, glucose-derived metabolites are important suppliers of other metabolic pathways that are not dedicated to energy production, and this deviation is partially controlled by PK. The levels of active PK determines the use of glucose derived compounds by OXPHOS or by the pentose phosphate pathway (PPP). Meaning that high activity levels of PK favors OXPHOS, whereas low levels favors PPP (Gui et al. 2013; Fukuda et al. 2015). PKM2 expression is upregulated by HIF-1α (Luo et al. 2011), and PKM2 activity is directly inhibited by reactive oxygen species (ROS), through the oxidation of cysteine residue 358; which will consequently inhibit glycolysis and deviate the glucose flux towards PPP (Anastasiou et al. 2011).

We cannot fail to recall that acetyl-CoA is a core molecule in carbon metabolism, being generated from the catabolism of glucose, lipids and amino acids (Lyssiotis and Cantley 2014). Afterwards, acetyl-CoA can be used in the synthesis of nucleotides, fatty acids and amino acids, or be further oxidized to sustain OXPHOS. The major part of acetyl-CoA used in Krebs cycle is glucose/pyruvate originated, however in highly glycolytic or hypoxic cancer cells, the major glycolysis endpoint is the production of lactate. In those cases, acetyl-CoA can be originated from fatty acids (FA) β-oxidation (Sect. 1.5), glutaminolysis (Sect. 1.3) or cancer cells manage to produce acetyl-CoA from acetate (Zhao et al. 2016; Mashimo et al. 2014) by the action of acyl-CoA synthetases 2 (ACSS2), whose relevance in cancer progression is extensively documented (Mashimo et al. 2014; Wen et al. 2019; Zhang et al. 2018a; Bidkhori et al. 2018), although some controversy persists (Sun et al. 2017). Nonetheless, in cancer, the increased glycolytic rate is mainly accounting to lactate production, which will be depicted in the next section.

1.1.2 Phosphate Pentose Pathway (PPP)

The PPP works in parallel to glycolysis, and it constitutes the first inter-pathways connected step of glucose metabolism (Ramos-Martinez 2017). The PPP occurs through two irreversible oxidative reactions followed by a series of reversible conversions comprising two biochemical branches: an oxidative and a non-oxidative branch. The PPP (Fig. 1.1) is pivotal for cancer cell survival and proliferation as it uses glucose-6-phosphate to generate pentose phosphates for amino acids and nucleotides synthesis (non-oxidative branch) and also NADPH (oxidative branch), crucial for FA synthesis and redox balance (Jin and Zhou 2019; Pavlova and Thompson 2016; Stincone et al. 2015; Patra and Hay 2014).

Several studies reinforce the crucial role of PPP in cancer by showing that PPP improves cell survival and proliferation (Patra and Hay 2014; Weber 2016). For PPP to be enhanced, the expression and activity of other glycolysis enzymes must be controlled. This is the case of PFK (mainly PFK1), which irreversibly phosphorylates fructose-6-phosphate into fructose-1,6-bisphosphate. So, there is a direct competition between PFK1 and glucose-6-phospate dehydrogenase (G6PD) – the limiting enzyme of PPP – as they depend on the availability of glucose-6-phosphate (Jin and Zhou 2019). Hence, the PFK1 activity is often suppressed by O-GlcNAcylation, and glucose 6 phosphate is directed to PPP, accounting for tumor growth (Yi et al. 2012). Thus, the increased G6PD expression is associated to cancer progression, autophagy and drug resistance (Mele et al. 2019). Importantly, there is a precise regulation of PPP during cell cycle in order to monetize the resources (Li et al. 2019a). According to this, the cell cycle clock elements, cyclin D3/CDK6 phosphorylates and inhibits PFK (Wang et al. 2017a), shifting the glucose-6-phosphate towards the PPP. The expression and activity of G6PD is regulated by PI3K pathway, whose intermediate AKT both activates G6PD expression and activity by phosphorylation and promotion of G6PD active dimers, while PTEN phosphatase, a PI3K pathway inhibitor, disrupts the G6PD active dimers (Hong et al. 2014). Wnt and NFkB pathways also activate the expression of G6PD, respectively by the activation of c-MYC and p65, being responsible for a more metastatic and chemoresistant cancer phenotype concomitant with PPP promotion (Yin et al. 2017; Gao et al. 2017). PPP, as an anti-oxidant pathway (Riganti et al. 2012), is regulated by the main redox balance promotor transcription factor, the NRF2, which directly induces the transcription of G6PD codifying gene (Kowalik et al. 2016).

1.1.3 Gluconeogenesis

The synthesis of glucose is not extensively explored in cancer context, including the synthesis of glucose from non-glucidic compounds, termed gluconeogenesis, which has recently received some attention. Gluconeogenesis pathway (Fig. 1.1) is almost a reversion of glycolysis, synthesizing glucose from glycerol, lactate, pyruvate, acetyl-CoA (also FA derived) or glucogenic amino acids, as alanine. However, the three irreversible steps of glycolysis impose gluconeogenesis to use four other enzymes: pyruvate carboxylase (PC); phosphoenolpyruvate carboxykinase (PKC); fructose 1,6-bisphosphatase (FBP), and glucose 6-phosphatase (G6PC) (Tsai et al. 2015). Because these enzymes are tissue specifically expressed, gluconeogenesis occurs predominantly in the liver and, in a lower rate, in the renal cortex and small intestine (Zhang et al. 2018a; Potts et al. 2018). PC converts pyruvate in oxaloacetate in mitochondria, which is subsequently converted into phosphoenolpyruvated in cytoplasm by PKC1 or in mitochondria by PKC2. Phosphoenolpyruvate is then included in the sequential reverse reactions of glycolysis. The conversion of fructose 1,6-bisphosphate into fructose 6-phosphate is catalyzed by the two FBP isoforms (FBP1/2). Finally, glucose 6-phosphate is converted into glucose by G6PC (Grasmann et al. 2019). The inhibition of this final step can redirect glucose 6-phosphate to PPP, making gluconeogenesis a supplier pathway for PPP in glucose depleted environment.

Gluconeogenesis can be activated under a panel of a metabolic remodeling due to nutrients scarcity, and its enzymes are under the control of signaling pathways crucial in cancer metabolism modulation as KRAS-dependent pathways, PIK3/mTOR pathway, Wnt pathway and HIF1 (as reviewed in Lao-On et al. (2018) and Wang and Dong (2019)). Being a way of supplying metabolic pathways accounting for cancer progression, makes gluconeogenesis itself a pro survival pathway relevant in certain cancer context as revised by Grasmann et al. (2019), pointing that the upregulation or *de novo* expression of certain gluconeogenesis-related enzymes are described in tumors, such as breast, colon, stomach, uterine cervix, liver and pancreas. Some controversy on the reflex of the action of these enzymes in cancer progression and tumor aggressiveness persists, however this can underlie the parallel roles of these enzymes in other metabolic pathways, since they are not exclusive of gluconeogenesis, as it will be referred along this chapter. Nevertheless, I believe much more is about to be discovered, in the near future, on gluconeogenesis role in cancer.

Glucose catabolism and anabolism are versatile and their intermediates can be quickly deviated to alternative pathways depending on the particular cancer cell demands and TME. This fact implies that in various situations the core pathways for energy production are not straightly sustained by glucose-derived compounds.

1.2 Lactate, the Usable "Waste" Product

In the line of increased rate of glycolysis in cancer cells, high levels of lactate are produced to cope with the increased levels of pyruvate production. Pyruvate is an endpoint of glycolysis that when gradually generated it is converted into acetyl-CoA and further conducted to Krebs (tricarboxylic acids- TCA) cycle or to the synthesis of amino acids or FA. However, a high glycolysis rate produces augmented levels of pyruvate, which cannot accumulate within the cell. Despite being a valuable metabolic intermediate and an anti-oxidant (Koprivica et al. 2019), pyruvate is an inhibitor of histone deacetylatases (McBrian et al. 2013; Hernández-Juárez et al. 2019) interfering with epigenetic remodeling and often working as an apoptosis inducer (Thangaraju et al. 2006; Zhang et al. 2019a). So, the conversion of pyruvate into lactate (Fig. 1.1) accomplishes two criteria: (1) it allows the maintenance of high glycolysis rate without cell injury, and (2) it is the fastest way of regenerating NAD^+ molecule, as NADH is converted from NAD^+ along glycolysis and further recovered upon the transformation of pyruvate into lactate (Hung and Yellen 2014). The availability of NAD^+ is crucial for the maintenance of the metabolic flow, since it is one of the main acceptors of electrons needed to hold chemical reactions (Hung and Yellen 2014; Hung et al. 2011; Lemire et al. 2008).

1.2.1 Lactate Synthesis and Transport

Lactate cannot accumulate in the cell otherwise the cytoplasmic pH would decrease and affect cell viability. Hence, the increased expression of monocarboxylate transporters (MCTs) capable of transporting lactate across the cell membrane is an usual phenomenon in cancer cells (Lopes-Coelho et al. 2017; Silva et al. 2016; Afonso et al. 2019; Gurrapu et al. 2015; Sanità et al. 2014; Baek et al. 2014; Doherty et al. 2014; Curry et al. 2013). So, lactate produced as a consequence of glycolysis is exported to the extracellular medium. Nevertheless, the role of lactate in cell metabolism is not finished in the metabolic network. At first lactate was considered a waste product, although it can be used as a valuable carbon source. The exported lactate can further be gradually imported and converted into pyruvate in order to serve as a substrate to sustain OXPHOS (Lopes-Coelho et al. 2017; Silva et al. 2016). The interconversion of pyruvate ↔ lactate is allowed by the action of lactate dehydrogenases (LDHs).

According to all the information stated before, we can find MCTs and LDHs as pivotal effectors in the production and the consumption of carbon through the use of lactate molecule as a reservoir. In humans, 14 *SLC16A* (solute carrier transporters family 16A) genes (encoding MCTs) are described but only 4 codify MCTs (MCT1-4) capable of transporting lactate (Halestrap 2013; Carneiro and Pellerin 2015). In cancer context, MCT1 and MCT4 are far the more studied. These transporters can transfer lactate in both ways across the cell membrane, but MCT1 is more associated to lactate import and MCT4 with lactate export (Lopes-Coelho et al. 2017; Silva et al. 2016; Boidot et al. 2012; Updegraff et al. 2018). LDHs work as tetramers and can be encoded by 4 different genes, LDH-A, LDH-B, LDH-C and LDH-D. Whereas the 3 first genes codify for enzymes that interconvert L-lactate and pyruvate, the former gene, encodes an enzyme that catalyzes the synthesis of D-lactate from endogenous and exogenous compounds that are not pyruvate (Monroe et al. 2019). So far, little is known about LDHD in cancer but it also seems to play a role as a prognostic factor (Wang et al. 2018b). *LDH-C*, encodes a testis specific enzyme that works as a homo-tetramer. Despite its tissue specific expression (Dodo et al. 2018), LDH-C can be *de novo* expressed in several cancer types (Kong et al. 2016; Tang and Goldberg 2009; Chen et al. 2014). *LDH-A* and *LDH-B* genes encode respectively the muscle (M) and heart (H) related LDH chains; they can be organized in homo or hetero tetramers. All these 3 LDH enzymes are capable of catalyzing the pyruvate ↔ lactate conversion, however LDH-C and LDH-A are more effective in converting pyruvate → lactate and LDH-B is more effective in lactate → pyruvate conversion (Valvona et al. 2016; Read et al. 2001; Hong et al. 2019). Therefore, MCT4 and LDHA-A work, in coordination, in cells that are producing high levels of lactate as a consequence of increased rates of glycolysis, so lactate in produced from pyruvate and must be exported to the extracellular medium. In turn, MCT1 and LDH-B can be considered metabolic partners, being expressed in cancer cells that are able to consume lactate, so lactate is imported and converted into pyruvate, which will be further canalized to diverse metabolic pathways (Fig. 1.1).

Based on the view of established partnerships between MCTs and LDHs, a symbiotic metabolism was described at first between cancer cells and then between cancer and stromal cells (Allen et al. 2016; Andersen et al. 2015). This symbiosis asset in the continuous sharing of nutrients adapted to the precise location of a cell within a tumor, managing which metabolic pathway is more adequate to each particular cell situation.

The acidified TME, due to the export of lactate, accounts for the improvement of migratory and invasive ability of cancer cells. In one hand, acidification contributes for the denaturation of extracellular matrix (ECM) proteins, facilitating the penetration and the matrix crossed-through by tumor cells (Gillies and Gatenby 2015). In the other hand, the acidified media stimulates the expression of genes associated to more invasive cell phenotypes by interfering with ECM remodeling, such as integrins, metalloproteinases (MMPs) and cancer progression related genes (Gillies and Gatenby 2015; Rohani et al. 2019; Li et al. 2018a; Paradise et al. 2011; Moellering et al. 2008).

Besides all these, TME contributes for the maintenance of the dynamics of glucose/lactate through the intervention of stromal cells. Some studies have pointed the role of stromal cells not only in the emission of signals that will control the metabolic switch but also by sharing organic compounds. This theme will be presented in more detail in Sect. 1.6.

1.3 Glutamine, the Main Substitute of Glucose

Glutamine catabolism is pointed out as important as glucose catabolism in cancer cells, being essential for mitochondrial metabolism. Glutamine provides anaplerotic carbons and oxaloacetate to supply the Krebs cycle, accounting for ATP and macromolecules synthesis (Mates et al. 2013; Wise and Thompson 2010; Reynolds et al. 2013, 2014). Glutaminolysis is widely addressed in cancer context and studies, associating the increase in glutaminolysis rate with carcinogenesis, as well as showing that its targeting impairs cancer cells proliferation (Wise and Thompson 2010; Lukey et al. 2013; Chen and Cui 2015). Furthermore, glutamine is considered the main Krebs cycle supplier upon cancer metabolic remodeling (Gaglio et al. 2011), being a preferential substitute of glucose.

1.3.1 Glutamine Transport and Glutaminolysis

The transport of glutamine is crucial and amino acids transporters capable of importing glutamine, such as $ATB^{0,+}$ (SLC6A14 gene), SNAT1 (SLC38A1 gene), ASCT2 (SLC1A5 gene), LAT1 (SLC7A5 gene) and LAT2 (SLC7A8 gene) (complete review in Bhutia and Ganapathy (2016)) are often upregulated in tumors (Ko et al. 2011; Bothwell et al. 2018; Rajasinghe et al. 2019; Feng et al. 2018; Bolzoni et al. 2016). Therefore, glutamine transporters are desirable therapeutic targets and some pre-clinical attempts have already been developed. LAT2 inhibitory experiments suggest that the abrogation of glutamine import can be a suitable strategy to disrupt chemoresistant cancer cell phenotypes (Feng et al. 2018). ASCT2 is unraveled as a good target to disturb cancer metabolism (Bolzoni et al. 2016; Wahi and Holst 2019; Giuliani et al. 2017; Wang et al. 2014a), and specific inhibitors are under investigation for future clinical application (Bröer et al. 2019). However, depending on the cancer type, the redundant expression of glutamine transporters (Bröer et al. 2018, 2019) can be a mechanism of resistance to a glutamine uptake-targeted therapy. The switch of these transporters from therapeutic targets to mediators of drug delivery can be a strategy to overlap the redundancy problem and take advantages of the cancer cells metabolic dependency on glutamine, as proposed by Sikder et al. (2017).

Once in the cytoplasm, glutamine is directly converted into glutamate (Fig. 1.2) by glutaminase isoenzymes (GLS1 and GSL2) (Mates et al.

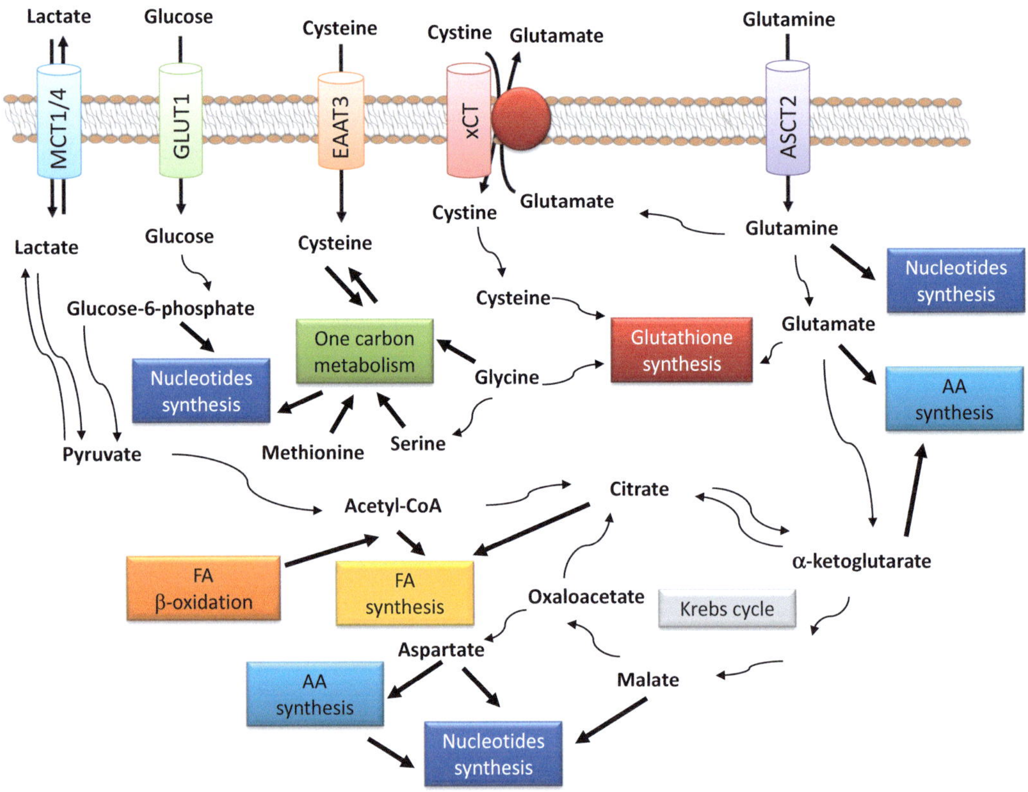

Fig. 1.2 Metabolic interplay between glucose, glutamine and cysteine. Glucose uptake and glycolysis is upregulated in most types of cancer. Glucose is taken up by specific transporters (e.g. GLUT1) and feeds, energy and biomass production. Increased levels of lactate production and secretion is a consequence of the augmented glycolysis rate. Afterwards, lactate is taken up and converted into pyruvate, following the intracellular metabolic course. The transport of lactate across the cell membrane is mediated mainly by MCT1 and MCT4. Glutamine originates glutamate (through glutaminolysis), which upon conversion into α-ketoglutarate, enters the Krebs cycle, replacing the glucose-derived compounds. Acetyl-CoA and citrate are pivotal intermediates between glucose-derived and glutamine-derived compounds and fatty acids metabolism. Glutamate, cysteine and glycine have a crucial role in GSH synthesis and redox control. Cysteine is taken up directly or as cystine, through the action of specific transporters and exchangers (e.g. respectively EAAT3 and xCT). Glutamine-derived glutamate sustains cystine entrance in the cell through xCT. Cysteine synthesis and degradation is linked to one-carbon metabolism, having glycine, methionine and serine as important substrates. The networking of glucose, glutamine and cysteine metabolism supports amino acids (AA), fatty acids (FA) and nucleotides syntheses

2013), whose expression is under the command of key transcription factors in cancer biology, as c-Myc and p53 (Gao et al. 2009; Wise et al. 2008; Li et al. 2019b; Kleszcz et al. 2018). GLS1 and GLS2 are inversely associated to cancer promotion, being GLS1 considered pro-tumoral (Xi et al. 2019; Xiang et al. 2019; Li et al. 2019c) and GLS2 tumor suppressive (Saha et al. 2019; Kuo et al. 2016). However, GLS2 tumor suppressor role is independent of its function in glutaminolysis, GSL2 contributes for the anti-cancer activity of p53, by inhibiting Rac1 and mediating the function of p53 in metastasis suppression (Zhang et al. 2016). Alternatively, GLS2 can also function as metastasis suppressor by interacting with Dicer and promoting the downregulation of Snail, a pro-metastatic transcription factor (Kuo et al. 2016). Anyway, Matés and colleagues in two very interesting reviews, depicted how to handle with the metabolic and molecular specificities of GLS isoenzymes in order to contribute for biomarkers identification, new drugs design, and personalized therapy (Matés et al. 2018; José et al. 2019).

Glutamate can be converted into α-ketoglutarate through oxidative deamination, by glutamate dehydrogenase 1 (GLDH1), or through transamination, by amino acids-specific transaminases. The first reaction occurs in the mitochondria and the second one can occur either in the cytoplasm or mitochondria. The former process produces in simultaneous α-ketoglutarate and nonessential amino acids, such as serine and aspartate (Fig. 1.2). This metabolic branch of glutamine is very important in cancer metabolism, since glutamine is a vital nitrogen source, and the activation of transaminases by MAPK pathway, the main regulator of glutamine metabolism, points that glutamine, in a metabolic remodeling scenario, is also canalized to the synthesis of other amino acids, important building blocks (Son et al. 2013).

Serine originated from glutamine plays an important role in cancer metabolism, being a good example of how the relevance of glutamine is not restricted to supplying Krebs cycles and all the pathways dependent on Krebs cycle intermediates. As very well revised by Amelio et al. (2014), serine is synthesized from glycolysis intermediates and glutamine-derived glutamate; and it fuels the synthesis of several other organic compounds, such as glycine and pyruvate (making the link between glutamine and lactate production). Serine can be converted into glycine under the action of serine hydroxymethyltransferase (SHMT), and afterwards these two amino acids through the one-carbon metabolism (methionine cycle plus folate cycle) can give rise to several other compounds to supply the epigenetic modulation (methyl and acetyl groups), nitrogen bases for nucleotides synthesis, scavenger molecules for free radicals elimination and proteins and lipids for biomass production (Fig. 1.2). Importantly, the methionine cycle can also be a source of cysteine, whose role in cancer metabolism will be depicted later on this chapter. Hence, serine holds an important part of glutamine's role as a carbon and nitrogen source. The relevance of serine in cancer metabolism is mirrored by the interest in developing SHMT inhibitors to treat cancer (Ducker et al. 2017; Marani et al. 2016).

Another example is asparagine (its physiologic and pathophysiologic role is revised by Lomelino et al. (2017)), whose conversion from aspartate is dependent on glutamine, and when asparagine synthetase (ASNS) is inhibited, cancer cells undergo an apoptotic cell death (Zhang et al. 2014). On the other hand, the supplementation of media with asparagine rescues the apoptosis induced by glutamine depletion (Zhang et al. 2014), indicating that asparagine can at least in part replace glutamine (Jiang et al. 2018). Importantly, tumors overexpressing ASNS are more resistant to therapy with asparaginase (Dufour et al. 2012) and with cytotoxic drugs (Cui et al. 2007). Accordingly, a therapeutic approach directed to glutamine and asparagine bioavailability was recently proposed (Jiang et al. 2018). However, cancer cells plasticity and survival stimuli, can also push cancer cells to undergo an adaptive process to overlap the glutamine depletion/metabolism inhibition; and recent metabolomics analysis presented this clearly (Biancur et al. 2017).

In the mitochondria, α-ketoglutarate continues the anaplerotic role of glutamine, entering the

Krebs cycle in which it will sequentially give rise to other different organic compounds, as fumarate, malate and citrate (Le et al. 2012; Li and Le 2018) by the action of different enzymes. As it will be mentioned in the FA-dedicated section, citrate is a key molecule in lipids synthesis, and in cancer it is of note that citrate is extensively glutamine-originated. Again, these findings highlight the minutious coordination between glucose and glutamine metabolism, in cancer. Whilst glucose is used to fulfill PPP and support cell proliferation, glutamine is used to support Krebs cycle, OXPHOS and amino acids and lipids syntheses. Another evidence of this synchronization is the fact that GLDH1 activity is increased when mitochondrial pyruvate uptake is impaired and the opposite is also observed (Yang et al. 2014; DeBerardinis et al. 2007). Furthermore, the glutamine derived-aspartate can re-enter the Krebs cycle by being further converted into oxaloacetate by aspartate transaminase (GOT1), which upregulation is described as being linked to GLHD1 downregulation (Son et al. 2013). Obviously, the Krebs cycle intervenient enzymes must also be deregulated in cancer, and isoenzymes of malic enzyme (ME1 and ME2) are upregulated in different cancer types (Liu et al. 2018; Sarfraz et al. 2018; Lu et al. 2018; Nakashima et al. 2018), being its pro-tumorigenic function validated in cancer models (Fernandes et al. 2018), which prompts their nomination as relevant therapeutic targets (Liu et al. 2018; Sarfraz et al. 2018). Interestingly, a direct ligation between ME1 and the PPP was disclosed, showing that ME1 forms a hetero-oligomer with 6-phosphogluconate dehydrogenase (6PGD) and increases the affinity of 6PGD by its substrate 6-phosphogluconate (Yao et al. 2017), encouraging the deviation of glycolysis intermediates to PPP and keeping Krebs cycle reliance on glutamine (Fig. 1.2). Citrate synthase (CS) is also described as being overexpressed in cancer mainly in metabolic stressful conditions, such as hypoxia, favoring the activation of the metabolic remodeling with glutamine reliance (Peng et al. 2019). Together with citrate, glutamine originated-acetyl-CoA is directed to lipid synthesis (Le et al. 2012; Metallo et al. 2011; Jiang et al. 2017), accounting for about 20% of lipogenic acetyl-CoA (Metallo et al. 2011).

1.3.2 Glutamine and Oxidative Stress and Glutaminolysis Regulation

Glutaminolysis also contributes for the maintenance of the redox state by supplying the synthesis of glutathione (GSH; a tripeptide of glutamyl – cysteinyl – glycine) (Fig. 1.2), since glutamine is directly converted into glutamate and contributes indirectly for glycine synthesis that can derive from serine or Krebs cycle intermediates (Lopes-Coelho et al. 2016; Bruntz et al. 2019). Accordingly, the regulation of GLS isoenzymes expression is sensitive to oxidative stress (Mates et al. 2013) and it is regulated by p53, the guardian of the genome, whose action is also regulated by reactive oxygen species (ROS) (Gào and Schöttker 2017; Zhang et al. 2017; Suzuki et al. 2010). Considering this biochemical connection between glutaminolysis and ROS scavenging, a study by Gregory et al. propose as an innovative therapeutic approach the simultaneous targeting of glutaminolysis and the redox state of cancer cells (Gregory et al. 2019). Moreover, different studies have been presenting inhibitors and strategies to trigger the inhibition of glutamine metabolism and treat cancer, as extensively reviewed (Hensley et al. 2013; Altman et al. 2016; Still and Yuneva 2017; Vanhove et al. 2019).

The main regulator of glutamine import and metabolism is the MAPK pathway and this is evidenced by the increased levels of glutamine-derived compounds in KRAS-mutated cancer cells (Gaglio et al. 2011; Carr et al. 2010; Bryant et al. 2014; Wang et al. 2019b; Galan-Cobo et al. 2019). The decoupling between glucose and glutamine, in order to perform an overall supply of cancer metabolism supportive of tumor growth, is also regulated by KRAS (Gaglio et al. 2011). In BRAF mutated melanoma, glutamine metabolism is pointed by Luebker and Koepsell as part of a putative chemoresistance mechanism to BRAF inhibitors (Luebker and Koepsell 2019).

The same was described for the development of resistance to PI3K inhibition in lung cancer (Koh et al. 2017). A recent study, unravel that scarcity of crucial amino acids, such as glutamine, in the TME accounts for the diminished function of immune cells, being a way of decreasing the immunotherapy efficacy (Ramapriyan et al. 2019). Therefore, the increased uptake of glutamine by cancer cells, decreasing its levels in TME, sustains cancer cells survival and immune evasion. The activation of some signaling pathways is also dependent on glutamine import, as demonstrated by a study showing that LAT2 inhibition and glutamine import disturbance, would affect the amino acids sensitive mTOR (Feng et al. 2018), which is a mediator of PI3K pathway, pivotal in cancer survival and metabolic remodeling. The established role of glutamine in cancer cells coping with oxidative stress and allowing the maintenance of the metabolic flow, corroborates the fact that NRF2 pathway, the major responsible for the control of the redox state, also plays a role in the regulation of glutamine metabolism (Galan-Cobo et al. 2019; Romero et al. 2017). In addition, a cooperation between NRF2 and MAPK, in KRAS-mutated cancer cells (Galan-Cobo et al. 2019; Romero et al. 2017), is also described and reinforces the orchestrating relevance of signaling networking in metabolic remodeling.

In cancer, as a counterpart of what is happening in the glucose metabolism, glutamine-derived organic compounds are often the direct substitutes of glucose-derived compounds, entering the core pathways for energy production (Fig. 1.2).

1.4 Cysteine, the Shield of Endogenous Metabolism

Cysteine, a semi-essential thiolic amino acid that interferes with cancer metabolic remodeling at three different levels: (1) in free radicals scavenging, free or as a component of glutathione (GSH); (2) in hydrogen sulfide (H_2S) production, and (3) as a carbon source for biomass and energy production. In the three fronts, cysteine contributes for cancer cell strongness and prosperity, facing the microenvironmental stresses and escaping from drugs aggression and cell injury.

Cysteine and GSH are scavengers of free radicals (mainly ROS) that will abrogate the mechanism of action of the majority of drugs used to treat cancer, as they are oxidative/alkylating agents. GSH is highly important as a detoxifying system, allowing the physiological maintenance of the metabolic flow (Ballatori et al. 2009; Wu et al. 2004; Wang and Ballatori 1998; Kalinina et al. 2014). However, in disease facing the action of cytotoxic drugs, GSH constitute a severe resistance mechanism (Lopes-Coelho et al. 2016; Colla et al. 2016; Zanotto-Filho et al. 2016; Lien et al. 2016; Harris et al. 2015; Traverso et al. 2013). These facts highlight the importance of cysteine in GSH-mediated resistance to cisplatin, as intracellular cysteine is a rate-limiting precursor for GSH synthesis (Okuno et al. 2003).

1.4.1 Cysteine Metabolism

Several studies show that cysteine plays a core role in cell metabolism, also because it is a relevant key player in carbon and sulfur metabolism. Cysteine catabolism generates H_2S (Bhattacharyya et al. 2013; Szabo et al. 2013; Sen et al. 2015; Panza et al. 2015; Gai et al. 2016; Pan et al. 2015) and it is dependent on the action of four enzymes: cystathionine β-synthase (CBS); cystathionine γ-lyase (CSE), and 3-mercapto-pyruvate sulfurtransferase (MpST), which works together with cysteine aminotransferase (CAT) (Wang 2012). The action of all these enzymes is associated with ATP production, as H_2S can serve as a donor for the electron transport chain (ETC) (Bhattacharyya et al. 2013; Szabo et al. 2013; Módis et al. 2013; Fu et al. 2012). From the degradation of cysteine, pyruvate, α-ketoglutarate, glutamate and serine are generated (Nunes and Serpa 2018), and can be diverted into other metabolic pathways, as Krebs cycle and the one carbon metabolism (Figs. 1.2 and 1.3).

Recently, an innovative way of looking at the H_2S contribution for ATP production, showed that sulfur can also function as an acceptor of

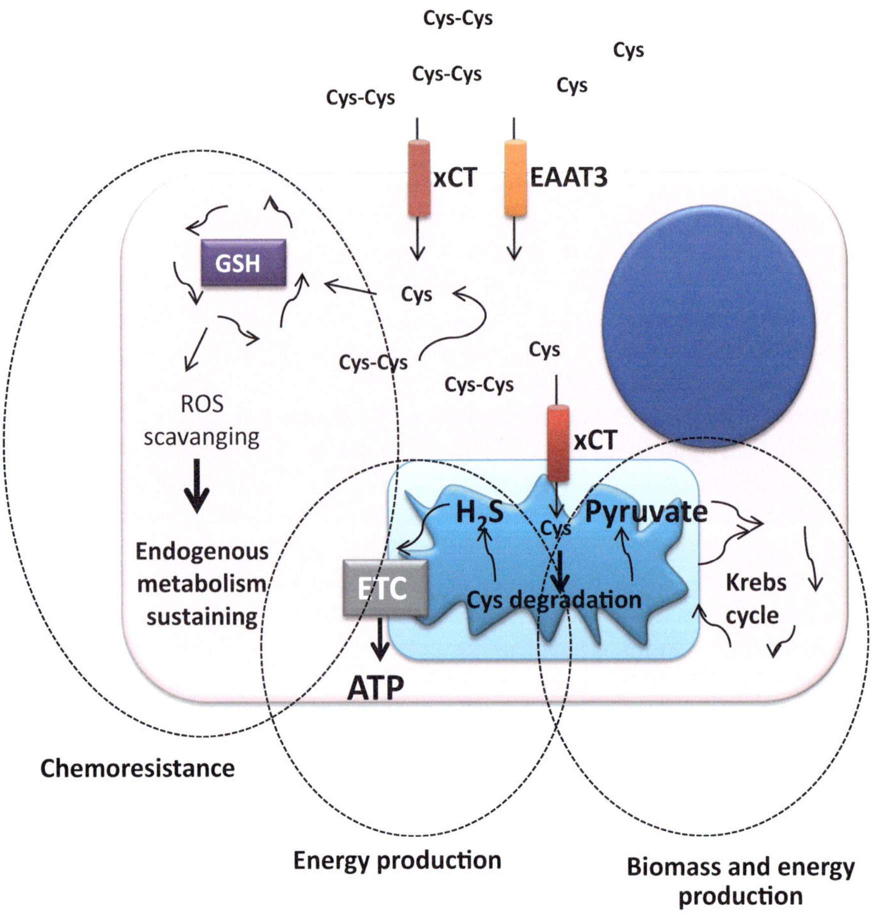

Fig. 1.3 Cysteine has a central role in cancer cell metabolism. The metabolic network of cysteine (cys), a thiolic amino acid that interferes with cancer metabolic remodeling and chemoresistance at three different levels: (1) free or as a component of glutathione, cysteine is a scavenger of free radicals (mainly reactive oxygen species- ROS) that will abrogate the mechanism of action of the majority of drugs used to treat cancer, as they are oxidative/alkylanting agents; (2) cysteine degradation gives rise to hydrogen sulphide (H_2S), which is an electron donor to electron transport chain (ETC) resulting in ATP production, and (3) cysteine is also a carbon source for biomass and energy production. The two former roles of cysteine contribute for cancer cell strongness and prosperity, facing the microenvironmental stresses and escaping from drugs aggression and cell injury. The uptake of cysteine is controlled by cystine (cys-Cys; e.g. xCT) or cysteine (cys; e.g. EAAT3) transporters

electrons in the end of ETC, substituting oxygen. But, in this case H_2S would be generated from persulfides metabolism and not directly from cysteine degradation (Fujii et al. 2019). Importantly, this study somehow changes the paradigm of H_2S role in metabolism, showing again that we are far from knowing all the secrets about metabolism. The role of H_2S as a signaling molecule is pivotal in cancer, since it regulates proliferation, bioenergetics and angiogenesis, both in a paracrine and autocrine fashion (Augsburger and Szabo 2018; Giuffrè and Vicente 2018).

The degradation of oxidized GSH (GSSG) will allow the recycling of its three components: glutamate, cysteine and glycine. This degradation occurs through the γ-glutamyl cycle. GSSG leaves the cell and at the external face of the cell membrane γ-glutamyl transpeptidase (GGT) releases glutamate (Umapathy et al. 2018) and the cysteinylglycine dipeptide that will be degraded by aminopeptidase N (APN), releasing cysteine and glycine (Hausheer et al. 2011). All the 3 amino acids can re-enter the cell in a process mediated by specific transporters. The cysteinylglycine dipeptide can also be taken up by cells in a process mediated by the peptide transporter 2 (PEPT2) and afterwards be degraded in the cytoplasm by dipeptidases (Frey et al. 2007).

Cysteine can also be *de novo* synthesized through the transsulfuration pathway (TSSP), deriving from methionine and serine (Fig. 1.2), which makes the synthesis of cysteine tightly linked to the methionine cycle, meaning that cysteine synthesis is dependent on the availability of the methionine cycle intermediates (Pérez-Miguelsanz et al. 2017). As mentioned in Sect. 1.3, serine can be glutamine-originated, making a link between glutamine and cysteine metabolism. As extensively revised by Combs and Nicola (Combs and De Nicola 2019), a step wise reactions transform methionine into homo-cysteine, which will be further converted into cystathionine through the condensation with serine catalyzed by CBS. Afterwards cystathionine is hydrolyzed by CSE, giving rise to cysteine, ammonia and α-ketoglutarate. Here, we can find a bond between TSSP and Krebs cycle through α-ketoglutarate.

Recent studies have pointed out cysteine as an important element in adaptation of cancer cells to stressful conditions, such as chemotherapy and hypoxia (Nunes et al. 2018a, b). In this context, the ability of cancer cells to deal with cysteine constitutes a mechanism accounting for the development of resistance to therapy. Cancer cells with a metabolic reliance on cysteine are more prone to adapt to metabolically damaging conditions like hypoxia, which accounts for their strength and consequent ability to develop chemoresistance (Nunes et al. 2018a). Furthermore, some studies associate the expression of enzymes involved in cysteine metabolism, as CBS and CSE, with malignancy related cell features and more aggressive cancer phenotypes (Bhattacharyya et al. 2013; Sen et al. 2015; Wang et al. 2019c; You et al. 2017; Turbat-Herrera et al. 2018; Alix-Panabières et al. 2017; Sekiguchi et al. 2016; Poisson et al. 2015). However, tumor-suppressive effects of CBS have also been reported in some cancer types, suggesting different roles of CBS in cancer, maybe dependent on the metabolic context (reviewed by Zhu et al. (2018)). In the sense of cysteine degradation and H_2S production in cancer, very low attention has been given to MpST. Nevertheless, emerging data based on assays using new pharmacological inhibitors and silencing approaches of MpST, support its role in cancer cell proliferation, bioenergetic and cell-signaling (Szabo et al. 2013).

1.4.2 Cysteine Transport

Looking at the entire flux of cysteine in biological cellular systems, the transport of cysteine into the cell is a crucial step that can also account to the development of resistance, considering the role of cysteine in oxidative stress control, as mentioned. Some transporters of cystine (oxidized form of cysteine) have been described and studied in cancer context, however, as some of these transporters mediate the transport of anionic form of cysteine and glutamate (Lo et al. 2008; Bianchi et al. 2014), several studies were developed on glutamate antiporter action (Fig. 1.3). Thus pointing out that the increased efflux of glutamate correlates with increased cancer aggressiveness and metastasis (Lo et al. 2008; Fazzari et al. 2015; Shiozaki et al. 2014; Stepulak et al. 2014; Koochekpour et al. 2012; Dornier et al. 2017), indicating a putative poor outcome in patients suffering from tumors with increased efflux of glutamate. For glutamate efflux to occur, the influx of cysteine is mandatory, supporting again the role of cysteine in cancer worst outcome. Thus, cancer cells in metabolic and phenotypical equilibrium can be disturbed by blocking cyst(e)ine/glutamate transporters (Lo et al. 2008; Drayton et al. 2014; Doxsee et al. 2007). Moreover, in cancer cell lines the expression levels of transport system xc- (xCT; SLC7A11 gene) was associated with intracellular GSH levels and cisplatin resistance (Okuno et al. 2003). In fact, xCT is considered the main transporter of cystine in cancer (Ji et al. 2018; Koppula et al. 2017; Lim et al. 2019) and its expression is regulated by the most relevant controller of cellular redox state,

NRF2 (Habib et al. 2015). In addition, its expression can also be under the command of PI3K/AKT/mTOR (Lien et al. 2017; Mossmann et al. 2018; Gu et al. 2017) and MAPK pathways, having a synergic functioning with the ATF4 transcription factor that is activated by endoplasmic reticulum stress (Lim et al. 2019). The pivotal role of xCT, in cyst(e)ine transport and cysteine metabolic reliance, is proved by the fact that cells with decreased expression of xCT increase the rate of TSSP to enhance cysteine endogenous synthesis (Lien et al. 2017; Kang et al. 2017).

Despite the uptake of cystine being the main form of cells acquiring cysteine, it is described that cancer cells uptake cysteine directly (Zhang et al. 2012), and it is mediated by cysteine transporters, which are overexpressed in different cancer types, such as the amino acid transporter 3 (EAAT3; SLC1A1 gene) (Pissimissis et al. 2009; Bianchi et al. 2012; Pedraz-Cuesta et al. 2015) and the alanine-serine-cysteine-transporter 2 (ASCT2; SLC1A5 gene) (Bothwell et al. 2018; Rajasinghe et al. 2019; Bolzoni et al. 2016). As mentioned in Sect. 1.3, regarding glutamine metabolism, ASCT2 was already elected as a putative therapeutic target in cancer (Wahi and Holst 2019). Unfortunately, the association between cysteine transport and the overexpression of these transporters in a cancer metabolic remodeling context was not yet established, since the performed studies main issue is the transport of other amino acids, as glutamine and glutamate.

1.4.3 Cysteine Interference in Cell Signaling

Cysteine and methionine are the only sulfur-containing proteinogenic amino acids. However, cysteine is the unique having a thiol group, allowing cysteine role as a nucleophile in the catalytic domain of certain enzymes and also the establishment of disulfide bonds that are crucial for protein stabilization and folding (Marino and Gladyshev 2012; Netto et al. 2007).

Besides its direct interference in metabolic pathways, cysteine has also an additional role in the regulation of cell functioning through the modulation of signal transduction. As mentioned, cysteine is the thiol component of GSH and the reversible thiolation, namely cysteinylation, of proteins was unraveled as a mechanism of regulating pivotal metabolic processes dependent on enzymes and transporters activity, signal transduction and gene expression (Traverso et al. 2013). Crucial signaling intervenients, such as RAS-GTPases, Jun N-terminal kinase (JNK)- 2, Activator protein 1 (AP-1), NFkB, PKC, caspases, thioredoxin and p53, are regulated by thiol oxidation (Traverso et al. 2013; Mieyal et al. 2008; Emmanuel et al. 2014; van Jaarsveld et al. 2015; Mackenzie et al. 2015; Zheng et al. 2015; Carduner et al. 2014; Al-Alem et al. 2013; Hajiahmadi et al. 2015; Su et al. 2013; Echevarría-Vargas et al. 2014; Yang-Hartwich et al. 2014; Wu et al. 2013a). In the case of p53, it can be inactivated by the oxidation of its cysteine residues, allowing the abrogation of DNA repair, cell cycle checkpoints and inhibition of apoptosis, contributing for carcinogenesis (Mieyal et al. 2008). In addition, many oncogenic mutations cause the insertion of a novel cysteine in the protein sequence, accounting for at least 12% of all activating mutations found in KRAS in cancer, and 88% of mutations in fibroblast growth factor receptor (FGFR) (Visscher et al. 2016). This fact points for a crucial role of cysteine in carcinogenesis, not only in a thiolation process but also in its aberrant inclusion in protein synthesis underlied by missense mutations.

Together, evidence supports that cancer cells dependence on cysteine might be due to its role on GSH-mediated and H_2S-mediated chemoresistance that promotes redox balance and allows the regulation of several metabolic processes vital to cancer cells survival in stressful and detrimental environments (Fig. 1.3).

1.5 Fatty Acids, the Versatile Macromolecules

Fatty acids (FA) are usable in all the intercrossed points of endogenous metabolism. They can be degraded through β-oxidation, giving rise to acetyl-CoA the intermediate capable of being used in Krebs cycle and redundantly in lipids synthesis (Fig. 1.4). In fact, FA are a valuable source of energy and biomass to sustain survival and proliferation, besides having a relevant role as precursors of signaling molecules (Infantino et al. 2014; Currie et al. 2013). The remodeling of FA metabolism is a part of the complex metabolic adjustment displayed by cancer cells to cope with the selective microenvironmental pressure and carry on cancer progression (Swierczynski et al. 2014).

1.5.1 Fatty Acids (FA) Synthesis

FA are obtained from dietary or *de novo* synthesized in the cells. The main precursor of acetyl-CoA used in FA synthesis is citrate mainly originated from Krebs cycle, which implies its export from mitochondria to cytoplasm. The transport of citrate across the inner mitochondrial membrane is mediated by the citrate transporter protein carrier (CTP; encoded by *SLC25A1* gene), which expression is augmented in several human cancer models *in vitro* and *in vivo*

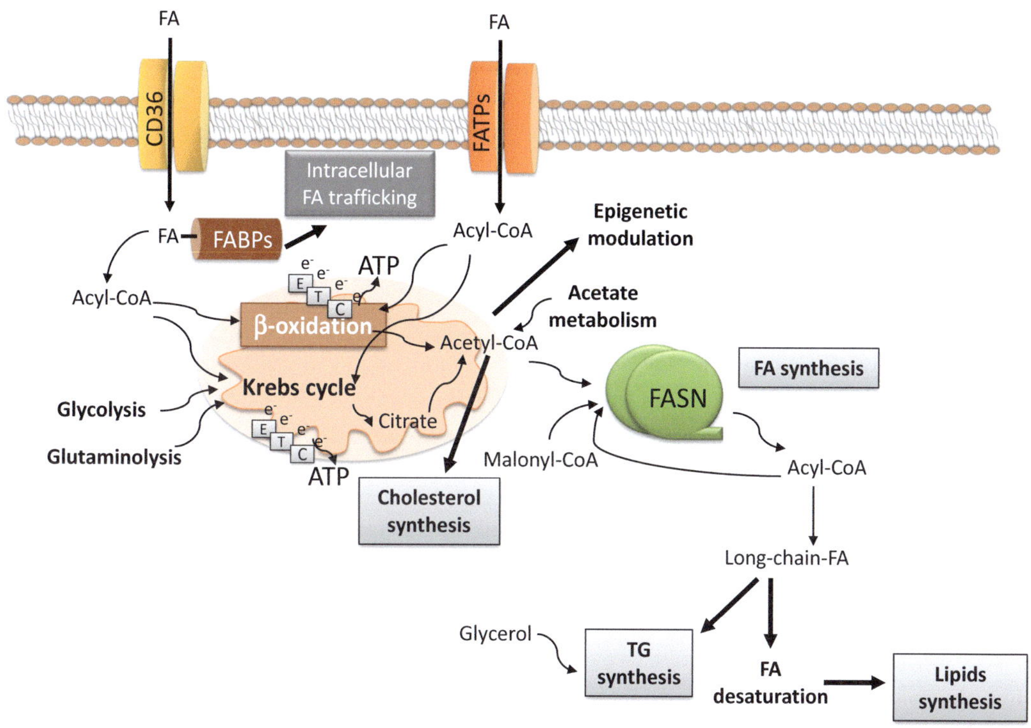

Fig. 1.4 Fatty acids (FA) synthesis and degradation can occur independent and simultaneously in a cancer cell. FA are a valuable source of energy and biomass, and FA consume depends on their uptake and degradation (β-oxidation). FA are taken up by cell membrane transporters (e.g CD36 and FATPs- FA transport proteins with acyl-CoA transferase activity) and their intracellular trafficking is mediated by FA binding proteins (FABPs). Acetyl-CoA results from FA degradation and can be deviated to the Krebs cycle, contributing for energy and biomass production, to the cholesterol synthesis (mevalonate pathway) or re-enter the FA synthesis. Acetyl-CoA can come from glucose, glutamine, FA or acetate metabolism. Acetyl-CoA has also an important role in epigenetic modulation, through the acetylation of proteins, namely histones. β-oxidation releases the same mount of electrons (e^-) to supply the electron transport chain (ETC) than glycolysis and Krebs cycle together. FA synthesis consists in a sequencial and repetitive adding of an acetyl-CoA molecule to an acyl-CoA (in the beginning malonyl-CoA). The elongation of FA is catalyzed by a unique enzyme, the FA synthase (FASN). Long acyl-CoA (long-chain-FA) are accumulated as triglycerides (TG) or serve as substrates for other lipids synthesis

(Catalina-Rodriguez et al. 2012). The expression of CTP seems to be regulated by mutant p53, being this way high CTP expression levels considered a poor prognosis marker (Kolukula et al. 2014). In addition, the activity and expression of CTP has been associated with resistance to therapy and the increased ability of cancer cells to adapt to metabolic stressful conditions, as hypoxia (Kolukula et al. 2014; Hlouschek et al. 2018). The relevance of FA synthesis in cancer cells is highlighted in deficient CTP models in which an alternative pathway to synthesize FA is followed due to the lack of mitochondrial citrate. Thus, α-ketoglutarate from Krebs cycle is trafficked to the cytosol and converted into citrate by isocitrate dehydrogenase 1 (IDH1) (Jiang et al. 2018).

The canonical FA synthesis has 3 limiting steps: (1) the conversion of citrate into acetyl-CoA; (2) the conversion of acetyl-CoA into malonyl-CoA, and (3) successive condensation reactions with acetyl-CoA.

In cytoplasm, ATP citrate lyase (ACLY) converts citrate and coenzyme A (CoA) to oxaloacetate and acetyl-CoA to feed FA synthesis, promoting a direct link between glucose, glutamine and FA metabolism, as the majority of citrate is Krebs cycle originated. The overexpression of ACLY is observed in a variety of cancer types (Zhou et al. 2013; Beckner et al. 2010; Wang et al. 2012a) and its cancer related usefulness is confirmed in cancer models, showing that ACLY downregulation suppresses cancer progression (Carrer et al. 2019).

As above-mentioned, acetyl-CoA can be alternatively produced from acetate under the action of ACSS2 (Carrer et al. 2019; Carrer and Wellen 2015). Despite ACLY being considered the central enzyme in acetyl-CoA synthesis (Wei et al. 2019; Zaidi et al. 2012), it is well known that both ACLY and ACSS2 are crucial in supplying histone acetylation (Carrer and Wellen 2015; Carrer et al. 2017; Wellen et al. 2009) and even in the post-translational stabilization by acetylation of ACLY itself (Lin et al. 2013). In addition, ACLY and ACSS2 expression and function have been linked to PI3K/AKT pathway activation, a core signaling pathway in cancer (Bidkhori et al. 2018; Hanai et al. 2012; Amrita Devi et al. 2015).

The carboxylation of acetyl-CoA to malonyl-CoA is catalyzed by acetyl-CoA carboxylases (ACC1 and ACC2); ACC1 is cytosolic and ACC2 is located in the mitochondrial membrane (Bourbeau and Bartberger 2015). ACC1 is responsible for the second rate limiting step of *de novo* FA synthesis by converting acetyl-CoA to malonyl-CoA, whereas ACC2 may be involved in the regulation of FA oxidation. This assumption is evidenced by the abrogation of FA synthesis when ACC1 is inhibited by phosphorylation upon the action of AMP-activated protein kinase (AMPK) (Lally et al. 2019; Zhang et al. 2019b).

The final limiting step of lipids synthesis is the successive malonyl-CoA/ acyl-CoA and acetyl-CoA condensation reactions that are catalyzed by the unique enzyme fatty acid synthase (FASN) in a NADPH dependent manner (Rudolph et al. 2012). The expression of FASN is upregulated in many types of cancers, being a metabolic marker of survival, proliferation and metastasis (Sun et al. 2019; Wang et al. 2019d). Equally important is the production of NADPH to support FASN activity. The main producer of NADPH is IDH1, thus IDH1-mutated gliomas are usually slow growing tumors, as the low levels of NADPH slows the rate of FASN activity and consequently cancer cells proliferation (Azar et al. 2018; Calvert et al. 2017). Whereas wild type IDH1 boosts tumor growth and therapy resistance (Calvert et al. 2017). Because *de novo* lipogenesis is critical for cell proliferation, the expression and activity of FA synthesis core enzymes (ACLY, ACC1 and FASN) are absolutely pivotal in cancer (Menendez et al. 2016; Menendez and Lupu 2007; Hopperton et al. 2014). Hence, their concerted expression must be tightly regulated and this synchronization is ensured by centralizing the expression of these genes under the control of the same signaling pathways and transcription factors; again, PI3K/AKT pathway plays a role (Zhu et al. 2018; Pattanayak et al. 2018; Li et al. 2016; Elhanati et al. 2013; Lin and Miner 2015; Porstmann et al. 2008).

Hypoxia drives metabolic remodeling in a large sense, and lipids metabolism is not an exception. FA synthesis is activated by hypoxia, using predominantly non-glucose-derived precursors, as hypoxia reduces the synthesis of acetyl-CoA from glucose (Munir et al. 2019); Hif-1 and AMPK are major regulators (Jia et al. 2019; Zhang et al. 2019b).

1.5.2 Fatty Acids (FA) β-oxidation

FA are a source of acetyl-CoA, thus FA catabolism is also important in cancer metabolic fitness, since it is a valuable way of sustaining energy production. FA degradation occurs through β-oxidation pathway in the mitochondria and peroxisome as well as during autophagy in lysosomes (Iershov et al. 2019; Singh and Cuervo 2012). In contrast to mitochondrial β-oxidation, the peroxisomal β-oxidation is coupled to hydrogen peroxide (H_2O_2) generation and residual ATP production (Liu et al. 2019).

Prior β-oxidation, FA must be converted into acyl-CoA by acyl-CoA synthetases (ACS), which are located on the endoplasmic reticulum, peroxisome and mitochondrial outer membrane (Yan et al. 2015). Depending on the length of FA they act on, ACS are classified into very long-chain (ACSVL), long-chain (ACSL), medium-chain (ACSM) and short-chain (ACSS) (Yan et al. 2015). Few is known on the role of ACS in cancer progression, however the studies presented so far on ACSL4 and ACSL5 are divisive. Some studies on breast cancer pointed ACSL4 and ACSL5 as good prognosis markers whereas other studies associate their expression with tumor growth promotion and hormone therapy resistance (Yan et al. 2017; Wu et al. 2013b). In human liver, about 16% and 21% of the palmitoyl-CoA synthetase activity (encoded by ACSL genes) is localized respectively in the peroxisome and in the mitochondria (Mashek et al. 2004). Moreover, ACS activity is related to increased survival of cancer cells and chemoresistance (Mashima et al. 2009; Sung et al. 2007). In addition, the expression of ACSL3 seems to be directly activated by peroxisome proliferator-activated receptor delta (PPARδ) (Cao et al. 2010), reinforcing the role of PPAR nuclear receptors in the regulation of lipids catabolism, being FA (e.g. oleic acid) their main ligands (Iershov et al. 2019). Some studies associate PPAR receptors to therapy resistance, both as promoters and as repressors. PPARα and co-activators are associated to increased radiotherapy resistance in nasopharyngeal cancer (Du et al. 2019), and PPARγ is associated to aggressiveness and chemoresistance in pancreatic cancer (Zhang et al. 2015). Nonetheless, the majority of studies, in different cancer models, stated the opposite for PPARγ and chemoresistance, showing that the activation of PPARγ mediates sensitivity to anti-cancer drugs (Xu et al. 2018b; Asukai et al. 2017; Wang et al. 2017b; Zhan et al. 2012a; Bräutigam et al. 2011). Interestingly, the anticancer effect of drugs functionalized with conjugated fatty acids (e.g. oleic acid) is PPARs mediated (Ricci et al. 2018).

The transfer of acyl-CoA molecules into mitochondria or peroxisome, to undergo β-oxidation, is mediated by carnitine palmitoyltransferases (CPT). In detail, acyl-CoA is converted to acyl-carnitine via CPT1 then acyl-carnitine is translocated to the mitochondria via carnitine acyl-carnitine translocase (CACT) and converted back to acyl-CoA by CPT2 (Fujiwara et al. 2018). In addition to CPT1 and CPT2, other carnitine acyltransferases may promote the transfer of medium-chain acyl-CoA across the cell membranes. As peroxisome only harbors the β-oxidation of long- and very long-chain FA, the medium chain FA must be transported to mitochondria. Carnitine octanoyltransferase plays a role in this process, as it catalyzes the reversible transfer of medium-chain acyl groups between coenzyme A and carnitine in the peroxisomes, followed by the transfer to mitochondria (Adeva-Andany et al. 2019). A certain controversy surrounds the meaning of the expression and activity of CPT1 and CTP2 in cancer. In breast cancer, some studies pointed the increased levels of CPT1 and CTP2 as part of a more aggressive cancer cells phenotype, thus poor prognosis (Reis et al. 2019), whereas other studies found a correlation between the expression of CPTs and estrogen receptor and consequently good prognosis

(Aiderus et al. 2018). In hepatocellular carcinoma, a poor prognosis type of tumor (Kitisin et al. 2011), the studies are more accordant, showing decreased β-oxidation rate together with decreased expression of CPT2 (Fujiwara et al. 2018; Lu et al. 2019; Lin et al. 2018). Therefore, it seems that the role of CPTs as prognostic factors are strongly dependent on the cancer cells organ of origin and, within this, on the molecular histotypes.

Finally, acyl-CoA dehydrogeneses (AD) catalyze the first of the four step FA breakdown pathway (Adeva-Andany et al. 2019), and the different AD enzymes have a FA length dependent affinity, being organized as short-chain AD (SCAD); medium-chain AD (MCAD), long-chain AD (LCAD) and very long-chain AD (VLCAD) (Bonito et al. 2016). AD convert acyl-CoA in 2-enoyl-CoA esters, which will be hydrated into 3-l-hydroxyacyl-coA, by enoyl-coA hydratase (ECH). Afterwards, 3-l-hydroxyacyl-CoA dehydrogenase forms a 3-ketoacyl-CoA intermediate that will be two carbons cleaved by 3-ketoacyl-CoA acetyltransferase/thiolase (ACAA), releasing an acetyl-CoA and an acyl-CoA ester with minus two carbons (Adeva-Andany et al. 2019); this newly generated acyl-CoA can re-enter the β-oxidation path. In β-oxidation, NADH and FADH2 are also produced, being electron deliverers to ETC (Adeva-Andany et al. 2019). Acetyl-CoA can be re-directed to other metabolic routes, as suppliers of Krebs cycle and redundantly FA synthesis.

FA β-oxidation is a motor for OXPHOS, as it supplies Krebs cycle with acetyl-CoA and ETC with NADH, being an efficient substitute of glycolysis in these metabolic routes (Fig. 1.4). This fact is evidenced by the disruption of OXPHOS in cancer cells upon mitochondrial MCAD inhibition, with consequent abrogation of β-oxidation (Yan et al. 2017). Again mitochondrial metabolic remodeling works in consonance with the whole metabolic readjustment in cancer cells. Glycolysis is no longer the direct supplier of Krebs cycle and OXPHOS but it can also be replaced by a mitochondrial resident pathway, β-oxidation, thus maximizing the existing resources.

Hypoxia also controls β-oxidation, being the main targets for this pathway inhibition the AD, responsible for the limiting step of β-oxidation and whose expression is abrogated by Hif-1α (Zhang et al. 2019a; Huang et al. 2014).

1.5.3 Fatty Acids (FA) Transport

A couple of decades ago, lipids were thought to cross the cell membrane by passive diffusion, nowadays different FA transporters are known, including CD36, plasma membrane-associated fatty acid-binding protein (FABP), and a family of fatty acid transport proteins (FATP1–6).

CD36 enhances cellular FA uptake, although the molecular mechanisms are not known. The background of CD36 function in low-density lipoprotein uptake together with the action of other FA transporters confound its real contribution in FA uptake (Xu et al. 2013). In fact, it seems that the effect of CD36 in enhancing the uptake of FA is a consequence of the CD36 action on intracellular metabolism, such as increased lipids esterification (Xu et al. 2013), which stimulates the need for the import of more FA. However, CD36 expression is often described in cancer, associated to more aggressive phenotypes (Zhang et al. 2019a; Wang et al. 2019d; Pan et al. 2019; Liang et al. 2018; Rozovski et al. 2018; Lopes-Coelho et al. 2018), being additionally proposed as a biomarker for metastatic disease (Enciu et al. 2018). An interplay between CD36 and MAPK and PI3K/AKT pathways was described, in hepatocellular and in uterine cervix carcinomas, pushing cancer progression and metastasis (Li et al. 2018b; Yang et al. 2018). In leukemia STAT3 plays a role in the activation of CD36 expression (Rozovski et al. 2018). Nevertheless, a study, dedicated to lung cancer, show an inverse association between CD36 expression and cancer progression (Sun et al. 2018). In some studies, the relevance of FABPs expression in stromal cells within the tumors for disease outcome is highlighted.

FABPs are intracellular FA carriers, counteracting the low solubility of FA in aqueous media, they are located close to the cytoplasmic side of the plasma membrane. After FA diffuse freely through the bilayer, FABP will bind them and promote their inter-compartment delivery. FABPs are expressed in a tissue-specific manner and despite their high homology and apparent functional redundancy, there are some structural specificities that highlight a certain specialization in FA transport (as reviewed in Storch and McDermott, (Storch and McDermott 2009)). In cancer, the expression of some FABPs, such as FABP4, FABP5 and FABP7, is associated with disease severity (Nagao et al. 2018; Guaita-Esteruelas et al. 2017) and outcome (Liu et al. 2011). The regulation of FABP expression is controlled by PI3K/AKT pathway (Lv et al. 2019) and Src oncoprotein (Hua et al. 2019).

FATP family has 6 members and it is not clear if they all mediate the plasma membrane crossing of lipids or if some of them only mediate the intracellular lipids delivery; and they are expressed in a tissue specific pattern (Anderson and Stahl 2013). FATP1/SLC27A1 and FATP4/SLC27A4 are the best studies and they are also able to convert FA in acyl-CoA, since all FATPs exhibit an acyl-CoA synthase activity (Zhan et al. 2012b). At a certain point the activity of FATP1 and FATP4 seem to be overlapped (Lin and Miner 2015). This feature simplifies the FA transport and intracellular delivery in cancer, because it exonerates the need for a partnership with other acyl-CoA synthase. Both FATP1 and FATP4 are insulin sensitive (reviewed by Fisher and Gertow (2005)). After expression, FATP1 is accumulated in the Golgi apparatus and upon a stimuli for its activation, such as insulin or adiponectin, FATP1 is translocated to the cell membrane in where it will exert its function in FA uptake (Wu et al. 2006; Stahl et al. 2002). The association of FATP1 and FATP4 with endoplasmic reticulum is also known and FATP1 is related to the production of lipids droplets, important in nuclear signaling and epigenetic modulation (Wu et al. 2018; Xu et al. 2012; Milger et al. 2006). The increased expression of FATPs is well documented in several types of cancer, such as melanoma, breast cancer and endometrial cancer, emphasizing the dependence of cancer cells on the uptake of FA from TME (Lopes-Coelho et al. 2018; Zhang et al. 2018b).

Interestingly and differently from what it was expected, β-oxidation and FA synthesis are not mutually exclusive, they are often active in simultaneous and independently of each other (Munir et al. 2019; Carracedo et al. 2013). This circuit of synthesis and degradation of FA is for sure important in cancer metabolic remodeling as a way of recycling and storing carbons in a transient fashion, maintaining the bond with other metabolic courses (Fig. 1.4).

1.6 Cancer and Non-cancerous Cells in TME, an Active Metabolic Symbiosis

The metabolic remodeling in a tumor niche is withstood by cancer cells in cooperation with non-cancerous cells, allocated to the same TME. Thus, cancer cells and stromal cells form a complex network of signaling and organic compounds that will sustain cell survival and tumor growth. In brief, it will be focused the role of the most well studied stromal cells contributing for tumor metabolic fitness to TME: cancer-associated fibroblasts (CAFs); cancer-associated adipocytes (CAAs), and tumor-associated macrophages (TAMs) (Fig. 1.5).

1.6.1 Cancer-Associated Fibroblasts (CAFs)

CAFs are the major component of the TME, and several studies stated their role in cancer cells survival, proliferation, migration, angiogenesis, epithelial to mesenchymal transition (EMT) and chemoresistance (Ko et al. 2011; Augsten 2014; Karagiannis et al. 2012; Tchou et al. 2012; Hwang et al. 2008; Pavlides et al. 2009; Witz 2009; Orimo and Weinberg 2006; Vangapandu et al. 2017). In some types of tumors, as breast, pancreas and lung tumors, cancer cells are embedded in a dense fibrotic tissue, called desmoplasia (Hwang et al. 2008; DeClerck 2012); this phenomenon indorses an intimacy between

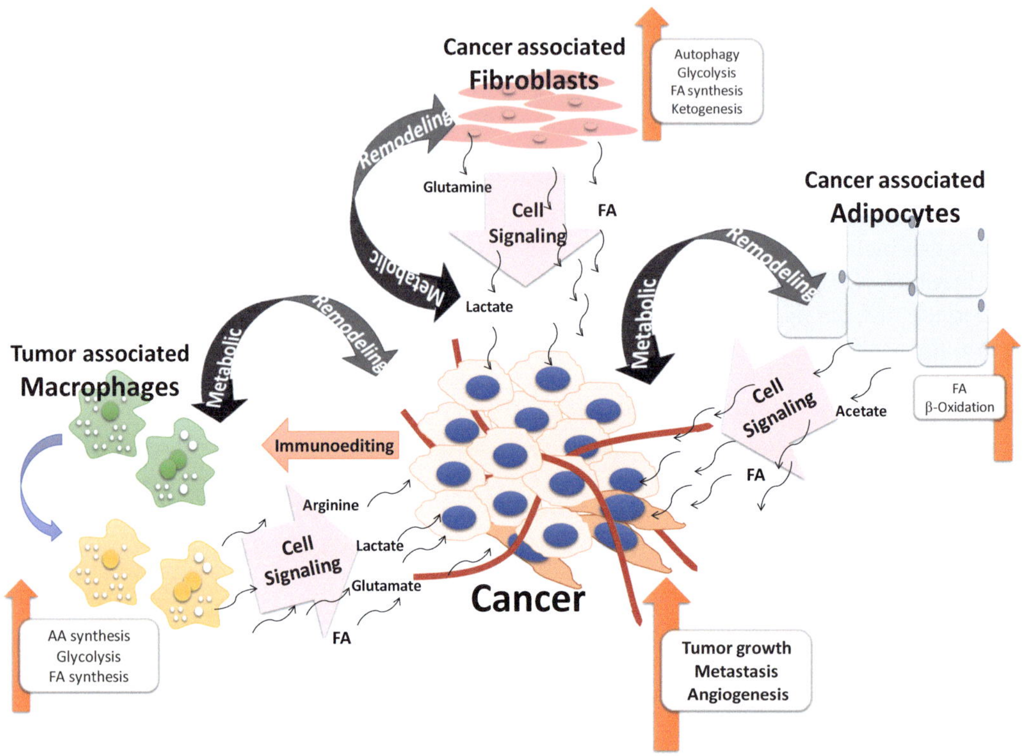

Fig. 1.5 Stromal non-cancerous cells in tumor microenvironment (TME) are pivotal players in cancer progression, through a metabolic symbiosis. Cancer associated fibroblasts (CAFs), cancer associated adipocytes (CAAs) and tumor associated macrophages (TAMs) are important TME players in cancer metabolic and signaling networking and sustainability. Cancer cells modulate and are modulated by stromal cells in order to establish an effective metabolic symbiosis. The most well known metabolic pathways altered in cancer associated stromal cells are: glycolysis, glutaminolysis, amino acids (AA) synthesis, ketogenesis and fatty acids (FA) synthesis and degradation (β-oxidation). Cancer-associated stromal cells often activate autophagy in order to increase the availibility of organic compounds to supply cancer cells, promoting tumor growth, angiogenesis and metastasis

fibroblasts and cancer cells. In the pro-tumorigenic activation, CAFs undergo a metabolic remodeling and cancer cells take advantage of the CAFs resulting metabolites.

The augmented proliferation observed in CAFs is accompanied by a glycolytic switch under the action of HIF-1α, tumor growth factor β (TGF-β), cytokines (e.g. IL6), platelet-derived growth factor (PDGF) and ROS (Ding et al. 2010; Valsecchi et al. 2016; Martinez-Outschoorn et al. 2010; Liu et al. 2016; Wang et al. 2016; Li et al. 2015); and this remodeling counts on the upregulation of glycolytic enzymes as PFK and HK1/2, fructose-2,6-bi-phosphatase (PFK1 regulator), as well as, glucose and lactate transporters (Zhang et al. 2015; Ando et al. 2010; Martinez-Outschoorn et al. 2011; Fiaschi et al. 2012; Pértega-Gomes et al. 2014). In CAFs, the continuous promotion of glycolysis and OXPHOS inhibition depend on HIF-1α stabilization, promoted by α-ketoglutarate resultant from IDH3a activity; allowing the HIF-1α-mediated upregulation of NADH dehydrogenase 1 that regenerates NAD^+ (Zhang et al. 2015). Afterwards, glycolysis-derived metabolites, as lactate, are transferred to cancer cells, serving as substrates for mitochondrial biogenesis and OXPHOS (Pavlides et al. 2009; Vangapandu et al. 2017; Marchiq and Pouysségur 2016; Yu et al. 2018).

As above-mentioned, the glutamine availability is essential for cancer cells mitochondrial metabolism and redox equilibrium. In some contexts glutamine catabolism is considered to be more important than glucose metabolism (Mates et al. 2013; Wise and Thompson 2010; Reynolds et al. 2013; Lukey et al. 2013; Chen and Cui 2015). CAFs are suppliers of glutamine to proliferating cancer cells (Eng et al. 2010; Knudsen et al. 2016); evidenced by the increased expression of glutaminase and glutamine transporters (e.g. SLC6A14) in cancer cells in the presence of CAFs, in a glutamine dependent pattern (Ko et al. 2011; Yang et al. 2016).

A study on the role of CAFs in the development of chemoresistance, in ovarian cancer, showed that the decrease in the nuclear accumulation of platinum in cancer cells was due to GSH and cysteine release by CAFs. They also demonstrated that $CD8^+$ T cells counteracted this resistance by changing GSH and cyst(e)ine metabolism in CAFs (Wang et al. 2016). Again, metabolic symbiosis between cancer cells and stromal cells, is a powerful weapon.

In lipids metabolism, an issue less addressed in this context, CAFs transfer lipids to neighboring cancer cells directly through FATPs (Lopes-Coelho et al. 2018) or mediated by microvesicles (Santi et al. 2013).

During metabolic reprogramming of CAFs, autophagy is upregulated (Qiao et al. 2016), being a way of sustaining the role of CAFs as suppliers (Mao et al. 2013), thus cancer cells, in nutrients scarcity, are fed by CAFs autophagy resulting products. Autophagy in CAFs is activated by TGF-β/SMAD, NFkB and mTOR/AMPK signaling pathways (Zhao et al. 2016; Liu et al. 2017; Guido et al. 2012; Thoen et al. 2011; Singh et al. 2009), in order to release lactate, glutamine and lipids.

1.6.2 Cancer-Associated Adipocytes (CAAs)

Adipocyte-secreted factors (e.g. adipokines, adipocytokines and insulin) act on cancer cells proliferation, local invasion, metastatic spread and resistance to therapy (Zhong et al. 2010; D'Esposito et al. 2012). In accordance, cancer cells express receptors for those signaling molecules (Wei et al. 2016; Yin et al. 2004), being prone to be modulated by the non-malignant cells in TME. Though, the FA provider function is the most well explored role of CAAs (Dirat et al. 2011; Wang et al. 2012b).

Cancer cells uptake FA to support the rapid growth and provide energy necessary for survival and metastatic behavior (D'Esposito et al. 2012; Wang et al. 2012b; Martinez-Outschoorn et al. 2014; Balaban et al. 2017; Nieman et al. 2011; Manabe et al. 2003; Carter and Church 2012; Grisouard et al. 2011; Shpilberg et al. 2015; Wang et al. 2015; Santander et al. 2015; Drew et al. 2015; Corbet et al. 2016); synchronized with the upregulation of FA transporters and β-oxidation relevant genes in order to fuel Krebs cycle (Balaban et al. 2017; Corbet et al. 2016; Corbet and Feron 2017; Wang et al. 2017c; Menard et al. 2016; Wen et al. 2017). CAAs and cancer cells reciprocal modulation is extremely important in this metabolic partnership. Hence, in tumor samples, the CAAs present delipidation with reduced expression of adipocytes differentiation markers (e.g. LIPE and FABP4) and in the tri-acyl glycerol content (Dirat et al. 2011; Wang et al. 2014b), undergoing a more fibroblastoid morphological and molecular makeover (Dirat et al. 2011; Wang et al. 2014b; Picon-Ruiz et al. 2016; Fujisaki et al. 2015), resembling the desmoplastic phenotype. Signaling feed-forward loops between CAAs and cancer cells, involving Src oncogene, proinflammatory cytokines and growth factors (e.g, TNFα and IL), are crucial for the maintenance of cancer progression (Picon-Ruiz et al. 2016). This pro-inflammatory environment seems to be also related to cachexia (progressive loss of muscle and adipose tissue) observed in advanced cancer patients (Tisdale 2002).

1.6.3 Tumor-Associated Macrophages (TAMs)

Macrophages result from the differentiation, in tissues, of extravasating monocytes and undergo specific differentiation according to the local tissue microenvironment (Gordon and Taylor 2005; Coffelt et al. 2009; Dijkgraaf et al. 2013). Two extreme stages of polarization have been described, M1 and M2. M1 macrophages are considered potent producers of pro-inflammatory cytokines and killers of microorganisms and tumor cells, whereas M2 cells are more prone to scavenge debris, and promote angiogenesis, tissue remodeling and repair (Mantovani et al. 2006; Mantovani and Sica 2010). Tumor-associated macrophages (TAMs) seem to have acquired features shared by M2 macrophages. However, their phenotypic plasticity (i.e. TAMs comprise several subpopulations) makes the M1/M2 nomenclature a not so accurate classification (Lee et al. 2013).

TAMs influence on cancer is present in all traits of carcinogenesis, including tumor initiation, growth, vascularization and metastasis. Each trait seems to be affected by a particular subpopulation, being recruited to strategic regions of the neoplasm, according to the chemokine expression pattern in TME (Lee et al. 2013). Chemokine (C-C motif) ligand 2 (CCL2):CCR2 is the main determinant axis of monocyte recruitment into tumors; and colony stimulating factor 1 (CSF-1):CSF1R is the major lineage regulator for macrophages (Lee et al. 2013; Pollard 2009). Both signaling molecules have been linked to TAMs accumulation in a wide panel of tumors (breast, ovarian, lung, glial cancers, melanoma and leukemia) and their overexpression was associated to more aggressive malignancies and consequent poor prognosis (Lee et al. 2013; Ueno et al. 2000; Lin et al. 2001; Toy et al. 2009; Mroczko et al. 2007).

Although little is known about the metabolic reprogramming of monocytes/macrophages in the context of carcinogenesis, this is another feature underlying their role as cancer "helpers" (Liu et al. 2016). TAMs have an increased aerobic glycolysis (Penny et al. 2016) via Akt/mTOR and HIF-1α stabilization (Tannahill et al. 2013; Wenes et al. 2016). The FA biosynthesis and uptake is also elevated and correlates with an enhanced pro-inflammatory, pro-tumorigenic profile of TAMs (Metallo et al. 2011; Wise et al. 2011). Glutamate and arginine metabolism are also upregulated in TAMs isolated from tumors, correlating with highly proliferative malignancies (Choi et al. 2015; Chang et al. 2001); which may be a way of sustaining glutamine demands of cancer cells. Additionally, the activation of ARG1 dependent polyamine synthesis pathway protects the tumor from nitric oxide-related cytotoxicity, allowing survival and proliferation (Chang et al. 2001). TAMs can also contribute for tumor progression by inducing cyclooxygenase-2 (COX-2) expression in cancer cells, through the IL1β-mediated stimulation of ROS – Src – MAPK signaling (Hou et al. 2011). A deeper knowledge on TAMs – cancer cells metabolic cross-talk is for sure a profitable niche in cancer research.

Despite all the mechanisms are not fully understood, the metabolic cooperation between non-cancerous stromal cells and cancer cells is an evident actively-functioning metabolic symbiosis (Fig. 1.5).

Highlights

If we look at each metabolic pathway as a carrousel, the cancer metabolism is a huge playground, which is fueled by the sharing of countless organic compounds that leap from one carrousel to another, enhancing the movement of the pivotal ones. In fact, it is very difficult to completely individualize the metabolic pathways, being the switch off of some of them adaptively compensated by the switch on of other ones.

However it is evident that whole the molecular consortium in a cancer cell makes a networking in the sense of sustaining, by signaling or feeding, cancer cells survival and proliferation towards a successful disease progression. This success is our main target in cancer research; and metabolism, as the basis of survival, assemble an enormous amount of information that need to be processed in order to find vital clues for new strategies to abrogate metabolic and signaling pathways and fight cancer. Nevertheless, it is important to keep in mind that a single target won't be enough to deadly disturb cancer, a concerted panel of targets and drugs must be applied. Adding the difficulty of finding the exact killing profile.

Acknowledgments The authors acknowledge iNOVA4Health – UID/Multi/04462/2013, a program financially supported by Fundação para a Ciência e Tecnologia/Ministério da Educação e Ciência, through national funds and co-funded by FEDER under the PT2020 Partnership Agreement.

References

Adeva-Andany MM et al (2019) Mitochondrial β-oxidation of saturated fatty acids in humans. Mitochondrion 46:73–90

Afonso J et al (2019) Clinical significance of metabolism-related biomarkers in non-Hodgkin lymphoma – MCT1 as potential target in diffuse large B cell lymphoma. Cell Oncol 42:303–318

Aiderus A et al (2018) Fatty acid oxidation is associated with proliferation and prognosis in breast and other cancers. BMC Cancer 18:805–805

Al-Alem LF et al (2013) Activation of the PKC pathway stimulates ovarian cancer cell proliferation, migration, and expression of MMP7 and MMP10. Biol Reprod 89:73–73

Alam MM et al (2016) A holistic view of cancer bioenergetics: mitochondrial function and respiration play fundamental roles in the development and progression of diverse tumors. Clin Transl Med 5:3–3

Alix-Panabières C et al (2017) Molecular portrait of metastasis-competent circulating tumor cells in colon cancer reveals the crucial role of genes regulating Energy metabolism and DNA repair. Clin Chem 63:700–713

Allen E et al (2016) Metabolic symbiosis enables adaptive resistance to anti-angiogenic therapy that is dependent on mTOR signaling. Cell Rep 15:1144–1160

Altman BJ et al (2016) From Krebs to clinic: glutamine metabolism to cancer therapy. Nat Rev Cancer 16:619–634

Amelio I et al (2014) Serine and glycine metabolism in cancer. Trends Biochem Sci 39:191–198

Amrita Devi K et al (2015) ATP Citrate Lyase (ACLY): a promising target for cancer prevention and treatment. Curr Drug Targets 16:156–163

Anastasiou D et al (2011) Inhibition of pyruvate kinase M2 by reactive oxygen species contributes to cellular antioxidant responses. Science 334:1278–1283

Andersen S et al (2015) Organized metabolic crime in prostate cancer: the coexpression of MCT1 in tumor and MCT4 in stroma is an independent prognosticator for biochemical failure. Urol Oncol Semin Orig Investig 33:338.e339–338.e317

Anderson CM, Stahl A (2013) SLC27 fatty acid transport proteins. Mol Asp Med 34:516–528

Ando M et al (2010) Interleukin 6 enhances Glycolysis through expression of the Glycolytic enzymes Hexokinase 2 and 6-Phosphofructo-2-kinase/Fructose-2,6-bisphosphatase-3. J Nippon Med Sch 77:97–105

Asukai K et al (2017) Micro-RNA-130a-3p regulates Gemcitabine resistance via PPARG in Cholangiocarcinoma. Ann Surg Oncol 24:2344–2352

Augsburger F, Szabo C (2018) Potential role of the 3-mercaptopyruvate sulfurtransferase (3-MST)—hydrogen sulfide (H2S) pathway in cancer cells. Pharmacol Res 104083

Augsten M (2014) Cancer-associated fibroblasts as another polarized cell type of the tumor microenvironment. Front Oncol 4:62

Azar S et al (2018) Cellular and molecular characterization of IDH1-mutated diffuse low grade gliomas reveals tumor heterogeneity and absence of EGFR/PDGFRα activation. Glia 66:239–255

Baek G et al (2014) MCT4 defines a glycolytic subtype of pancreatic cancer with poor prognosis and unique metabolic dependencies. Cell Rep 9:2233–2249

Balaban S et al (2017) Adipocyte lipolysis links obesity to breast cancer growth: adipocyte-derived fatty acids drive breast cancer cell proliferation and migration. Cancer Metab 5:1

Ballatori N et al (2009) Glutathione dysregulation and the etiology and progression of human diseases. Biol Chem 390:191–214

Beckner ME et al (2010) Identification of ATP citrate lyase as a positive regulator of glycolytic function in glioblastomas. Int J Cancer 126:2282–2295

Bhattacharyya S et al (2013) Cystathionine beta-synthase (CBS) contributes to advanced ovarian cancer progression and drug resistance. PLoS One 8:e79167–e79167

Bhutia YD, Ganapathy V (2016) Glutamine transporters in mammalian cells and their functions in physiology and cancer. Biochim Biophys Acta 1863:2531–2539

Bianchi MG et al (2012) Valproic acid induces the glutamate transporter excitatory amino acid transporter-3 in human oligodendroglioma cells. Neuroscience 227:260–270

Bianchi MG et al (2014) Changes in the expression of the glutamate transporter EAAT3/EAAC1 in health and disease. Cell Mol Life Sci 71:2001–2015

Biancur DE et al (2017) Compensatory metabolic networks in pancreatic cancers upon perturbation of glutamine metabolism. Nat Commun 8:15965–15965

Bidkhori G et al (2018) Metabolic network-based stratification of hepatocellular carcinoma reveals three distinct tumor subtypes. Proc Natl Acad Sci U S A 115:E11874–E11883

Boidot R et al (2012) Regulation of Monocarboxylate transporter MCT1 expression by p53 mediates inward and outward lactate fluxes in tumors. Cancer Res 72:939–948

Bolzoni M et al (2016) Dependence on glutamine uptake and glutamine addiction characterize myeloma cells: a new attractive target. Blood 128:667–679

Bonito CA et al (2016) Insights into medium-chain Acyl-CoA dehydrogenase structure by molecular dynamics simulations. Chem Biol Drug Des 88:281–292

Bothwell PJ et al (2018) Targeted suppression and knockout of ASCT2 or LAT1 in epithelial and mesenchy-

mal human liver cancer cells fail to inhibit growth. Int J Mol Sci 19:2093
Bourbeau MP, Bartberger MD (2015) Recent advances in the development of Acetyl-CoA Carboxylase (ACC) inhibitors for the treatment of Metabolic disease. J Med Chem 58:525–536
Bräutigam K et al (2011) Combined treatment with TRAIL and PPARγ ligands overcomes chemoresistance of ovarian cancer cell lines. J Cancer Res Clin Oncol 137:875–886
Bröer A et al (2018) Disruption of amino acid homeostasis by novel ASCT2 inhibitors involves multiple targets. Front Pharmacol 9:785–785
Bröer A et al (2019) Ablation of the ASCT2 (SLC1A5) gene encoding a neutral amino acid transporter reveals transporter plasticity and redundancy in cancer cells. J Biol Chem 294:4012–4026
Bruntz RC et al (2019) Inhibition of anaplerotic glutaminolysis underlies selenite toxicity in human lung cancer. Proteomics 0:1800486
Bryant KL et al (2014) KRAS: feeding pancreatic cancer proliferation. Trends Biochem Sci 39:91–100
Cabrera R et al (2011) The crystal complex of phosphofructokinase-2 of Escherichia coli with fructose-6-phosphate: kinetic and structural analysis of the allosteric ATP inhibition. J Biol Chem 286:5774–5783
Calvert AE et al (2017) Cancer-associated IDH1 promotes growth and resistance to targeted therapies in the absence of mutation. Cell Rep 19:1858–1873
Cao A et al (2010) Long chain acyl-CoA synthetase-3 is a molecular target for peroxisome proliferator-activated receptor delta in HepG2 hepatoma cells. J Biol Chem 285:16664–16674
Carduner L et al (2014) Cell cycle arrest or survival signaling through αv integrins, activation of PKC and ERK1/2 lead to anoikis resistance of ovarian cancer spheroids. Exp Cell Res 320:329–342
Carneiro L, Pellerin L (2015) Monocarboxylate transporters: new players in body weight regulation. Obes Rev 16:55–66
Carr EL et al (2010) Glutamine uptake and metabolism are coordinately regulated by ERK/MAPK during T lymphocyte activation. J Immunol 185:1037–1044
Carracedo A et al (2013) Cancer metabolism: fatty acid oxidation in the limelight. Nat Rev Cancer 13:227–232
Carrer A, Wellen KE (2015) Metabolism and epigenetics: a link cancer cells exploit. Curr Opin Biotechnol 34:23–29
Carrer A et al (2017) Impact of a high-fat diet on tissue Acyl-CoA and histone acetylation levels. J Biol Chem 292:3312–3322
Carrer A et al (2019) Acetyl-CoA metabolism supports multistep pancreatic tumorigenesis. Cancer Discov 9:416–435
Carter JC, Church FC (2012) Mature breast adipocytes promote breast cancer cell motility. Exp Mol Pathol 92:312–317
Catalina-Rodriguez O et al (2012) The mitochondrial citrate transporter, CIC, is essential for mitochondrial homeostasis. Oncotarget 3:1220–1235
Chang C-I et al (2001) Macrophage arginase promotes tumor cell growth and suppresses nitric oxide-mediated tumor cytotoxicity. Cancer Res 61:1100–1106
Chen L, Cui H (2015) Targeting glutamine induces apoptosis: a cancer therapy approach. Int J Mol Sci 16:22830–22855
Chen K et al (2014) Integrative metabolome and transcriptome profiling reveals discordant glycolysis process between osteosarcoma and normal osteoblastic cells. J Cancer Res Clin Oncol 140:1715–1721
Chen J et al (2019a) SIRT1 promotes GLUT1 expression and bladder cancer progression via regulation of glucose uptake. Hum Cell 32:193–201
Chen F et al (2019b) Extracellular vesicle-packaged HIF-1α-stabilizing lncRNA from tumour-associated macrophages regulates aerobic glycolysis of breast cancer cells. Nat Cell Biol 21:498–510
Cho H et al (2013) Overexpression of glucose transporter-1 (GLUT-1) predicts poor prognosis in epithelial ovarian cancer. Cancer Investig 31:607–615
Choi J et al (2015) Glioblastoma cells induce differential glutamatergic gene expressions in human tumor-associated microglia/macrophages and monocyte-derived macrophages. Cancer Biol Ther 16:1205–1213
Coffelt SB et al (2009) Tumor-associated macrophages: effectors of angiogenesis and tumor progression. Biochimica et Biophysica Acta (BBA) – Rev Cancer 1796:11–18
Colla R et al (2016) Glutathione-mediated antioxidant response and aerobic metabolism: two crucial factors involved in determining the multi-drug resistance of high-risk neuroblastoma. Oncotarget 7:70715–70737
Combs JA, De Nicola GM (2019) The non-essential amino acid cysteine becomes essential for tumor proliferation and survival. Cancers (Basel) 11:678
Corbet C, Feron O (2017) Cancer cell metabolism and mitochondria: nutrient plasticity for TCA cycle fueling. Biochimica et Biophysica Acta (BBA) – Rev Cancer 1868:7–15
Corbet C et al (2016) Acidosis drives the reprogramming of fatty acid metabolism in cancer cells through changes in mitochondrial and histone acetylation. Cell Metab 24:311–323
Cui H et al (2007) Enhanced expression of asparagine synthetase under glucose-deprived conditions protects pancreatic cancer cells from apoptosis induced by glucose deprivation and cisplatin. Cancer Res 67:3345–3355
Currie E et al (2013) Cellular fatty acid metabolism and cancer. Cell Metab 18:153–161
Curry JM et al (2013) Cancer metabolism, stemness and tumor recurrence: MCT1 and MCT4 are functional biomarkers of metabolic symbiosis in head and neck cancer. Cell Cycle (Georgetown Tex) 12:1371–1384
D'Esposito V et al (2012) Adipocyte-released insulin-like growth factor-1 is regulated by glucose and fatty acids and controls breast cancer cell growth in vitro. Diabetologia 55:2811–2822

DeBerardinis RJ et al (2007) Beyond aerobic glycolysis: transformed cells can engage in glutamine metabolism that exceeds the requirement for protein and nucleotide synthesis. Proc Natl Acad Sci U S A 104:19345–19350
DeClerck YA (2012) Desmoplasia: a response or a Niche? Cancer Discov 2:772–774
Dijkgraaf EM et al (2013) Chemotherapy alters monocyte differentiation to favor generation of cancer-supporting M2 macrophages in the tumor microenvironment. Cancer Res 73:2480–2492
Ding W et al (2010) Platelet-derived growth factor (PDGF)–PDGF receptor interaction activates bone marrow–derived mesenchymal stromal cells derived from chronic lymphocytic leukemia: implications for an angiogenic switch. Blood 116:2984–2993
Dirat B et al (2011) Cancer-associated adipocytes exhibit an activated phenotype and contribute to Breast cancer invasion. Cancer Res 71:2455–2465
Do SK et al (2019) Glucose transporter 3 gene variant is associated with survival outcome of patients with non-small cell lung cancer after surgical resection. Gene 703:58–64
Dodo M et al (2018) Lactate dehydrogenase C is required for the protein expression of a sperm-specific isoform of lactate dehydrogenase A. J Biochem 165:323–334
Doherty JR et al (2014) Blocking lactate export by inhibiting the myc target MCT1 Disables glycolysis and glutathione synthesis. Cancer Res 74:908–920
Dornier E et al (2017) Glutaminolysis drives membrane trafficking to promote invasiveness of breast cancer cells. Nat Commun 8:2255–2255
Doxsee DW et al (2007) Sulfasalazine-induced cystine starvation: potential use for prostate cancer therapy. Prostate 67:162–171
Drayton RM et al (2014) Reduced expression of miRNA-27a modulates cisplatin resistance in bladder cancer by targeting the cystine/glutamate exchanger SLC7A11. Clin Cancer Res 20:1990–2000
Drew BG et al (2015) Estrogen receptor (ER)α-regulated lipocalin 2 expression in adipose tissue links obesity with breast cancer progression. J Biol Chem 290:5566–5581
Du Q et al (2019) PGC1α/CEBPB/CPT1A axis promotes radiation resistance of nasopharyngeal carcinoma through activating fatty acid oxidation. Cancer Sci 110:2050–2062
Ducker GS et al (2017) Human SHMT inhibitors reveal defective glycine import as a targetable metabolic vulnerability of diffuse large B-cell lymphoma. Proc Natl Acad Sci U S A 114:11404–11409
Dufour E et al (2012) Pancreatic tumor sensitivity to plasma L-Asparagine starvation. Pancreas 41:940–948
Echevarría-Vargas IM et al (2014) Upregulation of miR-21 in cisplatin resistant ovarian cancer via JNK-1/c-Jun pathway. PLoS One 9:e97094–e97094
Elhanati S et al (2013) Multiple regulatory layers of SREBP1/2 by SIRT6. Cell Rep 4:905–912
Emmanuel C et al (2014) Genomic classification of serous ovarian cancer with adjacent borderline differentiates RAS pathway and TP53-mutant tumors and identifies NRAS as an oncogenic driver. Clin Cancer Res 20:6618–6630
Enciu A-M et al (2018) Targeting CD36 as biomarker for metastasis prognostic: how far from translation into clinical practice? Biomed Res Int 2018:7801202–7801202
Eng CH et al (2010) Ammonia derived from glutaminolysis is a diffusible regulator of autophagy. Sci Signal 3:ra31–ra31
Fazzari J et al (2015) Inhibitors of glutamate release from breast cancer cells; new targets for cancer-induced bone-pain. Sci Rep 5:8380–8380
Feng M et al (2018) LAT2 regulates glutamine-dependent mTOR activation to promote glycolysis and chemoresistance in pancreatic cancer. J Exp Clin Cancer Res 37:274–274
Fernandes LM et al (2018) Malic enzyme 1 (ME1) is pro-oncogenic in Apc(Min/+) mice. Sci Rep 8.14268–14268
Fiaschi T et al (2012) Reciprocal metabolic reprogramming through lactate shuttle coordinately influences tumor-stroma interplay. Cancer Res 72:5130–5140
Fisher RM, Gertow K (2005) Fatty acid transport proteins and insulin resistance. Curr Opin Lipidol 16:173–178
Frey IM et al (2007) Profiling at mRNA, protein, and metabolite levels reveals alterations in renal amino acid handling and glutathione metabolism in kidney tissue of Pept2−/− mice. Physiol Genomics 28:301–310
Fu M et al (2012) Hydrogen sulfide (H2S) metabolism in mitochondria and its regulatory role in energy production. Proc Natl Acad Sci U S A 109:2943–2948
Fujii S et al (2019) Persulfide synthases that are functionally coupled with translation mediate sulfur respiration in mammalian cells. Br J Pharmacol 176:607–615
Fujino M et al (2016) Expression of glucose transporter-1 is correlated with hypoxia-inducible factor 1α and malignant potential in pancreatic neuroendocrine tumors. Oncol Lett 12:3337–3343
Fujisaki K et al (2015) Cancer-mediated adipose reversion promotes cancer cell migration via IL-6 and MCP-1. Breast Cancer Res Treat 150:255–263
Fujiwara N et al (2018) CPT2 downregulation adapts HCC to lipid-rich environment and promotes carcinogenesis via acylcarnitine accumulation in obesity. Gut 67:1493–1504
Fukuda S et al (2015) Pyruvate Kinase M2 modulates esophageal squamous cell carcinoma chemotherapy response by regulating the pentose phosphate pathway. Ann Surg Oncol 22:1461–1468
Gaglio D et al (2011) Oncogenic K-Ras decouples glucose and glutamine metabolism to support cancer cell growth. Mol Syst Biol 7:523–523
Gai J-W et al (2016) Expression profile of hydrogen sulfide and its synthases correlates with tumor stage and grade in urothelial cell carcinoma of bladder. Urol Oncol Semin Orig Investig 34:166.e115–166.e120
Galan-Cobo A et al (2019) LKB1 and KEAP1/NRF2 pathways cooperatively promote metabolic reprogramming with enhanced glutamine dependence in

<em>KRAS</em>-mutant lung adenocarcinoma. Cancer Res 79:3251–3267
Gào X, Schöttker B (2017) Reduction-oxidation pathways involved in cancer development: a systematic review of literature reviews. Oncotarget 8:51888–51906
Gao P et al (2009) c-Myc suppression of miR-23a/b enhances mitochondrial glutaminase expression and glutamine metabolism. Nature 458:762–765
Gao Y et al (2017) TNFα-YAP/p65-HK2 axis mediates breast cancer cell migration. Oncogene 6:e383–e383
Gillies RJ, Gatenby RA (2015) Metabolism and its sequelae in cancer evolution and therapy. Cancer J 21:88–96
Giuffrè A, Vicente JB (2018) Hydrogen sulfide biochemistry and interplay with other gaseous mediators in mammalian physiology. Oxidative Med Cell Longev 2018:6290931–6290931
Giuliani N et al (2017) The potential of inhibiting glutamine uptake as a therapeutic target for multiple myeloma. Expert Opin Ther Targets 21:231–234
Gordon S, Taylor PR (2005) Monocyte and macrophage heterogeneity. Nat Rev Immunol 5:953–964
Grasmann G et al (2019) Gluconeogenesis in cancer cells – repurposing of a starvation-induced metabolic pathway? Biochimica et Biophysica Acta (BBA) – Rev Cancer 1872:24–36
Gregory MA et al (2019) Targeting glutamine metabolism and Redox state for Leukemia therapy. Clin Cancer Res 25:4079–4090
Grisouard J et al (2011) Targeting AMP-activated protein kinase in adipocytes to modulate obesity-related adipokine production associated with insulin resistance and breast cancer cell proliferation. Diabetol Metab Syndr 3:16–16
Gu Y et al (2017) mTORC2 regulates amino acid metabolism in cancer by phosphorylation of the cystine-glutamate antiporter xCT. Mol Cell 67:128–138. e127
Guaita-Esteruelas S et al (2017) Adipose-derived fatty acid-binding proteins plasma concentrations are increased in Breast cancer patients. Oncologist 22:1309–1315
Gui DY et al (2013) Allosteric regulation of PKM2 allows cellular adaptation to different physiological states. Science Signal 6:pe7–pe7
Guido C et al (2012) Mitochondrial fission induces glycolytic reprogramming in cancer-associated myofibroblasts, driving stromal lactate production, and early tumor growth. Oncotarget 3:798–810
Guppy M et al (2002) Contribution by different fuels and metabolic pathways to the total ATP turnover of proliferating MCF-7 breast cancer cells. Biochem J 364:309–315
Gurrapu S et al (2015) Monocarboxylate transporter 1 inhibitors as potential anticancer agents. ACS Med Chem Lett 6:558–561
Habib E et al (2015) Expression of xCT and activity of system xc(-) are regulated by NRF2 in human breast cancer cells in response to oxidative stress. Redox Biol 5:33–42
Hajiahmadi S et al (2015) Activation of A2b adenosine receptor regulates ovarian cancer cell growth: involvement of Bax/Bcl-2 and caspase-3. Biochem Cell Biol 93:321–329
Halestrap AP (2013) The SLC16 gene family – structure, role and regulation in health and disease. Mol Asp Med 34:337–349
Hamann I et al (2018) Expression and function of hexose transporters GLUT1, GLUT2, and GLUT5 in breast cancer—effects of hypoxia. FASEB J 32:5104–5118
Hanai J-I et al (2012) Inhibition of lung cancer growth: ATP citrate lyase knockdown and statin treatment leads to dual blockade of mitogen-activated protein kinase (MAPK) and phosphatidylinositol-3-kinase (PI3K)/AKT pathways. J Cell Physiol 227:1709–1720
Harris IS et al (2015) Glutathione and thioredoxin antioxidant pathways synergize to drive cancer initiation and progression. Cancer Cell 27:211–222
Hausheer FH et al (2011) Mechanistic study of BNP7787-mediated cisplatin nephroprotection: modulation of human aminopeptidase N. Cancer Chemother Pharmacol 67:381–391
Hensley CT et al (2013) Glutamine and cancer: cell biology, physiology, and clinical opportunities. J Clin Invest 123:3678–3684
Hernández-Juárez J et al (2019) Sodium-coupled monocarboxylate transporter is a target of epigenetic repression in cervical cancer. Int J Oncol 54:1613–1624
Hlouschek J et al (2018) The mitochondrial citrate carrier (SLC25A1) sustains redox homeostasis and mitochondrial metabolism supporting radioresistance of cancer cells with tolerance to cycling severe hypoxia. Front Oncol 8:170
Hong X et al (2014) PTEN antagonises Tcl1/hnRNPK-mediated G6PD pre-mRNA splicing which contributes to hepatocarcinogenesis. Gut 63:1635–1647
Hong SM et al (2019) Lactic acidosis caused by repressed lactate dehydrogenase subunit B expression down-regulates mitochondrial oxidative phosphorylation via the pyruvate dehydrogenase (PDH)-PDH kinase axis. J Biol Chem 294:7810–7820
Hopperton KE et al (2014) Fatty acid synthase plays a role in cancer metabolism beyond providing fatty acids for phospholipid synthesis or sustaining elevations in glycolytic activity. Exp Cell Res 320:302–310
Hou Z et al (2011) Macrophages induce COX-2 expression in breast cancer cells: role of IL-1β autoamplification. Carcinogenesis 32:695–702
Hua TNM et al (2019) Inhibition of oncogenic Src induces FABP4-mediated lipolysis via PPARγ activation exerting cancer growth suppression. EBioMedicine 41:134–145
Huang D et al (2014) HIF-1-mediated suppression of Acyl-CoA dehydrogenases and fatty acid oxidation is critical for cancer progression. Cell Rep 8:1930–1942
Hung YP, Yellen G (2014) Live-cell imaging of cytosolic NADH-NAD+ redox state using a genetically encoded fluorescent biosensor. Methods Mol Biol 1071:83–95

Hung YP et al (2011) Imaging cytosolic NADH-NAD(+) redox state with a genetically encoded fluorescent biosensor. Cell Metab 14:545–554

Hwang RF et al (2008) Cancer-associated stromal fibroblasts promote pancreatic tumor progression. Cancer Res 68:918–926

Iershov A et al (2019) The class 3 PI3K coordinates autophagy and mitochondrial lipid catabolism by controlling nuclear receptor PPARα. Nat Commun 10:1566–1566

Infantino V et al (2014) A key role of the mitochondrial citrate carrier (SLC25A1) in TNFα- and IFNγ-triggered inflammation. Biochim Biophys Acta 1839:1217–1225

Israelsen WJ et al (2013) PKM2 isoform-specific deletion reveals a differential requirement for pyruvate kinase in tumor cells. Cell 155:397–409

Iwamoto M et al (2014) Regulation of 18F-FDG accumulation in colorectal cancer cells with mutated KRAS. J Nucl Med 55:2038–2044

Iwasaki K et al (2015) Role of hypoxia-inducible factor-1α, carbonic anhydrase-IX, glucose transporter-1 and vascular endothelial growth factor associated with lymph node metastasis and recurrence in patients with locally advanced cervical cancer. Oncol Lett 10:1970–1978

Ji X et al (2018) xCT (SLC7A11)-mediated metabolic reprogramming promotes non-small cell lung cancer progression. Oncogene 37:5007–5019

Jia D et al (2019) Elucidating cancer metabolic plasticity by coupling gene regulation with metabolic pathways. Proc Natl Acad Sci 116:3909–3918

Jiang L et al (2017) Quantitative metabolic flux analysis reveals an unconventional pathway of fatty acid synthesis in cancer cells deficient for the mitochondrial citrate transport protein. Metab Eng 43:198–207

Jiang J et al (2018) Asparagine, a critical limiting metabolite during glutamine starvation. Mol Cell Oncol 5:e1441633–e1441633

Jin L, Zhou Y (2019) Crucial role of the pentose phosphate pathway in malignant tumors. Oncol Lett 17:4213–4221

José MM et al (2019) Metabolic reprogramming of cancer by chemicals that target glutaminase isoenzymes. Curr Med Chem 26:1–23

Kalinina EV et al (2014) Role of glutathione, glutathione transferase, and glutaredoxin in regulation of redox-dependent processes. Biochem Biokhimiia 79:1562–1583

Kang ES et al (2017) xCT deficiency aggravates acetaminophen-induced hepatotoxicity under inhibition of the transsulfuration pathway. Free Radic Res 51:80–90

Karagiannis GS et al (2012) Cancer-associated fibroblasts drive the progression of metastasis through both Paracrine and mechanical pressure on cancer tissue. Mol Cancer Res 10:1403–1418

Kim YH et al (2013) Factors Affecting 18F-FDG uptake by metastatic lymph nodes in gastric cancer. J Comput Assist Tomogr 37:815–819

Kim YH et al (2017) SLC2A2 (GLUT2) as a novel prognostic factor for hepatocellular carcinoma. Oncotarget 8:68381–68392

Kitisin K et al (2011) Presentation and outcomes of hepatocellular carcinoma patients at a Western centre. HPB (Oxford) 13:712–722

Kleszcz R et al (2018) The inhibition of c-MYC transcription factor modulates the expression of glycolytic and glutaminolytic enzymes in FaDu hypopharyngeal carcinoma cells. Adv Clin Exp Med Off Organ Wroclaw Med Univ 27:735–742

Knudsen ES et al (2016) Unique metabolic features of pancreatic cancer stroma: relevance to the tumor compartment, prognosis, and invasive potential. Oncotarget 7:78396–78411

Ko Y-H et al (2011) Glutamine fuels a vicious cycle of autophagy in the tumor stroma and oxidative mitochondrial metabolism in epithelial cancer cells. Cancer Biol Ther 12:1085–1097

Koh KX et al (2017) Acquired resistance to PI3K/mTOR inhibition is associated with mitochondrial DNA mutation and glycolysis. Oncotarget 8:110133–110144

Kolukula VK et al (2014) SLC25A1, or CIC, is a novel transcriptional target of mutant p53 and a negative tumor prognostic marker. Oncotarget 5:1212–1225

Kong L et al (2016) Expression of lactate dehydrogenase C in MDAMB231 cells and its role in tumor invasion and migration. Mol Med Rep 13:3533–3538

Koochekpour S et al (2012) Serum glutamate levels correlate with Gleason score and glutamate blockade decreases proliferation, migration, and invasion and induces apoptosis in prostate cancer cells. Clin Cancer Res 18:5888–5901

Koppula P et al (2017) The glutamate/cystine antiporter SLC7A11/xCT enhances cancer cell dependency on glucose by exporting glutamate. J Biol Chem 292:14240–14249

Koprivica I et al (2019) Ethyl pyruvate stimulates regulatory T cells and ameliorates type 1 diabetes development in mice. Front Immunol 9:3130–3130

Kowalik MA et al (2016) Metabolic reprogramming identifies the most aggressive lesions at early phases of hepatic carcinogenesis. Oncotarget 7:32375–32393

Kuo T-C et al (2016) Glutaminase 2 stabilizes dicer to repress snail and metastasis in hepatocellular carcinoma cells. Cancer Lett 383:282–294

Lally JSV et al (2019) Inhibition of acetyl-CoA carboxylase by phosphorylation or the inhibitor ND-654 suppresses lipogenesis and hepatocellular carcinoma. Cell Metab 29:174–182.e175

Lao-On U et al (2018) Roles of pyruvate carboxylase in human diseases: from diabetes to cancers and infection. J Mol Med 96:237–247

Le A et al (2012) Glucose-independent glutamine metabolism via TCA cycling for proliferation and survival in B cells. Cell Metab 15:110–121

Lee H-W et al (2013) Recruitment of monocytes/macrophages in different tumor microenvironments. Biochimica et Biophysica Acta (BBA) – Rev Cancer 1835:170–179

Lee J-H et al (2017) Stabilization of phosphofructokinase 1 platelet isoform by AKT promotes tumorigenesis. Nat Commun 8:949–949
Lemire J et al (2008) Mitochondrial lactate dehydrogenase is involved in oxidative-energy metabolism in human astrocytoma cells (CCF-STTG1). PLoS One 3:e1550–e1550
Li T, Le A (2018) Glutamine metabolism in cancer. In: Le A (ed) The heterogeneity of cancer metabolism. Springer, Cham, pp 13–32
Li H et al (2015) Cancer-associated fibroblasts provide a suitable microenvironment for tumor development and progression in oral tongue squamous cancer. J Transl Med 13:198
Li S et al (2016) Inhibition of mTOR complex 2 induces GSK3/FBXW7-dependent degradation of sterol regulatory element-binding protein 1 (SREBP1) and suppresses lipogenesis in cancer cells. Oncogene 35:642–650
Li S et al (2018a) Acidic pHe regulates cytoskeletal dynamics through conformational integrin β1 activation and promotes membrane protrusion. Biochim Biophys Acta (BBA) – Mol Basis Dis 1864:2395–2408
Li Q et al (2018b) HSCs-derived COMP drives hepatocellular carcinoma progression by activating MEK/ERK and PI3K/AKT signaling pathways. J Exp Clin Cancer Res 37:231–231
Li L et al (2019a) High developmental pluripotency-associated 4 expression promotes cell proliferation and glycolysis, and predicts poor prognosis in non-small-cell lung cancer. Mol Med Rep 20:445–454
Li J et al (2019b) miR-145 inhibits glutamine metabolism through c-myc/GLS1 pathways in ovarian cancer cells. Cell Biol Int 43:921–930
Li B et al (2019c) Targeting glutaminase 1 attenuates stemness properties in hepatocellular carcinoma by increasing reactive oxygen species and suppressing Wnt/beta-catenin pathway. EBioMedicine 39:239–254
Liang Y et al (2018) CD36 plays a critical role in proliferation, migration and tamoxifen-inhibited growth of ER-positive breast cancer cells. Oncogene 7:98–98
Lien EC et al (2016) Glutathione biosynthesis is a metabolic vulnerability in PI(3)K/Akt-driven breast cancer. Nat Cell Biol 18:572–578
Lien EC et al (2017) Oncogenic PI3K promotes methionine dependency in breast cancer cells through the cystine-glutamate antiporter xCT. Sci Signal 10:eaao6604
Lim JKM et al (2019) Cystine/glutamate antiporter xCT (SLC7A11) facilitates oncogenic RAS transformation by preserving intracellular redox balance. Proc Natl Acad Sci U S A 116:9433–9442
Lin M-H, Miner JH (2015) Fatty acid transport protein 1 can compensate for fatty acid transport protein 4 in the developing mouse epidermis. J Invest Dermatol 135:462–470
Lin EY et al (2001) Colony-stimulating factor 1 promotes progression of mammary tumors to malignancy. J Exp Med 193:727–740
Lin R et al (2013) Acetylation stabilizes ATP-citrate lyase to promote lipid biosynthesis and tumor growth. Mol Cell 51:506–518
Lin M et al (2018) Downregulation of CPT2 promotes tumorigenesis and chemoresistance to cisplatin in hepatocellular carcinoma. Onco Targets Ther 11:3101–3110
Liu R-Z et al (2011) Association of FABP5 expression with poor survival in triple-negative breast cancer: implication for retinoic acid therapy. Am J Pathol 178:997–1008
Liu F-L et al (2016) Autophagy is involved in TGF-β1-induced protective mechanisms and formation of cancer-associated fibroblasts phenotype in tumor microenvironment. Oncotarget 7:4122–4141
Liu D et al (2017) Comprehensive proteomics analysis reveals metabolic reprogramming of tumor-associated macrophages stimulated by the tumor microenvironment. J Proteome Res 16:288–297
Liu M et al (2018) Tumor-suppressing effects of microRNA-612 in bladder cancer cells by targeting malic enzyme 1 expression. Int J Oncol 52:1923–1933
Liu J et al (2019) Peroxisomal regulation of redox homeostasis and adipocyte metabolism. Redox Biol 24:101167–101167
Lo M et al (2008) The x cystine/glutamate antiporter: a potential target for therapy of cancer and other diseases. J Cell Physiol 215:593–602
Lomelino CL et al (2017) Asparagine synthetase: function, structure, and role in disease. J Biol Chem 292:19952–19958
Lopes-Coelho F et al (2016) HNF1β drives glutathione (GSH) synthesis underlying intrinsic carboplatin resistance of ovarian clear cell carcinoma (OCCC). Tumor Biol 37:4813–4829
Lopes-Coelho F et al (2017) Monocarboxylate transporter 1 (MCT1), a tool to stratify acute myeloid leukemia (AML) patients and a vehicle to kill cancer cells. Oncotarget 8:82803–82823
Lopes-Coelho F et al (2018) Breast cancer metabolic cross-talk: fibroblasts are hubs and breast cancer cells are gatherers of lipids. Mol Cell Endocrinol 462:93–106
Lu Y-X et al (2018) ME1 regulates NADPH homeostasis to promote gastric cancer growth and metastasis. Cancer Res 78:1972–1985
Lu X et al (2019) Metabolic profiling analysis upon acylcarnitines in tissues of hepatocellular carcinoma revealed the inhibited carnitine shuttle system caused by the downregulated carnitine palmitoyltransferase 2. Mol Carcinog 58:749–759
Luebker SA, Koepsell SA (2019) Diverse mechanisms of BRAF inhibitor resistance in melanoma identified in clinical and preclinical studies. Front Oncol 9:268–268
Lukey MJ et al (2013) Therapeutic strategies impacting cancer cell glutamine metabolism. Future Med Chem 5:1685–1700

Luo W et al (2011) Pyruvate kinase M2 is a PHD3-stimulated coactivator for hypoxia-inducible factor 1. Cell 145:732–744
Lv Q et al (2019) FABP5 regulates the proliferation of clear cell renal cell carcinoma cells via the PI3K/AKT signaling pathway. Int J Oncol 54:1221–1232
Lyssiotis CA, Cantley LC (2014) Acetate fuels the cancer engine. Cell 159:1492–1494
Mackenzie R et al (2015) Targeted deep sequencing of mucinous ovarian tumors reveals multiple overlapping RAS-pathway activating mutations in borderline and cancerous neoplasms. BMC Cancer 15:415–415
Madunić IV et al (2018) Sodium-glucose cotransporters: new targets of cancer therapy? Arhiv za higijenu rada i toksikologiju 69:278
Manabe Y et al (2003) Mature adipocytes, but not preadipocytes, promote the growth of breast carcinoma cells in collagen gel matrix culture through cancer–stromal cell interactions. J Pathol 201:221–228
Mantovani A, Sica A (2010) Macrophages, innate immunity and cancer: balance, tolerance, and diversity. Curr Opin Immunol 22:231–237
Mantovani A et al (2006) Role of tumor-associated macrophages in tumor progression and invasion. Cancer Metastasis Rev 25:315–322
Mao Y et al (2013) Stromal cells in tumor microenvironment and breast cancer. Cancer Metastasis Rev 32:303–315
Mao A et al (2019) KLF8 is associated with poor prognosis and regulates glycolysis by targeting GLUT4 in gastric cancer. J Cell Mol Med 23:5087–5097
Marani M et al (2016) A pyrazolopyran derivative preferentially inhibits the activity of human cytosolic serine hydroxymethyltransferase and induces cell death in lung cancer cells. Oncotarget 7:4570–4583
Marchiq I, Pouysségur J (2016) Hypoxia, cancer metabolism and the therapeutic benefit of targeting lactate/H+ symporters. J Mol Med 94:155–171
Marino SM, Gladyshev VN (2012) Analysis and functional prediction of reactive cysteine residues. J Biol Chem 287:4419–4425
Martinez-Outschoorn UE et al (2010) Autophagy in cancer associated fibroblasts promotes tumor cell survival. Cell Cycle 9:3515–3533
Martinez-Outschoorn UE et al (2011) Cancer cells metabolically "fertilize" the tumor microenvironment with hydrogen peroxide, driving the Warburg effect. Cell Cycle 10:2504–2520
Martinez-Outschoorn UE et al (2014) Catabolic cancer-associated fibroblasts transfer energy and biomass to anabolic cancer cells, fueling tumor growth. Semin Cancer Biol 25:47–60
Mashek DG et al (2004) Revised nomenclature for the mammalian long-chain acyl-CoA synthetase gene family. J Lipid Res 45:1958–1961
Mashima T et al (2009) Acyl-CoA synthetase as a cancer survival factor: its inhibition enhances the efficacy of etoposide. Cancer Sci 100:1556–1562
Mashimo T et al (2014) Acetate is a bioenergetic substrate for human glioblastoma and brain metastases. Cell 159:1603–1614
Mates JM et al (2013) Glutaminase isoenzymes as key regulators in metabolic and oxidative stress against cancer. Curr Mol Med 13:514–534
Matés JM et al (2018) Glutaminase isoenzymes in the metabolic therapy of cancer. Biochimica et Biophysica Acta (BBA) – Rev Cancer 1870:158–164
Mathupala SP et al (2009) Hexokinase-2 bound to mitochondria: cancer's stygian link to the "Warburg effect" and a pivotal target for effective therapy. Semin Cancer Biol 19:17–24
McBrian MA et al (2013) Histone acetylation regulates intracellular pH. Mol Cell 49:310–321
Mele L et al (2019) Glucose-6-phosphate dehydrogenase blockade potentiates tyrosine kinase inhibitor effect on breast cancer cells through autophagy perturbation. J Exp Clin Cancer Res 38:160–160
Menard JA et al (2016) Metastasis stimulation by hypoxia and acidosis-induced extracellular lipid uptake is mediated by proteoglycan-dependent endocytosis. Cancer Res 76:4828–4840
Menendez JA, Lupu R (2007) Fatty acid synthase and the lipogenic phenotype in cancer pathogenesis. Nat Rev Cancer 7:763
Menendez JA et al (2016) The metastasis inducer CCN1 (CYR61) activates the fatty acid synthase (FASN)-driven lipogenic phenotype in breast cancer cells. Oncoscience 3:242–257
Metallo CM et al (2011) Reductive glutamine metabolism by IDH1 mediates lipogenesis under hypoxia. Nature 481:380–384
Mieyal JJ et al (2008) Molecular mechanisms and clinical implications of reversible protein S-glutathionylation. Antioxid Redox Signal 10:1941–1988
Milger K et al (2006) Cellular uptake of fatty acids driven by the ER-localized acyl-CoA synthetase FATP4. J Cell Sci 119:4678–4688
Módis K et al (2013) Hydrogen sulfide-mediated stimulation of mitochondrial electron transport involves inhibition of the mitochondrial phosphodiesterase 2A, elevation of cAMP and activation of protein kinase A. Biochem Pharmacol 86:1311–1319
Moellering RE et al (2008) Acid treatment of melanoma cells selects for invasive phenotypes. Clin Exp Metastasis 25:411–425
Monroe GR et al (2019) Identification of human D lactate dehydrogenase deficiency. Nat Commun 10:1477–1477
Moon J-S et al (2011) Krüppel-like factor 4 (KLF4) activates the transcription of the gene for the platelet isoform of phosphofructokinase (PFKP) in breast cancer. J Biol Chem 286:23808–23816
Mossmann D et al (2018) mTOR signalling and cellular metabolism are mutual determinants in cancer. Nat Rev Cancer 18:744–757
Mroczko B et al (2007) Serum macrophage-colony stimulating factor levels in colorectal cancer patients cor-

relate with lymph node metastasis and poor prognosis. Clin Chim Acta 380:208–212
Munir R et al (2019) Lipid metabolism in cancer cells under metabolic stress. Br J Cancer 120:1090–1098
Nagao K et al (2018) Fatty acid binding protein 7 may be a marker and therapeutic targets in clear cell renal cell carcinoma. BMC Cancer 18:1114–1114
Nakashima C et al (2018) Expression of cytosolic malic enzyme (ME1) is associated with disease progression in human oral squamous cell carcinoma. Cancer Sci 109:2036–2045
Netto LES et al (2007) Reactive cysteine in proteins: protein folding, antioxidant defense, redox signaling and more. Comp Biochem Physiol Part C Toxicol Pharmacol 146:180–193
Nguyen XC et al (2008) High correlations between primary tumours and loco-regional metastatic lymph nodes in non-small-cell lung cancer with respect to glucose transporter type 1-mediated 2-deoxy-2-F18-fluoro-d-glucose uptake. Eur J Cancer 44:692–698
Nieman KM et al (2011) Adipocytes promote ovarian cancer metastasis and provide energy for rapid tumor growth. Nat Med 17:1498–1503
Nunes SC, Serpa J (2018) Glutathione in ovarian cancer: a double-edged sword. Int J Mol Sci 19:1882
Nunes SC et al (2018a) Cysteine boosters the evolutionary adaptation to CoCl(2) mimicked hypoxia conditions, favouring carboplatin resistance in ovarian cancer. BMC Evol Biol 18:97–97
Nunes SC et al (2018b) Cysteine allows ovarian cancer cells to adapt to hypoxia and to escape from carboplatin cytotoxicity. Sci Rep 8:9513–9513
Okuno S et al (2003) Role of cystine transport in intracellular glutathione level and cisplatin resistance in human ovarian cancer cell lines. Br J Cancer 88:951–956
Orimo A, Weinberg RA (2006) Stromal fibroblasts in cancer: a novel tumor-promoting cell type. Cell Cycle 5:1597–1601
Pan Y et al (2015) Radiation exposure promotes hepatocarcinoma cell invasion through epithelial mesenchymal transition mediated by H₂S/CSE pathway. Radiat Res 185(96–105):110
Pan J et al (2019) CD36 mediates palmitate acid-induced metastasis of gastric cancer via AKT/GSK-3β/β--catenin pathway. J Exp Clin Cancer Res 38:52–52
Panza E et al (2015) Role of the cystathionine γ lyase/hydrogen sulfide pathway in human melanoma progression. Pigment Cell Melanoma Res 28:61–72
Paradise RK et al (2011) Acidic extracellular pH promotes activation of integrin α(v)β(3). PLoS One 6:e15746–e15746
Park Y-Y et al (2013) Tat-activating regulatory DNA-binding protein regulates glycolysis in hepatocellular carcinoma by regulating the platelet isoform of phosphofructokinase through microRNA 520. Hepatology 58:182–191
Patra KC, Hay N (2014) The pentose phosphate pathway and cancer. Trends Biochem Sci 39:347–354
Pattanayak SP et al (2018) Bergapten inhibits liver carcinogenesis by modulating LXR/PI3K/Akt and IDOL/LDLR pathways. Biomed Pharmacother 108:297–308
Pavlides S et al (2009) The reverse Warburg effect: aerobic glycolysis in cancer associated fibroblasts and the tumor stroma. Cell Cycle 8:3984–4001
Pavlova NN, Thompson CB (2016) The emerging hallmarks of cancer metabolism. Cell Metab 23:27–47
Pedraz-Cuesta E et al (2015) The glutamate transport inhibitor DL-Threo-β-Benzyloxyaspartic acid (DL-TBOA) differentially affects SN38- and oxaliplatin-induced death of drug-resistant colorectal cancer cells. BMC Cancer 15:411–411
Penny HL et al (2016) Warburg metabolism in tumor-conditioned macrophages promotes metastasis in human pancreatic ductal adenocarcinoma. Oncoimmunology 5:e1191731–e1191731
Pérez-Miguelsanz J et al (2017) Betaine homocysteine S-methyltransferase emerges as a new player of the nuclear methionine cycle. Biochimica et Biophysica Acta (BBA) – Mol Cell Res 1864:1165–1182
Pértega-Gomes N et al (2014) A lactate shuttle system between tumour and stromal cells is associated with poor prognosis in prostate cancer. BMC Cancer 14:352
Picon-Ruiz M et al (2016) Interactions between adipocytes and breast cancer cells stimulate cytokine production and drive Src/Sox2/miR-302b–mediated malignant progression. Cancer Res 76:491–504
Pissimissis N et al (2009) The Glutamatergic system expression in human PC-3 and LNCaP prostate cancer cells. Anticancer Res 29:371–377
Poisson LM et al (2015) A metabolomic approach to identifying platinum resistance in ovarian cancer. J Ovarian Res 8:13–13
Pollard JW (2009) Trophic macrophages in development and disease. Nat Rev Immunol 9:259–270
Porstmann T et al (2008) SREBP activity is regulated by mTORC1 and contributes to Akt-dependent cell growth. Cell Metab 8:224–236
Potts A et al (2018) Cytosolic phosphoenolpyruvate carboxykinase as a cataplerotic pathway in the small intestine. Am J Physiol Gastrointest Liver Physiol 315:G249–G258
Qiao A et al (2016) Breast cancer-associated fibroblasts: their roles in tumor initiation, progression and clinical applications. Front Med 10:33–40
Rajasinghe LD et al (2019) Delta-tocotrienol modulates glutamine dependence by inhibiting ASCT2 and LAT1 transporters in non-small cell lung cancer (NSCLC) cells: a metabolomic approach. Metabolites 9:50
Ramapriyan R et al (2019) Altered cancer metabolism in mechanisms of immunotherapy resistance. Pharmacol Ther 195:162–171
Ramos-Martinez JI (2017) The regulation of the pentose phosphate pathway: remember Krebs. Arch Biochem Biophys 614:50–52
Read JA et al (2001) Structural basis for altered activity of M- and H-isozyme forms of human lactate dehydrogenase. Proteins 43:175–185

Reis LMD et al (2019) Dual inhibition of glutaminase and carnitine palmitoyltransferase decreases growth and migration of glutaminase inhibition–resistant triple-negative breast cancer cells. J Biol Chem 294:9342–9357

Reynolds MR et al (2013) Control of glutamine metabolism by the tumor suppressor Rb. Oncogene 33:556

Reynolds MR et al (2014) Control of glutamine metabolism by the tumor suppressor Rb. Oncogene 33:556–566

Ricci M et al (2018) PPARs are mediators of anti-cancer properties of superparamagnetic iron oxide nanoparticles (SPIONs) functionalized with conjugated linoleic acid. Chem Biol Interact 292:9–14

Riganti C et al (2012) The pentose phosphate pathway: an antioxidant defense and a crossroad in tumor cell fate. Free Radic Biol Med 53:421–436

Rodríguez-Enríquez S et al (2000) Substrate oxidation and ATP supply in AS-30D hepatoma cells. Arch Biochem Biophys 375:21–30

Rodríguez-Enríquez S et al (2006) Control of cellular proliferation by modulation of oxidative phosphorylation in human and rodent fast-growing tumor cells. Toxicol Appl Pharmacol 215:208–217

Rohani N et al (2019) Acidification of tumor at stromal boundaries drives transcriptome alterations associated with aggressive phenotypes. Cancer Res 79:1952–1966

Romero R et al (2017) Keap1 loss promotes Kras-driven lung cancer and results in dependence on glutaminolysis. Nat Med 23:1362–1368

Rozovski U et al (2018) STAT3-activated CD36 facilitates fatty acid uptake in chronic lymphocytic leukemia cells. Oncotarget 9:21268–21280

Rudolph MC et al (2012) Mammalian fatty acid synthase activity from crude tissue lysates tracing ^{13}C-labeled substrates using gas chromatography-mass spectrometry. Anal Biochem 428:158–166

Saha SK et al (2019) Multiomics analysis reveals that GLS and GLS2 differentially modulate the clinical outcomes of cancer. J Clin Med 8:355

Sanità P et al (2014) Tumor-stroma metabolic relationship based on lactate shuttle can sustain prostate cancer progression. BMC Cancer 14:154–154

Santander AM et al (2015) Paracrine interactions between adipocytes and tumor cells recruit and modify macrophages to the mammary tumor microenvironment: the role of Obesity and inflammation in breast adipose tissue. Cancers (Basel) 7:143–178

Santi A et al (2013) The effects of CA IX catalysis products within tumor microenvironment. Cell Commun Signal 11:81–81

Sarfraz I et al (2018) Malic enzyme 2 as a potential therapeutic drug target for cancer. IUBMB Life 70:1076–1083

Sato-Tadano A et al (2013) Hexokinase II in breast carcinoma: a potent prognostic factor associated with hypoxia-inducible factor-1α and Ki-67. Cancer Sci 104:1380–1388

Sawayama H et al (2019) Glucose transporter 1 regulates the proliferation and cisplatin sensitivity of esophageal cancer. Cancer Sci 110:1705–1714

Scheepers A et al (2004) The glucose transporter families SGLT and GLUT: molecular basis of normal and aberrant function. J Parenter Enter Nutr 28:364–371

Sekiguchi F et al (2016) Endogenous hydrogen sulfide enhances cell proliferation of human gastric cancer AGS cells. Biol Pharm Bull 39:887–890

Selvarajah B et al (2019) mTORC1 amplifies the ATF4-dependent de novo serine-glycine pathway to supply glycine during TGF-β₁–induced collagen biosynthesis. Sci Signal 12:eaav3048

Sen S et al (2015) Role of cystathionine β-synthase in human breast Cancer. Free Radic Biol Med 86:228–238

Shiozaki A et al (2014) xCT, component of cysteine/glutamate transporter, as an independent prognostic factor in human esophageal squamous cell carcinoma. J Gastroenterol 49:853–863

Shpilberg Y et al (2015) The direct and indirect effects of corticosterone and primary adipose tissue on MCF7 breast cancer cell cycle progression. In: Hormone molecular Biology and clinical investigation. De Gruyter, Berlin, p 91

Sikder MOF et al (2017) The Na+/Cl−-coupled, broad-specific, amino acid transporter SLC6A14 (ATB0,+): emerging roles in multiple diseases and therapeutic potential for treatment and diagnosis. AAPS J 20:12

Silva LS et al (2016) STAT3:FOXM1 and MCT1 drive uterine cervix carcinoma fitness to a lactate-rich microenvironment. Tumor Biol 37:5385–5395

Singh R, Cuervo AM (2012) Lipophagy: connecting autophagy and lipid metabolism. Int J Cell Biol 2012:282041–282041

Singh R et al (2009) Autophagy regulates lipid metabolism. Nature 458:1131–1135

Son J et al (2013) Glutamine supports pancreatic cancer growth through a KRAS-regulated metabolic pathway. Nature 496:101–105

Stahl A et al (2002) Insulin causes fatty acid transport Protein translocation and enhanced fatty Acid uptake in adipocytes. Dev Cell 2:477–488

Stepulak A et al (2014) Glutamate and its receptors in cancer. J Neural Transm (Vienna) 121:933–944

Still ER, Yuneva MO (2017) Hopefully devoted to Q: targeting glutamine addiction in cancer. Br J Cancer 116:1375–1381

Stincone A et al (2015) The return of metabolism: biochemistry and physiology of the pentose phosphate pathway. Biol Rev Camb Philos Soc 90:927–963

Storch J, McDermott L (2009) Structural and functional analysis of fatty acid-binding proteins. J Lipid Res 50(Suppl):S126–S131

Su Y et al (2013) Id1 enhances human ovarian cancer endothelial progenitor cell angiogenesis via PI3K/Akt and NF-κB/MMP-2 signaling pathways. J Transl Med 11:132–132

Sun L et al (2017) Decreased expression of acetyl-CoA synthase 2 promotes metastasis and predicts poor prognosis in hepatocellular carcinoma. Cancer Sci 108:1338–1346
Sun Q et al (2018) Hypermethylated CD36 gene affected the progression of lung cancer. Gene 678:395–406
Sun T et al (2019) Anoikis resistant mediated by FASN promoted growth and metastasis of osteosarcoma. Cell Death Dis 10:298–298
Sung YK et al (2007) Regulation of cell growth by fatty acid-CoA ligase 4 in human hepatocellular carcinoma cells. Exp Amp Mol Med 39:477
Suzuki S et al (2010) Phosphate-activated glutaminase (GLS2), a p53-inducible regulator of glutamine metabolism and reactive oxygen species. Proc Natl Acad Sci U S A 107:7461–7466
Swierczynski J et al (2014) Role of abnormal lipid metabolism in development, progression, diagnosis and therapy of pancreatic cancer. World J Gastroenterol 20:2279–2303
Szabo C et al (2013) Tumor-derived hydrogen sulfide, produced by cystathionine-β-synthase, stimulates bioenergetics, cell proliferation, and angiogenesis in colon cancer. Proc Natl Acad Sci U S A 110:12474–12479
Tang H, Goldberg E (2009) Homo sapiens Lactate Dehydrogenase c (Ldhc) gene expression in cancer cells is regulated by transcription factor Sp1, CREB, and CpG island methylation. J Androl 30:157–167
Tannahill GM et al (2013) Succinate is an inflammatory signal that induces IL-1β through HIF-1α. Nature 496:238–242
Tchou J et al (2012) Human breast cancer associated fibroblasts exhibit subtype specific gene expression profiles. BMC Med Genet 5:39
Thangaraju M et al (2006) SLC5A8 triggers tumor cell apoptosis through pyruvate-dependent inhibition of histone deacetylases. Cancer Res 66:11560–11564
Thoen LFR et al (2011) A role for autophagy during hepatic stellate cell activation. J Hepatol 55:1353–1360
Tisdale MJ (2002) Cachexia in cancer patients. Nat Rev Cancer 2:862–871
Toy EP et al (2009) Enhanced ovarian cancer tumorigenesis and metastasis by the macrophage colony-stimulating factor. Neoplasia (New York NY) 11:136–144
Traverso N et al (2013) Role of glutathione in cancer progression and chemoresistance. Oxidative Med Cell Longev 2013:972913–972913
Tsai W-W et al (2015) ATF3 mediates inhibitory effects of ethanol on hepatic gluconeogenesis. Proc Natl Acad Sci U S A 112:2699–2704
Turbat-Herrera EA et al (2018) Cystathione β-synthase is increased in thyroid malignancies. Anticancer Res 38:6085–6090
Ueno T et al (2000) Significance of macrophage chemoattractant protein-1 in macrophage recruitment, angiogenesis, and survival in human breast cancer. Clin Cancer Res 6:3282–3289
Umapathy A et al (2018) Functional characterisation of glutathione export from the rat lens. Exp Eye Res 166:151–159
Updegraff BL et al (2018) Transmembrane protease TMPRSS11B promotes lung cancer growth by enhancing lactate export and glycolytic metabolism. Cell Rep 25:2223–2233.e2226
Valsecchi R et al (2016) HIF-1α regulates the interaction of chronic lymphocytic leukemia cells with the tumor microenvironment. Blood 127:1987–1997
Valvona CJ et al (2016) The regulation and function of lactate dehydrogenase A: therapeutic potential in brain tumor. Brain Pathol 26:3–17
van Jaarsveld MTM et al (2015) miR-634 restores drug sensitivity in resistant ovarian cancer cells by targeting the Ras-MAPK pathway. Mol Cancer 14:196–196
Vangapandu HV et al (2017) The Stromal microenvironment modulates mitochondrial oxidative phosphorylation in chronic Lymphocytic leukemia cells. Neoplasia 19:762–771
Vanhove K et al (2019) Glutamine addiction and therapeutic strategies in lung cancer. Int J Mol Sci 20:252
Viale A, Corti D, Draetta GF (2015) Highlights from recent cancer literature. Cancer Res 75:3685–3686
Visscher M et al (2016) Covalent targeting of acquired cysteines in cancer. Curr Opin Chem Biol 30:61–67
Wahi K, Holst J (2019) ASCT2: a potential cancer drug target. Expert Opin Ther Targets 23:555–558
Wang R (2012) Physiological implications of hydrogen sulfide: a whiff exploration that blossomed. Physiol Rev 92:791–896
Wang W, Ballatori N (1998) Endogenous glutathione conjugates: occurrence and biological functions. Pharmacol Rev 50:335–356
Wang Z, Dong C (2019) Gluconeogenesis in cancer: function and regulation of PEPCK, FBPase, and G6Pase. Trends Cancer 5:30–45
Wang Y et al (2012a) Prognostic and therapeutic implications of increased ATP citrate lyase expression in human epithelial ovarian cancer. Oncol Rep 27:1156–1162
Wang Y-Y et al (2012b) Adipose tissue and breast epithelial cells: a dangerous dynamic duo in breast cancer. Cancer Lett 324:142–151
Wang Q et al (2014a) Targeting glutamine transport to suppress melanoma cell growth. Int J Cancer 135:1060–1071
Wang F et al (2014b) Mammary fat of breast cancer: gene expression profiling and functional characterization. PLoS One 9:e109742–e109742
Wang C et al (2015) Human adipocytes stimulate invasion of breast cancer MCF-7 cells by secreting IGFBP-2. PLoS One 10:e0119348–e0119348
Wang Q et al (2016) Autophagy protects ovarian cancer-associated fibroblasts against oxidative stress. Cell Cycle 15:1376–1385
Wang H et al (2017a) The metabolic function of cyclin D3-CDK6 kinase in cancer cell survival. Nature 546:426–430

Wang M et al (2017b) Uncoupling protein 2 downregulation by hypoxia through repression of peroxisome proliferator-activated receptor γ promotes chemoresistance of non-small cell lung cancer. Oncotarget 8:8083–8094

Wang YY et al (2017c) Mammary adipocytes stimulate breast cancer invasion through metabolic remodeling of tumor cells. JCI Insight 2:e87489–e87489

Wang S et al (2018a) KRAB-type zinc-finger proteins PITA and PISA specifically regulate p53-dependent glycolysis and mitochondrial respiration. Cell Res 28:572–592

Wang Y et al (2018b) Prognostic value of D-lactate dehydrogenase in patients with clear cell renal cell carcinoma. Oncol Lett 16:866–874

Wang L et al (2019a) Wnt1-inducible signaling protein 1 regulates laryngeal squamous cell carcinoma glycolysis and chemoresistance via the YAP1/TEAD1/GLUT1 pathway. J Cell Physiol 234:15941–15950

Wang X et al (2019b) Nf1 loss promotes Kras-driven lung adenocarcinoma and results in Psat1-mediated glutamate dependence. EMBO Mol Med 11:e9856

Wang L et al (2019c) I157172, a novel inhibitor of cystathionine gamma-lyase, inhibits growth and migration of breast cancer cells via SIRT1-mediated deacetylation of STAT3. Oncol Rep 41:427–436

Wang Y et al (2019d) Inhibition of fatty acid synthesis arrests colorectal neoplasm growth and metastasis: anti-cancer therapeutical effects of natural cyclopeptide RA-XII. Biochem Biophys Res Commun 512:819–824

Warburg O (1956) On the origin of cancer cells. Science 123:309–314

Weber GF (2016) Metabolism in cancer metastasis. Int J Cancer 138:2061–2066

Wei L et al (2016) Leptin promotes epithelial-mesenchymal transition of breast cancer via the upregulation of pyruvate kinase M2. J Exp Clin Cancer Res 35:166–166

Wei J et al (2019) An allosteric mechanism for potent inhibition of human ATP-citrate lyase. Nature 568:566–570

Wellen KE et al (2009) ATP-citrate lyase links cellular metabolism to histone acetylation. Science 324:1076–1080

Wen Y-A et al (2017) Adipocytes activate mitochondrial fatty acid oxidation and autophagy to promote tumor growth in colon cancer. Cell Death Dis 8:e2593–e2593

Wen H et al (2019) Glucose-derived acetate and ACSS2 as key players in cisplatin resistance in bladder cancer. Biochimica et Biophysica Acta (BBA) – Mol Cell Biol Lipids 1864:413–421

Wenes M et al (2016) Macrophage metabolism controls tumor blood vessel morphogenesis and metastasis. Cell Metab 24:701–715

Wise DR, Thompson CB (2010) Glutamine addiction: a new therapeutic target in cancer. Trends Biochem Sci 35:427–433

Wise DR et al (2008) Myc regulates a transcriptional program that stimulates mitochondrial glutaminolysis and leads to glutamine addiction. Proc Natl Acad Sci U S A 105:18782–18787

Wise DR et al (2011) Hypoxia promotes isocitrate dehydrogenase-dependent carboxylation of α-ketoglutarate to citrate to support cell growth and viability. Proc Natl Acad Sci U S A 108:19611–19616

Witz IP (2009) The tumor microenvironment: the making of a paradigm. Cancer Microenviron 2(Suppl 1):9–17

Wu G et al (2004) Glutathione metabolism and its implications for health. J Nutr 134:489–492

Wu Q et al (2006) FATP1 is an insulin-sensitive fatty acid transporter involved in diet-induced obesity. Mol Cell Biol 26:3455–3467

Wu R et al (2013a) Type I to type II ovarian carcinoma progression: mutant Trp53 or Pik3ca confers a more aggressive tumor phenotype in a mouse model of ovarian cancer. Am J Pathol 182:1391–1399

Wu X et al (2013b) Long chain fatty Acyl-CoA synthetase 4 is a biomarker for and mediator of hormone resistance in human breast cancer. PLoS One 8:e77060–e77060

Wu H et al (2018) Here, there, and everywhere: the importance of ER membrane contact sites. Science 361, eaan5835

Xi J et al (2019) GLS1 promotes proliferation in hepatocellular carcinoma cells via AKT/GSK3β/CyclinD1 pathway. Exp Cell Res 381:1–9

Xiang L et al (2019) Glutaminase 1 expression in colorectal cancer cells is induced by hypoxia and required for tumor growth, invasion, and metastatic colonization. Cell Death Dis 10:40–40

Xintaropoulou C et al (2018) Expression of glycolytic enzymes in ovarian cancers and evaluation of the glycolytic pathway as a strategy for ovarian cancer treatment. BMC Cancer 18:636–636

Xu N et al (2012) The FATP1-DGAT2 complex facilitates lipid droplet expansion at the ER-lipid droplet interface. J Cell Biol 198:895–911

Xu S et al (2013) CD36 enhances fatty acid uptake by increasing the rate of intracellular esterification but not transport across the plasma membrane. Biochemistry 52:7254–7261

Xu W et al (2018a) Crosstalk of protein kinase C ε with Smad2/3 promotes tumor cell proliferation in prostate cancer cells by enhancing aerobic glycolysis. Cell Mol Life Sci 75:4583–4598

Xu Y et al (2018b) miR-27b-3p is involved in doxorubicin resistance of human anaplastic thyroid cancer cells via targeting peroxisome proliferator-activated receptor gamma. Basic Clin Pharmacol Toxicol 123:670–677

Yan S-X et al (2013) Effect of antisense oligodeoxynucleotides glucose transporter-1 on enhancement of radiosensitivity of laryngeal carcinoma. Int J Med Sci 10:1375–1386

Yan S et al (2015) Long-chain acyl-CoA synthetase in fatty acid metabolism involved in liver and other diseases: an update. World J Gastroenterol 21:3492–3498

Yan X et al (2017) Eugenol inhibits oxidative phosphorylation and fatty acid oxidation via downregulation of c-Myc/PGC-1β/ERRα signaling pathway in MCF10A-ras cells. Sci Rep 7:12920–12920
Yang C et al (2014) Glutamine oxidation maintains the TCA cycle and cell survival during impaired mitochondrial pyruvate transport. Mol Cell 56:414–424
Yang L et al (2016) Targeting stromal glutamine synthetase in tumors disrupts tumor microenvironment-regulated cancer cell growth. Cell Metab 24:685–700
Yang P et al (2018) Dietary oleic acid-induced CD36 promotes cervical cancer cell growth and metastasis via up-regulation Src/ERK pathway. Cancer Lett 438:76–85
Yang H et al (2019) Roles of GLUT-1 and HK-II expression in the biological behavior of head and neck cancer. Oncotarget 10:3066–3083
Yang-Hartwich Y et al (2014) p53 protein aggregation promotes platinum resistance in ovarian cancer. Oncogene 34:3605
Yi W et al (2012) Phosphofructokinase 1 glycosylation regulates cell growth and metabolism. Science 337:975–980
Yin N et al (2004) Molecular mechanisms involved in the growth stimulation of Breast cancer cells by leptin. Cancer Res 64:5870–5875
Yin X et al (2017) ID1 promotes hepatocellular carcinoma proliferation and confers chemoresistance to oxaliplatin by activating pentose phosphate pathway. J Exp Clin Cancer Res 36:166–166
You J et al (2017) Cystathionine- γ-lyase promotes process of breast cancer in association with STAT3 signaling pathway. Oncotarget 8:65677–65686
Yu L et al (2018) Autophagy pathway: cellular and molecular mechanisms. Autophagy 14:207–215
Yu W et al (2019) SIRT6 promotes the Warburg effect of papillary thyroid cancer cell BCPAP through reactive oxygen species. Onco Targets Ther 12:2861–2868
Zaidi N et al (2012) ATP-citrate lyase: a key player in cancer metabolism. Cancer Res 72:3709–3714
Zanotto-Filho A et al (2016) Alkylating agent-induced NRF2 blocks endoplasmic reticulum stress-mediated Apoptosis via control of glutathione pools and protein thiol Homeostasis. Mol Cancer Ther 15:3000–3014
Zhan L et al (2012a) Regulatory role of KEAP1 and NRF2 in PPARγ expression and chemoresistance in human non-small-cell lung carcinoma cells. Free Radic Biol Med 53:758–768
Zhan T et al (2012b) Overexpressed FATP1, ACSVL4/FATP4 and ACSL1 increase the cellular fatty acid uptake of 3T3-L1 adipocytes but are localized on intracellular membranes. PLoS One 7:e45087–e45087
Zhang W et al (2012) Stromal control of cystine metabolism promotes cancer cell survival in chronic lymphocytic leukaemia. Nat Cell Biol 14:276–286
Zhang J et al (2014) Asparagine plays a critical role in regulating cellular adaptation to glutamine depletion. Mol Cell 56:205–218
Zhang D et al (2015) Metabolic reprogramming of cancer-associated fibroblasts by IDH3α downregulation. Cell Rep 10:1335–1348
Zhang C et al (2016) Glutaminase 2 is a novel negative regulator of small GTPase Rac1 and mediates p53 function in suppressing metastasis. elife 5:e10727–e10727
Zhang B et al (2017) IL-17A enhances microglial response to OGD by regulating p53 and PI3K/Akt pathways with involvement of ROS/HMGB1. Front Mol Neurosci 10:271–271
Zhang S et al (2018a) Acetyl-CoA synthetase 2 enhances tumorigenesis and is indicative of a poor prognosis for patients with renal cell carcinoma. Urol Oncol Semin Orig Investig 36:243.e249–243.e220
Zhang M et al (2018b) Adipocyte-derived lipids mediate melanoma progression via FATP proteins. Cancer Discov 8:1006–1025
Zhang X et al (2019a) LncRNA TINCR/microRNA-107/CD36 regulates cell proliferation and apoptosis in colorectal cancer via PPAR signaling pathway based on bioinformatics analysis. In Biol Chem, pp 663
Zhang Z-G et al (2019b) KDM5B promotes breast cancer cell proliferation and migration via AMPK-mediated lipid metabolism reprogramming. Exp Cell Res 379:182–190
Zhao H et al (2016) Tumor microenvironment derived exosomes pleiotropically modulate cancer cell metabolism. elife 5:e10250–e10250
Zhao W et al (2018) LINK-A promotes cell proliferation through the regulation of aerobic glycolysis in non-small-cell lung cancer. Onco Targets Ther 11:6071–6080
Zheng G-F et al (2015) Unfolded protein response mediated JNK/AP-1 signal transduction, a target for ovarian cancer treatment. Int J Clin Exp Pathol 8:6505–6511
Zhong J et al (2010) Temporal profiling of the secretome during adipogenesis in humans. J Proteome Res 9:5228–5238
Zhou Y et al (2013) ATP citrate lyase mediates resistance of colorectal cancer cells to SN38. Mol Cancer Ther 12:2782–2791
Zhou J et al (2019) Oncoprotein LAMTOR5 activates GLUT1 Via upregulating NF-κB in liver cancer. Open Med (Wars) 14:264–270
Zhu H et al (2018) Cystathionine β-Synthase in physiology and cancer. Biomed Res Int 2018:3205125–3205125

2 Tumor Microenvironment – Selective Pressures Boosting Cancer Progression

Sofia C. Nunes

Abstract

In 2018, 9.6 million deaths from cancer were estimated, being this disease the second leading cause of death worldwide. Notwithstanding all the efforts developed in prevention, diagnosis and new treatment approaches, chemoresistance seems to be inevitable, leading to cancer progression, recurrence and affecting the outcome of the disease. As more and more evidence support that cancer is an evolutionary and ecological process, this concept is rarely applied in the clinical context. In fact, cancer cells *emerge* and progress within an ecological niche – the tumor microenvironment – that is shared with several other cell types and that is continuously changing. Therefore, the tumor microenvironment imposes several selective pressures on cancer cells such as acidosis, hypoxia, competition for space and resources, immune predation and anti-cancer therapies, that cancer cells must be able to adapt to or will face extinction.

In here, the role of the tumor microenvironment selective pressures on cancer progression will be discussed, as well as the targeting of its features/components as strategies to fight cancer.

Keywords

Cancer · Evolution · Microenvironment · Metabolic selection

2.1 Cancer and Darwin's Theory of Evolution

"It may be said that natural selection is daily and hourly scrutinising, throughout the world, every variation, even the slightest; rejecting that which is bad, preserving and adding up all that is good; silently and insensibly working, whenever and wherever opportunity offers, at the improvement of each organic being in relation to its organic and inorganic conditions of life." (Darwin 1859)

Darwin's theory of Evolution by natural selection dramatically changed our vision of life, of species evolution, adaptation and extinction, and strikingly, the same ideas can be applied to cancer progression (e.g.(Cairns 1975; Nowell 1976; Crespi and Summers 2005; Merlo et al. 2006; Pepper et al. 2009; Greaves and Maley 2012; Nagraj et al. 2015; Ibrahim-Hashim et al. 2017; Maley et al. 2017; Fortunato et al. 2017; Leong

S. C. Nunes (✉)
CEDOC, Chronic Diseases Research Centre, NOVA Medical School | Faculdade de Ciências Médicas, Universidade NOVA de Lisboa, Lisbon, Portugal

Instituto Português de Oncologia de Lisboa Francisco Gentil (IPOLFG), Lisbon, Portugal

J. Serpa (ed.), *Tumor Microenvironment*, Advances in Experimental Medicine and Biology 1219,
https://doi.org/10.1007/978-3-030-34025-4_2

et al. 2018; Lacina et al. 2019)). In 1976, Nowell's introduced the clonal evolution theory of tumor populations, proposing that neoplasms have a clonal origin, and that the acquisition of genetic variability allows the sequential selection of more aggressive subclones, leading to tumor progression (Nowell 1976). Since then, cancer has been widely recognized as an evolutionary disease, albeit the integration of the Darwinian dynamics in clinical trials is still rare (Aktipis et al. 2011; Gallaher et al. 2017; Zhang et al. 2017).

Tumors are composed by several heterogeneous cells driven by genetic, transcriptomic, epigenetic, and/or phenotypic changes (reviewed in (Dagogo-Jack and Shaw 2017)), that coexist and compete with several other cells within an ecological context (Merlo et al. 2006) that is continuously changing. Hence, cancer cells must be able to adapt to these continuous changes or will face extinction.

The increasing number of cancer-related deaths, mainly due to metastatic disease (Rankin and Giaccia 2016; Lambert et al. 2017), undoubtedly show us the power of evolution in cancer progression. In fact, besides the gradual accumulation of mutations, there is also evidence for punctuated evolution in cancer (reviewed in (McGranahan and Swanton 2017)), where macro-evolutionary leaps can take place driving tumor evolution (reviewed in (Nagraj et al. 2015; McGranahan and Swanton 2017)).

In 2018, 9.6 million deaths from cancer were estimated (Bray et al. 2018), making this complex group of different diseases the second leading cause of death worldwide (Fitzmaurice et al. 2015). It is therefore urgent the introduction of evolutionary principles into clinical settings of cancer treatment, in order to be able to predict and avoid the evolutionary dynamics of cancer cells that will lead to poor outcomes of cancer patients.

In the next sections, I will discuss the role of the ecological environment of cancer cells – the tumor microenvironment – as a driving force of cancer progression.

2.2 Tumor Microenvironment: The Ecological Context of Cancer Cells

Albeit the existence of a high number of different types of cancer, it is widely accepted that the human tumors share eight key alterations in cell physiology: self-sufficiency in growth signals, evading growth suppressors, evasion of apoptosis, limitless replicative potential, sustained angiogenesis, tissue invasion and metastasis, energy metabolism reprogramming and evading immune destruction (Hanahan and Weinberg 2000, 2011). Furthermore, all cancer cells share another key feature: cancer cells *emerge* within a complex ecological niche, where several cell types coexist and several biomolecules and metabolites are secreted and shared. In fact, cancer cells coexist in a complex network of intracellular interactions – the tumor microenvironment – where the non-malignant microenvironmental metabolic interactions are repurposed by the malignant cells, allowing their survival and progression (Lyssiotis and Kimmelman 2017).

The tumor microenvironment is comprised by several cell types, by signaling molecules and by extracellular matrixes (reviewed in (Gupta et al. 2017)). The cellular fraction of the tumor microenvironment derive from the surrounding tissues and can have a hematopoietic (B cells, T cells, neutrophils, natural killer cells and macrophages) or mesenchymal (fibroblasts, adipocytes, endothelial cells, and pericytes) origin (reviewed in (Gupta et al. 2017)). Interestingly, the tumor-stroma ratio was reported as an independent prognostic predictor in patients with different types of cancer (e.g. (De Kruijf et al. 2011; Wang et al. 2012; Zhang et al. 2014; Liu et al. 2014; Chen et al. 2015b; Kemi et al. 2018; van Pelt et al. 2018; Karpathiou et al. 2019)).

The tumor microenvironment exerts several selective pressures including hypoxia, acidosis, competition for space and nutrients, cooperation and predation by the immune system and anticancer therapies (e.g.(Gatenby and Gillies 2004; Crespi and Summers 2005; Merlo et al. 2006; Gillies et al. 2012; Sun et al. 2016; Venkatesan

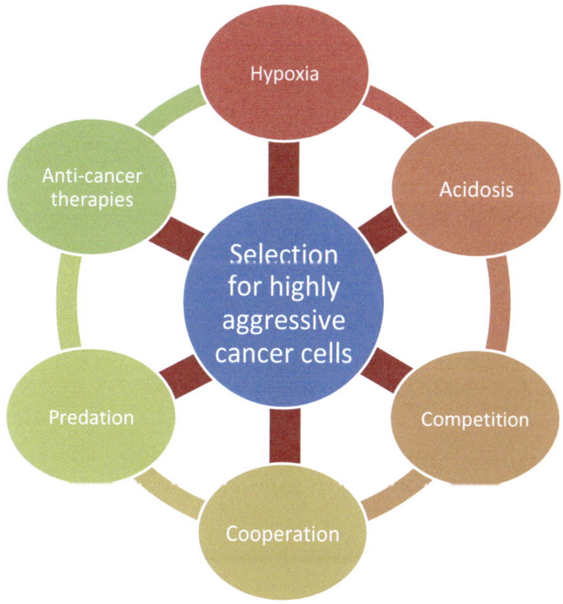

Fig. 2.1 Cancer cells are subject to several selective pressures within the tumor microenvironment, including hypoxia, acidosis, competition for space and resources, predation by the immune system and anti-cancer therapies. Cancer cells can also be selected to cooperate with other malignant and non-malignant cells. All this selective pressures lead to cancer cells evolution towards more aggressive and resistant phenotypes

et al. 2017)) (Fig. 2.1). Hence, cancer is an evolutionary and an ecological process (Merlo et al. 2006).

Interestingly, supporting the idea that cancer cells are subject to different selective pressures compared to the normal counterparts, Oven and Naugler have suggested that the same genes may experience different selection pressures within non-malignant and malignant tissues (Ovens and Naugler 2012). Moreover, Chen and colleagues, by using experimental evolution of a human breast cell-derived xenograft tumor in mice, have reported that cancer cells reversed multicellular evolution towards a unicellular state, by the knock down of the genetic constraints needed for the multicellularity maintenance (Chen et al. 2015a). The same group also reported a dominant convergent evolution toward an embryonic stem cell functional status in a large number of tumors that were grown in distinct tissue environments (Chen and He 2015). Therefore, a link between unicellularity, embryonic stem cells and cancer cells was proposed, all associated to fast proliferation leading to the formation of cell colonies, maximizing growth rate, and benefiting their own fitness (Chen and He 2015). These reports strongly support that cancer cells are subject to different selective pressures compared to the non-malignant cells. The existence of similar selective pressures within tumor microenvironments across all cancers can explain the convergence observed in cancer cell traits, albeit the unique genetic features of each cancer (Fortunato et al. 2017).

Therefore, the tumor microenvironment has been implicated not only in tumor growth, invasion, and metastasis but also in acquired drug resistance, mediated by myeloid cells, cancer-associated fibroblasts, mesenchymal stem cells and the interaction with the extracellular matrix (reviewed in (Son et al. 2017)), and radioresistance (Barker et al. 2015; Chen et al. 2018; Gu et al. 2018). Furthermore, certain conditions of the tumor microenvironment such as hypoxia (e.g. (Vaupel and Mayer 2007; Gillies et al. 2012; Semenza 2012; Paolicchi et al. 2016)) and acidosis (e.g. (Gillies et al. 2012; Pillai et al. 2019)) are intimately associated with chemoresistance and disease progression. In fact, the tumor microenvironment enables and sustains the hallmarks of cancer cells (Hanahan and Coussens 2012). Interestingly, Fouad and Aanei revisited the hallmarks of cancer and have defined seven hallmarks: selective growth and proliferative advantage, altered stress response, vascularization, invasion and metastasis, metabolic rewiring,

genomic instability through interplay with oncoproteins like c-MYC (Luoto et al. 2013). Koshiji et al. also reported a role of HIF-1α in hypoxia-induced genetic instability through the inhibition of MutSα expression, a DNA mismatch repair gene (Koshiji et al. 2005). Notably, HIF-1α targets include genes involved in angiogenesis, glucose metabolism, cell proliferation/viability, invasion and migration, thus, the upregulation of HIF-1α activates several hallmarks of cancer (Balamurugan 2016).

Besides HIF-1, evidence also supports a role of Nrf2 – a transcription factor with pivotal roles in the maintenance of oxidative homeostasis – in the response of cancer cells to hypoxia (reviewed in (Toth and Warfel 2017)). Interestingly, data also support a crosstalk between HIF-1 and Nrf2 in promoting tumor progression (reviewed in (Toth and Warfel 2017)).

Besides hypoxia, acidosis is also well known as a widespread tumor microenvironment condition that exerts a powerful selective pressure on cancer cells (Gillies et al. 2012). However, the studies have been focused mainly in hypoxia (reviewed in (Corbet and Feron 2017)).

Acidosis can result from hypoxia, since hypoxia selects for cells with a glycolytic phenotype (Gatenby and Gillies 2004; Gillies et al. 2012). The adaptation to intratumoral acidosis was already associated with selection of malignant traits, such as increased invasion and metastasis, chemoresistance (Gillies et al. 2012; Pillai et al. 2019), and escape from immune surveillance (reviewed in (Pillai et al. 2019)). Acidosis is known to promote upregulation of Na^+/H^+ exchangers as well as driving mutations in apoptotic pathways that decrease acid-mediated toxicity (reviewed in (Gatenby et al. 2007)). Moreover, Wojtkowiak and co-workers have reported that chronic autophagy represents a cellular adaptation to acidic microenvironments (Wojtkowiak et al. 2012). In addition, by chronically adapting pre-malignant cells in culture, Damaghi and colleagues have reported that these cells significantly upregulated lysosomal proteins, including the upregulation of the lysosomal-associated membrane protein 2 (LAMP2) in the plasma membrane, both *in vitro* and *in vivo* (Damaghi et al. 2015). Moreover, the authors have shown that the depletion of LAMP2 led to the increase of acidosis-mediated toxicity (Damaghi et al. 2015), hence showing a role of LAMP2 in the adaptation of cancer cells to acidosis. Importantly, the upregulation of HIF-1α under acidosis was also reported (reviewed in (Corbet and Feron 2017)).

It is important to highlight that several reports have already taken into account the combined effects of hypoxia and acidosis. In the context of breast cancer, Gatenby et al. have reported that the adaptation to hypoxia and acidosis may drive cancer invasiveness (Gatenby et al. 2007). Later, Alfarouk and colleagues have proposed tumor vascular density and blood flow resulting in fluctuations in substrates such as oxygen, and in metabolites, such as lactic acid, as a primary evolutionary force of cancer cells (Alfarouk et al. 2013). More recently, it has been reported that hypoxia and acidosis robustly impair the expression of inflammatory mediators in tumor cells (Riemann et al. 2017). In fact, Damgaci and co-workers have extensively reviewed the impact of hypoxia and acidosis on the immune system, and concluded that in general these environmental conditions function as 'immune suppressors' (Damgaci et al. 2018).

Interestingly, hypoxia can explain acidosis, by selecting glycolitic cells, and acidosis can further select cells with upregulated glycolysis and acidic resistance, hence selecting cells with growth advantage (Gatenby and Gillies 2004). The Warburg effect – the increased rate of glycolysis commonly observed in cancer cells even in the presence of oxygen – can be then explained by both intermittent hypoxia and acidosis and it was proposed to be required to the evolution of invasive human cancers as it confers a potent growth advantage (Gatenby and Gillies 2004).

In the next section, the role of competition, cooperation and predation among cancer cells and non-malignant cells within the tumor microenvironment will be discussed.

2.4 Competition, Cooperation and Predation Within the Tumor Microenvironment

It is well known that within the tumor microenvironment, cancer cells engage competitive processes for both space and nutrients with non-malignant cells as well as with other cancer cells.

In 1988, Miller et al. reported that when a mixture of two different cell lines was injected into syngeneic mice, the resulting tumors presented one dominant cell line being this effect possibly due to the production of a growth-inhibitory factor produced by the dominant line (Miller et al. 1988). These results clearly shown that cancer cells compete with other malignant cells in order to survive and proliferate. In fact, competition between tumor cells have been explored mainly in the context of clonal interference (reviewed in (Levayer 2019)).

Besides competition between the cancer clones, cancer cells also compete with the non-malignant ones, such as immune cells. In a mouse sarcoma model it has been reported that glucose consumption by tumor cells limits T cells activity, leading to their reduced mTOR activity, glycolytic capacity, and IFN-γ production, hence allowing tumor progression (Chang et al. 2015). Besides limiting glucose availability, cancer cells and tumor-associated macrophages also reduce tryptophan availability, impairing anti-tumor T effector cells, and indirectly supporting protumor regulatory T cells by releasing kynurenine (reviewed in (Lyssiotis and Kimmelman 2017)). Moreover, competition also exists among non-malignant cells, as tumor-associated macrophages, leading to arginine depletion in the tumor microenvironment, thus impairing anti-tumor T cell activity (reviewed in (Lyssiotis and Kimmelman 2017)).

It is important to highlight the review by Di Gregorio and co-workers on the role of competition in cellular fitness, where the authors explored its role both as a tumor suppressor and as a tumor promoter (Di Gregorio et al. 2016).

More recently, Levayer have reviewed the role of mechanical cell competition – a mechanical stress that promotes competitive interactions between cells – in tumor initiation and expansion (reviewed in (Levayer 2019)). During tumor initiation and tumor expansion, there is evidence for competition for space (reviewed in (Levayer 2019)). Therefore, via mechanical cell competition, cells presenting higher proliferative rates could compact and eliminate the neighbor cells which are more sensitive to compaction, which could further affect tumor progression and resistance to therapy (reviewed in (Levayer 2019)).

Whereas cancer is generally viewed as a 'breakdown of multicellular cooperation' (Aktipis et al. 2015), cancer cells also present cooperative interactions within the tumor microenvironment. Hence, in 2006, Axelrod and colleagues have proposed that the evolution of cooperation between heterogeneous cancer cells through the sharing of diffusible products enables tumor progression (Axelrod et al. 2006). In fact, in mouse models of breast cancer it was reported that cooperation between different clones can be pivotal for tumor maintenance (Cleary et al. 2014). In melanoma, cooperation among subpopulations of tumor cells was also reported to drive disease progression in a process the authors called 'cooperative invasion' (Chapman et al. 2014). The authors reported that the 'cooperative invasion' relied on the cooperation between the invasive and poorly invasive cells and that it was dependent on protease activity and fibronectin deposition (Chapman et al. 2014).

One more interesting example of cooperation between cancer cells was reported by Archetti et al., by applying evolutionary game theory to explain how heterogeneity among cancer cells can be maintained (Archetti et al. 2015). By using pancreatic cancer cells, the authors reported cooperation and stable heterogeneity for the production of insulin-like growth factor II (Archetti et al. 2015), a key growth factor associated with development and progression of several tumors (reviewed in (Denduluri et al. 2015)), that could have further implications for therapies targeting growth factors (Archetti et al. 2015). In addition, cooperative metabolism among the cellular components of the tumor microenvironment has been reported in several types of cancer (reviewed in (Lyssiotis and Kimmelman 2017)). For instance, the symbiotic metabolism of lactate and glucose allows the hypoxic cells within the tumor microenvironment to metabolize glucose while releasing lactate, that can be further metabolized by the normoxic cells, using lactate to support their energetic and biomass needs (Guillaumond et al. 2013; Allen et al. 2016; Pisarsky et al. 2016; Lyssiotis and Kimmelman 2017). Importantly, these symbiotic metabolic interactions support tumor metabolism, growth and resistance to angiogenesis inhibitors (Allen et al. 2016; Pisarsky et al. 2016; Lyssiotis and Kimmelman 2017).

More recently, Martín-Pardillos and colleagues have reported the existence of heterogeneity in stable cell lines and that the combination of different clones was able to enhance the malignant properties, supporting the cooperation among tumor clones mediated by secreted factors in the progression of the disease (Martín-Pardillos et al. 2018). In another interesting publication, Lopes-Coelho and co-workers reported that cancer associated fibroblasts cooperate with breast cancer cells by supplying fatty acids both *in vitro* and *in vivo* (Lopes-Coelho et al. 2018).

Besides competition and cooperation, cancer cells are also subject to predation by the immune system that imposes a high selective pressure on cancer cells. In fact, several studies were published considering the interactions between tumors and the immune system as prey-predator systems (Albano et al. 2007; Babbs 2012; Agarwal and Bhadauria 2013; Kartal 2014; Kaur and Ahmad 2014)). As Gonzalez and co-workers stated, "*it is currently accepted that an aberrant innate and adaptive immune response contributes to tumorigenesis by selecting aggressive clones, inducing immunosuppression, and stimulating cancer cell proliferation and metastasis*" (reviewed in (Gonzalez et al. 2018)). Therefore, the immune system has a key role not only in cancer prevention but also in cancer progression and therapy response, where the overall proportion and features of T cells within the tumor immune microenvironment is accepted as a key feature shaping tumor progression over time (reviewed in (Binnewies et al. 2018)).

Recently, it has been reported that the immune system restricts the levels of intratumor genomic heterogeneity, favoring clonal dominance (Milo et al. 2018). In another publication, the authors have reported similar findings, suggesting the "*duality tumor-immunity model of human cancer metastases* (that) *proposed that the development of tumor clones is linked to the intra-metastatic immune microenvironment via the immunoediting process*" (Angelova et al. 2018). Showing the complexity of the tumor immune microenvironment, in a patient with ovarian cancer, Jiménez-Sánchez et al. have reported the existence of multiple diverse tumor immune microenvironments in different metastases that could explain its further fate (Jiménez-Sánchez et al. 2017).

However, as the immune system is able to affect tumor cells composition, tumor cells can also directly (via tumor-derived cytokines and chemokines) and indirectly (via phenotypic changes in non-immune stromal cells within the local tumor microenvironment) affect immune cells (reviewed in (Binnewies et al. 2018)). In fact, it was proposed that the low microenvironment pH, through regulation of the secretion of lactic acid by cancer cells via metabolic reprogramming, is another key mechanism by which cancer cells can suppress the anti-cancer immune response (Choi et al. 2013; Gupta et al. 2017). Moreover, Korobeinikov and colleagues have hypothesized that cancer is able to avoid immune pressure due to the accumulation of mutations driving immune-resistance (Korobeinikova et al. 2017). The authors, by using mathematical models, have shown that cancer is able to persist if cancer cells are able to mutate rapidly in the presence of an insufficient immune response (Korobeinikova et al. 2017).

Interestingly, cancer cells itself are capable of cannibalism of other cells such as mesenchymal stem cells (Bartosh et al. 2016) and also T cells (Lugini et al. 2006). Recently Fais and Overholtzer reviewed the role of cannibalism and entosis in cancer progression, proposing the cannibalistic activity as a hallmark of cancer (Fais and Overholtzer 2018).

In the next section the role of anti-cancer therapies as selective pressures on cancer cells will be discussed.

2.5 Anti-cancer Therapies: Powerful Selective Pressures on Cancer Cells

Anti-cancer therapies can also exert strong selective pressures on cancer cells, selecting often for resistance (e.g. (Crespi and Summers 2005; Merlo et al. 2006; Gillies et al. 2012; Greaves and Maley 2012; Sun et al. 2016; Venkatesan et al. 2017)). For instance, recently, Niehr and co-workers have reported a role of treatment-induced clonal selection in the development of cisplatin resistance in the context of squamous cell carcinoma of the head and neck (Niehr et al. 2018), hence showing the selective power of anti-cancer therapies, leading to the selection of resistant cells, that will allow further tumor relapse.

In fact, resistance to treatment is a major problem in cancer management, where several mechanisms of drug (reviewed in (Mansoori et al. 2017)) and radiotherapy resistance (reviewed in (Wu et al. 2015; Tang et al. 2018)) were reported. Undoubtedly, resistance seems to be inevitable, altering the course and the outcome of the disease.

Besides strongly selecting drug resistant cancer cells, chemotherapy also induces mutagenesis through the generation of DNA damage such as double strand breaks and the consequent homologous recombination-based mutagenic break repair, via microhomologous mutagenic break repair that drives copy number alterations and other genome rearrangements, and also due to the limitation of DNA mismatch repair (reviewed in (Fitzgerald et al. 2017)). It is important to highlight that several conventional chemotherapeutic drugs such has platinum drugs (Marullo et al. 2013; Dasari and Tchounwou 2014) and paclitaxel (Alexandre et al. 2006) induce ROS generation. As already mentioned before, increased ROS are known to induce DNA damage which are linked to genomic instability (reviewed in (Gillies et al. 2012)). Furthermore, chemotherapy also drives competitive release (Gatenby et al. 2009; Enriquez-Navas et al. 2016; Gallaher et al. 2017; Zhang et al. 2017; Venkatesan et al. 2017). Therefore, chemotherapy increases heterogeneity within the tumor cells while eliminating competition with sensitive cells, thus influencing cancer evolution (reviewed in (Venkatesan et al. 2017)).

As already mentioned, conditions of the tumor microenvironment such as hypoxia and the consequent induction of ROS and acidosis (reviewed in (Gillies et al. 2012)), and also its cellular components were linked to both chemoresistance (reviewed in (Son et al. 2017; Senthebane et al. 2017)) and radioresistance (Barker et al. 2015; Chen et al. 2018; Gu et al. 2018), making the tumor microenvironment a promising target in cancer treatment.

In the next section some of the possible routes of tumor microenvironment targeting for cancer therapy will be discussed.

2.6 Tumor Microenvironment: Numerous Routes for Cancer Therapy

As already mentioned, evidence have been supporting a key role of the tumor microenvironment on resistance to anti-cancer treatments, including resistance to several drugs (reviewed in (Son et al. 2017; Senthebane et al. 2017)) and to radiotherapy (Barker et al. 2015; Son et al. 2017; Chen et al. 2018; Gu et al. 2018). Therefore, several strategies targeting the different features/components of the tumor microenvironment were already reported, including the inhibition of the recruitment and differentiation of macrophages, the activation of the immune system with anti-tumoral activity, the targeting of cancer-associated fibroblasts, hypoxia and acidosis, the impairment of the extracellular matrix, the targeting of tumor cell derived exosomes and chronic inflammation, and the avoidance of neovascularization (reviewed in (Roma-Rodrigues et al. 2019)).

Immunotherapy has recently become a powerful cancer treatment strategy, however, its efficacy has been often limited (Yu and Cui 2018; Datta et al. 2019). Recently, Yu and Cui have reviewed the role of the tumor microenvironment in immunotherapy and in immune response (Yu and Cui 2018). In fact, the simultaneous targeting of the tumor microenvironment was proposed to improve immunotherapy efficacy (Yu and Cui 2018; Datta et al. 2019). For instance, Odunsi have recently proposed the reprogramming of the tumor microenvironment along with T cells in ovarian cancer immunotherapy (Odunsi 2018). The epigenetic reprogramming of the tumor microenvironment with entinostat (a synthetic benzamide derivative histone deacetylase (HDAC) inhibitor) was also reported to improve immunotherapy (Hicks et al. 2018). Furthermore, it was reported that the reprogramming of cancer-associated fibroblasts with tumor-selective angiotensin blockers is able to improve cancer immunotherapy in mice models of primary and metastatic breast cancer (Chauhan et al. 2019). Hence, the combined targeting of the immune tumor microenvironment together with other components of the tumor microenvironment should bring promising results in cancer treatment.

As already mentioned, hypoxia and acidosis are both related to chemoresistance (reviewed in (Gillies et al. 2012)) and radioresistance (reviewed in (Tang et al. 2018)). Hence, the targeting of these conditions was widely explored. For instance, the targeting of hypoxia and hypoxia response have been proposed (reviewed in (Wigerup et al. 2016; Paolicchi et al. 2016)) and several clinical trials targeting the adaptation of cancer cells to hypoxia were already conducted, including the targeting of HIF1α, the inhibition of HIF targets involved in the regulation of acidosis and the targeting of mitochondria dysfunction, among other strategies (reviewed in (Paolicchi et al. 2016)).

The targeting of acidosis in cancer treatment was also proposed using four different strategies: by neutralizing the acid using buffers, by targeting metabolic vulnerabilities due to acidosis, by developing acid-activatable drugs and nanomedicines, and by inhibiting the metabolic processes that lead to acids production (Pillai et al. 2019). Interestingly, Pellegrini and colleagues have performed a drug screening assay on colorectal cancer cells chronically adapted to acidosis and identified several compounds with preferential cytotoxicity against acid-adapted cells (Pellegrini et al. 2018).

A key role of the extracellular matrix proteins in the response of esophageal cancer cells to chemotherapy was also reported, thus suggesting the targeting of these proteins as an effective therapeutic strategy against chemoresistant tumors (Senthebane et al. 2018).

The targeting of metabolic symbiosis among cancer cells and other malignant and non-malignant cells was also proposed. For instance, in the context of anti-angiogenic therapy, it was reported in a mouse model of breast cancer, that impairing glycolysis or disrupting the metabolic symbiosis improves nintedanib's (angiokinase inhibitor) efficacy (Pisarsky et al. 2016). Allen and co-workers found similar beneficial effects in disrupting metabolic symbiosis driven by angiogenesis inhibition in mice models of pancreatic neuroendocrine tumors, by the simultaneous inhibition of mTOR that leads to the upregulation of glucose transport in normoxic cells, thus decreasing glucose availability to the hypoxic cells and probably leading to toxic lactate accumulation (Allen et al. 2016).

Given the increasing evidence of the role of extracellular vesicles derived from both tumor and stromal cells in drug resistance, recently, the exploitation of vesicles molecular cargo together with the development of exogenous vesicles as drug vehicles were proposed has promising strategies in cancer treatment (Maacha et al. 2019).

Achard and colleagues have recently reviewed the role of oncolytic viruses in disrupting the immunosuppressive nature of the tumor microenvironment, thus enhancing the anti-tumor immune responses (Achard et al. 2018).

Other strategies targeting the cellular dynamics of competition within the tumor microenvironment were also proposed. In order to avoid the competitive release, several studies have reported the beneficial role of the adaptive therapy in several types of cancer (Gatenby et al. 2009; Enriquez-Navas et al. 2016; Gallaher et al. 2017;

Fig. 2.2 Routes for targeting the tumor microenvironment of cancer cells. These routes include the targeting of: hypoxia and ROS and related responses; acidosis; extracellular matrixes; tumor-derived exosomes; chronic inflammation; neovascularization; cancer-associated fibroblasts; cooperation and metabolic symbiosis; macrophages recruitment and differentiation. Other possibilities include the induction of competition between drug resistant and sensitive cancer cells and activation of antitumoral immune responses. Adapted from (Roma-Rodrigues et al. 2019)

Zhang et al. 2017). Therefore, the adaptive therapy aim to induce competition among the sensitive and the chemoresistant cancer cells within the tumor microenvironment, due to fitness costs of resistance in the absence of drugs, hence maintaining a stable population of chemosensitive cells that compete with the resistant cells (Gatenby et al. 2009).

In Fig. 2.2 the main routes for targeting the tumor microenvironment are summarized.

It is important to highlight that, given the pivotal role of the tumor microenvironment in chemoresistance, Jo and co-workers have reviewed the limitations of the *in vitro* cancer platforms generally used for drug screening, highlighting the need of *in vitro* cancer models that better mimic the *in vivo* physiology (Jo et al. 2018).

Together, evidences strongly support a key role of the tumor microenvironment in all stages of carcinogenesis, from cancer initiation to progression and resistance to therapy. Therefore, the malignant cells cannot be seen as individual entities but rather as entities living within a complex ecological niche with environment conditions that change over time. Thus, the targeting of the different features/components of the tumor microenvironment should be promising in the fight against cancer.

2.7 Final Remarks

More and more evidence support that cancer is an evolutionary disease where cancer cells are subject to several selective pressures imposed by the tumor microenvironment. Albeit the unique molecular traits among cancers, there is a widespread convergence in cancer cells traits, hence supporting that cancer cells are subject to common selective pressures. In fact, besides the shared hallmarks of cancer, cancer cells also coexist within a complex ecological niche, the tumor microenvironment, where several cellular and metabolic interactions take place.

Hypoxia, acidosis, competition for space and resources, predation by the immune system and anti-cancer treatments are common selective pressures observed within the tumor microenvironment, independent of the cancer type. These selective pressures drive the evolution of cancer cells, leading to cancer progression and resistance to anti-cancer treatments. Thus, the tumor microenvironment comprises several possible routes for damaging cancer cells. Therapeutic strategies that take into account both the complexity of the tumor microenvironment and the cancer cells adaptability and evolvability could undoubtedly bring new powerful strategies to fight this highly lethal group of diseases.

Acknowledgments The authors acknowledge iNOVA-4Health – UID/Multi/04462/2013, a program financially supported by Fundação para a Ciência e Tecnologia/Ministério da Educação e Ciência, through national funds and co-funded by FEDER under the PT2020 Partnership Agreement.

References

Achard C, Surendran A, Wedge M-E et al (2018) Lighting a fire in the tumor microenvironment using oncolytic immunotherapy. EBioMedicine 31:17–24. https://doi.org/10.1016/j.ebiom.2018.04.020

Agarwal M, Bhadauria AS (2013) A generalised prey-predator type model of immunogenic cancer with the effect of immunotherapy. Int J Eng Sci Technol 5:66–84. https://doi.org/10.4314/ijest.v5i1.6

Aktipis CA, Kwan VSY, Johnson KA et al (2011) Overlooking evolution: a systematic analysis of Cancer relapse and therapeutic resistance research. PLoS One 6:e26100. https://doi.org/10.1371/journal.pone.0026100

Aktipis CA, Boddy AM, Jansen G et al (2015) Cancer across the tree of life: cooperation and cheating in multicellularity. Philos Trans R Soc B Biol Sci 370:1–21. https://doi.org/10.1098/rstb.2014.0219

Albano G, Giorno V, Saturnino C (2007) A prey-predator model for immune response and drug resistance in tumor growth. In: Moreno Díaz R, Pichler FQAA (eds) Computer aided systems theory – EUROCAST 2007. Lecture notes in computer science, vol 4739. Springer, Berlin/Heidelberg, pp 171–178

Alexandre J, Batteux F, Nicco C et al (2006) Accumulation of hydrogen peroxide is an early and crucial step for paclitaxel-induced cancer cell death both in vitro and in vivo. Int J Cancer 119:41–48. https://doi.org/10.1002/ijc.21685

Alfarouk KO, Ibrahim ME, Gatenby RA, Brown JS (2013) Riparian ecosystems in human cancers. Evol Appl 6:46–53. https://doi.org/10.1111/eva.12015

Allen E, Ville PḾ, Warren CM et al (2016) Metabolic symbiosis enables adaptive resistance to anti-angiogenic therapy that is dependent on mTOR signaling. Cell Rep 15:1144–1160. https://doi.org/10.1016/j.celrep.2016.04.029

Angelova M, Mlecnik B, Vasaturo A et al (2018) Evolution of metastases in space and time under immune selection. Cell 175:751–765.e16. https://doi.org/10.1016/j.cell.2018.09.018

Archetti M, Ferraro DA, Christofori G (2015) Heterogeneity for IGF-II production maintained by public goods dynamics in neuroendocrine pancreatic cancer. Proc Natl Acad Sci U S A 112:1833–1838. https://doi.org/10.1073/pnas.1414653112

Axelrod R, Axelrod DE, Pienta KJ (2006) Evolution of cooperation among tumor cells. Proc Natl Acad Sci U S A 103:13474–13479. https://doi.org/10.1073/pnas.0606053103

Babbs CF (2012) Predicting success or failure of immunotherapy for cancer: insights from a clinically applicable mathematical model. Am J Cancer Res 2:204–213

Balamurugan K (2016) HIF-1 at the crossroads of hypoxia, inflammation, and cancer. Int J Cancer 138:1058–1066. https://doi.org/10.1002/ijc.29519

Barker HE, Paget JTE, Khan AA, Harrington KJ (2015) The tumour microenvironment after radiotherapy: mechanisms of resistance and recurrence. Nat Rev Cancer 15:409–425. https://doi.org/10.1038/nrc3958

Bartosh TJ, Ullah M, Zeitouni S et al (2016) Cancer cells enter dormancy after cannibalizing mesenchymal stem/stromal cells (MSCs). Proc Natl Acad Sci U S A 113:E6447–E6456. https://doi.org/10.1073/pnas.1612290113

Bergman A, Gligorijevic B (2015) Niche construction game cancer cells play. Eur Phys J Plus 130:203–215. https://doi.org/10.1140/epjp/i2015-15203-5

Bhandari V, Hoey C, Liu LY et al (2019) Molecular landmarks of tumor hypoxia across cancer types. Nat Genet 51:308–318. https://doi.org/10.1038/s41588-018-0318-2

Binnewies M, Roberts EW, Kersten K et al (2018) Understanding the tumor immune microenvironment (TIME) for effective therapy. Nat Med 24:541–550. https://doi.org/10.1038/s41591-018-0014-x

Bray F, Ferlay J, Soerjomataram I et al (2018) Global Cancer statistics 2018: GLOBOCAN estimates of incidence and mortality worldwide for 36 cancers in 185 countries. CA Cancer J Clin 68:394–424. https://doi.org/10.3322/caac.21492

Bristow RG, Hill RP (2008) Hypoxia and metabolism: hypoxia, DNA repair and genetic instability. Nat Rev Cancer 8:180–192. https://doi.org/10.1038/nrc2344

Cairns J (1975) Mutation selection and the natural history of cancer. Nature 255:197–200. https://doi.org/10.1038/255197a0

Chang C-H, Qiu J, O'Sullivan D et al (2015) Metabolic competition in the tumor microenvironment is a driver of cancer progression. Cell 162:1229–1241. https://doi.org/10.1016/j.cell.2015.08.016

Chapman A, del Ama LF, Ferguson J et al (2014) Heterogeneous tumor subpopulations cooperate to drive invasion. Cell Rep 8:688–695. https://doi.org/10.1016/j.celrep.2014.06.045

Chauhan VP, Chen IX, Tong R et al (2019) Reprogramming the microenvironment with tumor-selective angiotensin blockers enhances cancer immunotherapy. Proc Natl Acad Sci 116:10674–10680. https://doi.org/10.1073/pnas.1819889116

Chen H, He X (2015) The convergent cancer evolution toward a single cellular destination. Mol Biol Evol 33:4–12. https://doi.org/10.1093/molbev/msv212

Chen H, Lin F, Xing K, He X (2015a) The reverse evolution from multicellularity to unicellularity during carcinogenesis. Nat Commun 6:1–9. https://doi.org/10.1038/ncomms7367

Chen Y, Zhang L, Liu W, Liu X (2015b) Prognostic significance of the tumor-stroma ratio in epithelial ovarian cancer. Biomed Res Int 2015:1–8. https://doi.org/10.1155/2015/589301

Chen H, Xu L, Li L et al (2018) Inhibiting the CD8+ T cell infiltration in the tumor microenvironment after radiotherapy is an important mechanism of radioresistance. Sci Rep 8:1–10. https://doi.org/10.1038/s41598-018-30417-6

Choi SYC, Collins CC, Gout PW, Wang Y (2013) Cancer-generated lactic acid: a regulatory, immunosuppressive metabolite? J Pathol 230:350–355. https://doi.org/10.1002/path.4218

Cleary AS, Leonard TL, Gestl SA, Gunther EJ (2014) Tumour cell heterogeneity maintained by cooperating subclones in Wnt-driven mammary cancers. Nature 508:113–117. https://doi.org/10.1038/nature13187

Corbet C, Feron O (2017) Tumour acidosis: from the passenger to the driver's seat. Nat Rev Cancer 17:577–593. https://doi.org/10.1038/nrc.2017.77

Crespi B, Summers K (2005) Evolutionary biology of cancer. Trends Ecol Evol 20:545–552. https://doi.org/10.1016/j.tree.2005.07.007

Dagogo-Jack I, Shaw AT (2017) Tumour heterogeneity and resistance to cancer therapies. Nat Rev Clin Oncol 15:81–94. https://doi.org/10.1038/nrclinonc.2017.166

Damaghi M, Tafreshi NK, Lloyd MC et al (2015) Chronic acidosis in the tumour microenvironment selects for overexpression of LAMP2 in the plasma membrane. Nat Commun 6:1–13. https://doi.org/10.1038/ncomms9752

Damgaci S, Ibrahim-Hashim A, Enriquez-Navas PM et al (2018) Hypoxia and acidosis: immune suppressors and therapeutic targets. Immunology 154:354–362. https://doi.org/10.1111/imm.12917

Darwin C (1859) The origin of species (text). Pennsylvania State University 448 p. https://doi.org/10.5117/9781904633785

Dasari S, Tchounwou BP (2014) Cisplatin in cancer therapy: molecular mechanisms of action. Eur J Pharmacol 740:364–378. https://doi.org/10.1016/j.ejphar.2014.07.025

Datta M, Coussens LM, Nishikawa H et al (2019) Reprogramming the tumor microenvironment to improve immunotherapy: emerging strategies and combination therapies. Am Soc Clin Oncol Educ B 39:165–174. https://doi.org/10.1200/EDBK_237987

De Kruijf EM, Van Nes JGH, Van De Velde CJH et al (2011) Tumor-stroma ratio in the primary tumor is a prognostic factor in early breast cancer patients, especially in triple-negative carcinoma patients. Breast Cancer Res Treat 125:687–696. https://doi.org/10.1007/s10549-010-0855-6

Denduluri SK, Idowu O, Wang Z et al (2015) Insulin-like growth factor (IGF) signaling in tumorigenesis and the development of cancer drug resistance. Genes Dis 2:13–25. https://doi.org/10.1016/j.gendis.2014.10.004

Di Gregorio A, Bowling S, Rodriguez TA (2016) Cell competition and its role in the regulation of cell fitness from development to cancer. Dev Cell 38:621–634. https://doi.org/10.1016/j.devcel.2016.08.012

Enriquez-Navas PM, Kam Y, Das T et al (2016) Exploiting evolutionary principles to prolong tumor control in preclinical models of breast cancer. Sci Transl Med 8:1–9. https://doi.org/10.1126/scitranslmed.aad7842

Fais S, Overholtzer M (2018) Cell-in-cell phenomena in cancer. Nat Rev Cancer 18:758–766. https://doi.org/10.1038/s41568-018-0073-9

Fitzgerald DM, Hastings PJ, Rosenberg SM (2017) Stress-induced mutagenesis: implications in cancer and drug resistance. Ann Rev Cancer Biol 1:119–140. https://doi.org/10.1146/annurev-cancerbio-050216-121919

Fitzmaurice C, Dicker D, Pain A et al (2015) The global burden of cancer 2013. JAMA Oncol 1:505–527. https://doi.org/10.1001/jamaoncol.2015.0735

Fortunato A, Boddy A, Mallo D et al (2017) Natural selection in cancer biology: from molecular snowflakes to trait hallmarks. Cold Spring Harb Perspect Med 7:1–14. https://doi.org/10.1101/cshperspect.a029652

Fouad YA, Aanei C (2017) Revisiting the hallmarks of cancer. Am J Cancer Res 7:1016–1036

Gallaher JA, Enriquez-Navas PM, Luddy KA et al (2017) Spatial heterogeneity and evolutionary dynamics

modulate time to recurrence in continuous and adaptive cancer therapies. bioRxiv:1–21. https://doi.org/10.1101/128959
Gatenby RA, Gillies RJ (2004) Why do cancers have high aerobic glycolysis? Nat Rev Cancer 4:891–899. https://doi.org/10.1038/nrc1478
Gatenby RA, Smallbone K, Maini PK et al (2007) Cellular adaptations to hypoxia and acidosis during somatic evolution of breast cancer. Br J Cancer 97:646–653. https://doi.org/10.1038/sj.bjc.6603922
Gatenby RA, Silva AS, Gillies RJ, Frieden BR (2009) Adaptive therapy. Cancer Res 69:4894–4903. https://doi.org/10.1158/0008-5472.CAN-08-3658
Gillies RJ, Verduzco D, Gatenby RA (2012) Evolutionary dynamics of carcinogenesis and why targeted therapy does not work. Nat Rev Cancer 12:487–493. https://doi.org/10.1038/nrc3298
Gonzalez H, Hagerling C, Werb Z (2018) Roles of the immune system in cancer: from tumor initiation to metastatic progression. Genes Dev 32:1267–1284. https://doi.org/10.1101/gad.314617.118
Greaves M, Maley CC (2012) Clonal evolution in cancer. Nature 481:306–313. https://doi.org/10.1038/nature10762
Gu H, Huang T, Shen Y et al (2018) Reactive oxygen species-mediated tumor microenvironment transformation: the mechanism of radioresistant gastric Cancer. Oxidative Med Cell Longev 2018:1–8. https://doi.org/10.1155/2018/5801209
Guillaumond F, Leca J, Olivares O et al (2013) Strengthened glycolysis under hypoxia supports tumor symbiosis and hexosamine biosynthesis in pancreatic adenocarcinoma. Proc Natl Acad Sci 110:3919–3924. https://doi.org/10.1073/pnas.1219555110
Gupta S, Roy A, Dwarakanath BS (2017) Metabolic cooperation and competition in the tumor microenvironment: implications for therapy. Front Oncol 7:1–24. https://doi.org/10.3389/fonc.2017.00068
Hanahan D, Coussens LM (2012) Accessories to the crime: functions of cells recruited to the tumor microenvironment. Cancer Cell 21:309–322. https://doi.org/10.1016/j.ccr.2012.02.022
Hanahan D, Weinberg RA (2000) The hallmarks of cancer. Cell 100:57–70. https://doi.org/10.1007/s00262-010-0968-0
Hanahan D, Weinberg RA (2011) Hallmarks of cancer: the next generation. Cell 144:646–674. https://doi.org/10.1016/j.cell.2011.02.013
Hicks KC, Knudson KM, Jones FR, et al (2018, April 14–18) Abstract 1740: epigenetic reprogramming of the tumor microenvironment by entinostat increases tumor sensitivity to multivalent immunotherapy combinations with an IL-15 superagonist plus vaccine or immune checkpoint blockade. In: AACR annual meeting 2018. Chicago, p 1740
Ibrahim-Hashim A, Gillies RJ, Brown JS, Gatenby RA (2017) Coevolution of tumor cells and their microenvironment: "niche construction in Cancer" In: Ujvari B, Roche B, Thomas F (eds) Ecology and evolution of cancer. Elsevier Inc., pp 111–117
Jiménez-Sánchez A, Memon D, Pourpe S et al (2017) Heterogeneous tumor-immune microenvironments among differentially growing metastases in an ovarian cancer patient. Cell 170:927–938.e20. https://doi.org/10.1016/j.cell.2017.07.025
Jo Y, Choi N, Kim K et al (2018) Chemoresistance of cancer cells: requirements of tumor microenvironment-mimicking in vitro models in anti-cancer drug development. Theranostics 8:5259–5275. https://doi.org/10.7150/thno.29098
Karpathiou G, Vieville M, Gavid M et al (2019) Prognostic significance of tumor budding, tumor-stroma ratio, cell nests size, and stroma type in laryngeal and pharyngeal squamous cell carcinomas. Head Neck 41:1918–1927. https://doi.org/10.1002/hed.25629
Kartal S (2014) Mathematical modeling and analysis of tumor-immune system interaction by using Lotka-Volterra predator-prey like model with piecewise constant arguments. Period Eng Nat Sci 2:7–12. https://doi.org/10.21533/pen.v2i1.36
Kaur G, Ahmad N (2014) On study of immune response to tumor cells in prey-predator system. Int Sch Res Not 2014:1–8. https://doi.org/10.1155/2014/346597
Kemi N, Eskuri M, Herva A et al (2018) Tumour-stroma ratio and prognosis in gastric adenocarcinoma. Br J Cancer 119:435–439. https://doi.org/10.1038/s41416-018-0202-y
Korobeinikova A, Starkovc KE, Valle PA (2017) Modeling cancer evolution: evolutionary escape under immune system control. J Phys Conf Ser 811:012004–012012. https://doi.org/10.1088/1742-6596/811/1/012004
Koshiji M, To KKW, Hammer S et al (2005) HIF-1α induces genetic instability by transcriptionally downregulating MutSα expression. Mol Cell 17:793–803. https://doi.org/10.1016/j.molcel.2005.02.015
Lacina L, Čoma M, Dvořánková B et al (2019) Evolution of Cancer progression in the context of Darwinism. Anticancer Res 39:1–16. https://doi.org/10.21873/anticanres.13074
Lambert AW, Pattabiraman DR, Weinberg RA (2017) Emerging biological principles of metastasis. Cell 168:670–691. https://doi.org/10.1016/j.cell.2016.11.037
Leong SP, Aktipis A, Maley C (2018) Cancer initiation and progression within the cancer microenvironment. Clin Exp Metastasis 35:361–367. https://doi.org/10.1007/s10585-018-9921-y
Levayer R (2019) Solid stress, competition for space and cancer: the opposing roles of mechanical cell competition in tumour initiation and growth. Semin Cancer Biol:1–12. https://doi.org/10.1016/j.semcancer.2019.05.004
Li C, Little JB, Hu K (2001) Persistent genetic instability in Cancer cells induced by non-DNA-damaging stress exposures advances in brief persistent genetic instability in cancer cells induced by non-DNA-damaging. Cancer Res 61:428–432
Liu J, Liu J, Li J et al (2014) Tumor–stroma ratio is an independent predictor for survival in early cervical

carcinoma. Gynecol Oncol 132:81–86. https://doi.org/10.1016/j.ygyno.2013.11.003
Lopes-Coelho F, André S, Félix A, Serpa J (2018) Breast cancer metabolic cross-talk: fibroblasts are hubs and breast cancer cells are gatherers of lipids. Mol Cell Endocrinol 462:93–106. https://doi.org/10.1016/j.mce.2017.01.031
Lugini L, Matarrese P, Tinari A et al (2006) Cannibalism of live lymphocytes by human metastatic but not primary melanoma cells. Cancer Res 66:3629–3638. https://doi.org/10.1158/0008-5472.CAN-05-3204
Luoto KR, Kumareswaran R, Bristow RG (2013) Tumor hypoxia as a driving force in genetic instability. Genome Integr 4:1–15. https://doi.org/10.1186/2041-9414-4-5
Lyssiotis CA, Kimmelman AC (2017) Metabolic interactions in the tumor microenvironment. Trends Cell Biol 27:863–875. https://doi.org/10.1016/j.tcb.2017.06.003
Maacha S, Bhat AA, Jimenez L et al (2019) Extracellular vesicles-mediated intercellular communication: roles in the tumor microenvironment and anti-cancer drug resistance. Mol Cancer 18:1–16. https://doi.org/10.1186/s12943-019-0965-7
Maley CC, Aktipis A, Graham TA et al (2017) Classifying the evolutionary and ecological features of neoplasms. Nat Rev Cancer 17:605–619. https://doi.org/10.1038/nrc.2017.69
Mansoori B, Mohammadi A, Davudian S et al (2017) The different mechanisms of cancer drug resistance: a brief review. Adv Pharm Bull 7:339–348. https://doi.org/10.15171/apb.2017.041
Martín-Pardillos A, Valls-Chiva A, Serrano EB, et al (2018, April 14–18) Abstract 2183: clonal cooperation in cancer progression: a new paradigm in cancer. In: AACR annual meeting 2018. Chicago. p 2183
Marullo R, Werner E, Degtyareva N et al (2013) Cisplatin induces a mitochondrial-ros response that contributes to cytotoxicity depending on mitochondrial redox status and bioenergetic functions. PLoS One 8:1–15. https://doi.org/10.1371/journal.pone.0081162
McGranahan N, Swanton C (2017) Clonal heterogeneity and tumor evolution: past, present, and the future. Cell 168:613–628. https://doi.org/10.1016/j.cell.2017.01.018
Merlo LMF, Pepper JW, Reid BJ, Maley CC (2006) Cancer as an evolutionary and ecological process. Nat Rev Cancer 6:924–935. https://doi.org/10.1038/nrc2013
Miller BE, Miller FR, Wilburn D, Heppner GH (1988) Dominance of a tumor subpopulation line in mixed heterogeneous mouse mammary tumors. Cancer Res 48:5747–5753
Milo I, Bedora-Faure M, Garcia Z et al (2018) The immune system profoundly restricts intratumor genetic heterogeneity. Sci Immunol 3:1–14. https://doi.org/10.1126/sciimmunol.aat1435
Nagraj J, Mukherjee S, Chowdhury R (2015) Cancer: an evolutionary perspective. J Cancer Biol Res 3:1064–1068
Nakazawa MS, Keith B, Simon MC (2016) Oxygen availability and metabolic adaptations. Nat Rev Cancer 16:663–673. https://doi.org/10.1038/nrc.2016.84
Niehr F, Eder T, Pilz T et al (2018) Multilayered omics-based analysis of a head and neck Cancer model of cisplatin resistance reveals intratumoral heterogeneity and treatment-induced clonal selection. Clin Cancer Res 24:158–168. https://doi.org/10.1158/1078-0432.CCR-17-2410
Nowell PC (1976) The clonal evolution of tumor cell populations. Science 194:23–28. https://doi.org/10.1126/science.191.4224.241-a
Odunsi K (2018) Abstract IA22: reprogramming the tumor microenvironment and T cells for ovarian cancer immunotherapy. In: AACR special conference: addressing critical questions in ovarian Cancer research and treatment; October 1-4, 2017; Pittsburgh, PA. p IA22
Ovens K, Naugler C (2012) Preliminary evidence of different selection pressures on cancer cells as compared to normal tissues. Theor Biol Med Model 9:44–54. https://doi.org/10.1186/1742-4682-9-44
Paolicchi E, Gemignani F, Krstic-Demonacos M et al (2016) Targeting hypoxic response for cancer therapy. Oncotarget 7:13464–13478. https://doi.org/10.18632/oncotarget.7229
Pellegrini P, Serviss JT, Lundbäck T et al (2018) A drug screening assay on cancer cells chronically adapted to acidosis. Cancer Cell Int 18:1–15. https://doi.org/10.1186/s12935-018-0645-5
Pepper JW, Scott Findlay C, Kassen R et al (2009) Cancer research meets evolutionary biology. Evol Appl 2:62–70. https://doi.org/10.1111/j.1752-4571.2008.00063.x
Pillai SR, Damaghi M, Marunaka Y et al (2019) Causes, consequences, and therapy of tumors acidosis. Cancer Metastasis Rev 38:205–222. https://doi.org/10.1007/s10555-019-09792-7
Pisarsky L, Bill R, Fagiani E et al (2016) Targeting metabolic symbiosis to overcome resistance to anti-angiogenic therapy. Cell Rep 15:1161–1174. https://doi.org/10.1016/j.celrep.2016.04.028
Qian JJ, Akçay E (2018) Competition and niche construction in a model of cancer metastasis. PLoS One 13:1–20. https://doi.org/10.1371/journal.pone.0198163
Rankin EB, Giaccia AJ (2016) Hypoxic control of metastasis. Science 352:175–180. https://doi.org/10.1126/science.aaf4405
Reynolds TV, Rockwell S, Gazer PM (1996) Genetic instability induced by the tumor microenvironment. Cancer Res 56:5754–5757
Riemann A, Reime S, Thews O (2017) Tumor acidosis and hypoxia differently modulate the inflammatory program: measurements in vitro and in vivo. Neoplasia 19:1033–1042. https://doi.org/10.1016/j.neo.2017.09.005
Roma-Rodrigues C, Mendes R, Baptista PV, Fernandes AR (2019) Targeting tumor microenvironment for cancer therapy. Int J Mol Sci 20:1–31. https://doi.org/10.3390/ijms20040840

Semenza GL (2012) Hypoxia-inducible factors: mediators of cancer progression and targets for cancer therapy. Trends Pharmacol Sci 33:207–214. https://doi.org/10.1016/j.tips.2012.01.005

Senthebane DA, Rowe A, Thomford NE et al (2017) The role of tumor microenvironment in chemoresistance: to survive, keep your enemies closer. Int J Mol Sci 18:1–30. https://doi.org/10.3390/ijms18071586

Senthebane DA, Jonker T, Rowe A et al (2018) The role of tumor microenvironment in chemoresistance: 3D extracellular matrices as accomplices. Int J Mol Sci 19:1–32. https://doi.org/10.3390/ijms19102861

Son B, Lee S, Youn H et al (2017) The role of tumor microenvironment in therapeutic resistance. Oncotarget 8:3933–3945. https://doi.org/10.18632/oncotarget.13907

Sun D, Dalin S, Hemann MT et al (2016) Differential selective pressure alters rate of drug resistance acquisition in heterogeneous tumor populations. Sci Rep 6:1–13. https://doi.org/10.1038/srep36198

Tafani M, Sansone L, Limana F et al (2016) The interplay of reactive oxygen species, hypoxia, inflammation, and sirtuins in cancer initiation and progression. Oxidative Med Cell Longev 2016:1–18. https://doi.org/10.1155/2016/3907147

Tang L, Wei F, Wu Y et al (2018) Role of metabolism in cancer cell radioresistance and radiosensitization methods. J Exp Clin Cancer Res 37:1–15. https://doi.org/10.1186/s13046-018-0758-7

Toth RK, Warfel NA (2017) Strange bedfellows: nuclear factor, erythroid 2-like 2 (Nrf2) and hypoxia-inducible factor 1 (HIF-1) in tumor hypoxia. Antioxidants (Basel, Switzerland) 6:1–21. https://doi.org/10.3390/antiox6020027

van Pelt GW, Kjær-Frifeldt S, van Krieken JHJM et al (2018) Scoring the tumor-stroma ratio in colon cancer: procedure and recommendations. Virchows Arch 473:405–412. https://doi.org/10.1007/s00428-018-2408-z

Vaupel P, Mayer A (2007) Hypoxia in cancer: significance and impact on clinical outcome. Cancer Metastasis Rev 26:225–239. https://doi.org/10.1007/s10555-007-9055-1

Venkatesan S, Swanton C, Taylor BS, Costello JF (2017) Treatment-induced mutagenesis and selective pressures sculpt cancer evolution. Cold Spring Harb Perspect Med 7:1–16. https://doi.org/10.1101/cshperspect.a026617

Wang k, Ma W, Wang J et al (2012) Tumor-stroma ratio is an independent predictor for survival in esophageal squamous cell carcinoma. J Thorac Oncol 7:1457–1461. https://doi.org/10.1097/JTO.0b013e318260dfe8

Wigerup C, Påhlman S, Bexell D (2016) Therapeutic targeting of hypoxia and hypoxia-inducible factors in cancer. Pharmacol Ther 164:152–169. https://doi.org/10.1016/j.pharmthera.2016.04.009

Wojtkowiak JW, Rothberg JM, Kumar V et al (2012) Chronic autophagy is a cellular adaptation to tumor acidic pH microenvironments. Cancer Res 72:3938–3947. https://doi.org/10.1158/0008-5472.CAN-11-3881

Wu TS, Lin BR, Chang HH (2015) Radio resistance mechanisms of cancers: an overview and future perspectives. Biol Med s2:1–7. https://doi.org/10.4172/0974-8369.1000s2-003

Yu Y, Cui J (2018) Present and future of cancer immunotherapy: a tumor microenvironmental perspective. Oncol Lett 16:4105–4113. https://doi.org/10.3892/ol.2018.9219

Zhang X-L, Jiang C, Zhang Z-X et al (2014) The tumor-stroma ratio is an independent predictor for survival in nasopharyngeal cancer. Oncol Res Treat 37:480–484. https://doi.org/10.1159/000365165

Zhang J, Cunningham JJ, Brown JS, Gatenby RA (2017) Integrating evolutionary dynamics into treatment of metastatic castrate-resistant prostate cancer. Nat Commun 8:1–9. https://doi.org/10.1038/s41467-017-01968-5

Zhou J, Schmid T, Schnitzer S, Brüne B (2006) Tumor hypoxia and cancer progression. Cancer Lett 237:10–21. https://doi.org/10.1016/j.canlet.2005.05.028

Lactate and Lactate Transporters as Key Players in the Maintenance of the Warburg Effect

3

Andreia Pereira-Nunes, Julieta Afonso, Sara Granja, and Fátima Baltazar

Abstract

Reprogramming of energy metabolism is a key hallmark of cancer. Most cancer cells display a glycolytic phenotype, with increased glucose consumption and glycolysis rates, and production of lactate as the end product, independently of oxygen concentrations. This phenomenon, known as "Warburg Effect", provides several survival advantages to cancer cells and modulates the metabolism and function of neighbour cells in the tumour microenvironment. However, due to the presence of metabolic heterogeneity within a tumour, cancer cells can also display an oxidative phenotype, and corruptible cells from the microenvironment become glycolytic, cooperating with oxidative cancer cells to boost tumour growth. This phenomenon is known as "Reverse Warburg Effect". In either way, lactate is a key mediator in the metabolic crosstalk between cancer cells and the microenvironment, and lactate transporters are expressed differentially by existing cell populations, to support this crosstalk.

In this review, we will focus on lactate and on lactate transporters in distinct cells of the tumour microenvironment, aiming at a better understanding of their role in the acquisition and maintenance of the direct/reverse "Warburg effect" phenotype, which modulate cancer progression.

Keywords

Glycolysis · Lactate · Warburg effect · Reverse Warburg effect · Monocarboxylate transporters · Lactate shuttles · Cancer-associated fibroblasts · Endothelial cells · Immune cells

Andreia Pereira-Nunes, Julieta Afonso, Sara Granja and Fátima Baltazar have equally contributed to this chapter.

A. Pereira-Nunes · J. Afonso · S. Granja
F. Baltazar (✉)
Life and Health Sciences Research Institute (ICVS), School of Medicine, University of Minho, Braga, Portugal

ICVS/3B's – PT Government Associate Laboratory, Braga/Guimarães, Portugal
e-mail: fbaltazar@med.uminho.pt

3.1 Introduction

Cellular homeostasis is regulated by several coordinated mechanisms involving the production, utilization or transformation of energy to fulfil multiple biological activities. Adenosine triphosphate (ATP), the primary energetic molecule for mammalian cells, is produced through oxidative or non-oxidative metabolism of glucose

J. Serpa (ed.), *Tumor Microenvironment*, Advances in Experimental Medicine and Biology 1219,
https://doi.org/10.1007/978-3-030-34025-4_3

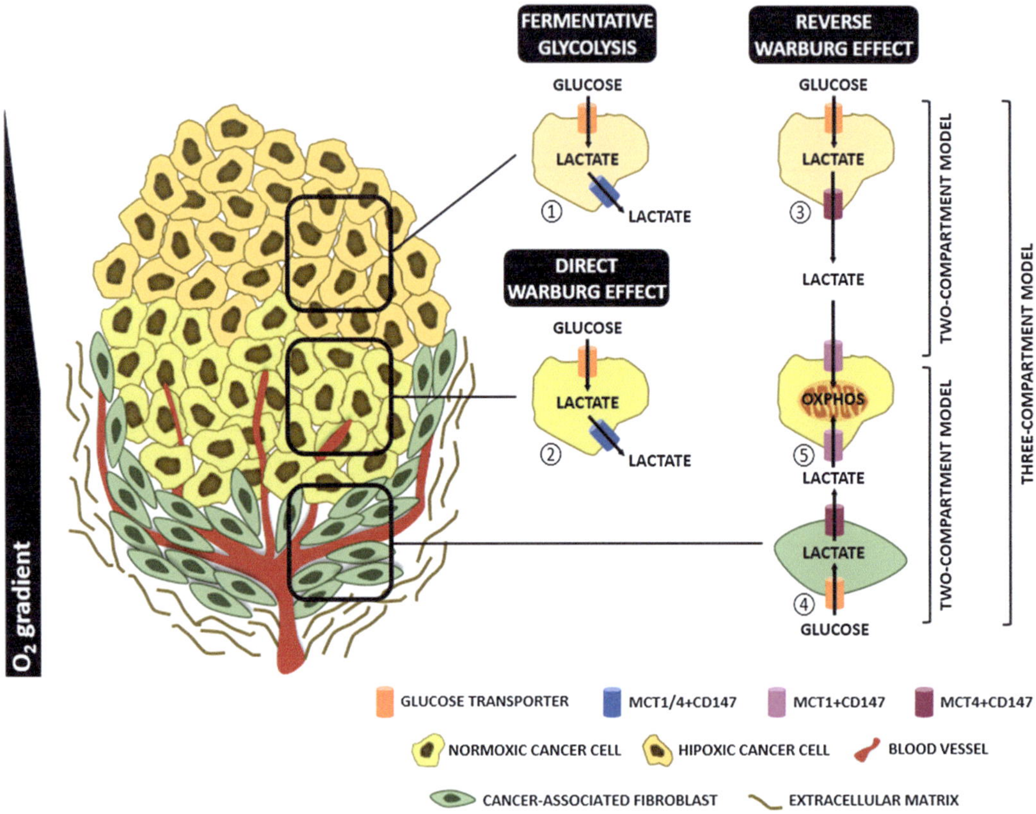

Fig. 3.1 Metabolic phenotypes occurring in the tumour microenvironment. In hypoxic conditions, cancer cells convert glucose to lactate through fermentative glycolysis ①. This phenotype also occurs in oxygenated regions, a phenomenon known as the Warburg effect ②. In either way, cancer cells need to export lactate to the extracellular milieu, in order to avoid intracellular acidosis. This occurs through monocarboxylate transporters 1 and 4 (chaperoned by CD147). Alternatively, glycolytic hypoxic cancer cells ③ and/or cancer-associated fibroblasts ④ are coupled to normoxic cancer cells ⑤ through a reverse Warburg effect, providing them with glycolysis-originating lactate that enters the Krebs cycle and boosts ATP production through OXPHOS (oxidative phosphorylation). Under these conditions, MCT4 is the preferred lactate exporter, while MCT1 is the preferred lactate importer

(Vander Heiden et al. 2009; Cheng and Ristow 2013). Conventional models of cellular energy dynamics stated that under oxygen availability, glucose is completely oxidized via respiration in the mitochondria originating CO_2 as the end product, while in periods of hypoxia, glucose is converted to lactate through fermentative glycolysis (Racker 1974) (Fig. 3.1).

Malignant solid tumours are usually heterogeneous, with both oxygenated and hypoxic areas, forcing cancer cells to adapt to the environment in order to survive (Whitaker-Menezes et al. 2011b). Thus, in the same tumour, it is possible to find both oxidative and glycolytic cells, which, in turn, will modulate the metabolism of co-existing cells in the tumour microenvironment to boost tumour growth and induce progression.

3.2 Direct and Reverse Warburg Effect

Cancer cells prefer to obtain their energy from fermentative rather than oxidative cell metabolism, regardless of oxygen availability. This observation was originally reported in 1924 by Otto Warburg, who identified lactate as a characteristic product released by cancer cells (Warburg

1956b). This phenomenon, then termed "Warburg effect" or "aerobic glycolysis" (Fig. 3.1), was recognized as a distinctive metabolic feature of many types of malignancies and, consequently, it was introduced as a new hallmark of cancer in 2011 (Hanahan and Weinberg 2011). The Warburg metabolic phenotype proved to be a useful tool not only for clinical detection of glucose-addicted tumours, but also to monitor tumour growth and dissemination by ^{18}F-deoxyglucose-Positron Emission Tomography (FDG-PET) imaging (Aide et al. 2017; Im et al. 2018).

Warburg-dependent cancer cells have an inefficient mechanism of energy production, since the amount of ATP produced by aerobic glycolysis (2 ATP per mol of glucose) is about 18-fold lower than the one obtained by oxidative phosphorylation (OXPHOS; 36 ATP per mol of glucose) (Vander Heiden et al. 2009; Cheng and Ristow 2013). To compensate for this metabolic inefficiency, the majority of cancer cells increase the rate of glucose uptake to support their metabolic demands. Indeed, the production of lactate from glucose was described to be 10–100 times faster than the complete oxidation of glucose, being the amount of ATP synthetized for a given period of time much higher (Shestov et al. 2014). The strong increase of glucose consumption has been associated with the generation of carbon skeletons needed for actively proliferating cells. A substantial fraction of these carbons accumulate as glycolytic intermediates that fuel anabolic pathways stemming from glycolysis, converging on protein, lipid and acid nucleic synthesis, thereby allowing cell growth. For instance, glucose-6-phosphate, a side product of glycolysis, can be deviated to the pentose phosphate pathway, that ultimately produces ribose-5-phosphate and NAPDH for nucleic acid and lipid synthesis (Langbein et al. 2006; Liberti and Locasale 2016). Glyceraldehyde-3-phosphate can give rise to phosphatidic acid necessary for lipid production. Moreover, amino acids such as serine and glycine, required for protein and DNA/RNA synthesis, can be produced from 3-phosphoglycerate, a pyruvate precursor (Mazurek et al. 2005). Additionally, in conditions of aerobic glycolysis, cancer cells can survive to fluctuating oxygen levels (Pouyssegur et al. 2006).

Beyond the cell-intrinsic functions described above, the Warburg effect has been commonly implicated in the tumour microenvironment (TME). Upregulation of glycolysis increases intracellular lactate and H^+ content and, consequently, decreases intracellular pH (pH_i). In order to maintain pH homeostasis and to avoid glycolysis inhibition due to a negative feedback mechanism, cancer cells export lactate and H^+ ions from the cell through monocarboxylate transporters (MCTs), acidifying the extracellular milieu (Parks et al. 2013). Low extracellular pH (pH_e) leads to H^+ diffusion, following the concentration gradients, from cancer cells to the surrounding stroma, a phenomenon that is harmful to normal cells (Gatenby and Gawlinski 1996; Gatenby et al. 2006). Cancer cells, due to their evolutionary biological capabilities, developed adaptive mechanisms to resist TME acidosis. These include the activation of anti-apoptotic proteins (*i.e.* Bcl-2 and GRP65) coupled with intracellular pathways, such as extracellular-signal related kinase 1/2 (ERK1/2) pathway, involved in cell proliferation, differentiation and survival (Riemann et al. 2011; Ryder et al. 2012); and the upregulation of pH_i-regulating systems, such as Na^+/H^+ exchangers (NHEs), Na^+/HCO_3^- cotransporters and carbonic anhydrases (CAs), that maintain an alkaline cytoplasm despite the acidic pH_e (Parks et al. 2013). Moreover, several studies reported that lactate secreted by cancer cells is strongly active within the TME, both as nutrient substrate and signalling molecule, contributing to carcinogenesis and to the modulation of the oncogenic process. Briefly, this monocarboxylic acid participates in the induction of angiogenesis (Beckert et al. 2006; Sonveaux et al. 2012), cell migration and cell invasion (Vegran et al. 2011; Baumann et al. 2009) and allows cancer cells to escape the immune system surveillance (Fischer et al. 2007; Husain et al. 2013a; Gottfried et al. 2006; Colegio et al. 2014).

The direct "Warburg-like" glycolytic phenotype of cancer cells (Fig. 3.1) stands nowadays as the main principle sustaining the emerging hallmark of energy metabolism reprogramming

(Hanahan and Weinberg 2011), but the metabolism of a malignant tumour is only partially explained by that theory. The first evidence showing that metabolic heterogeneity occurs within tumours came from the studies of Sonveaux et al. (Sonveaux et al. 2008). Using *in vitro* (human cervical cancer cell line) and *in vivo* (mouse xenografts of lung and colorectal carcinomas) models, the authors showed that glucose is used as a primary substrate only by hypoxic cancer cells; opposing the expectations, oxidative cancer cells used lactate as a prominent fuel for energy production through OXPHOS. Thus, lactate-producing glycolytic cancer cells fuelled oxidative cells with lactate, being this symbiotic association mediated by MCT1 (overexpressed by normoxic cancer cells; preferentially mediates lactate uptake) and MCT4 (overexpressed by hypoxic cancer cells; preferentially mediates lactate efflux) (Fig. 3.1). While lactate usage spared glucose for hypoxic cancer cells, genetic or pharmacological inhibition of MCT1 in pre-clinical models reverted the lactate-fuelled OXPHOS to glucose-dependent glycolysis, which hampered tumour growth due to glucose starvation, and promoted resistance to radiation (Sonveaux et al. 2008). Interestingly, this cancer-occurring lactate shuttle seems to be a cooption of normal physiological mechanisms that occur in skeletal muscle (from white to red fibres) (Park et al. 2015) and in brain (from astrocytes to neurons) (Machler et al. 2016). Therefore, and opposing the Warburg hypothesis on the dysfunctional mitochondria of cancer cells (Warburg 1956a), some tumours indeed exhibit high rates of OXPHOS (Whitaker-Menezes et al. 2011a), and aerobic glycolysis and OXPHOS may differently contribute to ATP production. This means that, in a single tumour mass, heterogeneous populations of cancer cells coexist and may even become metabolically coupled in a symbiotic interdependency, in order to cope with the demands of the TME (Wilde et al. 2017). Since Sonveaux's studies, several authors reported the occurrence of this metabolic symbiosis network between hypoxic and normoxic cancer cell populations. Using a mouse model of pancreatic cancer, Guillaumond et al. described hypoxia as a trigger of the "glycolytic" switch of cancer cells from OXPHOS to glycolysis, additionally demonstrating that lactate secreted from hypoxic cancer cells enhanced the growth of the normoxic compartment (Guillaumond et al. 2013). Lactate was also shown to be a carbon source for the TCA cycle in human lung tumours (Faubert et al. 2017). In the study by Allen et al., metabolic compartmentalization was induced by angiogenesis inhibitors, which made vessel-distal hypoxic cancer cells to increase glucose uptake, being the glycolysis-originating lactate imported by the cells near the blood vasculature; as those cells upregulated mTOR signalling due to increased lactate catabolism, mTOR inhibitors led to disruption of the metabolic symbiosis (Allen et al. 2016).

Sonveaux's hypothesis was later reinforced when Lisanti's group, using proteomic analysis and transcriptional profiling of caveolin-1 (Cav-1) null stromal cells, demonstrated that cancer cells are able to induce a "Warburg-like" metabolism in the surrounding stromal fibroblasts; these corruptible, cancer-associated fibroblasts (CAFs) become metabolic slaves of cancer cells, providing them with glycolysis-originated lactate and pyruvate that will further enter the Krebs' cycle, leading to effective ATP production via OXPHOS (Fig. 3.1). In fact, cancer-induced differentiation markers and glycolytic enzymes were upregulated under normoxia in the Cav-1 $^{(-/-)}$ stromal cells, which was further confirmed in human breast cancer tissues (Pavlides et al. 2009) and in a breast cancer mouse model; treatment of these xenografts with glycolytic blocking agents inhibited tumour growth (Bonuccelli et al. 2010). Cav-1 autophagic loss, frequent in CAFs (Martinez-Outschoorn et al. 2010a), hampers mitochondrial functionality and induces the glycolytic switch (Asterholm et al. 2012). This is thought to occur through HIF-1α (hypoxia inducible factor-1alpha) stabilization and NF-κβ (nuclear factor kappa-light-chain-enhancer of activated B cells) activation (Martinez-Outschoorn et al. 2010b). Moreover, Whitaker-Menezes et al. showed, using co-culture conditions, that breast cancer cells overexpressed MCT1, and were able to induce, through oxidative stress, MCT4 expression in CAFs; similar results were obtained when analysing MCT immunoexpression in breast cancer samples, corroborating the notion that a lactate shuttle exists between metabolically heterogeneous cell populations, namely epithelial and stromal cells (Whitaker-Menezes et al. 2011b). Moreover, it has been demonstrated that breast cancer cells generate 80% of their ATP through OXPHOS, which is in accordance with their metabolic dependence on CAF-originating lactate (Guppy et al. 2002).

This emerging metabolic phenotype of CAFs, coined as the "Reverse Warburg effect" (Pavlides et al. 2009) (Fig. 3.1), was further amplified with the "three-compartment model" described by Curry et al. (Curry et al. 2013) (Fig. 3.1). Using human samples of head and neck cancer and a large panel of metabolism biomarkers, the authors demonstrated the existence of three distinct cell populations: proliferative, mitochondrial-rich cancer cells; non-proliferative, mitochondrial-poor cancer cells; and non-proliferative, mitochondrial-poor stromal cells. The non-proliferative, catabolic cancer and stromal cells overexpressed MCT4, which was associated with a marked oxidative stress, an avidity for glucose and a poor clinical outcome. Conversely, the adjacent proliferative cancer cells overexpressed MCT1 and exhibited an anabolic oxidative metabolism, being specialized in the uptake of mitochondrial fuels paracrinally provided by the catabolic compartments. Since the first study demonstrating that metabolic heterogeneity exists within the tumour microenvironment (Sonveaux et al. 2008), several other reports showed, in different cancer models (reviewed in (Pinheiro et al. 2015c)), that cancer cells are able to create a compensatory ecosystem around them, in which CAFs, the most important components of the stromal compartment, are enslaved to generate metabolic fuels, namely lactate, that sustain tumour growth and progression by generating an enhanced anabolism and ATP production in cancer cells. Importantly, this metabolic flexibility has been shown to impact cancer patients' prognosis, which makes it an area of therapeutic intervention for which exciting results have already been obtained (Chen and Song 2018).

The metabolic phenotypes described above occur in each distinct TME in a context-dependent manner, but a common denominator exists between the direct Warburg effect and the reverse Warburg effect: lactate. Microenvironmental levels of this by-product of glycolysis are consistently elevated in malignancies as a result of increased glycolytic rates under hypoxic conditions or under energy metabolism reprogramming phenotypes, and upregulation of monocarboxylate transporters is an obligate condition for proper lactate exchange. As it will be described below, lactate not only fuels oxidative cancer cells, but is also the main source of microenvironmental acidosis. The pleiotropic effects of increased lactate concentrations contribute to the success of tumour growth and metastasis, while impairing therapeutic response and overall prognosis in cancer patients (San-Millan and Brooks 2017).

3.3 Role of Lactate and Lactate Transporters in Cancer Aggressiveness

3.3.1 Role of Lactate in Carcinogenesis

Microenvironmental stresses like hypoxia and nutrient limitations, exert selective pressure on cancer cells which imposes the glycolytic phenotype, resulting in the production of high amounts of lactate (Dhup et al. 2012; Wolf et al. 2010). As stated above, even though less efficient than oxidative phosphorylation (OXPHOS), lactate fermentation confers several competitive advantages to cancer cells, including rapid ATP production, redirection of glycolysis intermediates to biosynthetic pathways, and acidification of the tumour microenvironment, which sustains tumour growth and progression (Dhup et al. 2012).

During many years, lactate was considered a waste product of glycolysis, however, it has been more recently demonstrated that it plays an important role in malignancy. Lactate levels are up to 40-fold higher in glycolytic tumours and they are greatly associated with cancer aggressiveness (reviewed in (San-Millan and Brooks 2017)). Lactate has two main roles in cancer, as a metabolic fuel and as a signalling molecule. Lactate is a central metabolite in tumour cell symbiosis, being an important energy source for oxygenated tumour cells, in which it is converted back into pyruvate and oxidized in the mitochondria (Sonveaux et al. 2008). As most of the glucose is taken up by glycolytic cancer cells (e.g. hypoxic regions), oxidative cancer cells (normoxic regions) use lactate over glucose as fuel (Semenza 2008). The high glycolytic flux in cancer cells also allows to sustain proliferation, angiogenesis, immune escape, cell migration and metastasis, in which lactate production plays a key role (San-Millan and Brooks 2017).

The tumour-associated microenvironment has increasingly being recognized has playing a crucial role in carcinogenesis (Chen et al. 2015). Besides cancer cells, this tumour ecosystem comprises immune cells, non-malignant stromal cells, fibroblasts, as well as endothelial cells that constitute the vasculature of tumours. The typical acidic microenvironment of malignant tumours modulates the interaction and signalling among the cellular players, stimulating carcinogenesis. Lactate is the main player in the maintenance of this acidic phenotype and, by modulating the tumour microenvironment, contributes to several features of tumour progression (Hirschhaeuser et al. 2011), including cell migration and invasion, angiogenesis, and escape to immune surveillance.

Cell migration and invasion are essential elements in the carcinogenic process, however, the contribution of lactate to this process has been largely neglected in the literature (Dhup et al. 2012). It has been demonstrated that addition of exogenous lactate to cancer cells increases their motility and migration capacity (Goetze et al. 2011). One of the mechanisms appear to involve hyaluronan and its receptor CD44, since lactate stimulates the production of hyaluronan and expression of CD44, which function is associated with cell migration and invasion (Stern et al. 2002). Additionally, acidosis leads to activation of matrix metalloproteinases (MMPs), which results in extracellular matrix degradation, being involved in cell migration and invasion (Kato et al. 2007). High levels of lactate have been also associated with higher incidence of distant metastasis and patient poor prognosis in different types of cancer, however the mechanisms involved are not completely understood (Hirschhaeuser et al. 2011; Walenta and Mueller-Klieser 2004).

Lactate released from cancer cells is a promoter of angiogenesis, being a crucial signalling molecule in the cancer-endothelial cell crosstalk. Lactate stimulates the production of VEGF (vascular endothelial growth factor) and its receptor VEGFR2 in endothelial cells, through stabilization of HIF-1α (San-Millan and Brooks 2017; Sonveaux et al. 2012). However, this process is not exclusively dependent on HIF-1α expression. Vegran et al. demonstrated that lactate induces interleukin-8 (IL-8) production in endothelial cells by nuclear factor-kappa B (NFκB) stimulation, which incites new blood vessel formation and increased endothelial cell migration (Vegran et al. 2011). Additionally, lactate was reported to modulate angiogenesis through MYC stabilization, with upregulation of VEGF, IL-8 and CD31 levels under prolonged hypoxia, via ERK1/2 signalling (Lee et al. 2015).

Importantly, high lactate levels are associated with escape of immune surveillance. Cancer cell-generated lactate is described to inhibit dendritic and T cell activation, as well as natural killer cells (Fischer et al. 2007; Goetze et al. 2011; Wolf et al. 2010; Gottfried et al. 2006). In addition, lactate is known to stimulate polarization of resident macrophages to the M2 state, known as tumour-associated macrophages, which play an important pro-tumorigenic role (Carmona-Fontaine et al. 2013; Colegio et al. 2014). Further details on the role of lactate in the cancer cell immune escape are given below.

Tumour acidosis itself can also promote therapeutic resistance. In prostate (Hao et al. 2016), hepatocellular and cervical (Shimura et al. 2014) cancer models, radioresistance was associated with the glycolytic phenotype, as abrogation of glucose uptake and lactate production reverted acquired radioresistance. L- and D-lactate not only mediate DNA repair programs but also the resistance of cervical cancer cells to clinically used chemotherapeutics; inhibition of MCT activity supressed these pro-tumoural activities (Wagner et al. 2015). The lactate receptor GPR81 was also found to be involved in doxorubicin chemoresistance (Wagner et al. 2017). A recent study identified, in colorectal cancer cells, B7-H3, an immunoregulatory protein, as a novel mediator of the glycolytic phenotype and chemoresistance by interacting with hexokinase 2 (HK2), the first rate-limiting enzyme controlling glycolysis (Shi et al. 2019). Moreover, metabolic symbiosis through lactate shuttling among hypoxic and normoxic cancer cells was established as an escape mechanism to anti-angiogenic therapy (Allen et al. 2016).

3.3.2 Role of Lactate Transporters in Cancer Aggressiveness

Monocarboxylate transporters (MCTs) are transmembrane proteins responsible for the transport of lactate, and other metabolically important monocarboxylates, such as pyruvate, branched-chain oxoacids, and ketone bodies, across cell membranes (Halestrap and Wilson 2012). MCTs belong to the SLC16A gene family, with 14 members identified so far, however, only the first four, MCT1–4, catalyze the proton-coupled transport of monocarboxylates.

The expression pattern of each MCT isoform varies according to the metabolic requirements of each tissue [reviewed in (Pinheiro et al. 2012)]: MCT1 has an intermediate affinity for the substrates, being expressed in most human tissues; MCT2 is a high affinity transporter, being mainly present in tissues that use lactate as fuel (e.g. brain, cardiac and red skeletal muscle) and in tissues that use lactate as a gluconeogenic substrate (e.g. kidney and liver). MCT3 presents a more specific localization, being present in the retinal pigment and choroid plexus epithelia. MCT4 is a low affinity transporter and is present in tissues with high glycolytic activity, including white skeletal muscle fibres, astrocytes and white blood cells (Halestrap and Wilson 2012; Halestrap 2013).

During the last decades, the function of MCT1 and MCT4 in the maintenance of the metabolic phenotype of cancer cells, has been associated with cancer aggressiveness. For that reason, the role of MCTs in several cancer types, including expression analysis in human cancer tissues have been widely explored in the past years [reviewed in (Pinheiro et al. 2015c)]. MCT1 and MCT4 are upregulated in tumour cells when compared to adjacent normal tissue in a variety of human malignancies including lung, breast, head and neck, renal, brain, adrenal, melanomas, pancreatic, colorectal, ovarian, bladder and cervix [reviewed in (Granja et al. 2017)]. Interestingly, in liver and prostate cancer, MCT1 appears downregulated, while MCT4 is upregulated. On the other hand, fewer cancer types express MCT2, but overexpression has been reported in lung, brain, pancreas, colorectal and prostate cancer (Granja et al. 2017).

MCTs have been recognized as markers of poor survival in several tumour types. Especially MCT1 and MCT4 have been associated with some poor prognostic variables, advanced tumour staging, high grade tumours, shorter overall and disease-free survival, in a variety of human cancers, including adrenocortical tumours, breast, renal, prostate, head and neck, hepatocellular and pancreas [reviewed in (Pinheiro et al. 2015c)]. In contrast, MCT2 expression appears to be associated with favourable prognostic parameters, namely lower mitotic index, small tumour size, absence of metastasis and good prognosis in adrenocortical malignant tumours (Pinheiro et al. 2015b).

MCTs play a key role in the metabolic adaptations of cancer cells. On one hand, they mediate lactate efflux, essential for the maintenance of the hyper-glycolytic phenotype, and, on the other hand, they contribute to pH regulation, supporting the acid-resistant phenotype. Thus, considering the role of MCTs in cancer aggressiveness, inhibiting MCT activity will compromise intracellular pH homeostasis and will modulate the acidic tumour microenvironment. Therefore, these transporters are attractive targets in cancer therapy.

There are a number of MCT inhibitors described in the literature, with different affinities and specificities (Enerson and Drewes 2003). These include, aromatic compounds such as α-cyano-4-hydroxycinnamate (CHC), stilbene disulfonates, such as 4,4'-di-isothiocyanostilbene-2,2'-disulfonate (DIDS), and bioflavonoids such as quercetin. These compounds are not isoform specific but they display higher affinity for MCT1 and 2 at lower concentrations. Besides, these compounds are known to target other molecules: CHC inhibits the mitochondrial pyruvate carrier (Schell and Rutter 2013), and the stilbene disulfonates inhibit the chloride/bicarbonate exchanger AE1 (Halestrap and Wilson 2012). CHC is the most well studied classical MCT inhibitor. *In vitro*, CHC decreases lactate transport, cell proliferation, invasion and migration and increases cell death in different cancer models, including glioma, colorectal, cervical and breast cancer cells (Miranda-Gonçalves et al. 2013; Kumar et al. 2013; Sonveaux et al. 2008; Morais-Santos et al. 2014). *In vivo*, CHC decreases tumour growth, sensitizes to radiation, induces tumour necrosis and decreases invasion (Miranda-Gonçalves et al. 2013; Sonveaux et al. 2008; Colen et al. 2011).

A set of specific and high-affinity MCT1 inhibitors were more recently developed by AstraZeneca. One of the compounds is AR-C155858, which is active against MCT1 and MCT2 but not MCT4 (Ovens et al. 2010) and the other is AZD3965, a selective MCT1 inhibitor. AZD3965 has been tested *in vitro* and *in vivo* in different cancer models, leading to accumulation of intracellular lactate in cancer cells and decrease in tumour growth *in vivo* (Polanski et al. 2014; Beloueche-Babari et al. 2017). AZD3965 has already reached clinical trials, firstly in patients with advanced solid tumours (prostate and gastric) or lymphomas, mainly aiming to study safety, dose limiting toxicities, and determine the maximum tolerated dose (NCT01791595) (Halford et al. 2017). This trial is currently recruiting patients and there are still no results available about the efficacy of treatment. Importantly, inhibition of one MCT isoform can be compensated by another MCT isoform, which is the case of MCT1 and MCT4 (Le Floch et al. 2011). Thus, in some cancer models, to impair lactate efflux, it will be necessary to inhibit both isoforms. However, no commercial specific MCT4 inhibitors are available so far and the available studies impair MCT4 activity by gene deletion or downregulation.

MCT1-4 require the presence of a chaperone protein for membrane localization and activity. The chaperone protein for MCT1 and MCT4 is CD147, also known as basigin or EMMPRIN. CD147 major pro-tumoural action appears to be chaperoning MCTs (Le Floch et al. 2011) and several reports show that CD147 and MCTs are co-expressed in a variety of human tumour tissues [reviewed in (Granja et al. 2017)]. Therefore, targeting CD147 to inhibit MCTs appears to be a rational approach. In this context, CD147 silencing has been described to inhibit MCT1/MCT4 function, decrease lactate efflux (Slomiany et al. 2009) with consequent decrease in intracellular pH (Schneiderhan et al. 2009; Le Floch et al. 2011; Baba et al. 2008) and *in vivo* tumour malignant potential (Schneiderhan et al. 2009; Le Floch et al. 2011). Moreover, CD147 expression is also associated with poor prognosis cancer parameters such as tumour progression and chemoresistance (Xiong et al. 2014). Progress has been made with the development of CD147-directed monoclonal antibodies (Kasinrerk et al. 1999), however, there are still no clinical studies with inhibitors of CD147 in the clinical setting.

Table 3.1 shows a summary of *in vivo* studies targeting MCT1 and MCT4 in cancer models and the respective results. Different MCT inhibition approaches have been tried, and most of them show decrease in tumour growth. Importantly, there is also interest in the combined treatment of MCT inhibition with radio- and chemotherapy, as well as targeted therapy. However, most studies have been performed in mice xenografts, which may limit the conclusions due to the impairment of the immune system, which does not reflect the clinical setting.

Table 3.1 Studies targeting MCT1 and MCT4 in *in vivo* cancer models

Cancer type	Cancer model	Strategy for targeting lactate transport	Outcome	References
Bladder cancer	Orthotopic xenograft model	MCT4 siRNA	↓ tumour growth	(Todenhofer et al. 2018)
Breast cancer	Orthotopic syngraft mice model	AR-C155858 (MCT1/2 inhibitor)	No effect on tumour volume and weight	(Guan et al. 2018)
	Subcutaneous mice xenografts	MCT4 shRNA	↓ tumour growth ↑ tumour-free survival	(Andersen et al. 2018)
	Subcutaneous syngraft mice model	MCT1 CRISPR	Inhibition of migration, invasion, and spontaneous lung metastasis	(Payen et al. 2017)
	Orthotopic mice xenografts	MCT1 siRNA MCT4 siRNA	↓ tumour initiation ↓ tumour growth	(Morais-Santos et al. 2015)
	Subcutaneous mice xenografts	MCT1 overexpression	Sensitization to 3-BrPA treatment	(Birsoy et al. 2013)
Cervical cancer	Subcutaneous mice xenografts	7ACC2 (mitochondrial pyruvate carrier and MCT1 inhibitor)	Sensitization to radiotherapy	(Corbet et al. 2018)
	Subcutaneous mice xenografts	MCT1 and ASCT2 siRNA mixture (targeted nanoparticles)	↓ tumour growth	(Corbet et al. 2016)
	Subcutaneous mice xenografts	MCT1 shRNA	↓ tumour growth	(De Saedeleer et al. 2012)
Colorectal cancer	Subcutaneous mice xenografts	MCT4 and MCT1 siRNA	↓ tumour growth Additive with radiotherapy or chemotherapy	(Kim et al. 2018)
	Subcutaneous mice xenografts	CD147 downregulation	↓ tumour growth	(Li et al. 2013)
Gastric cancer	Subcutaneous mice xenografts	AR-C155858 (MCT1 inhibitor) MCT1 and MCT4 siRNA	↓ tumour growth and peritoneal dissemination Synergistic effect of MCT1/MCT4 silencing with radio- and chemotherapy	(Lee et al. 2016)
Glioma	Subcutaneous mice syngraft model	N,N-dialkyl carboxy coumarins (MCT1 inhibitors)	↓ tumour growth	(Gurrapu et al. 2016)
Lung cancer	Subcutaneous mice xenografts	AZD3965 (MCT1 inhibitor) MCT4 shRNA	↓ tumour growth in combination with tyrosine kinase inhibitors (TKIs) ↓ in vivo resistance to TKIs	(Apicella et al. 2018)
	Subcutaneous mice xenografts	AZD3965 (MCT1 inhibitor)	↓ tumour growth	(Polanski et al. 2014)
Lymphoma	B-cell lymphoma syngraft model (i.v. injection)	Indirect MCT1 downregulation by targeting *Myc* oncogene	↓ tumour growth	(Gan et al. 2016)
	Subcutaneous mice xenografts	AZD3965 (MCT1 inhibitor)	↓ tumour growth	(Beloueche-Babari et al. 2017)
Osteosarcoma	Subcutaneous and orthotopic mice xenografts	MCT1 shRNA or CHC (MCT1 inhibitor)	↓ tumour growth and enhancement of chemotherapy effect	(Zhao et al. 2014b)

(continued)

Table 3.1 (continued)

Cancer type	Cancer model	Strategy for targeting lactate transport	Outcome	References
Ovarian cancer	Subcutaneous mice xenografts	MCT1 shRNA	Reversal of cisplatin-resistance	(Yan et al. 2015)
Pancreatic carcinoma	Subcutaneous mice xenografts	miR-124 overexpression (MCT1 inhibition) MCT1 siRNA	↓ tumour growth for both approaches	(Wu et al. 2018a)
Prostate cancer	Subcutaneous mice xenografts	MCT4 siRNA (in nanoparticles)	↓ tumour growth	(Liu et al. 2018)
	Subcutaneous mice xenografts	MCT1 siRNA	↓ tumour growth	(Sanita et al. 2014)

↓ decreased; ↑ increased; 3-BrPA 3-bromopuruvate

3.4 Lactate Shuttles Between Cancer Cells and Stromal Cells

The replacement of the classical metabolomic studies in homotypic populations of cancer cells by the most recent studies involving the TME have brought new insights on the complicity between cancer cells and the surrounding stroma of cancer-associated cells, namely CAFs, endothelial cells (ECs) and immune cells. In fact, it is the metabolic cooperation between different cells of the TME that dictates the success of tumour growth, dissemination and metastasis. For that reason, the TME concept and the rewiring of energy metabolism have both been considered in the most recent versions on the hallmarks of cancer (Fouad and Aanei 2017; Hanahan and Weinberg 2011). That metabolic rewiring necessarily involves shuttling of metabolic intermediates, namely lactate, between cancer cells and the surrounding stroma, as it will be discussed in the next sections.

3.4.1 Fibroblasts

Fibroblasts are versatile, spindle-shaped connective tissue cells specialized in the secretion and resorption of extracellular matrix. Intrinsic survival programmes and cellular plasticity provide them with resilient adaptation abilities. Even in quiescent stages (non-dividing cells), fibroblasts remain metabolically active to preserve their self-integrity, exhibiting high glycolytic rates (Lemons et al. 2010). As a mirror of their versatility, they can become more oxidative under high lactate and pyruvate concentrations (McKay et al. 1983). Under proper signalling cues (growth factors, cytokines and mechanical stress), quiescent fibroblasts acquire biosynthetic, pro-inflammatory, contractile, and adhesive functions that enable them to transdifferentiate into myofibroblasts capable of mediating effective wound healing and take part of the inflammatory response, in a physiological self-limited repair program (McAnulty 2007). In the pathology of cancer, corresponding processes such as the "cancer wound" (development of the tumour) and tumour-promoting inflammation activate CAFs with enhanced proliferative and migratory skills, essential to mediate tumour initiation, desmoplastic reaction, angiogenesis, dissemination and metastasis. Due to their crucial roles in cancer pathology, these abundant microenvironmental stromal components are implicated in tumour recurrence, resistance to treatments and poor clinical outcomes and are, therefore, conspicuous targets for therapeutic intervention (Chen and Song 2018; LeBleu and Kalluri 2018; Rasanen and Vaheri 2010).

The recent advances in metabolic technology have highlighted the reciprocal interactions between cancer cells and CAFs as one of the most important metabolic crosstalks in that heterogeneous ecosystem. As previously mentioned, CAF, enslaved by the parasitic cancer cells, are able to enter into a reverse Warburg programme

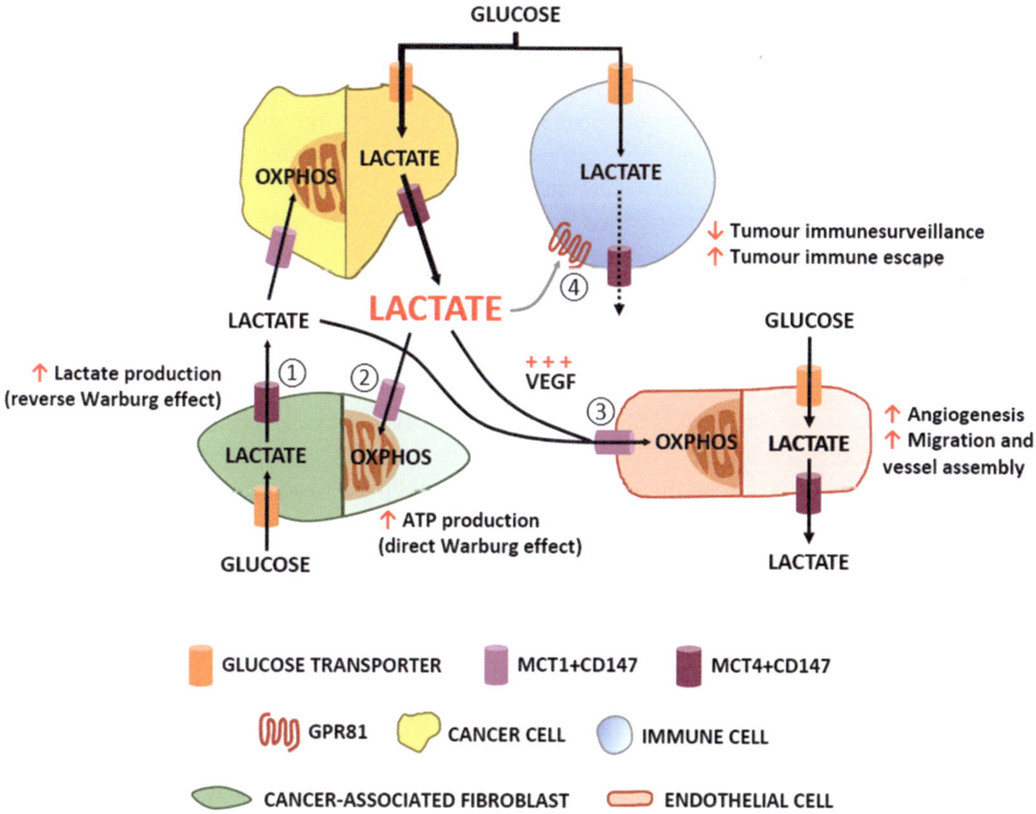

Fig. 3.2 Lactate shuttles between cancer cells and cancer-associated stromal cells. Similar to cancer cells, cancer-associated fibroblasts (CAF), endothelial cells and immune cells exhibit metabolic plasticity in order to adapt to varying microenvironmental conditions; these plastic abilities are hijacked by cancer cells to boost tumour expansion. ① Glycolytic CAF engage into a reverse Warburg effect with oxidative cancer cells, fuelling them with glycolysis-originating lactate to replenish the Krebs cycle and potentiate oxidative phosphorylation (OXPHOS). ② A direct Warburg effect may occur in some models, where oxidative CAF uptake lactate produced by glycolytic cancer cells in order to obtain energy for tumour growth support. ③ Activated endothelial cells (EC) are glycolysis-addicted, but under high microenvironmental acidosis they can switch into an oxidative phenotype, uptaking lactate to promote EC proliferation and angiogenesis, migration and vessel assembly. Increased VEGF secretion by cancer cells, CAF and macrophages further stimulates tumour angiogenesis. ④ Immune cells, dependent on glucose uptake for their proper activation and promotion of anti-tumour immune responses, are outcompeted by glucose-avid cancer cells. Microenvironmental acidosis blocks lactate export from immune cells, disturbing their proper metabolism, decreasing cytokine production and cytotoxic activity, and inducing activation of phenotypes that promote tumour immune escape (e.g. M2-like macrophages); conversely, lactate can act as a signalling molecule in these cells by interacting with lactate receptor GPR81, further impairing the glycolytic flux necessary for the cytolytic phenotype

that enables them with high expression of glycolytic enzymes and MCTs (Fig. 3.2). Since the original demonstrations of this phenotype by Lisantis's group (Pavlides et al. 2009; Curry et al. 2013), several additional studies of the CAF-cancer cells' lactate shuttles have been reported in other cancer models. Indeed, Shan et al. demonstrated increased glucose consumption and lactate production in pancreatic cancer cells-associated fibroblasts, when compared to normal fibroblasts; concurrently, lactate dehydrogenase (LDH) and pyruvate kinase m2 (PKM2) were also upregulated. In cancer cells, OXPHOS was increased, as well as MCT1 levels; MCT1 blockage disrupted the metabolic coupling and inhibited migration and invasion (Shan et al. 2017).

In non-Hodgkin lymphoma, MCT1 and MCT4 were highly expressed in neoplastic lymphocytes and in stromal cells (respectively), indicating a preference for OXPHOS and glycolytic phenotypes (respectively) in those cell populations (Gooptu et al. 2017); a recent study identified CAF-secreted pyruvate as responsible for survival of aerobic lymphoma cells (Sakamoto et al. 2019). In a co-culture system of oral squamous cell carcinoma (OSCC) cells and primary fibroblasts, MCT1 was upregulated in cancer cells, while glucose transporter 1 (GLUT1), hexokinase 2 (HK2), lactate dehydrogenase (LDH) and MCT4 were upregulated in CAF (Wu et al. 2018b). A similar phenotype in lung cancer cells and CAFs was observed in the study by Cruz-Bermúdez et al. (Cruz-Bermudez et al. 2019). Richter et al. demonstrated, in an "in vitro" co-culture system, that SDHB-mutated pheochromocytomas, normally larger and metastatic than wild-type tumours, depend on stromal lactate sources for continuous growth (Richter et al. 2018). The existence of a multi-compartment metabolic model has been associated with cancer aggressiveness and poor outcome in numerous malignancies (Afonso et al. 2016; Cruz-Bermudez et al. 2019; Pertega-Gomes et al. 2014; Zhao et al. 2014a); moreover, in the study by Afonso et al., this TME phenotype has allowed to discriminate prognosis among cisplatin-treated bladder cancer patients based on their tumour's expression profile (Afonso et al. 2016).

Recent studies are beginning to unravel the mechanisms underlying metabolic symbiosis among cancer cells and CAFs. Luo et al. demonstrated that metabolic interaction among non-small cell lung cancer cells and CAFs are physically mediated by unidirectional gap junctions, under the molecular control of connexin 43 (Luo et al. 2018). OSCC cells were shown to display high levels of IL-1β (interleukin-1β), that induced the same expression profile of co-cultured fibroblasts (upregulation of GLUT1, HK2, LDH and MCT4) when administered exogenously to monocultured fibroblasts; silencing of IL-1β in OSCC cells depleted stromal glycolysis in the co-culture system, which led the authors to suggest IL-1β as mediator of the metabolic reprogramming (Wu et al. 2018b). In another study, it was the antioxidant transcription factor Nrf2 (nuclear factor E2-related factor-2) that acted as key regulator of the metabolic shaping towards a reverse Warburg phenotype in malignant and pre-malignant colonic epithelial cells (Diehl et al. 2018). Late stage head and neck squamous cell carcinoma (HNSCC), which is mainly composed of CAFs, have been shown to secrete basic fibroblast growth factor (bFGF) in response to CAF-secreted hepatocyte growth factor (HGF); this facilitates tumour progression and increases extracellular lactate levels due to enhanced glycolytic activity. Concurrent inhibition of the Met and FGFR pathways significantly inhibited CAF-facilitated HNSCC proliferation *in vitro* and xenograft growth *in vivo*. Therefore, in this model, metabolic symbiosis seems to be mediated by reciprocal signalling between CAF and HNSCC involving bFGF and HGF (Kumar et al. 2018). Recently, a crucial role has been attributed to cancer-derived exosomes in enslaving CAFs to facilitate tumour development, with cancer cell-secreted exosomal microRNAs being implicated in the intercellular metabolic symbiosis (Yan et al. 2018; Dai et al. 2018; Rai et al. 2018). In the study by Rai et al., early and late-stage colorectal cancer cell-derived exosomes differentially activated fibroblasts into highly pro-proliferative and pro-angiogenic CAFs, or into pro-invasive CAFs, respectively; those two CAF populations displayed conserved secretion ability of extracellular matrix, oncogenic transformation and metabolic reprogramming, including upregulation of lactate transport (Rai et al. 2018). Interestingly, Zhao et al. demonstrated that miRNA contained in exomes derived from CAFs downregulated OXPHOS genes in pancreatic and prostate cancer cells; those exosomes also contained intact metabolites (glucose, amino acids, lipids, and TCA-cycle intermediates) that were able to replenish cancer cells in nutrient-deprivation settings, in order to maintain their rapid proliferation (Zhao et al. 2016). In fact, a similar phenotype had been described by Koukourakis et al. in preclinical and clinical studies with lung and colorectal cancer; in these studies, cancer cells and CAFs cooperate through a direct Warburg effect (Fig. 3.2), as cancer cells are shaped to reinforced anaerobic glycolytic activity and lactate extrusion, while CAFs display a functional Kreb's cycle and oxidize lactate to support proliferation and dissemination of the primary tumour (Koukourakis et al. 2007; Koukourakis et al. 2006; Koukourakis et al. 2017). The prevalence of the direct Warburg effect on those cancer models, opposed to the frequent occurrence of the reverse Warburg effect in others, denotes the biological complexity of each tumour entity that, in a search for sustained growth, hijacks CAFs in contrasting metabolic affairs, enslaving them in either a direct or reverse Warburg effect in order to optimize its own development. Therefore, careful should be undertaken when developing agents that exploit such metabolic phenotypes at cancer cell level or at CAF level.

3.4.2 Endothelial Cells

Even though cancer cell metabolism has been intensely studied over the past decades, endothelial cell metabolism only received attention only in the last years. ECs are essential in the expansion of the vascular network, and indeed ECs' metabolism has been proposed as a driving force of angiogenesis. In a normal physiological state ECs are in a quiescent state, however every time the formation of a new vessel is demanded, ECs become glycolysis-addicted (Fig. 3.2), increasing GLUT1 expression and converting high amounts of glucose into lactate (Eelen et al. 2015; Eelen et al. 2018). In particular, the endothelium in the neonatal brain expresses intensely MCT1 (Kishimoto et al. 2016; Gerhart et al. 1997; Mac and Nalecz 2003) and MCT2 (Perez-Escuredo et al. 2016).

In a growing tumour, hypoxia induces the recruitment of blood vessels to sustain tumour proliferation and growth. However, due to abnormal development, tumour vasculature presents a leaky, tortuous and disorganized vascularity leading to functionally defective vessels. As a result, tumour blood flow is inconstant, with poor oxygen perfusion, leading to tumour hypoxia regions, together with leaky vessels; this will facilitate intravasation of cancer cells and subsequent metastasis. Moreover, this hypoxic tumour microenvironment deprives cancer cells from nutrients and growth factors, which incites them to stimulate angiogenesis and consequently promotes formation of disorganized and non-funcional vessels, further aggravating tumour hypoperfusion (Potente et al. 2011).

As well described, hypoxia triggers the glycolytic switch not only in cancer cells but also in stromal cells, and this metabolic adaptation is known to stimulate angiogenesis (Pinheiro et al. 2015a). The most pronounced effect is probably exerted by lactate (Sonveaux et al. 2012; Porporato et al. 2012). Thus, understanding EC metabolism, namely the role of lactate as a key metabolic regulator in the interaction between cancer cells and the vasculature, became clearly needed. Lactate shuttling is implicated in the interplay of cancer cells with ECs (Doherty and Cleveland 2013) (Fig. 3.2). Extracellular lactate, secreted in part by cancer cells and CAFs, enhances vascular endothelial growth factor (VEGF) production and activity in fibroblasts (Trabold et al. 2003) and macrophages (Fig. 3.2) (Constant et al. 2000), mechanism proposed by the oxidation of lactate into pyruvate and a decrease in NAD^+ and in protein poly-ADP-ribosylation (Trabold et al. 2003; Constant et al. 2000; M. Xiong et al. 1998). In brain tumours, lactate induced HIF (hypoxia-inducible factor)-1 expression through pyruvate-mediated proline hydroxylation (PHD, prolyl hydroxylase domain) inhibition (Lu et al. 2005; Lu et al. 2002), resulting in increased VEGF production by tumour cells. Also, uptake of lactate, via MCT1, stimulated NF-κB and interleukin-8 (IL-8) activation, promoting EC migration and tube formation (Vegran et al. 2011; Hunt et al. 2007). Sonveaux and co-workers showed that lactate induces HIF-1α activity in normoxic ECs, via an increase in VEGFR2 expression. This occurs through competition of lactate with 2-oxoglutarate, a cofactor for PHD, inhibiting the degradation of HIF-1α in normoxia (Vegran et al. 2011). In counterpart, the pharmacological and genetic inhibition of MCT1 prevented lactate-induced EC migration, vascular sprouting and tube formation *in vitro* and, more importantly, blocked angiogenesis *in vivo* (Sonveaux et al. 2012). In a model of colon cancer xenograft co-injected with HUVECs (human umbilical vein endothelial cells), downregulation of MCT1 in ECs significantly delayed tumour growth (Vegran et al. 2011). Miranda-Gonçalves et al. described the role of MCTs in mediating tumour-EC crosstalk in malignant brain tumours. The *in vitro* results showed that HBMEC (human brain microvascular endothelial cells) expressed high levels of MCT4 and GLUT1, and MCTs downregulation led to a decrease in EC glycolytic phenotype, proliferation and vessel assembly capacity. Interestingly, by exposing ECs to cancer cell conditioned media (CM), it was shown that low levels of glucose and high levels of lactate promote a switch from a glycolytic into an oxidative phenotype (Miranda-Gonçalves et al. 2017). These findings suggest that in the TME, ECs rely on lactate as source of energy (Fig. 3.2). Moreover, exposure of brain ECs to glioma cell CM led to an increase in MCT1 but not MCT4 expression, suggesting that MCT1 was responsible for the uptake of lactate in ECs (Miranda-Gonçalves et al. 2017), as described by others (Sonveaux et al. 2012; Vegran et al. 2011).

The anti-angiogenic effect of MCT1 disruption documented above offers an additional rationale for the use of MCT inhibitors as potential anti-angiogenic therapy for cancer patients.

3.4.3 Immune Cells

The emerging role of immunometabolism as an important regulator of immune system function has been widely described in the last decade. In general, resting immune cells generate most of their energy from FAO (fatty acid oxidation) or using tricarboxylic acid (TCA) cycle, linked to the generation of ATP via OXPHOS (Pearce and Pearce 2013; Biswas 2015), to maintain their housekeeping functions. After activation, interferon-γ (IFN-γ) or LPS-stimulated macrophages (M1-like) and T cells, displaying increased demands for energy and biosynthetic precursors for proteins, lipids and nucleic acids, rapidly switch to aerobic glycolysis, with utilization of glucose and production of lactate (Newsholme et al. 1985; van der Windt and Pearce 2012; MacIver et al. 2013). Indeed, in an inflammation context, MCT4 is up-regulated in activated macrophages, and lactate efflux in required for the maintenance of high glycolysis and inflammatory response of macrophages (Tan et al. 2015). Likewise, exportation of lactate is essential for proper clonal expansion of activated T cells (Broer 2005). While MCT1 is overexpressed under lymphocyte activation, disruption of MCT1 activity resulted in decreased lymphocyte proliferation and increased intracellular lactate levels (Murray et al. 2005). Upregulation of glucose transporters (namely GLUT1) and glycolytic enzymes (namely LDH) also occurs during the glycolytic reprogramming of T cells (Macintyre et al. 2014; Cammann et al. 2016); these cells are dependent upon glucose for exponential growth, as they do not proliferate in a glucose-deprived media (Pearce et al. 2013).

In a nutrient-deprived tumour context, glucose-avid malignant cells, which present a highly glycolytic phenotype, easily succeed in the metabolic competition with immune cells (Fig. 3.2). In fact, metabolic restriction of T cells and a poor T cell tumour infiltrating response was observed in carcinoma tissues displaying upregulated glucose metabolism (Singer et al. 2011; Chang et al. 2015). Besides creating metabolic demanding environments that encroach on the metabolism and function of infiltrating immune cells, cancer cells also release immunosuppressive metabolites and by-products, forming a metabolic symbiosis with immune cells. Surprisingly, shuttling of metabolites has been described as a new route that cancer cells could use to evade the immune system (Wang et al. 2014). The mechanism by which lactate influences immunosuppression is not fully understood. However, it is thought that, on one hand, high lactate concentration exported by cancer cells block the export of lactate by glycolytic immune cells (Fig. 3.2) and, therefore, disturb their metabolism and function (Romero-Garcia et al. 2016); on the other hand, immune cells might take up lactate, which will impair the glycolytic flux necessary for the activated phenotype (Hargadon 2017) and act as signalling molecule (Fig. 3.2) (Romero-Garcia et al. 2016). In *in vitro* experiments, lactate could decrease the DNA binding activity of NF-κB, impairing maturation and differentiation of dendritic cells (DCs) (Puig-Kroger et al. 2003; Gottfried et al. 2006) and inducing a suppressor phenotype (Nasi et al. 2013). In addition, lactate derived from melanoma cells inhibited DC differentiation and suppressed IL-12 production (Hargadon 2017). Tumour-derived lactate has been shown to reduce immunosuppressive activity (Fig. 3.2), hampering lymphocyte proliferation and motility, cytokine production and cytotoxic activity (Fischer et al. 2007; Haas et al. 2015), decreasing the accumulation of myeloid-derived suppressor cells (MDSCs), activating natural killer (NK) cells (Husain et al. 2013a; Husain et al. 2013b) and strongly inhibiting TNF (tumour necrosis factor) secretion from human monocytes (Dietl et al. 2010). LDH- A-associated lactate accumulation in a mice melanoma model inhibited NFAT (nuclear factor of activated T cells) upregulation in T and NK cells and, consequently, decreased IFN-γ production (Brand et al. 2016). A state of anergy in human and murine tumour-infiltrating lymphocytes, as seen by reduced cytokine production and downregulation of IL2Rα (interleukin 2 receptor α) and TCR (T cell receptor), was also observed by Calcinotto et al. at microenvironmental pH of 6–6.5, condition that was rescued by rising pH to physiological levels (Calcinotto et al. 2012). Lactate can also influence the activation of macrophages into a pro-tumoural phenotype (M2-like) (Fig. 3.2) (Lin et al. 2017). In bone marrow-derived macrophages (BMDM), lactate inhibited the activation of a pro-inflammatory status in a GPR81-independent fashion (GPR81, G protein-coupled receptor 81, a cell-surface receptor for lactate) (Errea et al. 2016). In line with is, lactate derived from a pancreatic tumour cell line induced the polarization of THP1 (human monocytic cell line) into an M2-like phenotype (Ye et al. 2018). In a microfluidic device, lactate produced by bladder cancer cells reprogrammed tumour associated macrophages (TAMs) to an M2-like phenotype, while blockage of the lactate shuttle inhibited the acquisition of the M2-like phenotype (Zhao et al. 2015). In a murine model of Lewis lung carcinoma, Colegio and co-workers showed that tumour-derived lactate could induce VEGF expression and the protumoural M2-like polarization of TAMs by a mechanism mediated by HIF-1α and MCTs (Colegio et al. 2014). In a similar model, Seth et al. reported that myeloid-specific deletion of LDH-A resulted in accumulation of pro-inflammatory macrophages, diminished VEGF and PD-L1 (programmed death ligand 1, an immune checkpoint) expression by cancer cells, and increased $CD8^+$ T cell cytotoxicity; the authors suggested suppression of lactate-driven PD L1 expression as a putative mechanism for the increased anti-tumour activity (Seth et al. 2017). Recently, a study using human samples from OSCC described that macrophages within cancer cell nests or in the immediate adjacent stroma co-expressed CD163 (M2-like marker) and MCT4 (Bisetto et al. 2018).

As outlined above, metabolic competition occurs between cancer and immune cells in the heterogeneous tumour microenvironment. Coupled to glucose and amino acid deprivation, lactic acidosis, poor vascularization, low oxygen perfusion and high amounts of reactive oxygen species, strongly compromise the function of the immune system per se, while growth, dissemination and immune escape of the primary tumour is facilitated. For that reason, metabolic targeting of cancer cells seems to be a rational approach not only to compromise tumour progression but also to rescue anti-tumour immune function.

3.5 Summary and Future Directions

Cancer cells are especially known for their sweet tooth; they uptake glucose at overwhelming rates and convert a significant part into lactate. As emphasized above, even though lactate fermentation is less efficient than mitochondrial respiration, glucose addiction provides several survival advantages to cancer cells and allow them to develop a set of tools to thrive and conquer new environments. However, cancer cells are not alone in this battle; they rely on other cells also present in the malignant tumour ecosystem, which they control and dominate.

Lactate, previously considered as a waste product of glycolysis, is nowadays recognized as a key molecule in the crosstalk among the cells that constitute the malignant tumour. In the Warburg phenotype, lactate is produced by glycolytic cancer cells and can be used by neighbour oxidative cells as fuel, which includes cancer cells. This lactate can also serve as a signal molecule, which will modulate the function/activation of endothelial cells, fibroblasts, immune cells, as well as other stromal cells, boosting important malignant features, such as angiogenesis, capacity to invade, and escape the immune system. In the reverse Warburg effect phenotype model, cancer cells are oxidative and can use lactate produced from glycolytic stromal cells, also creating a symbiotic model which will boost tumour growth.

But why is it important to know cancer cell nutritional preferences? The knowledge of cancer cell metabolic phenotypes can be explored to improve diagnostic, prognostic and treatment of cancer patients. Cancer glucose addiction is already explored in the clinic as a diagnostic tool, with the use of a non-metabolizable glucose analogue (FDG-PET scan). Also, there are new tools being developed and tested, such as PET tracers for lactate, which allows to monitor MCT1-dependent lactate uptake in tumours *in vivo* (Van Hee et al. 2017).

The levels of lactate in the tumours, as well as the expression of proteins that participate in the metabolic phenotype of cancer cells can have prognostic value, such as the case of lactate transporters (MCTs) in different clinical cancer settings. Additionally, these molecules can be used as targets for cancer therapy, and some of these have already reached clinical trials, with promising preliminary results. MCTs are no exception and the activity of a specific MCT1 inhibitor (AZD3965) is now being investigated in cancer patients with positive expression of MCT1.

Thus, by pursuing the research on cancer metabolism, researchers can better understand this hallmark of cancer and exploit it to improve cancer patient's lives.

Acknowledgments This article has been developed under the scope of the project NORTE-01-0145-FEDER-000013, supported by the Northern Portugal Regional Operational Programme (NORTE 2020) under the Portugal Partnership Agreement, through the European Regional Development Fund (FEDER), and through the Competitiveness Factors Operational Programme (COMPETE) and by National funds, through the Foundation for Science and Technology (FCT), under the scope of the project POCI-01-0145-FEDER-007038. AP, JA and SG received fellowships from FCT, ref. SFRH/BD/148476/2019, SFRH/BPD/116784/2016 and SFRH/BPD/117858/2016, respectively.

Competing Interests The authors declare no conflict of interest.

References

Afonso J, Santos LL, Morais A, Amaro T, Longatto-Filho A, Baltazar F (2016) Metabolic coupling in urothelial bladder cancer compartments and its correlation to tumor aggressiveness. Cell Cycle 15(3):368–380. https://doi.org/10.1080/15384101.2015.1121329

Aide N, Lasnon C, Veit-Haibach P, Sera T, Sattler B, Boellaard R (2017) EANM/EARL harmonization strategies in PET quantification: from daily practice to multicentre oncological studies. Eur J Nucl Med Mol Imaging 44(Suppl 1):17–31. https://doi.org/10.1007/s00259-017-3740-2

Allen E, Mieville P, Warren CM, Saghafinia S, Li L, Peng MW et al (2016) Metabolic Symbiosis Enables Adaptive Resistance to Anti-angiogenic Therapy that Is Dependent on mTOR Signaling. Cell Rep 15(6):1144–1160. https://doi.org/10.1016/j.celrep.2016.04.029

Andersen AP, Samsoe-Petersen J, Oernbo EK, Boedtkjer E, Moreira JMA, Kveiborg M et al (2018) The net acid extruders NHE1, NBCn1 and MCT4 promote mammary tumor growth through distinct but overlapping mechanisms. Int J Cancer 142(12):2529–2542. https://doi.org/10.1002/ijc.31276

Apicella M, Giannoni E, Fiore S, Ferrari KJ, Fernandez-Perez D, Isella C et al (2018) Increased lactate secretion by cancer cells sustains non-cell-autonomous adaptive resistance to MET and EGFR targeted therapies. Cell Metab 28(6):848–865.e6. https://doi.org/10.1016/j.cmet.2018.08.006

Asterholm IW, Mundy DI, Weng J, Anderson RG, Scherer PE (2012) Altered mitochondrial function and metabolic inflexibility associated with loss of caveolin-1. Cell Metab 15(2):171–185. https://doi.org/10.1016/j.cmet.2012.01.004

Baba M, Inoue M, Itoh K, Nishizawa Y (2008) Blocking CD147 induces cell death in cancer cells through impairment of glycolytic energy metabolism. Biochem Biophys Res Commun 374(1):111–116. https://doi.org/10.1016/j.bbrc.2008.06.122

Baumann F, Leukel P, Doerfelt A, Beier CP, Dettmer K, Oefner PJ et al (2009) Lactate promotes glioma migration by TGF-beta2-dependent regulation of matrix metalloproteinase-2. Neuro-Oncology 11(4):368–380. https://doi.org/10.1215/15228517-2008-106

Beckert S, Farrahi F, Aslam RS, Scheuenstuhl H, Konigsrainer A, Hussain MZ et al (2006) Lactate stimulates endothelial cell migration. Wound Repair Regen 14(3):321–324. https://doi.org/10.1111/j.1743-6109.2006.00127.x

Beloueche-Babari M, Wantuch S, Casals Galobart T, Koniordou M, Parkes HG, Arunan V et al (2017) MCT1 inhibitor AZD3965 increases mitochondrial metabolism, facilitating combination therapy and noninvasive magnetic resonance spectroscopy. Cancer Res 77(21):5913–5924. https://doi.org/10.1158/0008-5472.CAN-16-2686

Birsoy K, Wang T, Possemato R, Yilmaz OH, Koch CE, Chen WW et al (2013) MCT1-mediated transport of a toxic molecule is an effective strategy for targeting glycolytic tumors. Nat Genet 45(1):104–108. https://doi.org/10.1038/ng.2471

Bisetto S, Whitaker-Menezes D, Wilski NA, Tuluc M, Curry J, Zhan T et al (2018) Monocarboxylate transporter 4 (MCT4) knockout mice have attenuated 4NQO induced carcinogenesis; a role for MCT4 in driving oral squamous cell cancer. Front Oncol 8:324. https://doi.org/10.3389/fonc.2018.00324

Biswas SK (2015) Metabolic reprogramming of immune cells in cancer progression. Immunity 43(3):435–449. https://doi.org/10.1016/j.immuni.2015.09.001

Bonuccelli G, Whitaker-Menezes D, Castello-Cros R, Pavlides S, Pestell RG, Fatatis A et al (2010) The reverse Warburg effect: glycolysis inhibitors prevent the tumor promoting effects of caveolin-1 deficient cancer associated fibroblasts. Cell Cycle 9(10):1960–1971. https://doi.org/10.4161/cc.9.10.11601

Brand A, Singer K, Koehl GE, Kolitzus M, Schoenhammer G, Thiel A et al (2016) LDHA-associated lactic acid production blunts tumor immunosurveillance by T and NK cells. Cell Metab 24(5):657–671. https://doi.org/10.1016/j.cmet.2016.08.011

Broer S (2005) Lactate transportation is required for lymphocyte activation. Nat Chem Biol 1(7):356–357. http://www.ncbi.nlm.nih.gov/pubmed/16370370

Calcinotto A, Filipazzi P, Grioni M, Iero M, De Milito A, Ricupito A et al (2012) Modulation of microenvironment acidity reverses anergy in human and murine tumor-infiltrating T lymphocytes. Cancer Res 72(11):2746–2756. https://doi.org/10.1158/0008-5472.CAN-11-1272

Cammann C, Rath A, Reichl U, Lingel H, Brunner-Weinzierl M, Simeoni L et al (2016) Early changes in the metabolic profile of activated CD8(+) T cells. BMC Cell Biol 17(1):28. https://doi.org/10.1186/s12860-016-0104-x.

Carmona-Fontaine C, Bucci V, Akkari L, Deforet M, Joyce JA, Xavier JB (2013) Emergence of spatial structure in the tumor microenvironment due to the Warburg effect. Proc Natl Acad Sci U S A 110(48):19402–19407. https://doi.org/10.1073/pnas.1311939110

Chang CH, Qiu J, O'Sullivan D, Buck MD, Noguchi T, Curtis JD et al (2015) Metabolic competition in the tumor microenvironment is a driver of cancer progression. Cell 162(6):1229–1241. https://doi.org/10.1016/j.cell.2015.08.016

Chen X, Song E (2018) Turning foes to friends: targeting cancer-associated fibroblasts. Nat Rev Drug Discov 18(2):99–115. https://doi.org/10.1038/s41573-018-0004-1

Chen F, Zhuang X, Lin L, Yu P, Wang Y, Shi Y et al (2015) New horizons in tumor microenvironment biology: challenges and opportunities. BMC Med 13:45. https://doi.org/10.1186/s12916-015-0278-7

Cheng Z, Ristow M (2013) Mitochondria and metabolic homeostasis. Antioxid Redox Signal 19(3):240–242. https://doi.org/10.1089/ars.2013.5255

Colegio OR, Chu NQ, Szabo AL, Chu T, Rhebergen AM, Jairam V et al (2014) Functional polarization of tumour-associated macrophages by tumour-derived lactic acid. Nature 513(7519):559–563. https://doi.org/10.1038/nature13490

Colen CB, Shen Y, Ghoddoussi F, Yu P, Francis TB, Koch BJ et al (2011) Metabolic targeting of lactate efflux by malignant glioma inhibits invasiveness and induces necrosis: an in vivo study. Neoplasia 13(7):620–632. https://www.ncbi.nlm.nih.gov/pubmed/21750656

Constant JS, Feng JJ, Zabel DD, Yuan H, Suh DY, Scheuenstuhl H et al (2000) Lactate elicits vascular endothelial growth factor from macrophages: a possible alternative to hypoxia. Wound Repair Regen 8(5):353–360

Corbet C, Ragelle H, Pourcelle V, Vanvarenberg K, Marchand-Brynaert J, Preat V et al (2016) Delivery of siRNA targeting tumor metabolism using non-covalent PEGylated chitosan nanoparticles: identification of an optimal combination of ligand structure, linker and grafting method. J Control Release 223:53–63. https://doi.org/10.1016/j.jconrel.2015.12.020

Corbet C, Bastien E, Draoui N, Doix B, Mignion L, Jordan BF et al (2018) Interruption of lactate uptake by inhibiting mitochondrial pyruvate transport unravels direct antitumor and radiosensitizing effects. Nat Commun 9(1):1208. https://doi.org/10.1038/s41467-018-03525-0.

Cruz-Bermudez A, Laza-Briviesca R, Vicente-Blanco RJ, Garcia-Grande A, Coronado MJ, Laine-Menendez S et al (2019) Cancer-associated fibroblasts modify lung cancer metabolism involving ROS and TGF-beta signaling. Free Radic Biol Med 130:163–173. https://doi.org/10.1016/j.freeradbiomed.2018.10.450

Curry JM, Tuluc M, Whitaker-Menezes D, Ames JA, Anantharaman A, Butera A et al (2013) Cancer metabolism, stemness and tumor recurrence: MCT1 and MCT4 are functional biomarkers of metabolic symbiosis in head and neck cancer. Cell Cycle 12(9):1371–1384. https://doi.org/10.4161/cc.24092

Dai G, Yao X, Zhang Y, Gu J, Geng Y, Xue F et al (2018) Colorectal cancer cell-derived exosomes containing miR-10b regulate fibroblast cells via the PI3K/Akt pathway. Bull Cancer 105(4):336–349. https://doi.org/10.1016/j.bulcan.2017.12.009

De Saedeleer CJ, Copetti T, Porporato PE, Verrax J, Feron O, Sonveaux P (2012) Lactate activates HIF-1 in oxidative but not in Warburg-phenotype human tumor cells. PLoS One 7(10):e46571. https://doi.org/10.1371/journal.pone.0046571

Dhup S, Dadhich RK, Porporato PE, Sonveaux P (2012) Multiple biological activities of lactic acid in cancer: influences on tumor growth, angiogenesis and metastasis. Curr Pharm Des 18(10):1319–1330. http://www.ncbi.nlm.nih.gov/pubmed/22360558

Diehl K, Dinges LA, Helm O, Ammar N, Plundrich D, Arlt A et al (2018) Nuclear factor E2-related factor-2 has a differential impact on MCT1 and MCT4 lactate carrier expression in colonic epithelial cells: a condition favoring metabolic symbiosis between colorectal cancer and stromal cells. Oncogene 37(1):39–51. https://doi.org/10.1038/onc.2017.299

Dietl K, Renner K, Dettmer K, Timischl B, Eberhart K, Dorn C et al (2010) Lactic acid and acidification inhibit TNF secretion and glycolysis of human monocytes. J Immunol 184(3):1200–1209. https://doi.org/10.4049/jimmunol.0902584

Doherty JR, Cleveland JL (2013) Targeting lactate metabolism for cancer therapeutics. J Clin Invest 123(9):3685–3692. https://doi.org/10.1172/JCI69741

Eelen G, de Zeeuw P, Simons M, Carmeliet P (2015) Endothelial cell metabolism in normal and diseased vasculature. Circ Res 116(7):1231–1244. https://doi.org/10.1161/CIRCRESAHA.116.302855

Eelen G, de Zeeuw P, Treps L, Harjes U, Wong BW, Carmeliet P (2018) Endothelial cell metabolism. Physiol Rev 98(1):3–58. https://doi.org/10.1152/physrev.00001.2017

Enerson BE, Drewes LR (2003) Molecular features, regulation, and function of monocarboxylate transporters: implications for drug delivery. J Pharm Sci 92(8):1531–1544. https://doi.org/10.1002/jps.10389

Errea A, Cayet D, Marchetti P, Tang C, Kluza J, Offermanns S et al (2016) Lactate inhibits the pro-inflammatory response and metabolic reprogramming in murine macrophages in a GPR81-independent manner. PLoS One 11(11):e0163694. https://doi.org/10.1371/journal.pone.0163694

Faubert B, Li KY, Cai L, Hensley CT, Kim J, Zacharias LG et al (2017) Lactate metabolism in human lung tumors. Cell 171(2):358–371.e9. https://doi.org/10.1016/j.cell.2017.09.019

Fischer K, Hoffmann P, Voelkl S, Meidenbauer N, Ammer J, Edinger M et al (2007) Inhibitory effect of tumor cell-derived lactic acid on human T cells. Blood 109(9):3812–3819. https://doi.org/10.1182/blood-2006-07-035972

Fouad YA, Aanei C (2017) Revisiting the hallmarks of cancer. Am J Cancer Res 7(5):1016–1036. http://www.ncbi.nlm.nih.gov/pubmed/28560055

Gan L, Xiu R, Ren P, Yue M, Su H, Guo G et al (2016) Metabolic targeting of oncogene MYC by selective activation of the proton-coupled monocarboxylate family of transporters. Oncogene 35(23):3037–3048. https://doi.org/10.1038/onc.2015.360

Gatenby RA, Gawlinski ET (1996) A reaction-diffusion model of cancer invasion. Cancer Res 56(24):5745–5753. http://www.ncbi.nlm.nih.gov/pubmed/8971186

Gatenby RA, Gawlinski ET, Gmitro AF, Kaylor B, Gillies RJ (2006) Acid-mediated tumor invasion: a multidisciplinary study. Cancer Res 66(10):5216–5223. https://doi.org/10.1158/0008-5472.CAN-05-4193

Gerhart DZ, Enerson BE, Zhdankina OY, Leino RL, Drewes LR (1997) Expression of monocarboxylate transporter MCT1 by brain endothelium and glia in adult and suckling rats. Am J Phys 273(1 Pt 1):E207–E213. https://doi.org/10.1152/ajpendo.1997.273.1.E207.

Goetze K, Walenta S, Ksiazkiewicz M, Kunz-Schughart LA, Mueller-Klieser W (2011) Lactate enhances motility of tumor cells and inhibits monocyte migration and cytokine release. Int J Oncol 39(2):453–463. https://doi.org/10.3892/ijo.2011.1055.

Gooptu M, Whitaker-Menezes D, Sprandio J, Domingo-Vidal M, Lin Z, Uppal G et al (2017) Mitochondrial and glycolytic metabolic compartmentalization in diffuse large B-cell lymphoma. Semin Oncol 44(3):204–217. https://doi.org/10.1053/j.seminoncol.2017.10.002

Gottfried E, Kunz-Schughart LA, Ebner S, Mueller-Klieser W, Hoves S, Andreesen R et al (2006) Tumor-derived lactic acid modulates dendritic cell activation and antigen expression. Blood 107(5):2013–2021. https://doi.org/10.1182/blood-2005-05-1795

Granja S, Tavares-Valente D, Queiros O, Baltazar F (2017) Value of pH regulators in the diagnosis, prognosis and treatment of cancer. Semin Cancer Biol 43:17–34. https://doi.org/10.1016/j.semcancer.2016.12.003

Guan X, Bryniarski MA, Morris ME (2018) In vitro and in vivo efficacy of the monocarboxylate transporter 1 inhibitor AR-C155858 in the murine 4T1 breast cancer tumor model. AAPS J 21(1):3. https://doi.org/10.1208/s12248-018-0261-2.

Guillaumond F, Leca J, Olivares O, Lavaut MN, Vidal N, Berthezene P et al (2013) Strengthened glycolysis under hypoxia supports tumor symbiosis and hexosamine biosynthesis in pancreatic adenocarcinoma. Proc Natl Acad Sci U S A 110(10):3919–3924. https://doi.org/10.1073/pnas.1219555110

Guppy M, Leedman P, Zu X, Russell V (2002) Contribution by different fuels and metabolic pathways to the total ATP turnover of proliferating MCF-7 breast cancer cells. Biochem J 364(Pt 1):309–315. http://www.ncbi.nlm.nih.gov/pubmed/11988105

Gurrapu S, Jonnalagadda SK, Alam MA, Ronayne CT, Nelson GL, Solano LN et al (2016) Coumarin carboxylic acids as monocarboxylate transporter 1 inhibitors: in vitro and in vivo studies as potential anticancer agents. Bioorg Med Chem Lett 26(14):3282–3286. https://doi.org/10.1016/j.bmcl.2016.05.054

Haas R, Smith J, Rocher-Ros V, Nadkarni S, Montero-Melendez T, D'Acquisto F et al (2015) Lactate regulates metabolic and pro-inflammatory circuits in control of T cell migration and effector functions. PLoS Biol 13(7):e1002202. https://doi.org/10.1371/journal.pbio.1002202
Halestrap AP (2013) The SLC16 gene family – structure, role and regulation in health and disease. Mol Asp Med 34(2–3):337–349. https://doi.org/10.1016/j.mam.2012.05.003
Halestrap AP, Wilson MC (2012) The monocarboxylate transporter family--role and regulation. IUBMB Life 64(2):109–119. https://doi.org/10.1002/iub.572
Halford SER, Jones P, Wedge S, Hirschberg S, Katugampola S, Veal G et al (2017) A first-in-human first-in-class (FIC) trial of the monocarboxylate transporter 1 (MCT1) inhibitor AZD3965 in patients with advanced solid tumours. J Clin Oncol 35(15_suppl):2516 2516. https://doi.org/10.1200/JCO.2017.35.15_suppl.2516
Hanahan D, Weinberg RA (2011) Hallmarks of cancer: the next generation. Cell 144(5):646–674. https://doi.org/10.1016/j.cell.2011.02.013
Hao J, Graham P, Chang L, Ni J, Wasinger V, Beretov J et al (2016) Proteomic identification of the lactate dehydrogenase A in a radioresistant prostate cancer xenograft mouse model for improving radiotherapy. Oncotarget 7(45):74269–74285. https://doi.org/10.18632/oncotarget.12368
Hargadon KM (2017) Strategies to improve the efficacy of dendritic cell-based immunotherapy for melanoma. Front Immunol 8:1594. https://doi.org/10.3389/fimmu.2017.01594
Hirschhaeuser F, Sattler UG, Mueller-Klieser W (2011) Lactate: a metabolic key player in cancer. Cancer Res 71(22):6921–6925. 71/22/6921 [pii]. https://doi.org/10.1158/0008-5472.CAN-11-1457
Hunt TK, Aslam RS, Beckert S, Wagner S, Ghani QP, Hussain MZ et al (2007) Aerobically derived lactate stimulates revascularization and tissue repair via redox mechanisms. Antioxid Redox Signal 9(8):1115–1124. https://doi.org/10.1089/ars.2007.1674
Husain Z, Huang Y, Seth P, Sukhatme VP (2013a) Tumor-derived lactate modifies antitumor immune response: effect on myeloid-derived suppressor cells and NK cells. J Immunol 191(3):1486–1495. https://doi.org/10.4049/jimmunol.1202702
Husain Z, Seth P, Sukhatme VP (2013b) Tumor-derived lactate and myeloid-derived suppressor cells: linking metabolism to cancer immunology. Oncoimmunology 2(11):e26383. https://doi.org/10.4161/onci.26383
Im HJ, Bradshaw T, Solaiyappan M, Cho SY (2018) Current methods to define metabolic tumor volume in positron emission tomography: which one is better? Nucl Med Mol Imaging 52(1):5–15. https://doi.org/10.1007/s13139-017-0493-6
Kasinrerk W, Tokrasinwit N, Phunpae P (1999) CD147 monoclonal antibodies induce homotypic cell aggregation of monocytic cell line U937 via LFA-1/ICAM-1 pathway. Immunology 96(2):184–192. https://www.ncbi.nlm.nih.gov/pubmed/10233694
Kato Y, Ozawa S, Tsukuda M, Kubota E, Miyazaki K, St-Pierre Y et al (2007) Acidic extracellular pH increases calcium influx-triggered phospholipase D activity along with acidic sphingomyelinase activation to induce matrix metalloproteinase-9 expression in mouse metastatic melanoma. FEBS J 274(12):3171–3183. https://doi.org/10.1111/j.1742-4658.2007.05848.x
Kim HK, Lee I, Bang H, Kim HC, Lee WY, Yun SH et al (2018) MCT4 expression is a potential therapeutic target in colorectal cancer with peritoneal carcinomatosis. Mol Cancer Ther 17(4):838–848. https://doi.org/10.1158/1535-7163.MCT-17-0535
Kishimoto A, Takahashi-Iwanaga H, Watanabe MM, Iwanaga T (2016) Differential expression of endothelial nutrient transporters (MCT1 and GLUT1) in the developing eyes of mice. Exp Eye Res 153:170–177. https://doi.org/10.1016/j.exer.2016.10.019
Koukourakis MI, Giatromanolaki A, Harris AL, Sivridis E (2006) Comparison of metabolic pathways between cancer cells and stromal cells in colorectal carcinomas: a metabolic survival role for tumor-associated stroma. Cancer Res 66(2):632–637. https://doi.org/10.1158/0008-5472.CAN-05-3260
Koukourakis MI, Giatromanolaki A, Bougioukas G, Sivridis E (2007) Lung cancer: a comparative study of metabolism related protein expression in cancer cells and tumor associated stroma. Cancer Biol Ther 6(9):1476–1479. http://www.ncbi.nlm.nih.gov/pubmed/17881895
Koukourakis MI, Kalamida D, Mitrakas AG, Liousia M, Pouliliou S, Sivridis E et al (2017) Metabolic cooperation between co-cultured lung cancer cells and lung fibroblasts. Lab Investig 97(11):1321–1331. https://doi.org/10.1038/labinvest.2017.79
Kumar A, Kant S, Singh SM (2013) Targeting monocarboxylate transporter by alpha-cyano-4-hydroxycinnamate modulates apoptosis and cisplatin resistance of Colo205 cells: implication of altered cell survival regulation. Apoptosis 18(12):1574–1585. https://doi.org/10.1007/s10495-013-0894-7
Kumar D, New J, Vishwakarma V, Joshi R, Enders J, Lin F et al (2018) Cancer-associated fibroblasts drive glycolysis in a targetable signaling loop implicated in head and neck squamous cell carcinoma progression. Cancer Res 78(14):3769–3782. https://doi.org/10.1158/0008-5472.CAN-17-1076
Langbein S, Zerilli M, Zur Hausen A, Staiger W, Rensch-Boschert K, Lukan N et al (2006) Expression of transketolase TKTL1 predicts colon and urothelial cancer patient survival: Warburg effect reinterpreted. Br J Cancer 94(4):578–585. https://doi.org/10.1038/sj.bjc.6602962
Le Floch R, Chiche J, Marchiq I, Naiken T, Ilc K, Murray CM et al (2011) CD147 subunit of lactate/H+ symporters MCT1 and hypoxia-inducible MCT4 is critical for energetics and growth of glycolytic tumors. Proc

Natl Acad Sci U S A 108(40):16663–16668. https://doi.org/10.1073/pnas.1106123108

LeBleu VS, Kalluri R (2018) A peek into cancer-associated fibroblasts: origins, functions and translational impact. Dis Model Mech 11(4). https://doi.org/10.1242/dmm.029447

Lee DC, Sohn HA, Park ZY, Oh S, Kang YK, Lee KM et al (2015) A lactate-induced response to hypoxia. Cell 161(3):595–609. https://doi.org/10.1016/j.cell.2015.03.011

Lee JY, Lee I, Chang WJ, Ahn SM, Lim SH, Kim HS et al (2016) MCT4 as a potential therapeutic target for metastatic gastric cancer with peritoneal carcinomatosis. Oncotarget 7(28):43492–43503. https://doi.org/10.18632/oncotarget.9523

Lemons JM, Feng XJ, Bennett BD, Legesse-Miller A, Johnson EL, Raitman I et al (2010) Quiescent fibroblasts exhibit high metabolic activity. PLoS Biol 8(10):e1000514. https://doi.org/10.1371/journal.pbio.1000514

Li R, Pan Y, He B, Xu Y, Gao T, Song G et al (2013) Downregulation of CD147 expression by RNA interference inhibits HT29 cell proliferation, invasion and tumorigenicity in vitro and in vivo. Int J Oncol 43(6):1885–1894. https://doi.org/10.3892/ijo.2013.2108

Liberti MV, Locasale JW (2016) The Warburg effect: how does it benefit cancer cells? Trends Biochem Sci 41(3):211–218. https://doi.org/10.1016/j.tibs.2015.12.001

Lin S, Sun L, Lyu X, Ai X, Du D, Su N et al (2017) Lactate-activated macrophages induced aerobic glycolysis and epithelial-mesenchymal transition in breast cancer by regulation of CCL5-CCR5 axis: a positive metabolic feedback loop. Oncotarget 8(66):110426–110443. https://doi.org/10.18632/oncotarget.22786

Liu Y, Ji X, Tong WWL, Askhatova D, Yang T, Cheng H et al (2018) Engineering multifunctional RNAi nanomedicine to concurrently target cancer hallmarks for combinatorial therapy. Angew Chem Int Ed Engl 57(6):1510–1513. https://doi.org/10.1002/anie.201710144

Lu H, Forbes RA, Verma A (2002) Hypoxia-inducible factor 1 activation by aerobic glycolysis implicates the Warburg effect in carcinogenesis. J Biol Chem 277(26):23111–23115. https://doi.org/10.1074/jbc.M202487200

Lu H, Dalgard CL, Mohyeldin A, McFate T, Tait AS, Verma A (2005) Reversible inactivation of HIF-1 prolyl hydroxylases allows cell metabolism to control basal HIF-1. J Biol Chem 280(51):41928–41939. https://doi.org/10.1074/jbc.M508718200

Luo M, Luo Y, Mao N, Huang G, Teng C, Wang H et al (2018) Cancer-associated fibroblasts accelerate malignant progression of non-small cell lung cancer via connexin 43-formed unidirectional gap junctional intercellular communication. Cell Physiol Biochem 51(1):315–336. https://doi.org/10.1159/000495232.

Mac M, Nalecz KA (2003) Expression of monocarboxylic acid transporters (MCT) in brain cells. Implication for branched chain alpha-ketoacids transport in neurons. Neurochem Int 43(4–5):305–309. http://www.ncbi.nlm.nih.gov/pubmed/12742073

Machler P, Wyss MT, Elsayed M, Stobart J, Gutierrez R, von Faber-Castell A et al (2016) In vivo evidence for a lactate gradient from astrocytes to neurons. Cell Metab 23(1):94–102. https://doi.org/10.1016/j.cmet.2015.10.010

Macintyre AN, Gerriets VA, Nichols AG, Michalek RD, Rudolph MC, Deoliveira D et al (2014) The glucose transporter Glut1 is selectively essential for CD4 T cell activation and effector function. Cell Metab 20(1):61–72. https://doi.org/10.1016/j.cmet.2014.05.004

MacIver NJ, Michalek RD, Rathmell JC (2013) Metabolic regulation of T lymphocytes. Annu Rev Immunol 31:259–283. https://doi.org/10.1146/annurev-immunol-032712-095956

Martinez-Outschoorn UE, Pavlides S, Whitaker-Menezes D, Daumer KM, Milliman JN, Chiavarina B et al (2010a) Tumor cells induce the cancer associated fibroblast phenotype via caveolin-1 degradation: implications for breast cancer and DCIS therapy with autophagy inhibitors. Cell Cycle 9(12):2423–2433. https://doi.org/10.4161/cc.9.12.12048

Martinez-Outschoorn UE, Trimmer C, Lin Z, Whitaker-Menezes D, Chiavarina B, Zhou J et al (2010b) Autophagy in cancer associated fibroblasts promotes tumor cell survival: Role of hypoxia, HIF1 induction and NFkappaB activation in the tumor stromal microenvironment. Cell Cycle 9(17):3515–3533. https://doi.org/10.4161/cc.9.17.12928.

Mazurek S, Boschek CB, Hugo F, Eigenbrodt E (2005) Pyruvate kinase type M2 and its role in tumor growth and spreading. Semin Cancer Biol 15(4):300–308. https://doi.org/10.1016/j.semcancer.2005.04.009

McAnulty RJ (2007) Fibroblasts and myofibroblasts: their source, function and role in disease. Int J Biochem Cell Biol 39(4):666–671. https://doi.org/10.1016/j.biocel.2006.11.005

McKay ND, Robinson B, Brodie R, Rooke-Allen N (1983) Glucose transport and metabolism in cultured human skin fibroblasts. Biochim Biophys Acta 762(2):198–204. http://www.ncbi.nlm.nih.gov/pubmed/6830872

Miranda-Gonçalves V, Honavar M, Pinheiro C, Martinho O, Pires MM, Pinheiro C et al (2013) Monocarboxylate transporters (MCTs) in gliomas: expression and exploitation as therapeutic targets. Neuro-Oncology 15(2):172–188. https://doi.org/10.1093/neuonc/nos298

Miranda-Gonçalves V, Bezerra F, Costa-Almeida R, Freitas-Cunha M, Soares R, Martinho O et al (2017) Monocarboxylate transporter 1 is a key player in glioma-endothelial cell crosstalk. Mol Carcinog 56(12):2630–2642. https://doi.org/10.1002/mc.22707

Morais-Santos F, Miranda-Goncalves V, Pinheiro S, Vieira AF, Paredes J, Schmitt FC et al (2014) Differential sensitivities to lactate transport inhibitors of breast cancer cell lines. Endocr Relat Cancer 21(1):27–38. https://doi.org/10.1530/ERC-13-0132.

Morais-Santos F, Granja S, Miranda-Goncalves V, Moreira AH, Queiros S, Vilaca JL et al (2015) Targeting lactate transport suppresses in vivo breast tumour growth. Oncotarget 6(22):19177–19189. https://doi.org/10.18632/oncotarget.3910

Murray CM, Hutchinson R, Bantick JR, Belfield GP, Benjamin AD, Brazma D et al (2005) Monocarboxylate transporter MCT1 is a target for immunosuppression. Nat Chem Biol 1(7):371–376. http://www.ncbi.nlm.nih.gov/pubmed/16370372

Nasi A, Fekete T, Krishnamurthy A, Snowden S, Rajnavolgyi E, Catrina AI et al (2013) Dendritic cell reprogramming by endogenously produced lactic acid. J Immunol 191(6):3090–3099. https://doi.org/10.4049/jimmunol.1300772

Newsholme EA, Crabtree B, Ardawi MS (1985) The role of high rates of glycolysis and glutamine utilization in rapidly dividing cells. Biosci Rep 5(5):393–400

Ovens MJ, Davies AJ, Wilson MC, Murray CM, Halestrap AP (2010) AR-C155858 is a potent inhibitor of monocarboxylate transporters MCT1 and MCT2 that binds to an intracellular site involving transmembrane helices 7-10. Biochem J 425(3):523–530. https://doi.org/10.1042/BJ20091515

Park JM, Josan S, Mayer D, Hurd RE, Chung Y, Bendahan D et al (2015) Hyperpolarized 13C NMR observation of lactate kinetics in skeletal muscle. J Exp Biol 218(Pt 20):3308–3318. https://doi.org/10.1242/jeb.123141

Parks SK, Chiche J, Pouyssegur J (2013) Disrupting proton dynamics and energy metabolism for cancer therapy. Nat Rev Cancer 13(9):611–623. https://doi.org/10.1038/nrc3579

Pavlides S, Whitaker-Menezes D, Castello-Cros R, Flomenberg N, Witkiewicz AK, Frank PG et al (2009) The reverse Warburg effect: aerobic glycolysis in cancer associated fibroblasts and the tumor stroma. Cell Cycle 8(23):3984–4001. https://doi.org/10.4161/cc.8.23.10238

Payen VL, Hsu MY, Radecke KS, Wyart E, Vazeille T, Bouzin C et al (2017) Monocarboxylate transporter MCT1 promotes tumor metastasis independently of its activity as a lactate transporter. Cancer Res 77(20):5591–5601. https://doi.org/10.1158/0008-5472.CAN-17-0764

Pearce EL, Pearce EJ (2013) Metabolic pathways in immune cell activation and quiescence. Immunity 38(4):633–643. https://doi.org/10.1016/j.immuni.2013.04.005

Pearce EL, Poffenberger MC, Chang CH, Jones RG (2013) Fueling immunity: insights into metabolism and lymphocyte function. Science 342(6155):1242454. https://doi.org/10.1126/science.1242454

Perez-Escuredo J, Van Hee VF, Sboarina M, Falces J, Payen VL, Pellerin L et al (2016) Monocarboxylate transporters in the brain and in cancer. Biochim Biophys Acta 1863(10):2481–2497. https://doi.org/10.1016/j.bbamcr.2016.03.013

Pertega-Gomes N, Vizcaino JR, Attig J, Jurmeister S, Lopes C, Baltazar F (2014) A lactate shuttle system between tumour and stromal cells is associated with poor prognosis in prostate cancer. BMC Cancer 14:352. https://doi.org/10.1186/1471-2407-14-352

Pinheiro C, Longatto-Filho A, Azevedo-Silva J, Casal M, Schmitt FC, Baltazar F (2012) Role of monocarboxylate transporters in human cancers: state of the art. J Bioenerg Biomembr 44(1):127–139. https://doi.org/10.1007/s10863-012-9428-1

Pinheiro C, Garcia EA, Morais-Santos F, Moreira MA, Almeida FM, Jube LF et al (2015a) Reprogramming energy metabolism and inducing angiogenesis: co-expression of monocarboxylate transporters with VEGF family members in cervical adenocarcinomas. BMC Cancer 15:835. https://doi.org/10.1186/s12885-015-1842-4

Pinheiro C, Granja S, Longatto-Filho A, Faria AM, Fragoso MC, Lovisolo SM et al (2015b) Metabolic reprogramming: a new relevant pathway in adult adrenocortical tumors. Oncotarget 6(42):44403–44421. https://doi.org/10.18632/oncotarget.5623

Pinheiro C, Morais-Santos F, Granja S, Miranda-Goncalves V, Afonso J, Amorim R et al (2015c) Targeting metabolic reprogramming as an anti-cancer strategy: aiming at monocarboxylate transporters. In: Atta-ur-Rahman, Choudhary MI (eds) Frontiers in anti-cancer drug discovery. Bentham Science Publishers, Sharjah, pp 3–65. https://doi.org/10.2174/97816810814961150600003

Polanski R, Hodgkinson CL, Fusi A, Nonaka D, Priest L, Kelly P et al (2014) Activity of the monocarboxylate transporter 1 inhibitor AZD3965 in small cell lung cancer. Clin Cancer Res 20(4):926–937. https://doi.org/10.1158/1078-0432.CCR-13-2270

Porporato PE, Payen VL, De Saedeleer CJ, Preat V, Thissen JP, Feron O et al (2012) Lactate stimulates angiogenesis and accelerates the healing of superficial and ischemic wounds in mice. Angiogenesis 15(4):581–592. https://doi.org/10.1007/s10456-012-9282-0

Potente M, Gerhardt H, Carmeliet P (2011) Basic and therapeutic aspects of angiogenesis. Cell 146(6):873–887. https://doi.org/10.1016/j.cell.2011.08.039

Pouyssegur J, Dayan F, Mazure NM (2006) Hypoxia signalling in cancer and approaches to enforce tumour regression. Nature 441(7092):437–443. https://doi.org/10.1038/nature04871

Puig-Kroger A, Pello OM, Muniz-Pello O, Selgas R, Criado G, Bajo MA et al (2003) Peritoneal dialysis solutions inhibit the differentiation and maturation of human monocyte-derived dendritic cells: effect of lactate and glucose-degradation products. J Leukoc Biol 73(4):482–492. https://doi.org/10.1189/jlb.0902451

Racker E (1974) History of the Pasteur effect and its pathobiology. Mol Cell Biochem 5(1–2):17–23. http://www.ncbi.nlm.nih.gov/pubmed/4279327

Rai A, Greening DW, Chen M, Xu R, Ji H, Simpson RJ (2018) Exosomes derived from human primary and metastatic colorectal cancer cells contribute to functional heterogeneity of activated fibroblasts by reprogramming their proteome. Proteomics:e1800148. https://doi.org/10.1002/pmic.201800148

Rasanen K, Vaheri A (2010) Activation of fibroblasts in cancer stroma. Exp Cell Res 316(17):2713–2722. https://doi.org/10.1016/j.yexcr.2010.04.032

Richter S, D'Antongiovanni V, Martinelli S, Bechmann N, Riverso M, Poitz DM et al (2018) Primary fibroblast co-culture stimulates growth and metabolism in Sdhb-impaired mouse pheochromocytoma MTT cells. Cell Tissue Res 374(3):473–485. https://doi.org/10.1007/s00441-018-2907-x

Riemann A, Schneider B, Ihling A, Nowak M, Sauvant C, Thews O et al (2011) Acidic environment leads to ROS-induced MAPK signaling in cancer cells. PLoS One 6(7):e22445. https://doi.org/10.1371/journal.pone.0022445

Romero-Garcia S, Moreno-Altamirano MM, Prado-Garcia H, Sanchez-Garcia FJ (2016) Lactate contribution to the tumor microenvironment: mechanisms, effects on immune cells and therapeutic relevance. Front Immunol 7:52. https://doi.org/10.3389/fimmu.2016.00052

Ryder C, McColl K, Zhong F, Distelhorst CW (2012) Acidosis promotes Bcl-2 family-mediated evasion of apoptosis: involvement of acid-sensing G protein-coupled receptor Gpr65 signaling to Mek/Erk. J Biol Chem 287(33):27863–27875. https://doi.org/10.1074/jbc.M112.384685

Sakamoto A, Kunou S, Shimada K, Tsunoda M, Aoki T, Iriyama C et al (2019) Pyruvate secreted from patient-derived cancer-associated fibroblasts supports survival of primary lymphoma cells. Cancer Sci 110(1):269–278. https://doi.org/10.1111/cas.13873

Sanita P, Capulli M, Teti A, Galatioto GP, Vicentini C, Chiarugi P et al (2014) Tumor-stroma metabolic relationship based on lactate shuttle can sustain prostate cancer progression. BMC Cancer 14:154. https://doi.org/10.1186/1471-2407-14-154

San-Millan I, Brooks GA (2017) Reexamining cancer metabolism: lactate production for carcinogenesis could be the purpose and explanation of the Warburg Effect. Carcinogenesis 38(2):119–133. https://doi.org/10.1093/carcin/bgw127.

Schell JC, Rutter J (2013) The long and winding road to the mitochondrial pyruvate carrier. Cancer Metab 1(1):6. https://doi.org/10.1186/2049-3002-1-6.

Schneiderhan W, Scheler M, Holzmann KH, Marx M, Gschwend JE, Bucholz M et al (2009) CD147 silencing inhibits lactate transport and reduces malignant potential of pancreatic cancer cells in in vivo and in vitro models. Gut 58(10):1391–1398. https://doi.org/10.1136/gut.2009.181412

Semenza GL (2008) Tumor metabolism: cancer cells give and take lactate. J Clin Invest 118(12):3835–3837. https://doi.org/10.1172/JCI37373.

Seth P, Csizmadia E, Hedblom A, Vuerich M, Xie H, Li M et al (2017) Deletion of lactate Dehydrogenase-A in myeloid cells triggers antitumor immunity. Cancer Res 77(13):3632–3643. https://doi.org/10.1158/0008-5472.CAN-16-2938

Shan T, Chen S, Chen X, Lin WR, Li W, Ma J et al (2017) Cancer-associated fibroblasts enhance pancreatic cancer cell invasion by remodeling the metabolic conversion mechanism. Oncol Rep 37(4):1971–1979. https://doi.org/10.3892/or.2017.5479

Shestov AA, Liu X, Ser Z, Cluntun AA, Hung YP, Huang L et al (2014) Quantitative determinants of aerobic glycolysis identify flux through the enzyme GAPDH as a limiting step. elife 3. https://doi.org/10.7554/eLife.03342

Shi T, Ma Y, Cao L, Zhan S, Xu Y, Fu F et al (2019) B7-H3 promotes aerobic glycolysis and chemoresistance in colorectal cancer cells by regulating HK2. Cell Death Dis 10(4):308. https://doi.org/10.1038/s41419-019-1549-6

Shimura T, Noma N, Sano Y, Ochiai Y, Oikawa T, Fukumoto M et al (2014) AKT-mediated enhanced aerobic glycolysis causes acquired radioresistance by human tumor cells. Radiother Oncol 112(2):302–307. https://doi.org/10.1016/j.radonc.2014.07.015

Singer K, Kastenberger M, Gottfried E, Hammerschmied CG, Buttner M, Aigner M et al (2011) Warburg phenotype in renal cell carcinoma: high expression of glucose-transporter 1 (GLUT-1) correlates with low CD8(+) T-cell infiltration in the tumor. Int J Cancer 128(9):2085–2095. https://doi.org/10.1002/ijc.25543

Slomiany MG, Grass GD, Robertson AD, Yang XY, Maria BL, Beeson C et al (2009) Hyaluronan, CD44, and emmprin regulate lactate efflux and membrane localization of monocarboxylate transporters in human breast carcinoma cells. Cancer Res 69(4):1293–1301. https://doi.org/10.1158/0008-5472.CAN-08-2491

Sonveaux P, Vegran F, Schroeder T, Wergin MC, Verrax J, Rabbani ZN et al (2008) Targeting lactate-fueled respiration selectively kills hypoxic tumor cells in mice. J Clin Invest 118(12):3930–3942. https://doi.org/10.1172/JCI36843.

Sonveaux P, Copetti T, De Saedeleer CJ, Vegran F, Verrax J, Kennedy KM et al (2012) Targeting the lactate transporter MCT1 in endothelial cells inhibits lactate-induced HIF-1 activation and tumor angiogenesis. PLoS One 7(3):e33418. https://doi.org/10.1371/journal.pone.0033418

Stern R, Shuster S, Neudecker BA, Formby B (2002) Lactate stimulates fibroblast expression of hyaluronan and CD44: the Warburg effect revisited. Exp Cell Res 276(1):24–31. https://doi.org/10.1006/excr.2002.5508

Tan Z, Xie N, Banerjee S, Cui H, Fu M, Thannickal VJ et al (2015) The monocarboxylate transporter 4 is required for glycolytic reprogramming and inflammatory response in macrophages. J Biol Chem 290(1):46–55. https://doi.org/10.1074/jbc.M114.603589

Todenhofer T, Seiler R, Stewart C, Moskalev I, Gao J, Ladhar S et al (2018) Selective Inhibition of the Lactate Transporter MCT4 Reduces Growth of Invasive Bladder Cancer. Mol Cancer Ther 17(12):2746–2755. https://doi.org/10.1158/1535-7163.MCT-18-0107

Trabold O, Wagner S, Wicke C, Scheuenstuhl H, Hussain MZ, Rosen N et al (2003) Lactate and oxygen constitute a fundamental regulatory mechanism in wound healing. Wound Repair Regen 11(6):504–509
van der Windt GJ, Pearce EL (2012) Metabolic switching and fuel choice during T-cell differentiation and memory development. Immunol Rev 249(1):27–42. https://doi.org/10.1111/j.1600-065X.2012.01150.x
Van Hee VF, Labar D, Dehon G, Grasso D, Gregoire V, Muccioli GG et al (2017) Radiosynthesis and validation of (+/−)-[18F]-3-fluoro-2-hydroxypropionate ([18F]-FLac) as a PET tracer of lactate to monitor MCT1-dependent lactate uptake in tumors. Oncotarget 8(15):24415–24428. https://doi.org/10.18632/oncotarget.14705
Vander Heiden MG, Cantley LC, Thompson CB (2009) Understanding the Warburg effect: the metabolic requirements of cell proliferation. Science 324(5930):1029–1033. https://doi.org/10.1126/science.1160809
Vegran F, Boidot R, Michiels C, Sonveaux P, Feron O (2011) Lactate influx through the endothelial cell monocarboxylate transporter MCT1 supports an NF-kappaB/IL-8 pathway that drives tumor angiogenesis. Cancer Res 71(7):2550–2560. https://doi.org/10.1158/0008-5472.can-10-2828
Wagner W, Ciszewski WM, Kania KD (2015) L- and D-lactate enhance DNA repair and modulate the resistance of cervical carcinoma cells to anticancer drugs via histone deacetylase inhibition and hydroxycarboxylic acid receptor 1 activation. Cell Commun Signal 13:36. https://doi.org/10.1186/s12964-015-0114-x
Wagner W, Kania KD, Blauz A, Ciszewski WM (2017) The lactate receptor (HCAR1/GPR81) contributes to doxorubicin chemoresistance via ABCB1 transporter up-regulation in human cervical cancer HeLa cells. J Physiol Pharmacol 68(4):555–564. http://www.ncbi.nlm.nih.gov/pubmed/29151072
Walenta S, Mueller-Klieser WF (2004) Lactate: mirror and motor of tumor malignancy. Semin Radiat Oncol 14(3):267–274. https://doi.org/10.1016/j.semradonc.2004.04.004
Wang T, Liu G, Wang R (2014) The intercellular metabolic interplay between tumor and immune cells. Front Immunol 5:358. https://doi.org/10.3389/fimmu.2014.00358
Warburg O (1956a) On respiratory impairment in cancer cells. Science 124(3215):269–270. http://www.ncbi.nlm.nih.gov/pubmed/13351639
Warburg O (1956b) On the origin of cancer cells. Science 123(3191):309–314. http://www.ncbi.nlm.nih.gov/pubmed/13298683
Whitaker-Menezes D, Martinez-Outschoorn UE, Flomenberg N, Birbe RC, Witkiewicz AK, Howell A et al (2011a) Hyperactivation of oxidative mitochondrial metabolism in epithelial cancer cells in situ: visualizing the therapeutic effects of metformin in tumor tissue. Cell Cycle 10(23):4047–4064. https://doi.org/10.4161/cc.10.23.18151
Whitaker-Menezes D, Martinez-Outschoorn UE, Lin Z, Ertel A, Flomenberg N, Witkiewicz AK et al (2011b) Evidence for a stromal-epithelial "lactate shuttle" in human tumors: MCT4 is a marker of oxidative stress in cancer-associated fibroblasts. Cell Cycle 10(11):1772–1783. https://doi.org/10.4161/cc.10.11.15659
Wilde L, Roche M, Domingo-Vidal M, Tanson K, Philp N, Curry J et al (2017) Metabolic coupling and the reverse Warburg effect in cancer: implications for novel biomarker and anticancer agent development. Semin Oncol 44(3):198–203. https://doi.org/10.1053/j.seminoncol.2017.10.004
Wolf A, Agnihotri S, Guha A (2010) Targeting metabolic remodeling in glioblastoma multiforme. Oncotarget 1(7):552–562. https://doi.org/10.18632/oncotarget.190
Wu DH, Liang H, Lu SN, Wang H, Su ZL, Zhang L et al (2018a) miR-124 suppresses pancreatic ductal adenocarcinoma growth by regulating monocarboxylate transporter 1-mediated cancer lactate metabolism. Cell Physiol Biochem 50(3):924–935. https://doi.org/10.1159/000494477
Wu J, Hong Y, Wu T, Wang J, Chen X, Wang Z et al (2018b) Stromal-epithelial lactate shuttle induced by tumorderived interleukin1beta promotes cell proliferation in oral squamous cell carcinoma. Int J Mol Med 41(2):687–696. https://doi.org/10.3892/ijmm.2017.3267.
Xiong M, Elson G, Legarda D, Leibovich SJ (1998) Production of vascular endothelial growth factor by murine macrophages: regulation by hypoxia, lactate, and the inducible nitric oxide synthase pathway. Am J Pathol 153(2):587–598. https://doi.org/10.1016/s0002-9440(10)65601-5
Xiong L, Edwards CK 3rd, Zhou L (2014) The biological function and clinical utilization of CD147 in human diseases: a review of the current scientific literature. Int J Mol Sci 15(10):17411–17441. https://doi.org/10.3390/ijms151017411
Yan C, Yang F, Zhou C, Chen X, Han X, Liu X et al (2015) MCT1 promotes the cisplatin-resistance by antagonizing Fas in epithelial ovarian cancer. Int J Clin Exp Pathol 8(3):2710–2718. http://www.ncbi.nlm.nih.gov/pubmed/26045776
Yan W, Wu X, Zhou W, Fong MY, Cao M, Liu J et al (2018) Cancer-cell-secreted exosomal miR-105 promotes tumour growth through the MYC-dependent metabolic reprogramming of stromal cells. Nat Cell Biol 20(5):597–609. https://doi.org/10.1038/s41556-018-0083-6
Ye H, Zhou Q, Zheng S, Li G, Lin Q, Wei L et al (2018) Tumor-associated macrophages promote progression and the Warburg effect via CCL18/NF-kB/VCAM-1 pathway in pancreatic ductal adenocarcinoma.

Cell Death Dis 9(5):453. https://doi.org/10.1038/s41419-018-0486-0.
Zhao Z, Han F, He Y, Yang S, Hua L, Wu J et al (2014a) Stromal-epithelial metabolic coupling in gastric cancer: stromal MCT4 and mitochondrial TOMM20 as poor prognostic factors. Eur J Surg Oncol 40(10):1361–1368. https://doi.org/10.1016/j.ejso.2014.04.005
Zhao Z, Wu MS, Zou C, Tang Q, Lu J, Liu D et al (2014b) Downregulation of MCT1 inhibits tumor growth, metastasis and enhances chemotherapeutic efficacy in osteosarcoma through regulation of the NF-kappaB pathway. Cancer Lett 342(1):150–158. https://doi.org/10.1016/j.canlet.2013.08.042
Zhao Y, Wang D, Xu T, Liu P, Cao Y, Wang Y et al (2015) Bladder cancer cells re-educate TAMs through lactate shuttling in the microfluidic cancer microenvironment. Oncotarget 6(36):39196–39210. https://doi.org/10.18632/oncotarget.5538
Zhao H, Yang L, Baddour J, Achreja A, Bernard V, Moss T et al (2016) Tumor microenvironment derived exosomes pleiotropically modulate cancer cell metabolism. elife 5:e10250. https://doi.org/10.7554/eLife.10250

Part I

Adaptive Metabolic Features Are Sustained by Tumor Microenvironment

Recycling the Interspecific Relations with Epithelial Cells: Bacteria and Cancer Metabolic Symbiosis

4

Sofia C. Nunes and Jacinta Serpa

Abstract

Several aspects of the human physiology are controlled by the microbiota that plays a key role in health and disease. In fact, microbial dysbiosis is associated with numerous diseases, including several types of cancer such as colon, gastric, esophageal, pancreatic, laryngeal, breast and gallbladder carcinomas.

Metabolic symbiosis between non-malignant cells and the resident microbita is crucial for the host homeostasis. However, cancer cells are able to repurpose the pre-existing metabolic symbiosis, being able to recycle those relations and also create novel metabolic symbiosis, leading to profound alterations on the local microenvironment.

In here we will explore some of these symbiotic metabolic interactions between bacteria and non-malignant cells in two different contexts: colon and uterine cervix. The way malignant cells are able to recycle these normal interactions and also create novel types of symbiotic metabolic relations will also be discussed.

The knowledge of these complex interactions and recycling mechanisms is of extreme importance for cancer treatment, as new therapeutic targets could be developed.

S. C. Nunes · J. Serpa (✉)
CEDOC, Chronic Diseases Research Centre, NOVA Medical School | Faculdade de Ciências Médicas, Universidade NOVA de Lisboa, Lisbon, Portugal

Instituto Português de Oncologia de Lisboa Francisco Gentil (IPOLFG), Lisbon, Portugal
e-mail: jacinta.serpa@nms.unl.pt

Keywords

Uterine cervix cancer · Colon cancer · Microflora · Metabolic symbiosis · Symbiosis bacteria · Epithelial cells · Lactate · Butyrate

4.1 Bacteria: Central Players in Humans' Health and Disease

"Eukaryotes presumably arose from prokaryotes (Margulis 1993) and have remained in close relationship with them ever since (Hickman 2005)" (Zilber-Rosenberg and Rosenberg 2008). In fact, humans have evolved in the presence of interactions with several other species. For instance, humans contain about 10^{14} microbes in the digestive tract and albeit the number of cells in the human microbiota has been considered 10 times higher than the number of cells in the human body, recent evidence showed that this ratio is variable and closer to 1 (Rosenberg and Zilber-Rosenberg 2016; Sender et al. 2016). Nonetheless, this ratio is impressive and our 'metagenome' is a combination of human and microbial genes that colonize us, allowing the

J. Serpa (ed.), *Tumor Microenvironment*, Advances in Experimental Medicine and Biology 1219,
https://doi.org/10.1007/978-3-030-34025-4_4

evolution of human traits that would be otherwise impossible (Turnbaugh et al. 2007). In fact, Gill and colleagues have found that our microbiome presents a significantly enriched metabolism of glycans, amino acids, and xenobiotics; methanogenesis; and 2-methyl-D- erythritol 4-phosphate pathway–mediated biosynthesis of vitamins and isoprenoids. The authors have then showed that our metabolism is a mixture of microbial and human features, considering humans 'superorganisms' (Gill S et al. 2006).

The human microbiota is central in both health and disease, and several studies reported the pivotal role of microbial symbiosis in several diseases including infection, liver diseases, metabolic disorders, respiratory diseases, mental or psychological diseases and autoimmune diseases (reviewed in Wang et al. 2017). Strikingly, microbial dysbiosis was shown to contribute to the etiology of several types of cancer, including colon, gastric, esophageal, pancreatic, laryngeal, breast and gallbladder carcinomas (Sheflin et al. 2014). Moreover, the International Agency for Research on Cancer (IARC) already published a list of microbes considered as Class 1 carcinogens (Humans 2012; Bhatt et al. 2017).

When analyzing the bacteria number on different organs, Sender and colleagues observed that the colon, a segment of the gut, presents a bacterial content at least two orders of magnitude higher compared to all other organs (Sender et al. 2016). In fact, the gut flora was already termed as the 'forgotten organ' (O'Hara and Shanahan 2006).

Given the central relevance of the microbiota, in 2008 the hologenome theory of evolution was proposed to animals and plants, proposing the holobiont (the host with all of its associated symbionts) as a unit of selection in evolution (Zilber-Rosenberg and Rosenberg 2008).

In the next section, we will discuss the role of tumor microenvironment in cancer initiation and progression, focusing on the role of bacteria on its modulation.

4.2 Cancer Initiation and Progression: The Role of Bacteria on the Modulation of the Tumor Microenvironment

In order to sustain a malignant neoplasm, the cells that undergo the malignant transformation must survive and proliferate within their microenvironment, allowing the tumor initiation and the further progression. The metabolic remodeling is a hallmark of cancer cells (Hanahan and Weinberg 2011; Serpa and Dias 2011), allowing the adaptation of these cells to hostile environments, namely to acidosis and hypoxia, and to environments where the availability of energy and biomass sources are scarce, allowing the further sustaining of cell survival and proliferation.

Each human organ has a particular microenvironment, which comprises several cell types and, in some cases, also symbiotic microorganisms. These biological partners continuously share organic compounds and signaling molecules that will impact cell proliferation and differentiation, accounting for the correct organ's function. For instance, the gut microbiota can modulate directly the gut epithelium or the immune system and pathological dysbiosis can lead to the production of high levels of toxins that will elicit both inflammation and tumorigenesis (Zou et al. 2018; Vivarelli et al. 2019). This clearly shows the profound effect of microbiota in the modulation of the local microenvironment. Impressively, evidence supports that gut microbiota also exert effects on distant microenvironments, as gut microbiome secretes several bioactive metabolites able to modulate breast cancer cells microenvironment (reviewed in Mikó et al. 2019).

In fact, human microbiota presents a dual role in tumor suppression and tumor promotion, where dysbiosis is profoundly associated with tumorigenesis (Schwabe and Jobin 2013; Parida and Sharma 2019; Saus et al. 2019). Nonetheless data on longitudinal cohort studies are required in order to directly evidence human commensal microbiome as crucial in the etiopathogenesis of cancer (Scott et al. 2019). In here, we do not intent to explore this duality since the specific role of gut microbiota in tumor promotion/suppression will be extensively explored in the chapter by Baffy. Instead, we do intent to explore how cancer cells are able to take advantage from the pre-existing metabolic symbiotic relations between microbiota and non-malignant cells.

In a malignant context, the cancer cells are able to repurpose the complex microenvironmental metabolic interactions in order to support the tumor metabolism and growth (Lyssiotis and Kimmelman 2017), being able to take advantage from the adaptations that the normal counterparts already possess.

In the next sections, we will explore some symbiotic metabolic interactions between bacteria and non-malignant cells in two different contexts: colon and uterine cervix, focusing on how the malignant cells are able to recycle the normal metabolic symbiosis and also create novel types of symbiotic metabolic relations.

4.3 Colon: The Organ with the Highest Bacterial Content

The colon is one of the pivotal parts of the gastrointestinal tract, with functions in the absorption of water, minerals, and nutrients (Arvelo et al. 2015). It also functions as a storage area for the waste material that forms the feces (Arvelo et al. 2015). The colon is comprised by four sections: the ascending, the transverse, the descending and the sigmoid colon (Arvelo et al. 2015). This organ is irregular and thick due to the longitudinal disposition of muscular fibers, presenting a scarce developed submucosa, but a very developed mucosa that harbors lymph tissue, the *Peyers' patches* (Arvelo et al. 2015). The mucosa is formed by multiple tubular invaginations called *'crypts of Lieberkühn'* along the surface of its epithelium, in which the epithelium regeneration occurs (Arvelo et al. 2015).

As already mentioned, the human gastrointestinal tract comprises a complex and dynamic population of microorganisms, with important roles not only on the host homeostasis but also in disease (Thursby and Juge 2017), being involved in essential human biological processes, including the modulation of the metabolic phenotype, regulation of epithelial development, and affecting innate immunity (Wang et al. 2017). In fact, the gut dysbiosis was already associated with several diseases, including several types of cancer, especially colorectal cancer (Kosumi et al. 2018), suggesting that the colon microbiota can have a profound systemic role on cancer initiation and progression that is not only confined to colorectal cancers.

The gastrointestinal microbiota includes Archaea, Bacteria and Eukarya and specifically the colon harbors *Bacteroidaceae*, *Prevotellaceae*, *Rikenellaceae*, *Lachnospiraceae* and *Ruminococcaceae* (Donaldson et al. 2015). The interactions between host-symbiotic gut bacteria provide several benefits to the host, including the metabolism of indigestible compounds and the supply of essential nutrients, defense against colonization by opportunistic pathogens, and contribution to the formation of intestinal architecture (Round and Mazmanian 2009; Wang et al. 2017). A good example of the pivotal role of gut bacteria on gut homeostasis is the production of butyrate in the colon microenvironment, via fermentation of dietary fibres that plays important roles in the regulation of tissue remodeling and metabolic homeostasis (Serpa and Dias 2011). Also, the dynamics of hydrogen sulfide (H_2S) production in the gastrointestinal tract have profound effects in this system, functioning as pro and anti-inflammatory, smooth muscle relaxant, prosecretory, and with pro- and anti-nociceptive activities (Linden 2014). Interestingly, evidence have also been supporting a role of gut bacteria-derived H_2S in the homeostasis of the circulatory system in mammals (Tomasova et al. 2016).

In fact, H_2S was proposed as another gasotransmitter, along with nitric oxide and carbon monoxide (Wang 2002), being involved in several biological processes including autophagy, cellular metabolism, stem cells fate regulation, inflammation, cell cycle and cell death, being crucial both in health and disease (Sen 2017). The natural metabolic symbiosis between non-malignant colonocytes and resident bacteria,

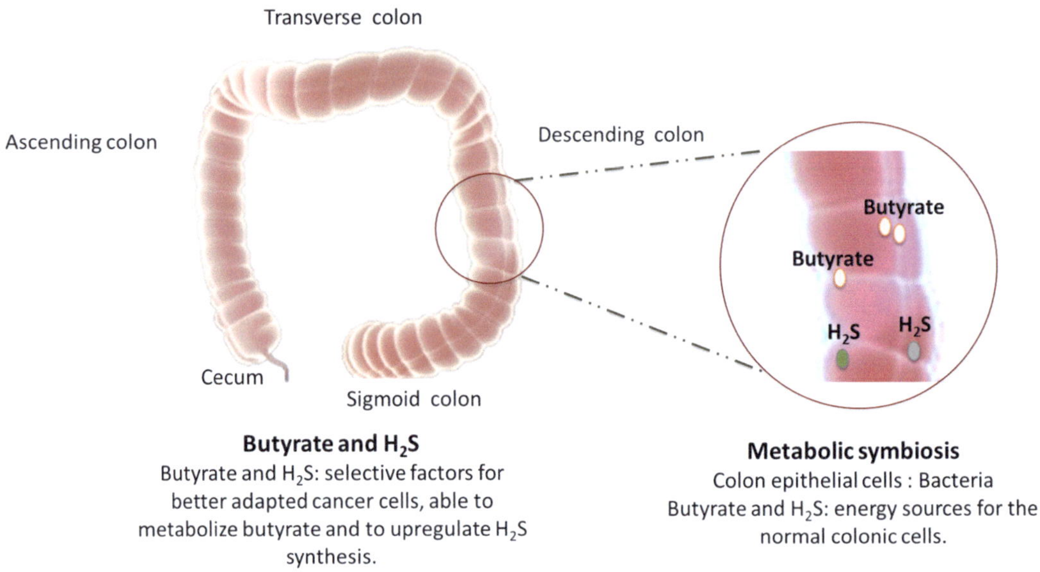

Fig. 4.1 Natural metabolic symbiosis between colonocytes and symbiotic bacteria
Within the colon, a natural metabolic symbiosis between non-malignant colonocytes and symbiotic bacteria takes place. Hence, bacteria produce butyrate via fermentation of dietary fibres, which is the main energy source for colonic epithelial cells, playing key roles in the maintenance of the gut microbiota stability and the integrity of intestinal epithelium (Serpa et al. 2010; Serpa and Dias 2011; Wu et al. 2018). This metabolite presents anti-inflammatory and tumor suppressor activities (reviewed in Wu et al. 2018). Other example of a metabolic symbiosis within the colon microenvironment is via the production of H_2S from both the colonic epithelial cells and from bacteria. H_2S has important functions on the host colonic mucosa metabolism, physiology, and physiopathology (reviewed in Blachier et al. 2019), being also used as an energy source for the colonic epithelial cells (Goubern et al. 2007)

mediated by both butyrate and H_2S are presented in the Fig. 4.1.

In the next section, we will explore how the recycling of non-malignant metabolic symbiosis by cancer cells can be advantageous for the malignant colonic cells and thus, leading to disease progression.

4.4 Colon Microbiota: From Healthy Interactions to Malignant Exploitation

As already mentioned, symbiotic bacteria within the colon produces several metabolites, including butyrate. Butyrate is the main energy source for colonic epithelial cells, playing key roles in the maintenance of the gut microbiota stability and the integrity of intestinal epithelium (Serpa et al. 2010; Serpa and Dias 2011; Wu et al. 2018). This metabolite presents anti-inflammatory and tumor suppressor activities, functioning as an histone deacetylase (HDAC) inhibitor, inhibiting cell proliferation and/or inducing apoptosis of cancer cells via several signaling pathways, including ERK2/MAPK, Wnt and p53 (reviewed in Wu et al. 2018). Butyrate was also reported to reduce neuropilin expression, suppressing angiogenesis, metastasis and colorectal cancer cells survival (reviewed in Wu et al. 2018). However, besides resistance through down-regulation of butyrate metabolism (reviewed in Wu et al. 2018), it was also reported that colon cancer cells are able to resemble the normal colonocytes and metabolize butyrate (Serpa et al. 2010; Serpa and Dias 2011). Therefore, Serpa and colleagues have reported that the chronic exposure of colon cancer cells to butyrate may result in the selection of more aggressive clones, where butyrate-resistant colon cancer cells retain the normal colonocytes ability to metabolize butyrate, exhibit a mesenchymal phenotype and are also more invasive (Serpa

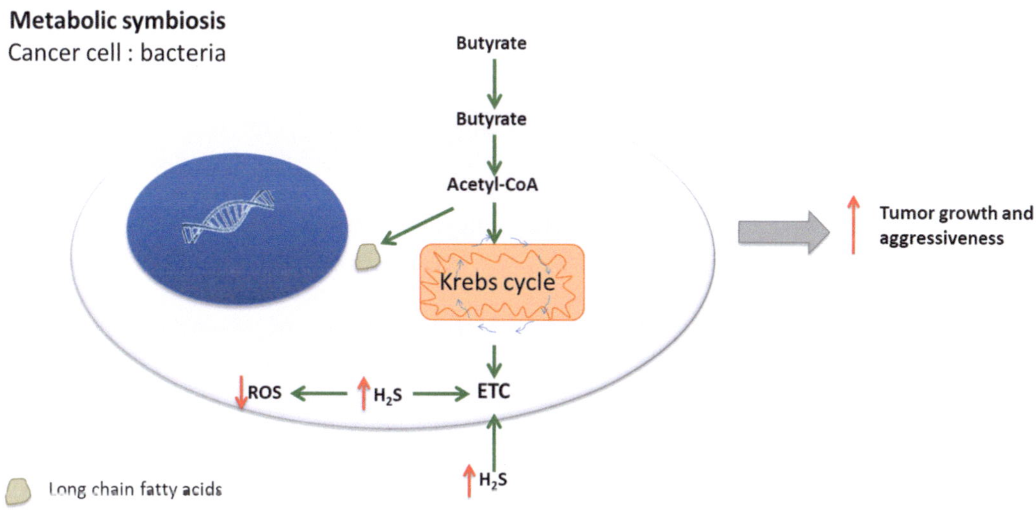

Fig. 4.2 Recycling of the metabolic interactions between non-malignant cells and bacteria in colon microenvironment by colon cancer cells
While butyrate is generally assumed to present tumor suppressor activities (reviewed in Wu et al. 2018), Serpa and colleagues have reported that the chronic exposure of colon cancer cells to butyrate may result in the selection of more aggressive clones, where butyrate-resistant colon cancer cells retain the normal colonocytes ability to metabolize butyrate, being capable to metabolize it to intermediates used in the Krebs cycle and also used as blocks in the synthesis of long chains fatty acids (Serpa et al. 2010). The normal metabolic symbiosis concerning H_2S synthesis can also be exploited in colorectal cancer. Hence, it was observed an increased H_2S production from both the host and from bacteria in colorectal cancer patients (reviewed in Blachier et al. 2019). Therefore, H_2S can be advantageous for colon tumor growth by promoting angiogenesis and vasorelaxation, and by stimulating bioenergetics (Szabo et al. 2013) and also by presenting antioxidant properties in the mitochondria (Hellmich et al. 2015)

et al. 2010; Serpa and Dias 2011). These results undoubtedly show that the transformed malignant cells can present a metabolic advantage when re-acquiring/recycling/retaining the metabolic features of the normal counterparts allowing, not only cancer cells survival, but also an increased aggressive phenotype, showing that the recycling of pre-existing features of the normal counterparts can be crucial for cancer cells survival and tumor progression (Fig. 4.2).

Other interesting example of the recycling of the metabolic symbiosis between non-malignant cells and bacteria by cancer cells concerns H_2S metabolism, which have key roles on the host colonic mucosa metabolism, physiology, and physiopathology (reviewed in Blachier et al. 2019). In the gastrointestinal tract, H_2S is produced by both the host and by the resident microbiota. In mammalian cells, H_2S is generated mainly via enzymatic pathways and from the metabolism of L-cysteine by the catalysis of three key enzymes: cystathionine β-synthase (CBS), cystathionine γ-lyase (CSE) and by 3-mercaptopyruvate sulphurtransferase (MpST) accompanied by cysteine aminotransferase (CAT) (Wang 2012). More recently, a new pathway generating H_2S from D-cysteine involving 3-mercaptopyruvate sulfurtransferase and D-amino acid oxidase was reported in mammalian cells (Shibuya et al. 2013). Specifically in the colon, evidence suggest that CBS is the major H_2S-synthesizing enzyme in this organ (reviewed in Guo et al. 2016; Blachier et al. 2019). Additionally, H_2S is also produced by the resident gut bacteria, mainly by cysteine degradation (Blachier et al. 2019). It is important to mention that bacteria present CBS, CSE and MpST homologs with similar function, but as Szabo alerted "*from the evolutionary standpoint, the correct way of stating would be to say that the bacterial enzymes have mammalian homologs*" (Szabo 2018). H_2S is also produced by colonic sulfate-reducing bacteria (Singh and Lin 2015; Guo et al. 2016) and by sulfite reduction (Tomasova et al. 2016).

The dynamics of H_2S production by both epithelial cells and by the intestinal microbiota can have profound effects on colonic mucosa inflammation and colorectal cancer progression (Blachier et al. 2019), being the increased H_2S generation induced by imbalances in the resident bacteria and also by the host possibly associated to colorectal cancer progression (Blachier et al. 2019).

In fact, the enzymes involved in H_2S generation were already reported to be highly expressed in several types of cancer, including colorectal cancer (reviewed in Cao et al. 2018; Szabo 2018). The fact that cancer cells upregulate these enzymes strongly supports the recycling of metabolic symbiosis between gut bacteria and normal colonocytes, where cancer cells take advantage from these interactions and additionally exacerbate this phenotype (Fig. 4.2). Supporting this recycling, Wu and colleagues have reported similar effects of H_2S in both non-malignant and malignant colon epithelial cells, inhibiting cells proliferation while inducing protective autophagy via the AMPK pathway (Wu et al. 2012). In fact, it was reported a dual effect of H_2S in cancer cells, where endogenous or low exogenous H_2S levels are advantageous for colon cancer cells, whereas the exposure to high levels is disadvantageous (Hellmich et al. 2015; Cao et al. 2018), exhibiting a bell-shaped curve effect (Hellmich et al. 2015). Therefore, it was reported that tumor-derived H_2S, generated by CBS activity, is advantageous for colon tumor growth by promoting angiogenesis and vasorelaxation, and by stimulating bioenergetics (Szabo et al. 2013). Moreover, physiological concentrations of H_2S were reported to counteract β-phenyethyl isothiocyanate- mediated apoptosis in colon cancer cells (Rose et al. 2005).

It worth to mention that there is evidence supporting that the endogenous or exogenous H_2S can affect intestinal microbiota (Wallace et al. 2017), suggesting that besides being able to recycle the normal metabolic symbiosis, cancer cells are also able to create new networks of metabolic symbiosis, as their upregulated H_2S synthesis could lead to dysbiosis. Interestingly, Johnson and colleagues have reported a direct correlation between biofilm formation on colon cancers and the upregulation of N1, N12-diacetylspermine, a polyamine metabolite that may enhance cancer growth, invasion and metastasis (Johnson et al. 2015), hence suggesting that cancer cells are also able to create and exploit new symbiotic relations with the resident microbiota.

Cao and colleagues have linked butyrate and H_2S synthesis, hence reporting that butyrate regulates endogenous H_2S production by inducing CBS expression in colon cancer cells, leading to cancer cell growth inhibition through different mechanisms (Cao et al. 2009). This information is apparently contradictory to the beneficial effects of H_2S in colon cancer cells but, as already mentioned, H_2S exhibit a bell-shape effect on colon cancer cells (Hellmich et al. 2015), which may explain these results. In fact, H_2S may be a promising strategy in cancer treatment, either through H_2S biosynthesis inhibition or through H_2S supplementation (Cao et al. 2018).

In the next section we will discuss the normal metabolic symbiosis between uterine cervix squamous epithelial cells and the resident bacteria and also the recycling of these interactions in a context of cervical cancer.

4.5 Uterine Cervix: An Ideal Acidic Microenvironment for Cancer Cells

Whereas the consensus of the existence of a complex gut microbiota in healthy human adults, for almost a century, the consensus was that a healthy uterine cavity is sterile (reviewed in Baker et al. 2018). However, uterine microbiota was reported in several mammals, including humans, nonetheless, it was estimated to be 100 and 10,000 times lower compared to the vaginal microbiome (reviewed in Baker et al. 2018).

Uterine cervix has a particular acidic microenvironment created by the biochemical collaboration of epithelial cells and symbiotic bacteria, mainly *Lactobacillus* sp. (Van Der Veer et al. 2019). It has been known for a long time that Lactobacilli are the predominant microorganisms found in the cervix and vagina, together with some skin and fecal contaminants (Wilson and Miles 1946; Corbishley 1977). In 1861, Albert Döderlein described the physiology of vaginal and cervix microbiota, considering the *Lactobacillus* genus the most prevalent group of bacteria which since then received his name Döderlein Bacilli (Thomas 1928). Despite a controversial study with new microbial isolation and identification techniques claiming that often *Lactobacillus* are not the predominant genus in vaginal/cervical microflora (Linhares et al. 2010), several other publications until today corroborate Döderlein's findings (Corbishley 1977; Vásquez et al. 2002; Liu et al. 2007; Ravel et al. 2011; Xiao and Liao 2012; Pendharkar et al. 2013; Wang et al. 2018; Kroon et al. 2018; Elovitz et al. 2019).

Because of the glycolytic phenotype of cancer cells, these cells may be especially capable to adapt to acidic environments, making the uterine cervix an ideal microenvironment for the newly transformed cells. In the next section, we will briefly describe the histological and metabolic properties of the uterine cervix in a healthy scenario.

4.6 Histological and Metabolic Features of the Uterine Cervix

The uterine cervix is formed by two components: the ectocervix that protrudes into vagina and the endocervix, a canal connecting the vagina to the uterus. The endocervical canal is lined by cells similar to endometrium, which covers the uterine cavity – a single layer of tall, columnar, mucus-secreting cells. Importantly, the ectocervix is directly exposed to the hostile acidic microenvironment of the vagina, being lined by a thick stratified squamous epithelium (Havens and Sullivan 2002; Brezinski 2006). Physiologically, the endocervical columnar epithelium in the portion more close to ectocervix undergoes metaplastic transformation to a mature squamous epithelium, being this transformation especially active during adolescence and pregnancy (Havens and Sullivan 2002; Reich et al. 2017). The metaplastic dynamics in this transformation zone occurs throughout female reproductive life, and it is regulated by environmental stressors such as hormones and pH (Braun and Anderson 2007; Reich et al. 2017). Although the physiological nature of the squamous metaplasia of the cervix, the newly differentiated squamous epithelium of the transformation zone is vulnerable to cell injury and damage (Braun and Anderson 2007), presenting an increased risk of oncogenesis, principally through the infection of Human Papillomavirus (HPV) (Havens and Sullivan 2002; Brezinski 2006), which is believed to be the main etiological factor for uterine cervix cancer (Gao et al. 2016). Interestingly, albeit the need of further elucidation, it was shown that the abundance of vaginal microbiota differs between patients with cervical cancer and healthy individuals, presenting different contents in *Mycoplasma genitalium*, aerobic lactobacilli, *Staphylococcus epidermidis*, Enterococci, *Escherichia coli*, and *Bacteriodes* species (Yang et al. 2018). Moreover, evidence has been supporting an important role of *Lactobacillus* in cervical cancer initiation and progression (Yang et al. 2018).

The squamous cells from both vagina and ectocervix are rich in glycogen and upon the physiological peeling process, the cell disintegration allows the glycogen release in the microenvironment. Glycogen is the main energy source for Döderlein bacilli, being firstly degraded into glucose that afterwards through lactic fermentation gives rise to lactic acid (lactate). Lactic acid is a crucial player in the control of chemical and physical conditions, being the main responsible for the acidification of the uterine cervix and vaginal microenvironment (Redondo-Lopez et al. 1990; Gupta 2011; Van Der Veer et al. 2019) (Fig. 4.3). It has been known for decades that cancer cells arising from the squamous mucosa of the cervix and vagina lose the ability to accumulate glycogen (Das and Chowdury 1978). However, it is also well established that oxidative cancer cells are able to use lactate as a carbon and energy source (reviewed in de Bari and Atlante 2018), including cervix cancer cells (Draoui and Feron 2011; Silva et al. 2016). Therefore, in the

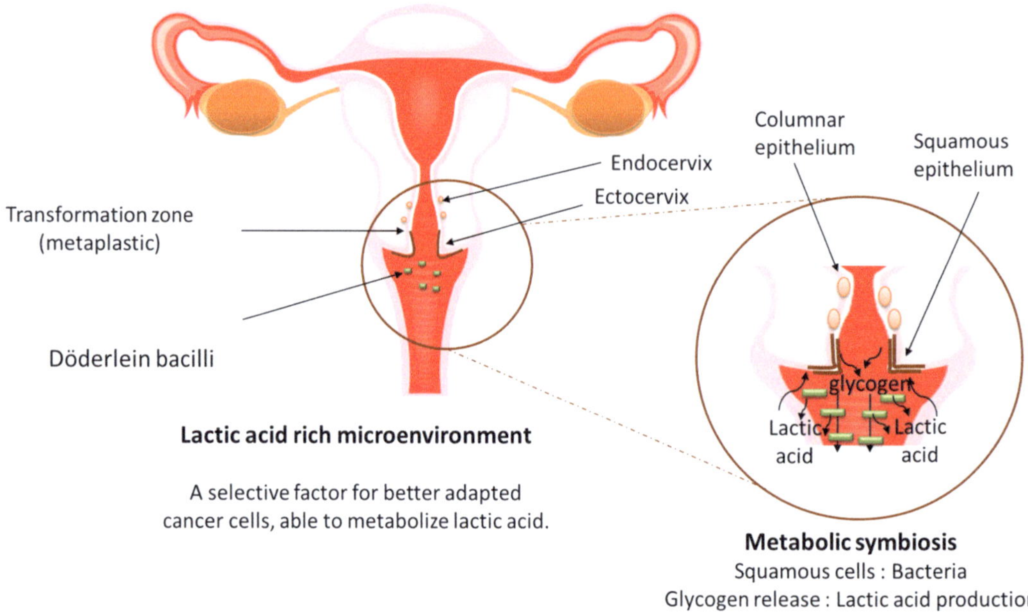

Fig. 4.3 Natural metabolic symbiosis between uterine cervix squamous cells and symbiotic microbiota
The squamous cells, from vagina and ectocervix are full of glycogen and upon the physiological peeling process the cell disintegration allows the glycogen release into microenvironment. Döderlein bacilli degrade glycogen at first into glucose and then into lactic acid through lactic fermentation. Lactic acid is the main responsible for the acidification of uterine cervix and vaginal microenvironment that can constitute a selective factor for cancer cells that are more able to metabolize lactic acid and carry cancer progression

next section, we will explore how the cervix cancer cells can exploit the normal metabolic symbiosis present between the non-malignant cells and the resident bacteria.

4.7 Microbiota-Derived Lactate in Uterine Cervix: The Role of Metabolic Symbiosis Recycling in Cancer Cells

The Warburg effect is the most well documented metabolic adaptation in cancer, proposing that cancer cells present increased rate of glycolysis even under normal oxygen concentrations. There is some debate about the selective advantages of glycolytic metabolism to proliferating tumor cells. Initial studies argued that tumor cells develop defects in mitochondrial function and that aerobic glycolysis is a necessary adaptation to the lack of ATP production through oxidative phosphorylation (Cairns et al. 2011). However, it was later observed that mitochondrial defects are rare (Frezza and Gottlieb 2009) and tumors retain the capacity of oxidative phosphorylation and consume oxygen at rates similar to those observed in normal tissues (Fantin et al. 2006; Moreno-Sánchez et al. 2007). Because the energetic yield of glycolysis is much lower than cellular respiration, glycolysis would occur at a very high rate, having the lactate production as a final event.

Alternatively, it has been proposed that glycolytic metabolism arises as an adaptation to hypoxic conditions during the early avascular phase of tumor development, as it allows ATP production under low levels of oxygen. Adaptation to the resulting acidified microenvironment that is caused by excessive lactate production may further drive the evolution of the glycolytic phenotype (Gatenby and Gillies 2004; Gillies et al. 2008). So, the traditional view of lactate as a mere excretion product of glucose accelerated metabolism was replaced by a 'functional' role of lactate, that can also be used as a carbon and energy source (Semenza 2008; Hirschhaeuser et al. 2011; Draoui et al. 2013; Silva et al. 2016).

Aerobic glycolysis was also reported to provide a biosynthetic advantage for tumor cells, and that a high flux of substrate through glycolysis allows for effective shunting of carbon to key subsidiary biosynthetic pathways (Vander Heiden et al. 2009; Lopes-Coelho et al. 2017; Lee et al. 2019). Moreover, glycolysis wouldn't be a predominantly energetic source but a biomass supplying source. Our group already published some studies showing that glycolysis mainly supplies phosphate pentose pathway (PPP) for nucleotide synthesis that will further support cell proliferation, both in acute myeloid leukemia (Lopes-Coelho et al. 2017; Lee et al. 2019) and in uterine cervix cancer (Silva et al. 2016).

Strikingly, our group has shown that squamous cell carcinomas originated from ectocervix cells are predominantly lactate consumers being the cells able to uptake lactate and convert it into pyruvate and Krebs cycle intermediates as well as into amino acids and fatty acids (Silva et al. 2016). However, adenocarcinoma cells originated from endocervix glandular cells, exhibit a glycolytic phenotype, being almost exclusively lactate producers by using glucose as a source (Silva et al. 2016). This observation is consistent with the fact that squamous ectocervix cells are naturally more exposed and adaptable to lactic acid metabolism than glandular endocervix cells, showing the ability of cancer cells to recycle the already existent symbiotic metabolic interactions.

Uterine cervix cancer is therefore another exceptional model in which the malignant cells recycle symbiotic metabolic interactions between non-malignant cells and bacteria and also create new networks of metabolic symbiosis in order to survive and proliferate. The Fig. 4.4 resumes this complex metabolic recycling.

Given the expanding evidence for the relevance of microbiota in cancer initiation, progression and treatment, in the next section we will briefly discuss the dual role of microbiota modulation in cancer treatment and also the effects of anti-cancer therapies in dysbiosis.

4.8 From the Role of Microbiota in Anticancer Therapies to Anticancer Therapies Role in Dysbiosis

Given the emergent key role of microbiota in carcinogenesis, the rational that the microbiota also impairs anticancer therapies have gained further attention.

Indeed, evidence has been supporting that microbiota modulate the effects of chemotherapy, radiotherapy (Bashiardes et al. 2017; Roy and Trinchieri 2017; Mikó et al. 2019; Wu et al. 2019b) and immunotherapy (Li et al. 2019; Wu et al. 2019b), including drug efficacy and toxicity sensitive modulation (Wu et al. 2019b). Alexander and colleagues reviewed the effects of gut microbiota in chemotherapy modulation and proposed the 'TIMER' mechanistic framework, suggesting that gut microbiota modulate chemotherapy through Translocation, Immunomodulation, Metabolism, Enzymatic degradation, and Reduced diversity and ecological variation (Alexander et al. 2017). Therefore, strategies targeting microbiota in an attempt to enhance the responses to treatment in cancer patients were already developed. Very recently, Ding and colleagues have reviewed the potential use of biotherapies as a strategy to modulate intestinal microbiota and improve colorectal cancer treatment (Ding et al. 2018). These strategies include the use of oral probiotics, prebiotics, antibiotics and other drugs and fecal microbiota transplantation (reviewed in Ding et al. 2018). Since *Lactobacillus rhamnosus* presents several anti-inflammatory properties, some ongoing clinical trials are testing its role on the prevention or enhancing the toxic effects associated with anticancer therapies, including colorectal cancer patients (Vivarelli et al. 2019). Moreover, 24 clinical trials of probiotic and/or synbiotic therapies were already published showing positive results for colorectal cancer patients (Ding et al. 2018). Besides this, McFadden and colleagues have reported that curcumin, a bioactive component derived from a rhizome of the *Curcuma longa* plant, was able to decrease colonic tumor burden and that this effect was associated with the modu-

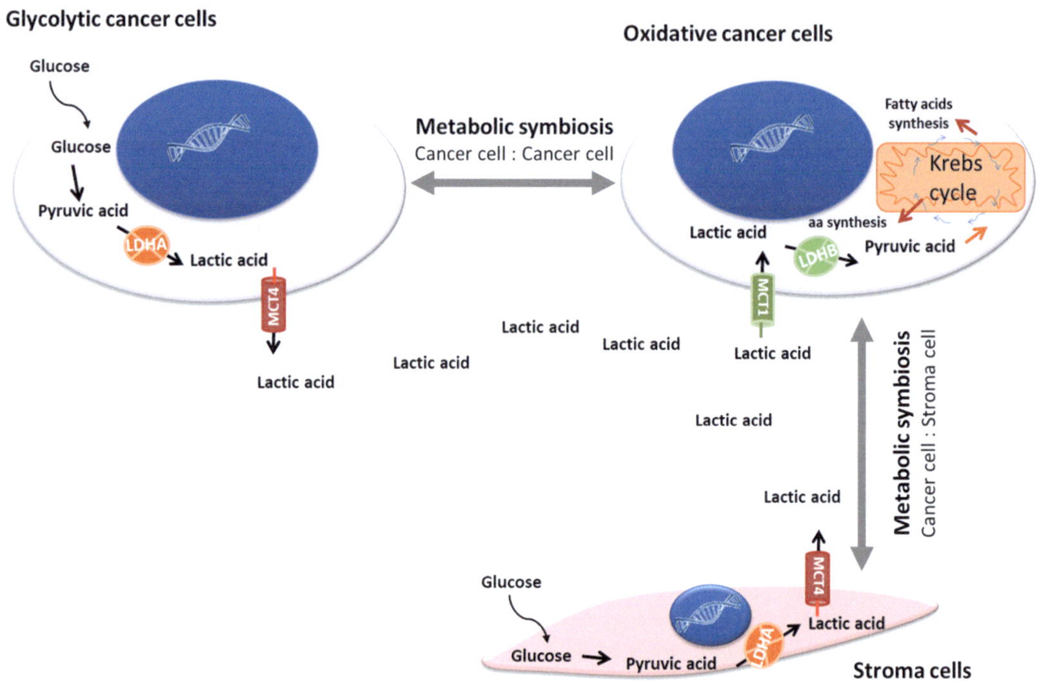

Fig. 4.4 Recycling of the metabolic interactions between non-malignant cells and bacteria in the uterine cervix microenvironment by malignant cells and the new metabolic symbiosis between cancer and stroma cells
The natural uterine cervix microenvironment is rich in lactate produced by Döderlein bacilli from epithelial glycogen; in malignant neoplasms, cancer cells decrease the production of glycogen as they increase their glucose demands but at the same time they produce high levels of lactate that will keep the acidity of the microenvironment. This way cancer cells keep the metabolic favorable conditions that will positively select cancer cells that are more prone to drive cancer progression. Furthermore, because the symbiosis between cancer cells and Lactobacilli is replaced by symbiosis between cancer cells, the bacterial density decreases in cancerous uterine cervix. Moreover, the pro-apoptotic effect exerted by Lactobacilli on cancer cells (Motevaseli et al. 2013) is also depleted by all this microenvironmental metabolic remodeling

lation of the colonic microbiota, where curcumin allowed the maintenance of colonic microbial diversity (McFadden et al. 2015).

In the context of uterine cervix cancer, radiotherapy is the main clinical approach to treatment. Unfortunately, few studies were performed relating the prevalence and role of uterine cervix and vaginal microbiota in therapy response (Choo et al. 1984; Gilstrap III et al. 1986; Gordon et al. 1989; Mubangizi et al. 2014). Interestingly, in an orthotopic model of cervical cancer, Colbert and colleagues have reported that the reduction of gut microbiome diversity and the incidence of vaginal *Lactobacillus* were associated with decreased activated CD8 T-cell infiltration and decreased response to radiation treatment (Colbert et al. 2018). Moreover, due to the possible role of bacteria in cervical cancer carcinogenesis, the manipulation of the cervical microbiome with antibiotics was already suggested (Lam et al. 2018). Also, Champer and colleagues alerted that research is needed in order to determine the benefits of probiotics in the modulation of the vaginal microbiome on the treatment of gynecological cancers (Champer et al. 2018).

Besides the ability of microbiota to modulate chemotherapy, radiotherapy and immunotherapy (Bashiardes et al. 2017; Roy and Trinchieri 2017; Mikó et al. 2019; Wu et al. 2019b), chemotherapy-induced dysbiosis (Montassier et al. 2015), radiation-induced dysbiosis (Gerassy-Vainberg et al. 2018) and immunotherapy-induced dysbiosis (Vétizou et al. 2015) were also reported. In an attempt to avoid this anticancer treatment-induced dysbiosis, Ichim and colleagues have recently identified a probiotic formulation able to counteract this effect on gut dysbiosis (Ichim et al. 2018). Also, fecal microbial transplantation was reported to counteract chemotherapy-induced gut dysbiosis in mice (Le Bastard et al. 2018).

Hence, the manipulation of the microbiota with approaches such as biotherapies seem to be promising in colon cancer management and probably also in cervical cancer most likely by re-establishing the normal microbiota microenvironment that reverts the malignant metabolic symbiosis to a non-malignant state, hence supporting the potent effects of cancer cells on the manipulation of the tumor microenvironment components and on the creation of novel symbiotic interactions. This leads to the rational that besides the key role of dysbiosis in carcinogenesis, carcinogenesis *per se* has also a key role on dysbiosis, hence being the cancer cells able to manipulate also the microbiome. The powerful effect of cancer cells in the manipulation of other cells is well reflected by the secretion of extracellular vesicles, being the cancer cells able to transfer the metastatic ability to other cells (reviewed in Kosaka et al. 2016). Outstandingly, in the context of colorectal cancer, a recent review was published discussing the intracellular communication mediated by extracellular vesicles between cancer cells, macrophages and the microbioma (Wu et al. 2019a). In fact, evidence suggests that not only cancer cells are able to internalize outer membrane vesicles derived from bacteria, but the extracellular vesicles derived from colorectal cancer cells are also able to modulate commensal bacteria in gut, possibly transferring the malignant features to microbiome (Wu et al. 2019a). This undoubtedly shows the complexity and the reciprocal influence between cancer cells and microbiota, where cancer cells are able not only to recycle non-malignant symbiotic interactions, but also create novel ones via the modulation of the microbiota, that ultimately will favor cancer cells survival and disease progression.

4.9 Highlights

Several aspects of the human physiology are controlled by the microbiota that plays a key role in health and disease. Metabolic symbiosis between non-malignant cells and the resident bacteria is crucial for the host homeostasis. However, the microbiota can present a dual effect on the host, being protective but being also harmful, when dysbiosis takes place. Under physiological conditions, each organ presents a unique microenvironment, where particular cellular metabolic interactions between the host and the microbiota take place. However, the same interactions can be recycled and exploited by cancer cells, allowing their survival and progression.

While tumorigenesis is intimately related with dysbiosis, cancer cells can also disturb the microbiota, and establish new symbiotic relations different from the healthy state, showing the complexity of the recycling of metabolic interactions by the malignant cells.

The modulation of the microbiota represents a promising strategy for cancer treatment, where new targets can be identified and new treatments implemented in order to fight this highly mortal group of diseases.

Acknowledgments The authors acknowledge iNOVA-4Health – UID/Multi/04462/2013, a program financially supported by Fundação para a Ciência e Tecnologia/Ministério da Educação e Ciência, through national funds and co-funded by FEDER under the PT2020 Partnership Agreement.

References

Alexander JL, Wilson ID, Teare J et al (2017) Gut microbiota modulation of chemotherapy efficacy and toxicity. Nat Rev Gastroenterol Hepatol 14:356–365. https://doi.org/10.1038/nrgastro.2017.20

Arvelo F, Sojo F, Cotte C (2015) Biology of colorectal cancer. Ecancermedicalscience 9:1–20. https://doi.org/10.1097/PPO.0b013e3181e076af

Baker JM, Chase DM, Herbst-Kralovetz MM (2018) Uterine microbiota: residents, tourists, or invaders? Front Immunol 9:208. https://doi.org/10.3389/fimmu.2018.00208

Bashiardes S, Tuganbaev T, Federici S, Elinav E (2017) The microbiome in anti-cancer therapy. Semin

Immunol 32:74–81. https://doi.org/10.1016/j.smim.2017.04.001
Bhatt AP, Redinbo MR, Bultman SJ (2017) The role of the microbiome in cancer development and therapy. CA Cancer J Clin 67:326–344. https://doi.org/10.3322/caac.21398
Blachier F, Beaumont M, Kim E (2019) Cysteine-derived hydrogen sulfide and gut health: a matter of endogenous or bacterial origin. Curr Opin Clin Nutr Metab Care 22:68–75. https://doi.org/10.1097/MCO.0000000000000526
Braun CA, Anderson CM (2007) Pathophysiology: functional alterations in human health, 6th edn. Lippincott Williams & Wilkins, Philadelphia
Brezinski M (2006) Optical coherence tomography: principles and applications. Academic, Amsterdam/Boston
Cairns R, Harris I, Mak T (2011) Regulation of cancer cell metabolism. Nat Rev Cancer 11:85–95. https://doi.org/10.1038/nrc2981
Cao Q, Zhang L, Yang G et al (2009) Butyrate-stimulated H2S production in colon cancer cells. Antioxid Redox Signal 12:1101–1109. https://doi.org/10.1089/ars.2009.2915
Cao X, Ding L, Xie Z et al (2018) A review of hydrogen sulfide synthesis, metabolism, and measurement: is modulation of hydrogen sulfide a novel therapeutic for cancer? Antioxid Redox Signal 16616:ars.2017.7058. https://doi.org/10.1089/ars.2017.7058
Champer M, Wong AM, Champer J et al (2018) The role of the vaginal microbiome in gynaecological cancer. BJOG 125:309–315. https://doi.org/10.1111/1471-0528.14631
Choo YC, Seto WH, Ma HK (1984) Cervical-vaginal bacterial flora in patients with cervical carcinoma treated with irradiation and febrile morbidity during intracavitary radium therapy. Aust N Z J Obstet Gynaecol 24:34–38. https://doi.org/10.1111/j.1479-828X.1984.tb03318.x
Colbert L, Mikkelson M, Delgado Medrano AY et al (2018) The gut and cervical microbiome promote immune activation and response to chemoradiation in cervical cancer. SSRN Electron J
Corbishley CM (1977) Microbial flora of the vagina and cervix. J Clin Pathol 30:745–748. https://doi.org/10.1136/jcp.30.8.745
Das D, Chowdury J (1978) The use of glycogen studies in the evaluation of treatment for carcinoma of the cervix uteri. Acta Cytol 21:578–587
de Bari L, Atlante A (2018) Including the mitochondrial metabolism of l-lactate in cancer metabolic reprogramming. Cell Mol Life Sci 75:2763–2776. https://doi.org/10.1007/s00018-018-2831-y
Ding C, Tang W, Fan X, Wu G (2018) Intestinal microbiota: a novel perspective in colorectal cancer biotherapeutics. Onco Targets Ther 11:4797–4810. https://doi.org/10.2147/OTT.S170626
Donaldson GP, Lee SM, Mazmanian SK (2015) Gut biogeography of the bacterial microbiota. Nat Rev Microbiol 14:20–32. https://doi.org/10.1038/nrmicro3552
Draoui N, Feron O (2011) Lactate shuttles at a glance: from physiological paradigms to anti-cancer treatments. Dis Model Mech 4:727–732. https://doi.org/10.1242/dmm.007724
Draoui N, Schicke O, Fernandes A et al (2013) Synthesis and pharmacological evaluation of carboxycoumarins as a new antitumor treatment targeting lactate transport in cancer cells. Bioorg Med Chem 21:7107–7117. https://doi.org/10.1016/j.bmc.2013.09.010
Elovitz MA, Gajer P, Riis V et al (2019) Cervicovaginal microbiota and local immune response modulate the risk of spontaneous preterm delivery. Nat Commun 10:1305. https://doi.org/10.1038/s41467-019-09285-9
Fantin VR, St-Pierre J, Leder P (2006) Attenuation of LDH-A expression uncovers a link between glycolysis, mitochondrial physiology, and tumor maintenance. Cancer Cell 9:425–434. https://doi.org/10.1016/j.ccr.2006.04.023
Frezza C, Gottlieb E (2009) Mitochondria in cancer: not just innocent bystanders. Semin Cancer Biol 19:4–11
Gao J, Fan L, Ma W, Xiao H (2016) Synergistic antitumor effect of a human papillomavirus DNA vaccine harboring E6E7 fusion gene and vascular endothelial growth factor receptor 2 gene. Microbiol Immunol 60:626–633. https://doi.org/10.1111/1348-0421.12408
Gatenby RA, Gillies RJ (2004) Why do cancers have high aerobic glycolysis? Nat Rev Cancer 4:891–899. https://doi.org/10.1038/nrc1478
Gerassy-Vainberg S, Blatt A, Danin-Poleg Y et al (2018) Radiation induces proinflammatory dysbiosis: transmission of inflammatory susceptibility by host cytokine induction. Gut 67:97–107. https://doi.org/10.1136/gutjnl-2017-313789
Gill S, Pop M, DeBoy R et al (2006) Metagenomic analysis of the human distal gut microbiome. Science 312:1355–1359. https://doi.org/10.1126/science.1124234.Metagenomic
Gillies RJ, Robey I, Gatenby RA (2008) Causes and consequences of increased glucose metabolism of cancers. J Nucl Med 49(Suppl 2):24S–42S. https://doi.org/10.2967/jnumed.107.047258
Gilstrap LC III, Gibbs RS, Michel TJ, Hauth JC (1986) Genital aerobic bacterial flora of women receiving radiotherapy for gynecologic malignancy. Gynecol Oncol 23:35–39. https://doi.org/10.1016/0090-8258(86)90112-5
Gordon AN, Martens M, LaPread Y, Faro S (1989) Response of lower genital tract flora to external pelvic irradiation. Gynecol Oncol 35:233–235. https://doi.org/10.1016/0090-8258(89)90050-4
Goubern M, Andriamihaja M, Nubel T et al (2007) Sulfide, the first inorganic substrate for human cells. FASEB J 21:1699–1706. https://doi.org/10.1096/fj.06-7407com
Guo FF, Yu TC, Hong J, Fang JY (2016) Emerging roles of hydrogen sulfide in inflammatory and neoplastic colonic diseases. Front Physiol 7:1–8. https://doi.org/10.3389/fphys.2016.00156

Gupta S (2011) A comprehensive textbook of obstetrics and gynecology, 1st edn. Jaypee Brothers Medical Publishers Ltd, New Delhi

Hanahan D, Weinberg RA (2011) Hallmarks of cancer: the next generation. Cell 144:646–674. https://doi.org/10.1016/j.cell.2011.02.013

Havens N, Sullivan C (2002) Manual of outpatient gynecology, 4th edn. Lippincott Williams & Wilkins, Philadelphia

Hellmich MR, Coletta C, Chao C, Szabo C (2015) The therapeutic potential of cystathionine β-synthetase/hydrogen sulfide inhibition in cancer. Antioxid Redox Signal 22:424–448. https://doi.org/10.1089/ars.2014.5933

Hirschhaeuser F, Sattler UGA, Mueller-Klieser W (2011) Lactate: a metabolic key player in cancer. Cancer Res 71:6921–6925

Humans IWG on the E of CR to (2012) Biological agents. Volume 100 B. A review of human carcinogens. IARC Monogr Eval Carcinog Risks Hum 100:1–441

Ichim TE, Kesari S, Shafer K (2018) Protection from chemotherapy- and antibiotic-mediated dysbiosis of the gut microbiota by a probiotic with digestive enzymes supplement. Oncotarget 9:30919–30935. https://doi.org/10.18632/oncotarget.25778

Johnson CH, Dejea CM, Edler D et al (2015) Metabolism links bacterial biofilms and colon carcinogenesis. Cell Metab 21:891–897. https://doi.org/10.1016/j.cmet.2015.04.011

Kosaka N, Yoshioka Y, Fujita Y, Ochiya T (2016) Versatile roles of extracellular vesicles in cancer. J Clin Invest 126:1163–1172. https://doi.org/10.1172/JCI81130

Kosumi K, Mima K, Baba H, Ogino S (2018) Dysbiosis of the gut microbiota and colorectal cancer: the key target of molecular pathological epidemiology. J Lab Precis Med 3:1–6. https://doi.org/10.1007/s11065-015-9294-9.Functional

Kroon SJ, Ravel J, Huston WM (2018) Cervicovaginal microbiota, women's health, and reproductive outcomes. Fertil Steril 110:327–336. https://doi.org/10.1016/j.fertnstert.2018.06.036

Lam KC, Vyshenska D, Hu J et al (2018) Transkingdom network reveals bacterial players associated with cervical cancer gene expression program. PeerJ 6:e5590. https://doi.org/10.7717/peerj.5590

Le Bastard Q, Ward T, Sidiropoulos D et al (2018) Fecal microbiota transplantation reverses antibiotic and chemotherapy-induced gut dysbiosis in mice. Sci Rep 8:6219. https://doi.org/10.1038/s41598-018-24342-x

Lee MH, Malloy CR, Corbin IR et al (2019) Assessing the pentose phosphate pathway using [2, 3-$^{13}C_2$]glucose. NMR Biomed 32(6):e4096. https://doi.org/10.1002/nbm.4096

Li W, Deng Y, Chu Q, Zhang P (2019) Gut microbiome and cancer immunotherapy. Cancer Lett 447:41–47. https://doi.org/10.1016/j.canlet.2019.01.015

Linden DR (2014) Hydrogen sulfide signaling in the gastrointestinal tract. Antioxid Redox Signal 20(5):818–830. https://doi.org/10.1089/ars.2013.5312. Epub 2013 May19

Linhares IM, Giraldo PC, Baracat EC (2010) New findings about vaginal bacterial flora. Rev Assoc Med Bras 56:370–374. https://doi.org/10.1590/S0104-42302010000300026

Liu JJ, Reid G, Jiang Y et al (2007) Activity of HIV entry and fusion inhibitors expressed by the human vaginal colonizing probiotic Lactobacillus reuteri RC-14. Cell Microbiol 9:120–130. https://doi.org/10.1111/j.1462-5822.2006.00772.x

Lopes-Coelho F, Nunes C, Gouveia-Fernandes S et al (2017) Monocarboxylate transporter 1 (MCT1), a tool to stratify acute myeloid leukemia (AML) patients and a vehicle to kill cancer cells. Oncotarget 8:82803–82823. https://doi.org/10.18632/oncotarget.20294

Lyssiotis CA, Kimmelman AC (2017) Metabolic interactions in the tumor microenvironment. Trends Cell Biol 27:863–875. https://doi.org/10.1016/j.tcb.2017.06.003

McFadden R-MT, Larmonier CB, Shehab KW et al (2015) The role of curcumin in modulating colonic microbiota during colitis and colon cancer prevention. Inflamm Bowel Dis 21:2483–2494. https://doi.org/10.1097/MIB.0000000000000522

Mikó E, Kovács T, Sebő É et al (2019) Microbiome—microbial metabolome—cancer cell interactions in breast cancer—familiar, but unexplored. Cell 8:pii: E293

Montassier E, Gastinne T, Vangay P et al (2015) Chemotherapy-driven dysbiosis in the intestinal microbiome. Aliment Pharmacol Ther 42:515–528. https://doi.org/10.1111/apt.13302

Moreno-Sánchez R, Rodríguez-Enríquez S, Marín-Hernández A, Saavedra E (2007) Energy metabolism in tumor cells. FEBS J 274:1393–1418. https://doi.org/10.1111/j.1742-4658.2007.05686.x

Motevaseli E, Shirzad M, Akrami SM et al (2013) Normal and tumour cervical cells respond differently to vaginal lactobacilli, independent of pH and lactate. J Med Microbiol 62:1065–1072. https://doi.org/10.1099/jmm.0.057521-0

Mubangizi L, Namusoke F, Mutyaba T (2014) Aerobic cervical bacteriology and antibiotic sensitivity patterns in patients with advanced cervical cancer before and after radiotherapy at a national referral hospital in Uganda. Int J Gynecol Obstet 126:37–40. https://doi.org/10.1016/j.ijgo.2014.01.013

O'Hara AM, Shanahan F (2006) The gut flora as a forgotten organ. EMBO Rep 7:688–693. https://doi.org/10.1038/sj.embor.7400731

Parida S, Sharma D (2019) The power of small changes: comprehensive analyses of microbial dysbiosis in breast cancer. Biochim Biophys Acta Rev Cancer 1871:392–405. https://doi.org/10.1016/j.bbcan.2019.04.001

Pendharkar S, Magopane T, Larsson P-G et al (2013) Identification and characterisation of vaginal lactobacilli from South African women. BMC Infect Dis 13:43. https://doi.org/10.1186/1471-2334-13-43

Ravel J, Gajer P, Abdo Z et al (2011) Vaginal microbiome of reproductive-age women. Proc Natl

Acad Sci 108:4680–4687. https://doi.org/10.1073/pnas.1002611107
Redondo-Lopez V, Cook RL, Sobel JD (1990) Emerging role of lactobacilli in the control and maintenance of the vaginal bacterial microflora. Rev Infect Dis 12:856–872. https://doi.org/10.1093/clinids/12.5.856
Reich O, Regauer S, McCluggage WG et al (2017) Defining the cervical transformation zone and Squamocolumnar junction. Int J Gynecol Pathol 36:517–522. https://doi.org/10.1097/pgp.0000000000000381
Rose P, Moore P-K, Ming S-H et al (2005) Hydrogen sulfide protects colon cancer cells from chemopreventative agent beta-phenylethyl isothiocyanate induced apoptosis. World J Gastroenterol 11:3990–3997. https://doi.org/10.3748/wjg.v11.i26.3990
Rosenberg E, Zilber-Rosenberg I (2016) Microbes drive evolution of animals and plants: the Hologenome concept. MBio 7:1–8. https://doi.org/10.1128/mbio.01395-15
Round JL, Mazmanian SK (2009) The gut microbiome shapes intestinal immune responses during health and disease. Nat Rev Immunol 9:313–323. https://doi.org/10.1038/nri2515
Roy S, Trinchieri G (2017) Microbiota: a key orchestrator of cancer therapy. Nat Rev Cancer 17:271
Saus E, Iraola-Guzmán S, Willis J et al (2019) Microbiome and colorectal cancer: roles in carcinogenesis and clinical potential. Mol Aspects Med S0098-2997:30032–30039. https://doi.org/10.1016/j.mam.2019.05.001
Schwabe RF, Jobin C (2013) The microbiome and cancer. Nat Rev Cancer 13:800
Scott AJ, Alexander JL, Merrifield CA et al (2019) International Cancer Microbiome Consortium consensus statement on the role of the human microbiome in carcinogenesis. Gut 68(9):1624–1632. https://doi.org/10.1136/gutjnl-2019-318556
Semenza GL (2008) Tumor metabolism: cancer cells give and take lactate. J Clin Invest 118:3835–3837
Sen N (2017) Functional and molecular insights of hydrogen sulfide signaling and protein sulfhydration. J Mol Biol 429:543–561. https://doi.org/10.1016/j.jmb.2016.12.015
Sender R, Fuchs S, Milo R (2016) Revised estimates for the number of human and bacteria cells in the body. PLoS Biol 14:e1002533. https://doi.org/10.1371/journal.pbio.1002533
Serpa J, Dias S (2011) Metabolic cues from the microenvironment act as a major selective factor for cancer progression and metastases formation. Cell Cycle 10:180–181. https://doi.org/10.4161/cc.10.2.14476
Serpa J, Caiado F, Carvalho T et al (2010) Butyrate-rich colonic microenvironment is a relevant selection factor for metabolically adapted tumor cells. J Biol Chem 285:39211–39223. https://doi.org/10.1074/jbc.M110.156026
Sheflin AM, Whitney AK, Weir TL (2014) Cancer-promoting effects of microbial dysbiosis. Curr Oncol Rep 16:1–9. https://doi.org/10.1007/s11912-014-0406-0
Shibuya N, Koike S, Tanaka M et al (2013) A novel pathway for the production of hydrogen sulfide from D-cysteine in mammalian cells. Nat Commun 4:1366–1367. https://doi.org/10.1038/ncomms2371
Silva LS, Goncalves LG, Silva F et al (2016) STAT3:FOXM1 and MCT1 drive uterine cervix carcinoma fitness to a lactate-rich microenvironment. Tumor Biol 37:5385–5395. https://doi.org/10.1007/s13277-015-4385-z
Singh S, Lin H (2015) Hydrogen sulfide in physiology and diseases of the digestive tract. Microorganisms 3:866–889. https://doi.org/10.3390/microorganisms3040866
Szabo C (2018) A timeline of hydrogen sulfide (H2S) research: from environmental toxin to biological mediator. Biochem Pharmacol 149:5–19. https://doi.org/10.1016/j.bcp.2017.09.010
Szabo C, Coletta C, Chao C et al (2013) Tumor-derived hydrogen sulfide, produced by cystathionine- β -synthase, stimulates bioenergetics, cell proliferation, and angiogenesis in colon cancer. PNAS Pharmacol 110:12474–12479. https://doi.org/10.1073/pnas.1306241110/-/DCSupplemental.www.pnas.org/cgi/doi/10.1073/pnas.1306241110
Thomas S (1928) Döderlein's Bacillus: Lactobacillus acidophilus. J Infect Dis 43:218–227
Thursby E, Juge N (2017) Introduction to the human gut microbiota. Biochem J 474:1823–1836. https://doi.org/10.1042/BCJ20160510
Tomasova L, Konopelski P, Ufnal M (2016) Gut bacteria and hydrogen sulfide: the new old players in circulatory system homeostasis. Molecules 21:1–18. https://doi.org/10.3390/molecules21111558
Turnbaugh PJ, Ley RE, Hamady M et al (2007) The human microbiome project: exploring the microbial part of ourselves in a changing world. Nature 449:804–810. https://doi.org/10.1038/nature06244
Van Der Veer C, Hertzberger RY, Bruisten SM et al (2019) Comparative genomics of human Lactobacillus crispatus isolates reveals genes for glycosylation and glycogen degradation: implications for in vivo dominance of the vaginal microbiota. Microbiome 7:1–14. https://doi.org/10.1186/s40168-019-0667-9
Vander Heiden MG, Cantley LC, Thompson CB (2009) Understanding the Warburg effect: the metabolic requirements of cell proliferation. Science 324:1029–1033. https://doi.org/10.1126/science.1160809
Vásquez A, Jakobsson T, Ahrné S et al (2002) Vaginal *Lactobacillus* flora of healthy Swedish women. J Clin Microbiol 40:2746–2749. https://doi.org/10.1128/JCM.40.8.2746
Vétizou M, Pitt JM, Daillère R et al (2015) Anticancer immunotherapy by CTLA-4 blockade relies on the gut microbiota. Science 350:1079–1084. https://doi.org/10.1126/science.aad1329
Vivarelli S, Salemi R, Candido S et al (2019) Gut microbiota and cancer: from pathogenesis to therapy. Cancers (Basel) 11:38. https://doi.org/10.3390/cancers11010038
Wallace JL, Motta J-P, Buret AG (2017) Hydrogen sulfide: an agent of stability at the microbiome-mucosa

interface. Am J Physiol Liver Physiol 314:G143–G149. https://doi.org/10.1152/ajpgi.00249.2017

Wang R (2002) Two's company, three's a crowd: can H2S be the third endogenous gaseous transmitter? FASEB J 16:1792–1798. https://doi.org/10.1096/fj.02-0211hyp

Wang R (2012) Physiological implications of hydrogen sulfide: a whiff exploration that blossomed. Physiol Rev 92:791–896. https://doi.org/10.1152/physrev.00017.2011

Wang B, Yao M, Lv L et al (2017) The human microbiota in health and disease. Engineering 3:71–82. https://doi.org/10.1016/J.ENG.2017.01.008

Wang Y, Zhang Y, Zhang Q et al (2018) Characterization of pelvic and cervical microbiotas from patients with pelvic inflammatory disease. J Med Microbiol 67:1519–1526

Wilson GS, Miles AA (1946) Topley and Wilson's principles of bacteriology and immunity. Topley Wilson's Princ Bacteriol Immun 1. https://doi.org/10.1136/jcp.38.2.239-a

Wu YC, Wang XJ, Yu L et al (2012) Hydrogen Sulfide lowers proliferation and induces protective autophagy in Colon epithelial cells. PLoS One 7:e37572

Wu X, Wu Y, He L et al (2018) Effects of the intestinal microbial metabolite butyrate on the development of colorectal cancer. J Cancer 9:2510–2517. https://doi.org/10.7150/jca.25324

Wu J, Li H, Xie H et al (2019a) The malignant role of exosomes in the communication among colorectal cancer cell, macrophage and microbiome. Carcinogenesis 40(5):601–610. https://doi.org/10.1093/carcin/bgy138

Wu X, Zhang T, Chen X et al (2019b) Microbiota transplantation: targeting cancer treatment. Cancer Lett 452:144–151. https://doi.org/10.1016/j.canlet.2019.03.010

Xiao BB, Liao QP (2012) Analysis of diversity of vaginal microbiota in healthy Chinese women by using DNA-fingerprinting. Beijing Da Xue Xue Bao 44:281–287

Yang X, Da M, Zhang W et al (2018) Role of Lactobacillus in cervical cancer. Cancer Manag Res 10:1219–1229. https://doi.org/10.2147/CMAR.S165228

Zilber-Rosenberg I, Rosenberg E (2008) Role of microorganisms in the evolution of animals and plants: the hologenome theory of evolution. FEMS Microbiol Rev 32:723–735. https://doi.org/10.1111/j.1574-6976.2008.00123.x

Zou S, Fang L, Lee MH (2018) Dysbiosis of gut microbiota in promoting the development of colorectal cancer. Gastroenterol Rep 6:1–12. https://doi.org/10.1093/gastro/gox031

Gut Microbiota and Cancer of the Host: Colliding Interests

5

Gyorgy Baffy

Abstract

Cancer develops in multicellular organisms from cells that ignore the rules of cooperation and escape the mechanisms of anti-cancer surveillance. Tumorigenesis is jointly encountered by the host and microbiota, a vast collection of microorganisms that live on the external and internal epithelial surfaces of the body. The largest community of human microbiota resides in the gastrointestinal tract where commensal, symbiotic and pathogenic microorganisms interact with the intestinal barrier and gut mucosal lymphoid tissue, creating a tumor microenvironment in which cancer cells thrive or perish. Aberrant composition and function of the gut microbiota (dysbiosis) has been associated with tumorigenesis by inducing inflammation, promoting cell growth and proliferation, weakening immunosurveillance, and altering food and drug metabolism or other biochemical functions of the host. However, recent research has also identified several mechanisms through which gut microbiota support the host in the fight against cancer. These mechanisms include the use of antigenic mimicry, biotransformation of chemotherapeutic agents, and other mechanisms to boost anti-cancer immune responses and improve the efficacy of cancer immunotherapy. Further research in this rapidly advancing field is expected to identify additional microbial metabolites with tumor suppressing properties, map the complex interactions of host-microbe 'transkingdom network' with cancer cells, and elucidate cellular and molecular pathways underlying the impact of specific intestinal microbial configurations on immune checkpoint inhibitor therapy.

Keywords

Holobiont · Immune checkpoint inhibitors · Dysbiosis · Biofilm · Butyrate paradox · Intestinal barrier · Transkingdom network · Fecal microbiota transplantation

5.1 Introduction

Multicellular life first evolved about 1 billion years ago (Maynard Smith and Szathmáry 1995). Multicellularity requires cooperation among cells to ensure division of labor, allocation of resources and maintenance of replication with effective mechanisms to control cell proliferation and suppression of cheating (Aktipis et al. 2015). Multicellular organisms co-evolved with their microbial environment from the very beginning. Microorganisms living inside and out of the

G. Baffy (✉)
Department of Medicine, VA Boston Healthcare System and Brigham and Women's Hospital, Harvard Medical School, Boston, MA, USA
e-mail: gbaffy@bwh.harvard.edu

J. Serpa (ed.), *Tumor Microenvironment*, Advances in Experimental Medicine and Biology 1219,
https://doi.org/10.1007/978-3-030-34025-4_5

multicellular host are termed microbiota, and the totality of their genetic information is termed microbiome, while the host-microbiota ecosystem is often referred to as 'super-organism' or 'holobiont' (Huitzil et al. 2018; Schwabe and Jobin 2013). Microbiota are involved in diverse interactions with each other while contributing key functions to host physiology, aptly described within the context of a 'transkingdom network' (Greer et al. 2016). The largest community of human microbiota resides in the gastrointestinal tract and contains over 1000 different bacterial species (Human Microbiome Project 2012). Gut microbiota support the host by maximizing dietary energy extraction, generating essential metabolites, assisting the biotransformation of xenobiotics, shaping innate and adaptive immunity, and protecting from the invasion of pathogenic microorganisms (Backhed et al. 2005; Lee and Hase 2014; Nicholson et al. 2012). Similarly, host-to-microbe effects related to nutrition, medications and other lifestyle or health factors are critical for the gut microbiota (Fischbach and Sonnenburg 2011; Foster et al. 2017; Zmora et al. 2019). Perturbations of the host-microbial relationship result in dysbiosis, which is defined as a loss of balance in microbiota composition and function and potentially results in disease phenotypes (Llorente and Schnabl 2015).

Cancer is an inherent feature of multicellular existence that develops when multicellular cooperation breaks down and cells reject cooperation to favor self-autonomy, evasion of growth control and programmed cell death, replicative immortality, and a limitless ability to invade and metastasize (Hanahan and Weinberg 2000, 2011). This process involves abandonment of multicellularity and a reverse evolution of cancer cells back to a unicellular state (Chen et al. 2015). Cancer-like phenomena characterized by abnormal cell proliferation and neoplastic growth have been observed in all seven branches of multicellular life (Aktipis et al. 2015). Cancer cells become increasingly robust during their emergence and induce reciprocal changes in the host and microbiota (Aktipis and Nesse 2013). Uncontrolled cancer growth eventually leads to demise of the host, making it a common existential threat for the entire holobiont. However, because natural selection primarily works at species level (Maynard Smith 1998), each individual microbial strain and the host itself are distinct entities in this conflict with potentially divergent selective pressures and it may be wrong to assume they are acting with common interests (Foster et al. 2017). Thus, while coexistence of host and microbiota may normally serve the entire holobiont's homeostasis, interactions between cancer and microbiota do not necessarily benefit the host. In fact, the range of host-microbial relationship extends from mutualistic (mutually beneficial) to parasitic (harmful to the host and beneficial to the parasite) (Wasielewski et al. 2016). A better understanding of the interplay between host, cancer and microbiota is therefore critical to take advantage of between-species cooperation and of the potential gains from modulating this relationship (Fig. 5.1). This review will focus on some recent insights into the complex role of gut microbiota as an essential partner of the host in facing the initiation, progression and prognosis of cancer.

5.2 Tumor Microenvironment: A Collective Affair

Tumor microenvironment includes several types of cells that modulate the growth, proliferation and dissemination of cancer cells (Leong et al. 2018). The holobiont has developed powerful mechanisms in this milieu to suppress tumorigenesis, which may explain why—against the mathematical odds—clinically apparent cancer is relatively rare (Aktipis and Nesse 2013). Cancer cells become members of a 'social microenvironment' that includes cellular elements of both the host immunosurveillance and commensal microbiota, embedded in a physicochemical microenvironment and threatened by adversities such as poor nutrient availability, hypoxia, low pH and redox stress (Sun et al. 2018). Local expansion and metastatic spreading of cancer cells occur through a web of key permissive and controlling factors within this microenvironment and has been the topic of several excellent reviews in

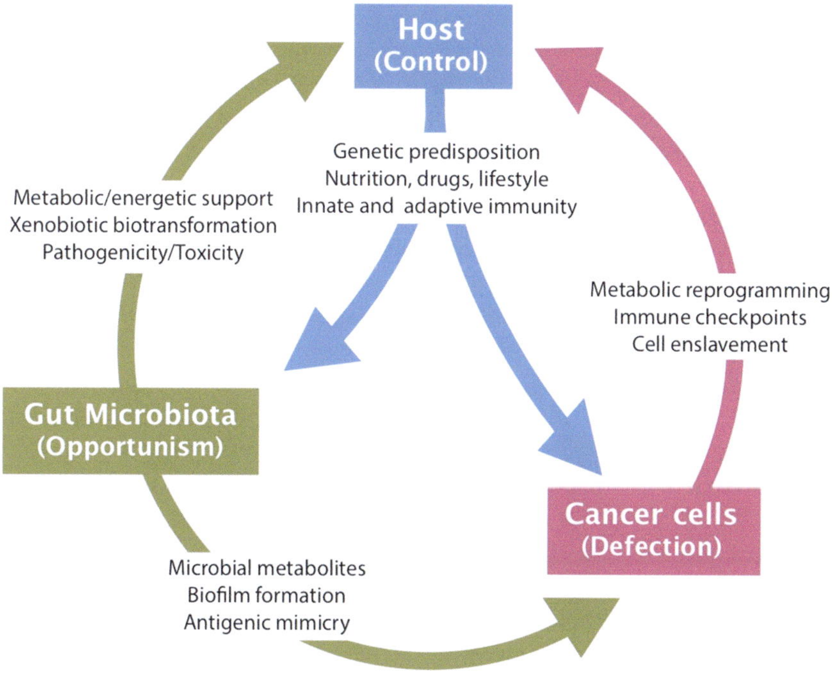

Fig. 5.1 Host—microbiota—cancer interplay
Schematic illustration of key interactions between host, microbiota and cancer, each multicellular community manifesting distinct behavior (shown in parentheses) in their triangular relationship. Genetic and environmental factors (e.g., diet and medications) may determine the success of host-mediated control over microbiota and tumorigenesis. Innate and adaptive immunity is essential in both cases, while additional structural (e.g., intestinal epithelial barrier) and functional elements (e.g., antimicrobial peptides, not shown), regulate the microbiota and sustain host homeostasis. Cancer cells seek defection with an aggressive agenda for limitless growth and proliferation at the host's expense, including coercion ('enslavement') of host cells into metabolic reprogramming (a.k.a. reverse Warburg effect in cancer-associated fibroblasts) and the use of immune checkpoints to stifle anti-cancer immune responses. Gut microbiota and cancer have an opportunistic relationship, since commensal microorganisms as individual species or in consortia respond to various selection pressures that may fortuitously assist cancer cells and act therefore not in the host's interest. Thus, providing unconventional nutrients to cancer cells, creating biofilms that impair the gut barrier and induce inflammation, or promoting genotoxicity in host cells may tilt the balance toward tumorigenesis in the gastrointestinal tract and beyond. However, eubiotic gut microbiota have also been shown to strengthen anti-cancer immunosurveillance through a variety of mechanisms such as synthesis of tumor suppressor metabolites (e.g., butyrate), cancer cell recognition via antigenic mimicry, or enhancement of anti-cancer chemotherapy and immunotherapy. Please see details in the main text.

recent years (Leong et al. 2018; Morgillo et al. 2018; Quail and Joyce 2013; Quante et al. 2013; Swartz et al. 2012).

5.2.1 Gut Microbiota

Microbial gene richness is a key feature of healthy gut microbiome, while diminishing bacterial diversity is often associated with disease (Le Chatelier et al. 2013). Metagenomic analysis of the feces by culture-independent methods found that most intestinal bacteria belong to 6 phyla: *Firmicutes, Bacteroidetes, Proteobacteria, Actinobacteria, Verrucomicrobia*, and *Fusobacteria* (Human Microbiome Project 2012). Similar to other living systems of multicellular aggregates, gut microbiota members interact through cooperation and competition. Cooperative activities between microbial species include the formation of biofilms, fermentation of complex substrates, and the use of supply chains such as exchange of metabolites and cross-feeding, whereby the metabolic end product of one strain becomes the nutrient for a different strain (Muller et al. 2018; Plichta et al. 2016). At the same time, microbial strains use diverse mechanisms to compete with each other for resources and adhesion sites, or produce antimicrobial substances to gain advantage (Foster et al. 2017). Gut microbial ecosystems analyzed in a large number of individuals fall into three major 'enterotypes' distinguished by variable microbial abundance and molecular functions (Arumugam et al. 2011). Interestingly, whereas microbial species composition differs from individual to individual, the overall distribution of expressed gene functions remains relatively constant (Plichta et al. 2016). A remarkable process regulating coexistence of microbial species in the gut is complementary silencing of genes that encode overlapping functions such as anaerobic fermentation, biosynthesis of short-chain fatty acids, and starch degradation (Plichta et al. 2016). Local stability analysis of microbiota networks indicates the importance of limiting positive feedbacks and weakening ecological interactions (Coyte et al. 2015). Cooperation can create dependency and increases the probability that perturbations rapidly spread and destabilize the system. Accordingly, high diversity is key to microbiota stability, characterized by competitiveness and weak interactions (Coyte et al. 2015).

5.2.2 Intestinal Barrier

Gut microbiota and the host are physically separated by a complex intestinal barrier (Goto and Kiyono 2012). The epithelial component is a single sheet of columnar cells tightly connected by intercellular complexes and covered by a thick mucous layer secreted by goblet cells (Marchiando et al. 2010). The endothelial or gut vascular barrier is another layer of tightly connected endothelial cells surrounded by pericytes and glial cells (Bouziat and Jabri 2015). Host self-defense is further enhanced via antimicrobial peptides secreted by Paneth cells at the crypt base (Dupont et al. 2014). Microbial metabolites or structural components of entire microorganisms residing in the gastrointestinal tract are sampled by the pattern recognition receptors of dendritic cells (Quante et al. 2013). Matured antigen-presenting dendritic cells migrate to mesenteric lymph nodes where they shape adaptive immunity by activating memory and naïve T cells (Steinman 2007). Depending on the context of microbial metabolites or byproducts, dendritic cells may turn naïve T cells into helper, effector (cytotoxic), or regulatory (Treg or suppressor) T cells. T_H17 cells, a special subset of $CD4^+$ helper T cells, protect the host from pathogenic microbes by strengthening tight junctions, stimulating the production of anti-microbial peptides, and by recruiting polymorphonuclear neutrophils (Garrett et al. 2010). Systemic immune responses are also affected by gut microbiota when T cells primed by dendritic cells circulate from local lymph nodes to distant sites or if microbial components and viable microorganisms enter the portal venous system. This response escalation is increasingly likely in the setting of dysbiosis characterized by reduced species diversity, enrichment of opportunistic pathogenic bacteria, and impaired gut barrier (Giannelli et al. 2014; Gopalakrishnan et al. 2018a).

5.2.3 Host Immunosurveillance

Augmented inflammation due to activation of innate and adaptive immune responses may promote cancer initiation and progression. Tumorigenic properties of dendritic cells, polymorphonuclear neutrophils, tumor-associated macrophages and myeloid-derived suppressor cells immature myeloid cells have been linked to the release of cytokines, growth factors, tissue-degrading enzymes and angiogenic mediators (Noonan et al. 2008; Quante et al. 2013). Several subsets of the adaptive immune system such as T_H17 cells have also been associated with inflammation and their increased presence of T_H17 cells in colorectal cancer predicts poor prognosis (Quante et al. 2013). The impact of Tregs on cancer is similarly controversial as they create a local anti-inflammatory milieu and mitigate tumorigenesis, but may simultaneously weaken anti-tumor immunosurveillance, often making their presence an ambivalent prognostic parameter (Quante et al. 2013). By contrast, cytotoxic $CD8^+$ T cells may specifically identify tumor differentiation antigens via their T cell receptors and destroy cancer cells, indicating that increased infiltration of the tumor tissue with these cells is a favorable prognostic sign (Reticker-Flynn and Engleman 2019).

However, anti-cancer immunosurveillance is not guaranteed by the presence of tumor-infiltrating lymphocytes, which are typically restrained by immune checkpoints consisting of a large and heterogeneous group of ligands and receptors preventing indiscriminate activation of the immune response (Restifo et al. 2012). Ligands that are able to activate immune checkpoints such as programmed cell death protein 1 (PD1) and cytotoxic T-lymphocyte-associated protein 4 (CTLA4) often become overexpressed in cancer cells and other components of the tumor microenvironment, thereby assisting evasion of immune-mediated destruction (Pardoll 2015). Tumor immunotherapy targeting immune checkpoints has been one of the fastest-developing and successful chapters of cancer research and, as discussed later, its clinical efficacy is remarkably determined by the composition and function of gut microbiota.

5.3 Gut Microbiota and Mechanisms of Tumor Promotion

In general, microbial contribution to tumorigenesis can be classified into three mechanisms: direct impact by modulating host cell proliferation and death, interference with the host innate and adaptive immune system, and altering food and drug metabolism or host biochemistry (Garrett 2015). Altogether, an estimated 20–30% of cancers are associated with chronic microbial infections (Garrett 2015; Yang et al. 2013). So far, however, only 10 microorganisms have been irrefutably designated as carcinogens (i.e., *bona fide* oncomicrobes), including *Helicobacter pylori* (gastric adenocarcinoma), hepatitis B and C viruses (liver cancer), *Schistosoma haematobium* (urinary bladder cancer), *Opisthorchis viverrini* and *Clonorchis sinensis* (cholangiocellular neoplasia), Epstein-Barr virus (nasopharyngeal carcinoma and lymphoma), human herpes virus type 8 (Kaposi's sarcoma), human T cell lymphotropic virus type 1 (lymphoma), and human papilloma viruses (cervical and anogenital cancer) (de Martel et al. 2012).

While only a few microorganisms have been specifically implicated in tumorigenesis, microbiota as a multicellular aggregate appears to have influence on the initiation and progression of cancer (Drewes et al. 2016; Garrett et al. 2010; Morgillo et al. 2018). Current view is that cancer is more likely to develop in dysbiosis that includes a marked decrease in both microbial diversity and community stability (Bhatt et al. 2017). Moreover, dysbiosis modulates the impact of anti-cancer therapy on host responses and adverse events (Gopalakrishnan et al. 2018a). Importantly, the tumorigenic impact of dysbiosis is not confined to colorectal cancer as dysbiosis has also been associated with other forms of cancers including breast, lung, urogenital tract and liver (Pope et al. 2017). In fact, gut microbiota, metabolites, and immune cells may exit the gut via the circulation and influence tumorigenesis at distant sites (Bhatt et al. 2017). Supporting this notion, metastases from patients with colorectal cancer continue to harbor bacteria, in particular *Fusobacterium* but also *Bacteroides*, *Selenomonas* and *Prevotella* species (Kroemer and Zitvogel 2018).

5.3.1 Microbiota-Associated Genotoxicity and Tumorigenesis

Many bacteria have developed competitive strategies, which include the ability to damage the genome of competing organisms. Some of these bacteria are part of the commensal human microbiota. Microbial-induced genotoxic mechanisms and the activation of related oncogenic signaling pathways also affect the host, potentially leading to cancer (Garrett 2015). For instance, *Salmonella typhi* strains possessing avrA, a virulence gene encoding acetyltransferase activity, establish chronic infection and activate epithelial β-catenin signaling, which has been associated with hepatobiliary and colorectal cancers (Dutta et al. 2000; Lu et al. 2017). Colibactin is a bacterial toxin, which is synthesized by the *pks* genomic island and found in members of the *Enterobacteriaceae* family such as group B2 *Escherichia coli* (Fais et al. 2018). Colibactin promotes colorectal tumorigenesis by inducing double-stranded DNA breaks and a senescence-associated secretory phenotype in intestinal epithelial cells (Fulbright et al. 2017; Schwabe and Jobin 2013). Analysis of the colonic mucosa of patients with colorectal cancer found significantly higher abundance of *pks*-harboring *E. coli* strains compared to healthy controls (Buc et al. 2013). Enterotoxigenic *Bacteroides fragilis* (ETBF), a gut commensal and opportunistic pathogen enriched in patients with colorectal cancer, secretes a zinc-dependent metalloprotease that cleaves and degrades E-cadherin (Sears 2009). This toxin enables nuclear translocation of β-catenin and transcription of Myc proto-oncogene, promoting cellular proliferation in the colon epithelium. In the $Apc^{Min/+}$ mouse model, ETBF facilitates colon tumorigenesis by triggering T_H17-mediated colitis and STAT3 activation (Wu et al. 2009).

5.3.2 Gut Microbiota Metabolism Favoring Tumorigenesis

Robust cancer phenotypes emerge from comprehensive reprogramming of macromolecular biosynthesis and energy metabolism (DeBerardinis et al. 2008). A prominent metabolic feature of cancer cells is the Warburg effect or reallocation of energy production from mitochondrial oxidative phosphorylation to the disproportionate use of glycolysis (Warburg et al. 1924). While the causes and rationale of the Warburg effect remain debated, it has been proposed that glycolytic ATP production can be scaled up with fewer regulatory constraints to match the vast energetic needs of rapidly proliferating cancer cells (Pfeiffer et al. 2001). In addition, diversion of substrates from the electron transport chain may diminish mitochondrial oxidative stress that is already substantial in cancer cells (Brand 1997). Also, glycolysis can be viewed as a versatile production line of precursors for almost all major biosynthetic routes including the pentose phosphate pathway, serine synthesis pathway, and *de novo* lipogenesis (Pavlova and Thompson 2016; Sun et al. 2018). Finally, excess lactate production gives cancer cells a competitive advantage by sustaining an acidic tumor microenvironment increasingly inhabitable for normal cells (DeBerardinis et al. 2008; Hsu and Sabatini 2008).

Cancer cells exploit all available resources to support limitless growth and proliferation in an increasingly harsh and nutrient-deprived microenvironment. Thus, cancer cells may engulf and digest apoptotic bodies or entire living cells and utilize cellular waste products such as lactate, acetate, branched-chain keto acids, ketone bodies and ammonia (Sun et al. 2018). Currently, it is impossible to tell how many gut microbial species actually provide benefits and how many may 'collude' with cancer cells while interacting via diffusible metabolites (Wegiel et al. 2018). Many dietary and digestive components are metabolized by bacteria in the gastrointestinal tract, yielding putative metabolites that have either oncogenic or tumor suppressing properties (Bhatt et al. 2017). Unconventional nutrients potentially made available by the microbiota for the energy metabolism of cancer cells include short-chain fatty acids, bile acids, polyamines, choline metabolites, indole derivatives and vitamins (Sun et al. 2018). This list is almost certainly incomplete since close to half of the metabolites found in human plasma have been estimated to originate in the microbiota (Martin et al. 2007). Moreover, many genes identified in gut microbiomes of participants in the Human Microbiome Project could not be characterized by standard annotation methods (Joice et al. 2014), leaving us with a possibly large number of small molecules that may influence tumorigenesis (Donia and Fischbach 2015; Foster et al. 2017).

Bile acids represent an important connection between microbial and host metabolism. Primary bile acids synthesized in the liver as cholic acid or chenodeoxycholic acid become conjugated with glycine or taurine and excreted into bile (Wahlstrom et al. 2016). Most conjugated bile acids are reabsorbed in the terminal ileum and return to the liver via the enterohepatic circulation, while a small amount is deconjugated by intestinal bacteria into secondary bile acids such as deoxycholic acid and lithocholic acid (Long et al. 2017). Secondary bile acids may contribute to colon tumorigenesis by triggering inflammation and promoting β-catenin and NF-κB signaling. In addition, intestinal deoxycholic acid was found to contribute to the development of liver cancer by inducing senescence-associated secretory phenotype in hepatic stellate cells and stimulating pro-inflammatory and tumor-promoting reactions in a mouse model of obesity-associated hepatocellular carcinoma (Yoshimoto et al. 2013). Consumption of a diet rich in saturated fat leads to dysbiosis with expansion of sulfur-reducing gut bacteria such as *Bilophila* and *Desulfovibrio*, which use sulfur as a terminal electron acceptor, primarily obtained from taurine-conjugated bile acids (Devkota et al. 2012; Wang 2012). Microbial-derived hydrogen sulfide (H_2S) has been implicated in the development of colitis and colorectal cancer based on its ability to impair the gut barrier and cause genotoxicity (Arkan 2017; Singh and Lin 2015).

Microbial contribution to amino acid metabolism has a significant impact on host immunosurveillance. A key example is tryptophan, an essential amino acid catabolized by both host and microbial enzymes and having derivatives with multiple biological functions. Endogenous tryptophan metabolites include kynurenine, serotonin and melatonin, whereas bacterial breakdown of tryptophan yields indole, indole acetate, indole propionate, skatole, and tryptamine (Gao et al. 2018). Tryptophan catabolism is an important effector system that modulates T cell responses and promotes immune tolerance (Fallarino et al. 2006). Not surprisingly, accelerated tryptophan catabolism has been reported by various tumors of the colon, breast, lung and brain due to higher expression of tryptophan-2, 3-dioxygenase and indoleamine-2, 3-dioxygenase (Sun et al. 2018). Endogenous and microbial derivatives of tryptophan activate the aryl hydrocarbon receptor (AhR), a xenobiotic sensor and regulator of inflammation and immunity (Zelante et al. 2013). AhR mediates activation of group 3 innate lymphoid cells (ILC3) resulting in enhanced secretion of IL-22, which dampens pro-inflammatory signals and protects the intestinal mucous barrier (Hernandez et al. 2018), but AhR activation also limits immunosurveillance by promoting apoptosis of effector T cells and creating therefore a cancer-permissive microenvironment (Grohmann et al. 2003).

5.3.3 Biofilm Formation and Tumorigenesis

Colon biofilms are dense consortia of approx. 100 bacterial strains embedded in a complex matrix in close proximity with, and partially invading, the intestinal mucosa (Dejea et al. 2014). Formation of biofilms has also been observed in conditions involving chronic mucosal inflammation beyond the colon (e.g., tonsillitis, otitis media, sinusitis, urethritis and vaginitis) (Costerton et al. 1999). Biofilms have been identified in about 15% of apparently healthy patients (Swidsinski et al. 2007), but they are almost universally present in patients with right-sided, surgically resected colorectal neoplasms and in paired biopsies of tumor-free mucosa (Dejea et al. 2014). Biofilms allow direct bacterial contact with colon epithelial cells, which is believed to trigger chronic inflammation conducive to tumorigenesis, characterized by diminished levels of E-cadherin, enhanced IL-6 and STAT3 activation, and increased crypt epithelial cell proliferation rates (Dejea et al. 2014). Notably, biofilm-positive tumor tissue indicated higher levels of acetylated polyamines compared to biofilm-negative tumors, which may explain how microbial biofilms contribute to colon cancer (Johnson et al. 2015).

Fusobacterium, a gram-negative anaerobic bacterial genus, is a main component of biofilms in various mucosal locations (Zhou et al. 2018). *F. nucleatum* produces several factors of fusobacterial virulence and associated functions (Wu et al. 2018). Thus, Fap2 engages the inhibitory receptor T cell immunoglobulin and ITIM domain (TIGIT) to silence anti-cancer cytotoxic activity of T cells and natural killer cells (Gur et al. 2015). In addition, FadA is a fusobacterial adhesin that binds to lectins and E-cadherin on the surface of host epithelial cells and activates β-catenin signaling, thus promoting cell proliferation (Rubinstein et al. 2013). *F. nucleatum* also binds to Toll-like receptor 4 on epithelial cells and activates the Myd88/NF-κB signaling pathway, promoting chemoresistance to colorectal cancer (Yu et al. 2017). Also, abundance of *F. nucleatum* has been correlated with enrichment of myeloid-derived suppressor cells and tumor-associated macrophages, both of which are known to weaken anti-cancer immunosurveillance (Wu et al. 2018).

5.4 Gut Microbiota and Mechanisms of Tumor Suppression

Recent research has identified a number of pathways through which commensal microbes support the host against cancer (Perez-Chanona and Trinchieri 2016; Pope et al. 2017). Thus, microbiota may beneficially influence a range of host immune functions including innate and adaptive immune responses, all related to anti-cancer surveillance (Pope et al. 2017). Also, accumulating evidence for the involvement of gut microbiota in cancer pathogenesis has opened new opportunities for preventive or therapeutic interventions. These include the administration of antibiotics, probiotics, prebiotics, and postbiotics (Bultman 2016; Zitvogel et al. 2015), but there are additional efforts to use specific intestinal microbial configurations for selectively manipulating the composition and function of gut microbiota (Comstock and Coyne 2003; Everard et al. 2013; Tanoue et al. 2019). However, caution is advised and some experts warn about the indiscriminate use of probiotics, particularly lactate-producing bacteria such as *Lactobacillus*, *Turicibacte*r and *Streptococcus* and anaerobic *Lachnospiraceae* and *Ruminococcaceae* that ferment complex carbohydrates and aromatic compounds, in cancer patients (Arkan 2017). It remains to be seen when and which probiotic to use to maximize benefits and minimize potential harm to those with increased susceptibility to cancer.

5.4.1 Tumor Suppressive Microbial Metabolites and Biotransformation

Carbohydrate fermentation by the gut microbiota yields large amounts of short-chain fatty acids, such as acetate, propionate and butyrate (Riscuta et al. 2018). Short-chain fatty acids are metabolized via β-oxidation and the tricarboxylic acid (TCA) cycle in the mitochondrial matrix, representing an energy source for normal and cancerous colonocytes. Importantly, breakdown of butyrate into acetyl-CoA also stimulates histone acetylase activity, allowing transcriptional activation of genes involved in cell growth and proliferation, therefore making the role of butyrate controversial in tumorigenesis (Donohoe et al. 2012). Microbial-derived short-chain fatty acids may also serve as carbon source for enhanced biosynthetic activity in cancer cells and as ligands for G protein coupled receptors such as GPR41 and GPR43 with an ambiguous role in intestinal inflammation (Ang and Ding 2016). Paradoxically, however, butyrate has mostly been found to inhibit cancer cell growth, referred to as the 'butyrate paradox', and the explanation appears to be related to metabolic reprogramming of cancer cells (Donohoe et al. 2012). Thus, mitochondrial breakdown of butyrate may become insufficient due to the Warburg effect with surplus butyrate accumulating in the nucleus where it functions as an inhibitor of histone deacetylase; an effect that can be experimentally reversed by preventing aerobic glycolysis in cancer cells (Donohoe et al. 2012). By contrast, cancer cells that retain the ability to metabolize butyrate are positively selected by the microenvironment and develop a more aggressive and invasive phenotype (Serpa et al. 2010).

There is accumulating evidence that intact commensal microbiota are required for optimal responses to cancer chemotherapy (Kroemer and Zitvogel 2018). Certain health conditions or repeated use of antibiotics may significantly, even if only temporarily, alter the composition and function of gut microbiota and loss of microbial diversity and dysbiosis have been linked to altered pharmacodynamics of anti-cancer agents with unfavorable clinical outcomes (Gopalakrishnan et al. 2018b). Disruption of gut microbiota by antibiotics in a variety of murine tumor models results in profoundly impaired responsiveness of mice to CpG-oligonucleotide immunotherapy and platinum-based chemotherapy (Iida et al. 2013). Platinum compounds such as oxaliplatin and cisplatin require the generation of intratumoral oxidative stress in order to exert DNA damage and apoptosis, an effect that was hampered in mice receiving antibiotics (Iida et al. 2013). Furthermore, oral gavage of bacterial endotoxin to antibiotic-treated mice restored responsiveness to CpG-oligonucleotide immunotherapy evidenced by immunogenic cell death also drives antitumor T cell responses (Iida et al. 2013). The same work also demonstrated that antibiotics reduce the therapeutic efficacy of platinum compounds against subcutaneously transplanted tumors and suggested that the efficacy of oxaliplatin depends on microbiota-based priming of myeloid cells for the release of reactive oxygen species that contribute to genotoxicity and tumor reduction (Iida et al. 2013). This problem is further compounded by chemotherapy-induced dysbiosis and breakdown of the intestinal epithelial barrier, compromising clinical outcomes in cancer patients (Galloway-Pena et al. 2017).

Biotransformation of pharmaceutical agents by the gut microbiota is not necessarily beneficial. One such example involves irinotecan (CPT-11), which is a chemotherapeutic drug often used in the treatment of metastatic colon cancer. Irinotecan is a prodrug, metabolized into the active topoisomerase I inhibitor SN38 and subsequently glucuronidated in the liver to form the inactive SN38-G, which is excreted with the bile into the GI tract. In the colon, SN38-G is converted back to the active form by commensal gut bacteria such as *Streptococcus agalactiae, Clostridium perfringens* or *Bacteroides fragilis* that also possess β-glucuronidase activity. Reactivated SN38 causes severe colitis and diarrhea in susceptible patients, and this adverse event may necessitate dose reduction or discontinuation of irinotecan therapy. Small-molecule inhibitors specific to bacterial β-glucuronidases have been developed to avoid this complication (Wallace et al. 2010). Subsequently, inhibitors of *E. coli* β-glucuronidase were shown to protect the host against gastrointestinal toxicity induced by irinotecan in mice (Pope et al. 2017).

5.4.2 Gut Microbiota and Anti-cancer Immunosurveillance

Host survival critically depends on timely recognition of tumor-associated antigens and the destruction of cells committed to tumorigenesis. There is increasing evidence that gut microbiota play an important role in this process through their ability to modulate anti-cancer immune responses and immunotherapy (Pope et al. 2017). Neoantigens related to malignant transformation often show sufficient similarity with microbial epitopes, and antigenic mimicry may therefore enhance the recognition of cancer cells (Zitvogel et al. 2016). Related to this concept, the 'cancer hygiene hypothesis' suggests that limited exposure to microbial antigens in highly industrialized societies may account for the increased incidence of certain cancers (Thorburn et al. 2014). Moreover, microbe-associated molecular patterns as danger signals may increase the overall 'vigor' of innate immune system through the summative impact of extraneous microbial products on pathogen recognition receptors and the generation of soluble mediators such as interferons and cytokines

(Zitvogel et al. 2016). Also, activation of T cell subsets implicated in anti-cancer immunosurveillance is impaired in germ-free mice and may be restored upon colonization with various intestinal microbial strains (Sommer and Backhed 2013).

These mechanisms are not necessarily restricted to gut-associated lymphoid tissue as translocation of microbial metabolites or entire microorganisms through a leaky gut-vascular barrier may allow systemic exposure and affects tumorigenesis at distal sites (Bhatt et al. 2017; Kroemer and Zitvogel 2018). In addition, the concept of a 'common mucosal immune system' based on animal models postulates that immune cells primed locally in the gut mucosa may travel to other mucosal or lymphoid sites, extending the impact of gut microbiota to the entire host (Wilson and Obradovic 2015).

As recently reported, long-term survivors of pancreatic ductal adenocarcinoma are characterized by high numbers of tumor-infiltrating $CD8^+$ T cells and tumor neoantigens cross-reacting with microbial-derived epitopes, suggesting that enhanced immune response due to antigenic mimicry may account for a more favorable prognosis in this cohort (Balachandran et al. 2017). Cross-reactive clones in these patients show selective loss on metastatic progression, further supporting the favorable impact of microbial homology while precise mechanisms of T cell clone selection and survival remain incompletely understood (Balachandran et al. 2017). The role of gut microbiota in eliciting anti-cancer immune responses was further evidenced by a recent work identifying a consortium of 11 bacterial strains derived from healthy human donor feces and able to induce accumulation of interferon-γ-producing $CD8^+$ T cells in the intestinal lamina propria of germ-free mice (Tanoue et al. 2019). Colonization with the 11-strain mixture protected these animals from dissemination of *Listeria monocytogenes* and suppressed the growth of syngeneic grafts of colon adenocarcinoma and melanoma, illustrating the profound potential of gut microbiota-based anti-cancer interventions (Tanoue et al. 2019).

5.4.3 Enhancement of Anti-cancer Immunotherapy by Gut Microbiota

A number of immune checkpoint inhibitor molecules have been developed and marketed in recent years, including monoclonal antibodies against CTLA4 (ipilimumab), PD1, (nivolumab), and PD1 ligand 1 or PDL1 (pembrolizumab) that proved highly efficient in several difficult-to-treat cancers (Fan et al. 2018; Kudo 2018; Robert et al. 2015). However, a truly remarkable impact of immune checkpoint inhibitors on these cancers is only observed in about 25% of patients while the remaining cases show limited or no response (Gopalakrishnan et al. 2018a; Sun et al. 2018). Intriguingly, abnormal composition of gut microbiota profoundly alters the efficacy of immune checkpoint inhibitor therapy and contributes to primary resistance (Bhatt et al. 2017). Anti-CTLA4 antibodies were ineffective when administered to subcutaneous tumors in germ-free or antibiotics-treated mice while CTLA4 blockade was restored by re-colonization with *Bacteriodes* and *Burkholderia*, bacterial strains that have markedly reduced abundance in response to anti-CTLA4 treatment in conventional mice (Vetizou et al. 2015). Moreover, fecal microbial transplantation (FMT) from melanoma patients treated with anti-CTLA4 and featuring abundance of *Bacteroides fragilis* indicated stronger anti-cancer properties when administered to the murine tumor model (Vetizou et al. 2015). In addition, repletion of gut microbiota with *B. fragilis* and *Burkholderia cepacia* ameliorated the mucosal toxicity of anti-CTLA4, indicating that microbiota composition may also improve therapeutic effectiveness by preventing adverse reactions (Vetizou et al. 2015).

The efficacy of PDL1 blockade was analyzed by using experimental tumor models in C57BL/6 mice obtained from different facilities and featuring distinct gut microbiota (Sivan et al. 2015). Metagenomic analysis revealed that abundance of *Bifidobacterium* spp. was associated with increased responsiveness to anti-cancer therapy and the response benefit was transferable between mouse strains by oral *Bifidobacterium* (Sivan et al. 2015). While *Bifidobacterium*-primed dendritic cells improved the function of tumor-specific $CD8^+$ T cells, authors postulated a more generic and antigen-independent effect based on the changes in innate immune functions observed in this experimental model (Sivan et al. 2015).

Another recent work has provided additional insights into the molecular and cellular mechanisms by which gut microbiota may influence the tumor microenvironment and enhance immunotherapies (Gopalakrishnan et al. 2018b). Patients with metastatic melanoma under treatment with anti-PD1 therapy were found to harbor significant differences in the composition of gut microbiota according to their response status. Thus, non-responders to immune checkpoint inhibition had increased abundance of *Bacteroidales* in the gut microbiota, while patients with prolonged progression-free survival in response to anti-PD1 therapy had a significantly higher diversity of bacteria in their gut microbiota as well as a higher relative abundance of *Clostridiales*, *Faecalibacterium* and *Ruminococcaceae* (Gopalakrishnan et al. 2018a). Moreover, tumor tissue infiltration with $CD8^+$ T cells was significantly more prominent among patients with abundance of *Faecalibacterium prausnitzii* and other *Firmicutes* in the gut microbiota (Gopalakrishnan et al. 2018a).

Another recent study aimed to assess the impact of dysbiosis associated with malignant disease or concomitant antibiotic use on primary resistance to PD1 blockade in patients with non-small cell lung cancer, renal cell carcinoma and urothelial carcinoma (Routy et al. 2018). Clinical review indicated that patients who received antibiotics had shorter progression-free survival, relapsed sooner, and responded poorly to immune checkpoint inhibitor therapy. Metagenomic sequencing indicated major differences in the composition of gut microbiota based on responsiveness to PD1 blockade. Improved clinical outcomes were similarly associated with increased abundance of *Akkermansia* and *Alistipes* species in this report (Routy et al. 2018). To establish causality between gut microbiota composition and responsiveness to anti-PD1 therapy, antibiotic-treated mice were given FMT from responder and non-responder cancer patients and then inoculated with tumor cells to assess the efficacy of immune checkpoint blockade in the xenografts. Importantly, tumor growth was delayed in mice that received FMT from responding patients, whereas FMT from non-responding patients had no such effect (Routy et al. 2018). Response to PD1 blockade was also restored by specific re-colonization of mice with *Akkermansia muciniphila* and *Enterococcus hirae*, bacterial strains associated with clinical benefits and shown to induce dendritic cells to secrete IL-12, which is involved in the immunogenicity of PD1 blockade in eubiosis (Routy et al. 2018).

Finally, strong correlation between commensal microbial composition and clinical response to immune checkpoint inhibitors in patients diagnosed with metastatic melanoma was recently demonstrated (Matson et al. 2018). Thus, integrated metagenomic analysis identified 10 microbial species differentially enriched in the intestines of responders vs. non-responders to anti-PDL1 or anti-CTLA4 therapy. Bacterial species more abundant in responders included *Bifidobacterium longum*, *Collinsella aerofaciens*, *Enterococcus faecium*, *Lactobacillus* spp. and *Veillonella parvula*. FMT from patients to germ-free mice recapitulated the patient phenotypes and animals reconstituted with responder microbiota had increased numbers of tumor antigen-specific $CD8^+$ T cells in their tumor microenvironments (Matson et al. 2018). Anti-PDL1 was highly effective in mice colonized with responder microbiota while it remained completely ineffective in mice receiving FMT from non-responder patients (Matson et al. 2018).

5.5 Conclusions

Tumorigenesis, the process of host cells opting for cheating over cooperation and breaking the rules of multicellular life, has been the source of much suffering. Advances in preventing cancer that may ultimately destroy the host are eagerly awaited. There is now hope that increasing knowledge about the human commensal microbial community, and in particular the gut microbiota, will give valuable insights into the evolution and ecological interactions of host-cancer-microbiota relationship and identify new molecular targets for preventing and treating cancer. There is already evidence for the importance of eubiosis, which supports the intestinal barrier and gut-associated lymphoid tissue by enhancing innate and adaptive immunity and creating a tumor microenvironment that becomes unwelcoming to cancer cells in the gastrointestinal tract and beyond. Additional research is needed to identify microbial metabolites and specific molecular mechanisms by which individual strains or well-defined consortia of gut microbiota can be utilized in the prevention and treatment of cancer.

References

Aktipis CA, Nesse RM (2013) Evolutionary foundations for cancer biology. Evol Appl 6:144–159. https://doi.org/10.1111/eva.12034

Aktipis CA, Boddy AM, Jansen G, Hibner U, Hochberg ME, Maley CC, Wilkinson GS (2015) Cancer across the tree of life: cooperation and cheating in multicellularity. Philos Trans R Soc Lond B Biol Sci 370:pii: 20140219. https://doi.org/10.1098/rstb.2014.0219

Ang Z, Ding JL (2016) GPR41 and GPR43 in obesity and inflammation – protective or causative? Front Immunol 7:28. https://doi.org/10.3389/fimmu.2016.00028

Arkan MC (2017) The intricate connection between diet, microbiota, and cancer: a jigsaw puzzle. Semin Immunol 32:35–42. https://doi.org/10.1016/j.smim.2017.08.009

Arumugam M et al (2011) Enterotypes of the human gut microbiome. Nature 473:174–180. https://doi.org/10.1038/nature09944

Backhed F, Ley RE, Sonnenburg JL, Peterson DA, Gordon JI (2005) Host-bacterial mutualism in the human intestine. Science 307:1915–1920. https://doi.org/10.1126/science.1104816

Balachandran VP et al (2017) Identification of unique neoantigen qualities in long-term survivors of pancreatic cancer. Nature 551:512–516. https://doi.org/10.1038/nature24462

Bhatt AP, Redinbo MR, Bultman SJ (2017) The role of the microbiome in cancer development and therapy. CA Cancer J Clin 67:326–344. https://doi.org/10.3322/caac.21398

Bouziat R, Jabri B (2015) Breaching the gut-vascular barrier. Science 350:742–743. https://doi.org/10.1126/science.aad6768

Brand K (1997) Aerobic glycolysis by proliferating cells: protection against oxidative stress at the expense of energy yield. J Bioenerg Biomembr 29:355–364

Buc E et al (2013) High prevalence of mucosa-associated E. coli producing cyclomodulin and genotoxin in colon cancer. PLoS One 8:e56964. https://doi.org/10.1371/journal.pone.0056964

Bultman SJ (2016) The microbiome and its potential as a cancer preventive intervention. Semin Oncol 43:97–106. https://doi.org/10.1053/j.seminoncol.2015.09.001

Chen H, Lin F, Xing K, He X (2015) The reverse evolution from multicellularity to unicellularity during carcinogenesis. Nat Commun 6:6367. https://doi.org/10.1038/ncomms7367

Comstock LE, Coyne MJ (2003) Bacteroides thetaiotaomicron: a dynamic, niche-adapted human symbiont. Bioessays 25:926–929. https://doi.org/10.1002/bies.10350

Costerton JW, Stewart PS, Greenberg EP (1999) Bacterial biofilms: a common cause of persistent infections. Science 284:1318–1322

Coyte KZ, Schluter J, Foster KR (2015) The ecology of the microbiome: networks, competition, and stability. Science 350:663–666. https://doi.org/10.1126/science.aad2602

de Martel C, Ferlay J, Franceschi S, Vignat J, Bray F, Forman D, Plummer M (2012) Global burden of cancers attributable to infections in 2008: a review and synthetic analysis. Lancet Oncol 13:607–615. https://doi.org/10.1016/S1470-2045(12)70137-7

DeBerardinis RJ, Lum JJ, Hatzivassiliou G, Thompson CB (2008) The biology of cancer: metabolic reprogramming fuels cell growth and proliferation. Cell Metab 7:11–20. https://doi.org/10.1016/j.cmet.2007.10.002

Dejea CM et al (2014) Microbiota organization is a distinct feature of proximal colorectal cancers. Proc Natl Acad Sci U S A 111:18321–18326. https://doi.org/10.1073/pnas.1406199111

Devkota S et al (2012) Dietary-fat-induced taurocholic acid promotes pathobiont expansion and colitis in Il10-/- mice. Nature 487:104–108. https://doi.org/10.1038/nature11225

Donia MS, Fischbach MA (2015) Small molecules from the human microbiota. Science 349:1254766. https://doi.org/10.1126/science.1254766

Donohoe DR, Collins LB, Wali A, Bigler R, Sun W, Bultman SJ (2012) The Warburg effect dictates the mechanism of butyrate-mediated histone acetylation and cell proliferation. Mol Cell 48:612–626. https://doi.org/10.1016/j.molcel.2012.08.033

Drewes JL, Housseau F, Sears CL (2016) Sporadic colorectal cancer: microbial contributors to disease prevention, development and therapy. Br J Cancer 115:273–280. https://doi.org/10.1038/bjc.2016.189

Dupont A, Heinbockel L, Brandenburg K, Hornef MW (2014) Antimicrobial peptides and the enteric mucus layer act in concert to protect the intestinal mucosa.

Gut Microbes 5:761–765. https://doi.org/10.4161/19490976.2014.972238
Dutta U, Garg PK, Kumar R, Tandon RK (2000) Typhoid carriers among patients with gallstones are at increased risk for carcinoma of the gallbladder. Am J Gastroenterol 95:784–787. https://doi.org/10.1111/j.1572-0241.2000.01860.x
Everard A et al (2013) Cross-talk between Akkermansia muciniphila and intestinal epithelium controls diet-induced obesity. Proc Natl Acad Sci U S A 110:9066–9071. https://doi.org/10.1073/pnas.1219451110
Fais T, Delmas J, Barnich N, Bonnet R, Dalmasso G (2018) Colibactin: more than a new bacterial toxin. Toxins (Basel) 10:pii: E151. https://doi.org/10.3390/toxins10040151
Fallarino F et al (2006) The combined effects of tryptophan starvation and tryptophan catabolites down-regulate T cell receptor zeta-chain and induce a regulatory phenotype in naive T cells. J Immunol 176:6752–6761
Fan CA, Reader J, Roque DM (2018) Review of immune therapies targeting ovarian cancer. Curr Treat Options Oncol 19:74. https://doi.org/10.1007/s11864-018-0584-3
Fischbach MA, Sonnenburg JL (2011) Eating for two: how metabolism establishes interspecies interactions in the gut. Cell Host Microbe 10:336–347. https://doi.org/10.1016/j.chom.2011.10.002
Foster KR, Schluter J, Coyte KZ, Rakoff-Nahoum S (2017) The evolution of the host microbiome as an ecosystem on a leash. Nature 548:43–51. https://doi.org/10.1038/nature23292
Fulbright LE, Ellermann M, Arthur JC (2017) The microbiome and the hallmarks of cancer. PLoS Pathog 13:e1006480. https://doi.org/10.1371/journal.ppat.1006480
Galloway-Pena JR, Jenq RR, Shelburne SA (2017) Can consideration of the microbiome improve antimicrobial utilization and treatment outcomes in the oncology patient? Clin Cancer Res 23:3263–3268. https://doi.org/10.1158/1078-0432.CCR-16-3173
Gao J et al (2018) Impact of the gut microbiota on intestinal immunity mediated by tryptophan metabolism. Front Cell Infect Microbiol 8:13. https://doi.org/10.3389/fcimb.2018.00013
Garrett WS (2015) Cancer and the microbiota. Science 348:80–86. https://doi.org/10.1126/science.aaa4972
Garrett WS, Gordon JI, Glimcher LH (2010) Homeostasis and inflammation in the intestine. Cell 140:859–870. https://doi.org/10.1016/j.cell.2010.01.023
Giannelli V, Di Gregorio V, Iebba V, Giusto M, Schippa S, Merli M, Thalheimer U (2014) Microbiota and the gut-liver axis: bacterial translocation, inflammation and infection in cirrhosis. World J Gastroenterol 20:16795–16810. https://doi.org/10.3748/wjg.v20.i45.16795
Gopalakrishnan V, Helmink BA, Spencer CN, Reuben A, Wargo JA (2018a) The influence of the gut microbiome on cancer, immunity, and cancer immunotherapy. Cancer Cell 33:570–580. https://doi.org/10.1016/j.ccell.2018.03.015
Gopalakrishnan V et al (2018b) Gut microbiome modulates response to anti-PD-1 immunotherapy in melanoma patients. Science 359:97–103. https://doi.org/10.1126/science.aan4236
Goto Y, Kiyono H (2012) Epithelial barrier: an interface for the cross-communication between gut flora and immune system. Immunol Rev 245:147–163. https://doi.org/10.1111/j.1600-065X.2011.01078.x
Greer R, Dong X, Morgun A, Shulzhenko N (2016) Investigating a holobiont: microbiota perturbations and transkingdom networks. Gut Microbes 7:126–135. https://doi.org/10.1080/19490976.2015.1128625
Grohmann U, Fallarino F, Puccetti P (2003) Tolerance, DCs and tryptophan: much ado about IDO. Trends Immunol 24:242–248
Gur C et al (2015) Binding of the Fap2 protein of Fusobacterium nucleatum to human inhibitory receptor TIGIT protects tumors from immune cell attack. Immunity 42:344–355. https://doi.org/10.1016/j.immuni.2015.01.010
Hanahan D, Weinberg RA (2000) The hallmarks of cancer. Cell 100:57–70
Hanahan D, Weinberg RA (2011) Hallmarks of cancer: the next generation. Cell 144:646–674. https://doi.org/10.1016/j.cell.2011.02.013
Hernandez P, Gronke K, Diefenbach A (2018) A catch-22: Interleukin-22 and cancer. Eur J Immunol 48:15–31. https://doi.org/10.1002/eji.201747183
Hsu PP, Sabatini DM (2008) Cancer cell metabolism: Warburg and beyond. Cell 134:703–707. https://doi.org/10.1016/j.cell.2008.08.021
Huitzil S, Sandoval-Motta S, Frank A, Aldana M (2018) Modeling the role of the microbiome in evolution. Front Physiol 9:1836. https://doi.org/10.3389/fphys.2018.01836
Human Microbiome Project C (2012) Structure, function and diversity of the healthy human microbiome. Nature 486:207–214. https://doi.org/10.1038/nature11234
Iida N et al (2013) Commensal bacteria control cancer response to therapy by modulating the tumor microenvironment. Science 342:967–970. https://doi.org/10.1126/science.1240527
Johnson CH et al (2015) Metabolism links bacterial biofilms and colon carcinogenesis. Cell Metab 21:891–897. https://doi.org/10.1016/j.cmet.2015.04.011
Joice R, Yasuda K, Shafquat A, Morgan XC, Huttenhower C (2014) Determining microbial products and identifying molecular targets in the human microbiome. Cell Metab 20:731–741. https://doi.org/10.1016/j.cmet.2014.10.003
Kroemer G, Zitvogel L (2018) Cancer immunotherapy in 2017: the breakthrough of the microbiota. Nat Rev Immunol 18:87–88. https://doi.org/10.1038/nri.2018.4
Kudo M (2018) Systemic therapy for hepatocellular carcinoma: latest advances. Cancers (Basel) 10. https://doi.org/10.3390/cancers10110412
Le Chatelier E et al (2013) Richness of human gut microbiome correlates with metabolic markers. Nature 500:541–546. https://doi.org/10.1038/nature12506

Lee WJ, Hase K (2014) Gut microbiota-generated metabolites in animal health and disease. Nat Chem Biol 10:416–424. https://doi.org/10.1038/nchembio.1535
Leong SP, Aktipis A, Maley C (2018) Cancer initiation and progression within the cancer microenvironment. Clin Exp Metastasis 35:361–367. https://doi.org/10.1007/s10585-018-9921-y
Llorente C, Schnabl B (2015) The gut microbiota and liver disease. Cell Mol Gastroenterol Hepatol 1:275–284. https://doi.org/10.1016/j.jcmgh.2015.04.003
Long SL, Gahan CGM, Joyce SA (2017) Interactions between gut bacteria and bile in health and disease. Mol Aspects Med 56:54–65. https://doi.org/10.1016/j.mam.2017.06.002
Lu R, Bosland M, Xia Y, Zhang YG, Kato I, Sun J (2017) Presence of Salmonella AvrA in colorectal tumor and its precursor lesions in mouse intestine and human specimens. Oncotarget 8:55104–55115. https://doi.org/10.18632/oncotarget.19052
Marchiando AM, Graham WV, Turner JR (2010) Epithelial barriers in homeostasis and disease. Annu Rev Pathol 5:119–144. https://doi.org/10.1146/annurev.pathol.4.110807.092135
Martin FP et al (2007) A top-down systems biology view of microbiome-mammalian metabolic interactions in a mouse model. Mol Syst Biol 3:112. https://doi.org/10.1038/msb4100153
Matson V et al (2018) The commensal microbiome is associated with anti-PD-1 efficacy in metastatic melanoma patients. Science 359:104–108. https://doi.org/10.1126/science.aao3290
Maynard Smith J (1998) The units of selection. Novartis Found Symp 213:203–211. discussion 211–207
Maynard Smith J, Szathmáry E (1995) The major transitions in evolution. W.H. Freeman Spektrum, Oxford/New York
Morgillo F et al (2018) Carcinogenesis as a result of multiple inflammatory and oxidative hits: a comprehensive review from tumor microenvironment to gut microbiota. Neoplasia 20:721–733. https://doi.org/10.1016/j.neo.2018.05.002
Muller EEL, Faust K, Widder S, Herold M, Martinez Abbas S, Wilmes P (2018) Using metabolic networks to resolve ecological properties of microbiomes. Curr Opin Syst Biol 8:73–80
Nicholson JK, Holmes E, Kinross J, Burcelin R, Gibson G, Jia W, Pettersson S (2012) Host-gut microbiota metabolic interactions. Science 336:1262–1267. https://doi.org/10.1126/science.1223813
Noonan DM, De Lerma Barbaro A, Vannini N, Mortara L, Albini A (2008) Inflammation, inflammatory cells and angiogenesis: decisions and indecisions. Cancer Metastasis Rev 27:31–40. https://doi.org/10.1007/s10555-007-9108-5
Pardoll D (2015) Cancer and the immune system: basic concepts and targets for intervention. Semin Oncol 42:523–538. https://doi.org/10.1053/j.seminoncol.2015.05.003
Pavlova NN, Thompson CB (2016) The emerging hallmarks of cancer metabolism. Cell Metab 23:27–47. https://doi.org/10.1016/j.cmet.2015.12.006
Perez-Chanona E, Trinchieri G (2016) The role of microbiota in cancer therapy. Curr Opin Immunol 39:75–81. https://doi.org/10.1016/j.coi.2016.01.003
Pfeiffer T, Schuster S, Bonhoeffer S (2001) Cooperation and competition in the evolution of ATP-producing pathways. Science 292:504–507. https://doi.org/10.1126/science.1058079
Plichta DR et al (2016) Transcriptional interactions suggest niche segregation among microorganisms in the human gut. Nat Microbiol 1:16152. https://doi.org/10.1038/nmicrobiol.2016.152
Pope JL, Tomkovich S, Yang Y, Jobin C (2017) Microbiota as a mediator of cancer progression and therapy. Transl Res 179:139–154. https://doi.org/10.1016/j.trsl.2016.07.021
Quail DF, Joyce JA (2013) Microenvironmental regulation of tumor progression and metastasis. Nat Med 19:1423–1437. https://doi.org/10.1038/nm.3394
Quante M, Varga J, Wang TC, Greten FR (2013) The gastrointestinal tumor microenvironment. Gastroenterology 145:63–78. https://doi.org/10.1053/j.gastro.2013.03.052
Restifo NP, Dudley ME, Rosenberg SA (2012) Adoptive immunotherapy for cancer: harnessing the T cell response. Nat Rev Immunol 12:269–281. https://doi.org/10.1038/nri3191
Reticker-Flynn NE, Engleman EG (2019) A gut punch fights cancer and infection. Nature 565:573–574. https://doi.org/10.1038/d41586-019-00133-w
Riscuta G, Xi D, Pierre-Victor D, Starke-Reed P, Khalsa J, Duffy L (2018) Diet, microbiome, and epigenetic changes in cancer. In: Dumitrescu RG, Verma M (eds) Cancer epigenetics for precision medicine. Methods in molecular biology. Humana Press, Clifton, pp 141–156
Robert C et al (2015) Pembrolizumab versus ipilimumab in advanced melanoma. N Engl J Med 372:2521–2532. https://doi.org/10.1056/NEJMoa1503093
Routy B et al (2018) Gut microbiome influences efficacy of PD-1-based immunotherapy against epithelial tumors. Science 359:91–97. https://doi.org/10.1126/science.aan3706
Rubinstein MR, Wang X, Liu W, Hao Y, Cai G, Han YW (2013) Fusobacterium nucleatum promotes colorectal carcinogenesis by modulating E-cadherin/beta-catenin signaling via its FadA adhesin. Cell Host Microbe 14:195–206. https://doi.org/10.1016/j.chom.2013.07.012
Schwabe RF, Jobin C (2013) The microbiome and cancer. Nat Rev Cancer 13:800–812. https://doi.org/10.1038/nrc3610
Sears CL (2009) Enterotoxigenic Bacteroides fragilis: a rogue among symbiotes. Clin Microbiol Rev 22:349–369. https://doi.org/10.1128/CMR.00053-08
Serpa J et al (2010) Butyrate-rich colonic microenvironment is a relevant selection factor for metabolically adapted tumor cells. J Biol Chem 285:39211–39223. https://doi.org/10.1074/jbc.M110.156026
Singh SB, Lin HC (2015) Hydrogen sulfide in physiology and diseases of the digestive tract.

Microorganisms 3:866–889. https://doi.org/10.3390/microorganisms3040866

Sivan A et al (2015) Commensal Bifidobacterium promotes antitumor immunity and facilitates anti-PD-L1 efficacy. Science 350:1084–1089. https://doi.org/10.1126/science.aac4255

Sommer F, Backhed F (2013) The gut microbiota – masters of host development and physiology. Nat Rev Microbiol 11:227–238. https://doi.org/10.1038/nrmicro2974

Steinman RM (2007) Lasker basic medical research award. Dendritic cells: versatile controllers of the immune system. Nat Med 13:1155–1159. https://doi.org/10.1038/nm1643

Sun L, Suo C, Li ST, Zhang H, Gao P (2018) Metabolic reprogramming for cancer cells and their microenvironment: beyond the Warburg effect. Biochim Biophys Acta Rev Cancer 1870:51–66. https://doi.org/10.1016/j.bbcan.2018.06.005

Swartz MA et al (2012) Tumor microenvironment complexity: emerging roles in cancer therapy. Cancer Res 72:2473–2480. https://doi.org/10.1158/0008-5472.CAN-12-0122

Swidsinski A et al (2007) Comparative study of the intestinal mucus barrier in normal and inflamed colon. Gut 56:343–350. https://doi.org/10.1136/gut.2006.098160

Tanoue T et al (2019) A defined commensal consortium elicits CD8 T cells and anti-cancer immunity. Nature 565:600–605. https://doi.org/10.1038/s41586-019-0878-z

Thorburn AN, Macia L, Mackay CR (2014) Diet, metabolites, and "western-lifestyle" inflammatory diseases. Immunity 40:833–842. https://doi.org/10.1016/j.immuni.2014.05.014

Vetizou M et al (2015) Anticancer immunotherapy by CTLA-4 blockade relies on the gut microbiota. Science 350:1079–1084. https://doi.org/10.1126/science.aad1329

Wahlstrom A, Sayin SI, Marschall HU, Backhed F (2016) Intestinal crosstalk between bile acids and microbiota and its impact on host metabolism. Cell Metab 24:41–50. https://doi.org/10.1016/j.cmet.2016.05.005

Wallace BD et al (2010) Alleviating cancer drug toxicity by inhibiting a bacterial enzyme. Science 330:831–835. https://doi.org/10.1126/science.1191175

Wang R (2012) Physiological implications of hydrogen sulfide: a whiff exploration that blossomed. Physiol Rev 92:791–896. https://doi.org/10.1152/physrev.00017.2011

Warburg O, Poesener K, Negelein E (1924) Über den Stoffwechsel der Carcinomzelle. Biochem Z 152:309–344

Wasielewski H, Alcock J, Aktipis A (2016) Resource conflict and cooperation between human host and gut microbiota: implications for nutrition and health. Ann N Y Acad Sci 1372:20–28. https://doi.org/10.1111/nyas.13118

Wegiel B, Vuerich M, Daneshmandi S, Seth P (2018) Metabolic switch in the tumor microenvironment determines immune responses to anti-cancer therapy. Front Oncol 8:284. https://doi.org/10.3389/fonc.2018.00284

Wilson HL, Obradovic MR (2015) Evidence for a common mucosal immune system in the pig. Mol Immunol 66:22–34. https://doi.org/10.1016/j.molimm.2014.09.004

Wu S et al (2009) A human colonic commensal promotes colon tumorigenesis via activation of T helper type 17 T cell responses. Nat Med 15:1016–1022. https://doi.org/10.1038/nm.2015

Wu C et al (2018) Forward genetic dissection of biofilm development by fusobacterium nucleatum: novel functions of cell division proteins FtsX and EnvC. MBio 9. https://doi.org/10.1128/mBio.00360-18

Yang T, Owen JL, Lightfoot YL, Kladde MP, Mohamadzadeh M (2013) Microbiota impact on the epigenetic regulation of colorectal cancer. Trends Mol Med 19:714–725. https://doi.org/10.1016/j.molmed.2013.08.005

Yoshimoto S et al (2013) Obesity-induced gut microbial metabolite promotes liver cancer through senescence secretome. Nature 499:97–101. https://doi.org/10.1038/nature12347

Yu T et al (2017) Fusobacterium nucleatum promotes chemoresistance to colorectal cancer by modulating autophagy. Cell 170:548–563.e516. https://doi.org/10.1016/j.cell.2017.07.008

Zelante T et al (2013) Tryptophan catabolites from microbiota engage aryl hydrocarbon receptor and balance mucosal reactivity via interleukin-22. Immunity 39:372–385. https://doi.org/10.1016/j.immuni.2013.08.003

Zhou Z, Chen J, Yao H, Hu H (2018) Fusobacterium and colorectal cancer. Front Oncol 8:371. https://doi.org/10.3389/fonc.2018.00371

Zitvogel L, Galluzzi L, Viaud S, Vetizou M, Daillere R, Merad M, Kroemer G (2015) Cancer and the gut microbiota: an unexpected link. Sci Transl Med 7:271ps271. https://doi.org/10.1126/scitranslmed.3010473

Zitvogel L, Ayyoub M, Routy B, Kroemer G (2016) Microbiome and anticancer immunosurveillance. Cell 165:276–287. https://doi.org/10.1016/j.cell.2016.03.001

Zmora N, Suez J, Elinav E (2019) You are what you eat: diet, health and the gut microbiota. Nat Rev Gastroenterol Hepatol 16:35–56. https://doi.org/10.1038/s41575-018-0061-2

Metabolic Plasticity of Tumor Cells: How They Do Adapt to Food Deprivation

6

Céline A. Schoonjans and Bernard Gallez

Abstract

Dysregulated metabolism is a key hallmark of cancer cells and an enticing target for cancer treatment. Since the last 10 years, research on cancer metabolism has moved from pathway attention to network consideration. This metabolic complexity continuously adapt to new constraints in the tumor microenvironment. In this review, we will highlight striking changes in cancer cell metabolism compared to normal cells. Understanding this tumor metabolic plasticity suggests potential new targets and innovative combinatorial treatments for fighting cancer.

Keywords

Metabolic reprogramming · Bioenergetics · Glycolysis · Mitochondria · Amino acid · Fatty acid · Hypoxia · Acidosis · Combined therapies

C. A. Schoonjans
Université catholique de Louvain (UCLouvain), Louvain Drug Research Institute, Brussels, Belgium

B. Gallez (✉)
Biomedical Magnetic Resonance, Brussels, Belgium
e-mail: bernard.gallez@uclouvain.be

6.1 Introduction

Cancer metabolism is under the spotlight since almost one century when Otto Warburg first observed an increase in glucose consumption in tumors in comparison with non-proliferating normal tissues (Warburg 1956a, b). Numerous studies on different tumor models and in patients confirmed that tumors present a high glucose consumption and lactate production regardless of oxygen availability (DeBerardinis and Chandel 2016). This phenomenon is called aerobic glycolysis and the resulting increased glycolytic flux was demonstrated to provide new biomass in order to fuel cancer cell growth. However, more recently, new investigation and new biochemical tools highlighted that tumors do not only depend on glucose as a major source of energy. New evidence demonstrated that mitochondrial activity still participates in tumor energy production. In order to survive and proliferate, cancer cells adapt their metabolism to acquire necessary nutrients from nutrient-poor environment or hostile environment like acidosis. Depending on cell type or microenvironment modification, tumor cells can rely on glucose, on specific amino acids (such as glutamine or serine) and/or on lipid metabolism (Fig. 6.1) (Corbet and Feron 2017a). These recent findings underscore the complexity of metabolic flexibility presented by cancer cells and the important role that this plasticity may play in cancer development.

J. Serpa (ed.), *Tumor Microenvironment*, Advances in Experimental Medicine and Biology 1219,
https://doi.org/10.1007/978-3-030-34025-4_6

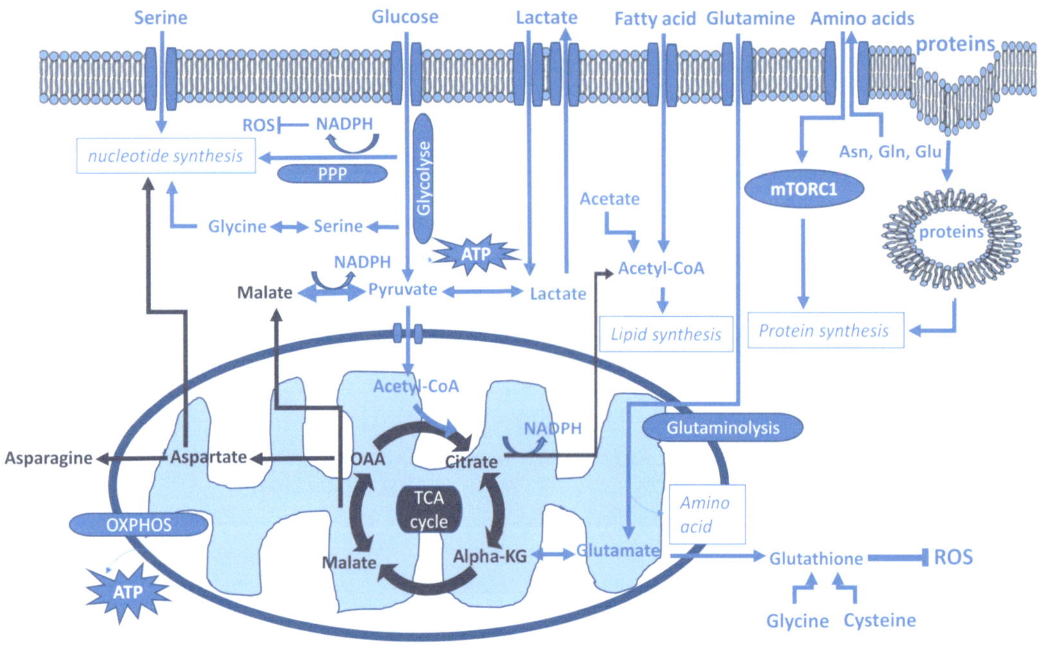

Fig. 6.1 Bioenergetics of cancer cells
Cancer cells can take up glucose, glutamine, serine, amino acids, fatty acids, acetate, lactate and extracellular proteins to support proliferation. Glucose metabolism through glycolysis produces ATP and carbon intermediates to sustain proliferation such as intermediates for the pentose phosphate pathway (PPP) and for de novo serine synthesis. PPP generates NADPH (defense against reactive species of oxygen, ROS) and is involved in nucleotide synthesis. Extracellular and de novo serine also participate in nucleotide synthesis. The end-product of glycolysis, pyruvate, can be converted into lactate or acetyl-coA. Lactate is exported out of the cells and acetyl-coA fuels tricarboxilic acid (TCA) cycle. In some cancer cells, lactate can be taken up and use as carbon substrate for pyruvate production. TCA cycle produces carbon intermediate for cellular proliferation such as aspartate, asparagine, glutamate and precursors for lipid synthesis. In the presence of oxygen, TCA cycle via oxidative phosphorylation (OXPHOS) produces ATP. TCA cycle can also be fueled by glutamate. Glutamate is the result of glutamine metabolism, glutaminolysis. Glutaminolysis is also involved in amino acid synthesis and in concert with glycine and cysteine, glutathione production (defense against ROS). Cancer cells can import fatty acids from extracellular and, via fatty acid oxidation (FA0), generate ATP and acetyl-CoA. Fatty acids can also be synthetized de novo via an acetyl-coA sources provide from citrate or acetate. Glutamine (Gln), glutamate (Glu) and asparagine (Asn) can act as exchange substrate for antiporter that allow the entrance of extracellular amino acids. The level of intercellular amino acid regulate mTORC1 activity, which in turn regulates protein synthesis. In case of depleted nutrient medium, cancer cells can import entire protein via micropinocytosis and through lysosomal degradation release free amino acids.

6.2 Bioenergetics of Cancer Cells

The metabolic needs of the proliferating cancer cells significantly differ from the quiescent cells. To support rapid proliferation, tumor cells must undergo metabolic reprogramming. Understanding the different metabolic strategies developed by the cancer cells to acquire necessary nutrients can build up promising strategies that may lead to food starvation or prevent their metabolic flexibility.

6.2.1 Aerobic Glycolysis and TCA Cycle Dependency

In most healthy cells, glucose is metabolized into pyruvate (glycolysis) and then oxidized in the mitochondria through tricarboxylic acid (TCA) cycle in order to produce energy via OXPHOS (Fig. 6.1). In the absence of oxygen, pyruvate is converted into lactate in the cytosol. Otto Warburg observed that this physiological response to hypoxia also occurs in cancer cells even in the presence of oxygen. This phenomenon is called aerobic glycolysis. Warburg first explained this phenomenon by an impaired mitochondrial function in tumor cells leading to a non-coupling of glycolysis to energy production from mitochondria (Warburg 1956a, b). However, the Warburg hypothesis has been now reassessed because only a few mutations of mitochondrial metabolic enzymes have been described, and this true only in some cancer cell lines (San-Millan and Brooks 2017; Laurenti and Tennant 2016). Moreover, recent studies have shown that various cancer cell types rely on mitochondrial respiration and that mitochondrial metabolism is necessary for cancer tumorigenesis (Weinberg et al. 2010). Nevertheless, Warburg's observation that cancer cells consume a large amount of glucose has been validated in many cancer patients. Today, it is clear that the exacerbated glycolytic flux is due to the loss of tumor suppressors and activation of oncogenes that modulate the expression or activity of proteins (transporters, enzymes, cofactors) involved in glycolysis (DeBerardinis and Chandel 2016). It has been demonstrated that one advantage of the resulting increase in glycolytic intermediates is to provide new precursors for anabolic pathways in order to fuel cancer cell growth, especially through the pentose phosphate pathway (PPP), a pivotal biosynthetic pathway branched to glycolysis (De Preter et al. 2016). PPP produces NADPH against oxidative stress and also produces pentose phosphate for nucleotide synthesis (Fig. 6.1). De Preter et al. demonstrated that reduction of PPP activity decreases cancer cells proliferation with a profound effect in Warburg-phenotype cancer cells. In addition, she found that inducing a switch of glycolytic cancer cells to a more oxidative phenotype led to a decreased proliferation (De Preter et al. 2016). This result confirmed the pivotal role of aerobic glycolysis in cancer proliferation.

The observation that mitochondria are still active in many cancers is explained by the fact that these organelles (which are at the crossroad of many essential metabolic pathways) can be fueled by other nutrients than glucose (DeBerardinis and Chandel 2016). Amino acids and fatty acids can supply substrates to TCA cycle in cancer cell lines (Fig. 6.1). Like glycolytic intermediates, TCA cycle intermediates also provide new precursors for anabolic pathways (Ahn and Metallo 2015). Due to tumor heterogeneity, different metabolic preferences could coexist within a tumor. Local microenvironment could influence which fuel is used by the cancer cells to sustain tumor growth (Boroughs and DeBerardinis 2015; Ahn and Metallo 2015). For instance, it has been demonstrated that lactate release by hypoxic tumors can be taken up by oxidative tumors and then oxidized into pyruvate for fueling TCA cycle (Corbet and Feron 2017a).

6.2.2 Amino Acid Metabolism

In addition to the increased glucose consumption, various cancer types exhibit an increased demand for specific amino acids and become dependent either on an exogenous supply or on an upregulated *de novo* synthesis (Lukey et al. 2017).

6.2.2.1 Glutamine Metabolism

This non-essential amino acid is the most abundant amino acid present in the plasma and is used by the cell for energy generation and biomass production (Altman et al. 2016). Glutamine enters into the cells via many transporters, such as SLC1A5 which is upregulated in several cancer types (Bhutia et al. 2015). Then, it can be exported via the efflux of glutamine through LAT1 antiporter (Pavlova and Thompson 2016). Glutamine can also be metabolized through glutaminolysis to generate glutamate and subsequently alpha-ketoglutarate that fuels the TCA cycle (DeBerardinis and Chandel 2016). Two different routes may be used by the TCA cycle to metabolize the alpha-ketoglutarate: the canonical oxidative decarboxylation pathway and the reductive carboxylation pathway. The oxidation of alpha-ketoglutarate produces energy, anaplerotic nutrients and pyruvate through malic enzyme (Fig. 6.1). On the other hand, the reduction of alpha-ketoglutarate leads to citrate which supports the fatty acid synthesis (Altman et al. 2016). In addition, glutamine via its conversion into glutamate is an important source of nitrogen, which supports the production of amino acids and the biosynthesis of nucleotides. Glutamate can also induce cysteine uptake by acting as an exchange substrate for the cysteine antiporter xCT (Pavlova and Thompson 2016). Glutamine metabolism is also involved in protein synthesis, mTORC1 regulation (a major positive regulator of cell growth), production of defense against reactive oxygen species (through glutathione synthesis and NADPH production) and autophagy.

As a consequence of these different reactions and pathways, glutamine is theoretically an ideal substrate for the development of cancer cells. Experimental studies have confirmed its pivotal role. In 2007, Yuneva et al. already described that some cancer cells were sensitive to glutamine depletion (Yuneva et al. 2007). Several experimental ^{13}C-NMR studies have shown that glutamine is used to replenish TCA cycle in cancer instead of glucose (DeBerardinis et al. 2007; Fan et al. 2013; Altman et al. 2016; Jones and Bianchi 2015). In different cancer cell lines, it has been demonstrated that glutamine supports OXPHOS and that OXPHOS remains the largest quantitative contributor for ATP production, a fact that is also true in hypoxia (Fan et al. 2013; Le et al. 2012). In glioblastoma, it has been shown that glutamine also serves as a carbon source for biosynthetic pathways. First, the decarboxylation of glutamine leads to pyruvate production through malic enzyme, consistent with NADPH production. Pyruvate can then be used as anaplerotic nutrient to replenish TCA cycle intermediates during citrate export while NADPH serves as an electron donor for reductive reaction in lipid synthesis. Second, glutamine decarboxylation feeds the production of oxaloacetate that supplies anabolism as oxaloacetate can be converted into aspartate, a required precursor for the synthesis of nucleotides (DeBerardinis et al. 2007). More recently, glutamine has been identified as a major source of citrate in cancer cells when the TCA cycle function is altered by hypoxia (Metallo et al. 2011). In these circumstances, glutamine preferentially undergoes reductive metabolism to produce citrate through the TCA cycle, then can be converted into acetyl-CoA for lipid synthesis (Zhang et al. 2014; Metallo et al. 2011; Wise et al. 2011; Mullen et al. 2011). Some studies reported circumstances where reductive and canonical metabolism of glutamine co-exist, depending on the tumor microenvironment (acidic pH) or mutations of oncogenes regulating the mitochondrial metabolism (McGuirk et al. 2013; Corbet et al. 2014). This ability to easily switch from one mode to another mode of glutamine metabolism offers to the tumor cells a metabolic plasticity advantage to rapidly adapt to new environmental constraints.

6.2.2.2 Serine Metabolism

Besides glutamine, the availability of other amino acids is also a limiting factor for cancer cell proliferation (Kory et al. 2018). Many cancer cells are dependent on serine, a non-essential amino acid involved in several metabolic processes that are crucial for cell growth (Yang and Vousden 2016). Serine can be synthesized *de novo* via a glycolysis-diverting pathway: the glycolytic intermediate 3-phosphoglycerate is converted into serine following a three-step enzymatic reaction (Amelio et al. 2014). Metabolic studies have shown that cancer cells may use as much as 50% of glucose-derived carbon in serine biosynthesis and downstream metabolism (Pavlova and Thompson 2016). Serine can also be imported from the extracellular medium via different transporters (Yang and Vousden 2016). After glucose and glutamine, serine is the third most consumed metabolite in mammalian cells (Frezza 2016; Hosios et al. 2016). Serine, through its conversion into glycine, is a major donor of one-carbon unit to the folate cycle which is essential for the synthesis of nucleic acids and contributes to NAPDH formation and methylation reaction (DNA, RNA, proteins and lipids). Of note, antifolate chemotherapies are currently used in cancer treatment. Serine is also involved in the production of cysteine and glutathione (Yang and Vousden 2016). In various cancer types, the first enzyme of *de novo* serine synthesis (PHGDH) was found to be overexpressed, opening new opportunities for cancer treatment (Yang and Vousden 2016).

6.2.2.3 Role of Other Amino Acids

To sustain their need for amino acids, cancer cells have developed means to either increase the import or the biosynthesis of these amino acids. By acting as an exchange substrate for the antiporter LAT1, glutamine is involved in the import of many essential amino acids through this antiporter (Pavlova and Thompson 2016). Fuchs et al. demonstrated that the expression of LAT1 was increased in several cancer types and that the expression of this antiporter was correlated with tumor growth (Fuchs and Bode 2005). The metabolism of serine and glutamine is involved in the biosynthesis of all non-essential amino acids (Zhang et al. 2017). These amino acids are essential for protein synthesis, and like glutamine and serine, they participate in other cellular functions. For instance, Krall et al. observed that asparagine level regulates the uptake of amino acids and mTORC1 activity. More specifically, asparagine regulates the serine uptake and influences the serine metabolism as well as nucleotides synthesis (Krall et al. 2016). The asparagine biosynthesis is the result of glutamine oxidation into aspartate which is metabolized into asparagine. Sullivan et al. demonstrated that a major role of glutamine oxidation for respiration was to provide access to electron acceptors to support aspartate biosynthesis. They demonstrated that aspartate supported cancer proliferation through its contribution to nucleotide synthesis (Sullivan et al. 2015). This data highlight the importance of the *glutamine-aspartate-asparagine* axis in cancer proliferation. In many tumor cells, some non-essential amino acids become essential because of the high metabolic demand (such as glutamine and serine) or mutations in specific metabolic enzymes (Geck and Toker 2016). For example, arginine auxotrophic tumors have been described. These tumors cannot synthesize arginine due to a lack of argininosuccinate synthase (Geck and Toker 2016). This feature is interesting for cancer therapy because healthy cells do not need extracellular sources of non-essential amino acids and starvation therapy is under consideration. In the absence of glutamine and other amino acids, cancer cells have developed a third way to obtain proper resources via the micropinocytosis of extracellular proteins, phagocytosis of apoptotic bodies and entosis living cells (Zhang et al. 2017; Pavlova and Thompson 2016). After lysosomal degradation, cancer cells recover free amino acids and therefore survive in unfavorable environment.

6.2.3 Fatty Acids Synthesis and Oxidation

Lipids are important building blocks for producing new cells. They are major components of membranes and they are also involved in lipidation reactions and cellular signaling (DeBerardinis and Chandel 2016). Fatty acids can be obtained from extracellular media or synthesized *de novo*. Fatty acid synthesis requires a source of acetyl-coA and the reducing factor NADPH. Depending on the microenvironment and the availability, glucose, glutamine or acetate are the major source of acetyl-coA while the NADPH mainly comes from PPP (DeBerardinis and Chandel 2016). While the *de novo* synthesis of fatty acids is low in most healthy cells, tumorigenesis is associated with high lipid synthesis (Pavlova and Thompson 2016). In the presence of abundant extracellular nutrients and oxygen, most tumors rely on *de novo* fatty acid production (Boroughs and DeBerardinis 2015). Moreover, the three major components involved in fatty acid synthesis are frequently upregulated in various cancer types (Pavlova and Thompson 2016). However, under some condition such as hypoxia, cancer cells are not able to synthesize fatty acids and need to import lipids from the extracellular environment. Fatty acids also supply energy, as mitochondrial fatty acid oxidation produces more than twice much ATP per mole than oxidation of glucose or amino acids (Boroughs and DeBerardinis 2015). To conclude, like for amino acid metabolism, cancer cells have developed various mechanisms to exploit lipid metabolism in order to sustain growth in unfavorable microenvironment.

6.3 Metabolic Reprogramming

Tumors adapt to different sources of energy to survive and proliferate. However, what dictates which pathway is used and when? It has been established that the metabolic phenotype of tumor is controlled by intrinsic genetic mutations and external response to microenvironment (Cairns et al. 2011). Intrinsic mutations lead to modification of metabolic pathways to optimize cancer proliferation (Obre and Rossignol 2015; Cairns et al. 2011). Here, we will describe illustrative examples of metabolic plasticity induced by the microenvironment, namely pH, oxygen and nutrient deprivation. We will also focus on metabolic changes that are induced by anti-cancer treatments.

6.3.1 Influence of Tumor Microenvironment on Tumor Plasticity: Illustrative Examples

Abnormal and heterogeneous microenvironment impose constraints on fast-growing tumors. Limited access to vasculature lead to local hypoxia and nutrient deprivation. The increase in metabolic needs lead to acidification of the tumor microenvironment. These constraints lead to profound cancer cells metabolic adaptation that need to be integrated together in order to develop efficient strategies to control tumor progression.

6.3.1.1 Hypoxia Enhanced Metabolic Plasticity

Hypoxia takes place when cancer cells have limited access to oxygen or when the balance between oxygen supply and consumption is disrupted. Several factors contribute to the occurrence of tumor hypoxia where co-exist chronic hypoxia (diffusion-limited hypoxia and hypoxemic hypoxia) and acute hypoxia (due to temporal fluctuations in red blood cell flux) (Bayer and Vaupel 2012). The main physiological response to hypoxia is the stabilization of the transcription factor HIF-1α (hypoxia-inducible factor 1 alpha), a common feature observed in hypoxic cancer cells (Nakazawa et al. 2016). HIF-1α activity upregulates almost all enzymes of glycolysis and glucose transporters thereby facilitating increased glycolytic flux (Eales et al. 2016). HIF-1α also prevents oxidative mitochondrial activity by controlling the glycolytic pyruvate fate. First, HIF-1α enhances pyruvate reduction into lactate (upregulation of LDHA) and increases lactate export (upregulation of MCT4). Second, HIF-1α prevents oxidation of pyruvate via the upregulation of PDK-1 and PDK-3 which inhibit the conversion of pyruvate into acetyl-CoA (Lu et al. 2008). In addition to the decreased mitochondrial respiration, the decrease in acetyl-CoA entering into TCA cycle affects other metabolic pathways as it decreases the level of TCA cycle intermediates required for biomass production and synthesis of non-essential amino acids. Glucose-derived acetyl-CoA is also the starting point for fatty acid synthesis (Nakazawa et al. 2016). Therefore, hypoxic cancer cells exhibit an increase in glutamine consumption to compensate the decrease in glucose contribution to TCA cycle (Nakazawa et al. 2016). As previously described, both oxidative and reductive glutamine metabolism can sustain TCA cycle. It has been shown that HIF-1α can shift glutamine metabolism from oxidative to reductive carboxylation to support lipid synthesis (Metallo et al. 2011).

There is increasing evidence that, under hypoxia, cancer cells rely on carbon sources other than glutamine and glucose (Eales et al. 2016). Under hypoxia, several cancer cells lines increased their uptake of extracellular lipids (Kamphorst et al. 2013). Hypoxia upregulates an enzyme involved in acetate metabolism (ACSS2, acetate to acetyl-CoA) (Schug et al. 2015). As a consequence, a large fraction of fatty acid carbon is derived from acetate (Kamphorst et al. 2014). Overall, specific mutations induced by hypoxia in acetate metabolism and/or fatty acid uptake influence the fuel choice. The acetate assimilation to maintain lipogenesis together with the increased uptake of fatty acid support the rapid proliferation of tumor cells under hypoxia. Serine metabolism under hypoxia was also investigated by Ye et al. They found that SHMT2, the enzyme that converts serine into glycine in mitochondria, was induced under hypoxia through HIF-1α. They also observed that depletion of this enzyme in hypoxic cells increased ROS levels, led to cell death *in vitro* and decreased tumor growth *in vivo*. They concluded that this enzyme was crucial to connect serine metabolism to mitochondrial redox control of cancer and to maintain cell survival (Ye et al. 2014).

Hypoxia is heterogeneous in solid tumors, a feature that can enhance metabolic cooperation between different microenvironment and cell types. For example, De Bock et al. demonstrated that tumor vascular endothelial cells rely on glycolysis rather than on oxidative phosphorylation for ATP production. This allows oxygen to permeate further into the tumor instead of being directly consumed by proximal endothelial cells (De Bock et al. 2013). Another example of metabolic cooperation is the symbiosis existing between hypoxic and oxygenated cancer cells within a tumor (Sonveaux et al. 2008): the lactate released by hypoxic highly glycolytic cancer cells is metabolized by normoxic cancer cells. As a consequence, glucose freely diffuses through the oxygenated tumor cell sheath to fuel glycolysis of distant, hypoxic tumor cells. Of note, this metabolic symbiosis can be disrupted by MCT1 inhibition (Sonveaux et al. 2008).

6.3.1.2 Interplay Between Cancer Metabolism and Extracellular pH

Aerobic glycolysis and hypoxia are responsible for this extracellular acidification (Corbet and Feron 2017b). The increase in glycolytic flux leads to the production of lactate and protons that need to be exported outside the cells to avoid cytoplasm acidity and to maintain cellular function. Therefore, cancer cells have developed different mechanisms to export protons outside the cell. It has been shown that HIF-1α upregulates plasma membrane ion pumps and transporters (Taddei et al. 2013). Lactate and H^+ are exported to the extracellular media through the MCT4 transporter. Protons can also be exported outside the cells by Na^+/H^+ exchangers and H^+-ATPases. Of note, oxygenated areas are also involved in extracellular acidification because CO_2 produced by cellular respiration diffuses outside the cell and is hydrated in the extracellular medium in bicarbonate and proton. The disorganized tumor vasculature prevents an efficient wash-out of H^+ ions released into the extracellular medium and contributes to its acidification (Corbet and Feron 2017b). Extracellular acidosis has been largely described promoting cancer cells migration, invasion and metastasis (Taddei et al. 2013). Acidosis can also play a role in tumor metabolism reprogramming (Corbet and Feron 2017b). In 2008, Chen et al. performed a genome-scale gene expression study in breast cancer to measure response to lactic acidosis: low-risk breast cancer patients exhibited a preference for aerobic respiration and a repression of glycolysis gene expression (Chen et al. 2008). In another study, Chen demonstrated that acidosis abolished stabilization of HIF-1α (Tang et al. 2012). In 2013, Lamonte et al. performed stable-isotope tracers of glucose, glutamine and palmitate under acidosis. They observed that acidosis redirected glucose away from glycolysis and increased respiratory metabolism (Lamonte et al. 2013). They also found that acidosis promoted an increase in glutamine uptake and consumption leading to the ATP generation. They measured an increase in fatty oxidation which, together with glutaminolysis, increased ROS production. Interestingly, acidosis also redirected glucose away from glycolysis towards the oxidative branch of the PPP to produce NADPH resulting in ROS neutralization (Lamonte et al. 2013). More recently, Corbet et al. described that acidosis is leading to a metabolic reprograming towards glutamine and fatty acid metabolism. They found that the main source of acetyl-CoA was fatty acid oxidation in acidic pH-adapted cancer cells. Acetyl-CoA fuels the TCA cycle and supports tumor cell respiration under acidosis. By mitochondrial protein hyperacetylation, acetyl-CoA also restrains complex I activity and ROS production (Corbet et al. 2014, 2016). Another recent study has shown that chronic acidosis increased acetate metabolism. It was found that extracellular acidic pH triggered activation of SREBP2 by stimulating nuclear translocation and promoter binding to its targets, namely ACSS2 that promotes conversion of acetate into acetyl-CoA (Kondo et al. 2017). In summary, in order to reduce proton production, cancer cells shift their metabolism away from glycolysis and rely on glutamine and fatty acid metabolism. Even if hypoxia and acidosis are often considered as a whole and often present similar adaptation, some metabolic reprogramming are specific to each hallmark. Moreover, different studies demonstrated areas of hypoxia and acidosis do not overlap in all tumors (Corbet and Feron 2017b). Consequently, these similarities and differences should be taken into account for the development of strategies targeting tumor metabolism.

6.3.1.3 How the Cancer Cell Escape to Nutrient Deprivation

Even if acidosis and hypoxia regulate cell preferences in nutrients, the presence of nutrients themselves is a limiting factor for cell function. While some tumors regions are well vascularized, some others are poorly perfused leading to decreased availability of oxygen and of exogenous nutrients. Nutrient availability also fluctuates during tumor development. It has been established that, when tumors become deprived of nutrients, they decrease their demand in ATP in order to keep an adequate ATP/ADP ratio to drive unfavorable reactions (DeBerardinis and Chandel 2016). The mTOR pathway, which controls cell proliferation, regulates this energy demand. Indeed, under nutrient-rich conditions, mTORC1 promotes cell growth by stimulating biosynthetic pathways and repressing autophagy. On the opposite, when amino acid level is low, mTORC1 is inhibited and cell proliferation is consequently reduced while autophagy is activated (Kim and Guan 2019; Rabanal-Ruiz et al. 2017). Moreover, Palm et al. demonstrated that the mTORC1 inhibition in a nutrient depleted microenvironment led to an increased micropinocytosis and lysosomal degradation resulting in free amino acids liberation (Palm et al. 2015). In response to specific nutrient limitations, cancer cells can shift their metabolism to other sources of energy. Polet et al. demonstrated that glutamine deprivation in leukemia cells led to the upregulation of the serine pathway independently from glucose metabolism. They observed that glutamine depleted medium induced the upregulation of serine metabolism enzymes (Polet et al. 2016). Another example of metabolic plasticity induced by specific food deprivation is provided by the glutamine metabolism that is increased when using a glucose depleted medium (Le et al. 2012).

Cancer cells may also benefit from a symbiosis with other cells in the microenvironment (Lyssiotis and Kimmelman 2017). By-products of one type of cell could serve as a carbon source for another type of cell. We already described two examples of metabolic symbiosis that occurs upon hypoxia: lactate/glucose between normoxic and hypoxic cancer cells and oxygen/glycolysis between endothelial cells and cancer cells (Sonveaux et al. 2008; De Bock et al. 2013). Independently from hypoxia, another symbiosis was observed in pancreatic cancer: pancreatic cancer-associated fibroblasts (CAFs) excrete alanine in response to interaction with pancreatic cancer cells. The alanine is taken up by pancreatic cancer cells and used to fuel TCA cycle. Moreover, alanine can outcompete glucose and glutamine to fuel metabolism and decrease tumor dependency to glucose and glutamine which are limited in the pancreatic tumor microenvironment (Sousa et al. 2016; Kamphorst et al. 2015). Comparably, it has been shown that ovarian CAFs produce glutamine for ovarian cancer cells and that the disruption of CAF glutamine production delays tumor growth (Yang et al. 2016). It has been hypothesized that CAFs consume glutamate and lactate, the waste products of ovarian cancer cells, in order to produce glutamine (Tajan and Vousden 2016). In a similar way, in glutamine-restricted microenvironment, astrocytes produce glutamine for glioblastoma (Tardito et al. 2015) and adipocytes produce glutamine for pancreatic cancer cells (Meyer et al. 2016). These results warrant further research on glutamine symbiosis in different cancer types as a potential niche for new therapies.

6.3.2 Influence of Treatment in Metabolic Reprogramming

With the advent of treatment targeting cancer metabolism, it is now crucial to evaluate if the treatment itself may promote changes in metabolism. The study of Polet et al. we described earlier is an example of metabolic adaptation to a treatment as they demonstrated that glutamine inhibition enhanced serine metabolism (Polet et al. 2016). Indeed, they also treat leukemia cells with pharmacological inhibitors of glutamine metabolism. They observed that these inhibitors decrease leukemia cells growth and that the addition of serine deprivation increased the anti-proliferative effect of these pharmacological inhibitors. Consequently, they demonstrated that treatment against one major metabolic pathway for cancer cell growth induces a metabolic reprogramming to another pathway that cancer cells exploit to sustain their proliferation. Other adaptations were observed in other cancer cell lines. For instance, the metabolic adaptation to anti-glutamine therapies has also been described in pancreatic cancer cells that are dependent on glutamine pathway (Biancur et al. 2017). When testing glutaminase inhibitors (GLSi, conversion of glutamine into glutamate), they observed effects only at an early stage, effects that were overcome in long-term proliferation assays. After GLSi treatment, they observed a decrease in glutamate derived from glutamine but concomitantly they also found an increase in glutamate not derived from glutamine. These results indicated an adaptive response to GLSi as an alternative pathway was used by treated pancreatic tumor cells. They identified multiple compensatory pathways that may explain the resistance to GLSi such as an alteration in metabolic enzymes involved in TCA cycle, lipid metabolism and amino acid metabolism. Finally, they demonstrated that combining inhibitors of these pathways with GLS inhibitors may have therapeutic utilities. They concluded that individual tumors may have unique kinetics or metabolic adaptations to the same metabolic perturbation (Biancur et al. 2017). This underlines the need to find biomarkers of response in order to perform personalized medicine. Another study identified a metabolic adaptation upon pharmacological glycolysis inhibition (Pusapati et al. 2016). They used the glucose analog 2-deoxy-D-glucose (2DG) that blocks the glycolysis by competitively inhibiting glucose-6-phosphate isomerase (the second enzyme of the glycolysis). The treatment was effective in blocking glycolysis but did not have significant therapeutic benefit *in vivo*. As expected, they observed a decrease in fructose 1,6-bisphosphate, a 2DG downstream glycolytic metabolite. Interestingly, they also observed an increase in PPP metabolites under 2DG treatment. PPP is fueled by glucose-6-phosphate which results from the conversion of glucose by the first enzyme of the glycolysis. They observed that 2DG induced a glucose-6-phosphate flux into the PPP. However, the PPP product, pentose 5-phosphate, re-entered in the glycolysis at a lower step by producing glyceraldehyde 3-phosphate. By redirecting the flux of the first step directly into PPP and re-injecting PPP products lower in glycolysis, resistant cancer cells can bypass the inhibition of 2DG. They demonstrated that cancer cells, that are resistant to 2DG, rely on mTORC1 for survival and that mTORC1 induced the expression of the rate-limiting enzyme in the PPP. Finally, they found that combining 2DG together with inhibition of mTORC1 signaling decreased tumor growth in cells that were resistant to 2DG.

Taken together, these three illustrative examples of metabolic reprogramming induced by pharmacological inhibitors highlight the high metabolic plasticity of cancer cells. These results prone the use of combined strategies to potentiate the effects of treatments targeting cancer metabolism.

6.4 Futures Directions: Understanding Metabolic Plasticity to Rationalize Combinatorial Treatments

This review highlights the ability of tumor cells to rapidly shift their metabolism and adapt to new microenvironment constraints such as nutrient deprivation, acidosis, hypoxia and pharmacological inhibitors. Therefore, treatments that target a single pathway or a single cancer hallmark seems to be old fashion. Nowadays, new therapies should be based on strategies such as combined therapies that are able to control tumor progression and their adaptation in order to prevent resistance. Chemotherapeutics agents together with surgery and radiotherapy are standard of care treatments to treat (or cure) most tumors. Despite significant results in early tumor regression, many tumors develop resistance to chemotherapy. Even if genetics and epigenetics are widely studied to understand the origins of resistance, the tumor microenvironment, such has hypoxia, has emerged as a key player in the development of these resistances (Agarwal and Kaye 2003; Senthebane et al. 2017) and metabolic modulators are increasingly considered in the treatment armamentarium.

As an illustrative example, we may cite the use of dichloroacetate (DCA), a metabolic modulator that has been widely studied in several combined therapies (Han et al. 2018). DCA inhibits Pyruvate Dehydrogenase Kinase (PDK) and consequently shifts metabolism from glycolysis to glucose oxidation through PDH reactivation. Therefore, DCA redirects pyruvate back into the mitochondria and reduces lactate production (Kankotia and Stacpoole 2014). In 2007, Bonnet et al. showed that DCA reduced cancer cell proliferation via the induction of oxidative stress and apoptosis in various cancer cell types but not in healthy cells (Bonnet et al. 2007). In a study of hypoxia-induced chemoresistance in gastric cancer, researchers observed that the levels of HIF-1α and PDK-1 were higher in-patients who showed recurrence after chemotherapy (Xuan et al. 2014). They found that the expression of PDK-1 was positively correlated with HIF-1α expression, the factor that regulates hypoxic responses in cancer cells. Moreover, they observed *in vitro* that PDK-1 expression was higher under hypoxia. Finally, they observed that DCA treatment was able to re-sensitize resistant gastric cancer cells through the alteration of glucose metabolism (Xuan et al. 2014). This study revealed that the metabolic reprogramming induced by hypoxia was involved in chemoresistance. The results suggest that metabolic change induces by hypoxia, such as higher PDK-1 expression, could be a marker of chemoresistance as well as a target for DCA therapy. DCA has also been studied in combination with bevacizumab, an antiangiogenic drug (Kumar et al. 2013). Kumar et al. showed that glioblastoma cells that were rendered resistant to antiangiogenic therapy (after long term exposure to bevacizumab) enhanced HIF-1α related gene expression and consequently presented a shift from mitochondrial respiration to glycolysis. They showed that DCA reversed this shift induced by bevacizumab which highlighted the plasticity of tumor metabolism in response to therapeutic agents. These two studies underline the ability of DCA to counteract drug resistance induces by HIF-1α and the benefit of combining a metabolic modulator with another anti-cancer agent.

DCA is also studied in combination with other metabolic agents. For instance, the combination of DCA with metformin, a metabolic drug that inhibits mitochondrial respiration, shows promising results in anti-cancer therapies. Metformin suppresses tumor growth in different cancers via the inhibition of complex 1 of the electron transport chain and therefore enhance ROS production and apoptosis induction (Ward et al. 2017). In return, metformin is leading to an excessive lactate production and glucose consumption. Therefore, in order to reduce this side effect of metformin, several studies sought to evaluate the potential benefit of combining metformin and DCA. Different studies on glioblastoma, ovarian cancer and breast cancer have shown that DCA enhanced cytotoxicity of metformin via the reduction of metformin-mediated lactate accumulation. Indeed, it has been demonstrated that DCA reversed metformin-induced glycolytic metabolism and that this co-treatment synergistically reduced tumor growth (Haugrud et al. 2014; Li et al. 2016; Ward et al. 2017). Another recent study combined DCA with arginase in auxotroph tumors to arginine (Verma et al. 2019). As explained earlier, some tumors are dependent on external sources of arginine. Arginase has been found to inhibit the growth of arginine auxotroph tumor. Verma et al. observed that this drug combination synergistically reduced tumor growth *in vitro* and *in vivo*. They observed an increase in genes involved in cell cycle and p53 signaling. Investigation is performed to understand the underlining mechanism of action leading to this synergy (Verma et al. 2019).

Using a similar approach, Bourdeau et al. studied the modulation of aerobic glycolysis via the inhibition of the enzyme that converts pyruvate into lactate (LDHA). They observed that some cancer cells were resistant to this inhibition or become rapidly resistant. They found that the resistance was due to a reduction of glycolysis and to an increase of OXPHOS. Therefore, they combined the LDHA inhibitor (GNE-140) with an OXPHOS inhibitor (phenformin) and observed that this combination re-sensitized the cell to GNE-140. Of note, this combination not only potentiated the effect of LDHA inhibitors but also prevented the emergence of cell resistance to LDHA inhibition (Boudreau et al. 2016).

These illustrative examples of combined therapies are mainly based on modulation of glycolysis or OXPHOS. Nevertheless, as described before, cancer cells can rely on other sources of energy and easily switch from one substrate to another due the metabolic plasticity. For instance, Imbert et al. have shown that targeting amino acid fluxes with pH regulators provides a promising therapeutic strategy (Imbert et al. 2018).

In the future, personalized medicine will benefit from specific biomarkers that will identify which sources of energy are used by cancer cells. Moreover, these biomarkers will tackle potential metabolic adaptation in the course of treatment. Overall, this will help to define the ideal combination of drugs that target metabolism in order to efficiently fight against cancer.

References

Agarwal R, Kaye SB (2003) Ovarian cancer: strategies for overcoming resistance to chemotherapy. Nat Rev Cancer 3(7):502–516. https://doi.org/10.1038/nrc1123

Ahn CS, Metallo CM (2015) Mitochondria as biosynthetic factories for cancer proliferation. Cancer Metab 3(1):1. https://doi.org/10.1186/s40170-015-0128-2

Altman BJ, Stine ZE, Dang CV (2016) From Krebs to clinic: glutamine metabolism to cancer therapy. Nat Rev Cancer 16(11):749. https://doi.org/10.1038/nrc.2016.114

Amelio I, Cutruzzola F, Antonov A, Agostini M, Melino G (2014) Serine and glycine metabolism in cancer. Trends Biochem Sci 39(4):191–198. https://doi.org/10.1016/j.tibs.2014.02.004

Bayer C, Vaupel P (2012) Acute versus chronic hypoxia in tumors: controversial data concerning time frames and biological consequences. Strahlenther Onkol 188(7):616–627. https://doi.org/10.1007/s00066-012-0085-4

Bhutia YD, Babu E, Ramachandran S, Ganapathy V (2015) Amino acid transporters in cancer and their relevance to "glutamine addiction": novel targets for the design of a new class of anticancer drugs. Cancer Res 75(9):1782–1788. https://doi.org/10.1158/0008-5472.CAN-14-3745

Biancur DE, Paulo JA, Malachowska B, Quiles Del Rey M, Sousa CM, Wang X, Sohn ASW, Chu GC, Gygi SP, Harper JW, Fendler W, Mancias JD, Kimmelman AC (2017) Compensatory metabolic networks in pancreatic cancers upon perturbation of glutamine metabolism. Nat Commun 8:15965. https://doi.org/10.1038/ncomms15965

Bonnet S, Archer SL, Allalunis-Turner J, Haromy A, Beaulieu C, Thompson R, Lee CT, Lopaschuk GD, Puttagunta L, Bonnet S, Harry G, Hashimoto K, Porter CJ, Andrade MA, Thebaud B, Michelakis ED (2007) A mitochondria-K+ channel axis is suppressed in cancer and its normalization promotes apoptosis and inhibits cancer growth. Cancer Cell 11(1):37–51. https://doi.org/10.1016/j.ccr.2006.10.020

Boroughs LK, DeBerardinis RJ (2015) Metabolic pathways promoting cancer cell survival and growth. Nat Cell Biol 17(4):351–359. https://doi.org/10.1038/ncb3124

Boudreau A, Purkey HE, Hitz A, Robarge K, Peterson D, Labadie S, Kwong M, Hong R, Gao M, Del Nagro C, Pusapati R, Ma S, Salphati L, Pang J, Zhou A, Lai T, Li Y, Chen Z, Wei B, Yen I, Sideris S, McCleland M, Firestein R, Corson L, Vanderbilt A, Williams S, Daemen A, Belvin M, Eigenbrot C, Jackson PK, Malek S, Hatzivassiliou G, Sampath D, Evangelista M, O'Brien T (2016) Metabolic plasticity underpins innate and acquired resistance to LDHA inhibition. Nat Chem Biol 12(10):779–786. https://doi.org/10.1038/nchembio.2143

Cairns RA, Harris IS, Mak TW (2011) Regulation of cancer cell metabolism. Nat Rev Cancer 11(2):85–95. https://doi.org/10.1038/nrc2981

Chen JL, Lucas JE, Schroeder T, Mori S, Wu J, Nevins J, Dewhirst M, West M, Chi JT (2008) The genomic analysis of lactic acidosis and acidosis response in human cancers. PLoS Genet 4(12):e1000293. https://doi.org/10.1371/journal.pgen.1000293

Corbet C, Feron O (2017a) Cancer cell metabolism and mitochondria: nutrient plasticity for TCA cycle fueling. Biochim Biophys Acta Rev Cancer 1868(1):7–15. https://doi.org/10.1016/j.bbcan.2017.01.002

Corbet C, Feron O (2017b) Tumour acidosis: from the passenger to the driver's seat. Nat Rev Cancer 17(10):577–593. https://doi.org/10.1038/nrc.2017.77
Corbet C, Draoui N, Polet F, Pinto A, Drozak X, Riant O, Feron O (2014) The SIRT1/HIF2alpha axis drives reductive glutamine metabolism under chronic acidosis and alters tumor response to therapy. Cancer Res 74(19):5507–5519. https://doi.org/10.1158/0008-5472.CAN-14-0705
Corbet C, Pinto A, Martherus R, Santiago de Jesus JP, Polet F, Feron O (2016) Acidosis drives the reprogramming of fatty acid metabolism in cancer cells through changes in mitochondrial and histone acetylation. Cell Metab 24(2):311–323. https://doi.org/10.1016/j.cmet.2016.07.003
De Bock K, Georgiadou M, Schoors S, Kuchnio A, Wong BW, Cantelmo AR, Quaegebeur A, Ghesquiere B, Cauwenberghs S, Eelen G, Phng LK, Betz I, Tembuyser B, Brepoels K, Welti J, Geudens I, Segura I, Cruys B, Bifari F, Decimo I, Blanco R, Wyns S, Vangindertael J, Rocha S, Collins RT, Munck S, Daelemans D, Imamura H, Devlieger R, Rider M, Van Veldhoven PP, Schuit F, Bartrons R, Hofkens J, Fraisl P, Telang S, Deberardinis RJ, Schoonjans L, Vinckier S, Chesney J, Gerhardt H, Dewerchin M, Carmeliet P (2013) Role of PFKFB3-driven glycolysis in vessel sprouting. Cell 154(3):651–663. https://doi.org/10.1016/j.cell.2013.06.037
De Preter G, Neveu MA, Danhier P, Brisson L, Payen VL, Porporato PE, Jordan BF, Sonveaux P, Gallez B (2016) Inhibition of the pentose phosphate pathway by dichloroacetate unravels a missing link between aerobic glycolysis and cancer cell proliferation. Oncotarget 7(3):2910–2920. https://doi.org/10.18632/oncotarget.6272
DeBerardinis RJ, Chandel NS (2016) Fundamentals of cancer metabolism. Sci Adv 2(5):e1600200. https://doi.org/10.1126/sciadv.1600200
DeBerardinis RJ, Mancuso A, Daikhin E, Nissim I, Yudkoff M, Wehrli S, Thompson CB (2007) Beyond aerobic glycolysis: transformed cells can engage in glutamine metabolism that exceeds the requirement for protein and nucleotide synthesis. Proc Natl Acad Sci U S A 104(49):19345–19350. https://doi.org/10.1073/pnas.0709747104
Eales KL, Hollinshead KE, Tennant DA (2016) Hypoxia and metabolic adaptation of cancer cells. Oncogene 5:e190. https://doi.org/10.1038/oncsis.2015.50
Fan J, Kamphorst JJ, Mathew R, Chung MK, White E, Shlomi T, Rabinowitz JD (2013) Glutamine-driven oxidative phosphorylation is a major ATP source in transformed mammalian cells in both normoxia and hypoxia. Mol Syst Biol 9:712. https://doi.org/10.1038/msb.2013.65
Frezza C (2016) Cancer metabolism: addicted to serine. Nat Chem Biol 12(6):389–390. https://doi.org/10.1038/nchembio.2086
Fuchs BC, Bode BP (2005) Amino acid transporters ASCT2 and LAT1 in cancer: partners in crime? Semin Cancer Biol 15(4):254–266. https://doi.org/10.1016/j.semcancer.2005.04.005
Geck RC, Toker A (2016) Nonessential amino acid metabolism in breast cancer. Adv Biol Regul 62:11–17. https://doi.org/10.1016/j.jbior.2016.01.001
Han CY, Patten DA, Richardson RB, Harper ME, Tsang BK (2018) Tumor metabolism regulating chemosensitivity in ovarian cancer. Genes Cancer 9(5–6):155–175. https://doi.org/10.18632/genesandcancer.176
Haugrud AB, Zhuang Y, Coppock JD, Miskimins WK (2014) Dichloroacetate enhances apoptotic cell death via oxidative damage and attenuates lactate production in metformin-treated breast cancer cells. Breast Cancer Res Treat 147(3):539–550. https://doi.org/10.1007/s10549-014-3128-y
Hosios AM, Hecht VC, Danai LV, Johnson MO, Rathmell JC, Steinhauser ML, Manalis SR, Vander Heiden MG (2016) Amino acids rather than glucose account for the majority of cell mass in proliferating mammalian cells. Dev Cell 36(5):540–549. https://doi.org/10.1016/j.devcel.2016.02.012
Imbert V, Nebout M, Mary D, Endou H, Wempe MF, Supuran CT, Winum JY, Peyron JF (2018) Co-targeting intracellular pH and essential amino acid uptake cooperates to induce cell death of T-ALL/LL cells. Leuk Lymphoma 59(2):460–468. https://doi.org/10.1080/10428194.2017.1339875
Jones W, Bianchi K (2015) Aerobic glycolysis: beyond proliferation. Front Immunol 6:227. https://doi.org/10.3389/fimmu.2015.00227
Kamphorst JJ, Cross JR, Fan J, de Stanchina E, Mathew R, White EP, Thompson CB, Rabinowitz JD (2013) Hypoxic and Ras-transformed cells support growth by scavenging unsaturated fatty acids from lysophospholipids. Proc Natl Acad Sci U S A 110(22):8882–8887. https://doi.org/10.1073/pnas.1307237110
Kamphorst JJ, Chung MK, Fan J, Rabinowitz JD (2014) Quantitative analysis of acetyl-CoA production in hypoxic cancer cells reveals substantial contribution from acetate. Cancer Metab 2:23. https://doi.org/10.1186/2049-3002-2-23
Kamphorst JJ, Nofal M, Commisso C, Hackett SR, Lu W, Grabocka E, Vander Heiden MG, Miller G, Drebin JA, Bar-Sagi D, Thompson CB, Rabinowitz JD (2015) Human pancreatic cancer tumors are nutrient poor and tumor cells actively scavenge extracellular protein. Cancer Res 75(3):544–553. https://doi.org/10.1158/0008-5472.CAN-14-2211
Kankotia S, Stacpoole PW (2014) Dichloroacetate and cancer: new home for an orphan drug? Biochim Biophys Acta 1846(2):617–629. https://doi.org/10.1016/j.bbcan.2014.08.005
Kim J, Guan KL (2019) mTOR as a central hub of nutrient signalling and cell growth. Nat Cell Biol 21(1):63–71. https://doi.org/10.1038/s41556-018-0205-1

Kondo A, Yamamoto S, Nakaki R, Shimamura T, Hamakubo T, Sakai J, Kodama T, Yoshida T, Aburatani H, Osawa T (2017) Extracellular acidic pH activates the sterol regulatory element-binding protein 2 to promote tumor progression. Cell Rep 18(9):2228–2242. https://doi.org/10.1016/j.celrep.2017.02.006

Kory N, Wyant GA, Prakash G, Uit de Bos J, Bottanelli F, Pacold ME, Chan SH, Lewis CA, Wang T, Keys HR, Guo YE, Sabatini DM (2018) SFXN1 is a mitochondrial serine transporter required for one-carbon metabolism. Science 362(6416):eaat9528. https://doi.org/10.1126/science.aat9528

Krall AS, Xu S, Graeber TG, Braas D, Christofk HR (2016) Asparagine promotes cancer cell proliferation through use as an amino acid exchange factor. Nat Commun 7:11457. https://doi.org/10.1038/ncomms11457

Kumar K, Wigfield S, Gee HE, Devlin CM, Singleton D, Li JL, Buffa F, Huffman M, Sinn AL, Silver J, Turley H, Leek R, Harris AL, Ivan M (2013) Dichloroacetate reverses the hypoxic adaptation to bevacizumab and enhances its antitumor effects in mouse xenografts. J Mol Med (Berl) 91(6):749–758. https://doi.org/10.1007/s00109-013-0996-2

Lamonte G, Tang X, Chen JL, Wu J, Ding CK, Keenan MM, Sangokoya C, Kung HN, Ilkayeva O, Boros LG, Newgard CB, Chi JT (2013) Acidosis induces reprogramming of cellular metabolism to mitigate oxidative stress. Cancer Metab 1(1):23. https://doi.org/10.1186/2049-3002-1-23

Laurenti G, Tennant DA (2016) Isocitrate dehydrogenase (IDH), succinate dehydrogenase (SDH), fumarate hydratase (FH): three players for one phenotype in cancer? Biochem Soc Trans 44(4):1111–1116. https://doi.org/10.1042/BST20160099

Le A, Lane AN, Hamaker M, Bose S, Gouw A, Barbi J, Tsukamoto T, Rojas CJ, Slusher BS, Zhang H, Zimmerman LJ, Liebler DC, Slebos RJ, Lorkiewicz PK, Higashi RM, Fan TW, Dang CV (2012) Glucose-independent glutamine metabolism via TCA cycling for proliferation and survival in B cells. Cell Metab 15(1):110–121. https://doi.org/10.1016/j.cmet.2011.12.009

Li B, Li X, Ni Z, Zhang Y, Zeng Y, Yan X, Huang Y, He J, Lyu X, Wu Y, Wang Y, Zheng Y, He F (2016) Dichloroacetate and metformin synergistically suppress the growth of ovarian cancer cells. Oncotarget 7(37):59458–59470. https://doi.org/10.18632/oncotarget.10694

Lu CW, Lin SC, Chen KF, Lai YY, Tsai SJ (2008) Induction of pyruvate dehydrogenase kinase-3 by hypoxia-inducible factor-1 promotes metabolic switch and drug resistance. J Biol Chem 283(42):28106–28114. https://doi.org/10.1074/jbc.M803508200

Lukey MJ, Katt WP, Cerione RA (2017) Targeting amino acid metabolism for cancer therapy. Drug Discov Today 22(5):796–804. https://doi.org/10.1016/j.drudis.2016.12.003

Lyssiotis CA, Kimmelman AC (2017) Metabolic interactions in the tumor microenvironment. Trends Cell Biol 27(11):863–875. https://doi.org/10.1016/j.tcb.2017.06.003

McGuirk S, Gravel SP, Deblois G, Papadopoli DJ, Faubert B, Wegner A, Hiller K, Avizonis D, Akavia UD, Jones RG, Giguere V, St-Pierre J (2013) PGC-1alpha supports glutamine metabolism in breast cancer. Cancer Metab 1(1):22. https://doi.org/10.1186/2049-3002-1-22

Metallo CM, Gameiro PA, Bell EL, Mattaini KR, Yang J, Hiller K, Jewell CM, Johnson ZR, Irvine DJ, Guarente L, Kelleher JK, Vander Heiden MG, Iliopoulos O, Stephanopoulos G (2011) Reductive glutamine metabolism by IDH1 mediates lipogenesis under hypoxia. Nature 481(7381):380–384. https://doi.org/10.1038/nature10602

Meyer KA, Neeley CK, Baker NA, Washabaugh AR, Flesher CG, Nelson BS, Frankel TL, Lumeng CN, Lyssiotis CA, Wynn ML, Rhim AD, O'Rourke RW (2016) Adipocytes promote pancreatic cancer cell proliferation via glutamine transfer. Biochem Biophys Rep 7:144–149. https://doi.org/10.1016/j.bbrep.2016.06.004

Mullen AR, Wheaton WW, Jin ES, Chen PH, Sullivan LB, Cheng T, Yang Y, Linehan WM, Chandel NS, DeBerardinis RJ (2011) Reductive carboxylation supports growth in tumour cells with defective mitochondria. Nature 481(7381):385–388. https://doi.org/10.1038/nature10642

Nakazawa MS, Keith B, Simon MC (2016) Oxygen availability and metabolic adaptations. Nat Rev Cancer 16(10):663–673. https://doi.org/10.1038/nrc.2016.84

Obre E, Rossignol R (2015) Emerging concepts in bioenergetics and cancer research: metabolic flexibility, coupling, symbiosis, switch, oxidative tumors, metabolic remodeling, signaling and bioenergetic therapy. Int J Biochem Cell Biol 59:167–181. https://doi.org/10.1016/j.biocel.2014.12.008

Palm W, Park Y, Wright K, Pavlova NN, Tuveson DA, Thompson CB (2015) The utilization of extracellular proteins as nutrients is suppressed by mTORC1. Cell 162(2):259–270. https://doi.org/10.1016/j.cell.2015.06.017

Pavlova NN, Thompson CB (2016) The emerging hallmarks of cancer metabolism. Cell Metab 23(1):27–47. https://doi.org/10.1016/j.cmet.2015.12.006

Polet F, Corbet C, Pinto A, Rubio LI, Martherus R, Bol V, Drozak X, Gregoire V, Riant O, Feron O (2016) Reducing the serine availability complements the inhibition of the glutamine metabolism to block leukemia cell growth. Oncotarget 7(2):1765–1776. https://doi.org/10.18632/oncotarget.6426

Pusapati RV, Daemen A, Wilson C, Sandoval W, Gao M, Haley B, Baudy AR, Hatzivassiliou G, Evangelista M, Settleman J (2016) mTORC1-dependent metabolic reprogramming underlies escape from glycolysis addiction in cancer cells. Cancer Cell 29(4):548–562. https://doi.org/10.1016/j.ccell.2016.02.018

Rabanal-Ruiz Y, Otten EG, Korolchuk VI (2017) mTORC1 as the main gateway to autophagy. Essays Biochem 61(6):565–584. https://doi.org/10.1042/EBC20170027

San-Millan I, Brooks GA (2017) Reexamining cancer metabolism: lactate production for carcinogenesis could be the purpose and explanation of the Warburg Effect. Carcinogenesis 38(2):119–133. https://doi.org/10.1093/carcin/bgw127

Schug ZT, Peck B, Jones DT, Zhang Q, Grosskurth S, Alam IS, Goodwin LM, Smethurst E, Mason S, Blyth K, McGarry L, James D, Shanks E, Kalna G, Saunders RE, Jiang M, Howell M, Lassailly F, Thin MZ, Spencer-Dene B, Stamp G, van den Broek NJ, Mackay G, Bulusu V, Kamphorst JJ, Tardito S, Strachan D, Harris AL, Aboagye EO, Critchlow SE, Wakelam MJ, Schulze A, Gottlieb E (2015) Acetyl-CoA synthetase 2 promotes acetate utilization and maintains cancer cell growth under metabolic stress. Cancer Cell 27(1):57–71. https://doi.org/10.1016/j.ccell.2014.12.002

Senthebane DA, Rowe A, Thomford NE, Shipanga H, Munro D, Mazeedi M, Almazyadi HAM, Kallmeyer K, Dandara C, Pepper MS, Parker MI, Dzobo K (2017) The role of tumor microenvironment in chemoresistance: to survive, keep your enemies closer. Int J Mol Sci 18(7):pii: E1586. https://doi.org/10.3390/ijms18071586

Sonveaux P, Vegran F, Schroeder T, Wergin MC, Verrax J, Rabbani ZN, De Saedeleer CJ, Kennedy KM, Diepart C, Jordan BF, Kelley MJ, Gallez B, Wahl ML, Feron O, Dewhirst MW (2008) Targeting lactate-fueled respiration selectively kills hypoxic tumor cells in mice. J Clin Invest 118(12):3930–3942. https://doi.org/10.1172/JCI36843

Sousa CM, Biancur DE, Wang X, Halbrook CJ, Sherman MH, Zhang L, Kremer D, Hwang RF, Witkiewicz AK, Ying H, Asara JM, Evans RM, Cantley LC, Lyssiotis CA, Kimmelman AC (2016) Pancreatic stellate cells support tumour metabolism through autophagic alanine secretion. Nature 536(7617):479–483. https://doi.org/10.1038/nature19084

Sullivan LB, Gui DY, Hosios AM, Bush LN, Freinkman E, Vander Heiden MG (2015) Supporting aspartate biosynthesis is an essential function of respiration in proliferating cells. Cell 162(3):552–563. https://doi.org/10.1016/j.cell.2015.07.017

Taddei ML, Giannoni E, Comito G, Chiarugi P (2013) Microenvironment and tumor cell plasticity: an easy way out. Cancer Lett 341(1):80–96. https://doi.org/10.1016/j.canlet.2013.01.042

Tajan M, Vousden KH (2016) The quid pro quo of the tumor/stromal interaction. Cell Metab 24(5):645–646. https://doi.org/10.1016/j.cmet.2016.10.017

Tang X, Lucas JE, Chen JL, LaMonte G, Wu J, Wang MC, Koumenis C, Chi JT (2012) Functional interaction between responses to lactic acidosis and hypoxia regulates genomic transcriptional outputs. Cancer Res 72(2):491–502. https://doi.org/10.1158/0008-5472.CAN-11-2076

Tardito S, Oudin A, Ahmed SU, Fack F, Keunen O, Zheng L, Miletic H, Sakariassen PO, Weinstock A, Wagner A, Lindsay SL, Hock AK, Barnett SC, Ruppin E, Morkve SH, Lund-Johansen M, Chalmers AJ, Bjerkvig R, Niclou SP, Gottlieb E (2015) Glutamine synthetase activity fuels nucleotide biosynthesis and supports growth of glutamine-restricted glioblastoma. Nat Cell Biol 17(12):1556–1568. https://doi.org/10.1038/ncb3272

Verma A, Lam YM, Leung YC, Hu X, Chen X, Cheung E, Tam KY (2019) Combined use of arginase and dichloroacetate exhibits anti-proliferative effects in triple negative breast cancer cells. J Pharm Pharmacol 71(3):306–315. https://doi.org/10.1111/jphp.13033

Warburg O (1956a) On respiratory impairment in cancer cells. Science 124(3215):269–270

Warburg O (1956b) On the origin of cancer cells. Science 123(3191):309–314

Ward NP, Poff AM, Koutnik AP, D'Agostino DP (2017) Complex I inhibition augments dichloroacetate cytotoxicity through enhancing oxidative stress in VM-M3 glioblastoma cells. PLoS One 12(6):e0180061. https://doi.org/10.1371/journal.pone.0180061

Weinberg F, Hamanaka R, Wheaton WW, Weinberg S, Joseph J, Lopez M, Kalyanaraman B, Mutlu GM, Budinger GR, Chandel NS (2010) Mitochondrial metabolism and ROS generation are essential for Kras-mediated tumorigenicity. Proc Natl Acad Sci U S A 107(19):8788–8793. https://doi.org/10.1073/pnas.1003428107

Wise DR, Ward PS, Shay JE, Cross JR, Gruber JJ, Sachdeva UM, Platt JM, DeMatteo RG, Simon MC, Thompson CB (2011) Hypoxia promotes isocitrate dehydrogenase-dependent carboxylation of alpha-ketoglutarate to citrate to support cell growth and viability. Proc Natl Acad Sci U S A 108(49):19611–19616. https://doi.org/10.1073/pnas.1117773108

Xuan Y, Hur H, Ham IH, Yun J, Lee JY, Shim W, Kim YB, Lee G, Han SU, Cho YK (2014) Dichloroacetate attenuates hypoxia-induced resistance to 5-fluorouracil in gastric cancer through the regulation of glucose metabolism. Exp Cell Res 321(2):219–230. https://doi.org/10.1016/j.yexcr.2013.12.009

Yang M, Vousden KH (2016) Serine and one-carbon metabolism in cancer. Nat Rev Cancer 16(10):650–662. https://doi.org/10.1038/nrc.2016.81

Yang L, Achreja A, Yeung TL, Mangala LS, Jiang D, Han C, Baddour J, Marini JC, Ni J, Nakahara R, Wahlig S, Chiba L, Kim SH, Morse J, Pradeep S, Nagaraja AS, Haemmerle M, Kyunghee N, Derichsweiler M, Plackemeier T, Mercado-Uribe I, Lopez-Berestein G, Moss T, Ram PT, Liu J, Lu X, Mok SC, Sood AK, Nagrath D (2016) Targeting stromal glutamine synthetase in tumors disrupts tumor microenvironment-regulated cancer cell growth. Cell Metab 24(5):685–700. https://doi.org/10.1016/j.cmet.2016.10.011

Ye J, Fan J, Venneti S, Wan YW, Pawel BR, Zhang J, Finley LW, Lu C, Lindsten T, Cross JR, Qing G, Liu

Z, Simon MC, Rabinowitz JD, Thompson CB (2014) Serine catabolism regulates mitochondrial redox control during hypoxia. Cancer Discov 4(12):1406–1417. https://doi.org/10.1158/2159-8290.CD-14-0250

Yuneva M, Zamboni N, Oefner P, Sachidanandam R, Lazebnik Y (2007) Deficiency in glutamine but not glucose induces MYC-dependent apoptosis in human cells. J Cell Biol 178(1):93–105. https://doi.org/10.1083/jcb.200703099

Zhang J, Ahn WS, Gameiro PA, Keibler MA, Zhang Z, Stephanopoulos G (2014) 13C isotope-assisted methods for quantifying glutamine metabolism in cancer cells. Methods Enzymol 542:369–389. https://doi.org/10.1016/B978-0-12-416618-9.00019-4

Zhang J, Pavlova NN, Thompson CB (2017) Cancer cell metabolism: the essential role of the nonessential amino acid, glutamine. EMBO J 36(10):1302–1315. https://doi.org/10.15252/embj.201696151

Multifaceted Oncogenic Role of Adipocytes in the Tumour Microenvironment

7

Yannasittha Jiramongkol and Eric W.-F. Lam

Abstract

Obesity has for decades been recognised as one of the major health concerns. Recently accumulated evidence has established that obesity or being overweight is strongly linked to an increased risk of cancer. However, it is still not completely clear how adipose tissue (fat), along with other stromal connective tissues and cells, contribute to tumour initiation and progression. In the tumour microenvironment, the adipose tissue cells, in particular the adipocytes, secrete a number of adipokines, including growth factors, hormones, collagens, fatty acids, and other metabolites as well as extracellular vesicles to shape and condition the tumour and its microenvironment. In fact, the adipocytes, through releasing these factors and materials, can directly and indirectly facilitate cancer cell proliferation, apoptosis, metabolism, angiogenesis, metastasis and even chemotherapy resistance. In this chapter, the multidimensional role played by adipocytes, a major and functional component of the adipose tissue, in promoting cancer development and progression within the tumour microenvironment will be discussed.

Y. Jiramongkol · E. W.-F. Lam (✉)
Department of Surgery and Cancer, Imperial College London, Hammersmith Hospital Campus, London, UK
e-mail: eric.lam@imperial.ac.uk

Keywords

Tumour microenvironment · Obesity · Adipocytes · Secretosomes · Fatty acids · Tumorigenesis and therapeutic resistance

7.1 Introduction

7.1.1 Obesity and Cancer

Changes in the environmental factors, diets, eating habits and daily lifestyles can cause an accumulation of adipose tissue in the body. This can ultimately lead to over-weight and the development of obesity, which has multiple detrimental health impacts. It is now evident that the accumulation of adipose tissue-fat, is strongly correlated with many diseases, including cardiovascular disease, type 2 diabetes, hypertension, dyslipidemia, liver disease and also cancer (Zhang and Scherer 2018; Amin et al. 2019). This has put obesity as one of the major health concerns, particularly with the steady rise in number of obese individuals (Calle and Kaaks 2004).

Obesity is defined as a body mass index (BMI) 30 kg/m^2 or above and is a pathological condition where there is excessive deposition of fat due to an imbalance between the dietary intake and energy output. The excessive energy is converted into lipids and stored primarily in the adipose tissue, ultimately leading to an increase in the mass

J. Serpa (ed.), *Tumor Microenvironment*, Advances in Experimental Medicine and Biology 1219,
https://doi.org/10.1007/978-3-030-34025-4_7

of the individual. This storage of lipids is an efficient way to put away energy for future use, because of their high caloric values. Lipids contain double the energy content (1 g = 38 Kj) as amino acids or glucose. Triglycerides (TG) are hydrophobic and insoluble in water, thus allowing the cell to store many of them as lipid droplets (Haczeyni et al. 2018). In fasting conditions, lipids are converted into energy by a process called lipolysis (Birsoy et al. 2013). The expansion of this energy-rich storage is predominantly due to an increase in volume (hypertrophy) of adipocytes (fat cells) rather than an increase in number (hyperplasia) of adipocytes (Sun et al. 2011). With increased levels of fat deposition, individuals with excessive body weight are known to have elevated risks of cancers, including head and neck (Wang et al. 2019) breast (Picon-Ruiz et al. 2017), prostate (Mistry et al. 2007), colon (Amemori et al. 2007), liver (Calle et al. 2003; Nair et al. 2002) and ovarian cancer (Olsen et al. 2013). Moreover, cancer survival outcomes are also influenced by obesity (Calle and Kaaks 2004). Although these observations are informative, the whole picture by which adipocytes in obese individuals contribute to cancer development is only beginning to unveil. In this chapter, we will discuss the potential mechanisms by which adipocytes can provide a favourable microenvironment for cancer initiation, progression and drug resistance.

7.1.2 The Tumour Microenvironment

Cancers are heterogeneous tissues made up of multiple components which include tumour cells and the stromal cells in the microenvironment. The stroma itself consists of connective tissues of different cell types, which function together to provide support for organs in our body (Bremnes et al. 2011). Recently, significant attention has been directed towards the interactions between the stromal and the cancer cells in the tumour microenvironment. The properties of stromal cells are known to be modified in cancer. On the other hand, these altered and sometimes deregulated stromal cells can also influence cancer progression in a positive feedforward mechanism (Hoy et al. 2017).

7.1.3 Adipose Tissue and Adipocytes

In our body, adipose tissues can broadly be classified into two main categories, the white adipose tissue (WAT) and the brown adipose tissue (BAT). The WAT is found predominantly at the subcutaneous, visceral organ and female mammary glands. Its main role is to store energy and regulate weight control. The BAT is found in supraclavicular regions and paracervical to control body temperature in response to the dietary intakes and the changes in the environmental temperature. In these adipose tissues, adipocytes are considered to be the major functional components (Duong et al. 2017) and account for over 20% of the adipose tissue cells (Suga et al. 2008). In humans, adipose tissues (fat) are found under the skin (subcutaneous fat), around internal organs (visceral fat), in bone marrow (yellow bone marrow), in muscles (muscle fat) and in the breast tissue (breast fat). This allows the adipocytes to be close to and crosstalk with many organs, such as the breast, prostate, colon and ovaries. For example, normal breast tissue is made up of mammary glands which are embedded in a stroma enriched with connective tissues (Hovey et al. 1999).

The stromal adipose tissues consist predominantly of adipocytes, but they also contain cells including preadipocytes, fibroblasts, vascular endothelial cells and immune cells, such as macrophages. These stromal adipocytes can contribute to the cancer cell development in a variety of ways (Bussard et al. 2016). Apart from being an energy provider, adipocytes also release into the tumour microenvironment various factors, such as adipokines, growth factors, hormones, collagens, fatty acids, extracellular vesicles and other metabolites, all of which can contribute to the cancer initiation, progression and therapeutic resistance (Park et al. 2011, 2013).

7.1.4 Cancer-Associated Adipocytes (CAAs)

In normal breast tissue, the stroma separates the mammary glands from the adipocytes. However, during tumour development, the breast tissue undergoes extracellular matrix remodelling, resulting in the adipocytes being in closer proximity to the mammary glands (Wang et al. 1975). Similar processes are also detected during the development of other solid tumours, including that of the ovarian and prostate cancer (Finley et al. 2009; Kristin et al. 2011). This close proximity between the adipocytes and the cancer cells has profound impacts on the adipocyte development (Dirat et al. 2011). In fact, adipocytes are transformed by proximal cancer cells into cancer-associated adipocytes (CAAs) to acquire an activated phenotype that contributes to cancer invasion and progression (Dirat et al. 2011). For instance, adipocytes cultivated with breast cancer cells exhibit an activated phenotype characterized by the overexpression of proteases, such as matrix metalloproteinase (MMP)-11, and proinflammatory cytokines, including interleukin (IL)-6 and IL-1β. In agreement, histological studies also showed that adipocytes situated close to the larger tumours and/or with enhanced local invasion express higher levels of the proinflammatory cytokine IL-6 (Dirat et al. 2011). Another study using an ovarian cancer and adipocyte co-culture system as well as a mouse model also showed that adipocytes promote homing, migration and invasion of ovarian cancer cells, through overexpressing adipokines, including IL-8 (Kristin et al. 2011). The study also revealed that co-culturing with adipocytes induce the ovarian cancer cells to express the fatty acid transporter FABP4, which has a key role in promoting ovarian cancer metastasis (Kristin et al. 2011).

To date, many studies have focused on determining the role of adipose tissues, in particular adipocytes, in cancer initiation and progression. However, this was only started recently in 1992, when co-transplantation studies using murine models showed that mammary carcinoma cells grow better with fat fragments. This finding suggests that adipose tissue plays an integral part in cancer development (Elliott et al. 1992). Later in 2003, another study revealed that only mature adipocytes could facilitate tumour growth in estrogen receptor positive (ER+) breast cancer cell lines using a collagen gel matrix culture system. This differentiates the mature adipocytes from pre-adipocytes in their contributions to the cancer progression, and highlights adipocytes as a key cancer promoting component in the tumour microenvironment (Manabe et al. 2003). Consistent with this finding, a further study in 2011 demonstrated that human adipocytes can promote the growth of ovarian cancer cells both *in vivo* and *in vitro* (Kristin et al. 2011). This finding narrows down further the mechanisms by which adipose tissue affects cancer growth. Likewise, similar observations were documented in the colon and prostate cancer and demonstrated a role for adipocytes (Tokuda et al. 2003; Aoki et al. 2007). Furthermore, another study also showed that CAAs contribute to breast cancer invasion (Dirat et al. 2011). However, recently in 2015, an *in vivo* study using breast tumours revealed that the estrogen receptor negative (ER-) breast tumours in close proximity to the adipose tissues have a low mitotic index (Han Suk et al. 2015). This observation is at odds with the results from other findings which suggested that CAAs contribute to cancer progression (Dirat et al. 2011; Kristin et al. 2011; Tokuda et al. 2003; Aoki et al. 2007). Nevertheless, the general role of the adipocytes in the cancer progression can be influenced by the tumour-type and other factors in the microenvironment.

These added layers of complexity indicate that despite plenty of evidence showing that tumour growth can be promoted by the presence of adipocytes, the complete picture of the relationships between adipocytes and cancer cells in the tumour is only beginning to be revealed.

7.2 Role of Adipocytes and Their Secreted Factors in Cancer Development

For years, adipocytes have been thought of as passive energy storage depots. However, subsequent research has revealed that adipocytes also act as the sources for endocrine and paracrine factors (Darimont et al. 1994; Amri et al. 1994). These secreted factors are also known as adipokines and they consist of cytokines, chemokines, hormones and other growth factors. Other adipocyte-derived materials include fatty acids and other metabolites, which play important parts in facilitate metabolic crosstalks between adipocytes and the cancer cells (Fig. 7.1). These factors function both locally and systemically to play distinct roles in the cancer proliferation, growth, invasion, angiogenesis and metabolism as well as therapeutic resistance (Duong et al. 2017).

7.2.1 Leptin

Leptin is a hormone that helps to control energy intake and the body weight (Zhang et al. 2005). It is produced mainly by the adipocytes, released into the bloodstream, and received by the hypothalamus in the brain. High leptin levels cause a reduction in energy/food intake, while low levels stimulate an increase in energy/food intake and fat storage. In obese individuals, the secretion of leptin is elevated due to the increase in adipose tissues, but the brain becomes insensitive to high leptin levels. This high levels of leptin have also been found to be accompanied by the overexpression of leptin receptors in many cancer cells, such as breast and ovarian cancer (Ishikawa et al. 2004; Uddin et al. 2009).

Leptin has also been shown to function as a growth factor for cancer cells (Endo et al. 2011). Consistent with this idea, the elevated levels of leptin have been found to promote cancer cell proliferation through the activation of the extracellular signal-regulated kinase1/2 (ERK1/2) and c-Jun N-terminal kinase (JNK) signalling pathways (Garofalo and Surmacz 2006). This concept is further supported by three-dimensional collagen gel co-culture studies using colon cancer cells and adipocytes from leptin-deficient mice, showing that the trophic effects induced by adipocytes are abolished in leptin-deleted mice (Aoki et al. 2007). In concordance, the mammary and colorectal tumour growth is retarded in obese mice-deficient for leptin or its receptor (Endo et al. 2011; Qiao et al. 2011). Similarly, the prostate tumours induced in leptin receptor-deficient mice are significantly smaller (Ribeiro et al. 2010). However, the prostate tumours induced in the leptin-deleted mice were significantly larger.

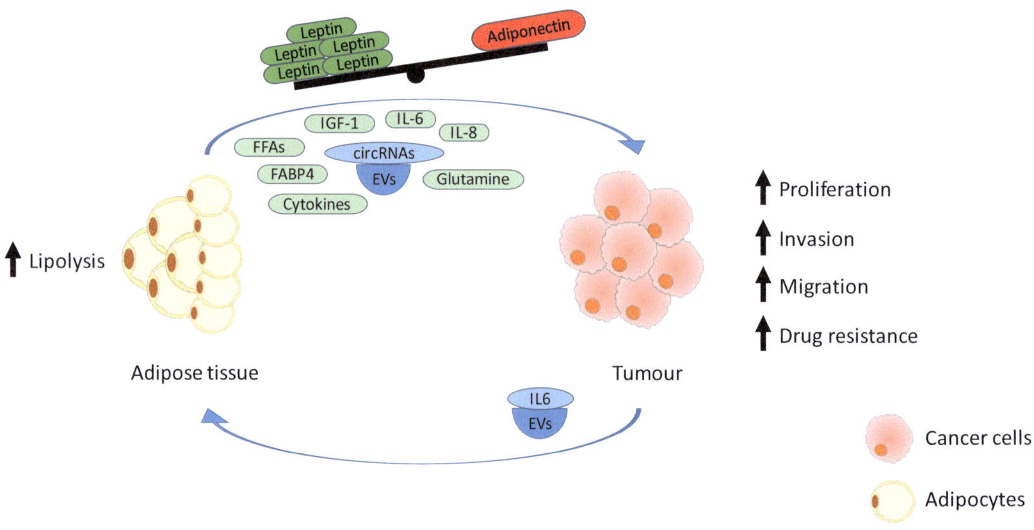

Fig. 7.1 Crosstalks between adipocytes and cancer cells Adipocytes secretes various adipokines and other factors to promote cancer progression. Specifically, the ratio of leptin and adiponectin plays an important role to the survival of cancer cells. At the same time, cancer cells drive adipocytes lipolysis to generate free fatty acids for themselves

This discrepancy may be attributed to the differences in the local tumour microenvironment and the fact that leptin is also a critical regulator of the development and activation of natural killer (NK) cells (Tian et al. 1959), which have the ability to detect and kill tumour cells (Wu and Lanier 2003). Nonetheless, in breast cancer cells, leptin has been shown to promote cancer cell proliferation in the tumour microenvironment via activating the phosphoinositide 3-kinase (PI3K)-Akt signalling pathway and pyruvate kinase M2 expression, which are important for cell proliferation and epithelial-mesenchymal transition (EMT) (Wei et al. 2016; Qiao et al. 2011). Consistently, the elevated proliferative effects of adipocytes on cancer cells are lowered when leptin expression is depleted using short hairpin (sh) RNA (Amy et al. 2015). Collectively, these findings show leptin produced by adipocytes play a critical role in stimulating cancer cell proliferation.

7.2.2 Adiponectin

Contrary to leptin, adiponectin is an adipokine whose plasma concentrations are significantly lower in obese individuals than in non-obese subjects (Arita et al. 1999). Exposure of cancer cells to adiponectin also causes the cancer cells to cease proliferation and undergo apoptosis, suggesting an anti-tumour role in adiponectin (Kang et al. 2005; Dieudonne et al. 2006; Ishikawa et al. 2007). In agreement, adiponectin depletion was shown to promote human breast cancer growth in nude mouse xenograft models, through activating the glycogen synthase kinase-3β/β-catenin signalling pathway (Wang et al. 2006). Subsequently, it was discovered that adiponectin also restricts cancer cell growth through activating the 5' AMP-activated protein kinase (AMPK) as well as inhibiting the AKT, ERK1/2, nuclear factor kappa-light-chain-enhancer of activated B cells (NF-kappaB) and Wnt pathways (Dalamaga et al. 2012). Interestingly, adipocytes from the tumour microenvironment are less differentiated and secrete significantly lower amounts of adiponectin than adipocytes from normal microenvironment (Fletcher et al. 2017). As both leptin and adiponectin are antagonistic adipokines secreted by the adipocytes, the ratio of these adipokines may influence the progression of the cancers. Indeed, the balance between leptin and adiponectin is often shifted in obese individuals (Al-Hamodi et al. 2014). These observations are further supported by clinical studies showing a positive correlation between leptin/adiponectin ratio and cancer risk (Ashizawa et al. 2010).

7.2.3 Insulin-Like Growth Factor 1 (IGF-1)

Another key cytokine secreted by adipocytes is insulin-like growth factor 1 (IGF-1). IGF-1 is strongly associated with cell proliferation and survival. As it is produced by adipocytes, the level of free IGF-1 is appropriately strongly associated with obesity (Nam et al. 1997). IGF-1 is an important regulator of energy metabolism and growth (Pollak 2008). The binding of the IGF-1 to its receptors activates primarily the PI3K-AKT and Mitogen-activated protein kinases (MAPK) pathways to promote cancer cell growth and progression (Pollak 2008). The concentration of circulating IGF-1 has also been linked to increased cancer risk (Shanmugalingam et al. 2016). Conversely, inhibition of IGF-1 receptor kinase activity limits the growth-promoting effect of adipocytes on cancer cells (D'Esposito et al. 2012). Thus, in summary, the secretion of IGF-1 by adipocytes has a direct role in stimulating the proliferation and survival of cancer cells in the tumour microenvironment and will have direct impact on cancer initiation and progression.

7.2.4 Vascular Endothelial Growth Factor (VEGF) and Cancer Angiogenesis

When a tumour grows beyond 1–2 mm^3, it requires extra blood vessels to supply nutrients, oxygen and growth factors for continuous proliferation and survival. In the absence of new blood vessel formation, the expanding tumours may become starved of oxygen (a condition termed hypoxia), nutrients and growth factors and undergo necrosis or even apoptosis (Muz et al. 2015). Thus, the formation of the new blood vessels, a phenomenon known as angiogenesis, is crucial for tumour expansion (Nishida et al. 2006). Angiogenesis is tightly regulated in the microenvironment by both the cancer and the stromal cells. The adipocytes can promote new vessels formation by secreting adipokines (Cao 2013). VEGF is a vital mediator of angiogenesis (the formation of new blood vessels) in cancer (Fig. 7.2) (Nishida et al. 2006), and it binds VEGF receptors which are expressed on the surface of vascular endothelial cells. In response to insulin, adipocytes secrete vascular endothelial growth factor A (VEGFA) to promote angiogenesis (Mick et al. 2002). Furthermore, leptin released by adipocytes also drives endothelial cell differentiation and proliferation (Gonzalez-Perez et al. 2010). Previous studies also showed that VEGF can also be upregulated by leptin in breast cancer cells through the transcription factors HIF-1 and NF-κB (Gonzalez-Perez et al. 2010). However, unlike leptin, the effects of adiponectin in angiogenesis are not clear-cut. Some studies showed adiponectin as a pro-angiogenesis factor, while other studies demonstrated that it has the opposite effect (Rei et al. 2004; Man et al. 2010).

7.2.5 Free Fatty Acids, Metabolites and the Warburg Effect

Cancer cells reprogramme their energy metabolism to promote their growth, survival, proliferation, and self-renewal. The hallmark of this altered cancer metabolism (also called Warburg effect) to yield the extra energy needed is to increase glucose uptake and generate energy through glucose to lactate fermentation in anaerobic glycolysis, even in the presence of sufficient oxygen (Warburg 1956). In other words, cancer cells prefer to produce adenosine triphosphate (ATP) via glycolysis instead of oxidative phosphorylation (OXPHO). The cancer-associated adipocytes have been shown to have increased secretion of inflammatory cytokines, such as TNFα, IL-6 and IL-1β, and MMP-11, which play

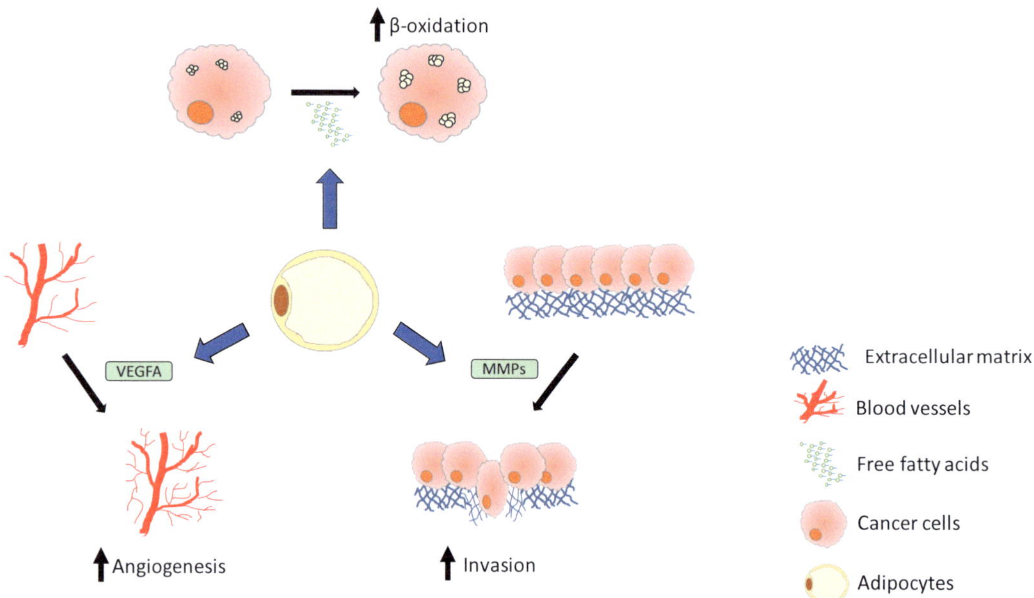

Fig. 7.2 Adipocytes and tumour microenvironment Cancer cells import adipocyte-secreted free fatty acids which resulting in larger lipid droplets and fat reservoir. In addition, adipocytes secrete VEGFA and MMPs to promote angiogenesis and cancer invasion respectively to enhance cancer progression

a key part in cancer energy metabolism by promoting metabolic switch in tumours (Ribeiro et al. 2012; Dali-Youcef et al. 2016; Dirat et al. 2011). In the tumour microenvironment, lactate and amino acids are the alternative external sources of energy available for cancer cells (Duong et al. 2017). Lately, the reverse Warburg effect has also been proposed, in which the cancer cells induce the CAFs to undergo aerobic glycolysis to produce the metabolic by-products, such as lactate and pyruvate, for the consumption of cancer cells. This induction also applies to the other stromal cells, including adipocytes, found in the tumour microenvironment (Pavlides et al. 2009). Specifically, under hypoxic conditions, lactate is secreted by these stromal cells through monocarboxylate transporters (MCTs) into the tumour microenvironment (Gonzalez-Perez et al. 2010). The secreted lactate is then imported and metabolised by the cancer cells to produce energy and other essential metabolites (Pavlides et al. 2009).

Nevertheless, many studies have now confirmed that free fatty acids (FFAs) are in fact the primary source of energy for cancer cells delivered from adipocytes. In terms of lipid and energy metabolism, the principal function of adipocytes is to store triglyceride and release fatty acids (FAs) for other cells and tissues when needed. *In vitro* and *in vivo* studies have shown that cancer cells can induce lipolysis and FA production in adipocytes. A recent study showed that CAAs have enhanced potentials to increase their lipid synthesis capacity and to breakdown the lipids by hydrolysis to release FAs, a process termed lipolysis (Balaban et al. 2017). Subsequently, the cancer cells will import the secreted FFAs from their microenvironment and use them for energy production or store them as triglycerides in lipid droplets (Young and Zechner 2013). Notably, the majority of the FFAs secreted by the adipocytes and imported by the cancer cells have been found to be long chain FAs, such as palmitic acid (Kwan et al. 2014). Interestingly, the FA transfer to cancer cells is further enhanced in "obese" adipocytes (FAs supplemented adipocytes) compared with normal adipocytes (Balaban et al. 2017). In addition to supplying the FAs, the adipocytes can stimulate the cancer cells to express higher levels of carnitine palmitoyltransferase 1A (CPT1A) and electron transport chain proteins to elevate their rates of fatty acid β-oxidation (FAO) (Fig. 7.2) (Balaban et al. 2017). Furthermore, the adipocytes also induce the cancer cells to release the stored FAs from their lipid droplets intracellularly through the adipose triglyceride lipase (ATGL)-dependent lipolytic pathway for FAO (Wang et al. 2017b).

In ovarian cancer, the uptake of FFAs is associated with an increased rate of FAO to generate a large amount of energy in the form of adenosine triphosphate (ATP) (Kristin et al. 2011). Similarly, FFAs imported have been shown to be used for FAO to generate energy to promote cell proliferation and migration in breast cancer (Balaban et al. 2017). However, recent breast cancer studies reported that the import of FFAs do not lead to the ATP production via FAO for cell proliferation or survival but cancer cell invasion (Wang et al. 2017b). In concordance with this, FFAs have been shown to enhance breast cancer cell migration through inducing the expression of plasminogen activator inhibitor-1 via SMAD4 (Byon et al. 2009). Moreover, a study following FA transfer from bone marrow adipocytes to metastatic prostate cancer cells also showed a shift in intracellular energy production from FAO to lactate production (Diedrich et al. 2016). The study also revealed that this shift to Warburg phenotype is mediated by the hypoxia-inducible factor 1α (HIF-1α) (Diedrich et al. 2016). Nevertheless, these findings collectively suggest that FFAs secreted by adipocytes into the tumour microenvironment could provide distinct effects under different conditions, but ultimately they all function to promote cancer progression and metabolism (Diedrich et al. 2016).

7.2.6 FABP4 (A-FABP)

Fatty acid binding protein 4 (FABP4) is a member of the FABP family of proteins involved in FA import, storage and export as well as cholesterol and phospholipid metabolism (Chmurzynska 2006; Furuhashi and Hotamisligil 2008). FABP4 expression itself is induced by high-fat diet (HFD) or obesity in adipocytes in the tumour microenvironment (Huang et al. 2017). Moreover, co-culture experiments also showed that adipocytes can also induce cancer cells to upregulate FABP4 expression (Kristin et al. 2011). FABP4 has many roles in cancer progression. FABP4 can enhance cancer progression by upregulating MMPs and stromal cell cytokine production (Huang et al. 2017). FABP4 also promotes cancer proliferation through inducing the expression of Forkhead box transcription M1 (FOXM1), a key transcription factor involved in driving cancer progression and drug resistance (Yao et al. 2018; Guaita-Esteruelas et al. 2017). The role of FABP4 in promoting cancer proliferation and migrating appears to be associated with its ability to facilitate FA metabolism in both adipocytes and cancer cells. A study showed that pharmaceutical inhibition of FABP4 in ovarian cancer cells lowered their lipid droplet accumulation and their metastatic and growth potentials. Likewise, FABP4-deletion in mice also impairs the metastatic tumour growth, indicating that FABP4 has a key role in cancer metastasis (Kristin et al. 2011). Moreover, FFAs secreted by adipocytes can elevate prostate cancer invasion, which can be reduced by pharmaceutical inhibition of the fatty acid transporter FABP4 (Herroon et al. 2013). A most recent study revealed that FABP4 also has a key role in mediating lipolysis (Hua et al. 2019). Besides FABP4, current research also showed that CAAs can also enhance lipid metabolism in breast cancer and melanoma cells by inducing the expression of another fatty acid transporter FATP1 (Lopes-Coelho et al. 2018; Zhang et al. 2018). Thus, in addition to being an energy source to fuel the Warburg effect in cancer, the adipocyte-derived FAs also function as a signalling molecule to drive the phenotypic changes in cancer to drive cancer progression within the tumour environment (Furuhashi and Hotamisligil 2008).

7.2.7 Matrix Metalloproteinases (MMPs)-Extracellular Matrix Remodelling

Apart from cancer growth and energy metabolism, adipocytes also support tumour progression by remodelling the extracellular matrix in the tumour microenvironment to facilitate cancer cell invasion and metastasis. Cancer cells can induce the adipocytes to secrete collagen VI which in turn promotes cancer cell survival in a paracrine positive feedback fashion (Petricoin Iii et al. 2005). Notably, a cleaved product of collagen VI, known as endotrophin, has also been shown to stimulate epithelial-mesenchymal transition (EMT) and cell metastasis (Park and Scherer 2012). To remodel the extracellular matrix, adipocytes also release degradation enzymes known as matrix metalloproteinases (MMPs) to facilitate cancer cell invasion and metastasis (Fig. 7.2) (Carine et al. 2003). Cancer cells can also induce the adipocytes to produce MMP-11 to facilitate the extracellular matrix remodelling and this enhances their invasion of the adipose tissue, highlighting the importance of MMP-11 in extracellular matrix remodelling and cancer invasion (Motrescu and Rio 2008). In support of this finding, small interfering ribonucleic acid (siRNA) mediated MMP-11 depletion causes a reduction in cancer metastasis. siRNA targeted against MMP-11 can restrict the of cancer cells to invade and metastasize to local lymph nodes (Jia et al. 2007). Indeed, high levels of MMP-11 expression are associated with cancer cell invasion and poor prognosis (Rouyer et al. 1994). Apart from secreting MMPs directly, adipocytes also release leptin to enhance the release of MMPs by the cancer cells to facilitate cancer invasion (Yeh et al. 2009).

After the local invasion, adipocytes also support cancer cells migration and seeding at the distant site. In fact, the adipose tissue is a preferential site for many metastatic cancers. This is true as adipocytes secrete many cytokines, including IL-8 in favour of the cancer cell survival (Kristin et al. 2011). An *in vitro* experiment using mice showed that inhibition of IL-8 receptors, particularly CXCR1 could reduce the homing of ovarian cancer cells towards adipocytes (Kristin et al. 2011). Similar observations were also found in the acute lymphoblastic leukaemia (ALL). For leukaemic cell migration, SDF-1 has been identified as a chemoattractant released by adipocytes. This leukaemic migration towards adipocytes could be blocked when SDF-1 receptors on the leukaemia cells were inhibited (Pramanik et al. 2012).

7.2.8 Inflammatory Cytokines-Cancer Metastasis

Metastasis is the spread and establishment of secondary cancer growths at a distal site from the primary cancer and the major cause of cancer deaths. The locations of future metastasis are predetermined microenvironments called 'premetastatic niches' (PMNs), and adipocytes also play an essential role in the homing of these metastasizing cancer cells (Peinado et al. 2017). Indeed, primary human omental adipocytes promote the homing, migration and invasion of metastasising ovarian cancer cells, through releasing adipokines including IL-6 and IL-8 (Kristin et al. 2011). The lipids, predominantly FFAs, released by resident adipocytes also serve as a source of energy to promote the growth and proliferation of the homing cancer cells (Kristin et al. 2011).

In terms of metastasis, adipocytes can induce the cancer cells to undergo an incomplete EMT, a crucial process involved in the development of an invasive and metastatic cell phenotype. In adipocyte-breast cancer co-culture studies, the cancer cells displayed reduced expression of the epithelial marker, E-cadherin, without a significant increase in mesenchymal marker expression (Dirat et al. 2011). However, subsequent experiments revealed that conditioned media from CAAs alone are enough to increase cancer cell invasiveness (Dirat et al. 2011). Experiments using the co-culture system or conditioned media with breast cancer cells also demonstrated that adipocytes have an ability to enhance the cancer cell proliferation, invasion and migration through Jak/STAT3 signalling pathway (Lapeire et al. 2014). Similar outcomes are also observed when the experiments are performed using xenograft models and 3D culture systems (Laetitia et al. 2013; Brian et al. 2014). Consistent findings were observed between adipocytes and prostate cancer cells (Abel 2012). To promote cancer invasion, adipocytes have been demonstrated to release the pro-inflammatory cytokine IL-6 into the tumour microenvironment. This level of IL-6 is directly correlated with the cancer aggressiveness, and its inhibition using an IL-6 antibody could suppress the invasiveness of the cancer cells (Dirat et al. 2011). Moreover, the secreted IL-6 can also lead to the local inflammation and activation of immune cells in the tumour microenvironment (Wright and Simone 2016). The IL-6-activated macrophages can further produce and secrete more inflammatory cytokines to promote further cancer metastasis (McNelis and Olefsky 2014). Apart from IL-6, leptin has also been identified as a promoter for breast cancer cell invasion. In fact, breast cancer-associated adipocytes also secrete proinflammatory cytokines, including IL-6, IL-8, IFNγ-inducible protein-10 (IP10, also called CXCL10), CCL2 (MCP1), and CCL5 (RANTES), to drive tumour-initiating cell abundance and metastatic progression (Picon-Ruiz et al. 2016).

7.2.9 Extracellular Vesicles

Tumour cells communicate with their microenvironment not only via soluble and secreted factors but also through extracellular vesicles (EVs). These EVs, which can be released by both cancer and stromal cells, are further classified into exosomes, microvesicles (MVs), and apoptotic bodies (ABs). There is a key function of extracellular vesicles in the establishment and maintenance of the tumour microenvironment (Han et al. 2017, 2019). These EVs facilitate bioactive cargo transfer between cancer and stromal cells in the tumour microenvironment and play essential roles in maintaining cell proliferation, evading growth suppression, resisting cell death, acquiring genomic instability and reprogramming stromal cell lineages and cancer cells, together contributing to the generation of a remodelled TME. For example, EVs secreted by cancer cells are involved in the transfer of IL-6 to activate the STAT3 signalling pathway to induce lipolysis and the generation of FFAs in adipocytes in lung cancer-adipocyte co-culture models (Hu et al. 2019). Conversely, Circular RNAs (circRNAs) secreted in exosomes by adipocytes promote the tumour growth by inhibiting deubiquitination in hepatocellular carcinoma (HCC) (Zhang et al. 2019).

7.3 Adipocytes Crosstalk with Other Stromal Cells in the Tumour Environment

Cancer-associated fibroblasts (CAFs) are known to play a vital role in cancer development and progression through the release of growth factors and chemokines and by involving in the remodelling of the extracellular matrix (Orimo et al. 2005; Buchsbaum and Oh 2016; Bochet et al. 2013). In the tumour microenvironment, the proximal localization of cancer cells to adipocytes can cause the adipocytes to undergo phenotypical changes to generate fibroblast-like cells termed adipocyte-derived fibroblasts (ADFs). These ADFs exhibit CAF-like phenotypes, including augmented fibronectin and collagen I secretion, enhanced migratory/invasive abilities, and enhanced expression of the CAF marker FSP-1 (Bochet et al. 2013). This may ultimately contribute to the number and the function of cancer-associated fibroblasts (CAFs) in the tumour environment. In addition, obese individuals also have more pre-adipocytes, macrophages and monocytes deposited in the adipose tissue. These changes to the microenvironment of obese individuals may promote cancer development (Wang et al. 2017a; Wang et al. 2017b). Adipose stem cells (ASCs) are known to alter the microenvironment and promote cancer progression. Accordingly, ASCs induce local inflammation through TGF-beta signalling pathway to recruit immune cells (Razmkhah et al. 2011). Moreover, ASCs can also promote angiogenesis by the platelet-derived growth factor BB/platelet-derived growth factor receptor-β (PDGF-BB/PDGFR-β) signalling pathway (Gehmert et al. 2010). Apart from the induction of inflammation and angiogenesis, co-culture and conditioned medium experiments revealed that ASCs also enhance EMT (Zimmerlin et al. 2011). In addition, ASCs can be differentiated into proliferation-promoting fibroblasts in a variety of cancers, including those of the breast, ovarian and lung, to accelerate cancer progression (Jotzu et al. 2010).

7.4 Therapeutic Resistance

Drug resistance is a major obstacle to effective cancer chemotherapy and is directly associated with limited therapeutic options and poor prognosis in cancer patients. Reduced or lack of drug responses are observed in obese cancer patients (Chen et al. 2012; Horowitz and Wright 2015), and there are plenty of examples to suggest that the cancer-associated adipocytes are involved in conferring resistance to therapeutic treatments. For instance, adipocytes secrete adipokines such as leptin and growth differentiation factor 15 (GDF15) to block the growth inhibitory action of trastuzumab in HER2-positive cancers (Griner et al. 2013). Adipocytes also produce leptin to promote melanoma drug resistance through the upregulation of pro-survival PI3K/Akt and MEK/ERK signalling pathways (Chi et al. 2014). A similar chemoprotective effect by adipocytes through leptin was also observed in colon cancer (Bartucci et al. 2010). Adipocytes impair the cytotoxic effects of the drug vincristine through elevating pro-survival signals such as Bcl-2 and Pim-2 in the leukaemic cells (Behan et al. 2009). Moreover, obesity-associated adipocytes also promote breast cancer chemotherapy resistance through releasing major vault protein (MVP), a suppressor of NF-κB signalling (Lehuede et al. 2019). L-asparaginase (ASNase) is a first-line therapy for acute lymphoblastic leukaemia (ALL) that breaks down the essential metabolic substrates asparagine and glutamine, which drive cancer metastasis (Luo et al. 2018). In this respect, adipocytes can release glutamine to the cancer microenvironment and provide the essential amino acid to the cancer cells directly to sustain proliferation and protein synthesis and lessen the cytotoxic effects of asparagine and glutamine shortage (Ehsanipour et al. 2013). Furthermore, another study reported that hypoxia conditions enhance the adipocyte-protection on breast cancer cells from chemotherapeutic toxicity (Rausch et al. 2017). Adipocytes in the bone marrow microenvironment can protect myeloma cells against chemotherapy through releasing adipokines, such as leptin and adipsin, to promote autophagy and therefore chemoresistance in the myeloma cells (Liu et al. 2015). In summary, the cancer-associated adipocytes can enhance cancer therapeutic resistance through upregulating the survival signalling pathways and DNA-repair activity (Shimizu et al. 2014) in the cancer cells and by providing metabolic substrates to boost cancer proliferation and survival.

7.5 Targeting Adipocyte Signalling and Other Therapeutic Perspectives

Recent data has shown that the mechanisms of energy metabolism are dysregulated in cancer. Because of the need to meet the high energy demands, tumour cells shift to the Warburg phenotype and preferentially import an excess of glucose and convert it to lactate for energy (ATP) production. At the same time, unlike their normal counterparts, the cancer cells are also less able to use lipids, like FAs, to generate ATP production via FAO (Diedrich et al. 2016; Wang et al. 2017b) because of the lower levels of FA metabolic enzymes in mitochondria of the cancer cells (Vidali et al. 1983). The low-carbohydrate, high-fat ketogenic diet is an example where our knowledge of cancer energy metabolism has been successfully translated into treatment strategies. The ketogenic diet has been designed to mimic the effects of glucose starvation and is based on the fact that cancer cells fail to adapt to glucose shortage and use lipids/fatty acids as alternative energy sources. On this diet, the cancer cells will rapidly consume intracellular energy reserves through glycolysis because of the Warburg effect. Therefore, the ketogenic diet has the potential to be an effective cancer diet therapy. In support, experiments with animal models showed that the ketogenic diet significantly reduced the growth of tumours and that the diet prolonged animal survival (Raphael Johannes et al. 2015; Kennedy et al. 2007). These findings are also well correlated with a reduction in the plasma glucose concentrations and cancer cellular proliferation markers in animal models (Martuscello et al. 2016). While the ketogenic diet works relatively well in the short term, the long-term health effects of the diet remain to be determined, as there is plenty of evidence to suggest that an accumulation of adipose tissue (fat) in the tumour microenvironment will accelerate cancer progression. Specifically, the cancer-associated adipocytes can modify the cancer cell phenotype to utilise alternative energetic nutrients to replace glucose, as well as essential metabolites, like FAs and glutamate, as energy sources to sustain cancer cell proliferation, survival and progression. In consequence, another potential strategy is to target the crosstalk between adipocytes and the cancer cells, in particular the fatty acid transporters, such as FABP4 which plays a pivotal role in fatty acid metabolism and transport in both the adipocytes and the cancer cells. Indeed, the highly specific FABP4 inhibitor, BMS309403, has been shown to be able decreased tumour cell proliferation and migration by downregulating HIF1 pathway in hepatocellular carcinoma (HCC) and reduce tumour growth in heterotopic and orthotopic xenografted mice models (Laouirem et al. 2019). In addition, prostate stromal cells can augment cancer cell invasiveness by secreting IL-8 and IL-6, which can be also abrogated by BMS309403. Thus, targeting the crosstalk between adipocytes and the cancer cells or vulnerabilities arisen as a result may provide novel strategies of cancer treatment and for overcoming drug resistance.

Moreover, FABP4 expression is frequently elevated in breast cancer and has been shown to be a potential good diagnostic and poor prognostic marker in patients with breast cancer (Cui et al. 2019).

7.6 Conclusion and Future Perspectives

Tumorigenesis and cancer progression require the interaction of tumour cells with the surrounding tissues in the tumour microenvironment. The contribution of adipose tissue, in particular adipocytes to the development of cancer has help us to establish a link between obesity and cancer. The proximal location of the cancer cells can dramatically modify the adipocytes to a pro-cancer phenotype. Over the past few decades the rates of obesity have risen at a dramatic rate globally (Chooi et al. 2019). Obesity and being overweight further condition the function of the cancer-associated adipocytes. Thus, targeting the crosstalks between adipocytes and the cancer cells or vulnerabilities arisen as a result of these interactions may provide novel strategies of cancer treatment and for overcoming drug resistance. Nevertheless, further work will be required to obtain a more complete understanding of the mechanisms by which adipocytes, in conjunction with other stomal cells, promote cancer initiation, progression and drug resistance (Table 7.1). This information will allow us to identify biomarkers for early cancer risk prediction and diagnosis as well as targets and opportunities for therapeutic intervention.

Table 7.1 Summary of cancer-associated adipocyte secretosome

Secretome	Targets	Descriptions
Leptin	ERK1/2	Cell proliferation
	JNK	Epithelial-mesenchymal transition
	PI3K/AKT	Angiogenesis
		Chemotherapeutic resistance
Adiponectin	GSK3β/β-catenin	Apoptosis
	AMPK	Restrict cell growth
Insulin-like growth factor (IGF)	PI3K/AKT	Cell growth
	MAPK	
Vascular endothelial growth factor (VEGF)	VEGFR	Angiogenesis
Free fatty acids (FFAs)	Lipid droplets	Source of energy
	SMAD4	Cell proliferation
		Migration
		Invasion
Fatty acid binding protein 4 (FABP4)	Free fatty acids (FFAs)	Cell proliferation
	FOXM1	Invasion
	FATP1	Metastasis
		Lipid storage
		Lipid metabolism
Matrix metalloproteinase (MMP)	Adhesion molecules	Extracellular matrix remodelling
		Invasion
		Metastasis
Interleukine-6 (IL-6)	JAK/STAT3	Invasion
		Local inflammation
		Immune cells activation
Extracellular vesicles (EVs)	Bioactive cargo	Cell proliferation
		Resist cell death
		Reprogramme stromal cells

References

Abel M (2012) Human periprostatic adipose tissue promotes prostate cancer aggressiveness in vitro. Cambridge Scholars Publishing, Newcastle upon Tyne

Al-Hamodi Z, Al-Habori M, Al-Meeri A, Saif-Ali R (2014) Association of adipokines, leptin/adiponectin ratio and C-reactive protein with obesity and type 2 diabetes mellitus. Diabetol Metab Syndr 6:99

Amemori S, Ootani A, Aoki S, Fujise T, Shimoda R, Kakimoto T, Shiraishi R, Sakata Y, Tsunada S, Iwakiri R, Fujimoto K (2007) Adipocytes and preadipocytes promote the proliferation of colon cancer cells in vitro. Am J Physiol Gastrointest Liver Physiol 292:G923–G929

Amin MN, Hussain MS, Sarwar MS, Rahman Moghal MM, Das A, Hossain MZ, Chowdhury JA, Millat MS, Islam MS (2019) How the association between obesity and inflammation may lead to insulin resistance and cancer. Diabetes Metab Syndr 13:1213–1224

Amri EZ, Ailhaud G, Grimaldi PA (1994) Fatty acids as signal transducing molecules: involvement in the differentiation of preadipose to adipose cells. J Lipid Res 35:930–937

Amy LS, Jason FO, Brandi AB, Lyndsay VR, Dorothy TP, Tucker HA, Claire L, Annie CB, Maria FD, Shijia Z, Jeffrey MG, Matthew EB, Bruce AB (2015) Leptin produced by obese adipose stromal/stem cells enhances proliferation and metastasis of estrogen receptor positive breast cancers. Breast Cancer Res 17:112

Aoki S, Tsunada S, Ootani A, Shimoda R, Sakata Y, Kakimoto T, Fujise T, Iwakiri R, Fujimoto K, Amemori S, Shiraishi R (2007) Adipocytes and preadipocytes promote the proliferation of colon cancer cells in vitro. Am J Physiol 292:G923

Arita Y, Kihara S, Ouchi N, Takahashi M, Maeda K, Miyagawa J-I, Hotta K, Shimomura I, Nakamura T, Miyaoka K, Kuriyama H, Nishida M, Yamashita S, Okubo K, Matsubara K, Muraguchi M, Ohmoto Y, Funahashi T, Matsuzawa Y (1999) Paradoxical decrease of an adipose-specific protein, adiponectin, in obesity. Biochem Biophys Res Commun 257:79–83

Ashizawa N, Yahata T, Quan J, Adachi S, Yoshihara K, Tanaka K (2010) Serum leptin–adiponectin ratio and endometrial cancer risk in postmenopausal female subjects. Gynecol Oncol 119:65–69

Balaban S, Shearer RF, Lee LS, Van Geldermalsen M, Schreuder M, Shtein HC, Cairns R, Thomas KC, Fazakerley DJ, Grewal T, Holst J, Saunders DN, Hoy AJ (2017) Adipocyte lipolysis links obesity to breast cancer growth: adipocyte-derived fatty acids drive breast cancer cell proliferation and migration. Cancer Metab 5:1

Bartucci M, Svensson S, Ricci-Vitiani L, Dattilo R, Biffoni M, Signore M, Ferla R, De Maria R, Surmacz E (2010) Obesity hormone leptin induces growth and interferes with the cytotoxic effects of 5-fluorouracil in colorectal tumor stem cells. Endocr Relat Cancer 17:823–833

Behan JW, Yun JP, Proektor MP, Ehsanipour EA, Arutyunyan A, Moses AS, Avramis VI, Louie SG, Butturini A, Heisterkamp N, Mittelman SD (2009) Adipocytes impair leukemia treatment in mice. Cancer Res 69:7867–7874

Birsoy K, Festuccia WT, Laplante M (2013) A comparative perspective on lipid storage in animals. J Cell Sci 126:1541–1552

Bochet L, Lehuede C, Dauvillier S, Wang YY, Dirat B, Laurent V, Dray C, Guiet R, Maridonneau-Parini I, Le Gonidec S, Couderc B, Escourrou G, Valet P, Muller C (2013) Adipocyte-derived fibroblasts promote tumor progression and contribute to the desmoplastic reaction in breast cancer. Cancer Res 73:5657–5668

Bremnes RM, Donnem T, Al-Saad S, Al-Shibli K, Andersen S, Sirera R, Camps C, Marinez I, Busund LT (2011) The role of tumor stroma in cancer progression and prognosis: emphasis on carcinoma-associated fibroblasts and non-small cell lung cancer. J Thorac Oncol 6:209–217

Brian GR, Jeffrey MG, Mei S, Muralidharan A, Ryan KJ, Trivia PF, Majdouline A, Eduardo AL, Paul LF, Robert K, Ernest SC (2014) Human adipose tissue-derived stromal/stem cells promote migration and early metastasis of triple negative breast cancer xenografts. PLoS One 9:e89595

Buchsbaum RJ, Oh SY (2016) Breast cancer-associated fibroblasts: where we are and where we need to go. Cancers (Basel) 8(2):19

Bussard KM, Mutkus L, Stumpf K, Gomez-Manzano C, Marini FC (2016) Tumor-associated stromal cells as key contributors to the tumor microenvironment. Breast Cancer Res 18:84

Byon CH, Hardy RW, Ren C, Ponnazhagan S, Welch DR, Mcdonald JM, Chen Y (2009) Free fatty acids enhance breast cancer cell migration through plasminogen activator inhibitor-1 and SMAD4. Lab Invest 89:1221–1228

Calle EE, Kaaks R (2004) Overweight, obesity and cancer: epidemiological evidence and proposed mechanisms. Nat Rev Cancer 4:579–591

Calle EE, Rodriguez C, Walker-Thurmond K, Thun MJ (2003) Overweight, obesity, and mortality from cancer in a prospectively studied cohort of U.S adults. N Engl J Med 348:1625–1638

Cao Y (2013) Angiogenesis and vascular functions in modulation of obesity, adipose metabolism, and insulin sensitivity. Cell Metab 18:478–489

Carine C, Bernard M, Marie-Noëlle M, Stéphanie B, Patrick A, Emmanuel Van O, Sophie T-D (2003) Matrix metalloproteinases are differentially expressed in adipose tissue during obesity and modulate adipocyte differentiation. J Biol Chem 278:11888–11896

Chen S, Chen C-M, Zhou Y, Zhou R-J, Yu K-D, Shao Z-M (2012) Obesity or overweight is associated with worse pathological response to neoadjuvant chemotherapy among Chinese women with breast cancer. PLoS One e41380:7

Chi M, Chen J, Ye Y, Tseng H-Y, Lai F, Tay KH, Jin L, Guo ST, Jiang CC, Zhang XD (2014) Adipocytes contribute to resistance of human melanoma cells to chemotherapy and targeted therapy. Curr Med Chem 21:1255–1267

Chmurzynska A (2006) The multigene family of fatty acid-binding proteins (FABPs): function, structure and polymorphism. J Appl Genet 47:39–48

Chooi YC, Ding C, Magkos F (2019) The epidemiology of obesity. Metabolism 92:6–10

Cui Y, Song M, Kim SY (2019) Prognostic significance of fatty acid binding protein-4 in the invasive ductal carcinoma of the breast. Pathol Int 69:68–75

D'esposito V, Passaretti F, Hammarstedt A, Liguoro D, Terracciano D, Molea G, Canta L, Miele C, Smith U, Beguinot F, Formisano P (2012) Adipocyte-released insulin-like growth factor-1 is regulated by glucose and fatty acids and controls breast cancer cell growth in vitro. Diabetologia 55:2811–2822

Dalamaga M, Diakopoulos KN, Mantzoros CS (2012) The role of adiponectin in cancer: a review of current evidence. Endocr Rev 33:547–594

Dali-Youcef N, Hnia K, Blaise S, Messaddeq N, Blanc S, Postic C, Valet P, Tomasetto C, Rio M-C (2016) Matrix metalloproteinase 11 protects from diabesity and promotes metabolic switch. Sci Rep 6:25140

Darimont C, Vassaux G, Ailhaud G, Negrel R (1994) Differentiation of preadipose cells: paracrine role of prostacyclin upon stimulation of adipose cells by angiotensin-II. Endocrinology 135:2030–2036

Diedrich JD, Rajagurubandara E, Herroon MK, Mahapatra G, Huttemann M, Podgorski I (2016) Bone marrow adipocytes promote the Warburg phenotype in metastatic prostate tumors via HIF-1α activation. Oncotarget 7:64854–64877

Dieudonne M-N, Bussiere M, Dos Santos E, Leneveu M-C, Giudicelli Y, Pecquery R (2006) Adiponectin mediates antiproliferative and apoptotic responses in human MCF7 breast cancer cells. Biochem Biophys Res Commun 345:271–279

Dirat B, Bochet L, Dabek M, Daviaud D, Dauvillier S, Majed B, Wang YY, Meulle A, Salles B, Le Gonidec S, Garrido I, Escourrou G, Valet P, Muller C (2011) Cancer-associated adipocytes exhibit an activated phenotype and contribute to breast cancer invasion. Cancer Res 71:2455–2465

Duong MN, Geneste A, Fallone F, Li X, Dumontet C, Muller C (2017) The fat and the bad: mature adipocytes, key actors in tumor progression and resistance. Oncotarget 8:57622–57641

Ehsanipour EA, Sheng X, Behan JW, Wang X, Butturini A, Avramis VI, Mittelman SD (2013) Adipocytes cause leukemia cell resistance to L-asparaginase via release of glutamine. Cancer Res 73:2998–3006

Elliott BE, Tam SP, Dexter D, Chen ZQ (1992) Capacity of adipose tissue to promote growth and metastasis of a murine mammary carcinoma: effect of estrogen and progesterone. Int J Cancer 51:416–424

Endo H, Hosono K, Uchiyama T, Sakai E, Sugiyama M, Takahashi H, Nakajima N, Wada K, Takeda K, Nakagama H, Nakajima A (2011) Leptin acts as a growth factor for colorectal tumours at stages subsequent to tumour initiation in murine colon carcinogenesis. Gut 60:1363–1371

Finley D, Calvert V, Inokuchi J, Lau A, Narula N, Petricoin E, Zaldivar F, Santos R, Tyson D, Ornstein D (2009) Periprostatic adipose tissue as a modulator of prostate cancer aggressiveness. J Urol 182:1621–1627

Fletcher SJ, Sacca PA, Pistone-Creydt M, Colo FA, Serra MF, Santino FE, Sasso CV, Lopez-Fontana CM, Caron RW, Calvo JC, Pistone-Creydt V (2017) Human breast adipose tissue: characterization of factors that change during tumor progression in human breast cancer. J Exp Clin Cancer Res 36:26

Furuhashi M, Hotamisligil GS (2008) Fatty acid-binding proteins: role in metabolic diseases and potential as drug targets. Nat Rev Drug Discov 7:489–503

Garofalo C, Surmacz E (2006) Leptin and cancer. J Cell Physiol 207:12–22

Gehmert S, Gehmert S, Prantl L, Vykoukal J, Alt E, Song Y-H (2010) Breast cancer cells attract the migration of adipose tissue-derived stem cells via the PDGF-BB/PDGFR-β signaling pathway. Biochem Biophys Res Commun 398:601–605

Gonzalez-Perez RR, Xu Y, Guo S, Watters A, Zhou W, Leibovich SJ (2010) Leptin upregulates VEGF in breast cancer via canonic and non-canonical signalling pathways and NFκB/HIF-1α activation. Cell Signal 22:1350–1362

Griner SE, Wang KJ, Joshi JP, Nahta R (2013) Mechanisms of adipocytokine-mediated trastuzumab resistance in HER2-positive breast cancer cell lines. Curr Pharmacogen Personal Med 11:31–41

Guaita-Esteruelas S, Bosquet A, Saavedra P, Guma J, Girona J, Lam EW, Amillano K, Borras J, Masana L (2017) Exogenous FABP4 increases breast cancer cell proliferation and activates the expression of fatty acid transport proteins. Mol Carcinog 56:208–217

Haczeyni F, Bell-Anderson KS, Farrell GC (2018) Causes and mechanisms of adipocyte enlargement and adipose expansion. Obes Rev 19:406–420

Han Suk R, Han-Byoel L, Wonshik H, Dong-Young N, Hyeong-Gon M (2015) Reduced proliferation in breast cancer cells contacting the neighboring adipocytes in human breast cancer tissues. Breast Cancer Res 17:90

Han L, Xu J, Xu Q, Zhang B, Lam EW, Sun Y (2017) Extracellular vesicles in the tumor microenvironment: therapeutic resistance, clinical biomarkers, and targeting strategies. Med Res Rev 37:1318–1349

Han L, Lam EW, Sun Y (2019) Extracellular vesicles in the tumor microenvironment: old stories, but new tales. Mol Cancer 18:59

Herroon MK, Rajagurubandara E, Hardaway AL, Powell K, Turchick A, Feldmann D, Podgorski I (2013) Bone marrow adipocytes promote tumor growth in bone via FABP4-dependent mechanisms. Oncotarget 4:2108

Horowitz NS, Wright AA (2015) Impact of obesity on chemotherapy management and outcomes in women with gynecologic malignancies. Gynecol Oncol 138:201–206

Hovey RC, Mcfadden TB, Akers RM (1999) Regulation of mammary gland growth and morphogenesis by the mammary fat pad: a species comparison. J Mammary Gland Biol Neoplasia 4:53–68

Hoy AJ, Balaban S, Saunders DN (2017) Adipocyte–tumor cell metabolic crosstalk in breast cancer. Trends Mol Med 23:381–392

Hu W, Ru Z, Zhou Y, Xiao W, Sun R, Zhang S, Gao Y, Li X, Zhang X, Yang H (2019) Lung cancer-derived extracellular vesicles induced myotube atrophy and adipocyte lipolysis via the extracellular IL-6-mediated STAT3 pathway. Biochim Biophys Acta Mol Cell Biol Lipids 1864:1091–1102

Hua TNM, Kim MK, Vo VTA, Choi JW, Choi JH, Kim HW, Cha SK, Park KS, Jeong Y (2019) Inhibition of oncogenic Src induces FABP4-mediated lipolysis via PPARgamma activation exerting cancer growth suppression. EBioMedicine 41:134–145

Huang M, Narita S, Inoue T, Koizumi A, Saito M, Tsuruta H, Numakura K, Satoh S, Nanjo H, Sasaki T, Habuchi T (2017) Fatty acid binding protein 4 enhances prostate cancer progression by upregulating matrix metalloproteinases and stromal cell cytokine production. Oncotarget 8:111780–111794

Ishikawa M, Kitayama J, Nagawa H (2004) Enhanced expression of leptin and leptin receptor (OB-R) in human breast cancer. Clin Cancer Res 10:4325–4331

Ishikawa M, Kitayama J, Yamauchi T, Kadowaki T, Maki T, Miyato H, Yamashita H, Nagawa H (2007) Adiponectin inhibits the growth and peritoneal metastasis of gastric cancer through its specific membrane receptors AdipoR1 and AdipoR2. Cancer Sci 98:1120–1127

Jia L, Wang S, Cao J, Zhou H, Wei W, Zhang J (2007) siRNA targeted against matrix metalloproteinase 11 inhibits the metastatic capability of murine hepatocarcinoma cell Hca-F to lymph nodes. Int J Biochem Cell Biol 39:2049–2062

Jotzu C, Alt E, Welte G, Li J, Hennessy BT, Devarajan E, Krishnappa S, Pinilla S, Droll L, Song Y-H (2010) Adipose tissue-derived stem cells differentiate into carcinoma-associated fibroblast-like cells under the influence of tumor-derived factors. Anal Cell Pathol (Amst) 33:61–79

Kang JH, Lee YY, Yu BY, Yang B-S, Cho K-H, Yoon DK, Roh YK (2005) Adiponectin induces growth arrest and apoptosis of MDA-MB-231 breast cancer cell. Arch Pharm Res 28:1263–1269

Kennedy AR, Pissios P, Otu H, Roberson R, Xue B, Asakura K, Furukawa N, Marino FE, Liu FF, Kahn BB, Libermann TA, Maratos-Flier E (2007) A high-fat, ketogenic diet induces a unique metabolic state in mice. Am J Physiol Endocrinol Metab 292:E1724–E1739

Kristin MN, Hilary AK, Carla VP, Andras L, Rebecca B-G, Marion RZ, Iris LR, Mark SC, Gordon BM, Gökhan SH, Yamada SD, Marcus EP, Katja G, Ernst L (2011) Adipocytes promote ovarian cancer metastasis and provide energy for rapid tumor growth. Nat Med 17:1498–1503

Kwan HY, Fu X, Liu B, Chao X, Chan CL, Cao H, Su T, Tse AKW, Fong WF, Yu Z-L (2014) Subcutaneous adipocytes promote melanoma cell growth by activating the Akt signaling pathway: role of palmitic acid. J Biol Chem 289:30525–30537

Laetitia D, Charlotte L, Virginie D, Alice D, Hermine B, Ali M, Odile D, Marie-Paule V, Florence C-C (2013) Reciprocal interactions between breast tumor and its adipose microenvironment based on a 3D adipose equivalent model. PLoS One 8:e66284

Laouirem S, Sannier A, Norkowski E, Cauchy F, Doblas S, Rautou PE, Albuquerque M, Garteiser P, Sognigbe L, Raffenne J, Van Beers BE, Soubrane O, Bedossa P, Cros J, Paradis V (2019) Endothelial fatty liver binding protein 4: a new targetable mediator in hepatocellular carcinoma related to metabolic syndrome. Oncogene 38:3033–3046

Lapeire L, Hendrix A, Lambein K, Van Bockstal M, Braems G, Van Den Broecke R, Limame R, Mestdagh P, Vandesompele J, Vanhove C, Maynard D, Lehuede C, Muller C, Valet P, Gespach CP, Bracke M, Cocquyt V, Denys H, De Wever O (2014) Cancer-associated adipose tissue promotes breast cancer progression by paracrine oncostatin M and Jak/STAT3 signaling. Cancer Res 74:6806–6819

Lehuede C, Li X, Dauvillier S, Vaysse C, Franchet C, Clement E, Esteve D, Longue M, Chaltiel L, Le Gonidec S, Lazar I, Geneste A, Dumontet C, Valet P, Nieto L, Fallone F, Muller C (2019) Adipocytes promote breast cancer resistance to chemotherapy, a process amplified by obesity: role of the major vault protein (MVP). Breast Cancer Res 21:7

Liu Z, Xu J, He J, Liu H, Lin P, Wan X, Navone NM, Tong Q, Kwak LW, Orlowski RZ, Yang J (2015) Mature adipocytes in bone marrow protect myeloma cells against chemotherapy through autophagy activation. Oncotarget 6:34329–34341

Lopes-Coelho F, Andre S, Felix A, Serpa J (2018) Breast cancer metabolic cross-talk: fibroblasts are hubs and breast cancer cells are gatherers of lipids. Mol Cell Endocrinol 462:93–106

Luo M, Brooks M, Wicha MS (2018) Asparagine and Glutamine: Co-conspirators Fueling Metastasis. Cell Metab 27:947–949

Man K, Ng KTP, Xu A, Cheng Q, Lo CM, Xiao JW, Sun BS, Lim ZXH, Cheung JS, Wu EX, Sun CKW, Poon RTP, Fan ST (2010) Suppression of liver tumor growth and metastasis by adiponectin in nude mice through inhibition of tumor angiogenesis and downregulation of rho kinase/IFN-inducible protein 10/matrix metalloproteinase 9 signaling. Clin Cancer Res 16:967–977

Manabe Y, Toda S, Miyazaki K, Sugihara H (2003) Mature adipocytes, but not preadipocytes, promote the growth of breast carcinoma cells in collagen gel matrix culture through cancer–stromal cell interactions. J Pathol 201:221–228

Martuscello RT, Vedam-Mai V, Mccarthy DJ, Schmoll ME, Jundi MA, Louviere CD, Griffith BG, Skinner CL, Suslov O, Deleyrolle LP, Reynolds BA (2016) A supplemented high-fat low-carbohydrate diet for the treatment of glioblastoma. Clin Cancer Res 22:2482–2495

Mcnelis JC, Olefsky JM (2014) Macrophages, immunity, and metabolic disease. Immunity 41:36–48

Mick GJ, Wang X, Mccormick K (2002) White adipocyte vascular endothelial growth factor: regulation by insulin. Endocrinology 143:948–953

Mistry T, Digby JE, Desai KM, Randeva HS (2007) Obesity and prostate cancer: a role for adipokines. Eur Urol 52:46–53

Motrescu ER, Rio MC (2008) Cancer cells, adipocytes and matrix metalloproteinase 11: a vicious tumor progression cycle. Biol Chem 389:1037–1041

Muz B, De La Puente P, Azab F, Azab AK (2015) The role of hypoxia in cancer progression, angiogenesis, metastasis, and resistance to therapy. Hypoxia (Auckland, NZ) 3:83–92

Nair S, Mason A, Eason J, Loss G, Perrillo RP (2002) Is obesity an independent risk factor for hepatocellular carcinoma in cirrhosis? Hepatology 36:150–155

Nam SY, Lee EJ, Kim KR, Cha BS, Song YD, Lim SK, Lee HC, Huh KB (1997) Effect of obesity on total and free insulin-like growth factor (IGF)-1, and their relationship to IGF-binding protein (BP)-1, IGFBP-2, IGFBP-3, insulin, and growth hormone. Int J Obes Relat Metab Disord 21:355–359

Nishida N, Yano H, Nishida T, Kamura T, Kojiro M (2006) Angiogenesis in cancer. Vasc Health Risk Manag 2:213–219

Olsen CM, Nagle CM, Whiteman DC, Ness R, Pearce CL, Pike MC, Rossing MA, Terry KL, Wu AH, Australian Cancer Study, Australian Ovarian Cancer Study Group, Risch HA, Yu H, Doherty JA, Chang-Claude J, Hein R, Nickels S, Wang-Gohrke S, Goodman MT, Carney ME, Matsuno RK, Lurie G, Moysich K, Kjaer SK, Jensen A, Hogdall E, Goode EL, Fridley BL, Vierkant RA, Larson MC, Schildkraut J, Hoyo C, Moorman P, Weber RP, Cramer DW, Vitonis AF, Bandera EV, Olson SH, Rodriguez-Rodriguez L, King M, Brinton LA, Yang H, Garcia-Closas M, Lissowska J, Anton-Culver H, Ziogas A, Gayther SA, Ramus SJ, Menon U, Gentry-Maharaj A, Webb PM, Ovarian Cancer Association Consortium (2013) Obesity and risk of ovarian cancer subtypes: evidence from the Ovarian Cancer Association Consortium. Endocr Relat Cancer 20:251–262

Orimo A, Gupta PB, Sgroi DC, Arenzana-Seisdedos F, Delaunay T, Naeem R, Carey VJ, Richardson AL, Weinberg RA (2005) Stromal fibroblasts present in invasive human breast carcinomas promote tumor growth and angiogenesis through elevated SDF-1/CXCL12 secretion. Cell 121:335–348

Park J, Scherer PE (2012) Adipocyte-derived endotrophin promotes malignant tumor progression. J Clin Invest 122:4243–4256

Park J, Euhus DM, Scherer PE (2011) Paracrine and endocrine effects of adipose tissue on cancer development and progression. Endocr Rev 32:550–570

Park J, Morley TS, Scherer PE (2013) Inhibition of endotrophin, a cleavage product of collagen VI, confers cisplatin sensitivity to tumours. EMBO Mol Med 5:935–948

Pavlides S, Whitaker-Menezes D, Castello-Cros R, Flomenberg N, Witkiewicz AK, Frank PG, Casimiro MC, Wang C, Fortina P, Addya S, Pestell RG, Martinez-Outschoorn UE, Sotgia F, Lisanti MP (2009) The reverse Warburg effect: aerobic glycolysis in cancer associated fibroblasts and the tumor stroma. Cell Cycle 8:3984–4001

Peinado H, Zhang H, Matei IR, Costa-Silva B, Hoshino A, Rodrigues G, Psaila B, Kaplan RN, Bromberg JF, Kang Y, Bissell MJ, Cox TR, Giaccia AJ, Erler JT, Hiratsuka S, Ghajar CM, Lyden D (2017) Pre-metastatic niches: organ-specific homes for metastases. Nat Rev Cancer 17:302–317

Petricoin III EF, Liotta L, Dadachova E, Pestell RG, Lisanti MP, Bonaldo P, Scherer PE (2005) Adipocyte-derived collagen VI affects early mammary tumor progression in vivo, demonstrating a critical interaction in the tumor/stroma microenvironment. J Clin Investig 115:1163–1176

Picon-Ruiz M, Pan C, Drews-Elger K, Jang K, Besser AH, Zhao D, Morata-Tarifa C, Kim M, Ince TA, Azzam DJ, Wander SA, Wang B, Ergonul B, Datar RH, Cote RJ, Howard GA, El-Ashry D, Torne-Poyatos P, Marchal JA, Slingerland JM (2016) Interactions between adipocytes and breast cancer cells stimulate cytokine production and drive Src/Sox2/miR-302b-mediated malignant progression. Cancer Res 76:491–504

Picon-Ruiz M, Morata-Tarifa C, Valle-Goffin JJ, Friedman ER, Slingerland JM (2017) Obesity and adverse breast cancer risk and outcome: mechanistic insights and strategies for intervention. CA Cancer J Clin 67:378–397

Pollak M (2008) Insulin and insulin-like growth factor signalling in neoplasia. Nat Rev Cancer 8:915–928

Pramanik R, Sheng X, Ichihara B, Heisterkamp N, Mittelman SD (2012) Adipose tissue attracts and protects acute lymphoblastic leukemia cells from chemotherapy. Leuk Res 37:503–509

Qiao Z, Sarah MD, Jinling Z, Erinn D-K, Jeremy R, Stephen DH, Nathan AB, Ofer R (2011) Leptin deficiency suppresses MMTV-Wnt-1 mammary tumor growth in obese mice and abrogates tumor initiating cell survival. Endocr Relat Cancer 18:491–503

Raphael Johannes M, Sepideh A-G, René Gunther F, Johannes Adalbert M, Roland L, Daniel N, Wolfgang S, Barbara K (2015) Inhibition of neuroblastoma tumor growth by ketogenic diet and/or calorie

restriction in a CD1-Nu mouse model. PLoS One 10:e0129802
Rausch LK, Netzer NC, Hoegel J, Pramsohler S (2017) The linkage between breast cancer, hypoxia, and adipose tissue. Front Oncol 7:211
Razmkhah M, Jaberipour M, Erfani N, Habibagahi M, Talei A-R, Ghaderi A (2011) Adipose derived stem cells (ASCs) isolated from breast cancer tissue express IL-4, IL-10 and TGF-β1 and upregulate expression of regulatory molecules on T cells: do they protect breast cancer cells from the immune response? Cell Immunol 266:116–122
Rei S, Noriyuki O, Shinji K, Kaori S, Tohru F, Kenneth W (2004) Adiponectin stimulates angiogenesis in response to tissue ischemia through stimulation of AMP-activated protein kinase signaling. J Biol Chem 279:28670–28674
Ribeiro AM, Andrade S, Pinho F, Monteiro JD, Costa M, Lopes C, Aguas AP, Monteiro MP (2010) Prostate cancer cell proliferation and angiogenesis in different obese mice models. Int J Exp Pathol 91:374
Ribeiro RJT, Monteiro CPD, Cunha VFPM, Azevedo ASM, Oliveira MJ, Monteiro R, Fraga AM, Principe P, Lobato C, Lobo F, Morais A, Silva V, Sanches-Magalhaes J, Oliveira J, Guimaraes JT, Lopes CMS, Medeiros RM (2012) Tumor cell-educated periprostatic adipose tissue acquires an aggressive cancer-promoting secretory profile. Cell Physiol Biochem 29:233–240
Rouyer N, Wolf C, Chenard MP, Rio MC, Chambon P, Bellocq JP, Basset P (1994) Stromelysin-3 gene expression in human cancer: an overview. Invasion Metastasis 14:269–275
Shanmugalingam T, Bosco C, Ridley AJ, Van Hemelrijck M (2016) Is there a role for IGF-1 in the development of second primary cancers? Cancer Med 5: 3353–3367
Shimizu I, Yoshida Y, Suda M, Minamino T (2014) DNA damage response and metabolic disease. Cell Metab 20:967–977
Suga H, Matsumoto D, Inoue K, Shigeura T, Eto H, Aoi N, Kato H, Abe H, Yoshimura K (2008) Numerical measurement of viable and nonviable adipocytes and other cellular components in aspirated fat tissue. Plast Reconstr Surg 122:103–114
Sun K, Kusminski CM, Scherer PE (2011) Adipose tissue remodeling and obesity. J Clin Invest 121:2094–2101
Tian Z, Sun R, Wei H, Gao B (1959) Impaired natural killer (NK) cell activity in leptin receptor deficient mice: leptin as a critical regulator in NK cell development and activation. Biochem Biophys Res Commun 298:297–302
Tokuda Y, Satoh Y, Fujiyama C, Toda S, Sugihara H, Masaki Z (2003) Prostate cancer cell growth is modulated by adipocyte-cancer cell interaction. BJU Int 91:716–720
Uddin S, Bu R, Ahmed M, Abubaker J, Al-Dayel F, Bavi P, Al-Kuraya KS (2009) Overexpression of leptin receptor predicts an unfavorable outcome in middle eastern ovarian cancer. Mol Cancer 8:74
Vidali S, Aminzadeh S, Lambert B, Rutherford T, Sperl W, Kofler B, Feichtinger RG (1983) Mitochondria: the ketogenic diet—a metabolism-based therapy. Cell Biochem Funct 63:55–59
Wang Y-Y, Lehuede C, Laurent V, Dirat B, Dauvillier S, Bochet L, Le Gonidec S, Escourrou G, Valet P, Muller C (1975) Adipose tissue and breast epithelial cells: a dangerous dynamic duo in breast cancer. Cancer Lett 324:142–151
Wang Y, Lam JB, Lam KSL, Liu J, Lam MC, Hoo RLC, Wu D, Cooper GJS, Xu A (2006) Adiponectin modulates the glycogen synthase kinase-3β/β-catenin signaling pathway and attenuates mammary tumorigenesis of MDA-MB-231 cells in nude mice. Cancer Res 66:11462–11470
Wang M, Zhao J, Zhang L, Wei F, Lian Y, Wu Y, Gong Z, Zhang S, Zhou J, Cao K, Li X, Xiong W, Li G, Zeng Z, Guo C (2017a) Role of tumor microenvironment in tumorigenesis. J Cancer 8:761–773
Wang YY, Attane C, Milhas D, Dirat B, Dauvillier S, Guerard A, Gilhodes J, Lazar I, Alet N, Laurent V, Le Gonidec S, Biard D, Herve C, Bost F, Ren GS, Bono F, Escourrou G, Prentki M, Nieto L, Valet P, Muller C (2017b) Mammary adipocytes stimulate breast cancer invasion through metabolic remodeling of tumor cells. JCI Insight 2:e87489
Wang K, Yu XH, Tang YJ, Tang YL, Liang XH (2019) Obesity: an emerging driver of head and neck cancer. Life Sci 233:116687
Warburg O (1956) On the origin of cancer cells. Science (NY) 123:309–314
Wei L, Li K, Pang X, Guo B, Su M, Huang Y, Wang N, Ji F, Zhong C, Yang J, Zhang Z, Jiang Y, Liu Y, Chen T (2016) Leptin promotes epithelial-mesenchymal transition of breast cancer via the upregulation of pyruvate kinase M2. J Exp Clin Cancer Res 35:166–110
Wright C, Simone N (2016) Obesity and tumor growth: inflammation, immunity, and the role of a ketogenic diet. Curr Opin Clin Nutr Metab Care 19:294–299
Wu J, Lanier LL (2003) Natural killer cells and cancer. Adv Cancer Res 90:127–156
Yao S, Fan LY, Lam EW (2018) The FOXO3-FOXM1 axis: a key cancer drug target and a modulator of cancer drug resistance. Semin Cancer Biol 50:77–89
Yeh W-L, Lu D-Y, Lee M-J, Fu W-M (2009) Leptin induces migration and invasion of glioma cells through MMP-13 production. Glia 57:454–464
Young SG, Zechner R (2013) Biochemistry and pathophysiology of intravascular and intracellular lipolysis. Genes Dev 27:459–484
Zhang Z, Scherer PE (2018) Adipose tissue: the dysfunctional adipocyte—a cancer cell's best friend. Nat Rev Endocrinol 14:132–134
Zhang F, Chen Y, Heiman M, Dimarchi R (2005) Leptin: structure, function and biology. Vitam Horm 71:345–372

Zhang M, Di Martino JS, Bowman RL, Campbell NR, Baksh SC, Simon-Vermot T, Kim IS, Haldeman P, Mondal C, Yong-Gonzales V, Abu-Akeel M, Merghoub T, Jones DR, Zhu XG, Arora A, Ariyan CE, Birsoy K, Wolchok JD, Panageas KS, Hollmann T, Bravo-Cordero JJ, White RM (2018) Adipocyte-derived lipids mediate melanoma progression via FATP proteins. Cancer Discov 8:1006–1025

Zhang H, Deng T, Ge S, Liu Y, Bai M, Zhu K, Fan Q, Li J, Ning T, Tian F, Li H, Sun W, Ying G, Ba Y (2019) Exosome circRNA secreted from adipocytes promotes the growth of hepatocellular carcinoma by targeting deubiquitination-related USP7. Oncogene 38:2844–2859

Zimmerlin L, Donnenberg AD, Rubin JP, Basse P, Landreneau RJ, Donnenberg VS (2011) Regenerative therapy and cancer: in vitro and in vivo studies of the interaction between adipose-derived stem cells and breast cancer cells from clinical isolates. Tissue Eng A 17:93–106

Endothelial Cells (ECs) Metabolism: A Valuable Piece to Disentangle Cancer Biology

8

Filipa Lopes-Coelho, Filipa Martins, and Jacinta Serpa

Abstract

Effective therapies to fight cancer should not be focused specifically on cancer cells, but it should consider the various components of the TME. Non-cancerous cells cooperate with cancer cells by sharing signaling and organic molecules, accounting for cancer progression. Most of the anti-angiogenic therapy clinically approved for the treatment of human diseases relies on targeting vascular endothelial growth factor (VEGF) signaling pathway. Unexpectedly and unfortunately, the results of anti-angiogenic therapies in the treatment of human diseases are not so effective, showing an insufficient efficacy and resistance.

This chapter will give some insights on showing that targeting endothelial cell metabolism is a missing piece to revolutionize cancer therapy. Only recently endothelial cell (EC) metabolism has been granted as an important inducer of angiogenesis. Metabolic studies in EC demonstrated that targeting EC metabolism can be an alternative to overcome the failure of anti-angiogenic therapies. Hence, it is urgent to increase the knowledge on how ECs alter their metabolism during human diseases, in order to open new therapeutic perspectives in the treatment of pathophysiological angiogenesis, as in cancer.

Keywords

Metabolic remodeling · Tumor microenvironment (TME) · Angiogenesis · Endothelial differentiation · Cancer progression · Cancer therapy

8.1 From a Quiescent ECs to a Functional Blood Vessel: A Brief Overview of Angiogenesis

Angiogenesis is the formation of new blood vessels from a pre-existing one (Bergers and Benjamin 2003). ECs are their major cellular component, creating a highly branched and tree-like tubular network, essential for oxygen and nutrient supply of peripheral tissues (Adams and Alitalo 2007). Blood vessels, besides the continuous supply of nutrients and oxygen, control the systemic pH, temperature, homeostasis and mediate immune responses (Wilting and Chao 2015). During human life, angiogenesis plays an

F. Lopes-Coelho · F. Martins · J. Serpa (✉)
CEDOC, Chronic Diseases Research Centre, NOVA Medical School | Faculdade de Ciências Médicas, Universidade NOVA de Lisboa, Lisbon, Portugal

Instituto Português de Oncologia de Lisboa Francisco Gentil (IPOLFG), Lisbon, Portugal
e-mail: jacinta.serpa@nms.unl.pt

J. Serpa (ed.), *Tumor Microenvironment*, Advances in Experimental Medicine and Biology 1219,
https://doi.org/10.1007/978-3-030-34025-4_8

essential role in physiological and pathophysiologic situations (Choi et al. 2014), including cancer, diabetes, atherosclerosis and chronic inflammation (Sliwinska et al. 2018).

During embryogenesis, the assembly of new endothelial cells into a primitive vascular plexus is called vasculogenesis (Domingues et al. 2015). Additionally, new vessels can grow through the recruitment of endothelial progenitor cells (EPCs) from the bone marrow, a process called neovasculogenesis (Domingues et al. 2015; Jung and Kleinheinz 2013; Hillen and Griffioen 2007).

After birth, angiogenesis is a fundamental process for growth, development and wound healing. The formation of new blood vessels, designated as sprouting angiogenesis, is a process with several steps initiated by pro-angiogenic stimuli that leads to the activation of endothelial tip cells (Domingues et al. 2015). These cells are motile and guide the growing sprout, being followed by the stalk cells, that proliferate and elongate to form the new blood vessel. At the same time, the basement membrane that surrounds the capillaries is degraded by matrix metalloproteinases (MMP), allowing the formation of the new tube (Domingues et al. 2015). Then, quiescent phalanx cells line the newly established vessel, regulating vascular homeostasis and barrier function. Also, mural cells (cells supporting the vessels structure, as smooth muscle cells and pericytes) are recruited to stabilize the new connections (Potente et al. 2011).

The balance between pro- and anti-angiogenic factors determines the level of ongoing angiogenesis (Hillen and Griffioen 2007). Among the positive regulators, VEGF-A (commonly called VEGF) is the most well-studied, being ubiquitously detected in all tissues/tumors undergoing angiogenesis (Stacker and Achen 2013). VEGF is also released by tumor cells and once it binds to its tyrosine kinases receptors (VEGFRs), the receptor dimerizes and becomes activated, triggering an intracellular signaling cascade (Blanco and Gerhardt 2013). The VEGF/VEGFR2 signaling pathway activates downstream molecules that mediate ECs proliferation, migration, differentiation, tube formation and permeability (Blanco and Gerhardt 2013). For example, rat sarcoma virus GTPase (Ras), rous sarcoma oncogene homolog kinase (Src), and phosphatidylinositol kinase (PI3K) pathways are activated by VEGFR2 (Kowanetz and Ferrara 2006).

8.1.1 Angiogenesis in Cancer: A Chaotic Blood Vessel Network

The term "tumor angiogenesis" was first proposed in 1971 by Judah Folkman (1971). For the first time, he described the notion that the formation of new vessels was necessary for tumor growth (Zetter 2008) and suggested a relationship between neo-angiogenesis and the malignancy of a tumor (Gimbrone et al. 1972), since these neo-vessels could be used as a route for neoplastic cells metastasize.

During tumor progression, uncontrolled proliferation of tumor cells leads to a point where tumor grows into a more hypoxic microenvironment, with an extreme need of accessing to nutrients and oxygen. Tumor cells start to release pro-angiogenic factors, as VEGF, leading to the activation of an "angiogenic switch" that stimulates the proliferation and migration of ECs to form new blood vessels, accounting for tumor growth (Sliwinska et al. 2018; Domingues et al. 2015). However, the newly formed vessels are structurally and functionally abnormal. This aberrant vasculature is characterized by excessive vessel branching, leakiness, enlarged, distorted and tortuous vessels and with less mural cells (Rohlenova et al. 2017). Dysfunctional vessels with weak ECs junctions and with increased leakiness facilitate the intravasation of cancer cells and reduce the efficacy of anti-cancer drugs delivery.

Folkman hypothesized that anti-angiogenic therapy would stimulate tumor regression by reducing tumor vasculature and consequently it will starve the tumor to death. Decades after this statement, anti-angiogenic strategies were developed and, antibodies such as bevacizumab (Zetter 2008), an inhibitor of VEGF signaling, were available for therapy (Wong et al. 2017). So far, these strategies have failed, in part, because the precise molecular mechanisms of cancer neoangiogenesis still remain unclear. Therefore, instead of targeting tumor vasculature, new findings in restoring tumor vessel normalization would improve drug delivery (Draoui et al. 2017), which in combination with chemotherapy could efficiently improve cancer therapy and impair metastasis (Rohlenova et al. 2017).

8.2 ECs Metabolism: A Driving Force of Angiogenesis

In adults, ECs are essentially in a quiescent state however upon a pro-angiogenic stimuli ECs are activated to form a new blood vessel or to repair the existing ones. This highly regulated mechanism is a 3-step process: sprouting, proliferation and maturation; which are carried out respectively by 3 EC subtypes–tip, stalk and phalanx cells (Potente et al. 2011; De Bock et al. 2013; Schoors et al. 2015). These ECs subtypes differ in energy, biomass and redox requirements, meaning that they differ in cellular metabolism (Potente et al. 2011; De Bock et al. 2013; Schoors et al. 2015). Perturbed ECs metabolism had been already linked to pathophysiological vascularization, as in cancer (Folkman 1971; Hanahan and Weinberg 2011).

ECs metabolism is an emerging field in the study of ECs biology and the most studied metabolic pathways rely on glucose, amino acids and fatty acids (FA) metabolism, which are key compounds in energy and biomass production.

8.2.1 Glycolysis, the Main Carburetor in ECs Metabolism

Glucose metabolism seems to be the major energy source of ECs with faster ATP kinetics, accounting for 85% of the ATP produced, which is vital for ECs activation and rapid vascularization (De Bock et al. 2013). It is estimated that in human umbilical ECs (HUVECs) the glycolysis is 200-fold increased in comparison to oxidative phosphorylation (OXPHOS) (De Bock et al. 2013), resulting in decreased reactive oxygen species (ROS) generation and saving oxygen to be transferred to the surrounding perivascular cells (De Bock et al. 2013; Eelen et al. 2015; Ghesquière et al. 2014; Helmlinger et al. 2002). Besides ATP production, glycolysis is fundamental during cytoskeleton remodeling upon filopodia and lamelipodia formation. Due to their tiny structures, filopodia and lamedipodia have no space for mitochondria to support the highly energetic requirements. Though, glycolytic machinery localizes close to these structures and supports energetically their formation and function (De Bock et al. 2013; Eelen et al. 2015; Ghesquière et al. 2014). On another hand, lactate, resultant from glycolysis, is often considered a pro-angiogenic signaling molecule (Hunt et al. 2007; Ruan and Kazlauskas 2013; Végran et al. 2011), since it is capable of activating PI3K/AKT pathway and stimulating an autocrine NF-kB/IL-8 pathway, both essential for ECs organization, migrating and tube formation (Hunt et al. 2007; Ruan and Kazlauskas 2013; Végran et al. 2011).

Pro-angiogenic stimuli, as VEGF, increases the glycolytic flux of ECs through the increased expression of glucose transporter-1 (GLUT1), pyruvate:lactate converting enzymes as lactate dehydrogenase isoform-A (LDHA) and glycolytic enzymes as bifunctional 6-phosphofructo-2-kinase/fructose-2,6-bisphosphatase-3 (PFKFB3) and hexokinase-2 (HK2) (De Bock et al. 2013; Hunt et al. 2007; Yu et al. 2017; Parra-Bonilla et al. 2010; Peters et al. 2009) (Fig. 8.1).

PFKFB3, the most studied glycolytic enzyme in ECs, generates fructose-2,6-biphosthase, an activator of the rate-limiting glycolytic enzyme phosphofructokinase-1 (PFK1). *In vitro* and *in vivo*, PFKFB3 was upregulated upon treatment with pro-angiogenic factors and its silencing abolished both ECs proliferation and migration (Xu et al. 2014). In addition, the pharmacological inhibition of PFKFB3, using the small molecule 3-(3-pyridinyl)-1-(4-pyridinyl)-2-propen-1-one (3PO), promotes ECs quiescence and reduces partially and transiently glycolysis *in vivo* (Schoors et al. 2014a).

PFKFB3 kinase activity is essential for the rapid ATP production required by tip cells and

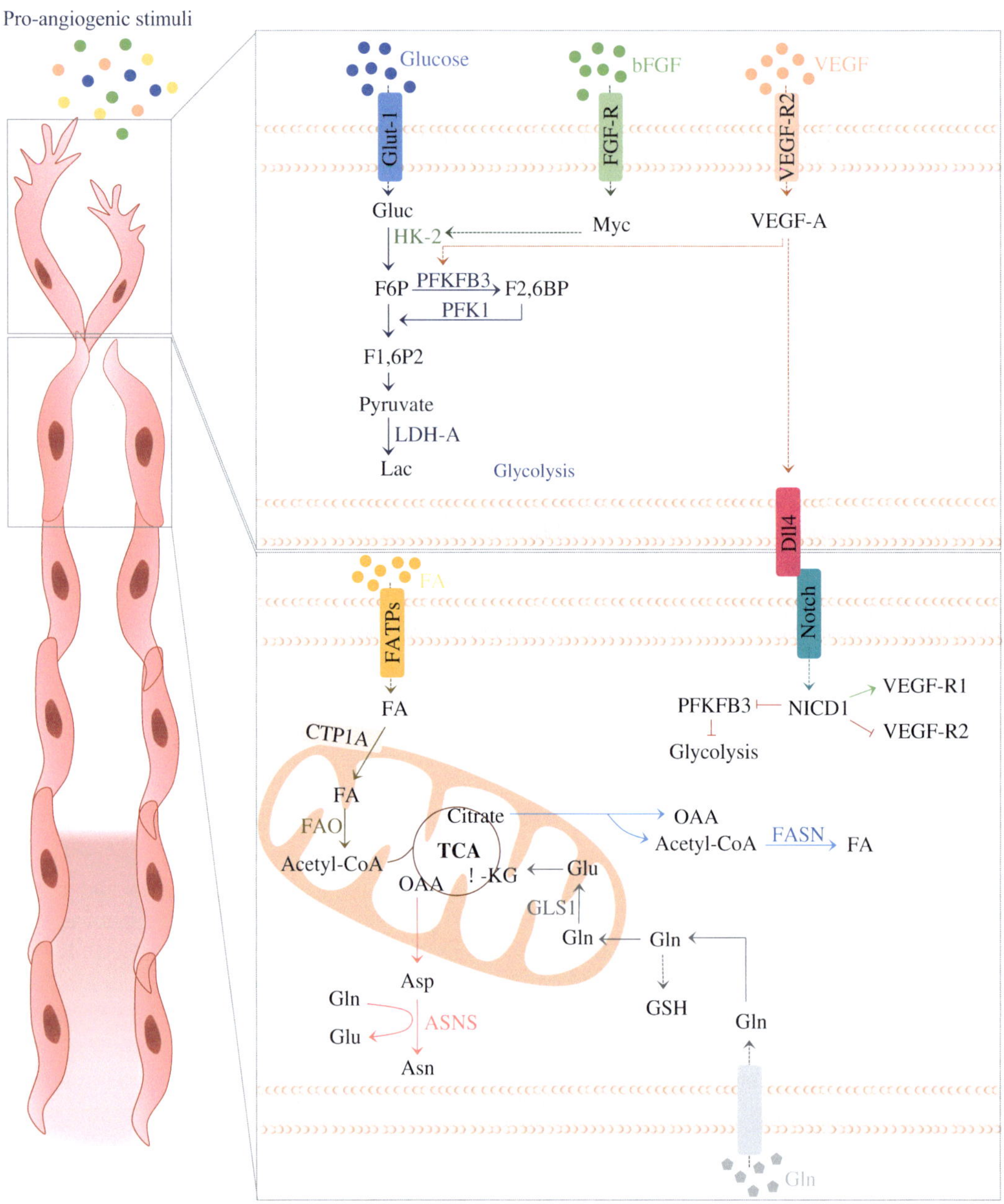

Fig. 8.1 **Sprouting angiogenesis is mediated by pro-angiogenic stimuli, such as vascular endothelial growth factor-A (VEGF-A) and basic fibroblast growth factor (bFGF), leading to the formation of new blood vessels**. VEGF-A and bFGF bind respectively to vascular endothelial growth factor receptor-2 (VEGFR2) and fibroblast growth factor receptor (FGF-R) activating endothelial tip cells (Domingues et al. 2015; Potente et al. 2011), highly motile cells with filopodia that guide the capillary sprout towards the pro-angiogenic source (Domingues et al. 2015; Hillen and Griffioen 2007; Potente et al. 2011; Stacker and Achen 2013). Tip cells are followed by stalk cells, highly proliferative cells that elongate to form the new blood vessel (Domingues et al. 2015; Hillen and Griffioen 2007; Potente et al. 2011; Stacker and Achen 2013). During sprouting angiogenesis, the justacrine Delta-Notch signaling pathway is essential for the establishment of tip-stalk fate. VEGF-A stimulates the production of the ligand delta-like-4 (Dll4) by tip cells that binds to the Notch receptor on neighboring stalk cells, activating Notch1 intracellular domain (NICD1) that suppresses VEGFR2 and inhibits migration of stalk cells (Potente et al. 2011; Stacker and Achen 2013; Blanco and Gerhardt 2013). Tip cells metabolism rely essentially on glycolysis,

for the control of directional migration by filopodia and lamellipodia, during vessel sprouting (De Bock et al. 2013; Schoors et al. 2014a, b). Moreover, stalk cells are transformed into a tip cell upon PFKFB3 activation, followed by an increased glycolytic flux (De Bock et al. 2013). Also, basic fibroblast growth factor (bFGF) signaling pathway had been linked with glycolysis, since bFGF promotes the expression of HK2, mediated by c-MYC oncogene (Yu et al. 2017).

In ECs, glucose consumption seems to be in the same range as cancer cells (De Bock et al. 2013). *In vivo*, PKFB3 inhibition decreases cancer cell invasion, intravasion and metastasis mediated by a lower glycolytic flux of ECs and increased expression of N-cadherin, responsible for the maintenance of pericytes in a more quiescent and adhesive state (Cantelmo et al. 2016). Interestingly, tumors implanted in mice with compromised endothelial PFKFB3 grew slower (Xu et al. 2014) and exhibited an improved chemotherapy response (Cantelmo et al. 2016).

These evidences in cancer context show that targeting ECs metabolism is a promising alternative to overcome the failure of anti-angiogenic therapy. The pivotal role of ECs in cancer progression will be discussed forward in this chapter.

8.2.2 Fatty Acid β-oxidation (FAO), a Valuable Energy and Biomass Source

In contrast to the tip cells, stalk cells were more dependent on FAO for vessel sprout elongation (De Bock et al. 2013; Schoors et al. 2015). ECs are able to use FA as carbon source to sustain tricarboxilic acid (TCA) cycle and nucleotides synthesis (Schoors et al. 2015) (Fig. 8.1). Metabolic studies showed that in ECs, FAO is not critical for the maintenance of energy homeostasis but contrarily to other cell types, ECs rely more on FA metabolism to fulfill nucleotides cell needs, essential for ECs proliferation (De Bock et al. 2013; Schoors et al. 2014a, 2015).

FAO blockage has a colossal impact in vessel sprouting. *In vitro* silencing of carnitine palmitoyltransferase 1A (CPT1A)- that transports acyl-CoA compounds (activated FA) into mitochondria- impairs FAO, leading to decreased nucleotides synthesis, which inhibits ECs proliferation and results in vascular sprouting defects (Schoors et al. 2015). So, the genetic or pharmacological *in vivo* loss of CPT1A decreases the number of branch points and radial expansion of vascular networks (Schoors et al. 2015).

FA transporter CD36 and FA binding proteins (FABPs) have been linked to some ECs features, but this dynamics is not fully understood

Fig. 8.1 (continued) because filopodia and lamedipodia do not have mitochondria (De Bock et al. 2013; Eelen et al. 2015; Ghesquière et al. 2014). Pro-angiogenic stimuli increases the glucose influx through glucose transporter-1 (GLUT-1) and bFGF activates Myc that activates hexokinase-2 (HK-2) expression. HK-2 converts glucose (gluc) to fructose-6-phosphate (F6P). VEGF-A leads into 6-phosphofructo-2-kinase/fructose-2,6-bisphosphatase-3 (PFKFB3) expression, which converts F6P into fructose-2,6-biphosphate (F2,6BP), an activator of phosphofructokinase-1 (PFK1) (De Bock et al. 2013; Hunt et al. 2007; Yu et al. 2017; Parra-Bonilla et al. 2010; Peters et al. 2009). PFK-1 generates fructose-1,6-bisphosphate (F1,6P2) that is converted into pyruvate and afterwards to lactate (lac) by lactate dehydrogenase-A (LDH-A) (Hunt et al. 2007; Ruan and Kazlauskas 2013; Végran et al. 2011). On stalk cells, NICD1 suppresses PFKFB3 leading to glycolysis inhibition and cells start to use fatty acids (FA) and glutamine (Gln) to sustain tricarboxylic acid (TCA) cycle, producing ATP, carbon dioxide, precursors of certain amino acids and NADH (De Bock et al. 2013; Schoors et al. 2015). FA are imported via fatty acid transporters (FATPs) and transferred to mitochondria via carnitine palmitoyltransferase I (CTP1A), undergoing fatty acid β-oxidation (FAO) (Schoors et al. 2015; Gerbod-Giannone et al. 2019; Cho 2012; Silverstein and Febbraio 2009; Elmasri et al. 2009). Gln enters the cell through several amino acids transporters or it is synthesized through the conversion of aspartate (Asp) into asparagine (Asn) by asparagine synthetase (ASNS). Citrate, from TCA, supplies FA production via fatty acid synthetase (FASN) (Elmasri et al. 2009, 2012; Huang et al. 2017; Kim et al. 2017)

(Gerbod-Giannone et al. 2019; Cho 2012; Silverstein and Febbraio 2009; Elmasri et al. 2009). In aorta rings assay, FABP4 loss impairs proliferation and migration, increases ECs apoptosis and affects angiogenic sprouting (Elmasri et al. 2009, 2012). ECs also express FA synthase (FASN) the key enzyme in FA synthesis, however it is not yet established if FA synthesis in ECs is essential for generating lipids for membrane, for signaling purposes or if it is irrelevant, since ECs are actively exposed to FA from the blood stream (Elmasri et al. 2009, 2012).

8.2.3 Glutamine Anabolism and Catabolism, a Crucial Network in ECs Survival

Recently the relevance of glutamine metabolism has been addressed in ECs, being unraveled the essential role of glutaminolysis in cell proliferation and vascular expansion. In HUVECs about 30% of the TCA carbons are derived from glutamine (Huang et al. 2017; Kim et al. 2017), being considered the major supplier of carbons for TCA cycle of ECs (Kim et al. 2017). Thus, in proliferating ECs, glutamine is an anaplerotic source of carbons, replenishing the TCA cycle via glutaminase-1 (GLS-1), and effectively supporting amino acids and nucleotide synthesis (Huang et al. 2017; Kim et al. 2017) (Fig. 8.1).

Moreover, glutamine is essential for redox signaling, via glutathione (GSH) synthesis, as glutamine is the main source of glutamate, a component of glutathione tripeptide (Fig. 8.1). In ECs, under glutamine-deprivation, GSH levels decrease, leading to an increased susceptibility to ROS-induced damage (Huang et al. 2017; Kim et al. 2017; DeBerardinis and Cheng 2010).

GLS-1 deletion or pharmacological inhibition impairs ECs proliferation, vessel sprouting and induces a senescent phenotype, supporting an ECs-activating role of glutaminolysis (Huang et al. 2017; Kim et al. 2017; Peyton et al. 2018; Eelen et al. 2018a). In glutamine scarcity *in vitro*, asparagine and α-ketoglutarate are glutamine sources, rescuing the proliferative defects of glutamine-deprived ECs (Huang et al. 2017), though deficits in asparagine synthetase (ASNS) also impair ECs proliferation (Huang et al. 2017).

Besides the lack of precise information on the relevance of glutamine metabolism during angiogenesis, in physiological and pathophysiological conditions; GLS-1 and ASNS can be attractive new targets for novel anti-angiogenic strategies directed to ECs metabolism.

8.2.4 Pentose Phosphate Pathway (PPP), Crucial but Dubious for ECs Function

The PPP does not involve glucose oxidation, since its primary role is anabolic rather than catabolic, but glycolytic intermediates can be diverted to PPP. For instance, glucose-6-phosphate (G6P), and other glycolytic intermediates, can be converted in PPP providing ribose units essential for nucleotide synthesis and NADPH, fundamental for ROS scavenging (Riganti et al. 2012; Peiró et al. 2016).

PPP is divided in two branches, the oxidative PPP (ox-PPP) and the non-oxidative PPP (non-ox-PPP). Glucose-6-phosphate dehydrogenase (G6PD) is a rate-limiting enzyme of ox-PPP, responsible for the conversion of G6P in ribulose-5-phosphate (R5P), whereas transkelotase (TK) is responsible for R5P and NADPH generation via non-ox-PPP (Zang et al. 2000; Vizán et al. 2009). Targeting G6PD and TK decreases PPP and leads to reduced VEGF-induced proliferation, migration and tube forming capacity of ECs (Vizán et al. 2009; Leopold et al. 2003). Importantly, assays using *in vitro* and *in vivo* ECs and vascular models showed that upon increased ROS the accelerated rate of PPP correlates with glutathione turnover and ROS scavenging (Kaczara et al. 2018), indicating that PPP is a suitable target to act on vascular remodeling. However, some studies on vascular damage associated to hyperglycemia, defend that the over-activation of PPP is an underlying pathophysiological mechanism, since it increases ROS generation through the activation of NADPH-oxidase (Peiró et al. 2016). In sum, it is notorious that the redox balance plays a role in

ECs and vascular remodeling and PPP is a relevant metabolic pathway thereon. However more studies are needed to ascertain the favorable or deleterious role of PPP in ECs functioning and vascular networking, namely in cancer.

8.2.5 Hexosamine Biosynthesis Pathway (HBP), an Unexplored but Promising Pathway in ECs

HBP uses glutamine, glucose, acetyl-CoA and uridine to mediate *O*-glycosylation and *N*-glycosylation, whose deregulation has been already linked to diseases, as diabetes and cancer (Slawson et al. 2010; Chiaradonna et al. 2018). Although the precise mechanism of HBP in ECs metabolism is not yet established, HBP has a nutrient sensing role (Chiaradonna et al. 2018), being expectable to have a fundamental function in ECs sprouting into avascular and nutrient-scare tissues. Moreover, a study regarding the metabolic remodeling in toxic ECs dysfunction showed that HBP was altered, which would affect proteins glycosylation in particular membrane proteins that are essential for ECs function, as I-CAM and V-CAM (Zhong et al. 2017). Considering cancer cells, the expression of the HBP rate-limiting enzyme, glutamine fructose-6-phosphate amidotransferase (GFAT), is usually upregulated, accounting for cancer associated aberrant glycosylation (Hanover et al. 2018; Li et al. 2017). Besides being deeply related to augmented glucose availability (Vasconcelos-Dos-Santos et al. 2017; Phoomak et al. 2017), increased HBP rate is also described as part of the profile of highly invasive cancer cells (Vasconcelos-Dos-Santos et al. 2017; Phoomak et al. 2017; de Queiroz et al. 2019) and of cancer cells undergoing EMT (Carvalho-cruz et al. 2018; Zhang et al. 2019; Lucena et al. 2016). Furthermore, inhibition of HBP in cancer cells drives differentiation and death (Asthana et al. 2018). As above mentioned, hyper-activated HBP cancer cell phenotypes rely on glucose, which is a common feature among ECs. Also, the increased invasion/ migration ability is very important in new vessels formation, thus HBP will be for sure part of the altered metabolic panel in ECs activation upon cancer angiogenesis.

8.2.6 Oxidative Phosphorylation (OXPHOS), a Surrogated Pathway in ECs

OXPHOS is an efficient pathway for energy production, but although ECs have functional mitochondria, they rely minimally on OXPHOS for ATP generation, using glucose derived compounds, due to the afore mentioned specialized glycolytic and lactate producing phenotype (De Bock et al. 2013). Besides ATP production, OXPHOS also provides metabolites to support cell proliferation, which can have an indispensable role during ECs proliferation. ECs increase their oxygen consumption upon cell activation during angiogenesis and mitochondrial inhibition, leading to the death of these proliferating ECs (Blecha et al. 2017; Coutelle et al. 2014; Rohlena et al. 2011; Don et al. 2003; Orecchioni et al. 2015). Nevertheless, the exact effect of OXPHOS and which are the most relevant compounds to sustain it remains to be fully understood in ECs.

8.2.7 Reactive Oxygen Species (ROS), an Essential Motor in Angiogenesis

ROS are highly reactive molecules that derive from incomplete reduction of molecular oxygen. This group includes free radicals, such as superoxide (O_2^-) and non-radicals, as hydrogen peroxide (H_2O_2) (Santoro 2018).

In ECs, superoxide-generating enzymes (NOX; NADPH oxidases) are a major source of ROS (Manuneedhi Cholan et al. 2017). These trans-membrane proteins transport electrons converting oxygen to superoxide. In vascular cells there are four NOX isoforms (NOX-1, NOX-2, NOX-4 and NOX-5) but only NOX-4 preferentially produces H_2O_2 (Manuneedhi Cholan et al. 2017; Panieri and Santoro 2015).

High levels of ROS can promote oxidative stress and be detrimental for ECs, however recent evidences support that physiological levels of ROS can be pro-angiogenic (Manuneedhi Cholan et al. 2017; Kim and Byzova 2014; Shafique et al. 2017), stimulating essential stages during vascular formation: proliferation, migration, sprouting and tubule formation (Manuneedhi Cholan et al. 2017). For example, NOX-2 can lead to VEGFR activation and, therefore, increase EC proliferation (Chen et al. 2014). Moreover, NOX-4-derived H_2O_2 can stimulate NOX-2.22 and can stabilize HIF1α, leading to an induction of VEGF expression that promotes tumor angiogenesis (Helfinger et al. 2016).

VEGF signaling pathway is a target of ROS in ECs. Besides VEGF stimulation of ROS production through the activation of NOXs (Ushio-Fukai and Nakamura 2008), ROS promote VEGFR2 dimerization and autophosphorylation, leading to the activation of signal transduction and consequently angiogenesis promotion (Kim and Byzova 2014; Colavitti et al. 2002). Thus, ROS act both up- and downstream to VEGF/VEGFR2 interaction.

Several studies have identified VEGF-independent mechanisms of ROS-induced angiogenesis (Kim and Byzova 2014; Huang and Nan 2019; Kim et al. 2013). The CEP/TLR2/MyD88 axis is one of those mechanisms, that consists in the generation of new lipid oxidation products (CEP, ω-carboxyethylpyrrole), inducing angiogenesis via the TLR2/MyD88 pathway. This pathway activates Rac1, a key factor that promotes cell migration and adhesion (West et al. 2010). Another mechanism occurs in response to ROS accumulation, which selectively activates ataxia-telangiectasia mutated kinase (ATM) an activator of DNA repair. ATM activation decreases p38α phosphorylation, the most prominent isoform of p38, promoting ECs proliferation and survival (Okuno et al. 2012).

Tumor-associated angiogenesis is more affected by oxidative stress than physiological angiogenesis, since substantial amounts of ROS are produced by cells in TME, such tumor and inflammatory cells (Kim and Byzova 2014; Huang and Nan 2019). Taking this into account, ROS-scavenging strategies could be applied in the treatment of diseases with pathological angiogenesis, as vascular diseases and cancer (Prieto-Bermejo and Hernández-Hernández 2017). For instance, the use of dietary antioxidants such as vitamin C and E (Nespereira 2003) could be an option, however it is difficult to determine the exact dosage since they are nonspecific and can also affect physiological angiogenesis. Another option could be pharmacological inhibitors of NADPH oxidase. However, *in vivo* studies using such inhibitors are scarce (Coso et al. 2012).

Therefore, oxidative-stress induced angiogenesis needs to be further studied and the molecular mechanisms underlying the effect of the use of antioxidant strategies should be explored.

8.3 Tumor Microenvironment (TME) Commands the Regulation of ECs Metabolism

The metabolic adaptation of cells is highly dependent on and regulated by the microenvironment. In the context of cancer, this is true for cancerous and non-cancerous cells. Thus, the sharing of signaling molecules and the bioavailability of organic compounds in the TME certainly play a decisive role in the regulation of the metabolic remodeling of ECs.

8.3.1 Metabolic Remodeling of ECs During Tumorigenesis, a Driving Force for Cancer Progression

Cancer angiogenesis is characterized by structurally and functionally defective vessels, with poor mural cells coverage. This chaotic network characterized by a loss of ECs organization triggers a perturbed blood flow that facilitates metastasis (DeClerck 2010; Jain 2014; Chang et al. 2002). In TME the high availability of VEGF, bFGF and other pro-angiogenic molecules prompts the modulation of ECs metabolism and promotes the growth of new blood vessels to fulfill cancer cells' needs (Folkman 1971; Hanahan and Weinberg 2011; Li et al. 2018). In this context, ECs must adapt their metabolism to meet minimal energy requirements to survive in areas with nutrient and oxygen scarcity (Potente et al. 2011).

Currently, it is well established that cancer cells contribute to a highly pro-angiogenic microenvironment by secreting growth factors and cytokines, as VEGF, bFGF, angiopoietins (Ang), hepatocyte growth factor (HGF), epidermal growth factor (EGF) and platelet-derived growth factor (PDGF) (Senger et al. 1983). In comparison to normal ECs (NECs), tumor ECs (TECs) are more resistant to apoptosis and are more reactive to paracrine signaling, inducing cell proliferation and migration to support tumor neo-angiogenesis (Senger et al. 1983; Bussolati et al. 2003).

The capacity of ECs to adapt to fluctuations in oxygen reveal that ECs metabolism is modulated and foments a more aggressive cancer phenotype (Hausenloy and Yellon 2011). Hence, in hypoxic environments, ECs increase their migratory capacity and tube forming potential (Martinive et al. 2006).

In tumorigenesis, cancer cells functioning promotes angiogenesis. In addition to VEGF production, the increased release of lactate by tumor cells stimulates angiogenesis (San-Millán and Brooks 2017; Goel and Mercurio 2013), and this was proved by different ways. Oxamate, a LDHA inhibitor reduces angiogenesis (Hunt et al. 2008) and lactate produced by cancer cells stimulates the pro-angiogenic NFκB/IL8 axis and HIF1α stabilization, leading to the activation of TECs (Hunt et al. 2007; Végran et al. 2011; Sonveaux et al. 2012). In fact, these signaling and phenotypic changes in TECs are concomitant with the remodeling of the metabolic players profile. *In vitro* studies showed that ECs exposed to lactate rich glioblastoma cells-conditioned media increase the expression of monocarboxylate transporter 1 (MCT1) (Miranda-Gonçalves et al. 2017), a very important lactate transporter described in different cancer models as relevant in the uptake of lactate to sustain OXPHOS (Silva et al. 2015; Lopes-Coelho et al. 2017). In the context of a TME-driven metabolic symbiosis, cancer cells and non-cancerous cells share relevant metabolic compounds and in a certain point, ECs can benefit from organic molecules released by cancer cells. A good example would be lactate secreted as a consequence of the high glycolytic rate. Nevertheless, the metabolic alterations of ECs, during tumor growth, do not seem to be extensively studied.

Despite being more responsive to paracrine and autocrine pro-angiogenic signaling, TECs are also more resistant to chemotherapy than NECs (Bussolati et al. 2003; Amin et al. 2006; Kurosu et al. 2011; Matsuda et al. 2010; Yamamoto et al. 2012). TECs associated to high-metastatic melanoma present higher proliferative and invasive rates in comparison to TECs associated to low-metastatic melanoma and NECs (Ohga et al. 2012).

Upon cancer cell stimuli, ECs activation leads to a hyperglycolytic phenotype. In fact, LDHA inhibition in ECs impair endothelial cell growth, indicating that glycolysis is fundamental for ECs functioning (Parra-Bonilla et al. 2010). TECs, in comparison to NECs, present an increased expression of enzymes and transporters that will sustain the hyper-glycolytic TECs phenotype, favoring the production of glucose-dependent biomass (Cantelmo et al. 2016).

TECs express high levels of GLUT-1 and PFKFB3, the main glycolytic activator in ECs (Xu et al. 2014; Cantelmo et al. 2016; Yeh et al. 2007; Trenti et al. 2017). The crucial role of PFKFB3 in ECs metabolism is documented in an *in vivo* model, in which the deletion of PFKFB3 in ECs decreases vessel perfusion of tumors (Xu et al. 2014). In hypoxic TME, GLUT-1 expression in TECs is regulated by VEGF (Yeh et al. 2007). In the same way PFKFB3 is also upregulated by hypoxia and by pro-inflammatory cytokines, as IL-1β and TNF-α, which also have a role in angiogenesis (Cantelmo et al. 2016; Yeh et al. 2007; Trenti et al. 2017).

Unfortunately, few studies have been developed concerning ECs metabolism in particular in cancer context. Most data on cancer metabolism are related to the metabolic adaptation of cancer cells, although non-cancerous cells are starting to be studied in TME, ECs are not deeply explored.

Since ECs adapt their metabolism, in part, by increasing glycolysis to support energy demands for proliferation and sprouting (Pfeiffer et al. 2001), targeting PFKFB3 seems to be a good alternative strategy to fight angiogenesis. Mice treated with high doses of 3PO showed the disintegration of vessels in tumors due to TECs' decreased proliferation and increased cell death (Conradi et al. 2017). On another hand, low doses of 3PO induced tumor vessel stability and decreased cancer cell invasion, vessels intravasion and metastasis and improved the delivery of chemotherapeutic agents, not affecting blood vessels of heathy tissues (Cantelmo et al. 2017). This phenomenon appears to be dose dependent and it seems that may be more beneficial to normalize the hyper-glycolytic phenotype in ECs, without eliminating the glycolytic flux.

In recent years, it has been highlighted that tumor vessel stabilization could be a strategy to improve anti-cancer drug delivery (Conradi et al. 2017). In this context, tackling glycolysis seems to be an attractive therapeutic approach, but this must be tightly controlled.

The role of PFKFB3 in cancer promotion is not resumed to ECs. PFKFB3 is expressed in various tumors and in hepatocellular carcinomas it has been correlated with advanced stages and poor prognosis (Shi et al. 2018; Peng et al. 2018). PFKFB3 inhibition blocks glucose consumption in cancer cells and reduces tumor growth *in vivo* (Shi et al. 2018). In breast cancer, PFKFB3 expression is associated with poor overall survival and their *in vitro* inhibition decreases the release of VEGF by cancer cells, inhibiting the angiogenic switch in TECs (Peng et al. 2018). In melanoma and pancreatic cancer *in vivo* models with haplo-deficient PFKFB3 in ECs, PFKFB3 depletion does not affect tumor growth but decreases invasion, intravasion and metastasis by inducing tumor vessel normalization through the reduction of glycolysis in TECs and through the maintenance of pericytes in a more quiescent and adhesive state (Cantelmo et al. 2016). Moreover, cisplatin, a conventional alkylating drug used in cancer therapy, induces PFKFB3 acetylation in lysine residues, leading to PFKFB3 accumulation and glycolysis promotion. Accordingly, a xenograft cancer model shows that the inhibition of PFKFB3 sensitizes cells to cisplatin treatment (Li et al. 2018).

As above mentioned, PFKFB3 pharmacological inhibition is difficult to adopt as it produces opposite effects on cancer cells and TECs; high doses of 3PO inhibits tumor growth and metastasis (Conradi et al. 2017), whereas low doses of 3PO induces vessel stabilization, improving chemotherapy efficacy (Cantelmo et al. 2016). The concomitant inhibition of HK, PFK and LDHA has also been tested in breast cancer using and epigallocatechin-3-gallate (EGCG, a polyphenol), which is capable of reducing breast cancer cell proliferation and increasing cell death. The effect is related to HIF1α downregulation and glycolysis suppression (Wei et al. 2018). This anti-tumoral effect of EGCG can also be related to ECs disturbance, as it was demonstrated that the pro-EGCG (a more stable form of EGCG) avoids the activation of ECs by downregulating the production and release of VEGF by tumor cells (Wang et al. 2013, 2018). The use of this polyphenol, besides targeting cancer cells, can suppress TEC activation, not only by decreasing VEGF release by cancer cells, but also by acting on glucose metabolism of TECs, though this is not known yet.

Regarding glutamine metabolism in ECs, its regulation relies on the action of TGF-β1, leading to the activation protein phosphatase 2A (PP2A)-mediated Raf-MEK-ERK signaling (Guo et al. 2016). This signaling pathway has been characterized as playing a significant role in the occurrence and development of cancer, being an attractive target for the development of new anticancer drugs (Li et al. 2016; Hilger et al. 2002). Moreover, increased endoglin expression, a component of the TGF-β-receptor complex, has been correlated to the activation of endothelium at the tumor edges (Miller et al. 1999). Treatments against this signaling pathway is a novel therapeutic perspective that would fight cancer not only by acting on cancer cells but also by interfering with TECs and the angiogenic process.

8.3.2 Hypoxic TME – Key Regulators of EC Metabolic Remodeling

During cancer angiogenesis, hypoxia activates the expression of HIF1α that will be responsible for the expression of VEGF amongst other pro-angiogenic stakeholders (e.g. VEGFR, PDGF, ANGPT1 and 2), promoting ECs proliferation, sprouting and vascular remodeling (Ceradini et al. 2004; Liu et al. 1995). Experimental models prove the orchestrating role of HIF1α in cancer angiogenesis by demonstrating that the loss of HIF1α abrogates the VEGF-mediated autocrine loop, impairing the highly proliferative and migratory behavior of ECs that culminates with the inhibition of blood vessel growth in solid tumors (Tang et al. 2004). Additionally, it was shown that the mechanism through which salinomycin (SAL) decreases the VEGF-A release by breast cancer cells *in vivo* involves the interference with HIF1α/VEGF signaling (Dewangan et al. 2019). Moreover, breast cancer cells expressing ALDH1A1 (stemness marker) express higher levels of HIF1α and VEGF, potentiating ECs proliferation, migration and tube formation (Ciccone et al. 2018).

Hypoxia is also a crucial regulator of the expression of genes related to glucose transport and glucose/pyruvate/lactate metabolism (De Bock et al. 2013; Weigand et al. 2012; Semenza 2003) that will account for the angiogenic switch underlied by the higher glycolytic flux in TECs (Cantelmo et al. 2016). In HUVECs exposed to conditions mimicking hypoxia, HIF1α induces the upregulation of genes related to glucose transport, such as *SLC2A1* (the gene coding glucose transporter protein type 1, GLUT-1), hexose metabolism (e.g. *HK-2*, which encodes hexokinase 2) and genes responsive to hypoxia, such as *ALDOC*, gene for aldolase C (Weigand et al. 2012). In ECs, hypoxia also leads to an increased expression of PFKFB3 (Xu et al. 2014) as well as of pyruvate dehydrogenase kinase-1 (PDK1), an enzyme that inhibits PDH (pyruvate dehydrogenase), responsible for the conversion of pyruvate into acetyl-CoA (Wu et al. 2017).

Again, targeting ECs metabolism triggers a panoply of events that will for sure affect tumor viability.

8.3.3 The Influence of Non-cancerous Cells of TME in ECs Metabolic Alterations – Partners in Crime

TME is not a static environment and the crosstalk between stromal and cancer cells are essential to modulate the metabolism and behavior of the cancerous and non-cancerous cells within the tumor. This whole network must be considered for the improvement of new cancer therapies.

Unfortunately, very few studies have been developed in order to understand the crosstalk between ECs and malignant and non-malignant cells in TME that will favor cancer progression. However, we can integrate some other studies on cancer metabolism and try to find a clue on what can happen in ECs metabolic remodeling by TME. For instance, in certain subsets of leukemia, asparagine becomes an essential amino acid, ensuring the clinical effectiveness of the asparaginase therapy. However, some mechanisms of resistance have been developed by leukemia cells (Hinze et al. 2019; Lee et al. 2019), namely the development of an alternative asparagine synthetic pathways using glutamate as a substrate. In this context, fatty-tissue adipocytes contribute for resistance to asparaginase in leukemia cells by secreting glutamine that will be used by cancer cells to produce glutamate and consequently asparagine (Ehsanipour et al. 2013). Moreover, it has been described that cancer associated fibroblasts (CAFs), by increasing autophagy, contribute to a glutamine-rich microenvironment that promotes glutamine catabolism and decreases glutamine synthesis in cancer cells (Ko et al. 2011). Perhaps, glutamine released by adipocytes and CAFs not only serves as a source of glutamine to fuel cancer, but also feeds TECs.

Controversially, hypoxic tumor-associated macrophages (TAMs) are capable of upregulating the expression of REDD1 (a negative regulator of mTOR), increasing the glycolytic flux. The upregulated glycolysis in TAMs leads to a competitive state of TECs and TAMs for glucose, promoting in TECs the formation of quiescent vascular junctions (characteristic of functional vessels) and the stabilization of tumor vasculature, reverting the abnormal and chaotic neo-angiogenesis and metastasis (Wenes et al. 2016) and simultaneously contributing for a more effective drug delivery.

Thereby, a better knowledge on the way TME interferes with ECs metabolism promoting angiogenesis will pave the path on the way of designing new drugs targeting ECs and disrupting the vascular network from which tumor survival depends on.

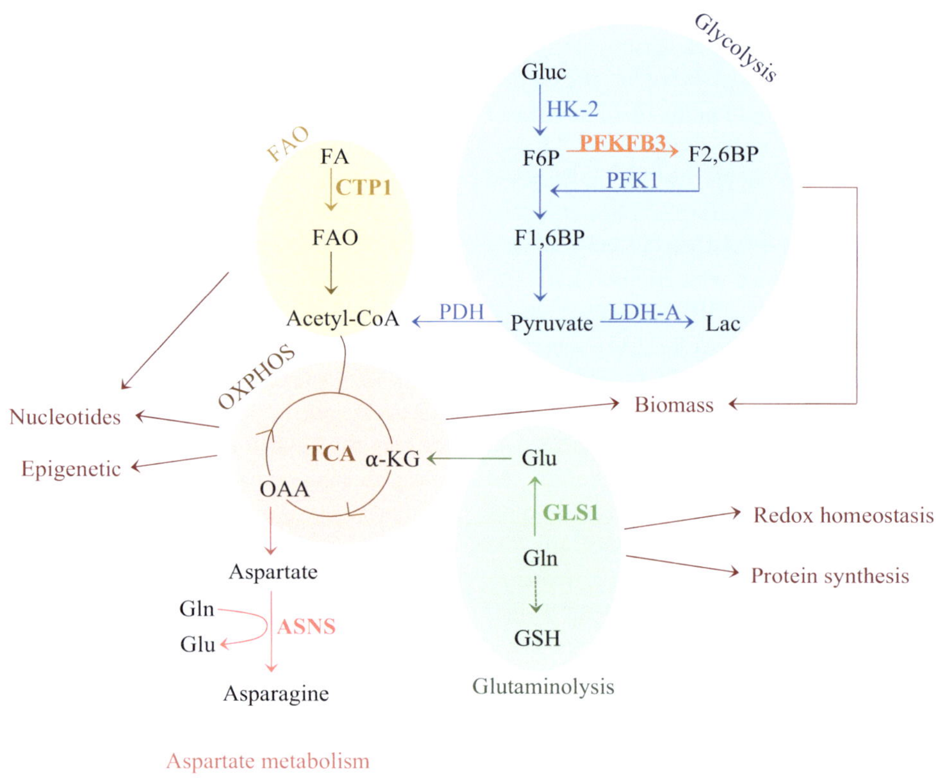

Fig. 8.2 Alternatives for the conventional anti-angiogenic therapies are urgent and targeting EC metabolism can contribute for this improvement. Regarding ECs, and since PFKFB3 inhibition promotes ECs quiescence by abrogating glycolysis (Xu et al. 2014; Schoors et al. 2014a) with the consequent reduction of tumor growth and vessel perfusion (Xu et al. 2014; Shi et al. 2018); it is tempting to target PFKFB3 in order to tackle neoangiogenesis. Moreover, targeting other enzymes involved in cellular metabolism (glycolysis: HK-2, PFK-1, LDH-A; FAO: CPT1A; glutamine metabolism: GLS-1 and ASNS), (Schoors et al. 2015) (Huang et al. 2017; Kim et al. 2017; Peyton et al. 2018; Eelen et al. 2018b) (Huang et al. 2017) also affects cancer cells survival, tackling not only ECs but also cancer cells metabolism
ASNS asparagine synthetase, *CTP1* carnitine palmitoyl-transferase I, *FA* fatty acids, *FAO* fatty acid β-oxidation, *F1,6P* fructose-1,6-bisphosphate, *F2,6BP* fructose-2,6-biphosphate, *F6P* fructose-6-phosphate, *Gln* glutamine, *Glu* glutamate, *Gluc* glucose, *GLS1* glutaminase-1, *Gluc* glucose, *GSH* glutathione, *HK-2* hexokinase-2, *Lac* lactate, *LDH-A* lactate dehydrogenase-A, *OAA* oxaloacetic acid, *OXPHOS* Oxidative phosphorylation, *PFK-1* phosphofructokinase-1, *PFKFB3* 6-phosphofructo-2-kinase/fructose-2,6-bisphosphatase-3, *TCA* tricarboxylic acid, *α-KG* alpha-ketoglutarate

8.4 Final Remarks and Future Perspectives

During tumor growth, it is pivotal to increase the supply of oxygen and nutrients and neo-angiogenesis takes care of this. In the past years, anti-angiogenic therapeutic strategies focusing on the inhibition of VEGF signaling seemed to be a promising strategy to treat solid tumors in general. However, it failed in the majority of cases, demonstrating that we do not fully know the ECs biology and the cancer angiogenesis process.

The role of ECs metabolism in angiogenesis started to be explored recently and in line with this rationale, metabolism seems a very attractive target (Fig. 8.2), in part, because the metabolic remodeling of ECs underlies the pro-angiogenic phenotypical remodeling activated by TME. Since ECs metabolism drives angiogenesis, a deeper knowledge on metabolic dynamics in healthy and pathological angiogenesis will for sure open new perspectives on a next generation anti-angiogenic therapy to fight cancer.

Acknowledgments The authors acknowledge iNOVA-4Health – UID/Multi/04462/2013, a program financially supported by Fundação para a Ciência e Tecnologia/Ministério da Educação e Ciência, through national funds and co-funded by FEDER under the PT2020 Partnership Agreement.

References

Adams RH, Alitalo K (2007) Molecular regulation of angiogenesis and lymphangiogenesis. Nat Rev Mol Cell Biol 8:464–478. https://doi.org/10.1038/nrm2183

Amin DN, Hida K, Bielenberg DR, Klagsbrun M (2006) Tumor endothelial cells express epidermal growth factor receptor (EGFR) but not ErbB3 and are responsive to EGF and to EGFR kinase inhibitors. Cancer Res 66:2173–2180. https://doi.org/10.1158/0008-5472.CAN-05-3387

Asthana A, Ramakrishnan P, Vicioso Y et al (2018) Hexosamine biosynthetic pathway inhibition leads to AML cell differentiation and cell death. Mol Cancer Ther 17:2226–2237. https://doi.org/10.1158/1535-7163.mct-18-0426

Bergers G, Benjamin LE (2003) Tumorigenesis and the angiogenic switch. Nat Rev Cancer 3:401–410. https://doi.org/10.1038/nrc1093

Blanco R, Gerhardt H (2013) VEGF and notch in tip and stalk cell selection. Cold Spring Harb Perspect Med 3:1–19

Blecha J, Novais SM, Rohlenova K et al (2017) Antioxidant defense in quiescent cells determines selectivity of electron transport chain inhibition-induced cell death. Free Radic Biol Med 112:253–266. https://doi.org/10.1016/j.freeradbiomed.2017.07.033

Bussolati B, Deambrosis I, Russo S et al (2003) Altered angiogenesis and survival in human tumor-derived endothelial cells. FASEB J 17:1159–1161

Cantelmo AR, Conradi LC, Brajic A et al (2016) Inhibition of the glycolytic activator PFKFB3 in endothelium induces tumor vessel normalization, impairs metastasis, and improves chemotherapy. Cancer Cell 30:968–985. https://doi.org/10.1016/j.ccell.2016.10.006

Cantelmo AR, Pircher A, Kalucka J, Carmeliet P (2017) Vessel pruning or healing: endothelial metabolism as a novel target? Expert Opin Ther Targets 21:239–247. https://doi.org/10.1080/14728222.2017.1282465

Carvalho-cruz P, Alisson-Silva F, Todeschini AR, Dias WB (2018) Cellular glycosylation senses metabolic changes and modulates cell plasticity during epithelial to mesenchymal transition. Dev Dyn 247:481–491. https://doi.org/10.1002/dvdy.24553

Ceradini DJ, Kulkarni AR, Callaghan MJ et al (2004) Progenitor cell trafficking is regulated by hypoxic gradients through HIF-1 induction of SDF-1. Nat Med 10:858–864. https://doi.org/10.1038/nm1075

Chang YS, di Tomaso E, McDonald DM et al (2002) Mosaic blood vessels in tumors: frequency of cancer cells in contact with flowing blood. Proc Natl Acad Sci 97:14608–14613. https://doi.org/10.1073/pnas.97.26.14608

Chen L, Hou X, Xiao J et al (2014) Both hydrogen peroxide and transforming growth factor beta 1 contribute to endothelial Nox4 mediated angiogenesis in endothelial Nox4 transgenic mouse lines. Biochim Biophys Acta Mol basis Dis 1842:2489–2499. https://doi.org/10.1016/j.bbadis.2014.10.007

Chiaradonna F, Ricciardiello F, Palorini R (2018) The nutrient-sensing Hexosamine biosynthetic pathway as the hub of Cancer metabolic rewiring. Cells 7:pii: E53. https://doi.org/10.3390/cells7060053

Cho S (2012) CD36 as a therapeutic target for endothelial dysfunction in stroke. Curr Pharm Des 18:3721–3730. https://doi.org/10.2174/138161212802002760

Choi BH, Pagano M, Da W (2014) Plk1 protein phosphorylates phosphatase and tensin homolog (PTEN) and regulates its mitotic activity during the cell cycle. J Biol Chem 289:14066–14074. https://doi.org/10.1074/jbc.M114.558155

Ciccone V, Terzuoli E, Donnini S et al (2018) Stemness marker ALDH1A1 promotes tumor angiogenesis via retinoic acid/HIF-1α/VEGF signalling in MCF-7 breast cancer cells\. J Exp Clin Cancer Res 37:311. https://doi.org/10.1186/s13046-018-0975-0

Colavitti R, Pani G, Bedogni B et al (2002) Reactive oxygen species as downstream mediators of Angiogenic signaling by vascular endothelial growth factor Receptor-2/KDR. J Biol Chem 277:3101–3108. https://doi.org/10.1074/jbc.M107711200

Conradi LC, Brajic A, Cantelmo AR et al (2017) Tumor vessel disintegration by maximum tolerable PFKFB3

blockade. Angiogenesis 20:599–613. https://doi.org/10.1007/s10456-017-9573-6
Coso S, Harrison I, Harrison CB et al (2012) NADPH oxidases as regulators of tumor angiogenesis: current and emerging concepts. Antioxid Redox Signal 16:1229–1247. https://doi.org/10.1089/ars.2011.4489
Coutelle O, Hornig-Do HT, Witt A et al (2014) Embelin inhibits endothelial mitochondrial respiration and impairs neoangiogenesis during tumor growth and wound healing. EMBO Mol Med 6:624–639. https://doi.org/10.1002/emmm.201303016
De Bock K, Georgiadou M, Schoors S et al (2013) Role of PFKFB3-driven glycolysis in vessel sprouting. Cell 154:651–663. https://doi.org/10.1016/j.cell.2013.06.037
de Queiroz RM, Oliveira IA, Piva B et al (2019) Hexosamine biosynthetic pathway and glycosylation regulate cell migration in melanoma cells. Front Oncol 17:2226–2237. https://doi.org/10.3389/fonc.2019.00116
DeBerardinis RJ, Cheng T (2010) Q's next: the diverse functions of glutamine in metabolism, cell biology and cancer. Oncogene 29:313–324. https://doi.org/10.1038/onc.2009.358
DeClerck K (2010) The role of hypoxia and acidosis in promoting metastasis and resistance to chemotherapy. Front Biosci 15:213–225. https://doi.org/10.2741/3616
Dewangan J, Srivastava S, Mishra S et al (2019) Salinomycin inhibits breast cancer progression via targeting HIF-1α/VEGF mediated tumorangiogenesis in vitro and in vivo. Biochem Pharmacol. pii:S0006-:30161–3
Domingues G, Fernandes SG, Serpa J (2015) Chapter 3: Dynamics of VEGF-A and its receptors in Cancer vascularization – an overview In: Understand cancer: research and treatment. iConcept Press–ISBN 978-1-922227-386
Don AS, Kisker O, Dilda P et al (2003) A peptide trivalent arsenical inhibits tumor angiogenesis by perturbing mitochondrial function in angiogenic endothelial cells. Cancer Cell 3:497–509. https://doi.org/10.1016/S1535-6108(03)00109-0
Draoui N, De Zeeuw P, Carmeliet P (2017) Angiogenesis revisited from a metabolic perspective : role and therapeutic implications of endothelial cell metabolism. Open Biol 7(12). pii: 170219. https://doi.org/10.1098/rsob.170219
Eelen G, de Zeeuw P, Simons M, Carmeliet P (2015) Endothelial cell metabolism in normal and diseased vasculature. Circ Res 116:1231–1244. https://doi.org/10.1161/CIRCRESAHA.116.302855
Eelen G, de Zeeuw P, Treps L et al (2018a) Endothelial cell metabolism. Physiol Rev 98:3–58. https://doi.org/10.1152/physrev.00001.2017
Eelen G, Dubois C, Cantelmo AR et al (2018b) Role of glutamine synthetase in angiogenesis beyond glutamine synthesis. Nature 561:63–69. https://doi.org/10.1038/s41586-018-0466-7
Ehsanipour EA, Sheng X, Behan JW et al (2013) Adipocytes cause leukemia cell resistance to l-asparaginase via release of glutamine. Cancer Res 73:2998–3006. https://doi.org/10.1158/0008-5472.CAN-12-4402
Elmasri H, Karaaslan C, Teper Y et al (2009) Fatty acid binding protein 4 is a target of VEGF and a regulator of cell proliferation in endothelial cells. FASEB J 23:3865–3873. https://doi.org/10.1096/fj.09-134882
Elmasri H, Ghelfi E, Yu CW et al (2012) Endothelial cell-fatty acid binding protein 4 promotes angiogenesis: role of stem cell factor/c-kit pathway. Angiogenesis 15:457–468. https://doi.org/10.1007/s10456-012-9274-0
Folkman J (1971) Tumor angiogenesis: therapeutic implications. N Engl J Med 285(21):1182–1186
Gerbod-Giannone MC, Dallet L, Naudin G et al (2019) Involvement of caveolin-1 and CD36 in native LDL endocytosis by endothelial cells. Biochim Biophys Acta – Gen Subj 1863:830–838. https://doi.org/10.1016/j.bbagen.2019.01.005
Ghesquière B, Wong BW, Kuchnio A, Carmeliet P (2014) Metabolism of stromal and immune cells in health and disease. Nature 511:167–176. https://doi.org/10.1038/nature13312
Gimbrone MA, Leapman SB, Cotran RS, Folkman J (1972) Tumor dormancy in vivo by prevention of neovascularization. J Exp Med 136:261–276
Goel HL, Mercurio AM (2013) VEGF targets the tumour cell. Nat Rev Cancer 13:871–882. https://doi.org/10.1038/nrc3627
Guo Y, Deng Y, Li X et al (2016) Glutaminolysis was induced by TGF-β1 through PP2Ac regulated Raf-MEK-ERK signaling in endothelial cells. PLoS One 11:e0162658. https://doi.org/10.1371/journal.pone.0162658
Hanahan D, Weinberg RA (2011) Hallmarks of cancer: the next generation. Cell 144:646–674. https://doi.org/10.1016/j.cell.2011.02.013
Hanover JA, Chen W, Bond MR (2018) O-GlcNAc in cancer: an Oncometabolism-fueled vicious cycle. J Bioenerg Biomembr 50:155–173. https://doi.org/10.1007/s10863-018-9751-2
Hausenloy DJ, Yellon DM (2011) The therapeutic potential of ischemic conditioning: an update. Nat Rev Cardiol 8:619–629. https://doi.org/10.1038/nrcardio.2011.85
Helfinger V, Henke N, Harenkamp S et al (2016) The NADPH oxidase Nox4 mediates tumour angiogenesis. Acta Physiol 216:435–446. https://doi.org/10.1111/apha.12625
Helmlinger G, Endo M, Ferrara N et al (2002) Formation of endothelial cell networks. Nature 405:139–141. https://doi.org/10.1038/35012132
Hilger RA, Scheulen ME, Strumberg D (2002) The Ras-Raf-MEK-ERK pathway in the treatment of cancer. Onkologie 25:511–518. https://doi.org/10.1159/000068621
Hillen F, Griffioen AW (2007) Tumour vascularization: sprouting angiogenesis and beyond. Cancer

Metastasis Rev 26:489–502. https://doi.org/10.1007/s10555-007-9094-7
Hinze L, Pfirrmann M, Karim S et al (2019) Synthetic lethality of Wnt pathway activation and asparaginase in drug-resistant acute leukemias. Cancer Cell 35:664–676. https://doi.org/10.1016/j.ccell.2019.03.004
Huang Y-J, Nan G-X (2019) Oxidative stress-induced angiogenesis. J Clin Neurosci 63:13–16. https://doi.org/10.1016/j.jocn.2019.02.019
Huang H, Vandekeere S, Kalucka J et al (2017) Role of glutamine and interlinked asparagine metabolism in vessel formation. EMBO J 36:2334–2352. https://doi.org/10.15252/embj.201695518
Hunt TK, Aslam RS, Beckert S et al (2007) Aerobically derived lactate stimulates revascularization and tissue repair via redox mechanisms. Antioxid Redox Signal 9:1115–1124. https://doi.org/10.1089/ars.2007.1674
Hunt TK, Aslam R, Hussain Z, Beckert S (2008) Lactate, with oxygen, incites angiogenesis. In: Advances in experimental medicine and Biology – in oxygen transport to tissue XXIX. Springer, New York, pp 73–80
Jain RK (2014) Antiangiogenesis strategies revisited: from starving tumors to alleviating hypoxia. Cancer Cell 26:605–622. https://doi.org/10.1016/j.ccell.2014.10.006
Jung S, Kleinheinz J (2013) Angiogenesis – the key to regeneration. In: Regenerative medicine and tissue engineering. IntechOpen, Rijeka, pp 453–473
Kaczara P, Proniewski B, Lovejoy C et al (2018) CORM-401 induces calcium signalling, NO increase and activation of pentose phosphate pathway in endothelial cells. FEBS J 285:1346–1358. https://doi.org/10.1111/febs.14411
Kim Y-W, Byzova TV (2014) Oxidative stress in angiogenesis and vascular disease. Blood 123:625–631. https://doi.org/10.1182/blood-2013-09-512749
Kim Y-W, West XZ, Byzova TV (2013) Inflammation and oxidative stress in angiogenesis and vascular disease. J Mol Med 91:323–328. https://doi.org/10.1007/s00109-013-1007-3
Kim B, Li J, Jang C, Arany Z (2017) Glutamine fuels proliferation but not migration of endothelial cells. EMBO J 36. https://doi.org/10.15252/embj.201796436
Ko YH, Lin Z, Flomenberg N et al (2011) Glutamine fuels a vicious cycle of autophagy in the tumor stroma and oxidative mitochondrial metabolism in epithelial cancer cells: implications for preventing chemotherapy resistance. Cancer Biol Ther 12:1085–1097. https://doi.org/10.4161/cbt.12.12.18671
Kowanetz M, Ferrara N (2006) Vascular endothelial growth factor signaling pathways: therapeutic perspective. Clin Cancer Res 12:5018–5023. https://doi.org/10.1158/1078-0432.CCR-06-1520
Kurosu T, Ohga N, Hida Y et al (2011) HuR keeps an angiogenic switch on by stabilising mRNA of VEGF and COX-2 in tumour endothelium. Br J Cancer 104:819–829. https://doi.org/10.1038/bjc.2011.20
Lee J, Kang S, Wang X et al (2019) HAP1 loss confers L-asparaginase resistance in ALL by downregulating the calpain-1-Bid-caspase-3/12 pathway. Blood 133:2222–2232. https://doi.org/10.1182/blood-2018-12-890236
Leopold JA, Walker J, Scribner AW et al (2003) Glucose-6-phosphate dehydrogenase modulates vascular endothelial growth factor-mediated angiogenesis. J Biol Chem 278:32100–32106. https://doi.org/10.1074/jbc.M301293200
Li L, Zhao GD, Shi Z et al (2016) The Ras/Raf/MEK/ERK signaling pathway and its role in the occurrence and development of HCC (Review). Oncol Lett 12:3045–3050. https://doi.org/10.3892/ol.2016.5110
Li L, Shao M, Peng P et al (2017) High expression of GFAT1 predicts unfavorable prognosis in patients with hepatocellular carcinoma. Oncotarget 8:19205–19217. https://doi.org/10.18632/oncotarget.15164
Li FL, Liu JP, Bao RX et al (2018) Acetylation accumulates PFKFB3 in cytoplasm to promote glycolysis and protects cells from cisplatin-induced apoptosis. Nat Commun 9:508. https://doi.org/10.1038/s41467-018-02950-5
Liu Y, Cox SR, Morita T, Kourembanas S (1995) Hypoxia regulates vascular endothelial growth factor gene expression in endothelial cells: identification of a 5′ enhancer. Circ Res 77:638–643. https://doi.org/10.1161/01.RES.77.3.638
Lopes-Coelho F, Nunes C, Gouveia-Fernandes S et al (2017) Monocarboxylate transporter 1 (MCT1), a tool to stratify acute myeloid leukemia (AML) patients and a vehicle to kill cancer cells. Oncotarget 8:82803–82823. https://doi.org/10.18632/oncotarget.20294
Lucena MC, Carvalho-Cruz P, Donadio JL et al (2016) Epithelial mesenchymal transition induces aberrant glycosylation through hexosamine biosynthetic pathway activation. J Biol Chem 291:12917–12929. https://doi.org/10.1074/jbc.M116.729236
Manuneedhi Cholan P, Cartland S, Kavurma M (2017) NADPH oxidases, angiogenesis, and peripheral artery disease. Antioxidants 6:56. https://doi.org/10.3390/antiox6030056
Martinive P, Defresne F, Bouzin C et al (2006) Preconditioning of the tumor vasculature and tumor cells by intermittent hypoxia: implications for anticancer therapies. Cancer Res 66:11736–11744. https://doi.org/10.1158/0008-5472.CAN-06-2056
Matsuda K, Ohga N, Hida Y et al (2010) Isolated tumor endothelial cells maintain specific character during long-term culture. Biochem Biophys Res Commun 394:947–954. https://doi.org/10.1016/j.bbrc.2010.03.089
Miller DW, Graulich W, Karges B et al (1999) Elevated expression of endoglin, a component of the TGF-β-receptor complex, correlates with proliferation of tumor endothelial cells. Int J Cancer 81:568–572. https://doi.org/10.1002/(SICI)1097-0215(19990517)81:4<568::AID-IJC11>3.0.CO;2-X
Miranda-Gonçalves V, Bezerra F, Costa-Almeida R et al (2017) Monocarboxylate transporter 1 is a key player in glioma-endothelial cell crosstalk. Mol Carcinog 56:2630–2642. https://doi.org/10.1002/mc.22707

Nespereira B (2003) Vitamins C and E downregulate vascular VEGF and VEGFR-2 expression in apolipoprotein-E-deficient mice. Atherosclerosis 171:67–73. https://doi.org/10.1016/j.atherosclerosis.2003.08.009

Ohga N, Ishikawa S, Maishi N et al (2012) Heterogeneity of tumor endothelial cells: comparison between tumor endothelial cells isolated from high- and low-metastatic tumors. Am J Pathol 180:1294–1307. https://doi.org/10.1016/j.ajpath.2011.11.035

Okuno Y, Nakamura-Ishizu A, Otsu K et al (2012) Pathological neoangiogenesis depends on oxidative stress regulation by ATM. Nat Med 18:1208–1216. https://doi.org/10.1038/nm.2846

Orecchioni S, Reggiani F, Talarico G et al (2015) The biguanides metformin and phenformin inhibit angiogenesis, local and metastatic growth of breast cancer by targeting both neoplastic and microenvironment cells. Int J Cancer 136:E534–E544. https://doi.org/10.1002/ijc.29193

Panieri E, Santoro MM (2015) ROS signaling and redox biology in endothelial cells. Cell Mol Life Sci 72:3281–3303. https://doi.org/10.1007/s00018-015-1928-9

Parra-Bonilla G, Alvarez DF, Al-Mehdi A-B et al (2010) Critical role for lactate dehydrogenase a in aerobic glycolysis that sustains pulmonary microvascular endothelial cell proliferation. Am J Physiol Cell Mol Physiol 299:L513–L522. https://doi.org/10.1152/ajplung.00274.2009

Peiró C, Romacho T, Azcutia V et al (2016) Inflammation, glucose, and vascular cell damage: the role of the pentose phosphate pathway. Cardiovasc Diabetol 15:82. https://doi.org/10.1186/s12933-016-0397-2

Peng F, Li Q, Sun JY et al (2018) PFKFB3 is involved in breast cancer proliferation, migration, invasion and angiogenesis. Int J Oncol 52:945–954. https://doi.org/10.3892/ijo.2018.4257

Peters K, Kamp G, Berz A et al (2009) Changes in human endothelial cell energy metabolic capacities during in vitro cultivation. The role of aerobic glycolysis and proliferation. Cell Physiol Biochem 24:483–492. https://doi.org/10.1159/000257490

Peyton KJ, Liu XM, Yu Y et al (2018) Glutaminase-1 stimulates the proliferation, migration, and survival of human endothelial cells. Biochem Pharmacol 156:204–214. https://doi.org/10.1016/j.bcp.2018.08.032

Pfeiffer T, Schuster S, Bonhoeffer S (2001) Cooperation and competition in the evolution of ATP-producing pathways. Science 292(80):504–507. https://doi.org/10.1126/science.1058079

Phoomak C, Vaeteewoottacharn K, Silsirivanit A et al (2017) High glucose levels boost the aggressiveness of highly metastatic cholangiocarcinoma cells via O-GlcNAcylation. Sci Rep 7:43842. https://doi.org/10.1038/srep43842

Potente M, Gerhardt H, Carmeliet P (2011) Basic and therapeutic aspects of angiogenesis. Cell 146:873–887. https://doi.org/10.1016/j.cell.2011.08.039

Prieto-Bermejo R, Hernández-Hernández A (2017) The importance of NADPH oxidases and redox signaling in angiogenesis. Antioxidants 6:32. https://doi.org/10.3390/antiox6020032

Riganti C, Gazzano E, Polimeni M et al (2012) The pentose phosphate pathway: an antioxidant defense and a crossroad in tumor cell fate. Free Radic Biol Med 53:421–436. https://doi.org/10.1016/j.freeradbiomed.2012.05.006

Rohlena J, Dong L-F, Kluckova K et al (2011) Mitochondrially targeted α-Tocopheryl succinate is antiangiogenic: potential benefit against tumor angiogenesis but caution against wound healing. Antioxid Redox Signal 15:2923–2935. https://doi.org/10.1089/ars.2011.4192

Rohlenova K, Veys K, Miranda-Santos I et al (2017) Endothelial cell metabolism in health and disease. Trends Cell Biol:1–13. https://doi.org/10.1016/j.tcb.2017.10.010

Ruan GX, Kazlauskas A (2013) Lactate engages receptor tyrosine kinases Axl, Tie2, and vascular endothelial growth factor receptor 2 to activate phosphoinositide 3-kinase/AKT and promote angiogenesis. J Biol Chem 288:21161–21172. https://doi.org/10.1074/jbc.M113.474619

San-Millán I, Brooks GA (2017) Reexamining cancer metabolism: lactate production for carcinogenesis could be the purpose and explanation of the Warburg effect. Carcinogenesis 38:119–133. https://doi.org/10.1093/carcin/bgw127

Santoro MM (2018) Fashioning blood vessels by ROS signalling and metabolism. Semin Cell Dev Biol 80:35–42. https://doi.org/10.1016/j.semcdb.2017.08.002

Schoors S, De Bock K, Cantelmo AR et al (2014a) Partial and transient reduction of glycolysis by PFKFB3 blockade reduces pathological angiogenesis. Cell Metab 19:37–48. https://doi.org/10.1016/j.cmet.2013.11.008

Schoors S, Cantelmo AR, Georgiadou M et al (2014b) Incomplete and transitory decrease of glycolysis: a new paradigm for anti-angiogenic therapy? Cell Cycle 13:16–22. https://doi.org/10.4161/cc.27519

Schoors S, Bruning U, Missiaen R et al (2015) Fatty acid carbon is essential for dNTP synthesis in endothelial cells. Nature 520:192–197. https://doi.org/10.1038/nature14362

Semenza GL (2003) Targeting HIF-1 for cancer therapy. Nat Rev Cancer 3:721–732. https://doi.org/10.1038/nrc1187

Senger DR, Galli SJ, Dvorak AM et al (1983) Tumor cells secrete a vascular permeability factor that promotes accumulation of ascites fluid. Science 219(80):983–985. https://doi.org/10.1126/science.6823562

Shafique E, Torina A, Reichert K et al (2017) Mitochondrial redox plays a critical role in the paradoxical effects of NAPDH oxidase-derived ROS on coronary endothelium. Cardiovasc Res 113:234–246. https://doi.org/10.1093/cvr/cvw249

Shi WK, Zhu XD, Wang CH et al (2018) PFKFB3 blockade inhibits hepatocellular carcinoma growth by impairing DNA repair through AKT article. Cell Death Dis 9:428. https://doi.org/10.1038/s41419-018-0435-y

Silva LS, Goncalves LG, Silva F et al (2015) STAT3:FOXM1 and MCT1 drive uterine cervix carcinoma fitness to a lactate-rich microenvironment.

Tumor Biol 37:5385–5395. https://doi.org/10.1007/s13277-015-4385-z
Silverstein RL, Febbraio M (2009) CD36, a scavenger receptor involved in immunity, metabolism, angiogenesis, and behavior. Sci Signal 2. https://doi.org/10.1126/scisignal.272re3
Slawson C, Copeland RJ, Hart GW (2010) O-GlcNAc signaling: a metabolic link between diabetes and cancer? Trends Biochem Sci 35(35):547–555. https://doi.org/10.1016/j.tibs.2010.04.005
Sliwinska PN, Alitalo K, Allen E et al (2018) Consensus guidelines for the use and interpretation of angiogenesis assays. Springer, New York
Sonveaux P, Copetti T, de Saedeleer CJ et al (2012) Targeting the lactate transporter MCT1 in endothelial cells inhibits lactate-induced HIF-1 activation and tumor angiogenesis. PLoS One 7:e33418. https://doi.org/10.1371/journal.pone.0033418
Stacker SA, Achen MG (2013) The VEGF signaling pathway in cancer: the road ahead. Chin J Cancer 32:297–302. https://doi.org/10.5732/cjc.012.10319
Tang N, Wang L, Esko J et al (2004) Loss of HIF-1α in endothelial cells disrupts a hypoxia-driven VEGF autocrine loop necessary for tumorigenesis. Cancer Cell 6:485–495. https://doi.org/10.1016/j.ccr.2004.09.026
Trenti A, Tedesco S, Boscaro C et al (2017) The glycolytic enzyme PFKFB3 is involved in estrogen-mediated angiogenesis via GPER1. J Pharmacol Exp Ther 361:398–407. https://doi.org/10.1124/jpet.116.238212
Ushio-Fukai M, Nakamura Y (2008) Reactive oxygen species and angiogenesis: NADPH oxidase as target for cancer therapy. Cancer Lett 266:37–52. https://doi.org/10.1016/j.canlet.2008.02.044
Vasconcelos-Dos-Santos A, Loponte HFBR, Mantuano NR et al (2017) Hyperglycemia exacerbates colon cancer malignancy through hexosamine biosynthetic pathway. Oncogene 6:e306. https://doi.org/10.1038/oncsis.2017.2
Végran F, Boidot R, Michiels C et al (2011) Lactate influx through the endothelial cell monocarboxylate transporter MCT1 supports an NF-kB/IL-8 pathway that drives tumor angiogenesis. Cancer Res 71:2550–2560. https://doi.org/10.1158/0008-5472.CAN-10-2828
Vizán P, Sánchez-Tena S, Alcarraz-Vizán G et al (2009) Characterization of the metabolic changes underlying growth factor angiogenic activation: identification of new potential therapeutic targets. Carcinogenesis 30:946–952. https://doi.org/10.1093/carcin/bgp083
Wang CC, Xu H, Man GCW et al (2013) Prodrug of green tea epigallocatechin-3-gallate (Pro-EGCG) as a potent anti-angiogenesis agent for endometriosis in mice. Angiogenesis 16:59–69. https://doi.org/10.1007/s10456-012-9299-4
Wang J, Man GCW, Chan TH et al (2018) A prodrug of green tea polyphenol (–)-epigallocatechin-3-gallate (Pro-EGCG) serves as a novel angiogenesis inhibitor in endometrial cancer. Cancer Lett 412:10–20. https://doi.org/10.1016/j.canlet.2017.09.054
Wei R, Mao L, Xu P et al (2018) Suppressing glucose metabolism with epigallocatechin-3-gallate (EGCG) reduces breast cancer cell growth in preclinical models. Food Funct 9:5682–5696. https://doi.org/10.1039/c8fo01397g
Weigand JE, Boeckel J-N, Gellert P, Dimmeler S (2012) Hypoxia-induced alternative splicing in endothelial cells. PLoS One 7:e42697. https://doi.org/10.1371/journal.pone.0042697
Wenes M, Shang M, Di Matteo M et al (2016) Macrophage metabolism controls tumor blood vessel morphogenesis and metastasis. Cell Metab 24:701–715. https://doi.org/10.1016/j.cmet.2016.09.008
West XZ, Malinin NL, Merkulova AA et al (2010) Oxidative stress induces angiogenesis by activating TLR2 with novel endogenous ligands. Nature 467:972–976. https://doi.org/10.1038/nature09421
Wilting J, Chao TI (2015) Integrated vascular anatomy. In: PanVascular medicine, 2nd edn. Springer, New York, pp 193–241
Wong BW, Marsch E, Treps L et al (2017) Endothelial cell metabolism in health and disease : impact of hypoxia. EMBO J 36:1–17. https://doi.org/10.15252/embj.201696150
Wu D, Huang R-T, Hamanaka RB et al (2017) HIF-1α is required for disturbed flow-induced metabolic reprogramming in human and porcine vascular endothelium. elife 6:1–26. https://doi.org/10.7554/eLife.25217
Xu Y, An X, Guo X et al (2014) Endothelial PFKFB3 plays a critical role in angiogenesis. Arterioscler Thromb Vasc Biol 34:1231–1239. https://doi.org/10.1161/ATVBAHA.113.303041
Yamamoto K, Ohga N, Hida Y et al (2012) Biglycan is a specific marker and an autocrine angiogenic factor of tumour endothelial cells. Br J Cancer 106:1214–1223. https://doi.org/10.1038/bjc.2012.59
Yeh W-L, Lin C-J, Fu W-M (2007) Enhancement of glucose transporter expression of brain endothelial cells by vascular endothelial growth factor derived from glioma exposed to hypoxia. Mol Pharmacol 73:170–177. https://doi.org/10.1124/mol.107.038851
Yu P, Wilhelm K, Dubrac A et al (2017) FGF-dependent metabolic control of vascular development. Nature 545:224–228. https://doi.org/10.1038/nature22322
Zang Z, Apse K, Pang J, Stanton RC (2000) High glucose inhibits glucose-6-phosphate dehydrogenase via cAMP in aortic endothelial cells. J Biol Chem 275:40042–40047. https://doi.org/10.1074/jbc.M007505200
Zetter BR (2008) The scientific contributions of M. Judah Folkman to cancer research. Nat Rev Cancer 8:647–654. https://doi.org/10.1038/nrc2458
Zhang N, Zhu T, Yu K et al (2019) Elevation of O-GlcNAc and GFAT expression by nicotine exposure promotes epithelial-mesenchymal transition and invasion in breast cancer cells. Cell Death Dis 10:343. https://doi.org/10.1038/s41419-019-1577-2
Zhong Q, Li X, Nong Q et al (2017) Metabolic profiling in association with vascular endothelial cell dysfunction following non-toxic cadmium exposure. Int J Mol Sci 18:E1905. https://doi.org/10.3390/ijms18091905

Monocytes and Macrophages in Cancer: Unsuspected Roles

9

Sofia Gouveia-Fernandes

Abstract

The behavior of cancer is undoubtedly affected by stroma. Macrophages belong to this microenvironment and their presence correlates with reduced survival in most cancers. After a tumor-induced "immunoediting", these monocytes/macrophages, originally the first line of defense against tumor cells, undergo a phenotypic switch and become tumor-supportive and immunosuppressive.

The influence of these tumor-associated macrophages (TAMs) on cancer is present in all traits of carcinogenesis. These cells participate in tumor initiation and growth, migration, vascularization, invasion and metastasis. Although metastasis is extremely clinically relevant, this step is always reliant on the angiogenic ability of tumors. Therefore, the formation of new blood vessels in tumors assumes particular importance as a limiting step for disease progression.

Herein, the once unsuspected roles of macrophages in cancer will be discussed and their importance as a promising strategy to treat this group of diseases will be reminded.

S. Gouveia-Fernandes (✉)
CEDOC, Chronic Diseases Research Centre, NOVA Medical School | Faculdade de Ciências Médicas, Universidade NOVA de Lisboa, Lisbon, Portugal

Keywords

Tumor stroma · Immunoediting · Tumor-associated macrophages (TAM) · Metastasis · Tumor vascularization

9.1 Introduction

Tumors have increasingly been recognized as organic systems (Bloch and Harel 2016) whose complexity cannot be ignored. Therefore, the study of the biology of a tumor must consider every cell type within it and its surroundings, the tumor microenvironment (Hanahan and Weinberg 2011; Lopes-Coelho et al. 2018). Among the several cell types that compose and surround a tumor mass are hematopoietic cells. These are recruited to most solid tumors and monocytes/macrophages can abundantly populate them. Furthermore, several studies have suggested a causal relationship between macrophage high density and poor disease prognosis (Lopes-Coelho et al. 2018; Komohara et al. 2011; Lee et al. 2013; Leek et al. 1996; Steidl et al. 2010; Zhu et al. 2017; Lissbrant et al. 2000; Salvesen and Akslen 1999; Freire Valls et al. 2019; Wei et al. 2019; Koukourakis et al. 1998; Pogoda et al. 2016; Zhu et al. 2008; Komohara et al. 2012).

Monocytes and macrophages belong to the myeloid lineage of leukocytes. Macrophages result from the differentiation, in tissues, of

J. Serpa (ed.), *Tumor Microenvironment*, Advances in Experimental Medicine and Biology 1219,
https://doi.org/10.1007/978-3-030-34025-4_9

extravasating monocytes and undergo specific differentiation according to the local tissue microenvironment. These cells are critical to our innate and acquired immune response (Coffelt et al. 2009; Dijkgraaf et al. 2013; Prenen and Mazzone 2019).

Macrophages are present in almost every tissue and are mostly known for playing a critical role in injury emergence and resolution of infection. They are the first line of defense against anything that expresses signatures on its surface different from the molecules present on host cells. Also, they are involved in the maintenance of tissue homeostasis through remodeling and repair, they secrete a wide array of immunomodulatory cytokines and are able to present antigens, and, are characterized by the ability to engulf invading pathogens or dying/dead cells, cell debris and cancer cells (Prenen and Mazzone 2019; De Palma and Lewis 2011; Mills and Ley 2014; Aras and Zaidi 2017; Karlmark et al. 2012).

Tumor cells are targets of immune surveillance. An antitumor response can be launched by a series of events. Here, monocytes/macrophages may be the first line of defense by stimulating dendritic cells and natural killer cells, which ultimately activate the cytotoxic lymphoid system against transformed cells (Mills and Ley 2014; Karlmark et al. 2012; Lamagna et al. 2006; Dunn et al. 2004). However, tumor cells are often capable of escaping the immune machinery, giving place to the neoplastic progression by promoting tumor vascularization and its spreading. When tumor cells escape the immune surveillance and cancer is installed, macrophages seem to quit their original purposes as body defenders and they start to function as cancer supporters. These macrophages are commonly referred to as tumor-associated macrophages (TAMs) (Aras and Zaidi 2017; Lamagna et al. 2006; Qian and Pollard 2010; Condeelis and Pollard 2006). "Immunoediting" is the term proposed to explain the modulation of monocytes/macrophages roles from an anti-tumor response to a pro-malignant function (Dunn et al. 2002).

9.2 "Immunoediting" in Cancer

Macrophages are conventionally classified into M1 and M2 subtypes according to the local tissue, their differentiation status and functional role in the immune system (Aras and Zaidi 2017). M1 macrophages are considered as potent pro-inflammatory, cytotoxic and anti-tumorigenic agents, whereas M2 cells are mostly involved in anti-inflammatory functions, angiogenesis and tissue repair (Mills and Ley 2014; Mantovani et al. 2006; Mantovani and Sica 2010). Although TAMs seem to have acquired features shared by M2 macrophages, this binary classification is now considered oversimplified for failing to fully account for the complexity of the macrophage activation process (Lee et al. 2013; Sica and Bronte 2007).

Hematopoietic cells are recruited to most solid tumors and TAMs can abundantly populate the tumor mass (Condeelis and Pollard 2006; Pollard 2004). Remarkably, the vast majority of the studies on the subject suggests a causal relationship between macrophage high density and poor patient prognosis. This association has been well documented in several cancer types, such as beast carcinoma (Leek et al. 1996), thyroid (Ryder et al. 2008) and prostate cancer (Lissbrant et al. 2000), endometrial carcinomas (Salvesen and Akslen 1999), colorectal cancer (Freire Valls et al. 2019; Wei et al. 2019), kidney (Komohara et al. 2011) and lung cancer (Koukourakis et al. 1998; Pogoda et al. 2016), hepatocellular carcinoma (Zhu et al. 2008) and Hodgkin's lymphoma (Steidl et al. 2010). Furthermore, the clodronate-induced depletion of macrophages, in several neoplastic contexts, has resulted in reduced tumor growth and vascularization, which unequivocally demonstrates TAMs intervention in the establishment of the malignant potential (Gazzaniga et al. 2007; Halin et al. 2009; Kimura et al. 2007; Robinson-Smith et al. 2007; Zeisberger et al. 2006). Accordingly, TAMs suppression, combined with dendritic cell immunotherapy, in malignant mesothelioma mice models,

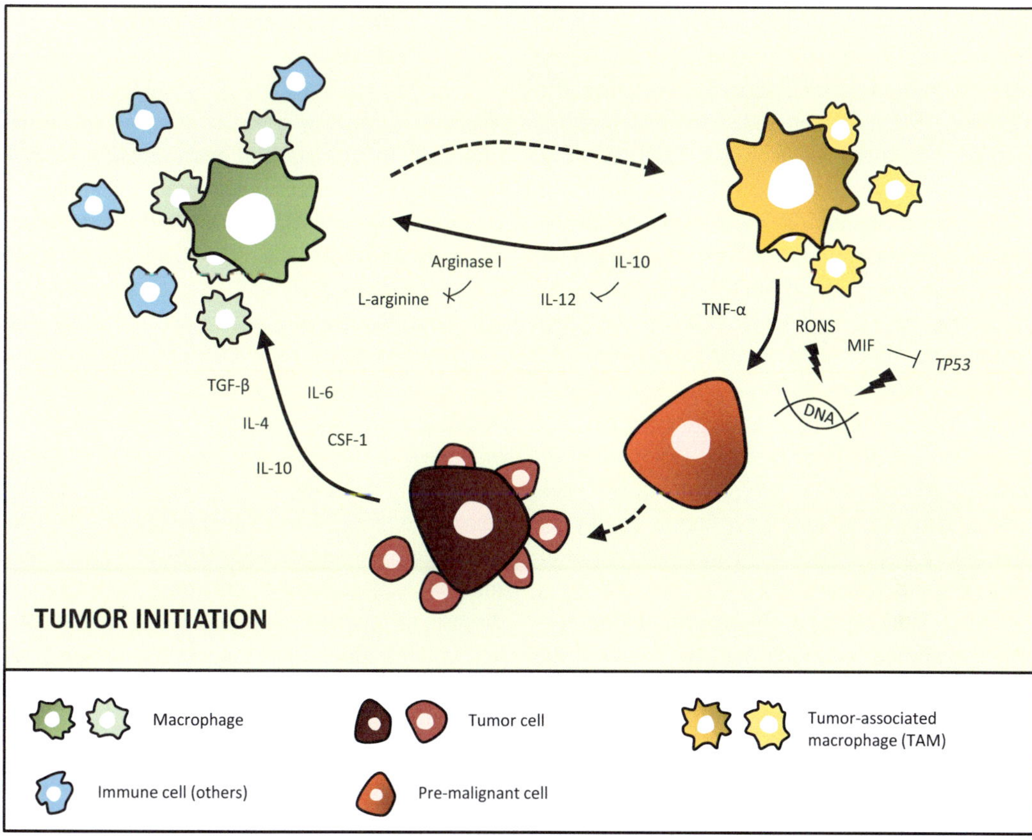

Fig. 9.1 Tumor-associated macrophages (TAMs) are present in the initial oncogenic events and tumor growth. Immunoediting and inflammation seem to mark the initiation of cancer. This immunosuppressive microenvironment includes a mixture of cytokines produced by tumor cells, including: TGF-β (transforming growth factor-beta), known to suppress immune surveillance (Bloch and Harel 2016; Hanahan and Weinberg 2011; Lopes-Coelho et al. 2018), CSF-1 (macrophage colony-stimulating factor-1), which acts by blocking the maturation of monocytes into dendritic cells (Komohara et al. 2011), and IL (interleukin)-4, IL-6 and IL-10, important players in the modulation of macrophages into a pro-tumor phenotype (Lee et al. 2013; Leek et al. 1996; Steidl et al. 2010; Zhu et al. 2017). By releasing IL-10, which prevents the release of pro-inflammatory IL-12 (Lissbrant et al. 2000), and arginase I, which, by depleting L-arginine, causes the impairment of T cells functions (Salvesen and Akslen 1999; Freire Valls et al. 2019; Wei et al. 2019), TAMs themselves are able to suppress immune surveillance. TAMs also generate an oncogenic microenvironment as a result of mutagenic events in the surrounding cells, in consequence of the release of RONS (reactive oxygen and nitrogen species) (Koukourakis et al. 1998; Pogoda et al. 2016; Zhu et al. 2008) and cytokines associated to mutagenic events, including TNF-α (tumor necrosis factor-alpha) and MIF (macrophage migration inhibitory factor), which is known to suppress *TP53* (Komohara et al. 2012)

enhanced anti-tumor immunity and survival (Dammeijer et al. 2017).

An immunosuppressive microenvironment created within the tumor seems to "educate" macrophages toward the tumor's own benefit (Condeelis and Pollard 2006; Pollard 2004) (Fig. 9.1). A cytokine mixture is produced by tumor cells, mostly preventing the immune response against themselves (Pollard 2004). TGF-β1 (transforming growth factor-beta 1) plays an important role in mediating the mechanism of tumor evasion from immune response and it is upregulated in many tumors (Yang et al. 2010; Derynck and Zhang 2003). TGF-β1 is known to suppress immune surveillance, by impacting proliferation, differentiation and survival of multiple immune cell lineages (Yang et al. 2010; Li et al. 2006; Gorelik and Flavell 2002), acting as a potent pro-oncogenic. TGF-β1 signaling blockade has generated an immune response capable of tumor rejection (Gorelik and Flavell 2001) and mutations in *TGF-β1* and TGF-β1 pathway-related genes have been reported as determinant in cancer (Yang et al. 2010; Akhurst and Derynck 2001). Besides TGF-β1, some evidence has also suggested the implication of IL (interleukin)-4, IL-6, IL-10 and in the reprogramming of "ordinary" macrophages into a pro-tumor phenotype (Komohara et al. 2008; Qiu et al. 2011; Sica et al. 2006; Wu and Watabe 2017).

CSF-1 (macrophage colony-stimulating factor-1) is another cytokine that, by blocking the maturation of monocytes into dendritic cells (the other pathway of differentiation of monocytes), impairs an anti-tumor response and promotes the development of immunosuppressive trophic TAMs (Menetrier-Caux et al. 1998).

TAMs become themselves able to suppress defense responses of other immune cells in the tumor by secreting interleukin IL-10, which, in turn, prevents the release of pro-inflammatory cytokine IL-12 by TAMs (Sica et al. 2000) and also by preventing the maturation of monocyte into dendritic cells (Allavena et al. 1998).

Arginase I is another product of TAMs capable of tumor immune surveillance evasion, by depleting the amino acid L-arginine with the consequent impairing of T cells functions (Rodriguez et al. 2007; Rodriguez et al. 2003; Bak et al. 2008).

This deviation of monocytes/macrophages functions due to the surrounding microenvironment resembles the one that occurs during the embryonic development. However, in cancer, this intervention is not controlled, due to the loss of positional identity by tumor cells in consequence of intrinsic mutations, and malignancy progresses, now even aided by monocytes/macrophages (Condeelis and Pollard 2006; Pollard 2004).

9.3 Cancer, an Inflammatory Disease

There is a growing appreciation that inflammation is the root of cause of many cancers and this has already been proposed as another hallmark of cancer (Hanahan and Weinberg 2011; Mantovani 2010; Mantovani et al. 2008; Coussens et al. 2013). Approximately 25% of cancers worldwide are caused by chronic inflammation (Mantovani 2010; Mantovani et al. 2008). This association is well documented, for instance, in *Helicobacter pylori* infection and gastric cancer, *Haemophilus influenzae*, asbestos and cigarette smoke and lung cancer, *Schistosoma hematobium* and bladder cancer, hepatitis virus related hepatocellular carcinoma and in genetic conditions that cause continuous inflammatory disorders as Crohn's disease, or prostatitis and pancreatitis (Condeelis and Pollard 2006; Sica et al. 2008a; Brown et al. 2017; El-Serag 2012; Parsonnet et al. 1991). Supporting this association is the reduction of cancer risk by treatment with anti-inflammatory drugs (Balkwill et al. 2005; Zhang et al. 2018; Roxburgh and McMillan 2014). Importantly, even when cancer origin is not etiologically related to inflammation, an inflammatory component is always present in the tumor microenvironment (Mantovani et al. 2008), this being ultimately a booster to cancer progression.

Macrophages are key regulators between inflammation and cancer (Fig. 9.1) and work in concert with other immune cells (Balkwill et al. 2005; Sica et al. 2008b). TAMs generate an oncogenic microenvironment by producing high levels of reactive oxygen and nitrogen species (RONS), which, in turn, react with DNA (deoxyribonucleic acid), resulting in mutagenic events in epithelial and other surrounding cells (Pollard 2004; Maeda and Akaike 1998; Fulton et al. 1984).

Some cytokines, produced by macrophages and other immune cells, such as TNF-α (tumor necrosis factor-alpha) and MIF (macrophage migration inhibitory factor), which suppresses *TP53* in tumor cells, also contribute to the generation of chromosomal abnormalities (Hudson et al. 1999). These mutations may be permanent by successive cell replication, marking the initiation of cancer. This replication is additionally promoted by macrophages by producing cytokines and growth factors, along with other immune cells (Qian and Pollard 2010; Lin and Karin 2007; Karin et al. 2006).

The increase in aerobic glycolysis (Penny et al. 2016) and fatty acid biosynthesis and uptake are other metabolic adjustments that could explain the pro-inflammatory and pro-tumorigenic profile of TAMs (Arts et al. 2016; Metallo et al. 2011).

Taken together, evidence introduces monocytes/macrophages as pivotal players in tumorigenesis, by promoting tumor progression, induced by the immunosuppressive tumor microenvironment and, remarkably, even by direct involvement in primary oncogenic events by responding and mediating inflammation. This transcriptional/functional program appears to be mediated by the upregulation and activation of several transcription factors, including NF-kB (nuclear factor kappa-light-chain-enhancer of activated B cells) and STAT (signal transducer and activator of transcription) proteins (Coffelt et al. 2009; Wu and Watabe 2017; Karin and Greten 2005; Pahl 1999).

9.4 Monocytes Recruitment

TAMs derive from circulating monocytes which are selectively attracted within the tumor microenvironment by locally produced chemotactic factors (Sica et al. 2008b).

CSF-1 and chemokine ligand 2 (CCL2) are important players in the recruitment of monocytes that become TAMs in the tumor microenvironment, a process called intraepithelial neoplasia (Alahari et al. 2015).

CSF-1, a cytokine whose effect is mediated by CSF-1 tyrosine kinase receptor (CSF-1R), is the major lineage regulator for macrophages (Qian and Pollard 2010; Pollard 2009). This cytokine is overexpressed by tumor cells in several malignancies, including breast (Lin et al. 2002; Lin et al. 2007; Lin et al. 2001; Kacinski 1997), ovarian (Kacinski 1997; Toy et al. 2009), liver (Zhu et al. 2008), prostate (Ide et al. 2002) and colorectal cancer (Mroczko et al. 2007), and has been associated with TAMs accumulation and more aggressive cancers and consequent poor prognosis. The ablation of CSF-1:CSF-1R interaction has resulted in the reduction of monocytes/macrophages recruitment or function and subsequent impairment of tumor growth and spreading (Dammeijer et al. 2017; Lin et al. 2002; Lin et al. 2001; Kubota et al. 2009; Xu et al. 2013; Oguma et al. 2008; Zabuawala et al. 2010; Abraham et al. 2010; Hung et al. 2014; DeNardo et al. 2011). On opposite, the induction of CSF-1 overexpression has been shown to accelerate tumor progression (Lin et al. 2001).

Chemokine (C-C motif) ligand 2 (CCL2):CCR2 is also a very important determinant of monocytes recruitment into tumors. CCL2 (also known as monocyte chemoattractant protein (MCP)-1 or small inducible cytokine A2 (SCYA2)) expression has been positively associated with TAMs accumulation in a broad panel of cancers, including breast (Ueno et al. 2000), ovarian (Negus et al. 1997), lung (Arenberg et al. 2000) and glial (Leung et al. 1997). This ligand-receptor pair is mostly implicated in the recruitment of monocytes into tumor epithelial regions (Lee et al. 2013; Negus et al. 1997; Willenborg et al. 2012). In fact, the induction of CCL2 expression has resulted in an increased accumulation of TAMs, in melanoma tumors (Bottazzi et al. 1992). Accordingly, the interruption of CCL2:CCR2 axis was associated with a reduced recruitment of monocytes/macrophages and a consequent decline in tumor burden (Gazzaniga et al. 2007; Qian et al. 2011).

CD62L (CD62 ligand):CD62R along with CX3CL1 (C-X3-C motif ligand 1):CX3CR1 are also important axis implicated in the recruitment of monocytes into the tumor. CD62L, also known as L-selectin, and CX3CL1, also called fractalkine, are both chemokines and adhesion molecules that mediate the recruitment of monocytes into the perivascular region of the tumor. They attract and arrest leukocytes, including monocytes, to the sites of inflamed endothelium in the tumor (Lee et al. 2013).

Some tumor cells, along with naïve T cells, also secrete CCL5 (chemokine (C-C motif) ligand 5), also known as RANTES (regulated on activation, normal T cell expressed and secreted), which stimulates monocyte migration into the tumor, namely through CCR1 (C-C chemokine receptor 1). This receptor recognizes innumerous monocytes-attractant chemokines, leading to the continuous recruitment of monocytes into the tumor (Lamagna et al. 2006; Locati et al. 2002).

SDF-1 (stromal cell-derived factor – 1)/CXCL12 (C-X-C motif chemokine 12):CXCR4 (chemokine receptor type 4) is another axis strongly associated with tumor progression, whose interaction is also known for retaining the recruited cells around blood vessels (Lamagna et al. 2006; Grunewald et al. 2006). Interferon-γ (IFN- γ) can also induce monocytes/macrophages recruitment into the tumor microenvironment (Sun et al. 2014). Endothelins (ET-1, ET-2, ET-3) comprise a group of small vasoconstrictor peptides that also act as chemoattractants of monocytes into hypoxic areas of tumors (Lamagna et al. 2006).

Hypoxia, in solid tumors, is recurrent and arises from the incapacity of the growing vasculature to accompany the high rate of tumor cells proliferation. In an attempt to meet inner tumor cells (under hypoxia) metabolic needs, a hypoxia-responsive transcriptional adjustment occurs (Semenza 2003). Low levels of tissue oxygenation induce TAM differentiation of macrophages (Erler et al. 2009) and hypoxic tumor regions commonly detain the highest density of macrophages. TAMs in the inner areas of the tumor seem to play crucial roles in all traits of tumor progression (Lewis and Murdoch 2005; Murdoch et al. 2004; Ohno et al. 2004). VEGF (vascular endothelial growth factor) overexpression by tumor cells is one of the most recognized adjustments induced by HIF (hypoxia inducible factor)-1α, whose activation is one of the most documented responses to hypoxia (Semenza 2003; Pages and Pouyssegur 2005). That growth factor has been correlated to the presence of macrophages within the tumor (Leek et al. 2000; Cursiefen et al. 2004). In fact, VEGF has been shown to function as a chemoattractant for monocytes via the activation of tyrosine kinase VEGF receptor 1 (VEGFR-1), also known as FLT-1 (fms related tyrosine kinase 1) (Barleon et al. 1996; Sawano et al. 2001). Moreover, VEGF is still able to induce the expression of CXCL12 (Leek et al. 2000; Barleon et al. 1996). Besides VEGF, HIF1-1α is also implicated in the upregulation of *bFGF* (*basic fibroblast growth factor*), *IL-8*, *COX* (*cyclooxygenase*) *2*, *MMP* (*matrix metalloproteinase*)*-7*, *MMP-9*, *MMP-12* and *ANG* (*angiopoietin*), genes implicated in macrophages reprogramming and malignant transition (Murdoch and Lewis 2005; Murdoch et al. 2008).

Activin was also shown to promote skin carcinogenesis, by increasing the number of skin macrophages via attraction of blood monocytes, which was prevented by the depletion of CCR2-positive monocytes. Activin was even implicated in the reprogramming of macrophages, which resembled the phenotype of TAMs (Antsiferova et al. 2017).

9.5 TAMs and Cancer Progression

Once the right microenvironment is set, TAMs will continue to serve tumor needs by enhancing tumor migration, vascularization, invasion and metastasis. This last step is extremely clinically relevant once the vast majority of cancer patients die from the spreading of tumors. However, in order to reach distant sites, the tumor needs to acquire angiogenic ability; hence, the formation of new blood vessels in tumors assumes particular importance as a limiting step for disease progression (Hanahan and Weinberg 2011; Hanahan and Folkman 1996; Folkman and Hanahan 1991).

Most of the chemokines/growth factors involved in monocytes recruitment is also instrumental in the regulation of TAMs phenotype (Fig. 9.2). This cooperation has an obvious impact on patient survival.

Although all traits of carcinogenesis are related, each one seems to be affected by a particular subpopulation, displaying a more suitable phenotype for each function and recruited to strategic regions of the neoplasm, according to the chemokine expression pattern in the microenvironment (Lee et al. 2013).

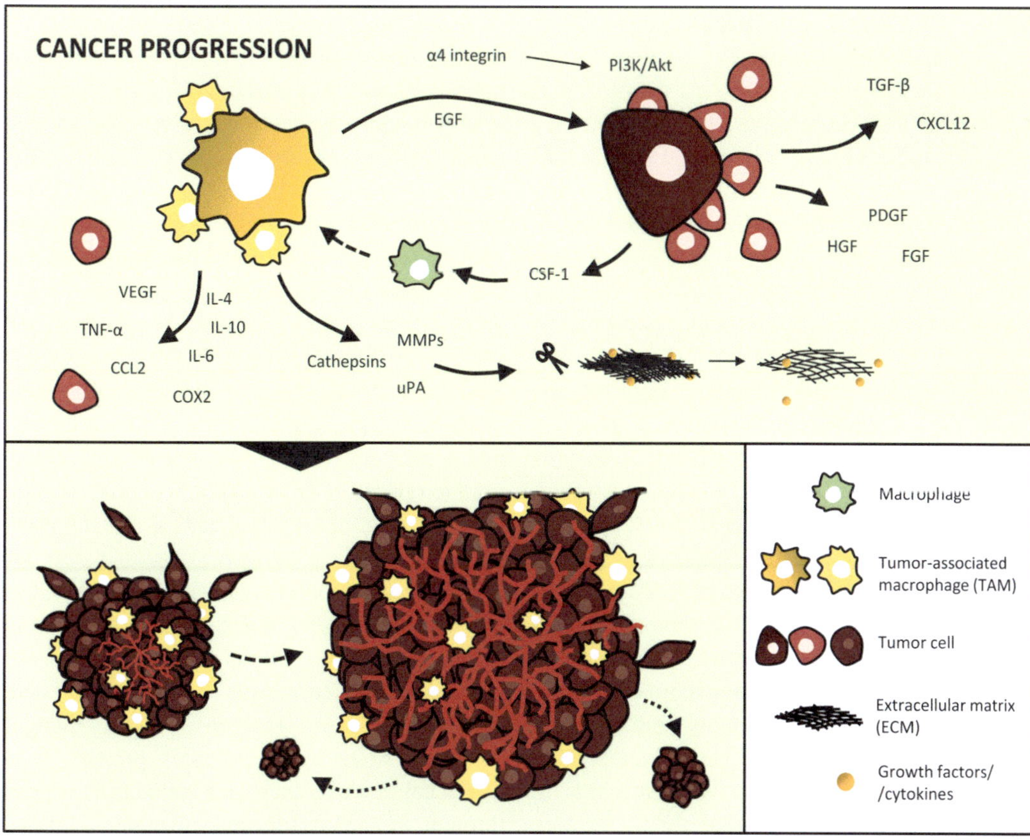

Fig. 9.2 Tumor-associated macrophages (TAMs) are important players in the progression of cancer: migration, invasion and metastasis. Once set the right microenvironment, a paracrine interaction between TAMs and cancer cells is crucial for the success of tumor progression. The production of CSF-1 (macrophage colony-stimulating factor-1) by tumor cells attracts monocytes/macrophages, which, as a response, release EGF (epidermal growth factor), leading to the co-migration of tumor cells and macrophages (Coffelt et al. 2009; Dijkgraaf et al. 2013; Prenen and Mazzone 2019; De Palma and Lewis 2011; Mills and Ley 2014). The increased survival of tumor cells through PI3K/Akt (phosphoinositide 3 kinase/protein kinase B) signaling is another achievement that counts on the influence of TAMs through the expression of integrin α4, which engages VCAM1 (vascular cell adhesion protein 1) on tumor cells (Aras and Zaidi 2017). TAMs are also known to secrete TNF-α (tumor necrosis factor-alpha), an important enhancer of invasive phenotype (Zhu et al. 2017; Karlmark et al. 2012), along with IL (interleukin)-4, IL-6, IL-10, CCL2 (chemokine ligand 2), COX2 (cyclooxygenase 2) and VEGF (vascular endothelial growth factor) (Lamagna et al. 2006; Dunn et al. 2004; Qian and Pollard 2010; Condeelis and Pollard 2006). By releasing proteases, such as MMPs (matrix metalloproteinases), cathepsins and uPA (urokinase plasminogen activator), TAMs are even implicated in EMC (extracellular matrix) proteolysis, allowing the creation of free space, which facilitates cells migration and escape, and the release of growth factors and other cytokines (Coffelt et al. 2009; Mills and Ley 2014; Karlmark et al. 2012; Dunn et al. 2002; Mantovani et al. 2006). This microenvironment still includes tumor cells-derived TGF-β (transforming growth factor-beta), CXCL12 (C-X-C motif chemokine 12), PDGF (platelet-derived growth factor), HGF (hepatocyte growth factor) and FGF (fibroblast growth factor), important promoters of cancer progression as well (Coffelt et al. 2009; Mantovani and Sica 2010; Sica and Bronte 2007)

9.5.1 TAMs in Tumor Invasion and Metastasis

A paracrine interaction between monocytes/macrophages and cancer cells is crucial for the success of tumor progression (Fig. 9.2). This cooperation starts with the migration and motility of both cell types. One of the mechanisms involved starts with the production of CSF-1 by tumor cells, which attracts monocytes/macrophages. In response, they secrete EGF (epidermal growth factor), whose signaling leads to co-migration of tumor cells and macrophages towards blood vessels where macrophages produce VEGF to enhance vessel permeability (Qian and Pollard 2010; Goswami et al. 2005; Wyckoff et al. 2004; Zeng et al. 2019; Nielsen and Schmid 2017). Indeed, the depletion of TAMs with a CSF-1R kinase inhibitor in combination with dendritic cell immunotherapy was suggested as a good strategy for mesothelioma treatment, by improving anti-tumor immunity and survival (Dammeijer et al. 2017). Additionally, the CSF-1:EGF loop was proved to be reinforced by CXCL12 stimulation (Rigo et al. 2010). Besides CSF-1:EGF axis, other pathways implicated in the initiation of the metastatic behavior have been pointed out. TAMs-derived TNF-α is an enhancer of tumor cells invasive phenotype while inducing MIF and extracellular matrix metalloproteinase inducer (EMMRIPN), which will both act on TAMs. Also IL-4, another cytokine produced by tumor cells or CD4+ T cells, is known to play an important role in the promotion of an invasive phenotype (Wu and Watabe 2017; Gocheva et al. 2010), along with tumor cells-derived CXCL12, FGF, HGF (hepatocyte growth factor), PDGF (platelet-derived growth factor) and macrophages-derived TGF-β signaling pathways (Coffelt et al. 2009; Qian and Pollard 2010; Rigo et al. 2010). TAMs have been shown to promote migration and metastasis in several malignant contexts, such as brain, renal and gastric cancer, by secreting other cytokines, such as IL-6 (Wei et al. 2019; Kovaleva et al. 2016), IL-10 (Wu and Watabe 2017; Wang et al. 2018), TNF-α (Kovaleva et al. 2016) and CCL2 (Wei et al. 2019; Kovaleva et al. 2016).

Extracellular matrix (ECM) proteolysis is another important trait in tumor cells migration and invasion that is also under TAMs influence. Once set within the stroma, TAMs are capable of producing several proteases, such as cathepsins, MMPs (MMP-2, MMP-7, MMP-9, MMP-19) and serine proteases (as urokinase plasminogen activator (uPA)), as well as TGF-β and IL-6 that will degrade ECM, thereby creating "free" space and facilitating the tumor cells escape (Qian and Pollard 2010; Mantovani et al. 2006; Nielsen and Schmid 2017; Gocheva et al. 2010; Yan et al. 2016). By degrading ECM, these proteases may additionally uncover cryptic sites of ECM, allowing the release of growth factors and other plausible inducers of tumor spreading (Polette et al. 2004).

Ets2 transcription factor was also identified as a central driver of a transcriptional program in TAMs that acts by promoting lung metastasis of breast tumors (Zabuawala et al. 2010). These cells also express integrin α4 that engages VCAM1 (vascular cell adhesion protein 1) on tumor cells, which increases tumor cell survival through PI3K/Akt (phosphoinositide 3 kinase/protein kinase B) signaling (Chen et al. 2011). TAMs have still been identified as contributors to tumor progression by inducing COX2 as a consequence of IL1b-mediated stimulation of ROS-Src-MAPK signaling (Hou et al. 2011).

Intriguingly, evidence has also implicated macrophages in another phenomenon: cell fusion. Macrophages have been observed to fuse with cancer cells. Cell fusion, which plays a well-recognized physiological role during development, is now emerging as an event that may explain, partially, the metastatic conversion of cancer cells. The data supports the novel possibility that tumor cells, by fusing with macrophages, can acquire the physical behavior attributed to migratory macrophages, including navigation through circulation, (mesenchymal traits) while still carrying oncogenic, tumor-derived genetic information (Powell et al. 2011; Martin-Padura et al. 2012; Kemeny et al. 2016). Cell fusion was also presented as a mechanism behind the development of radioresistance and tumor recurrence (Lindstrom et al. 2017).

9.5.2 TAMs in Tumor Angiogenesis

Due to intensive proliferation and expansion of tumor mass, oxygen demand is surpassed by oxygen supply, leading to tumor hypoxia. Therefore, for the success of tumor spreading, vascularization is mandatory. Despite of tumor cells self-sufficiency in growth signals, insensitivity to antigrowth signals, evasion of apoptosis and limitless replicative potential, tumors are not able to grow beyond 2–3 mm^3 and cannot metastasize, through hematogenous route, unless they become vascularized (Folkman 1985).

In order to sustain survival and to expand in size, incipient neoplasias must develop angiogenic ability, which seems to be acquired in a discrete step during tumor development via an "angiogenic switch" (Hanahan and Folkman 1996; Folkman and Hanahan 1991; Verbridge et al. 2009). Tumor vessels play an essential role in supplying tumor cells with nutrients, oxygen and immune cells, and also in the removal of waste products, enabling tumors to grow beyond the limitations of passive diffusion. In addition, and very importantly, newly formed vessels also afford the possibility of primary tumor to invade adjacent tissues, and circulate, through bloodstream, to distant sites, where they may form metastases (Nishida et al. 2006; Kawaguchi 2005; Sporn 1996). Thus, tumor vascularization represents a determining step in cancer progression.

Angiogenesis is defined as the *de novo* formation of blood vessels, in adults, through the proliferation of endothelial cells and it is the type of vascularization most widely studied in cancer (Kovacic et al. 2008; George et al. 2011).

Angiogenic switch is induced by several factors. Hypoxia induces the activation of a number of intracellular signaling pathways such as the major HIF, the PI3K/AKT/mTOR (mammalian target of rapamycin) and the NF-kB pathways (Prenen and Mazzone 2019; Keith et al. 2011). Several pro-angiogenic regulators are implicated, such as FGFs, thymidine phosphorylase (TP), TGF-β, TNF-α, PDGFs, angiopoietins, IL-8 and VEGFs (Nishida et al. 2006; Ferrara et al. 2003). VEGF assumes particular relevance, being the only growth factor observed almost ubiquitously at sites of angiogenesis and representing a critical rate-limiting step in this process (Ferrara et al. 2003; Robinson and Stringer 2001; Gerber and Ferrara 2003; Cebe-Suarez et al. 2006; Koong et al. 2000).

TAMs accumulate preferentially in the hypoxic regions within the tumors, mainly due to necrosis, and also at the surrounding blood vessels (Leek et al. 1996; Prenen and Mazzone 2019; Ohno et al. 2004; Knowles et al. 2004). Moreover, high TAMs infiltration is positively correlated with microvessel density, tumor stage and increased tumor angiogenesis in different malignancies, including melanoma (Torisu et al. 2000), breast (Bingle et al. 2006), endometrial (Salvesen and Akslen 1999), cervix (Jiang et al. 2016), gastric (Wu et al. 2012) and lung cancers (Takanami et al. 1999). Altogether, there seems to be an evident role of TAMs in promoting tumor vascularization.

Under this stressful environment, macrophages seem to undergo a transcriptional adaptation, at least partially induced by HIF-1α activation, which triggers the expression of angiogenic activators, as VEGF, IL-8, FGF, PDGF and CXCL12 receptor, CXCR4, whose influence relays mainly on the recruitment of more macrophages, promotion of endothelial cells proliferation, migration and survival, and vascular permeability (Metinko et al. 1992; Harmey et al. 1998; Kuwabara et al. 1995; Ceradini et al. 2004; Kioi et al. 2010) (Fig. 9.3). A recent study concluded that VEGFR1+ metastasis-associated macrophages contribute to metastatic angiogenesis and influence colorectal cancer patient outcome (Freire Valls et al. 2019).

The TSC2 (tuberous sclerosis complex 2)-mTOR-STAT3 pathway is also involved in the regulation of macrophages-induced tumor angiogenesis by increasing IL-10 and decreasing IL-12 (Chen et al. 2012).

A study on the interaction between ovarian cancer cell and TAMs has revealed this interaction would promote angiogenesis *in vitro*, by favoring the migration and tube formation of endothelial cells. In this case, underlying this enhancement was the increase in the expression of IL-8, regulated in part through the NFkβ pathway (Wang et al. 2013). Another work focusing on glioblastoma, demonstrated that, in this context, CECR1 (cat eye syndrome critical region protein 1) produced by TAMs regulated the crosstalk between macrophages and pericytes via paracrine PDGFB–PDGFRβ signaling, promoting pericyte recruitment and migration, and tumor angiogenesis. This signaling was related to the expression of periostin by pericytes (Zhu et al. 2017).

Activin seems to be another important player in carcinogenesis, at least in skin cancer, by increasing the number of skin macrophages via attraction of blood monocytes, which was prevented by the depletion of CCR2-positive monocytes. Activin induced the expression of genes that promote tumor cell proliferation and also the migration and proliferation of endothelial cells, and even by creating the space that could be filled by new blood capillaries (Antsiferova et al. 2017).

In a breast tumor context, hyaluronan, a glycosaminoglycan usually present in the ECM was described as an important inductor of the angiogenic behavior of TAMs, as a result of the increase in the expression of angiogenic factors VEGF, IL-8, FGF2 and MMP2 (Spinelli et al. 2019).

Besides the secretion of angiogenesis activators by TAMs themselves, TAMs may also contribute to tumor vascularization by making those mediators bioavailable, through the release of ECM-degrading enzymes, such as MMPs and uPA (Lamagna et al. 2006; Qian and Pollard 2010; Mantovani et al. 2006; Zajac et al. 2013). Moreover, ECM modulation may even minimize the stress applied by ECM on endothelial cells (Ingber 1992). MMP-9, MMP-19 and uPA are abundantly secreted by TAMs, under the action of CCL2 and CCL5 (Giraudo et al. 2004; Robinson et al. 2002). The degradation of ECM and the sustaining basement membrane by these enzymes facilitates migration and proliferation of endothelial cells and may even create the space that could be filled by new blood capillaries (Moldovan 2002).

Intriguingly, a new subset of monocytes/macrophages related to tumor vascularization has been identified (Venneri et al. 2007) (Fig. 9.3). Although recruited to tumor in lower numbers than TAMs, these monocytes seem to exert more potent pro-angiogenic functions, in a paracrine manner, through the release of pro-angiogenic factors (De Palma et al. 2005; Ribatti 2009).

These monocytes express Tie2, a marker normally restricted to endothelial cells. Tie2-expressing monocytes (TEMs) do not express CCR-2. Thus, TEMs might be attracted to tumors in a CCL-2:CCR-2 independent manner (Ribatti 2009). These monocytes appear to be recruited by Ang-2, an endothelial cell-derived ligand of Tie2 (Venneri et al. 2007; Atanasov et al. 2018; Wang et al. 2017). In addition, rebastinib inhibition of angiopoietin/Tie2 signaling impaired tumor progression mediated by TEM and angipoietin/Tie2-dependent angiogenesis (Harney et al. 2017). An interaction between Tie2 and Ang-1 also seems to take part in oral cancer metastasis (Kitajima et al. 2018). CSF-1 appears to boost Tie2-expressing monocyte differentiation and their recruitment as well (Forget et al. 2014).

TEMs have been found on several human tumors, including kidney, colon, ovary, pancreas, lung, mouth, breast and liver, where angiogenesis is correlated to tumor progression (Venneri et al. 2007; Atanasov et al. 2018; Wang et al. 2017; Kitajima et al. 2018; Forget et al. 2014; Matsubara et al. 2013; Ji et al. 2013; Yang et al. 2018; Roodhart et al. 2013). Within the tumor, TEMs are confined to perivascular sites and to hypoxic regions (Venneri et al. 2007; De Palma et al. 2005; Matsubara et al. 2013; De Palma et al. 2003; Lewis et al. 2007). In a study regarding bone marrow-derived cells involvement in tumor regrowth after chemotherapy, Roodhart et al. (Roodhart et al. 2013) observed, *in vivo*, an influx of cells from bone marrow into the tumor that was accompanied by a significant increase in tumor angiogenesis. Two specific populations ($Gr1^{+}/CD11b^{+}$ and $Tie2^{high}/CD31^{low}$) were located in the tumors perivascular areas.

IGF1-IGF1R signaling in ovarian cancer was disclosed as an important promoter of angiogenesis and metastasis, both *in vitro* and *in vivo*, by modulating the interaction between TEMs and endothelial cells (Wang et al. 2017).

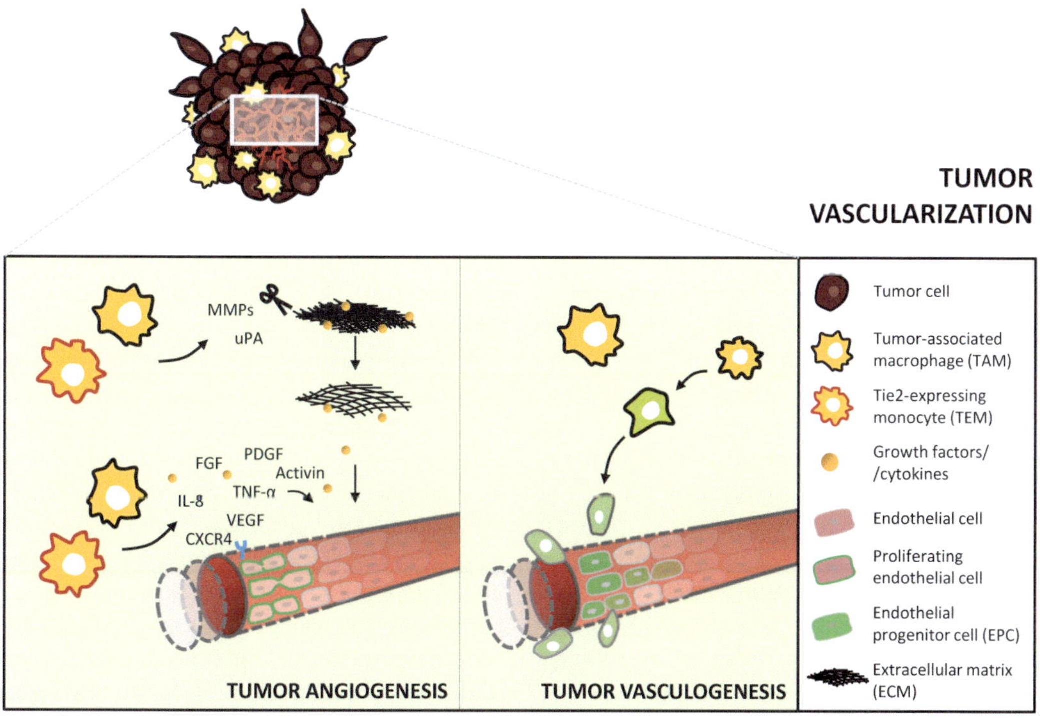

Fig. 9.3 Tumor-associated macrophages (TAMs) greatly influence tumor vascularization, acting on several fronts. TAMs appear to be implicated in both tumor angiogenesis and neovasculogenesis. In angiogenesis, TAMs act as important angiogenesis mediators by secreting several angiogenic activators, such as VEGF (vascular endothelial growth factor), IL (interleukin)-8, FGF (fibroblast growth factor), PDGF (platelet-derived growth factor), TNF-α (tumor necrosis factor-alpha), activin and CXCL12 (C-X-C motif chemokine 12) receptor, CXCR4 (chemokine receptor type 4), which promote endothelial cells proliferation, migration and survival, and vascular permeability (Pollard 2004; Ryder et al. 2008; Gazzaniga et al. 2007; Halin et al. 2009; Kimura et al. 2007; Robinson-Smith et al. 2007). Additionally, the release of ECM (extracellular matrix)-degrading enzymes, as MMPs (matrix metalloproteinases) (MMP-2, MMP-9, MMP-19) and uPA (urokinase plasminogen activator) by TAMs makes some angiogenesis regulators available and allows the creation of space for endothelial cells proliferation and migration (Coffelt et al. 2009; Dunn et al. 2002; Zeisberger et al. 2006; Dammeijer et al. 2017; Yang et al. 2010; Derynck and Zhang 2003). Tie2-expressing monocytes (TEM) are another subset of blood mononuclear cells strongly implicated in tumor angiogenesis (Li et al. 2006; Gorelik and Flavell 2002). Recent evidence is now including monocytes/macrophages in a new trait of vascularization, neovasculogenesis. These cells have been observed to undergo a transdifferentiation process into endothelial-like cells (ELCs), assuming, presumably, a more direct role in neovessels formation by integrating their walls (Gorelik and Flavell 2001; Akhurst and Derynck 2001; Komohara et al. 2008; Qiu et al. 2011; Sica et al. 2006; Wu and Watabe 2017; Menetrier-Caux et al. 1998; Sica et al. 2000; Allavena et al. 1998; Rodriguez et al. 2007; Rodriguez et al. 2003)

Gabrusiewicz and co-workers (Gabrusiewicz et al. 2014) have identified TEMs as pivotal players in the development of an invasive glioma phenotype resistant to anti-VEGF therapy, within the several myeloid populations that have been associated to this pattern of relapse, upon studies on mice models and human malignant glioma surgical specimens. These cells were even found as major sources of MMP-9 secretion (Gabrusiewicz et al. 2014). The ablation of these monocytes subset has remarkably reduced vascularization in the tumor, even causing its regression (De Palma et al. 2003, 2005). Besides, TEMs elimination has not affected the overall number of TAMs, suggesting they are a distinct monocytes subset, primarily responsible for promoting angiogenesis, and in a remarkable way given its lower number in comparison to TAMs (De Palma et al. 2003, 2005).

9.5.3 TAMs and Neovasculogenesis in Cancer

In the context of tumor vascularization, two types of blood vessels formation have been pointed out, angiogenesis and neovasculogenesis. Even though these two events serve the same purpose, they are quite distinct, although they are often globally referred solely as "tumor angiogenesis". Angiogenesis refers to the formation of new capillaries from pre-existing vessels, by their sprouting or splitting (Kovacic et al. 2008; George et al. 2011), whereas neovasculogenesis, also termed postnatal/adult vasculogenesis, comprises the *de novo* formation of a primary vascular plexus from endothelial progenitor cells (EPCs) (Kovacic et al. 2008; George et al. 2011). Nevertheless, the relative contribution of angiogenesis and neovasculogenesis in cancer is still under debate. Furthermore, even the origin of neovascular endothelial cells remains controversial.

EPCs are defined as a minor population of mononuclear non-endothelial cells capable of proliferating, migrating and differentiating into endothelial cell lineage, but have not yet acquired characteristics of mature endothelial cells (Ribatti 2007; Fadini et al. 2008; Medina et al. 2010). Asahara and co-workers (Asahara et al. 1997) were the first to isolate putative EPCs from human peripheral blood on the basis of cell surface expression of CD34 and VEGFR-2 markers, observing experimentally EPCs differentiation into endothelial cells. Since then, increasing knowledge on EPCs has emerged. Although some questions persist regarding the precise panel of cell surface markers defining EPCs (Hirschi et al. 2008; Timmermans et al. 2009; Yoder 2012), the combinations of $CD133^{+}CD34^{+}VEGFR\text{-}2^{+}$, $CD34^{+}VEGFR\text{-}2^{+}$, or $CD114^{+}CD34^{low}$ are now widely used to define or select cells expressing properties attributed to EPCs (Schmidt-Lucke et al. 2010; Romagnani et al. 2005; Peichev et al. 2000). Most circulating EPCs are believed to reside in the bone marrow in close association with hematopoietic stem cells and the stroma. EPCs circulating in the peripheral blood correspond, thus, to cells derived from the bone marrow, not yet incorporated into the vessel wall (Ribatti et al. 2005; Nolan et al. 2007). A strong correlation has also been described between EPCs number and tumor growth and progression in several neoplastic contexts (Gao et al. 2009; Yu et al. 2007; Monestiroli et al. 2001; Shaked et al. 2005; Real et al. 2011). Moreover, increased circulating levels of EPCs have been detected in cancer patients (Sakamori et al. 2012; Mancuso et al. 2001; Ahn et al. 2010a; Pircher et al. 2008; Richter-Ehrenstein et al. 2007).

Monocytes are in intimate contact with endothelium. The prevailing knowledge on their intervention in tumor vascularization recognizes monocytes/macrophages as indirect mediators of the process. However, increasing evidence regarding neovasculogenesis in cancer has now introduced the blood mononuclear cell population as a much closer system related to endothelial cells, by intervening directly in vascularization both in physiological and pathophysiological conditions (Domingues et al. 2015). Research has emerged by the increased awareness of adult vasculogenesis and by observations that include the often co-localization of new capillaries and monocytes/macrophages and the presence of angiogenic factor receptors on monocytes, previously considered to be expressed exclusively on endothelial cells (Barleon et al. 1996; Sawano et al. 2001; Moldovan 2002) (Fig. 9.3).

Several studies have now suggested a link between blood mononuclear cells population and endothelial cells that goes beyond the stimulation of capillary growth by secretion of angiogenic factors. Nineteen years ago, Fernandez and co-workers (Fernandez Pujol et al. 2000) demonstrated the capacity of CD14-positive cells from the normal peripheral blood to transdifferentiate into endothelial-like cells (ELCs) in the presence of endothelial growth factors. This transformation was achieved by sequential events starting from small attached mononuclear cells, then converted into adherent caudated or oval cells, capable of proliferating, which, upon culturing on three-dimensional fibrin gels, built network-like structures. That transdifferentiation process was accompanied by a clear expression of endothelial cell markers, including von Willebrand factor (vWF), CD144 (VE-cadherin), CD105 (endoglin), acetylated low-density lipoprotein (AC-LDL)-receptor, CD36 (thrombospondin receptor), VEGF receptors FLT-1 and, at a lower extent, KDR (kinase insert domain receptor, also named VEGF receptor 2) (Fernandez Pujol et al. 2000). Peripheral blood mononuclear cells collected from humans were shown to be enriched in EPCs after addition of VEGF, FGF-2, insulin-like growth factor (IGF) and EGF to the culture medium for 7–10 days. Afterwards, these cells contributed to the formation of new vessels in ischemic limbs in mice (Kalka et al. 2000). Urbich and colleagues (Urbich et al. 2003) have also demonstrated that EPCs have distinct monocytic features and can be cultured from CD14-positive cells. In other studies, monocytes cultured under angiogenic conditions also displayed an EPC phenotype with expression of specific surface markers and even formed cord-like structures (Schmeisser et al. 2001; Rohde et al. 2006). The incorporation of bone marrow-derived cells exhibiting characteristics of macrophages has been observed in brain vascularization (Hao et al. 2008). Moreover, CXCL12:CXCR4 axis was show to be crucial for the recruitment of bone marrow-derived cells to the pre-metastatic niches (Kaplan et al. 2007; Psaila and Lyden 2009).

Another study has introduced pleiotrophin as an important mediator of monocytes/macrophages transdifferentiation (Sharifi et al. 2006). This cytokine is expressed by monocytes/macrophages in ischemic tissues and it is usually known to promote neovascularization through stimulation of local endothelial cells proliferation. The authors, aimed at investigating an eventual autocrine interaction between pleiotrophin and monocytes/macrophages themselves and have found this cytokine was able to induce the transdifferentiation of monocytes into functional endothelial cells: its expression has led to a downregulation of monocytic cell markers and the upregulation of endothelium markers, along with the formation of tube-like structures, *in vitro*, upon fibrin gel culturing, under treatment of monocytic cells with pleiotrophin. *In vivo* assays have confirmed the integration of these cells into the neovasculature of chicken embryos and murine ischemic hindlimb models.

Pro-neovasculogenic and angiogenic effects by monocytic endothelial cells precursors have also been reported in the context of systemic sclerosis, both *in vitro* and *in vivo* (Yamaguchi et al. 2010).

The role of monocytes/macrophages as integrators of the neovasculature was still emphasized by the efforts of Kim and co-workers (Kim et al. 2009), who have described an elegant series of experiments in order to investigate the contribution of circulating cells in the establishment of neovasculature. For that, they have parabiosed a transgenic green fluorescent protein (GFP) mouse with a wild-type mouse, so that a common circulation was achieved. The contribution of GFP cells was assessed in acute (wound healing), sub-acute (implanted gel foam fragments) and chronic (subcutaneous tumors) phases of neovascularization. The staining of a panel of markers have informed on the origin of the cells composing the neoformed (GFP labeled) vasculature. The authors have concluded that the cell type incorporated in the neovessels was monocytes/macrophages.

The concept of circulating precursors of neovascular endothelial cells, derived from the bone marrow, has become a common and exciting topic in modern vascular biology. However, that is still far from gathering the consensus of the scientific community. Strikingly, unlike the outcomes by Kim and colleagues (Kim et al. 2009), Purhonen et al. (2008), aiming at studying the mobilization and differentiation of stem and precursor cells from the adult bone marrow during vascularization and tumor growth, have described a series of experiments based on 3D confocal microscopy and genetic labeling of bone marrow transplant and parabiosis systems, whose outcomes showed that bone marrow-derived cells, which include monocytic cells, do not incorporate into the luminal lining of blood vessels, in their models, and that tumor growth was independent of those cells (Purhonen et al. 2008). A previous study reported that “EPC” monocytes do not differentiate into endothelial cells, serving, instead, a proangiogenic function through paracrine secretion of angiogenic factors at sites of angiogenesis (Rehman et al. 2003).

Despite some controversy on ECs precursors' origin and their actual involvement in tumor neovasculogenesis and the lack of a full characterization of monocytes/macrophages-derived ELCs, we believe the emerging data on the ability of monocytes/macrophages to transdifferentiate into ELCs presents as an exciting field worthy of more investigation.

One of the new emerging paradigms in tumor vascularization is "vascular mimicry", which describes the *de novo* formation of perfusable, matrix-rich, vasculogenic-like networks in aggressive malignant tumors (Kirschmann et al. 2012; Liu et al. 2016).

More recently, Barnett and colleagues proved that macrophages form non-endothelial vessels in both tumor and angiogenesis *in vivo* models, a process dependent on HIF-1α (Barnett et al. 2016).

Three categories of bone marrow-derived cells have been suggested as contributors in tumor vascularization: endothelial progenitor cells, which serve as structural components of blood vessels (neovasculogenesis); myeloid progenitors subsets that can differentiate into endothelial-like cells and incorporate lumenally into the tumor neovessels (neovasculogenesis) and another wide group of cells from the monocytic lineage which act as angiogenesis regulators, not being part of tumor vasculature (Ribatti 2009). Indeed, evidence *in vitro* and *in vivo* has proved monocytic populations as cells able to transdifferentiate into cells that, at least, mimic endothelial cells phenotype (ELCs). Nevertheless, in both observations, these cells are capable of presenting epithelial markers and of disposing themselves into a vascular structure, which leads us to reinforce their classification as vasculogenic monocytes/macrophages.

9.6 Monocytes/Macrophages in Resistance to Cancer Therapy and Cancer Relapse

Some studies have suggested that chemotherapy induces the production of monocytes recruitment factors by cancer cells, enhancing macrophages infiltration, which explains, at least partially, tumors chemoresistance. DeNardo and co-authors (DeNardo et al. 2011) have shown, in a mouse breast tumor model, that chemotherapy increased the expression of CSF-1 in tumor cells, inducing the recruitment of a high number of monocytes expressing its receptor CSF-1R. The blockade of CSF-1:CSF-1R signaling markedly enhanced the ability of paclitaxel to improve the survival of the mice, by slowing the growth of both primary and metastatic tumors accompanied by a decrease in vessel density and by increasing the number of cytotoxic T cells. Indeed, the researchers pioneered the prognostic value of the inverse correlation between the number of TAMs and cytotoxic T cells: patients bearing breast tumors and displaying high amounts of TAMs and low numbers of cytotoxic T cells faced a worse prognosis.

In hepatocellular carcinoma, a high expression of CSF-1 in peritumoral liver tissue was associated with disease recurrence and poor survival after hepatectomy, which also highlighted the importance of peritumoral tissue (Zhu et al. 2008).

TAMs also play an important part in cancer relapse after irradiation. This function was clear in a study focusing on head and neck cancer in mice (Ahn et al. 2010b). After irradiation of tumors, intense recruitment of monocytes was observed as well as restored angiogenesis, which the researchers have linked to increased infiltration of TAMs. By inhibiting monocytes recruitment, the authors observed the inhibition of tumor growth and invasion of the post-irradiated tumors. In agreement, other researchers have demonstrated the upregulation of VEGF from TAMs following radiotherapy in human patients (McDonnell et al. 2003). Moreover, the accumulation of tyrosine kinase with immunoglobulin-like and EGF-like domains 2 (Tie2)-expressing monocytes, a subset of very potent pro-angiogenic monocytes (see "TAMs and Neovasculogenesis in Cancer"), has been strongly associated with anti-VEGF therapy-induced glioma invasion (Gabrusiewicz et al. 2014).

Cell fusion was also presented as a mechanism behind the development of radioresistance and tumor recurrence (Lindstrom et al. 2017).

9.7 Concluding Remarks

The importance of non-malignant cells within the tumor seems now to be properly recognized. Besides their role as tumor-supporters, those arise as very exciting targets due to their genome stability, as normal cells. Indeed, one of the main flops of conventional tumor cells-directed chemotherapy is the development of resistance by cancer cells, greatly due to their genome instability. Non-malignant cells are much more stable, and then much less likely to acquire chemoresistance, which makes them very tempting targets (Pollard 2004).

Overall, increasing evidence has highlighted the strong involvement of monocytes/macrophages in tumor progression. The paracrine interaction between TAMs and tumor cells is pivotal throughout all traits of cancer. Therefore, the molecular mediators in this inter-dependent communication represent important targets to consider for therapeutic purposes, which should encompass tumor-specific approaches so that the homeostatic functions of monocytes/macrophages in the normal body physiology is safeguarded.

Despite the obvious clinical relevance of metastasis, since the majority of cancer patients die from the spreading of tumors, prior angiogenesis is of no less significance, since it represents a limiting step of cancer progression, and, thus, ultimately, the formation of metastases. Anti-angiogenic therapy already focuses on non-malignant cells, by acting on tumor vasculature in an attempt to starve the tumor. However, this approach has not been as successful as initially idealized. Strikingly, a new concept has emerged in anti-angiogenic therapy, which supports the normalization of the abnormal neovasculature, instead of destroying it, in order to improve the chemotherapy proper delivery (Jain 2005; Goel et al. 2012). Indeed, the secretion of angiogenic activators by TAMs is recognized as being persistent, which contributes to an excessive vascularization, composed of abnormal and hypo-perfused blood vessels (De Palma and Lewis 2011; Stockmann et al. 2008). Interestingly, the depletion of pro-angiogenic TAMs (not vasculogenic), located at sites away from blood capillaries, has skewed their role to an angiostatic function, thereby normalizing the vessels and enhancing the efficacy of chemotherapy (DeNardo et al. 2011). Either way, monocytes/macrophages are certainly important targets to consider.

It is our belief that monocytes/macrophages, as stable non-malignant tumor supportive cells, represent powerful therapeutic targets, either as TAMs and/or as vasculature structure components. The strategies to treat cancer considering TAMs as therapeutic targets could include the direct targeting of TAMs by reducing monocyte recruitment, the repression of their support, namely on tumor angiogenesis, and the reconversion to their original immune-suppression and cytotoxicity functions. These are challenging tasks, but should, for sure, be considered among the efforts to beat cancer.

References

Abraham D, Zins K, Sioud M, Lucas T, Schafer R, Stanley ER, Aharinejad S (2010) Stromal cell-derived CSF-1 blockade prolongs xenograft survival of CSF-1-negative neuroblastoma. Int J Cancer 126(6):1339–1352. https://doi.org/10.1002/ijc.24859

Ahn JB, Rha SY, Shin SJ, Jeung HC, Kim TS, Zhang X, Park KH, Noh SH, Roh JK, Chung HC (2010a) Circulating endothelial progenitor cells (EPC) for tumor vasculogenesis in gastric cancer patients. Cancer Lett 288(1):124–132. https://doi.org/10.1016/j.canlet.2009.06.031

Ahn GO, Tseng D, Liao CH, Dorie MJ, Czechowicz A, Brown JM (2010b) Inhibition of Mac-1 (CD11b/CD18) enhances tumor response to radiation by reducing myeloid cell recruitment. Proc Natl Acad Sci U S A 107(18):8363–8368. https://doi.org/10.1073/pnas.0911378107

Akhurst RJ, Derynck R (2001) TGF-beta signaling in cancer--a double-edged sword. Trends Cell Biol 11(11):S44–S51. https://doi.org/10.1016/S0962-8924(01)02130-4

Alahari SV, Dong S, Alahari SK (2015) Are macrophages in tumors good targets for novel therapeutic approaches? Mol Cells 38(2):95–104. https://doi.org/10.14348/molcells.2015.2298

Allavena P, Piemonti L, Longoni D, Bernasconi S, Stoppacciaro A, Ruco L, Mantovani A (1998) IL-10 prevents the differentiation of monocytes to dendritic cells but promotes their maturation to macrophages. Eur J Immunol 28(1):359–369. https://doi.org/10.1002/

(SICI)1521-4141(199801)28:01<359::AID-IMMU359>3.0.CO;2-4

Antsiferova M, Piwko-Czuchra A, Cangkrama M, Wietecha M, Sahin D, Birkner K, Amann VC, Levesque M, Hohl D, Dummer R, Werner S (2017) Activin promotes skin carcinogenesis by attraction and reprogramming of macrophages. EMBO Mol Med 9(1):27–45. https://doi.org/10.15252/emmm.201606493

Aras S, Zaidi MR (2017) TAMeless traitors: macrophages in cancer progression and metastasis. Br J Cancer 117(11):1583–1591. https://doi.org/10.1038/bjc.2017.356

Arenberg DA, Keane MP, DiGiovine B, Kunkel SL, Strom SR, Burdick MD, Iannettoni MD, Strieter RM (2000) Macrophage infiltration in human non-small-cell lung cancer: the role of CC chemokines. Cancer Immunol Immunother 49(2):63–70

Arts RJ, Plantinga TS, Tuit S, Ulas T, Heinhuis B, Tesselaar M, Sloot Y, Adema GJ, Joosten LA, Smit JW, Netea MG, Schultze JL, Netea-Maier RT (2016) Transcriptional and metabolic reprogramming induce an inflammatory phenotype in non-medullary thyroid carcinoma-induced macrophages. Oncoimmunology 5(12):e1229725. https://doi.org/10.1080/2162402X.2016.1229725

Asahara T, Murohara T, Sullivan A, Silver M, van der Zee R, Li T, Witzenbichler B, Schatteman G, Isner JM (1997) Isolation of putative progenitor endothelial cells for angiogenesis. Science 275(5302):964–967

Atanasov G, Potner C, Aust G, Schierle K, Dietel C, Benzing C, Krenzien F, Bartels M, Eichfeld U, Schmelzle M, Bahra M, Pascher A, Wiltberger G (2018) TIE2-expressing monocytes and M2-polarized macrophages impact survival and correlate with angiogenesis in adenocarcinoma of the pancreas. Oncotarget 9(51):29715–29726. https://doi.org/10.18632/oncotarget.25690

Bak SP, Alonso A, Turk MJ, Berwin B (2008) Murine ovarian cancer vascular leukocytes require arginase-1 activity for T cell suppression. Mol Immunol 46(2):258–268. https://doi.org/10.1016/j.molimm.2008.08.266

Balkwill F, Charles KA, Mantovani A (2005) Smoldering and polarized inflammation in the initiation and promotion of malignant disease. Cancer Cell 7(3):211–217. https://doi.org/10.1016/j.ccr.2005.02.013

Barleon B, Sozzani S, Zhou D, Weich HA, Mantovani A, Marme D (1996) Migration of human monocytes in response to vascular endothelial growth factor (VEGF) is mediated via the VEGF receptor flt-1. Blood 87(8):3336–3343

Barnett FH, Rosenfeld M, Wood M, Kiosses WB, Usui Y, Marchetti V, Aguilar E, Friedlander M (2016) Macrophages form functional vascular mimicry channels in vivo. Sci Rep 6:36659. https://doi.org/10.1038/srep36659

Bingle L, Lewis CE, Corke KP, Reed MW, Brown NJ (2006) Macrophages promote angiogenesis in human breast tumour spheroids in vivo. Br J Cancer 94(1):101–107. https://doi.org/10.1038/sj.bjc.6602901

Bloch N, Harel D (2016) The tumor as an organ: comprehensive spatial and temporal modeling of the tumor and its microenvironment. BMC Bioinf 17(1):317. https://doi.org/10.1186/s12859-016-1168-5

Bottazzi B, Walter S, Govoni D, Colotta F, Mantovani A (1992) Monocyte chemotactic cytokine gene transfer modulates macrophage infiltration, growth, and susceptibility to IL-2 therapy of a murine melanoma. J Immunol 148(4):1280–1285

Brown JM, Recht L, Strober S (2017) The promise of targeting macrophages in cancer therapy. Clin Cancer Res 23(13):3241–3250. https://doi.org/10.1158/1078-0432.CCR-16-3122

Cebe-Suarez S, Zehnder-Fjallman A, Ballmer-Hofer K (2006) The role of VEGF receptors in angiogenesis; complex partnerships. Cell Mol Life Sci 63(5):601–615. https://doi.org/10.1007/s00018-005-5426-3

Ceradini DJ, Kulkarni AR, Callaghan MJ, Tepper OM, Bastidas N, Kleinman ME, Capla JM, Galiano RD, Levine JP, Gurtner GC (2004) Progenitor cell trafficking is regulated by hypoxic gradients through HIF-1 induction of SDF-1. Nat Med 10(8):858–864. https://doi.org/10.1038/nm1075

Chen Q, Zhang XH, Massague J (2011) Macrophage binding to receptor VCAM-1 transmits survival signals in breast cancer cells that invade the lungs. Cancer Cell 20(4):538–549. https://doi.org/10.1016/j.ccr.2011.08.025

Chen W, Ma T, Shen XN, Xia XF, Xu GD, Bai XL, Liang TB (2012) Macrophage-induced tumor angiogenesis is regulated by the TSC2-mTOR pathway. Cancer Res 72(6):1363–1372. https://doi.org/10.1158/0008-5472.CAN-11-2684

Coffelt SB, Hughes R, Lewis CE (2009) Tumor-associated macrophages: effectors of angiogenesis and tumor progression. Biochim Biophys Acta 1796(1):11–18. https://doi.org/10.1016/j.bbcan.2009.02.004

Condeelis J, Pollard JW (2006) Macrophages: obligate partners for tumor cell migration, invasion, and metastasis. Cell 124(2):263–266. https://doi.org/10.1016/j.cell.2006.01.007

Coussens LM, Zitvogel L, Palucka AK (2013) Neutralizing tumor-promoting chronic inflammation: a magic bullet? Science 339(6117):286–291. https://doi.org/10.1126/science.1232227

Cursiefen C, Chen L, Borges LP, Jackson D, Cao J, Radziejewski C, D'Amore PA, Dana MR, Wiegand SJ, Streilein JW (2004) VEGF-A stimulates lymphangiogenesis and hemangiogenesis in inflammatory neovascularization via macrophage recruitment. J Clin Invest 113(7):1040–1050. https://doi.org/10.1172/JCI20465

Dammeijer F, Lievense LA, Kaijen-Lambers ME, van Nimwegen M, Bezemer K, Hegmans JP, van Hall T, Hendriks RW, Aerts JG (2017) Depletion of tumor-associated macrophages with a CSF-1R kinase inhibitor enhances antitumor immunity and survival induced by DC immunotherapy. Cancer Immunol Res

5(7):535–546. https://doi.org/10.1158/2326-6066.CIR-16-0309

De Palma M, Lewis CE (2011) Cancer: macrophages limit chemotherapy. Nature 472(7343):303–304. https://doi.org/10.1038/472303a

De Palma M, Venneri MA, Roca C, Naldini L (2003) Targeting exogenous genes to tumor angiogenesis by transplantation of genetically modified hematopoietic stem cells. Nat Med 9(6):789–795. https://doi.org/10.1038/nm871

De Palma M, Venneri MA, Galli R, Sergi Sergi L, Politi LS, Sampaolesi M, Naldini L (2005) Tie2 identifies a hematopoietic lineage of proangiogenic monocytes required for tumor vessel formation and a mesenchymal population of pericyte progenitors. Cancer Cell 8(3):211–226. https://doi.org/10.1016/j.ccr.2005.08.002

DeNardo DG, Brennan DJ, Rexhepaj E, Ruffell B, Shiao SL, Madden SF, Gallagher WM, Wadhwani N, Keil SD, Junaid SA, Rugo HS, Hwang ES, Jirstrom K, West BL, Coussens LM (2011) Leukocyte complexity predicts breast cancer survival and functionally regulates response to chemotherapy. Cancer Discov 1(1):54–67. https://doi.org/10.1158/2159-8274.CD-10-0028

Derynck R, Zhang YE (2003) Smad-dependent and Smad-independent pathways in TGF-beta family signalling. Nature 425(6958):577–584. https://doi.org/10.1038/nature02006

Dijkgraaf EM, Heusinkveld M, Tummers B, Vogelpoel LT, Goedemans R, Jha V, Nortier JW, Welters MJ, Kroep JR, van der Burg SH (2013) Chemotherapy alters monocyte differentiation to favor generation of cancer-supporting M2 macrophages in the tumor microenvironment. Cancer Res 73(8):2480–2492. https://doi.org/10.1158/0008-5472.CAN-12-3542

Domingues G, Gouveia-Fernandes S, Salgado D, et al (2015) Monocytes/macrophages in cancer, from tumor aggressors to vascular components – a new insight for anti-angiogenic therapy. In: EACR-AACR-SIC special conference on anticancer drug action and drug resistance from cancer biology to the clinic, pp 98–99

Dunn GP, Bruce AT, Ikeda H, Old LJ, Schreiber RD (2002) Cancer immunoediting: from immunosurveillance to tumor escape. Nat Immunol 3(11):991–998. https://doi.org/10.1038/ni1102-991ni1102-991

Dunn GP, Old LJ, Schreiber RD (2004) The immunobiology of cancer immunosurveillance and immunoediting. Immunity 21(2):137–148. https://doi.org/10.1016/j.immuni.2004.07.017

El-Serag HB (2012) Epidemiology of viral hepatitis and hepatocellular carcinoma. Gastroenterology 142(6):1264–1273.e1261. https://doi.org/10.1053/j.gastro.2011.12.061

Erler JT, Bennewith KL, Cox TR, Lang G, Bird D, Koong A, Le QT, Giaccia AJ (2009) Hypoxia-induced lysyl oxidase is a critical mediator of bone marrow cell recruitment to form the premetastatic niche. Cancer Cell 15(1):35–44. https://doi.org/10.1016/j.ccr.2008.11.012

Fadini GP, Baesso I, Albiero M, Sartore S, Agostini C, Avogaro A (2008) Technical notes on endothelial progenitor cells: ways to escape from the knowledge plateau. Atherosclerosis 197(2):496–503. https://doi.org/10.1016/j.atherosclerosis.2007.12.039

Fernandez Pujol B, Lucibello FC, Gehling UM, Lindemann K, Weidner N, Zuzarte ML, Adamkiewicz J, Elsasser HP, Muller R, Havemann K (2000) Endothelial-like cells derived from human CD14 positive monocytes. Differentiation 65(5):287–300

Ferrara N, Gerber HP, LeCouter J (2003) The biology of VEGF and its receptors. Nat Med 9(6):669–676. https://doi.org/10.1038/nm0603-669

Folkman J (1985) Tumor angiogenesis. Adv Cancer Res 43:175–203

Folkman J, Hanahan D (1991) Switch to the angiogenic phenotype during tumorigenesis. Princess Takamatsu Symp 22:339–347

Forget MA, Voorhees JL, Cole SL, Dakhlallah D, Patterson IL, Gross AC, Moldovan L, Mo X, Evans R, Marsh CB, Eubank TD (2014) Macrophage colony-stimulating factor augments Tie2-expressing monocyte differentiation, angiogenic function, and recruitment in a mouse model of breast cancer. PLoS One 9(6):e98623. https://doi.org/10.1371/journal.pone.0098623

Freire Valls A, Knipper K, Giannakouri E, Sarachaga V, Hinterkopf S, Wuehrl M, Shen Y, Radhakrishnan P, Klose J, Ulrich A, Schneider M, Augustin HG, Ruiz de Almodovar C, Schmidt T (2019) VEGFR1(+) metastasis-associated macrophages contribute to metastatic angiogenesis and influence colorectal cancer patient outcome. Clin Cancer Res 25(18):5674–5685. https://doi.org/10.1158/1078-0432.CCR-18-2123

Fulton AM, Loveless SE, Heppner GH (1984) Mutagenic activity of tumor-associated macrophages in Salmonella typhimurium strains TA98 and TA 100. Cancer Res 44(10):4308–4311

Gabrusiewicz K, Liu D, Cortes-Santiago N, Hossain MB, Conrad CA, Aldape KD, Fuller GN, Marini FC, Alonso MM, Idoate MA, Gilbert MR, Fueyo J, Gomez-Manzano C (2014) Anti-vascular endothelial growth factor therapy-induced glioma invasion is associated with accumulation of Tie2-expressing monocytes. Oncotarget 5(8):2208–2220. https://doi.org/10.18632/oncotarget.1893

Gao D, Nolan D, McDonnell K, Vahdat L, Benezra R, Altorki N, Mittal V (2009) Bone marrow-derived endothelial progenitor cells contribute to the angiogenic switch in tumor growth and metastatic progression. Biochim Biophys Acta 1796(1):33–40. https://doi.org/10.1016/j.bbcan.2009.05.001

Gazzaniga S, Bravo AI, Guglielmotti A, van Rooijen N, Maschi F, Vecchi A, Mantovani A, Mordoh J, Wainstok R (2007) Targeting tumor-associated macrophages and inhibition of MCP-1 reduce angiogenesis and tumor growth in a human melanoma xenograft. J Invest Dermatol 127(8):2031–2041. https://doi.org/10.1038/sj.jid.5700827

George AL, Bangalore-Prakash P, Rajoria S, Suriano R, Shanmugam A, Mittelman A, Tiwari RK (2011) Endothelial progenitor cell biology in disease and tissue regeneration. J Hematol Oncol 4:24. https://doi.org/10.1186/1756-8722-4-24

Gerber HP, Ferrara N (2003) The role of VEGF in normal and neoplastic hematopoiesis. J Mol Med (Berl) 81(1):20–31. https://doi.org/10.1007/s00109-002-0397-4

Giraudo E, Inoue M, Hanahan D (2004) An amino-bisphosphonate targets MMP-9-expressing macrophages and angiogenesis to impair cervical carcinogenesis. J Clin Invest 114(5):623–633. https://doi.org/10.1172/JCI22087

Gocheva V, Wang HW, Gadea BB, Shree T, Hunter KE, Garfall AL, Berman T, Joyce JA (2010) IL-4 induces cathepsin protease activity in tumor-associated macrophages to promote cancer growth and invasion. Genes Dev 24(3):241–255. https://doi.org/10.1101/gad.1874010

Goel S, Wong AH, Jain RK (2012) Vascular normalization as a therapeutic strategy for malignant and nonmalignant disease. Cold Spring Harb Perspect Med 2(3):a006486. https://doi.org/10.1101/cshperspect.a006486

Gorelik L, Flavell RA (2001) Immune-mediated eradication of tumors through the blockade of transforming growth factor-beta signaling in T cells. Nat Med 7(10):1118–1122. https://doi.org/10.1038/nm1001-1118

Gorelik L, Flavell RA (2002) Transforming growth factor-beta in T-cell biology. Nat Rev Immunol 2(1):46–53. https://doi.org/10.1038/nri704

Goswami S, Sahai E, Wyckoff JB, Cammer M, Cox D, Pixley FJ, Stanley ER, Segall JE, Condeelis JS (2005) Macrophages promote the invasion of breast carcinoma cells via a colony-stimulating factor-1/epidermal growth factor paracrine loop. Cancer Res 65(12):5278–5283. https://doi.org/10.1158/0008-5472.CAN-04-1853

Grunewald M, Avraham I, Dor Y, Bachar-Lustig E, Itin A, Jung S, Chimenti S, Landsman L, Abramovitch R, Keshet E (2006) VEGF-induced adult neovascularization: recruitment, retention, and role of accessory cells. Cell 124(1):175–189. https://doi.org/10.1016/j.cell.2005.10.036

Halin S, Rudolfsson SH, Van Rooijen N, Bergh A (2009) Extratumoral macrophages promote tumor and vascular growth in an orthotopic rat prostate tumor model. Neoplasia 11(2):177–186

Hanahan D, Folkman J (1996) Patterns and emerging mechanisms of the angiogenic switch during tumorigenesis. Cell 86(3):353–364. https://doi.org/10.1016/s0092-8674(00)80108-7

Hanahan D, Weinberg RA (2011) Hallmarks of cancer: the next generation. Cell 144(5):646–674. https://doi.org/10.1016/j.cell.2011.02.013

Hao Q, Liu J, Pappu R, Su H, Rola R, Gabriel RA, Lee CZ, Young WL, Yang GY (2008) Contribution of bone marrow-derived cells associated with brain angiogenesis is primarily through leukocytes and macrophages. Arterioscler Thromb Vasc Biol 28(12):2151–2157. https://doi.org/10.1161/ATVBAHA.108.176297

Harmey JH, Dimitriadis E, Kay E, Redmond HP, Bouchier-Hayes D (1998) Regulation of macrophage production of vascular endothelial growth factor (VEGF) by hypoxia and transforming growth factor beta-1. Ann Surg Oncol 5(3):271–278

Harney AS, Karagiannis GS, Pignatelli J, Smith BD, Kadioglu E, Wise SC, Hood MM, Kaufman MD, Leary CB, Lu WP, Al-Ani G, Chen X, Entenberg D, Oktay MH, Wang Y, Chun L, De Palma M, Jones JG, Flynn DL, Condeelis JS (2017) The selective Tie2 inhibitor rebastinib blocks recruitment and function of Tie2(Hi) macrophages in breast cancer and pancreatic neuroendocrine tumors. Mol Cancer Ther 16(11):2486–2501. https://doi.org/10.1158/1535-7163.MCT-17-0241

Hirschi KK, Ingram DA, Yoder MC (2008) Assessing identity, phenotype, and fate of endothelial progenitor cells. Arterioscler Thromb Vasc Biol 28(9):1584–1595. https://doi.org/10.1161/ATVBAHA.107.155960

Hou Z, Falcone DJ, Subbaramaiah K, Dannenberg AJ (2011) Macrophages induce COX-2 expression in breast cancer cells: role of IL-1beta autoamplification. Carcinogenesis 32(5):695–702. https://doi.org/10.1093/carcin/bgr027

Hudson JD, Shoaibi MA, Maestro R, Carnero A, Hannon GJ, Beach DH (1999) A proinflammatory cytokine inhibits p53 tumor suppressor activity. J Exp Med 190(10):1375–1382

Hung JY, Horn D, Woodruff K, Prihoda T, LeSaux C, Peters J, Tio F, Abboud-Werner SL (2014) Colony-stimulating factor 1 potentiates lung cancer bone metastasis. Lab Investig 94(4):371–381. https://doi.org/10.1038/labinvest.2014.1

Ide H, Seligson DB, Memarzadeh S, Xin L, Horvath S, Dubey P, Flick MB, Kacinski BM, Palotie A, Witte ON (2002) Expression of colony-stimulating factor 1 receptor during prostate development and prostate cancer progression. Proc Natl Acad Sci U S A 99(22):14404–14409. https://doi.org/10.1073/pnas.222537099

Ingber DE (1992) Extracellular matrix as a solid-state regulator in angiogenesis: identification of new targets for anti-cancer therapy. Semin Cancer Biol 3(2):57–63

Jain RK (2005) Normalization of tumor vasculature: an emerging concept in antiangiogenic therapy. Science 307(5706):58–62. https://doi.org/10.1126/science.1104819

Ji J, Zhang G, Sun B, Yuan H, Huang Y, Zhang J, Wei X, Zhang X, Hou J (2013) The frequency of tumor-infiltrating Tie-2-expressing monocytes in renal cell carcinoma: its relationship to angiogenesis and progression. Urology 82(4):974.e979–974.e913. https://doi.org/10.1016/j.urology.2013.05.026

Jiang S, Yang Y, Fang M, Li X, Yuan X, Yuan J (2016) Co-evolution of tumor-associated macrophages and tumor neo-vessels during cervical cancer invasion. Oncol Lett 12(4):2625–2631. https://doi.org/10.3892/ol.2016.5014

Kacinski BM (1997) CSF-1 and its receptor in breast carcinomas and neoplasms of the female reproductive tract. Mol Reprod Dev 46(1):71–74. https://doi.org/10.1002/(SICI)1098-2795(199701)46:1<71::AID-MRD11>3.0.CO;2-6

Kalka C, Masuda H, Takahashi T, Kalka-Moll WM, Silver M, Kearney M, Li T, Isner JM, Asahara T (2000) Transplantation of ex vivo expanded endothelial progenitor cells for therapeutic neovascularization. Proc Natl Acad Sci U S A 97(7):3422–3427. https://doi.org/10.1073/pnas.070046397

Kaplan RN, Psaila B, Lyden D (2007) Niche-to-niche migration of bone-marrow-derived cells. Trends Mol Med 13(2):72–81. https://doi.org/10.1016/j.molmed.2006.12.003

Karin M, Greten FR (2005) NF-kappaB: linking inflammation and immunity to cancer development and progression. Nat Rev Immunol 5(10):749–759. https://doi.org/10.1038/nri1703

Karin M, Lawrence T, Nizet V (2006) Innate immunity gone awry: linking microbial infections to chronic inflammation and cancer. Cell 124(4):823–835. https://doi.org/10.1016/j.cell.2006.02.016

Karlmark KR, Tacke F, Dunay IR (2012) Monocytes in health and disease – minireview. Eur J Microbiol Immunol (Bp) 2(2):97–102. https://doi.org/10.1556/EuJMI.2.2012.2.1

Kawaguchi T (2005) Cancer metastasis: characterization and identification of the behavior of metastatic tumor cells and the cell adhesion molecules, including carbohydrates. Curr Drug Targets Cardiovasc Haematol Disord 5(1):39–64

Keith B, Johnson RS, Simon MC (2011) HIF1alpha and HIF2alpha: sibling rivalry in hypoxic tumour growth and progression. Nat Rev Cancer 12(1):9–22. https://doi.org/10.1038/nrc3183

Kemeny LV, Kurgyis Z, Buknicz T, Groma G, Jakab A, Zanker K, Dittmar T, Kemeny L, Nemeth IB (2016) Melanoma cells can adopt the phenotype of stromal fibroblasts and macrophages by spontaneous cell fusion in vitro. Int J Mol Sci 17(6). https://doi.org/10.3390/ijms17060826

Kim SJ, Kim JS, Papadopoulos J, Wook Kim S, Maya M, Zhang F, He J, Fan D, Langley R, Fidler IJ (2009) Circulating monocytes expressing CD31: implications for acute and chronic angiogenesis. Am J Pathol 174(5):1972–1980. https://doi.org/10.2353/ajpath.2009.080819

Kimura YN, Watari K, Fotovati A, Hosoi F, Yasumoto K, Izumi H, Kohno K, Umezawa K, Iguchi H, Shirouzu K, Takamori S, Kuwano M, Ono M (2007) Inflammatory stimuli from macrophages and cancer cells synergistically promote tumor growth and angiogenesis. Cancer Sci 98(12):2009–2018. https://doi.org/10.1111/j.1349-7006.2007.00633.x

Kioi M, Vogel H, Schultz G, Hoffman RM, Harsh GR, Brown JM (2010) Inhibition of vasculogenesis, but not angiogenesis, prevents the recurrence of glioblastoma after irradiation in mice. J Clin Invest 120(3):694–705. https://doi.org/10.1172/JCI40283

Kirschmann DA, Seftor EA, Hardy KM, Seftor RE, Hendrix MJ (2012) Molecular pathways: vasculogenic mimicry in tumor cells: diagnostic and therapeutic implications. Clin Cancer Res 18(10):2726–2732. https://doi.org/10.1158/1078-0432.CCR-11-3237

Kitajima D, Kasamatsu A, Nakashima D, Miyamoto I, Kimura Y, Endo-Sakamoto Y, Shiiba M, Tanzawa H, Uzawa K (2018) Evidence for critical role of Tie2/Ang1 interaction in metastatic oral cancer. Oncol Lett 15(5):7237–7242. https://doi.org/10.3892/ol.2018.8212

Knowles H, Leek R, Harris AL (2004) Macrophage infiltration and angiogenesis in human malignancy. Novartis Found Symp 256:189–200; discussion 200–184, 259–169

Komohara Y, Ohnishi K, Kuratsu J, Takeya M (2008) Possible involvement of the M2 anti-inflammatory macrophage phenotype in growth of human gliomas. J Pathol 216(1):15–24. https://doi.org/10.1002/path.2370

Komohara Y, Hasita H, Ohnishi K, Fujiwara Y, Suzu S, Eto M, Takeya M (2011) Macrophage infiltration and its prognostic relevance in clear cell renal cell carcinoma. Cancer Sci 102(7):1424–1431. https://doi.org/10.1111/j.1349-7006.2011.01945.x

Komohara Y, Horlad H, Ohnishi K, Fujiwara Y, Bai B, Nakagawa T, Suzu S, Nakamura H, Kuratsu J, Takeya M (2012) Importance of direct macrophage-tumor cell interaction on progression of human glioma. Cancer Sci 103(12):2165–2172. https://doi.org/10.1111/cas.12015

Koong AC, Denko NC, Hudson KM, Schindler C, Swiersz L, Koch C, Evans S, Ibrahim H, Le QT, Terris DJ, Giaccia AJ (2000) Candidate genes for the hypoxic tumor phenotype. Cancer Res 60(4):883–887

Koukourakis MI, Giatromanolaki A, Kakolyris S, O'Byrne KJ, Apostolikas N, Skarlatos J, Gatter KC, Harris AL (1998) Different patterns of stromal and cancer cell thymidine phosphorylase reactivity in non-small-cell lung cancer: impact on tumour neoangiogenesis and survival. Br J Cancer 77(10):1696–1703. https://doi.org/10.1038/bjc.1998.280

Kovacic JC, Moore J, Herbert A, Ma D, Boehm M, Graham RM (2008) Endothelial progenitor cells, angioblasts, and angiogenesis--old terms reconsidered from a current perspective. Trends Cardiovasc Med 18(2):45–51. https://doi.org/10.1016/j.tcm.2007.12.002

Kovaleva OV, Samoilova DV, Shitova MS, Gratchev A (2016) Tumor associated macrophages in kidney cancer. Anal Cell Pathol (Amst) 2016:9307549. https://doi.org/10.1155/2016/9307549

Kubota Y, Takubo K, Shimizu T, Ohno H, Kishi K, Shibuya M, Saya H, Suda T (2009) M-CSF inhibition selectively targets pathological angiogenesis and lymphangiogenesis. J Exp Med 206(5):1089–1102. https://doi.org/10.1084/jem.20081605

Kuwabara K, Ogawa S, Matsumoto M, Koga S, Clauss M, Pinsky DJ, Lyn P, Leavy J, Witte L, Joseph-Silverstein J et al (1995) Hypoxia-mediated induction of acidic/basic fibroblast growth factor and platelet-derived growth factor in mononuclear phagocytes stimulates growth of hypoxic endothelial cells. Proc Natl Acad Sci U S A 92(10):4606–4610

Lamagna C, Aurrand-Lions M, Imhof BA (2006) Dual role of macrophages in tumor growth and angiogenesis. J Leukoc Biol 80(4):705–713. https://doi.org/10.1189/jlb.1105656

Lee HW, Choi HJ, Ha SJ, Lee KT, Kwon YG (2013) Recruitment of monocytes/macrophages in different tumor microenvironments. Biochim Biophys Acta 1835(2):170–179. https://doi.org/10.1016/j.bbcan.2012.12.007

Leek RD, Lewis CE, Whitehouse R, Greenall M, Clarke J, Harris AL (1996) Association of macrophage infiltration with angiogenesis and prognosis in invasive breast carcinoma. Cancer Res 56(20):4625–4629

Leek RD, Hunt NC, Landers RJ, Lewis CE, Royds JA, Harris AL (2000) Macrophage infiltration is associated with VEGF and EGFR expression in breast cancer. J Pathol 190(4):430–436. https://doi.org/10.1002/(SICI)1096-9896(200003)190:4<430::AID-PATH538>3.0.CO;2-6

Leung SY, Wong MP, Chung LP, Chan AS, Yuen ST (1997) Monocyte chemoattractant protein-1 expression and macrophage infiltration in gliomas. Acta Neuropathol 93(5):518–527

Lewis C, Murdoch C (2005) Macrophage responses to hypoxia: implications for tumor progression and anti-cancer therapies. Am J Pathol 167(3):627–635. https://doi.org/10.1016/S0002-9440(10)62038-X

Lewis CE, De Palma M, Naldini L (2007) Tie2-expressing monocytes and tumor angiogenesis: regulation by hypoxia and angiopoietin-2. Cancer Res 67(18):8429–8432. https://doi.org/10.1158/0008-5472.CAN-07-1684

Li MO, Wan YY, Sanjabi S, Robertson AK, Flavell RA (2006) Transforming growth factor-beta regulation of immune responses. Annu Rev Immunol 24:99–146. https://doi.org/10.1146/annurev.immunol.24.021605.090737

Lin WW, Karin M (2007) A cytokine-mediated link between innate immunity, inflammation, and cancer. J Clin Invest 117(5):1175–1183. https://doi.org/10.1172/JCI31537

Lin EY, Nguyen AV, Russell RG, Pollard JW (2001) Colony-stimulating factor 1 promotes progression of mammary tumors to malignancy. J Exp Med 193(6):727–740

Lin EY, Gouon-Evans V, Nguyen AV, Pollard JW (2002) The macrophage growth factor CSF-1 in mammary gland development and tumor progression. J Mammary Gland Biol Neoplasia 7(2):147–162

Lin EY, Li JF, Bricard G, Wang W, Deng Y, Sellers R, Porcelli SA, Pollard JW (2007) Vascular endothelial growth factor restores delayed tumor progression in tumors depleted of macrophages. Mol Oncol 1(3):288–302. https://doi.org/10.1016/j.molonc.2007.10.003

Lindstrom A, Midtbo K, Arnesson LG, Garvin S, Shabo I (2017) Fusion between M2-macrophages and cancer cells results in a subpopulation of radioresistant cells with enhanced DNA-repair capacity. Oncotarget 8(31):51370–51386. https://doi.org/10.18632/oncotarget.17986

Lissbrant IF, Stattin P, Wikstrom P, Damber JE, Egevad L, Bergh A (2000) Tumor associated macrophages in human prostate cancer: relation to clinicopathological variables and survival. Int J Oncol 17(3):445–451. https://doi.org/10.3892/ijo.17.3.445

Liu Q, Qiao L, Liang N, Xie J, Zhang J, Deng G, Luo H (2016) The relationship between vasculogenic mimicry and epithelial-mesenchymal transitions. J Cell Mol Med 20(9):1761–1769. https://doi.org/10.1111/jcmm.12851

Locati M, Deuschle U, Massardi ML, Martinez FO, Sironi M, Sozzani S, Bartfai T, Mantovani A (2002) Analysis of the gene expression profile activated by the CC chemokine ligand 5/RANTES and by lipopolysaccharide in human monocytes. J Immunol 168(7):3557–3562

Lopes-Coelho F, Gouveia-Fernandes S, Serpa J (2018) Metabolic cooperation between cancer and non-cancerous stromal cells is pivotal in cancer progression. Tumour Biol 40(2):1010428318756203. https://doi.org/10.1177/1010428318756203

Maeda H, Akaike T (1998) Nitric oxide and oxygen radicals in infection, inflammation, and cancer. Biochemistry (Mosc) 63(7):854–865

Mancuso P, Burlini A, Pruneri G, Goldhirsch A, Martinelli G, Bertolini F (2001) Resting and activated endothelial cells are increased in the peripheral blood of cancer patients. Blood 97(11):3658–3661

Mantovani A (2010) Molecular pathways linking inflammation and cancer. Curr Mol Med 10(4):369–373. https://doi.org/10.2174/156652410791316968

Mantovani A, Sica A (2010) Macrophages, innate immunity and cancer: balance, tolerance, and diversity. Curr Opin Immunol 22(2):231–237. https://doi.org/10.1016/j.coi.2010.01.009

Mantovani A, Schioppa T, Porta C, Allavena P, Sica A (2006) Role of tumor-associated macrophages in tumor progression and invasion. Cancer Metastasis Rev 25(3):315–322. https://doi.org/10.1007/s10555-006-9001-7

Mantovani A, Allavena P, Sica A, Balkwill F (2008) Cancer-related inflammation. Nature 454(7203):436–444. https://doi.org/10.1038/nature07205

Martin-Padura I, Marighetti P, Gregato G, Agliano A, Malazzi O, Mancuso P, Pruneri G, Viale A, Bertolini F (2012) Spontaneous cell fusion of acute leukemia cells and macrophages observed in cells with leukemic potential. Neoplasia 14(11):1057–1066

Matsubara T, Kanto T, Kuroda S, Yoshio S, Higashitani K, Kakita N, Miyazaki M, Sakakibara M, Hiramatsu N, Kasahara A, Tomimaru Y, Tomokuni A, Nagano H, Hayashi N, Takehara T (2013) TIE2-expressing monocytes as a diagnostic marker for hepatocellular carcinoma correlates with angiogenesis. Hepatology 57(4):1416–1425. https://doi.org/10.1002/hep.25965

McDonnell CO, Bouchier-Hayes DJ, Toomey D, Foley D, Kay EW, Leen E, Walsh TN (2003) Effect of neoadjuvant chemoradiotherapy on angiogenesis in oesophageal cancer. Br J Surg 90(11):1373–1378. https://doi.org/10.1002/bjs.4338

Medina RJ, O'Neill CL, Sweeney M, Guduric-Fuchs J, Gardiner TA, Simpson DA, Stitt AW (2010) Molecular analysis of endothelial progenitor cell (EPC) subtypes reveals two distinct cell populations with different identities. BMC Med Genet 3:18. https://doi.org/10.1186/1755-8794-3-18

Menetrier-Caux C, Montmain G, Dieu MC, Bain C, Favrot MC, Caux C, Blay JY (1998) Inhibition of the differentiation of dendritic cells from CD34(+) progenitors by tumor cells: role of interleukin-6 and macrophage colony-stimulating factor. Blood 92(12):4778–4791

Metallo CM, Gameiro PA, Bell EL, Mattaini KR, Yang J, Hiller K, Jewell CM, Johnson ZR, Irvine DJ, Guarente L, Kelleher JK, Vander Heiden MG, Iliopoulos O, Stephanopoulos G (2011) Reductive glutamine metabolism by IDH1 mediates lipogenesis under hypoxia. Nature 481(7381):380 384. https://doi.org/10.1038/nature10602

Metinko AP, Kunkel SL, Standiford TJ, Strieter RM (1992) Anoxia-hyperoxia induces monocyte-derived interleukin-8. J Clin Invest 90(3):791–798. https://doi.org/10.1172/JCI115953

Mills CD, Ley K (2014) M1 and M2 macrophages: the chicken and the egg of immunity. J Innate Immun 6(6):716–726. https://doi.org/10.1159/000364945

Moldovan NI (2002) Role of monocytes and macrophages in adult angiogenesis: a light at the tunnel's end. J Hematother Stem Cell Res 11(2):179–194. https://doi.org/10.1089/152581602753658394

Monestiroli S, Mancuso P, Burlini A, Pruneri G, Dell'Agnola C, Gobbi A, Martinelli G, Bertolini F (2001) Kinetics and viability of circulating endothelial cells as surrogate angiogenesis marker in an animal model of human lymphoma. Cancer Res 61(11):4341–4344

Mroczko B, Groblewska M, Wereszczynska-Siemiatkowska U, Okulczyk B, Kedra B, Laszewicz W, Dabrowski A, Szmitkowski M (2007) Serum macrophage-colony stimulating factor levels in colorectal cancer patients correlate with lymph node metastasis and poor prognosis. Clin Chim Acta 380(1–2):208–212. https://doi.org/10.1016/j.cca.2007.02.037

Murdoch C, Lewis CE (2005) Macrophage migration and gene expression in response to tumor hypoxia. Int J Cancer 117(5):701–708. https://doi.org/10.1002/ijc.21422

Murdoch C, Giannoudis A, Lewis CE (2004) Mechanisms regulating the recruitment of macrophages into hypoxic areas of tumors and other ischemic tissues. Blood 104(8):2224–2234. https://doi.org/10.1182/blood-2004-03-1109

Murdoch C, Muthana M, Coffelt SB, Lewis CE (2008) The role of myeloid cells in the promotion of tumour angiogenesis. Nat Rev Cancer 8(8):618–631. https://doi.org/10.1038/nrc2444

Negus RP, Stamp GW, Hadley J, Balkwill FR (1997) Quantitative assessment of the leukocyte infiltrate in ovarian cancer and its relationship to the expression of C-C chemokines. Am J Pathol 150(5):1723–1734

Nielsen SR, Schmid MC (2017) Macrophages as key drivers of cancer progression and metastasis. Mediat Inflamm 2017:9624760. https://doi.org/10.1155/2017/9624760

Nishida N, Yano H, Nishida T, Kamura T, Kojiro M (2006) Angiogenesis in cancer. Vasc Health Risk Manag 2(3):213–219

Nolan DJ, Ciarrocchi A, Mellick AS, Jaggi JS, Bambino K, Gupta S, Heikamp E, McDevitt MR, Scheinberg DA, Benezra R, Mittal V (2007) Bone marrow-derived endothelial progenitor cells are a major determinant of nascent tumor neovascularization. Genes Dev 21(12):1546–1558. https://doi.org/10.1101/gad.436307

Oguma K, Oshima H, Aoki M, Uchio R, Naka K, Nakamura S, Hirao A, Saya H, Taketo MM, Oshima M (2008) Activated macrophages promote Wnt signalling through tumour necrosis factor-alpha in gastric tumour cells. EMBO J 27(12):1671–1681. https://doi.org/10.1038/emboj.2008.105

Ohno S, Ohno Y, Suzuki N, Kamei T, Koike K, Inagawa H, Kohchi C, Soma G, Inoue M (2004) Correlation of histological localization of tumor-associated macrophages with clinicopathological features in endometrial cancer. Anticancer Res 24(5C):3335–3342

Pages G, Pouyssegur J (2005) Transcriptional regulation of the Vascular Endothelial Growth Factor gene--a concert of activating factors. Cardiovasc Res 65(3):564–573. https://doi.org/10.1016/j.cardiores.2004.09.032

Pahl HL (1999) Activators and target genes of Rel/NF-kappaB transcription factors. Oncogene 18(49):6853–6866. https://doi.org/10.1038/sj.onc.1203239

Parsonnet J, Friedman GD, Vandersteen DP, Chang Y, Vogelman JH, Orentreich N, Sibley RK (1991) Helicobacter pylori infection and the risk of gastric carcinoma. N Engl J Med 325(16):1127–1131. https://doi.org/10.1056/NEJM199110173251603

Peichev M, Naiyer AJ, Pereira D, Zhu Z, Lane WJ, Williams M, Oz MC, Hicklin DJ, Witte L, Moore MA, Rafii S (2000) Expression of VEGFR-2 and AC133 by circulating human CD34(+) cells identifies a population of functional endothelial precursors. Blood 95(3):952–958

Penny HL, Sieow JL, Adriani G, Yeap WH, See Chi Ee P, San Luis B, Lee B, Lee T, Mak SY, Ho YS, Lam KP, Ong CK, Huang RY, Ginhoux F, Rotzschke O, Kamm RD, Wong SC (2016) Warburg metabolism in tumor-conditioned macrophages promotes metastasis in human pancreatic ductal adenocarcinoma. Oncoimmunology 5(8):e1191731. https://doi.org/10.1080/2162402X.2016.1191731

Pircher A, Kahler CM, Skvortsov S, Dlaska M, Kawaguchi G, Schmid T, Gunsilius E, Hilbe W (2008) Increased numbers of endothelial progenitor cells in peripheral blood and tumor specimens in non-small cell lung cancer: a methodological challenge and an ongoing debate on the clinical relevance. Oncol Rep 19(2):345–352

Pogoda K, Pyszniak M, Rybojad P, Tabarkiewicz J (2016) Monocytic myeloid-derived suppressor cells as a potent suppressor of tumor immunity in non-small cell lung cancer. Oncol Lett 12(6):4785–4794. https://doi.org/10.3892/ol.2016.5273

Polette M, Nawrocki-Raby B, Gilles C, Clavel C, Birembaut P (2004) Tumour invasion and matrix metalloproteinases. Crit Rev Oncol Hematol 49(3):179–186. https://doi.org/10.1016/j.critrevonc.2003.10.008

Pollard JW (2004) Tumour-educated macrophages promote tumour progression and metastasis. Nat Rev Cancer 4(1):71–78. https://doi.org/10.1038/nrc1256nrc1256

Pollard JW (2009) Trophic macrophages in development and disease. Nat Rev Immunol 9(4):259–270. https://doi.org/10.1038/nri2528

Powell AE, Anderson EC, Davies PS, Silk AD, Pelz C, Impey S, Wong MH (2011) Fusion between Intestinal epithelial cells and macrophages in a cancer context results in nuclear reprogramming. Cancer Res 71(4):1497–1505. https://doi.org/10.1158/0008-5472.CAN-10-3223

Prenen H, Mazzone M (2019) Tumor-associated macrophages: a short compendium. Cell Mol Life Sci 76(8):1447–1458. https://doi.org/10.1007/s00018-018-2997-3

Psaila B, Lyden D (2009) The metastatic niche: adapting the foreign soil. Nat Rev Cancer 9(4):285–293. https://doi.org/10.1038/nrc2621

Purhonen S, Palm J, Rossi D, Kaskenpaa N, Rajantie I, Yla-Herttuala S, Alitalo K, Weissman IL, Salven P (2008) Bone marrow-derived circulating endothelial precursors do not contribute to vascular endothelium and are not needed for tumor growth. Proc Natl Acad Sci U S A 105(18):6620–6625. https://doi.org/10.1073/pnas.0710516105

Qian BZ, Pollard JW (2010) Macrophage diversity enhances tumor progression and metastasis. Cell 141(1):39–51. https://doi.org/10.1016/j.cell.2010.03.014

Qian BZ, Li J, Zhang H, Kitamura T, Zhang J, Campion LR, Kaiser EA, Snyder LA, Pollard JW (2011) CCL2 recruits inflammatory monocytes to facilitate breast-tumour metastasis. Nature 475(7355):222–225. https://doi.org/10.1038/nature10138

Qiu B, Zhang D, Wang C, Tao J, Tie X, Qiao Y, Xu K, Wang Y, Wu A (2011) IL-10 and TGF-beta2 are overexpressed in tumor spheres cultured from human gliomas. Mol Biol Rep 38(5):3585–3591. https://doi.org/10.1007/s11033-010-0469-4

Real C, Remedio L, Caiado F, Igreja C, Borges C, Trindade A, Pinto-do OP, Yagita H, Duarte A, Dias S (2011) Bone marrow-derived endothelial progenitors expressing Delta-like 4 (Dll4) regulate tumor angiogenesis. PLoS One 6(4):e18323. https://doi.org/10.1371/journal.pone.0018323

Rehman J, Li J, Orschell CM, March KL (2003) Peripheral blood "endothelial progenitor cells" are derived from monocyte/macrophages and secrete angiogenic growth factors. Circulation 107(8):1164–1169

Ribatti D (2007) The discovery of endothelial progenitor cells. An historical review. Leuk Res 31(4):439–444. https://doi.org/10.1016/j.leukres.2006.10.014

Ribatti D (2009) The paracrine role of Tie-2-expressing monocytes in tumor angiogenesis. Stem Cells Dev 18(5):703–706. https://doi.org/10.1089/scd.2008.0385

Ribatti D, Nico B, Crivellato E, Vacca A (2005) Endothelial progenitor cells in health and disease. Histol Histopathol 20(4):1351–1358

Richter-Ehrenstein C, Rentzsch J, Runkel S, Schneider A, Schonfelder G (2007) Endothelial progenitor cells in breast cancer patients. Breast Cancer Res Treat 106(3):343–349. https://doi.org/10.1007/s10549-007-9505-z

Rigo A, Gottardi M, Zamo A, Mauri P, Bonifacio M, Krampera M, Damiani E, Pizzolo G, Vinante F (2010) Macrophages may promote cancer growth via a GM-CSF/HB-EGF paracrine loop that is enhanced by CXCL12. Mol Cancer 9:273. https://doi.org/10.1186/1476-4598-9-273

Robinson CJ, Stringer SE (2001) The splice variants of vascular endothelial growth factor (VEGF) and their receptors. J Cell Sci 114(Pt 5):853–865

Robinson SC, Scott KA, Balkwill FR (2002) Chemokine stimulation of monocyte matrix metalloproteinase-9 requires endogenous TNF-alpha. Eur J Immunol 32(2):404–412. https://doi.org/10.1002/1521-4141(200202)32:2<404::AID-IMMU404>3.0.CO;2-X

Robinson-Smith TM, Isaacsohn I, Mercer CA, Zhou M, Van Rooijen N, Husseinzadeh N, McFarland-Mancini MM, Drew AF (2007) Macrophages mediate inflammation-enhanced metastasis of ovarian tumors in mice. Cancer Res 67(12):5708–5716. https://doi.org/10.1158/0008-5472.CAN-06-4375

Rodriguez PC, Zea AH, DeSalvo J, Culotta KS, Zabaleta J, Quiceno DG, Ochoa JB, Ochoa AC (2003) L-arginine consumption by macrophages modulates the expression of CD3 zeta chain in T lymphocytes. J Immunol 171(3):1232–1239

Rodriguez PC, Quiceno DG, Ochoa AC (2007) L-arginine availability regulates T-lymphocyte cell-cycle progression. Blood 109(4):1568–1573. https://doi.org/10.1182/blood-2006-06-031856

Rohde E, Malischnik C, Thaler D, Maierhofer T, Linkesch W, Lanzer G, Guelly C, Strunk D (2006) Blood monocytes mimic endothelial progenitor cells. Stem Cells 24(2):357–367. https://doi.org/10.1634/stemcells.2005-0072

Romagnani P, Annunziato F, Liotta F, Lazzeri E, Mazzinghi B, Frosali F, Cosmi L, Maggi L, Lasagni L, Scheffold A, Kruger M, Dimmeler S, Marra F, Gensini G, Maggi E, Romagnani S (2005) CD14+CD34low cells with stem cell phenotypic and functional features are the major source of circulating endothelial progenitors. Circ Res 97(4):314–322. https://doi.org/10.1161/01.RES.0000177670.72216.9b

Roodhart JM, He H, Daenen LG, Monvoisin A, Barber CL, van Amersfoort M, Hofmann JJ, Radtke F, Lane TF, Voest EE, Iruela-Arispe ML (2013) Notch1 regulates angio-supportive bone marrow-derived cells in mice: relevance to chemoresistance. Blood 122(1):143–153. https://doi.org/10.1182/blood-2012-11-459347

Roxburgh CS, McMillan DC (2014) Cancer and systemic inflammation: treat the tumour and treat the host. Br J Cancer 110(6):1409–1412. https://doi.org/10.1038/bjc.2014.90

Ryder M, Ghossein RA, Ricarte-Filho JC, Knauf JA, Fagin JA (2008) Increased density of tumor-associated macrophages is associated with decreased survival in advanced thyroid cancer. Endocr Relat Cancer 15(4):1069–1074. https://doi.org/10.1677/ERC-08-0036

Sakamori Y, Masago K, Ohmori K, Togashi Y, Nagai H, Okuda C, Kim YH, Ichiyama S, Mishima M (2012) Increase in circulating endothelial progenitor cells predicts response in patients with advanced non-small-cell lung cancer. Cancer Sci 103(6):1065–1070. https://doi.org/10.1111/j.1349-7006.2012.02249.x

Salvesen HB, Akslen LA (1999) Significance of tumour-associated macrophages, vascular endothelial growth factor and thrombospondin-1 expression for tumour angiogenesis and prognosis in endometrial carcinomas. Int J Cancer 84(5):538–543. https://doi.org/10.1002/(SICI)1097-0215(19991022)84:5<538::AID-IJC17>3.0.CO;2-B

Sawano A, Iwai S, Sakurai Y, Ito M, Shitara K, Nakahata T, Shibuya M (2001) Flt-1, vascular endothelial growth factor receptor 1, is a novel cell surface marker for the lineage of monocyte-macrophages in humans. Blood 97(3):785–791

Schmeisser A, Garlichs CD, Zhang H, Eskafi S, Graffy C, Ludwig J, Strasser RH, Daniel WG (2001) Monocytes coexpress endothelial and macrophagocytic lineage markers and form cord-like structures in Matrigel under angiogenic conditions. Cardiovasc Res 49(3):671–680. https://doi.org/10.1016/s0008-6363(00)00270-4

Schmidt-Lucke C, Fichtlscherer S, Aicher A, Tschope C, Schultheiss HP, Zeiher AM, Dimmeler S (2010) Quantification of circulating endothelial progenitor cells using the modified ISHAGE protocol. PLoS One 5(11):e13790. https://doi.org/10.1371/journal.pone.0013790

Semenza GL (2003) Angiogenesis in ischemic and neoplastic disorders. Annu Rev Med 54:17–28. https://doi.org/10.1146/annurev.med.54.101601.152418

Shaked Y, Bertolini F, Man S, Rogers MS, Cervi D, Foutz T, Rawn K, Voskas D, Dumont DJ, Ben-David Y, Lawler J, Henkin J, Huber J, Hicklin DJ, D'Amato RJ, Kerbel RS (2005) Genetic heterogeneity of the vasculogenic phenotype parallels angiogenesis; Implications for cellular surrogate marker analysis of antiangiogenesis. Cancer Cell 7(1):101–111. https://doi.org/10.1016/j.ccr.2004.11.023

Sharifi BG, Zeng Z, Wang L, Song L, Chen H, Qin M, Sierra-Honigmann MR, Wachsmann-Hogiu S, Shah PK (2006) Pleiotrophin induces transdifferentiation of monocytes into functional endothelial cells. Arterioscler Thromb Vasc Biol 26(6):1273–1280. https://doi.org/10.1161/01.ATV.0000222017.05085.8e

Sica A, Bronte V (2007) Altered macrophage differentiation and immune dysfunction in tumor development. J Clin Invest 117(5):1155–1166. https://doi.org/10.1172/JCI31422

Sica A, Saccani A, Bottazzi B, Polentarutti N, Vecchi A, van Damme J, Mantovani A (2000) Autocrine production of IL-10 mediates defective IL-12 production and NF-kappa B activation in tumor-associated macrophages. J Immunol 164(2):762–767. https://doi.org/10.4049/jimmunol.164.2.762

Sica A, Schioppa T, Mantovani A, Allavena P (2006) Tumour-associated macrophages are a distinct M2 polarised population promoting tumour progression: potential targets of anti-cancer therapy. Eur J Cancer 42(6):717–727. https://doi.org/10.1016/j.ejca.2006.01.003

Sica A, Larghi P, Mancino A, Rubino L, Porta C, Totaro MG, Rimoldi M, Biswas SK, Allavena P, Mantovani A (2008a) Macrophage polarization in tumour progression. Semin Cancer Biol 18(5):349–355. https://doi.org/10.1016/j.semcancer.2008.03.004

Sica A, Allavena P, Mantovani A (2008b) Cancer related inflammation: the macrophage connection. Cancer Lett 267(2):204–215. https://doi.org/10.1016/j.canlet.2008.03.028

Spinelli FM, Vitale DL, Icardi A, Caon I, Brandone A, Giannoni P, Saturno V, Passi A, Garcia M, Sevic I, Alaniz L (2019) Hyaluronan preconditioning of monocytes/macrophages affects their angiogenic behavior and regulation of TSG-6 expression in a tumor type-specific manner. FEBS J 286(17):3433–3449. https://doi.org/10.1111/febs.14871

Sporn MB (1996) The war on cancer. Lancet 347(9012):1377–1381

Steidl C, Lee T, Shah SP, Farinha P, Han G, Nayar T, Delaney A, Jones SJ, Iqbal J, Weisenburger DD, Bast MA, Rosenwald A, Muller-Hermelink HK, Rimsza LM, Campo E, Delabie J, Braziel RM, Cook JR, Tubbs RR, Jaffe ES, Lenz G, Connors JM, Staudt LM, Chan WC, Gascoyne RD (2010) Tumor-associated macrophages and survival in classic Hodgkin's lymphoma. N Engl J Med 362(10):875–885. https://doi.org/10.1056/NEJMoa0905680

Stockmann C, Doedens A, Weidemann A, Zhang N, Takeda N, Greenberg JI, Cheresh DA, Johnson RS (2008) Deletion of vascular endothelial growth factor in myeloid cells accelerates tumorigenesis. Nature 456(7223):814–818. https://doi.org/10.1038/nature07445

Sun T, Yang Y, Luo X, Cheng Y, Zhang M, Wang K, Ge C (2014) Inhibition of tumor angiogenesis by interferon-gamma by suppression of tumor-associated macrophage differentiation. Oncol Res 21(5):227–235. https://doi.org/10.3727/096504014X13890370410285

Takanami I, Takeuchi K, Kodaira S (1999) Tumor-associated macrophage infiltration in pulmonary adenocarcinoma: association with angiogenesis and poor prognosis. Oncology 57(2):138–142. https://doi.org/10.1159/000012021

Timmermans F, Plum J, Yoder MC, Ingram DA, Vandekerckhove B, Case J (2009) Endothelial progenitor cells: identity defined? J Cell Mol Med 13(1):87–102. https://doi.org/10.1111/j.1582-4934.2008.00598.x

Torisu H, Ono M, Kiryu H, Furue M, Ohmoto Y, Nakayama J, Nishioka Y, Sone S, Kuwano M (2000) Macrophage infiltration correlates with tumor stage and angiogenesis in human malignant melanoma: possible involvement of TNFalpha and IL-1alpha. Int J Cancer 85(2):182–188. https://doi.org/10.1002/(SICI)1097-0215(20000115)85:2<182::AID-IJC6>3.0.CO;2-M

Toy EP, Azodi M, Folk NL, Zito CM, Zeiss CJ, Chambers SK (2009) Enhanced ovarian cancer tumorigenesis and metastasis by the macrophage colony-stimulating factor. Neoplasia 11(2):136–144

Ueno T, Toi M, Saji H, Muta M, Bando H, Kuroi K, Koike M, Inadera H, Matsushima K (2000) Significance of macrophage chemoattractant protein-1 in macrophage recruitment, angiogenesis, and survival in human breast cancer. Clin Cancer Res 6(8):3282–3289

Urbich C, Heeschen C, Aicher A, Dernbach E, Zeiher AM, Dimmeler S (2003) Relevance of monocytic features for neovascularization capacity of circulating endothelial progenitor cells. Circulation 108(20):2511–2516. https://doi.org/10.1161/01.CIR.0000096483.29777.50

Venneri MA, De Palma M, Ponzoni M, Pucci F, Scielzo C, Zonari E, Mazzieri R, Doglioni C, Naldini L (2007) Identification of proangiogenic TIE2-expressing monocytes (TEMs) in human peripheral blood and cancer. Blood 109(12):5276–5285. https://doi.org/10.1182/blood-2006-10-053504

Verbridge SS, Choi NW, Zheng Y, Brooks DJ, Stroock AD, Fischbach C (2009) Oxygen-controlled three-dimensional cultures to analyze tumor angiogenesis. Tissue Eng Part A 16(7):2133–2141. https://doi.org/10.1089/ten.TEA.2009.0670

Wang X, Zhao X, Wang K, Wu L, Duan T (2013) Interaction of monocytes/macrophages with ovarian cancer cells promotes angiogenesis in vitro. Cancer Sci 104(4):516–523. https://doi.org/10.1111/cas.12110

Wang X, Zhu Q, Lin Y, Wu L, Wu X, Wang K, He Q, Xu C, Wan X (2017) Crosstalk between TEMs and endothelial cells modulates angiogenesis and metastasis via IGF1-IGF1R signalling in epithelial ovarian cancer. Br J Cancer 117(9):1371–1382. https://doi.org/10.1038/bjc.2017.297

Wang F, Li B, Wei Y, Zhao Y, Wang L, Zhang P, Yang J, He W, Chen H, Jiao Z, Li Y (2018) Tumor-derived exosomes induce PD1(+) macrophage population in human gastric cancer that promotes disease progression. Oncogene 7(5):41. https://doi.org/10.1038/s41389-018-0049-3

Wei C, Yang C, Wang S, Shi D, Zhang C, Lin X, Liu Q, Dou R, Xiong B (2019) Crosstalk between cancer cells and tumor associated macrophages is required for mesenchymal circulating tumor cell-mediated colorectal cancer metastasis. Mol Cancer 18(1):64. https://doi.org/10.1186/s12943-019-0976-4

Willenborg S, Lucas T, van Loo G, Knipper JA, Krieg T, Haase I, Brachvogel B, Hammerschmidt M, Nagy A, Ferrara N, Pasparakis M, Eming SA (2012) CCR2 recruits an inflammatory macrophage subpopulation critical for angiogenesis in tissue repair. Blood 120(3):613–625. https://doi.org/10.1182/blood-2012-01-403386

Wu SY, Watabe K (2017) The roles of microglia/macrophages in tumor progression of brain cancer and metastatic disease. Front Biosci (Landmark Ed) 22:1805–1829. https://doi.org/10.2741/4573

Wu H, Xu JB, He YL, Peng JJ, Zhang XH, Chen CQ, Li W, Cai SR (2012) Tumor-associated macrophages promote angiogenesis and lymphangiogenesis of gastric cancer. J Surg Oncol 106(4):462–468. https://doi.org/10.1002/jso.23110

Wyckoff J, Wang W, Lin EY, Wang Y, Pixley F, Stanley ER, Graf T, Pollard JW, Segall J, Condeelis J (2004) A paracrine loop between tumor cells and macrophages is required for tumor cell migration in mammary tumors. Cancer Res 64(19):7022–7029. https://doi.org/10.1158/0008-5472.CAN-04-1449

Xu J, Escamilla J, Mok S, David J, Priceman S, West B, Bollag G, McBride W, Wu L (2013) CSF1R signaling blockade stanches tumor-infiltrating myeloid cells and improves the efficacy of radiotherapy in prostate cancer. Cancer Res 73(9):2782–2794. https://doi.org/10.1158/0008-5472.CAN-12-3981

Yamaguchi Y, Okazaki Y, Seta N, Satoh T, Takahashi K, Ikezawa Z, Kuwana M (2010) Enhanced angiogenic potency of monocytic endothelial progenitor cells in patients with systemic sclerosis. Arthritis Res Ther 12(6):R205. https://doi.org/10.1186/ar3180

Yan D, Wang HW, Bowman RL, Joyce JA (2016) STAT3 and STAT6 signaling pathways synergize to promote cathepsin secretion from macrophages via IRE1alpha activation. Cell Rep 16(11):2914–2927. https://doi.org/10.1016/j.celrep.2016.08.035

Yang L, Pang Y, Moses HL (2010) TGF-beta and immune cells: an important regulatory axis in the tumor microenvironment and progression. Trends Immunol 31(6):220–227. https://doi.org/10.1016/j.it.2010.04.002

Yang WJ, Hao YX, Yang X, Fu XL, Shi Y, Yue HL, Yin P, Dong HL, Yu PW (2018) Overexpression of Tie2 is associated with poor prognosis in patients with gastric cancer. Oncol Lett 15(5):8027–8033. https://doi.org/10.3892/ol.2018.8329

Yoder MC (2012) Human endothelial progenitor cells. Cold Spring Harb Perspect Med 2(7):a006692. https://doi.org/10.1101/cshperspect.a006692

Yu D, Sun X, Qiu Y, Zhou J, Wu Y, Zhuang L, Chen J, Ding Y (2007) Identification and clinical significance of mobilized endothelial progenitor cells in tumor vasculogenesis of hepatocellular carcinoma. Clin Cancer Res 13(13):3814–3824. https://doi.org/10.1158/1078-0432.CCR-06-2594

Zabuawala T, Taffany DA, Sharma SM, Merchant A, Adair B, Srinivasan R, Rosol TJ, Fernandez S, Huang K, Leone G, Ostrowski MC (2010) An ets2-driven transcriptional program in tumor-associated macrophages promotes tumor metastasis. Cancer Res 70(4):1323–1333. https://doi.org/10.1158/0008-5472.CAN-09-1474

Zajac E, Schweighofer B, Kupriyanova TA, Juncker-Jensen A, Minder P, Quigley JP, Deryugina EI (2013) Angiogenic capacity of M1- and M2-polarized macrophages is determined by the levels of TIMP-1 complexed with their secreted proMMP-9. Blood 122(25):4054–4067. https://doi.org/10.1182/blood-2013-05-501494

Zeisberger SM, Odermatt B, Marty C, Zehnder-Fjallman AH, Ballmer-Hofer K, Schwendener RA (2006) Clodronate-liposome-mediated depletion of tumour-associated macrophages: a new and highly effective antiangiogenic therapy approach. Br J Cancer 95(3):272–281. https://doi.org/10.1038/sj.bjc.6603240

Zeng XY, Xie H, Yuan J, Jiang XY, Yong JH, Zeng D, Dou YY, Xiao SS (2019) M2-like tumor-associated macrophages-secreted EGF promotes epithelial ovarian cancer metastasis via activating EGFR-ERK signaling and suppressing lncRNA LIMT expression. Cancer Biol Ther 20(7):956–966. https://doi.org/10.1080/15384047.2018.1564567

Zhang Z, Chen F, Shang L (2018) Advances in antitumor effects of NSAIDs. Cancer Manag Res 10:4631–4640. https://doi.org/10.2147/CMAR.S175212

Zhu XD, Zhang JB, Zhuang PY, Zhu HG, Zhang W, Xiong YQ, Wu WZ, Wang L, Tang ZY, Sun HC (2008) High expression of macrophage colony-stimulating factor in peritumoral liver tissue is associated with poor survival after curative resection of hepatocellular carcinoma. J Clin Oncol 26(16):2707–2716. https://doi.org/10.1200/JCO.2007.15.6521

Zhu C, Chrifi I, Mustafa D, van der Weiden M, Leenen PJM, Duncker DJ, Kros JM, Cheng C (2017) CECR1-mediated cross talk between macrophages and vascular mural cells promotes neovascularization in malignant glioma. Oncogene 36(38):5356–5368. https://doi.org/10.1038/onc.2017.145

Part II

Microenvironment and Metabolic Signalling: The Way Cancer Cells Know How to Survive

Wnt Signaling: Paths for Cancer Progression

10

Filipa Carreira-Barbosa and Sofia C. Nunes

Abstract

The Wnt signaling pathways are well known for having several pivotal roles during embryonic development. However, the same developmental signaling pathways also present key roles in cancer initiation and progression. In this chapter, several issues regarding the roles of both canonical and non-canonical Wnt signaling pathways in cancer will be explored, mainly concerning their role in the maintenance of cancer *stemness*, in the metabolism reprograming of cancer cells and in the modulation of the tumor microenvironment. The role of Wnt signaling cascades in the response of cancer cells to anti-cancer treatments will be also discussed, as well as its potential therapeutic targeting during cancer treatment. Collectively, increasing evidence has been supporting pivotal roles of Wnt signaling in several features of cancer biology, however; a lot is still to be elucidated.

F. Carreira-Barbosa (✉)
Faculdade de Ciências da Universidade de Lisboa (FCUL-UL), Lisbon, Portugal

S. C. Nunes
CEDOC, Chronic Diseases Research Centre, NOVA Medical School | Faculdade de Ciências Médicas, Universidade NOVA de Lisboa, Lisbon, Portugal

Instituto Português de Oncologia de Lisboa Francisco Gentil (IPOLFG), Lisbon, Portugal

Keywords

Cancer · Metabolism reprogramming · Resistance · Stemness · Tumor microenvironment · Wnt

10.1 Canonical Wnt Signaling: From Embryonic Development to Cancer Promotion

Several Wnt signal transduction pathways have been identified so far. The best known is the canonical Wnt signaling pathway that initially was found to specify segment polarity in *Drosophila* and mediate axis formation in *Xenopus* (reviewed in (Gilbert and Barresi 2019)). Currently, it is well known that the canonical Wnt signaling pathway has crucial roles not only during embryogenesis but also during adult tissue homeostasis, having biological functions in stem cell renewal, cell proliferation and cell differentiation (reviewed in (Steinhart and Angers 2018)).

The canonical pathway transduces the Wnt ligand and signal via attachment to its transmembrane receptor Frizzled (Fz). Sequentially there is activation of the cytoplasmic protein Dishevelled (Dsh). This leads to blocking of the breakdown of the complex containing APC, axin and GSK-3, which in turn permit the stabilization of β-catenin

J. Serpa (ed.), *Tumor Microenvironment*, Advances in Experimental Medicine and Biology 1219,
https://doi.org/10.1007/978-3-030-34025-4_10

and its consequent translocation to the nucleus. Once stabilized, β-catenin binds to proteins of the lymphoid enhancer factor/T-cell factor (LEF/TCF) family that is implicated in the transcriptional regulation of several target genes (Cadigan and Nusse 1997; Huelsken and Birchmeier 2001). These target genes include, among others, genes involved in cellular proliferation and transformation as *c-MYC*, *c-Jun*, *CCND1*, *EGFR*, CD44, CD133 and leucine-rich repeat-containing receptor 5 (*LG45*) (reviewed in (Jeong et al. 2018)).

The canonical Wnt signaling pathway was already implicated in different types of cancer, including colorectal, hepatocarcinoma, medulloblastoma, ovarian cancer, and breast cancer (Fior and Zilhão 2019). Elements of the Wnt/β-catenin signaling are generally mutated in tumors (Fior and Zilhão 2019). Additionally, loss of functional axin or other mutations that stabilize β-catenin expression were also reported (Zurawel et al. 1998; Palacios and Gamallo 1998; Satoh et al. 2000; Laurent-Puig and Zucman-Rossi 2006; Bao et al. 2012; Stewart et al. 2014). Furthermore, expanding evidence has been supporting the existence of a cross-talk between microRNAS and Wnt/β-catenin signaling, leading to carcinogenesis, cancer metastasis, and drug-resistance (reviewed in (Peng et al. 2017)). It is also important to highlight that a synergistic cooperation between the Wnt/β-catenin and RAS-ERK pathways was also reported in colorectal cancer, leading to the stabilization of β-catenin and RAS, thus driving tumorigenesis (reviewed in (Jeong et al. 2018)).

These discoveries have launched scientists to create inhibitors of the Wnt/β-catenin pathway for therapeutic procedure to treat cancer, even though the large majority of these are still at the preclinical trials (Anastas and Moon 2013). This therapy in cancer treatment is, nevertheless, altered by the answer of particular tumor types to Wnt/β-catenin pathway. For example in few cancers, like, melanoma and prostate cancer, patients with higher rates of active β-catenin signaling showed an improved response. This kind of results urges the necessity for a deepest knowledge of the characteristics of Wnt pathway in cancer.

This pathway is implicated in different tumors, as already explained, and has a central role for anticancer treatments, with different important inhibitors at several levels of clinical progress. The advanced task of immunotherapies in cancer and new progress into the Wnt- pathway in cancer-related immune-regulation will allow the development of novel therapeutics. The case of cancers that have upregulated Wnt/β-catenin pathway like colorectal cancer have been analyzed as objects for Wnt inhibition. But, Wnt inhibitors may have a broader feature in cancers such as melanoma, lung, and renal cancers where immunotherapy has come to the front position. There are a few questions to be answered in the situation of Wnt pathway in immunomodulation but previously this issue must go to clinical trials (reviewed in (Pai et al. 2017)).

Collectively, data has been supporting a pivotal role of Wnt signaling in carcinogenesis, cellular proliferation, adhesion, migration, invasion, angiogenesis, progression, survival, epithelial-to-mesenchymal transition and chemoresistance (reviewed in (Niiro et al. 2018)). Besides the canonical pathway, also non-canonical Wnt pathways were reported to have a role in cancer (Fior and Zilhão 2019). In the next section, we will discuss the role of non-canonical pathways in cancer initiation and progression.

10.2 Non-canonical Wnt Pathways in Cancer

Evidence also strongly support a role of non-canonical Wnt pathways in cancer, where several regulators and downstream effectors of Wnt were already reported.

The potential regulators of the non-canonical Wnt pathway comprise the small GTPases of the Rho family. Rho, Rac, and Cdc42 are involved in vertebrate non-canonical Wnt signaling (Habas et al. 2001; Choi and Han 2002; Penzo-Mendèz et al. 2003). In cancer the small GTPase Rac1 and Rho combined activate ROCK (Rho kinase) and JNK, which are reported to have a role in the reorganization of the cytoskeleton and/or transcriptional activities, via for instance ATF2 (activating transcription factor 2) (Zhan et al. 2017).

In cancer, there is also activation of transcription of YAP/TAZ-dependent. Wnt/PCP pathway controls then actin cytoskeletal dynamics, directional cell movement and JNK- or YAP/TAZ-dependent transcription (reviewed in (Katoh 2017)).

In cancer, non-canonical Wnt pathway via ROR1 and ROR2 can activate the PI3K-AKT and YAP pathways. Wnt cascades are connected to therapeutic resistance and relapse of human cancers in part through PI3K-AKT cascade (reviewed in (Katoh 2017)).

The JAK/STAT (Janus kinase and signal transducer and activator of transcription) pathway has been proposed to have a possible role in mesendermal cell polarization/migration and germ-layer separation and tumor development, functioning as a modulator of cell movements. Various JAK and STAT homologs have been recognized and are expressed and/or mutated during cancer development (reviewed in (Gilbert and Barresi 2019)).

A study experiment suggests that a downstream target of STAT3 may be secretory molecules that are capable of non-cell-autonomously activating Dsh-RhoA in the adjacent cells, thereby modulating the Planar Cell Polarity (PCP) pathway (Miyagi et al. 2004).

Upstream regulators of JAK/STAT pathway are being identified, for instance the activation of Stat3 may depend on the activity of the canonical Wnt/β-catenin pathway (Yamashita et al. 2002).

Collectively, evidence strongly supports an association of the Wnt signaling with several other pathways with pivotal roles in cancer biology.

In the next section we will discuss the role of the main non-canonical signaling pathways in cancer, the Wnt-Calcium and the Planar Cell Polarity Pathway (PCP).

10.2.1 Wnt-Calcium Pathway in Cancer

There is a pathway regulating intracellular calcium levels, the so called Wnt/Ca^{2+} pathway. There are several pieces of evidence proposing that another possible bifurcation of the Wnt pathway controls intracellular Ca^{2+} levels and could be regulating cancer (reviewed in (Sherwood 2015)). This pathway is triggered by the ligand Wnt5A, which is the most popular Wnt ligand that has been found to activate this signaling cascade in cancer cells, and its expression is linked with equally tumor-suppressive and pro-oncogenic tasks, varying on the tumor type. For instance, higher Wnt5A expression is linked with a good patient prognosis in breast and colon cancers (Lejeune et al. 1995; Dejmek et al. 2005) yet weak survival in melanoma and gastric cancer (Kurayoshi et al. 2006; Da Forno et al. 2008), giving the context-dependent disposition of Wnt pathway in carcinogenesis.

The receptor frizzled FZD2 was implicated in the degradation of the guanine nucleotide binding protein (G-protein), in various group of amino acids beta/gamma subunits G-protein alpha-t2, leading to Ca^{2+} to be discharged into the cytoplasm and stimulating the neuronal differentiation. Calcium activates CaMK II and Calmodulin, increasing the phosphorylation of Tcf/Lef (T-cell factor and lymphoid enhancer factor) thus blocking the canonical Wnt signaling (Sheldahl et al. 2003).

The Wnt/Ca^{2+} pathway is necessary for the regulation of differential cell adhesiveness (Winklbauer et al. 2001). More specifically interfering with *fz7* function gives rise to a failure of proper separation of the cell layers, and this function could be mediated via PKC in a G-protein-dependent manner. Thus Fz7 might regulate the adhesive characteristics of cells by activation of Ca^{2+} pathway (Winklbauer et al. 2001). Another study proposes that the Wnt/Ca^{2+} pathway is Dsh-dependent and also Pk1 has the ability to stimulate calcium flux, suggesting that the Wnt/Ca^{2+} and PCP pathways overlap to some extend (Veeman et al. 2003; Sheldahl et al. 2003). Additionally, full-length Dsh is able to activate calcium signaling by the calcium flux suggesting the promiscuous role of Dsh (Sheldahl et al. 2003).

Choi and Han have proposed a Dsh-independent Wnt/Ca $^{2+}$ pathway that activates PKC by regulating the activity of the p21 GTPase, Cdc42 (Choi and Han 2002). Fascinatingly, PKC has been proposed to act as a complex with Dsh and is necessary for the translocation of Dsh to the cell membrane in response to Fz7 (Kinoshita et al. 2003).

In addition to the intracellular role of this pathway, Ca^{2+}-release into the extracellular space might play a role in the cell-cell-communication involving and cell movements (Slusarski et al. 1997; Tada and Concha 2001; Wallingford et al. 2001). It is possible that Wnts regulates Ca^{2+} waves as their frequency is lowered when a dominant negative *fz8* is over-expressed which inhibits both canonical and non-canonical Wnt pathways. Together, these results suggest a permissive role for an intracellular Ca^{2+} signal (Wallingford et al. 2001).

In conclusion, it is established for the last 10 years that Wnt ligands lead to the discharge of intracellular Ca^{2+} to trigger Ca^{2+} – dependent enzymes such as phosphatase, calcineurin (Calcin), protein kinase C (PKC), and calmodulin-dependent kinase II (CamKII) to control various outcomes in animal tissues. PKC and CamKII regulate cell adhesion, migration, and differentiation, which are mediated by the transcription factor nuclear factor of triggered T cells (NFAT). Alternatively, calcineurin actuates nemo-like kinase (NLK) to phosphorylate TCF transcription factors and blocks canonical Wnt signaling (Ishitani et al. 2003). This Ca^{2+} pathway has been most intensely coupled with cancer initiation and its development (reviewed in (Sherwood 2015)).

10.2.2 Planar Cell Polarity Pathway in Cancer

The PCP is a signaling pathway downstream of the Wnt receptor *frizzled* (*fz*), which encodes a serpentine receptor (Vinson et al. 1989) and has been shown to be involved in the correct orientation of eye ommatidia and the polarized growth of sensory bristles in the thorax and also in the wing hairs in the wing epithelium of *Drosophila* and during development of embryo especially in gastrulation (Adler and Lee 2001; Adler 2002). Wnt/PCP signaling cascade is not directly implicated in transcriptional regulation, but on the other hand, FZ activation leads to amplified Rac and Rho small GTPase activity that has an outcome in cytoskeletal restructuring and consequently changes the cell polarity and migration (reviewed in (Sherwood 2015)).

Several PCP players are abnormally expressed in tumors, leading to cancer cell proliferation and migration. For instance, overexpression of VANGL-1, a central player of the PCP signaling cascade, is coupled with an increased risk of relapse in breast cancer patients and its knockdown diminished breast cancer cell migration (reviewed in (Corda and Sala 2017)).

Other regulators of the PCP such as Celsr1, Prickle1, Fzd3, Fzd7, Dvl2, Dvl3 and casein kinase 1 (CK1)-ε were discovered to be upregulated in B lymphocytes of patients with chronic lymphocytic leukemia. PCP stimulation in these individuals leads a worse prognosis and at cellular level is involved in transendothelial cell migration (reviewed in (Corda and Sala 2017)).

The critical roles of PCP and non-canonical Wnt pathways in oncogenic process are still a subject of study. Strikingly, concordant with its role in cell migration coordination during development, data has been supporting a role of Wnt/PCP pathway in cell migration, invasion, or metastasis in several types of cancer (Carreira-Barbosa et al. 2009, reviewed in (VanderVorst et al. 2019)). For instance, exosomes produced from fibroblast in the tumor microenvironment can increase motility and protrusive activity of breast cancer cells by the Wnt/PCP pathway (Luo et al. 2019b). In recent experiments, it was study whether signals coming from the cancer-associated stroma could control PCP. The exosomes secreted by cancer- associated fibroblasts can lead breast cancer cells to have a motile phenotype by activating PCP key players like FZD6, DVL1, DVL2 VANGL1, PK1, PLK4, AURKB and stimulating the autocrine discharge of the non-canonical ligand Wnt11 (Carreira-Barbosa et al. 2009; Corda and Sala 2017; Luo et al. 2019b; Mo et al. 2019). Interestingly, evidence suggests a role of Wnt/PCP signaling not only in single cancer cells migration but also in collective cell migration (reviewed in (VanderVorst et al. 2019)).

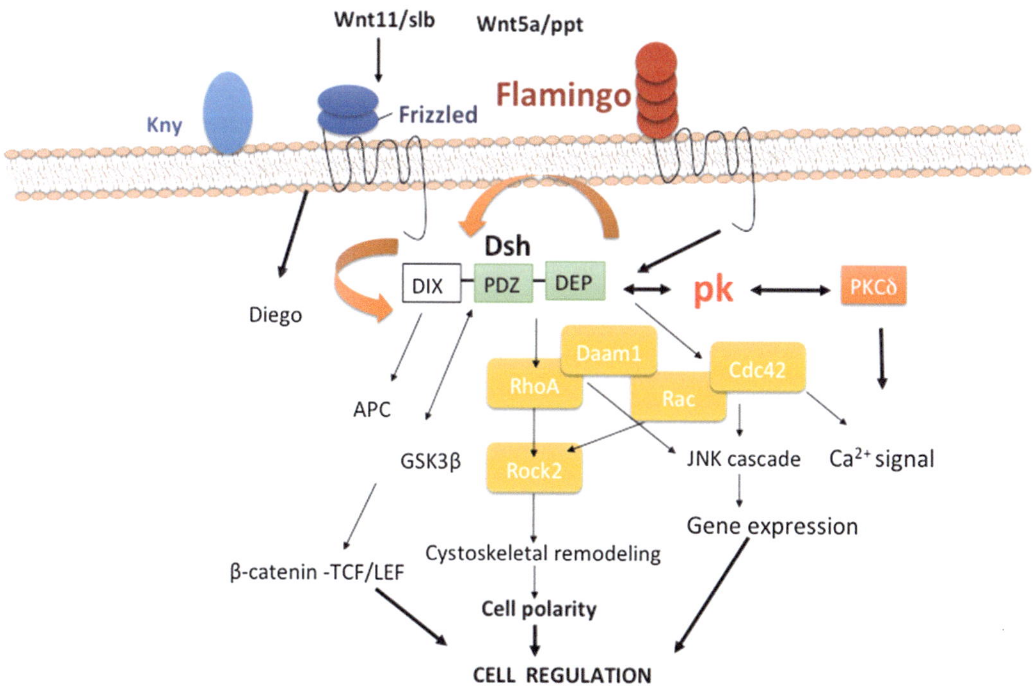

Fig. 10.1 Summary of Wnt signaling pathways. Wnt signals are transduced by various downstream pathways in a cell microenvironment-dependent manner. Canonical Wnt signaling through Frizzled (FZD) is transduced by the Wnt/β- catenin, while non-canonical Wnt pathways through FZD is transduced by Wnt/PCP (planar cell polarity), and Wnt/Ca^{2+} pathways (reviewed in (Corda and Sala 2017; Mo et al. 2019))

In conclusion these experiments propose that key player PCP molecules are not only pivotal in controlling the migration of cancer cells and the development of metastasis, but also can be pivotal for tumorigenesis (reviewed in (Zhan et al. 2017)).

The main Wnt signaling pathways are summarized in the Fig. 10.1. In the next section we will discuss the role of Wnt signaling pathways in cancer cells *stemness*, in metabolic reprogramming, on tumor microenvironment modulation and in resistance to anti-cancer treatments.

10.3 Wnt Signaling and Maintenance of Cancer Cells Stemness

The pivotal role of cancer stem cells (CSC) in cancer biology is well known. Those small populations of cells within tumors are implicated in tumor recurrence after anti-cancer therapies, resistance to therapies and metastasis (reviewed in (Rycaj and Tang 2015; Batlle and Clevers 2017; Moharil et al. 2017)). The tumorigenic factor of CSCs has been proved by xenotransplantation in immune-deficient mice. These cells have *stemness* traits that undergo reprograming and are not committed to specific cell fate, being recognized as stem-like cells (Maccalli et al., 2019).

The classical Wnt/β-catenin pathway has been implicated in self-renewal of stem cells and propagation or differentiation of progenitor cells, and non-canonical Wnt pathways were connected with preservation of stem cells, directional cell movement or down-regulation of the canonical Wnt pathway (reviewed in (Katoh 2017)). Expanding evidence has been supporting a role of both canonical and non-canonical Wnt pathways also in the expansion and evolution of CSCs, where the canonical pathway was associated to the maintenance and expansion of CSCs, whereas the non-canonical pathways were reported to induce invasion, survival and metastasis of CSCs (reviewed in (Katoh 2017)). For instance, it was reported that a novel Wnt co-activator ASPM (abnormal spindle-like microcephaly associated) activates prostate cancer stemness and expansion by enhancing Wnt – Dvl-3 – β-catenin pathway (Pai et al. 2019).

These cells have high level of plasticity and heterogeneity because of the interaction with the tumor microenvironment (reviewed in (Murphy and Weaver 2016)). For these reasons it is difficult to target CSCs/cancer initiating cells (CICs) with immunotherapy and additionally recognition by immune components of these cells needs to be increased by combination approaches to develop their susceptibility (reviewed in (Katoh 2017)). Also, future studies are needed in order to assess the function of CSCs/CICs as prognostic and predictive biomarkers of response of therapy for cancer patients (Maccalli et al. 2019). These markers have been used either to isolate stem-like cells from neoplastic tissues or to localize these cells within tumor tissues. However, the lack of standardized assays in order to isolate CSCs/CICs and the high level of plasticity were not conclusive (Maccalli et al. 2019).

As already mentioned, CSCs are intimately associated with resistance to therapies and metastasis (reviewed in (Batlle and Clevers 2017; Moharil et al. 2017)). The identification of therapeutic elements targeting CSCs/CICs is necessary in order to achieve the complete eradication of malignancy. With this goal, an extensive characterization of genomic, epigenetic, phenotypic, and immunological profile of CSCs/CICs could lead to a better understand of the functions involved in their biological characteristics, specially the Wnt canonical and non-canonical pathways. The capacity of maintaining long telomeres by the action of TERT, which expression is directly regulated by β-catenin binding to its promoter region, involves telomerase activity with Wnt pathway (Saretzki 2014). The stimulation of the Wnt antagonist DKK1 as well as action with the anti-Fzd antibody OMP18R5 paused PDAC development (Zhan et al. 2017; Fior and Zilhão 2019). In sum, the goal was to create better therapeutic molecules for cancer patients, but this is going to be addressed in the Wnt targeting section.

10.4 Wnt Signaling: A Driver of Metabolism Reprograming in Cancer Cells

The disclosure that tumor cells have specific modifications of metabolic signaling cascades has modified the perception and the knowledge of cancer. Almost 90 years ago, the first conclusions about tumor-exclusive modifications in cellular levels of energy were published (Cori and Cori 1925; Warburg et al. 1927; Potter et al. 2016). But only recently, after the genome era, the understanding of pathogenicity of cancer was impressively increased.

Metabolism reprograming is then a hallmark of cancer cells (Hanahan and Weinberg 2011; Ward and Thompson 2012). The Warburg effect is the best characterized metabolic phenotype observed in cancer cells, proposing that these cells present an increased rate of glycolysis even under normal oxygen concentrations due to defective respiration (Warburg 1956). However, expanding evidence has been supporting that mitochondrial oxidative phosphorylation is intact in most tumors (Rodríguez-Enríquez et al. 2000, 2006; Guppy et al. 2002; Viale et al. 2015; Alam et al. 2016).

The major pathways that have been studied are glucose and lipid metabolism, the lactate cascade, the PI3K-Akt-mTOR pathway and Myc signaling. Additionally, these metabolic regulations might have a function in the metabolism of cancer cells (Cramer and Schmitt 2016).

Recently, a role of Wnt pathways in metabolic reprogramming of cancer cells has been proposed, by altering the pathways mentioned above (reviewed in (Mo et al. 2019)).

Therefore, as already mentioned, Wnt signaling cascade is involved in tumor metabolic reprogramming through TCF/LEF pathway, c-myc pathway and Akt-mTOR cascade. One of the canonical cascade it is when, TCF/LEF is triggered, the expression of MCT-1, CYC1, and ATP synthase is enhanced leading to an increased aerobic glycolysis and intracellular lactate secretion, leading also to the secretion of factors like VEGF, thus developing tumor angiogenesis. Also, the classical Wnt pathway can be implicated in tumor metabolic reprogramming by over-overexpression of c-Myc. Overactivation of c-Myc's generally happens in cancer, which is a central player as a transcriptional element of oncogenic growth factor cascades (reviewed in (Mo et al. 2019)).

Myc's regulation leads to the production of proteins that control energy production, anabolic signaling pathways, protein synthesis, by this means enhancing the expression of genes such as GLUT-1, PDH, PFK1, HK, LDH, PKM2. One example comes from Vergara and colleagues, that reported that the Wnt signaling pathway via β-catenin-transcriptional regulation of Myc and its target genes have a role in mitochondria biogenesis and lipid metabolism in breast cancer cells (Vergara et al. 2017).

The conserved element in evolution, the famous Ser/Thr protein kinase target of rapamycin mTORC1 is also triggered by Wnt cascade. This pathway is involved in tumor metabolic reprogramming by overactivation of Akt -mTOR. Wnt pathway enhances mTORC1 through the PI3K-phosphoinositide-dependent kinase 1 (PDK1)-AKT pathway (reviewed in (Mo et al. 2019)).

Other targets of Wnt signaling with important roles in the metabolic reprograming of cancer cells were also reported. For instance, Senni and colleagues have disclosed a role of β-catenin in metabolic reprograming towards fatty acid oxidation in hepatocellular carcinomas, through *Pparα* induction (Senni et al. 2019).

Very recently, Tejeda-Muñoz and colleagues have reported that besides regulating endocytosis, the canonical Wnt signaling has also a role in nutrient uptake by engulfment of extracellular fluids through macropinocytosis in HeLa cells (Tejeda-Muñoz et al. 2019). Interestingly, a role of Wnt signaling in the activation of Keap1/NRF2 via TCF7L1 was also reported in the context of gastric cancer (Zhang et al. 2019). It is important to highlight that besides playing pivotal roles in cellular redox homeostasis regulation, evidence also suggest a role of NRF2 in metabolic reprogramming during cellular oxidative stress (reviewed in (Hayes and Dinkova-Kostova 2014)).

Whereas expanding evidence suggests a role of canonical Wnt signaling in metabolism reprograming and endocytosis, the opposite was also reported. Therefore Albrecht and co-workers have shown that methionine availability regulates canonical Wnt signaling in HeLa and HEK293T cells, suggesting a role of one-carbon metabolism in the regulation of Wnt signaling and endocytosis in cancer cells (Albrecht et al. 2019).

Collectively, data have been supporting a complex and key role of both canonical and non-canonical Wnt pathways in the metabolism reprogramming of cancer cells.

10.5 Wnt Signaling as a Modulator of the Tumor Microenvironment

The tumor microenvironment is composed by signaling molecules, extracellular matrixes, and by several cell types including fibroblasts, adipocytes, endothelial cells, pericytes, B cells, T cells, neutrophils, natural killer cells and macrophages (reviewed in (Gupta et al. 2017)). These important ecological niches of cancer cells have been implicated in tumor growth, invasion, metastasis and resistance to anti-cancer therapies (Barker et al. 2015; Son et al. 2017; Chen et al. 2018; Gu et al. 2018).

Expanding evidence supports that the Wnt signaling cascade is implicated in the modulation of the tumor microenvironment.

Hypoxia is a common feature of the tumor microenvironment that is responsible for tumor progression and resistance to therapy (reviewed in (Vaupel and Mayer 2007; Semenza 2012; Gillies et al. 2012)). Interestingly, a crosstalk between the Wnt signaling and hypoxia-inducible factors (HIF) was already suggested both *in vitro* and *in vivo* experiments (Bogaerts et al. 2014; Xu et al. 2017; Lyou et al. 2018). For instance, Lyou and colleagues have reported a role of hypoxia inducible factor 1 alpha subunit (HIF-1α) in the regulation of Wnt signaling through the regulation of the transcription factors LEF1 and TCF1 in human colon cancer cells (Lyou et al. 2018). Importantly, the authors' data also suggested a crosstalk between the HIF-1α and Wnt signaling pathways in the regulation of the metabolic targets expression (Lyou et al. 2018).

Interestingly, not only a role of HIF-1α in the regulation of Wnt signaling, was reported, but also the opposite. Therefore, Xiang et al. found a role of Wnt signaling in HIF-1α regulation (Xiang et al. 2018). The authors have shown that the β-catenin transcriptional partner TCF7L2 induced aerobic glycolysis in pancreatic cancer cells by suppressing Egl-9 family hypoxia inducible factor 2 (EGLN2), that consequently led to HIF-1α upregulation (Xiang et al. 2018).

The Wnt non-canonical pathway modulates adipocytes within the tumor microenvironment. Therefore, it was revealed that these pathways can facilitate adipocyte de-differentiation within the tumor microenvironment (Vona-Davisa and Gibsona 2013). Furthermore, it was also unraveled a role of Wnt in the function of adipose tissue in melanoma microenvironment and progression (Zoico et al. 2018).

Recently, Castañeda-Patlán and colleagues have reviewed the role of Wnt signaling in antitumor immune responses, stating that both canonical and non-canonical pathways mediate tumor immune tolerance (reviewed in (Castañeda-Patlán et al. 2018). In fact, evidence supports that Wnt signaling pathways modulate dendritic cells, CD4 T regulatory cells, cytotoxic CD8+ T cells and NK cells functions (reviewed in (Castañeda-Patlán et al. 2018). Corroborating this, Luke and colleagues have recently published the paper entitled "Wnt/β-catenin pathway activation correlates with immune exclusion across human cancers" (Luke et al. 2019).

Interestingly, in a cardiac microenvironment, the non-canonical Wnt pathway triggers monocytes after myocardial infarction (Meyer et al. 2017). Moreover, in primary mammary tumors, Ojalvo and colleagues suggested a role of Wnt signaling in mediating the activity of invasive tumor-associated macrophages (TAMs) – that probably have roles in tumor progression by promoting metastasis and angiogenesis – through TAM-derived Wnt7b (Ojalvo et al. 2010).

Very recently, Frenquelli and co-workers have disclosed a role of ROR2 – a receptor of a non-canonical Wnt pathway – in the adhesion of multiple myeloma cells to the bone marrow microenvironment mediated by PI3K/AKT and mTOR axes (Frenquelli et al. 2019).

Furthermore, both canonical and non-canonical Wnt signaling pathways are pivotal in angiogenesis in several organs, both in physiologic and pathological conditions (reviewed in (Olsen et al. 2017)), being also implicated in cancer angiogenesis (e.g. (Reis et al. 2012; Pate et al. 2014)).

Taken together, expanding evidence strongly support a key role of Wnt signaling pathways in the modulation of the different components/features of the tumor microenvironment – the niche of cancer cells that is pivotal for cancer initiation and progression – in several cancer types. In the next section we will discuss the role of Wnt signaling in chemo and radio-resistance.

10.6 Wnt Signaling Drives Resistance to Anti-cancer Therapies

Drug resistance is intimately associated to the poor outcome of cancer patients, baring cancer treatment.

Several studies have discovered that processes underlying the induction of drug resistance are intricate and may rely on cell and microenvironment landscape (Kozovska et al. 2014; Freimund et al. 2018; Wang et al. 2018). So, it is crucial to get data about the process of drug resistance and understand which are the dominant resistance cascades in a specific tumor group that could indicate candidates for targets in a clinical treatment environment (Mehta and Siddik 2009).

Importantly, both genetic and epigenetic modifications in canonical and non-canonical Wnt pathways were already associated to the development of drug resistance in several types of cancer (Su et al. 2010; Vangipuram et al. 2012; Fan et al. 2014; Wickström et al. 2015; Staal et al. 2016; Tan et al. 2018; Wils and Bijlsma 2018). For instance, Vangipuram and colleagues have reported, in the context of a neuroblastoma cell line, that an increased activity of the Wnt pathway via β-catenin and p-GSK3β (S-9) is able to confer Doxorubicin resistance to cancer-stem-like cells (Vangipuram et al. 2012). Wickström and colleagues have shown that Wnt/β-catenin pathway regulates the expression of the DNA repair enzyme O6-methylguanine-DNA methyltransferase (MGMT) in several types of cancer and that the inhibition of Wnt activity leads to MGMT downregulation while restoring chemosensitivity to DNA-alkylating drugs in mouse models (Wickström et al. 2015). Furthermore, Fan and co-workers related the urothelial cancer-associated 1 (UCA1) long non-coding RNA to cisplatin resistance of bladder cancer cells by enhancing the expression of Wnt6 (Fan et al. 2014). More recently, Tan and colleagues, in the context of gliomas, reported that TRIM14 upregulation is able to confer chemoresistance to temozolomide both *in vitro* and *in vivo* by stabilizing dishevelled (Dvl2) that afterwards activates the canonical Wnt signaling (Tan et al. 2018). Furthermore, preliminary results supports the use of Wnt/CTNNB1 mutations (the gene that encodes β-catenin) as biomarkers for the prediction of resistance to immune checkpoint inhibitors in hepatocellular carcinomas (Pinyol et al. 2019), hence supporting a role of Wnt signaling also in immunotherapy resistance.

Wnt signaling pathways were also reported to have a role in radioresistance (e.g. (Zhao et al. 2018; Luo et al. 2019a)). For instance, Zhao and colleagues reported a role of Wnt signaling in radioresistance by promoting DNA damage repair in esophageal squamous cell carcinoma (Zhao et al. 2018). In addition, Luo and co-workers have recently found that in nasopharyngeal carcinoma, forkhead box O 3a (FOXO3a) knockdown promotes radioresistance both in *in vitro* and *in vivo* models by inducing epithelial-mesenchymal transition and activating the canonical Wnt/β-catenin pathway (Luo et al. 2019a). Radioresistance mediated by long non-coding RNAs that activate the Wnt/β-catenin signaling pathway was also described in head and neck squamous cell carcinoma (Han et al. 2018).

Collectively, data has been supporting a key role of Wnt signaling in chemo and radioresistance in several types of cancer. In the next section we will discuss some strategies aiming to target Wnt signaling for cancer treatment.

10.7 Targeting Wnt Signaling in Cancer

As evidence strongly supports a key role of aberrant Wnt signaling pathways in cancer initiation and progression, their targeting was already suggested, as it was already referred in previous sections.

For instance, as data supports that β-catenin is the main cause of malfunction of Wnt pathway in cancer, various protein knockdowns mechanisms were developed. Besides, more information of the 300–400 Wnt enhanced genes gave contribution to targeted therapy (Morgan et al. 2017).

Therefore, present strategies aim the Wnt targeting in different tumor subgroups or with particular mutational landscape, including aberrant Wnt cascade activation in human cancers leading to CSCs survival, bulk-tumor development and invasion/metastasis (Zhan et al. 2017). Anti-FZD mAb, anti-ROR1 mAb, anti-RSPO3 mAb, PORCN inhibitors and β-catenin inhibitors are examples of Wnt cascade-targeted treatments in clinical trials (reviewed in (Katoh and Katoh 2017)).

Wnt pathway-targeted therapy can be combined with tyrosine kinase blockers or immune checkpoint inhibitors. Omics study is essential for therapeutic evaluation of Wnt pathway-targeted therapy (reviewed in (Katoh 2017)). It is important to highlight, that recently it has been reported that MEK inhibitors increase Wnt activity and induce stem cell plasticity both in *in vitro* and *in vivo* models of colorectal cancer, proposing a combined therapy of MEK and Wnt inhibitors as a promising strategy in this type of cancer (Zhan et al. 2019). Similar pathway interactions with Wnt signaling activation were found after BRAF targeting in both *BRAF* mutant colorectal cancer (Chen et al. 2019) and melanoma (Biechele et al. 2012). These findings support that the combination of different targeted therapies including Wnt targeting should be advantageous in the avoidance of drug resistance.

Wnt signaling cascades interact with other pathways including the Notch and Sonic Hedgehog, giving precious information about clinical trials in different cancers with inhibitors of these pathways. They may influence the normal Wnt dependent stem cell population, mainly in regions of fast turnover like hair follicles and the gastro-intestinal tract (reviewed in (Krishnamurthy and Kurzrock 2018)).

There are potential side effects of long-term usage with Wnt antibodies and other treatments like small molecule inhibitors that will need to be examined in different model organisms (reviewed in (Macheda and Stacker 2008)). A great improvement in this therapeutic approach is the consideration of the proteins in non-canonical Wnt signaling, as they are likely to be less critical in adult tissues, mainly proteins comprised in the PCP pathway (reviewed in (Macheda and Stacker 2008)).

Very recently, Harb and colleagues have reviewed the potential of canonical and non-canonical Wnt targeting for cancer treatment, stating that although initial clinical trials are showing promising results, no Wnt-specific drugs have been approved so far for clinical use (Harb et al. 2019).

10.8 Final Remarks

More and more evidence supports a critical role of Wnt signaling in cancer. Both canonical and non-canonical Wnt cascades are implicated in several aspects pivotal for cancer progression, namely the maintenance of cancer cells *stemness*, the acquisition of metabolic adaptations supporting cancer cells survival and proliferation and resistance to therapies.

Several questions regarding the role of Wnt signaling in cancer and its therapeutic targeting are still unclear. How fast all those different questions and issues that are proposed in this chapter can be answered will depend on the development or adaptation of new techniques. As many of these techniques are already available (although not necessarily in cancer research), one can expect significant progress in understanding the molecular and cellular mechanisms that regulate cancer development under Wnt canonical and non-canonical pathway in the near future. This would ultimately allow the development of new promising treatment strategies in the fight against cancer.

Acknowledgments The authors acknowledge iNOVA-4Health – UID/Multi/04462/2013, a program financially supported by Fundação para a Ciência e Tecnologia/Ministério da Educação e Ciência, through national funds and co-funded by FEDER under the PT2020 Partnership Agreement.

References

Adler PN (2002) Planar signaling and morphogenesis in drosophila. Dev Cell 2:525–535. https://doi.org/10.1016/S1534-5807(02)00176-4

Adler PN, Lee H (2001) Frizzled signaling and cell–cell interactions in planar polarity. Curr Opin Cell Biol 13:635–640. https://doi.org/10.1016/S0955-0674(00)00263-5

Alam MM, Lal S, FitzGerald KE, Zhang L (2016) A holistic view of cancer bioenergetics: mitochondrial function and respiration play fundamental roles in the development and progression of diverse tumors. Clin Transl Med 5:1–14. https://doi.org/10.1186/s40169-016-0082-9

Albrecht LV, Bui MH, De Robertis EM (2019) Canonical Wnt is inhibited by targeting one-carbon metabolism through methotrexate or methionine deprivation. Proc Natl Acad Sci 116:2987–2995. https://doi.org/10.1073/pnas.1820161116

Anastas JN, Moon RT (2013) WNT signalling pathways as therapeutic targets in cancer. Nat Rev Cancer 13:11–26. https://doi.org/10.1038/nrc3419

Bao R, Christova T, Song S et al (2012) Inhibition of tankyrases induces axin stabilization and blocks Wnt signalling in breast cancer cells. PLoS One 7:1–9. https://doi.org/10.1371/journal.pone.0048670

Barker HE, Paget JTE, Khan AA, Harrington KJ (2015) The tumour microenvironment after radiotherapy: mechanisms of resistance and recurrence. Nat Rev Cancer 15:409–425. https://doi.org/10.1038/nrc3958

Batlle E, Clevers H (2017) Cancer stem cells revisited. Nat Med 23:1124–1134. https://doi.org/10.1038/nm.4409

Biechele TL, Kulikauskas RM, Toroni RA et al (2012) Wnt/β-catenin signaling and AXIN1 regulate apoptosis mediated by inhibition of BRAFV600E kinase in human melanoma. Sci Signal 5:1–26. https://doi.org/10.1126/scisignal.2002274.Wnt/

Bogaerts E, Heindryckx F, Vandewynckel YP et al (2014) The roles of transforming growth factor-β, Wnt, Notch and hypoxia on liver progenitor cells in primary liver tumours (review). Int J Oncol 44:1015–1022. https://doi.org/10.3892/ijo.2014.2286

Cadigan KM, Nusse R (1997) Wnt signaling: a common theme in animal development. Genes Dev 11:3286–3305. https://doi.org/10.1101/gad.11.24.3286

Carreira-Barbosa F, Kajita M, Morel V et al (2009) Flamingo regulates epiboly and convergence/extension movements through cell cohesive and signalling functions during zebrafish gastrulation. Development 136:383–392. https://doi.org/10.1242/dev.026542

Castañeda-Patlán MC, Fuentes-García G, Robles-Flores M (2018) Wnt signaling as a master regulator of immune tolerance in a tumor microenvironment. In: Ray S (ed) Cell signalling-thermodynamics and molecular control. IntechOpen

Chen H, Xu L, Li L et al (2018) Inhibiting the CD8+ T cell infiltration in the tumor microenvironment after radiotherapy is an important mechanism of radioresistance. Sci Rep 8:1–10. https://doi.org/10.1038/s41598-018-30417-6

Chen G, Gao C, Gao X et al (2019) Wnt/β-catenin pathway activation mediates adaptive resistance to BRAF inhibition in colorectal cancer. Mol Cancer Ther 17:806 813. https://doi.org/10.1158/1535-7163.MCT-17-0561.Wnt/

Choi S-C, Han J-K (2002) Xenopus Cdc42 regulates convergent extension movements during gastrulation through Wnt/Ca2+ signaling pathway. Dev Biol 244:342–357. https://doi.org/10.1006/dbio.2002.0602

Corda G, Sala A (2017) Non-canonical WNT/PCP signalling in cancer: Fzd6 takes centre stage. Oncogene 6:1–6. https://doi.org/10.1038/oncsis.2017.69

Cori CA, Cori GT (1925) The carbohydrate metabolism of tumours. J Biol Chem 65:397–405

Cramer T, Schmitt C (eds) (2016) Metabolism in cancer. Springer

Da Forno PD, Pringle JH, Hutchinson P et al (2008) WNT5A expression increases during melanoma progression and correlates with outcome. Clin Cancer Res 14:5825–5832. https://doi.org/10.1158/1078-0432.CCR-07-5104

Dejmek J, Dejmek A, Säfholm A et al (2005) Wnt-5a protein expression in primary dukes B colon cancers identifies a subgroup of patients with good prognosis. Cancer Res 65:9142–9146. https://doi.org/10.1158/0008-5472.CAN-05-1710

Fan Y, Shen B, Tan M et al (2014) Long non-coding RNA UCA1 increases chemoresistance of bladder cancer cells by regulating Wnt signaling. FEBS J 281:1750–1758. https://doi.org/10.1111/febs.12737

Fior R, Zilhão R (eds) (2019) Molecular and cell biology of cancer. Springer

Freimund AE, Beach JA, Christie EL, Bowtell DDL (2018) Mechanisms of drug resistance in high-grade serous ovarian cancer. Hematol Oncol Clin North Am 32:983–996. https://doi.org/10.1016/j.hoc.2018.07.007

Frenquelli M, Caridi N, Antonini E et al (2019) The WNT receptor ROR2 drives the interaction of multiple myeloma cells with the microenvironment through AKT activation. Leukemia:1–14. https://doi.org/10.1038/s41375-019-0486-9

Gilbert SF, Barresi MJF (2019) Developmental biology, 12th edn. Sinauer Associates, Inc, Sunderland

Gillies RJ, Verduzco D, Gatenby RA (2012) Evolutionary dynamics of carcinogenesis and why targeted therapy does not work. Nat Rev Cancer 12:487–493. https://doi.org/10.1038/nrc3298

Gu H, Huang T, Shen Y et al (2018) Reactive oxygen species-mediated tumor microenvironment transformation: the mechanism of radioresistant gastric cancer. Oxidative Med Cell Longev 2018:1–8. https://doi.org/10.1155/2018/5801209

Guppy M, Leedman P, Zu X, Russell V (2002) Contribution by different fuels and metabolic pathways to the total ATP turnover of proliferating MCF-7 breast cancer cells. Biochem J 364:309–315. https://doi.org/10.1042/bj3640309

Gupta S, Roy A, Dwarakanath BS (2017) Metabolic cooperation and competition in the tumor microenvironment: implications for therapy. Front Oncol 7:1–24. https://doi.org/10.3389/fonc.2017.00068

Habas R, Kato Y, He X (2001) Wnt/Frizzled activation of Rho regulates vertebrate gastrulation and requires a novel formin homology protein Daam1. Cell 107:843–854. https://doi.org/10.1016/S0092-8674(01)00614-6

Han P, Ji X, Zhang M, Gao L (2018) Upregulation of lncRNA LINC00473 promotes radioresistance of HNSCC cells through activating Wnt/beta-catenin signaling pathway. Eur Rev Med Pharmacol Sci 22:7305–7313. https://doi.org/10.26355/eurrev_201811_16267

Hanahan D, Weinberg RA (2011) Hallmarks of cancer: the next generation. Cell 144:646–674

Harb J, Lin PJ, Hao J (2019) Recent development of Wnt signaling pathway inhibitors for cancer therapeutics. Curr Oncol Rep 21:1–9. https://doi.org/10.1007/s11912-019-0763-9

Hayes JD, Dinkova-Kostova AT (2014) The Nrf2 regulatory network provides an interface between redox and intermediary metabolism. Trends Biochem Sci 39:199–218. https://doi.org/10.1016/j.tibs.2014.02.002

Huelsken J, Birchmeier W (2001) New aspects of Wnt signaling pathways in higher vertebrates. Curr Opin Genet Dev 11:547–553. https://doi.org/10.1016/S0959-437X(00)00231-8

Ishitani T, Kishida S, Hyodo-Miura J et al (2003) The TAK1-NLK mitogen-activated protein kinase cascade functions in the Wnt-5a/Ca2+ pathway to antagonize Wnt/β-catenin signaling. Mol Cell Biol 23:131–139. https://doi.org/10.1128/MCB.23.1.131-139.2003

Jeong W-J, Ro EJ, Choi K-Y (2018) Interaction between Wnt/β-catenin and RAS-ERK pathways and an anti-cancer strategy via degradations of β-catenin and RAS by targeting the Wnt/β-catenin pathway. NPJ Precis Oncol 2:1–10. https://doi.org/10.1038/s41698-018-0049-y

Katoh M (2017) Canonical and non-canonical WNT signaling in cancer stem cells and their niches: cellular heterogeneity, omics reprogramming, targeted therapy and tumor plasticity (review). Int J Oncol 51:1357–1369. https://doi.org/10.3892/ijo.2017.4129

Katoh M, Katoh M (2017) Molecular genetics and targeted therapy of WNT-related human diseases (review).

Int J Mol Med 40:587–606. https://doi.org/10.3892/ijmm.2017.3071

Kinoshita N, Iioka H, Miyakoshi A, Ueno N (2003) PKCδ is essential for dishevelled function in a noncanonical Wnt pathway that regulates Xenopus convergent extension movements. Genes Dev 17:1663–1676. https://doi.org/10.1101/gad.1101303

Kozovska Z, Gabrisova V, Kucerova L (2014) Colon cancer: cancer stem cells markers, drug resistance and treatment. Biomed Pharmacother 68:911–916. https://doi.org/10.1016/j.biopha.2014.10.019

Krishnamurthy N, Kurzrock R (2018) Targeting the Wnt/beta-catenin pathway in cancer: update on effectors and inhibitors. Cancer Treat Rev 62:50–60. https://doi.org/10.1016/j.ctrv.2017.11.002

Kurayoshi M, Oue N, Yamamoto H et al (2006) Expression of Wnt-5a is correlated with aggressiveness of gastric cancer by stimulating cell migration and invasion. Cancer Res 66:10439–10448. https://doi.org/10.1158/0008-5472.CAN-06-2359

Laurent-Puig P, Zucman-Rossi J (2006) Genetics of hepatocellular tumors. Oncogene 25:3778–3786. https://doi.org/10.1038/sj.onc.1209547

Lejeune S, Huguet EL, Hamby A et al (1995) Wnt5a cloning, expression, and up-regulation in human primary breast cancers. Clin Cancer Res 1:215–222

Luke JJ, Bao R, Sweis RF et al (2019) WNT/β-catenin pathway activation correlates with immune exclusion across human cancers. Clin Cancer Res 25:3074–3083. https://doi.org/10.1158/1078-0432.CCR-18-1942

Luo M, Wu C, Guo E et al (2019a) FOXO3a knockdown promotes radioresistance in nasopharyngeal carcinoma by inducing epithelial-mesenchymal transition and the Wnt/β-catenin signaling pathway. Cancer Lett 455:26–35. https://doi.org/10.1016/j.canlet.2019.04.019

Luo Y, Barrios-Rodiles M, Gupta GD et al (2019b) Atypical function of a centrosomal module in WNT signalling drives contextual cancer cell motility. Nat Commun 10:1–20. https://doi.org/10.1038/s41467-019-10241-w

Lyou Y, Chen G, Waterman M (2018) Abstract 385: HIF1A can regulate Wnt signaling in human colon cancer cells. In: AACR annual meeting 2018 April 14–18. Chicago, pp 385–385

Maccalli C, Todaro M, Ferrone S (eds) (2019) Cancer stem cell resistance to targeted therapy. Springer

Macheda ML, Stacker SA (2008) Importance of Wnt signaling in the tumor stroma microenvironment. Curr Cancer Drug Targets 8:454–465

Mehta K, Siddik ZH (eds) (2009) Drug resistance in cancer cells. Springer

Meyer IS, Jungmann A, Dieterich C et al (2017) The cardiac microenvironment uses non-canonical WNT signaling to activate monocytes after myocardial infarction. EMBO Mol Med 9:1279–1293. https://doi.org/10.15252/emmm.201707565

Miyagi C, Yamashita S, Ohba Y et al (2004) STAT3 noncell-autonomously controls planar cell polarity during zebrafish convergence and extension. J Cell Biol 166:975–981. https://doi.org/10.1083/jcb.200403110

Mo Y, Wang Y, Zhang L et al (2019) The role of Wnt signaling pathway in tumor metabolic reprogramming. J Cancer 10:3789–3797. https://doi.org/10.7150/jca.31166

Moharil R, Dive A, Khandekar S, Bodhade A (2017) Cancer stem cells: an insight. J Oral Maxillofac Pathol 21:463–463. https://doi.org/10.4103/jomfp.JOMFP_132_16

Morgan R, Ankrah R, El-Tanani S et al (2017) Wnt Signaling as a therapeutic target in cancer and metastasis. In: Introduction to cancer metastasis. pp 375–394

Murphy K, Weaver C (2016) Janeway'S immunobiology, 9th edn. Garland Science/Taylor & Francis Group, LLC, New York

Niiro E, Morioka S, Iwai K et al (2018) Potential signaling pathways as therapeutic targets for overcoming chemoresistance in mucinous ovarian cancer (review). Biomed Rep:215–223. https://doi.org/10.3892/br.2018.1045

Ojalvo LS, Whittaker CA, Condeelis JS, Pollard JW (2010) Gene expression analysis of macrophages that facilitate tumor invasion supports a role for Wnt-signaling in mediating their activity in primary mammary tumors. J Immunol 184:702–712. https://doi.org/10.1038/mp.2011.182.doi

Olsen JJ, Pohl SO-G, Deshmukh A et al (2017) The role of Wnt signalling in angiogenesis. Clin Biochem Rev 38:131–142. https://doi.org/10.1161/CIRCRESAHA.116.310498

Pai SG, Carneiro BA, Mota JM et al (2017) Wnt/beta-catenin pathway: modulating anticancer immune response. J Hematol Oncol 10:1–12. https://doi.org/10.1186/s13045-017-0471-6

Pai VC, Hsu C-C, Chan T-S et al (2019) ASPM promotes prostate cancer stemness and progression by augmenting Wnt−Dvl-3−β-catenin signaling. Oncogene 38:1340–1353. https://doi.org/10.1038/s41388-018-0497-4

Palacios J, Gamallo C (1998) Mutations in the β-catenin gene (CTNNB1) in endometrioid ovarian carcinomas. Cancer Res 58:1344–1347

Pate KT, Stringari C, Sprowl-Tanio S et al (2014) Wnt signaling directs a metabolic program of glycolysis and angiogenesis in colon cancer. EMBO J 33:1454–1473. https://doi.org/10.15252/embj.201488598

Peng Y, Zhang X, Feng X et al (2017) The crosstalk between microRNAs and the Wnt/β-catenin signaling pathway in cancer. Oncotarget 8:14089–14106. https://doi.org/10.18632/oncotarget.12923

Penzo-Mendèz A, Umbhauer M, Djiane A et al (2003) Activation of Gβγ signaling downstream of Wnt-11/Xfz7 regulates Cdc42 activity during Xenopus gastrulation. Dev Biol 257:302–314. https://doi.org/10.1016/S0012-1606(03)00067-8

Pinyol R, Sia D, Llovet JM (2019) Immune exclusion-WNT/CTNNB1 class predicts resistance to immunotherapies in HCC. Clin Cancer Res 25:2021–2023. https://doi.org/10.1158/1078-0432.CCR-18-3778

Potter M, Newport E, Morten KJ (2016) The Warburg effect: 80 years on. Biochem Soc Trans 44:1499–1505. https://doi.org/10.1042/bst20160094

Reis M, Czupalla CJ, Ziegler N et al (2012) Endothelial Wnt/β-catenin signaling inhibits glioma angiogenesis and normalizes tumor blood vessels by inducing PDGF-B expression. J Exp Med 209:1611–1627. https://doi.org/10.1084/jem.20111580

Rodríguez-Enríquez S, Torres-Márquez ME, Moreno-Sánchez R (2000) Substrate oxidation and ATP supply in AS-30D hepatoma cells. Arch Biochem Biophys 375:21–30. https://doi.org/10.1006/abbi.1999.1582

Rodríguez-Enríquez S, Vital-González PA, Flores-Rodríguez FL et al (2006) Control of cellular proliferation by modulation of oxidative phosphorylation in human and rodent fast-growing tumor cells. Toxicol Appl Pharmacol 215:208–217. https://doi.org/10.1016/j.taap.2006.02.005

Rycaj K, Tang DG (2015) Cell of origin of cancer versus cancer stem cells: assays and interpretations. Cancer Res 75:4003–4011. https://doi.org/10.1158/0008-5472.CAN-15-0798

Saretzki G (2014) Extra-telomeric functions of human telomerase: cancer, mitochondria and oxidative stress. Curr Pharm Des 20:6386–6403. https://doi.org/10.2174/1381612820666140630095606

Satoh S, Daigo Y, Furukawa Y et al (2000) AXIN1 mutations in hepatocellular carcinomas, and growth suppression in cancer cells by virus-mediated transfer of AXIN1. Nat Genet 24:245–250. https://doi.org/10.1038/73448

Semenza GL (2012) Hypoxia-inducible factors: mediators of cancer progression and targets for cancer therapy. Trends Pharmacol Sci 33:207–214. https://doi.org/10.1016/j.tips.2012.01.005

Senni N, Savall M, Cabrerizo Granados D et al (2019) Β-catenin-activated hepatocellular carcinomas are addicted to fatty acids. Gut 68:322–334. https://doi.org/10.1136/gutjnl-2017-315448

Sheldahl LC, Slusarski DC, Pandur P et al (2003) Dishevelled activates Ca2+ flux, PKC, and CamKII in vertebrate embryos. J Cell Biol 161:769–777. https://doi.org/10.1083/jcb.200211094

Sherwood V (2015) WNT signaling: an emerging mediator of cancer cell metabolism? Mol Cell Biol 35:2–10. https://doi.org/10.1128/MCB.00992-14

Slusarski DC, Corces VG, Moon RT (1997) Interaction of Wnt and a Frizzled homologue triggers G-protein-linked phosphatidylinositol signalling. Nature 390:410–413. https://doi.org/10.1038/37138

Son B, Lee S, Youn H et al (2017) The role of tumor microenvironment in therapeutic resistance. Oncotarget 8:3933–3945. https://doi.org/10.18632/oncotarget.13907

Staal FJT, Famili F, Perez LG, Pike-Overzet K (2016) Aberrant Wnt signaling in leukemia. Cancers (Basel) 8:1–15. https://doi.org/10.3390/cancers8090078

Steinhart Z, Angers S (2018) Wnt signaling in development and tissue homeostasis. Development 145:1–8. https://doi.org/10.1242/dev.146589

Stewart DJ, Chang DW, Ye Y et al (2014) Wnt signaling pathway pharmacogenetics in non-small cell lung cancer. Pharmacogenomics J 14:509–522. https://doi.org/10.1038/tpj.2014.21

Su H-Y, Lai H-C, Lin Y-W et al (2010) Epigenetic silencing of SFRP5 is related to malignant phenotype and chemoresistance of ovarian cancer through Wnt signaling pathway. Int J Cancer 127:555–567. https://doi.org/10.1002/ijc.25083

Tada M, Concha ML (2001) Vertebrate gastrulation: calcium waves orchestrate cell movements. Curr Biol 11:R470–R472. https://doi.org/10.1016/S0960-9822(01)00284-6

Tan Z, Song L, Wu W et al (2018) TRIM14 promotes chemoresistance in gliomas by activating Wnt/β-catenin signaling via stabilizing Dvl2. Oncogene 37:5403–5415. https://doi.org/10.1038/s41388-018-0344-7

Tejeda-Muñoz N, Albrecht LV, Bui MH, De Robertis EM (2019) Wnt canonical pathway activates macropinocytosis and lysosomal degradation of extracellular proteins. Proc Natl Acad Sci 116:10402–10411. https://doi.org/10.1073/pnas.1903506116

VanderVorst K, Dreyer CA, Konopelski SE et al (2019) Wnt/PCP signaling contribution to carcinoma collective cell migration and metastasis. Cancer Res 79:1719–1729. https://doi.org/10.1158/0008-5472.CAN-18-2757

Vangipuram SD, Buck SA, Lyman WD (2012) Wnt pathway activity confers chemoresistance to cancer stem-like cells in a neuroblastoma cell line. Tumor Biol 33:2173–2183. https://doi.org/10.1007/s13277-012-0478-0

Vaupel P, Mayer A (2007) Hypoxia in cancer: significance and impact on clinical outcome. Cancer Metastasis Rev 26:225–239. https://doi.org/10.1007/s10555-007-9055-1

Veeman MT, Slusarski DC, Kaykas A et al (2003) Zebrafish prickle, a modulator of noncanonical Wnt/Fz signaling, regulates gastrulation movements. Chem Biol 10:53–60. https://doi.org/10.1016/s0960-9822(03)00240-9

Vergara D, Stanca E, Guerra F et al (2017) Β-catenin knockdown affects mitochondrial biogenesis and lipid metabolism in breast cancer cells. Front Physiol 8:1–17. https://doi.org/10.3389/fphys.2017.00544

Viale A, Corti D, Draetta GF (2015) Tumors and mitochondrial respiration: a neglected connection. Cancer Res 75:3685–3686. https://doi.org/10.1158/0008-5472.CAN-15-0491

Vinson CR, Conover S, Adler PN (1989) A Drosophila tissue polarity locus encodes a protein containing seven potential transmembrane domains. Nature 338:263–264. https://doi.org/10.1038/338263a0

Vona-Davisa L, Gibsona LF (2013) Adipocytes as a critical component of the tumor microenvironment. Leuk Res 37:483–484. https://doi.org/10.1016/j.leukres.2013.01.007

Wallingford JB, Ewald AJ, Harland RM, Fraser SE (2001) Calcium signaling during convergent extension in Xenopus. Curr Biol 11:652–661. https://doi.org/10.1016/S0960-9822(01)00201-9

Wang L, Liu L, Chen Y et al (2018) Correlation between adenosine triphosphate (ATP)-binding cassette transporter G2 (ABCG2) and drug resistance of esophageal cancer and reversal of drug resistance by artesunate. Pathol – Res Pract 214:1467–1473. https://doi.org/10.1016/j.prp.2018.08.001

Warburg O (1956) On the origin of cancer cells. Science 123:309–314. https://doi.org/10.1126/science.123.3191.309

Warburg O, Wind F, Negelein E (1927) The metabolism of tumors in the body. J Gen Physiol 8:519–530. https://doi.org/10.1085/jgp.8.6.519

Ward PS, Thompson CB (2012) Metabolic reprogramming: a cancer hallmark even Warburg did not anticipate. Cancer Cell 21:297–308. https://doi.org/10.1016/j.ccr.2012.02.014

Wickström M, Dyberg C, Milosevic J et al (2015) Wnt/β--catenin pathway regulates MGMT gene expression in cancer and inhibition of Wnt signalling prevents chemoresistance. Nat Commun 6:1–10. https://doi.org/10.1038/ncomms990

Wils LJ, Bijlsma MF (2018) Epigenetic regulation of the hedgehog and Wnt pathways in cancer. Crit Rev Oncol Hematol 121:23–44. https://doi.org/10.1016/j.critrevonc.2017.11.013

Winklbauer R, Medina A, Swain RK, Steinbeisser H (2001) Frizzled-7 signalling controls tissue separation during Xenopus gastrulation. Nature 413:856–860. https://doi.org/10.1038/35101621

Xiang J, Hu Q, Qin Y et al (2018) TCF7L2 positively regulates aerobic glycolysis via the EGLN2/HIF-1α axis and indicates prognosis in pancreatic cancer. Cell Death Dis 9:1–14. https://doi.org/10.1038/s41419-018-0367-6

Xu W, Zhou W, Cheng M et al (2017) Hypoxia activates Wnt/β-catenin signaling by regulating the expression of BCL9 in human hepatocellular carcinoma. Sci Rep 7:1–13

Yamashita S, Miyagi C, Carmany-Rampey A et al (2002) Stat3 controls cell movements during zebrafish gastrulation. Dev Cell 2:363–375. https://doi.org/10.1016/S1534-5807(02)00126-0

Zhan T, Rindtorff N, Boutros M (2017) Wnt signaling in cancer. Oncogene 36:1461–1473. https://doi.org/10.1038/onc.2016.304

Zhan T, Ambrosi G, Wandmacher AM et al (2019) MEK inhibitors activate Wnt signalling and induce stem cell plasticity in colorectal cancer. Nat Commun 10:2197. https://doi.org/10.1038/s41467-019-09898-0

Zhang B, Wu J, Cai Y et al (2019) TCF7L1 indicates prognosis and promotes proliferation through activation of Keap1/NRF2 in gastric cancer. Acta Biochim Biophys Sin Shanghai 51:375–385. https://doi.org/10.1093/abbs/gmz015

Zhao Y, Yi J, Tao L et al (2018) Wnt signaling induces radioresistance through upregulating HMGB1 in esophageal squamous cell carcinoma. Cell Death Dis 9:1–15. https://doi.org/10.1038/s41419-018-0466-4

Zoico E, Darra E, Rizzatti V et al (2018) Role of adipose tissue in melanoma cancer microenvironment and progression. Int J Obes 42:344–352. https://doi.org/10.1038/ijo.2017.21

Zurawel RH, Chiappa SA, Allen C, Raffel C (1998) Sporadic medulloblastomas contain oncogenic β-catenin mutations. Cancer Res 58:896–899

Melanoma Metabolism: Cell Survival and Resistance to Therapy

11

Rafael Luís, Cheila Brito, and Marta Pojo

Abstract

Cutaneous melanoma is one of the most aggressive types of cancer, presenting the highest potential to form metastases, both locally and distally, which are associated with high death rates of melanoma patients. A high somatic mutation burden is characteristic of these tumours, with most common oncogenic mutations occurring in the *BRAF*, *NRAS* and *NF1* genes. These intrinsic oncogenic pathways contribute to the metabolic switch between glycolysis and oxidative phosphorylation metabolisms of melanoma, facilitating tumour progression and resulting in a high plasticity and adaptability to unfavourable conditions. Moreover, melanoma microenvironment can influence its own metabolism and reprogram several immune cell subset functions, enabling melanoma to evade the immune system. The knowledge of the biology, molecular alterations and microenvironment of melanoma has led to the development of new targeted therapies and the improvement of patient care. In this work, we reviewed the impact of melanoma metabolism in the resistance to BRAF and MEK inhibitors and immunotherapies, emphasizing the requirement to evaluate metabolic alterations upon development of novel therapeutic approaches. Here we summarized the current understanding of the impact of metabolic processes in melanomagenesis, metastasis and microenvironment, as well as the involvement of metabolic pathways in the immune modulation and resistance to targeted and immunocheckpoint therapies.

Keywords

Melanoma · Metabolic profile · Microenvironment · Immunotherapy · Targeted therapy · Oncogenic mutations

11.1 Introduction

The skin is the largest organ of the human body and represents the interface between the organism and the environment. It is comprised by two different layers, the dermis and the epidermis. The dermis, constitutes the lower layer of skin and contains connective tissue, blood vessels, nerves and appendages such as sweat glands (Tobin 2017). On the other hand, the epidermis constitutes the upper layer of skin and includes: skin-resident dendritic antigen presenting cells termed Langerhans cells (Bandarchi et al. 2010);

Authors Rafael Luís and Cheila Brito have equally contributed to this chapter.

R. Luís · C. Brito · M. Pojo (✉)
Unidade de Investigação em Patobiologia Molecular (UIPM), Instituto Português de Oncologia de Lisboa Francisco Gentil E.P.E, Lisbon, Portugal

J. Serpa (ed.), *Tumor Microenvironment*, Advances in Experimental Medicine and Biology 1219,
https://doi.org/10.1007/978-3-030-34025-4_11

keratinocytes, responsible for the production of important structural and catalytic proteins, essential for the structure and homeostasis of the skin (Naves et al. 2017); and melanocytes, which are originated from the neural crest stem cells during the foetal development, remaining contained in the skin, eyes, mucosal epithelia, meninges of the brain and spinal cord (Smith et al. 2016). Melanocytes are specialized cells with an important role related to skin pigmentation, displaying pivotal photo-protection and thermoregulation functions through the production of melanin (Cichorek et al. 2013).

Melanoma derives from the malignant transformation of melanocytes localized predominantly in the skin (also referred as cutaneous melanoma) representing approximately 90% of the all cases (Leonardi et al. 2018). Melanoma could also be found, in a lower frequency, in other organs that contain melanocytes, for example: mucosal melanoma arises from melanocytes in the mucous membranes; and uveal melanoma from melanocytes residing in ocular stroma (Kuk et al. 2016). Besides melanoma, there are benign neoplasms also derived from melanocytes, designated as melanocytic nevi (Bastian 2014). Among these lesions were identified atypical nevi characterized by an unusual morphology and a not well-defined border – dysplastic nevi. There is still no agreement on the appearance of dysplastic nevi as precursor lesions for melanoma development. Although, it is commonly accepted that these skin malignancies constitute risk factors for melanoma (Goldstein and Tucker 2013).

Cutaneous melanoma represents less than 5% of all skin cancer, however is the most common lethal type of skin malignancies (Ali et al. 2013; Siegel et al. 2019), accounting for around 60–75% of deaths related with skin neoplasms (Bandarchi et al. 2010; Potrony et al. 2015; Siegel et al. 2019).

In the last decades, cutaneous melanoma incidence has been rising rapidly worldwide, while its associated mortality rate stabilized, mainly due to early detection and improved treatments administered to patients (Ali et al. 2013; Guy et al. 2015). Despite significant efforts to raise awareness and to inform the public, melanoma continues to increase among adolescents mainly affecting young women worldwide (Weir et al. 2011). After the age of 40 the incidence pattern changes and melanoma is mostly diagnosed in men (Garbe and Leiter 2009). The incidence of melanoma is also influenced by the ethnic groups and geographical locations, mostly due to the variation in sun exposure and skin phenotype (Whiteman et al. 2016; Weller and Castellsague 2017). Australia and New Zealand have the highest rates of melanoma incidence (40.4 and 27.5; 35.8 and 31.1 cases per 100,000 men and women respectively) according to the recent analysis of global cancer statistics (Bray et al. 2018). This incidence is really elevated when compared with the incidence of melanoma in men and women in United Sates (14.9 and 11.0 cases per 100,000 men and women, respectively) (Bray et al. 2018).

Although early-stage melanoma could be treated with surgery leading to a 5 years-survival rate up to 90% (Allemani et al. 2018), advanced tumours with deeper lesions have a worse prognosis, due these tumour cells' predisposition to invade and to resist to anti-cancer therapies (Lo and Fisher 2014). In this perspective, over the past years several adjuvant therapies were developed mainly targeted therapies and immunotherapies, to improve melanoma patient survival. Presently, the United States Food and Drug Administration (FDA) approved Vemurafenib, Dabrafenib, Trametinib and Cobimetinib (*BRAF* and MEK kinase inhibitors), as the main targeted therapies, for melanoma patients diagnosed with a $BRAF^{V600E}$ mutation and Ipilimumab (α-CTLA-4); Pembrolizumab and Nivolumab (α-PD-1) as the most common immunotherapies administered in patients with melanoma (reviewed in Domingues et al. 2018). All these therapies successfully impacted the overall survival (OS) and progression-free survival (PFS) of metastatic melanoma patients, when comparing chemo and radiotherapies with an OS of 6–8 months and a PFS of 1.9 months (Gogas et al. 2007), with BRAF and MEK inhibitors showing a median OS between 6 and 15.9 months and median PFS between 4.8 and 6.8 months (Hauschild et al. 2012; Sosman et al. 2012; Batus et al. 2013; Domingues et al. 2018). However, only half of the patients reveal a positive response to these targeted therapies (Eroglu and Ribas 2016). In this context, immunocheckpoint therapies were developed and further improved OS and PFS, showing a median OS of 10.1 months and median PFS between 5.9 and 11.5 months (reviewed in Domingues et al. 2018). Despite progresses in melanoma treatment, heterogeneity leads to a high variability of responses which contribute for the elevated capacity of tumour cells to change and adapt to the microenvironment conditions and to acquire distinct energy sources to induce tumour proliferation, progression and metastasis (Fischer et al. 2019). It is increasingly considered that metabolism plays an active role in development of resistance to the available therapies, influencing the efficiency of treatments.

11.2 Melanoma Mutations Reprogram Metabolic Profile and Sustain Cell Survival

In the last decade melanoma was described with a much higher somatic mutation burden relative to other types of cancers (Lawrence et al. 2013), which is related with a signature of ultraviolet radiation (UVR) (Hodis et al. 2012). Most of these alterations are described as oncogenic or drive mutations which are implicated in the development and progression of melanoma due to the constitutive activation of oncogenic pathways (Fig. 11.1). Around 50–65% of all melanoma cases demonstrate mutations in the RAS/RAF/MEK/ERK mitogen-activated protein kinase (MAPK) signal transduction pathway, which is essential for melanoma cells survival (Mcarthur and Ribas 2012; Parmenter et al. 2014; Shah and Dronca 2015).

The most frequent oncogenic mutation in melanoma is detected in the v-Raf murine sarcoma viral oncogene homolog (*BRAF*), identified in 40–50% of all melanoma cases (Davies et al. 2002; The Cancer Genome Atlas Network 2015). The second most prevalent mutation is found in the neuroblastoma RAS viral oncogene homolog (*NRAS*) which is mutated in approximately 30% (The Cancer Genome Atlas Network 2015). More recently the study of Cancer Genome Network disclosed a new frequent mutated gene the neurofibromin 1 (*NF1*) which is present in 14% of all melanomas (The Cancer Genome Atlas Network 2015). Interestingly these genes are important players in the MAPK pathway, which represents one of the master pathways of oncogenic signalling (Zhang and Liu 2006).

A high frequency of mutations in *BRAF* has also been found in benign and dysplastic nevi, further implicating them as a neoplastic event during the transformation of melanocytes (Pollock et al. 2003; Vredeveld et al. 2012). Most *BRAF* mutations result from substitutions at residue 600, namely valine-to-glutamic acid (V600E) (72%), valine-to-lysine (V600K) (17%) and valine-to-arginine (V600R) (5%), while *NRAS* mutations occur primarily due to substitutions at residue 61, specifically glutamine-to-lysine (Q61K) (41.5%) and glutamine-to-arginine (Q61R) (8.8%) (Heppt et al. 2017). Moreover, *BRAF* and *NRAS* mutations were described as mutually exclusive in melanoma cells (Sensi et al. 2006). Interestingly, activation of the key oncogenic pathway RAS/RAF/MEK/ERK seems to further promote a glycolytic metabolism rather than oxidative phosphorylation, more specifically through *BRAF* mutations that have been shown to regulate metabolic reprogramming in melanoma cells (Haq et al. 2013).

The constitutive activation of the MAPK signalling pathway induces glycolytic metabolism in melanoma cells, since it increases the transcription of the hypoxia inducible factor 1α (*HIF1α*) (Parmenter et al. 2014). In response to both hypoxic stress and oncogenic signals, the $BRAF^{V600E}$ mutant increases the expression of *HIF1α* (Kumar et al. 2007). A recent study demonstrates that the inhibition of HIF1α through $ZnSO_4$ treatment induced an anti-proliferative and anti-metastatic effect *in vivo* in melanoma cells (Burián et al. 2019). According to these results the anti-tumoral effect regarding $ZnSO_4$ is dependent on the downregulation of HIF1α. As such, HIF1α allows the adaptation of the cell to the hypoxic conditions through the alteration of the central carbon metabolism. This transcription

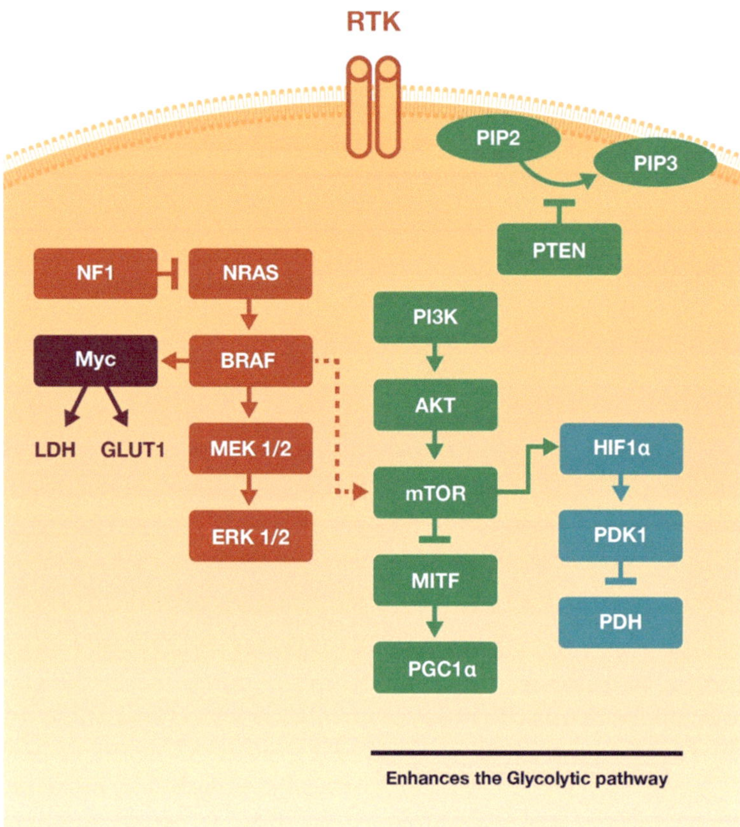

Fig. 11.1 The molecular drivers and signaling pathways determinant for the regulation of melanoma cells metabolism. The MAPK pathway is marked in orange and the PI3K/AKT/mTOR pathway is marked in green. MAPK pathway can regulate the transcription of key players for glucose uptake and metabolism such as GLUT1 and LDH, through the activation of Myc (purple). Both pathways regulate several factors, such as HIF1α, Myc, MITF, mTOR, PGC1α, that regulate the metabolic balance between the glycolytic pathway and oxidative phosphorylation. The crosstalk between these signaling pathways introduces a metabolic plasticity that contributes for the adaptation of melanoma cells to the microenvironment under stress conditions

factor is crucial for the expression of several key glycolytic genes important for hypoxia adaptation and glycolysis stimulation. Additionally, HIF1α suppresses mitochondrial oxidative phosphorylation by trans-activating the gene encoding pyruvate dehydrogenase kinase 1 (*PDK1*) (Kim et al. 2006). PDK1 is the enzyme that phosphorylates pyruvate dehydrogenase (PDH), resulting in its inactivation (Holness and Sugden 2003). PDH establishes the association between glycolysis and the tricarboxylic acid (TCA) cycle, because it is in the mitochondrial matrix converting pyruvate into acetyl coenzyme A (acetyl-CoA), pivotal for the TCA cycle. The re-activation of PDH is obtained by de-phosphorylation of the PHD-E1α subunit, exerted by the pyruvate dehydrogenase phosphatases (PDP1 and PDP2) (Cesi et al. 2017). The upregulation of *HIF1α* induces the activation of PDK1 and subsequently the suppression of PDH activity, decreasing mitochondrial respiration and oxidative phosphorylation (Kim et al. 2006; Papandreou et al. 2006). Recently, it was reported that Ku80, a protein involved in non-homolog end joining repair pathway, is positively correlated with the PDK1 protein in melanoma and both proteins were upregulated in tissues of melanoma patients (Nemoz et al. 2018; Liu et al. 2019). Ku80 is able of binding to the PDK1 promoter, activating its transcription, showing a pro-tumoral effect in melanoma both *in vitro* and *in vivo* (Liu et al. 2019). Additionally, STAT3, a well-known transcription factor involved in oncogenesis, can bind to the PDK1 promoter and positively regulate its expression, constitutively activating the large family of AGC protein kinases, important for many tumoral cell survival (Picco et al. 2019). The combined treatment between BRAF/MEK inhibitors and a PDK1 inhibitor supressed melanoma growth *in vitro*, as well as *in vivo*, evidencing that this therapeutic approach could be efficient in this context (Scortegagna et al. 2015). However, PDK1 holds a pivotal physiological role in non-malignant cells, which indicates that apart from the survival benefit conferred by this treatment its inhibition could have several side effects (Emmanouilidi and Falasca 2017).

In addition to *HIF-1α*, avian myelocytomatosis viral oncogene homolog (*MYC*) transcription is also enhanced by MAPK signal transduction pathway, increasing the glycolytic rate. *MYC* is an oncogene which can activate key players for glucose uptake and metabolism such as, the glucose transporter 1 (GLUT1) and lactate dehydrogenase (LDH) (Zeller et al. 2003; Dang et al. 2009). The two proteins referred are crucial for the maintenance of the glycolytic pathway, being GLUT1 behind glucose transport and LDH responsible for the conversion of pyruvate into lactate (Stine et al. 2016). Additionally, the elevated expression of *GLUT1* in melanoma metastasis was significantly associated with poor prognosis of melanoma patients (Koch et al. 2015). Another study confirmed these previous results showing the decrease of *GLUT1* expression was correlated with a better overall survival in melanoma patients (Yan et al. 2016). The overexpression of this glucose transporter enhances the metastatic behaviour of melanoma cells in murine models, being directly associated with melanoma progression (Koch et al. 2015). Furthermore, it has been noticed the presence of higher levels of GLUT1 in melanoma tissues compared to melanocytic nevi (Slominski et al. 2014). It was also established that *GLUT1* upregulation is correlated with melanoma ulceration and thickness (Yan et al. 2016). Altogether these results showed that GLUT1 could be a putative therapeutic target in melanoma.

Besides GLUT1, LDH also had a relevant role in the context of melanoma, since it was one of the earliest serologic biomarkers studied due to its importance as a measurable marker to monitor melanoma progression and outcome (Weinstein et al. 2014). It was previously shown that high levels of circulating LDH are correlated with shorter overall survival in patients with advanced melanoma, although LDH is a negative predictor of response to therapy (Palmer et al. 2011). All proteins mentioned enhance the glycolytic pathway in melanoma. Even though melanoma proliferating cells generally metabolize glucose into lactate regardless of oxygen levels, around 25% of all the pyruvate produced enters into the mitochondria of these cells (Scott et al. 2011).

This metabolic reprogramming is accompanied by the suppression of microphthalmia transcription factor *(MITF)* and peroxisome proliferator-activated receptor γ coactivator 1 (*PGC1α)* (Haq et al. 2013). *MITF* is a direct regulator of *PGC1α* expression (Haq and Fisher 2011) and encodes a protein that regulates melanocyte development and it is responsible for pigment cell-specific transcription of the melanogenesis enzyme genes (Hsiao and Fisher 2015). On the other hand, *PGC1α* regulates mitochondrial biogenesis and oxidative phosphorylation, controlling the mitochondrial production of reactive oxygen species (ROS), its suppression decreases the mitochondrial oxidative metabolism rate. Melanomas expressing high levels of *PGC1α* are usually associated with shorter overall survival in patients with metastatic melanomas, while decreased levels of this protein are correlated with a more invasive phenotype of primary melanomas (Vazquez et al. 2013).

In non-malignant cells, *MITF* nuclear transcription is promoted by mammalian target of rapamycin (mTOR), which is capable of enhance *PGC1α* expression, which consequently can induce the expression of oxidative metabolism related genes (McQuade and Vashisht Gopal 2015). On the other hand, in melanoma cells, $BRAF^{V600E}$ mutation enhances *HIF1α* expression through the activation of mTOR and suppression of *MITF* (Abildgaard and Guldberg 2015). Consequently, PGC1α expression is downregulated by MITF (Nasti et al. 2016). Thus, most melanomas demonstrate an increase in *HIF1α* expression (Kuphal et al. 2010) and a decrease in *PGC1α* transcription (Vazquez et al. 2013), which ensure the glycolytic flux. In sum, it was verified an interaction between HIF1α, mTOR, MITF and PGC1α in $BRAF^{V600E}$ mutated melanomas (Fig. 11.1). Overall, these evidences support the fact that RAS/RAF/MEK/ERK pathway is intrinsically involved in the switch between the oxidative metabolism and glycolytic synthesis.

In addition to MAPK pathway, it is widely accepted that phosphoinositide-3-kinase–protein kinase B-PI3K/AKT/mTOR pathway also has an important role in melanoma development. The later regulates many aspects of cell growth and survival, in physiological as well as in pathological conditions (Porta et al. 2014). This signalling begins with the activation of the tyrosine kinase receptors (RTKs), which induce the phosphorylation of phosphatidylinositol-4,5-bisphosphate (PIP2) in phosphatidylinositol-3,4,5-triphosphate (PIP3) through the activation of PI3K (Lai et al. 2015). PIP3 is stored in the intracellular medium near the plasmatic membrane of cells, leading to the recruitment of AKT. After activated, AKT phosphorylates several effector proteins such as mTOR and murine double minute 2 (Mdm2) (Vara et al. 2004). A study of Land and Tee, reported that mTOR can stimulate the transcriptional activity of *HIF1α* under hypoxic conditions (Land and Tee 2007). mTOR mediates the expression of several genes involved in the glycolytic metabolism and its activation is closely linked to malignant melanocytic lesions compared to the benign neoplasms (Karbowniczek et al. 2008). This regulator was also described as a growth promoting factor for melanoma cells (Karbowniczek et al. 2008).

As previously mentioned, AKT is a pivotal regulator of the PI3K/AKT/mTOR pathway, its activation promotes the expression and activity of glucose transporters such as GLUT1, as well as, the phosphorylation of PDK1 during hypoxia which ultimately induces glycolysis (Li et al. 2002; Chae et al. 2016). Furthermore, it was verified that phosphorylated AKT is mainly expressed in metastatic melanoma (77%), compared to primary melanoma (49%), dysplastic nevi (43%) and normal nevi (17%) (Dai et al. 2005). The expression of this phosphorylated protein increases melanoma invasion and is associated with poor outcomes (Dai et al. 2005).

This pathway ensures many physiological functions vital for melanoma cells survival and it can be constitutively activated by *NRAS* mutations, which also has a determinant role in the MAPK pathway (Muñoz-Couselo et al. 2017). Additionally, *BRAF* mutations and Phosphatase and Tensin Homolog (*PTEN*) inactivation can also enhance the PI3K/AKT/mTOR pathway, promoting mTOR signalling and *HIF1α* transcription, consequently driving glycolytic machinery synthesis (Land and Tee 2007; Kwong and Davies 2013). *PTEN* is a tumour suppressor gene with an opposite role to PI3K, regulating the levels of PIP3 in the cell. The study of the genetic profile of several cell lines of melanoma showed that the most of them (60%) contain a hemizygous deletion in the *PTEN* locus and 20–30% have *PTEN* loss through homozygous deletion or mutation (Zhou et al. 2000; Pollock et al. 2002). Therefore, there is a well-known crosstalk between RAS/RAF/MEK/ERK and PI3K/AKT/mTOR pathways in tumour cells, being NRAS and mTOR the most common mediators of both pathways, with predominant effect in melanoma context (Mendoza et al. 2011).

Additionally, in melanoma there is another important player that could regulate both RAS/RAF/MEK/ERK and PI3K/AKT/mTOR signalling pathways. *NF1* mutations were the third most common molecular alteration identified in melanoma. *NF1* is a tumour suppressor gene included in the family of RAS GTPase-activating proteins (GAP) that negatively regulates RAS (Martin et al. 1990). RAS proteins are activated when bound to GTP; conversely, hydrolysis of GTP to GDP, which is accelerated by GAPs, inactivates RAS (Ratner and Miller 2015). *NF1* loss hyperactivates NRAS protein, which consequently induces the activation of both RAS/RAF/MEK/ERK and PI3K/AKT/mTOR signalling pathways. Mutations in *NF1* are associated with an increased risk of melanoma-associated death (Kvist et al. 2017). *NF1*-mutated melanoma is characterized by a higher thickness and commonly by the manifestation of a second neoplasia (Guillot et al. 2001). As such, *NF1* gene is an important player in the context of melanoma metabolism, since its loss can regulate both previously mentioned signalling pathways, which are pivotal for the metabolic switch occurring in melanoma cells (Nissan et al. 2014; Kiuru and Busam 2017).

Overall, the most common molecular alterations in melanoma *BRAF*, *NRAS* and *NF1* contribute for its adaptation to the microenvironment under stress conditions, which allowed melanoma cells survival and aggressiveness. Therefore, the activation of RAS/RAF/MEK/ERK and PI3K/AKT/mTOR pathways is mainly driven by the above-mentioned mutations which are pointed as the main causes of melanocyte malignant transformation. Both pathways regulate several factors, such as HIF1α, Myc, MITF, mTOR, PGC1α, that control the metabolic balance between the glycolytic pathway and oxidative phosphorylation, which are central for carbon metabolism. Summarily, molecular alterations contribute to melanoma pathogenesis and heterogeneity through the introduction of metabolic plasticity, which should be further assessed to improve our understanding about the impact of these metabolic pathways in melanoma treatments.

11.3 The Impact of Melanoma Metabolism in the Microenvironment and Therapy Resistance

The interplay of metabolic reprogramming is a key feature in melanoma, impacting the whole tumour microenvironment. The pH dysregulation that occurs as a consequence of glycolysis is an important factor associated with tumour cell metabolism plasticity and invasiveness (Payen et al. 2016). While normal cells usually show an intracellular pH of 7.2 and an extracellular pH of 7.4, tumour cells demonstrate a higher intracellular pH (>7.2) and a lower extracellular pH (6.7–7.1) (Andreucci et al. 2018). Hence, these cells must adapt to a more acidic extracellular medium in order to survive. This is possible since tumour cells are extremely plastic even in terms of cellular energetics gaining a selective advantage under unfavourable environments (Andreucci et al. 2018). In this context, Na^+/H^+ exchanger (NHE) are ions channels and transporters responsible by the efflux of protons for the extracellular medium, causing its acidification. It was reported that the inhibition of the Na^+/H^+ exchanger channels induces a drastic impairment in melanoma cell motility and alters the morphology of cells (Stock et al. 2005). In addition, the Na^+/H^+ exchanger isoform 1, NHE1, makes part of the focal contacts, being in an ideal position to interact with integrins. Thus, NHE1 is important for the contact cell-matrix, its loss induces the impairment of cell polarity and reduced the directionality in cell migration (Denker and Barber 2002). Cell-matrix interactions or also called focal adhesions are large stable complexes assemblies which comprise integrins, focal adhesion kinase (FAK), talin, vinculin, paxillin and other proteins attached to the actin filament network (Friedl and Wolf 2003; Brakebusch 2003). Integrins and their downstream signalling are regulators of angiogenesis, immune response and the stromal context of the metastatic niche, with several different integrins being involved in the organ-specific metastasis of malignant melanoma (reviewed in Huang and Rofstad 2018). Some of the best studied examples of this interaction are integrin α4β1, which has been implicated in the formation of melanoma metastasis in the lymph nodes (Garmy-Susini et al. 2013) and integrins α2β1 and αvβ3, which are associated with lung melanoma metastasis (Pickarski et al. 2015; Bartolomé et al. 2017), however the mechanisms underlying their impact on metastasis location remain to be elucidated. With this information, one cannot exclude that in melanoma, NHE1 might provide an acidic microenvironment which influences the focal adhesion and the efficacy of the interaction between integrins and collagen. In fact, acidosis strengths the focal adhesions, promoting integrin-collagen bound in the lamellipodia of melanoma cells, which can increase their migration (Payen et al. 2016). Thus, migration and morphology of melanoma cells can be influenced by the interactions between integrin and collagen. Additionally, extracellular acidification impairs cellular junctions through the upregulation of several cytokines, pro-angiogenic factors and structural proteins (Rofstad et al. 2006). For example, matrix metalloproteases (MMPs) which are included in a family of zinc dependent endopeptidases, are involved in extracellular matrix degradation in both physiological and pathological conditions (Hofmann et al. 2000). Specifically in human melanoma cell lines, MMP-2 and MMP-9 are upregulated and associated with a more invasive phenotype (Rofstad et al. 2006), suggesting the association between the microenvironment acidification and tumour progression.

As previously referred, the overexpression of *HIF1α* and *MYC* is associated with the manifestation of a glycolytic behaviour in melanoma cells (Dang et al. 2009; Parmenter et al. 2014). This leads to an upregulation of monocarboxylate 4 (*MCT4*) to promote the secretion of the lactate by-product into the melanoma microenvironment (Payen et al. 2016; Pinheiro et al. 2016). The lactate secreted into the melanoma microenvironment is uptaken by endothelial cells via MCT1, which promotes signalling via HIF1α, interleukin-8 (IL-8), interleukin-10 (IL-10) and nuclear factor kappa-light-chain-enhancer of activated B cells (NFκB), culminating in the upregulation of growth factors including vascular endothelial growth factor receptor 2 (VEGFR2) and basic fibroblast growth factor (bFGF) (Sonveaux et al. 2012; Payen et al. 2016). MCTs have been described as upregulated in many types of cancer and this increase in their expression is associated with poor prognosis (reviewed in Fisel et al. 2018). *MCT4* and *GLUT1* showed significantly increased levels of expression in melanoma metastatic tumours compared with melanoma primary tumours and benign nevi and the upregulation of these genes was associated with a shorter overall survival (Pinheiro et al. 2016). In fact, glucose uptake through GLUT1 transporter and efflux of lactate by MCT4 contributes for the invasive phenotype and aggressiveness of melanoma cells (Koch et al. 2015). However, to our knowledge, no studies investigating a putative interaction between MCTs and the MAPK pathway, namely with *BRAF* and *NRAS* mutations, which also impacts *GLUT1* have been performed. Overall, lactate profoundly impacts the tumour microenvironment by facilitating angiogenesis, however critical components such as immune cells are also affected, possibly altering metabolic interactions between melanoma cells and immune cells (Romero-Garcia et al. 2016).

Indeed, melanoma is one of the most immunogenic types of tumours and its immunomodulatory mechanisms disable recognition and targeting by the immune system resulting in immune suppression and escape (Passarelli et al. 2017). The microenvironment of melanoma contains several types of immune cells, such as mast cells, neutrophils, dendritic cells (DC), B lymphocytes, T lymphocytes and macrophages (Passarelli et al. 2017). The later are the most abundant cells interfering with the melanoma microenvironment and are designated tumour-associated macrophages (TAM) (Kakizaki et al. 2015). These TAM can impact directly the tumour microenvironment in two opposing processes: if they present an M2-like or pro-tumoral phenotype, they are capable of remodelling the extracellular matrix, increasing tumour initiation and growth, promoting angiogenesis, and supressing anti-tumour activity through the production of growth factors, cytokines and chemokines (Mantovani and Sica 2010; Fujimura et al. 2018); if they are polarized towards a M1 macrophage or anti-tumoral phenotype, they are involved in the production of reactive oxygen and nitrogen species, phagocytosis of tumoral cells and activation of T cells by production of pro-inflammatory cytokines and chemokines (Pathria et al. 2019). Lactate has the ability to supress pro-inflammatory response and polarize M1 macrophages towards an M2 phenotype, promoting tumour development (Romero-Garcia et al. 2016). Since LDH is pivotal for the transformation of pyruvate into lactate, this phenomenon could explain why high levels of LDH are correlated with shorter overall survival of patients with advanced melanoma and why LDH is a negative predictor of response to therapy, as previously mentioned (Palmer et al. 2011). Nevertheless, macrophages are not the only subset of immune cells affected by metabolic changes.

Lactate is also capable of preventing the maturation of dendritic cells and, consequently, increases the amount of immunosuppressive cytokine IL-10 in the tumour microenvironment, further downregulating an anti-tumoral immune response (Nasi et al. 2013). This interaction is particularly relevant in the context of melanoma, since IL-10 expression has been correlated with melanoma growth and development of metastatic competence (Itakura et al. 2011). Nevertheless, to our knowledge, research to employ IL-10 as serological biomarker in melanoma are limited. Mature DCs locally mediate the efficiency of the response of the immune system and the ability of other cells, such as T lymphocytes, to orchestrate a cytotoxic effect (Nussenzweig 2010). DCs circulate in the peripheral blood and upon recognition of an antigen, they migrate towards the lymph nodes to perform their antigen-presenting functions (Nussenzweig 2010). Nevertheless, to mature and efficiently execute their cross-priming functions, DCs require an interplay between the T-cell receptor (TCR) and the major histocompatibility complex (MHC) molecules, binding of CD80/CD86 expressed by DCs with CD28 expressed by T cells, cytokine-mediated signalling and a chemokine profile which promotes migration to the lymph nodes (Tucci et al. 2014). In the context of melanoma, the maturation and stimulation of DCs is compromised by the enrichment of the tumour milieu in VEGF and IL-10 produced by the melanoma cells, that survive at the expense of DCs, which suffer from an immature or tolerogenic phenotype. Interestingly, blockade of the MAPK signalling pathway restores the normal co-stimulation of DCs by abolishing the effects of melanoma cells on the CD80 and CD86 expression (Ott et al. 2013).

This intrinsic defect of DCs to prime the immune system against melanoma cells through antigen presentation mainly impacts T cells. $CD8^+$ T cells play a key role in the immunity to melanoma as they can eliminate these cells upon TCR mediated recognition of specific antigenic peptides presented by DCs on the surface of target cells. Upon recognition, TCR and other signalling molecules become clustered at the centre of the contact area between the T cell and the tumour cells, initiating a cascade of T lymphocyte effector functions (Durgeau et al. 2018). These functions can be of a direct nature, through the exocytosis of granules containing perforin and granzymes into the target, or indirect, mediated by the secretion of pro-inflammatory cytokines, including interferon γ (IFNγ) and tumour necrosis alpha (TNF-α) (Durgeau et al. 2018). This TCR signalling is of such importance that strategies to employ it against cancer have been actively sought. Currently, TCR-engineered T cells that can recognize specific cancer antigens and induce potent immune responses are promising targets for use as a T cell therapy. In fact, in the last decade some promising candidate antigens have been subjected to clinical trials. TCR-engineered T cells targeting melanoma antigen recognized by T cells 1 (MART-1), were shown to result in a durable engraftment in 15 patients with metastatic melanoma (Morgan et al. 2006) and other study revealed that they led to an objective cancer regression in 30% of the patients subjected to the clinical trial (Johnson et al. 2009). Furthermore, a 2014 study involving TCR transgenic T cells targeting MART-1 showed evidence of melanoma regression in 9 out of 13 treated metastatic melanoma patients (Chodon et al. 2014). However, melanoma-associated antigen-A3 (MAGE-A3), that also appeared as a potential candidate and led to metastatic melanoma regression, was unsuccessful since severe adverse events were predominant and TCR-mediated inflammatory response resulted in neuronal destruction due to the previously unrecognized MAGE-A12 (Stewart et al. 2013). The most recent clinical trial involving this approach was based on New York oesophageal squamous cell carcinoma-1 (NY-ESO-1) reactive T cells where 11 out of 20 melanoma patients refractory to standard treatments revealed an objective clinical response, with overall 3- and 5-year survival rates of 33% (Robbins et al. 2015). Even though the adoptive transfer of autologous T cells encoding a TCR against melanoma epitopes seems to be a promising therapy, other immune cell function can impair the function of T cells. For instance, the deficient antigen presentation by DCs compromises the entire $CD8^+$ T cell cytotoxic function, contributing further for the downregulation of anti-tumoral responses in favour of pro-tumoral responses. Additionally, lactate excreted by melanoma cells can also directly impact $CD8^+$ T lymphocyte function. The high concentrations of lactate present in the tumour microenvironment prevent the adequate secretion of lactate from T cells, further inhibiting their function and keeping these cells in an anergic state (Romero-Garcia et al. 2016; Scott and Cleveland 2016). Moreover, $CD8^+$ T lymphocytes must compete against melanoma's highly glycolytic metabolism for the access to glucose and, if unable to compete, no glucose uptake by $CD8^+$ T cells occurs, and cytotoxic functions are again inhibited (Chang et al. 2015). Altogether, it is established that lactate supresses $CD8^+$ T cell anti-tumoral functions through the different mechanisms summarized above. Nevertheless, in the context of melanoma, lactate is known to also modulate the immune system towards an immunosuppressive state by permitting regulatory T cells (Treg) to perform their functions (Gerriets et al. 2015).

Contrarily to cytotoxic T cells, Tregs are less dependent on glycolysis and typically display oxidative phosphorylation metabolism for their energy production (Gerriets et al. 2015). Hence, while the tumour microenvironment promotes suppression of cytotoxic T cells, Tregs exhibit metabolic adaptations that enable them to thrive in low glucose and high lactate environments. In particular, Treg transcription factor forkhead box P3 (FOXP3) regulates their metabolism, mediating the transcriptional suppression of MYC signalling and, consequently, repressing glycolysis, whereas oxidative phosphorylation is enhanced (Gerriets et al. 2016). Thus, in the glucose-deprived and lactate-enriched melanoma microenvironment, Tregs acquire resistance to the suppressive effects of lactate which allows for a dominance of these subsets of cells. Tregs promote immunosuppression through IL-10, transforming growth factor beta (TGF-β) and indolamine 2,3-dioxygenase (IDO) overproduction, enabling melanoma to hijack a physiologic mechanism of self-tolerance (Munn and Mellor 2007; Gerriets et al. 2016).

Even though lactate is a major player in melanoma cell survival by influencing both the tumour metabolism and its microenvironment. However, another metabolic pathway involved in melanoma's modulation of the immune system is the tryptophan metabolism (Brody et al. 2009). When activated effector T helper-1 (Th1) cells secrete IFNγ, IDO is produced by tumour cells. Then, IDO will metabolize tryptophan, producing kynurenine during this process (Brody et al. 2009). Kynurenine binds to the aryl hydrocarbon receptor (AHR), which induces differentiation and activation of $FOXP3^+$ Tregs, and polarization of DCs and macrophages towards their corresponding immunosuppressive features (Munn et al. 2016). Since Tregs produce IDO by themselves, this pathway acts a positive feedback, perpetuating the cycle of immunosuppression in the tumour microenvironment. Moreover, IDO starves effector T cells of tryptophan, suppressing their proliferation and inducing the apoptosis of these cells (Fuchs et al. 2011), as indicated by studies that correlated low levels of tryptophan in the serum with a poor prognosis of melanoma patients (Mezrich et al. 2010; Munn et al. 2016). All this data suggested that tryptophan metabolism can be a key factor for the immune modulation in the context of melanoma, which led to development of phase II and phase III clinical trials in order to evaluate a treatment combination of pembrolizumab (α-PD-1) and the IDO inhibitor epacadostat in metastatic melanoma patients (Active, not recruiting – NCT02752074).

Along with lactate, TGF-β, VEGF and IDO, melanoma's immune escape is also dependent on the production of adenosine. Intracellular adenosine is involved with several biological processes including energy metabolism, nucleic acid metabolism and methionine cycle (Fredholm et al. 2001). On the other hand, extracellular adenosine is thought to be originated from ATP in the extracellular compartment, where the surface nucleosidase CD39 hydrolyses ATP/ADP to AMP and CD73 converts AMP to adenosine (reviewed in Ohta 2016). Extracellular adenosine has been implicated in intercellular signalling through G protein-coupled adenosine receptors on the surface of cells, with physiological impacts in the nervous, cardiovascular and immunological systems (Fredholm et al. 2001). Namely, immune cells predominantly express A2A adenosine receptor (A2AR). The stimulation of this receptor is associated with immunosuppressive signalling by induction of FOXP3 expression (Ohta and Sitkovsky 2014) and recruitment of Treg cells and myeloid-derived suppressor cells (MDSCs) (Umansky et al. 2014). Simultaneously A2AR stimulation leads to inhibition of proliferation and cytokine production by Th cells, inhibition of the cytotoxic functions from $CD8^+$ T cells, as well as, natural killer (NK) cells. The impairment of macrophage/DC antigen presentation and induction of anti-inflammatory cytokine production and neutrophil oxidative burst are also effects resulting from A2AR stimulation (Sitkovsky et al. 2004; Gabrilovich and Hurwitz 2014). Besides, A2B adenosine receptor (A2BR) has been shown to be play a major role in the adenosine-dependent differentiation of M2 macrophages, which become activated in the presence of a stimuli of A2BR, expressing arginase, IDO and TGF-β (Csóka et al. 2012). Specifically, in melanoma microenvironment, the presence of high levels of extracellular adenosine strongly restrain immune functions through paracrine signals that favour the initiation and progression of melanoma and the evasion of antitumor immune response (Umansky et al. 2014) (Fig. 11.2).

One certainty regarding melanoma is that the metabolic reprogramming that occurs in tumour cells has an accentuated impact in the immune system and, consequently, in immunotherapy. Since melanoma is particularly resistant to the traditional radio and chemotherapies, the use of kinase inhibitors and immunocheckpoint therapies (target therapies) has been rising in recent years to treat metastatic melanoma (Kalal et al. 2017). As previously explained, the inhibition of *BRAF*^{V600E} derived from the administration of vemurafenib, dabrafenib, trametinib and cobimetinib can suppress the expression of glycolytic enzymes and regulators such as HIF1α and MYC, inducing a reduction in glucose consumption (Parmenter et al. 2014). Interestingly, the resistance to *BRAF* inhibitors is mediated by the oxidative metabolism, since the treatment with vemurafenib enhances the mitochondrial respiration rate and ROS production (reviewed in Corazao-Rozas et al. 2013). In these conditions, the overexpression of *PGC1α*, the key cofactor in mitochondrial genesis, is enough to restore mitochondrial activity (Vazquez et al. 2013). Furthermore, *in vitro* studies combining vemurafenib and an inhibitor of mitochondrial respiration increased the *BRAF* inhibitor related cell death, confirming that mitochondrial activity is used by melanoma cells as a mechanism against drugs (Zaal and Berkers 2018). It was also suggested that this increase in oxidative metabolism is correlated with a metabolic switch that makes cells more dependent on glutamine rather than glucose (Dhomen et al. 2015). Moreover, studies using melanoma cell lines revealed evidence that MAPK inhibitors combined with a glutaminase inhibitor could be a viable therapeutic approach for melanoma, which demonstrate resistance developed via reactivation of oxidative metabolism (Dhomen et al. 2015). In addition, there are several compensatory mechanisms that could occur in MAPK signalling pathway and even in the PI3K/AKT pathway that can potentiate *BRAF* inhibitors resistance (McCubrey et al. 2015), such as the overactivation of downstream kinases by oncogenic mutations in RAS or MEK (reviewed in Griffin et al. 2017).

Even though combined *BRAF* and MEK inhibitor therapies lead to a better clinical response, the adverse effects resulting from this treatment are a major concern. Moreover, resistance to these therapies remains an obstacle (Wang et al. 2018), as the most common form of resistance results from the reactivation of the signalling pathway which is the target of the therapy (Van Allen et al. 2014). Considering all the currently available information, the combination between target therapies and metabolic inhibitors

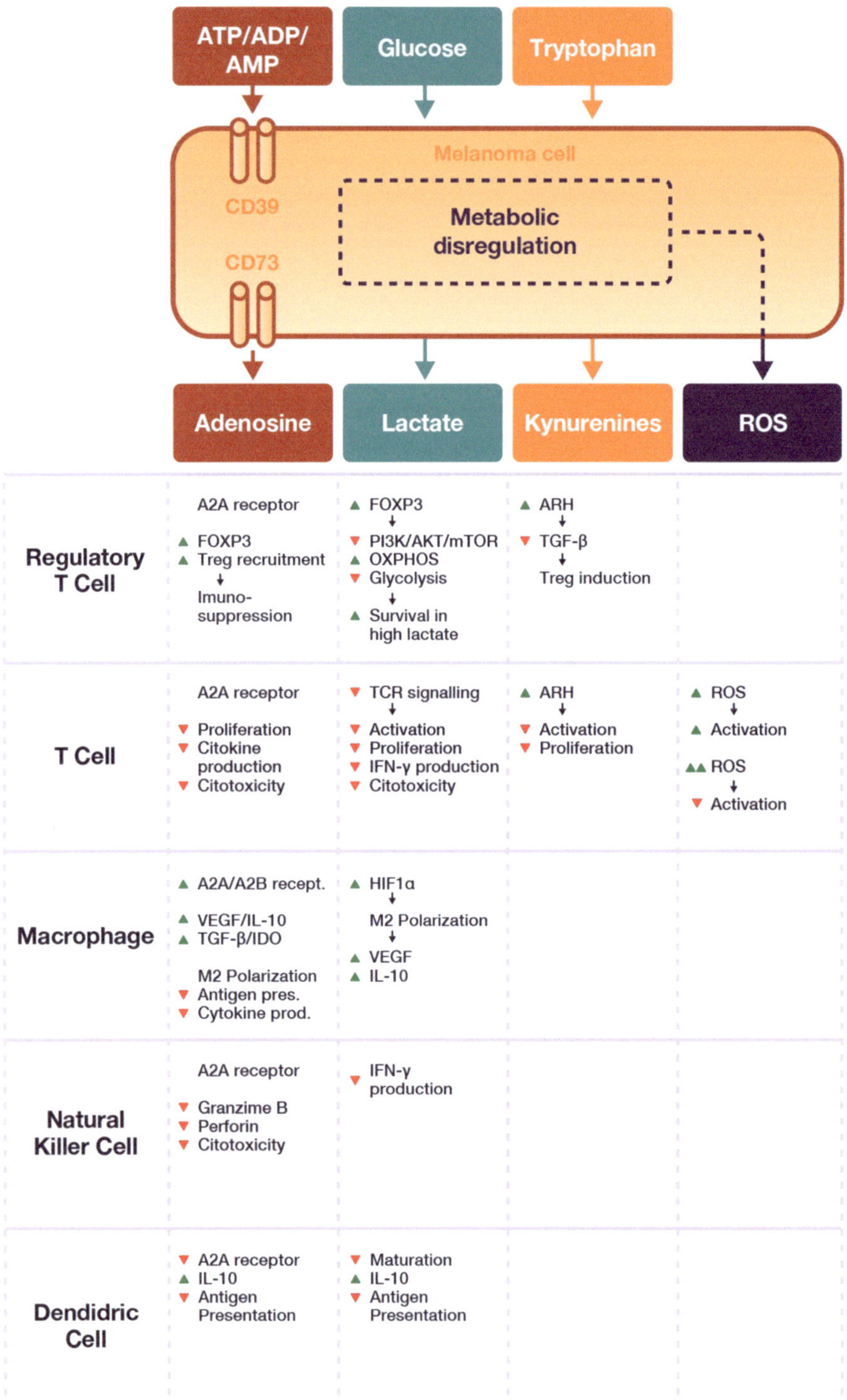

Fig. 11.2 Immunomodulatory effects of the presence of metabolites in melanoma microenvironment. In red, it is depicted the adenosine metabolism, where ATP/ADP are hydrolysed to AMP by CD39 and converted to adenosine by CD73. The presence of adenosine in the tumour microenvironment induces Treg differentiation, T cell and NK cell anergy, as well as macrophage and dendritic cell differentiation to their corresponding immunosuppressive state.

could be a viable therapeutic opportunity. Nevertheless, alternative therapies to answer the challenges presented by melanoma have been developed in the form of immunotherapies. For instance, ipilimumab is an antibody that targets the cytotoxic T-lymphocyte-associated protein 4 (CTLA-4), inhibiting its signalling on T cells (Wolchok et al. 2010). It has been shown that CTLA-4 expression increases significantly after $CD4^+$ and $CD8^+$ T cell activation (Alegre et al. 2000). It competes with CD28 to bind with CD80 and CD86, which are present on the surface of macrophages and other antigen presenting cells (Walunas et al. 1996). Due to the fact that CTLA-4 has a higher affinity for CD80 and CD86 than CD28, it will result in a more favourable binding of these molecules when compared with the later (Walunas et al. 1996). As a response, inhibitory signals are transmitted to $CD8^+$ T cells, suppressing their functions. Moreover, CTLA-4 signalling impairs signalling through AKT, which is mediated by protein phosphatase 2A (PP2A) inhibition (Alegre et al. 2000). As this pathway promotes glycolysis, CTLA-4 indirectly promotes a metabolism based on oxidative phosphorylation that is averse to activation and function of effector T cells, and therefore promotes immunosuppression (Walunas et al. 1996). Moreover, it has been described that treatment with ipilimumab enhances IFN-γ production (Gao et al. 2017). Since this pro-inflammatory cytokine plays a major role in the activation of macrophages and antigen presenting cells, as well as in the activation of the innate immune system and regulation of T helper cells (Tau and Rothman 1999), a decrease in IFN-γ may be associated with primary resistance to α-CTLA-4 treatment. In fact, patients identified as non-responders to ipilimumab have melanomas with genomic defects in IFN-γ pathway genes (Gao et al. 2017), further streghtening this hypothesis.

Pembrolizumab and Nivolumab are also blocking antibodies, both targeting programmed cell death protein-1 (PD-1). Similarly to CTLA-4, PD-1 is expressed on the surface of T cells, while its ligand PD-L1, is constitutively expressed on the surface of DCs, macrophages and tumour cells (Pardoll 2012). Activation of PD-1 occurs upon binding to PD-L1, leading to the impairment of effector T cell function and anti-tumour response (Lee et al. 2016). Nevertheless, a study showed that the growth of B16.F10 melanoma cell lines, is delayed in mice deficient for PD-L1 suggesting that PD L1 expression on non-melanoma cells in wildtype mice plays a role in the inhibition of anti-tumour immunity (Juneja et al. 2017). Besides, research by Lin et al. showed that α-PD-1 treatment is effective at reducing tumour growth in mice bearing a PD-L1-deficient B16.F10 melanoma, implying that host but not melanoma-derived PD-L1 is pivotal for PD-L1 therapy (Lin et al. 2018). Additionally, immunofluorescence microscopy revealed a positive correlation between expression of PD-L1 on DCs/macrophages and the efficacy of treatments with α-PD-1 in locally advanced and metastatic melanoma patients, indicating that the host DCs and macrophage-derived PD-L1 is indispensable for the therapeutic efficacy of α-PD-1 (Lin et al. 2018). Even though PD-1 signalling inhibits the same pathway as CTLA-4, PI3K function is suppressed instead of AKT (Walunas et al. 1996). Thereafter, PD-1 promotes oxidative phosphorylation, culminating in the same processes as CTLA-4 blockade. Furthermore, the binding of PD-1 to PD-L1, leads to the suppression of glycolysis and to the upregulation of fatty acid oxidation

Fig. 11.2 (continued) In blue, it is represented the glycolytic pathway which results in the excretion of lactate into the microenvironment. Tregs can perform oxidative phosphorylation and survive in this environment, while, T cells compete for glucose with melanoma cells, culminating in the inhibition of T cell function. The lactate-rich microenvironment also promotes M2 macrophage polarization, while reducing IFN-γ production by NK cells and antigen presentation by DCs. The tryptophan metabolic pathway is represented in orange, with the presence of kynurenines in the microenvironment as a by-product result in Treg differentiation and T cell proliferation inhibition. All the molecular instability, along with highly altered microenvironment contribute to the metabolic dysregulation of the melanoma cell, that triggers a substantial increase in ROS production which inhibits T cell proliferation

(Patsoukis et al. 2015), which is a major alternative metabolic pathway involved in melanoma cell survival, proliferation and progression (Aiderus et al. 2018; Li et al. 2018; Zhang et al. 2018). In these conditions, there is a metabolic switch that leads T cells to use a lipidic source of energy instead of glucose as the main metabolic pathway to obtain energy. T cells that receive PD-1 signals are unable to perform glycolysis, glutaminolysis or even to metabolize amino acids resorting to an elevated rate of fatty acid oxidation. This metabolic pathway compromises their functionality and triggers T cells to remain in a hyporesponsive or anergic state (Patsoukis et al. 2015, Chang et al. 2015). Checkpoint inhibitors recover PI3K-AKT pathway signalling in effector T cells, enabling their activation through adaptation to a glycolytic phenotype and triggering anti-tumoral responses (Chang et al. 2015). Tumour cell death will then result in the release of glucose to the tumour microenvironment, promoting further activation of immune cells (Chang et al. 2015). Recently, an approach combining α-PD-1 and α-CTLA4 therapy has been explored and treatment showed superior efficacy when compared to individual administration of each drug (reviewed in Seidel et al. 2018). Still, there was also an increase in severe side effects (Seidel et al. 2018), which may explain the lack of research assessing the impact of this therapy on melanoma metabolism.

11.4 Current Challenges and Future Trends

The metabolic heterogeneity of melanoma cells contributes for their singular ability to change and adapt to the microenvironment conditions and to acquire distinct energy sources to induce tumour progression and metastasis. However, the molecular mechanisms and specific immunological and microenvironmental cues regarding metastatic dissemination are not well understood. The lack of knowledge related to metastasis initiation remains the main obstacle to the development of efficient therapies for this malignancy. Therefore, it is urgent to develop new biomarkers of metastasis, prognosis and response to therapy to better monitor the progression of melanoma.

Hitherto, the mechanisms by which melanoma metastasis develop preferentially in specific organs remain to be elucidated. Chemokines and their respective receptors have been implicated as potential melanoma metastasis mediators (Richmond et al. 2009). Nevertheless, one must not ignore the possibility of extravasation and subsequent proliferation of the metastasis that are possibly regulated by different mechanisms. The differential metabolism and putative signalling cascades of each organ, may promote an ideal microenvironment where the metastatic melanoma cells can thrive, conditioning the metastasis location to the target organ rather than the remaining. Alternatively, the metabolic reprogramming of melanoma cells can be determinant to the site of metastasis since the metastatic cell may migrate to the organ according to the availability of resources for proliferation.

Currently, the development of an efficient and effective therapy is of the utmost importance to attenuate the negative effects of melanoma, since, 5-year survival rate of metastatic melanoma patients is around 23% (based on data from the United States; Siegel et al. 2019); while the global 5-year survival rate of localized and regional melanoma patients is estimated to be 90% (Allemani et al. 2018). Immunocheckpoint therapies have an important role on melanoma metabolic reprogramming by enhancing T cells' ability to compete for resources against melanoma cells. However, despite the recent successes on the development of targeted therapies in the clinical field, recurrence rates remain high and resistance to target inhibitor therapies and immunotherapy is a looming concern, with survival of the affected patients being only extended several months (Haq et al. 2013). Additional investigation is needed to define the putative interactions between the metabolic and oncogenic signaling pathways, while evaluating how these pathways cooperate with microenvironment to promote melanoma progression and metastasis.

Targeting metabolic pathways seems to be a promising approach, although the toxicities inherent to these therapies must be overcome, while accounting for melanoma's metabolic flexibility. With the information currently available and summarized on this chapter, the way to achieve the desired therapy possibly involves the combination between targeted therapies or immunotherapies with efficient and non-toxic metabolic reprogramming drugs. The drugs presently accessible are only employed in the treatment of stage III and IV melanomas, which correspond to the most aggressive types. Stage IV melanomas presumably possess a high rate of alterations which might render them resistant to several therapeutic strategies, including simultaneous targeting of the immune system and the metabolism. Consequently, it would be interesting to focus research towards the development of a metabolic-based drug therapy centred on stage II and III melanomas, since the failure of therapeutic approaches could be influenced by their late administration. Additionally, the rate of alterations should be considerably lower in earlier stages, allowing for a more efficient targeting and remodulation that could halt tumour progression, positively impacting the prognosis of these melanoma patients.

Furthermore, it will be crucial to uncover biomarkers that could select the group of patients who will benefit the most from the administration of metabolic inhibitors. Still, there would be the obstacle of metastasis with different metabolic profiles due to their different destinations. In this context, future research should focus on the relationship between alterations in the primary melanoma and its microenvironmental niche that might represent adaptations to a specific organ, allowing the prediction of the location of the metastasis. Moreover, it is pivotal to aim for correlations between the molecular alterations of the primary melanoma and its plasticity with chemotactic molecules and factors, that might be determining the migration of the metastatic cells towards the signalling location. In spite of these hypothesis, exosomes must be considered as putative mediators of melanoma metastasis due to their involvement in the induction of malignant transformation in non-malignant cells by interchange of determined molecules and factors (reviewed in Couto et al. 2018). Exosomes are mediators of cell-cell communication and the ones derived from tumour cells can be uptake by non-malignant cells, conferring molecules and factors that induce the malignant transformation. Besides, they induce the recruitment of bone marrow-derived endothelial progenitor cells to the lungs, leading to the formation of pulmonary pre-metastatic niches (Couto et al. 2018).

Hence, it is crucial that further endeavours aim to identify exosome association with metastasis molecular signatures to uncover effective melanoma progression biomarkers.

As such, to pave the way to target melanoma, research must focus on a combination of metabolism, physiology and therapy to ultimately develop therapies that allow an enhancement of the prognosis and survival of metastatic melanoma patients.

Acknowledgments The authors would like to thank the support of Liga Portuguesa Contra o Cancro, Núcleo Regional Sul (LPCC-NRS). The authors acknowledge iNOVA4Health – UID/Multi/04462/2013, a program financially supported by Fundação para a Ciência e Tecnologia/Ministério da Educação e Ciência, through national funds and co-funded by FEDER under the PT2020 Partnership Agreement. Sérgio Abreu was responsible for the design of figures.

References

Abildgaard C, Guldberg P (2015) Molecular drivers of cellular metabolic reprogramming in melanoma. Trends Mol Med 21:164–171. https://doi.org/10.1016/j.molmed.2014.12.007

Aiderus A, Black MA, Dunbier AK (2018) Fatty acid oxidation is associated with proliferation and prognosis in breast and other cancers. BMC Cancer 18:1–15. https://doi.org/10.1186/s12885-018-4626-9

Alegre ML, Frauwirth KA, Thompson CB et al (2000) Inhibition of CTLA-4 function by the regulatory subunit of serine/threonine phosphatase 2A. J Immunol 168:5070–5078. https://doi.org/10.4049/jimmunol.168.10.5070

Ali Z, Yousaf N, Larkin J (2013) Melanoma epidemiology, biology and prognosis. Eur J Cancer Suppl 11:81–91. https://doi.org/10.1016/j.ejcsup.2013.07.012

Allemani C, Matsuda T, Di Carlo V et al (2018) Global surveillance of trends in cancer survival 2000–14 (CONCORD-3): analysis of individual records for

37 513 025 patients diagnosed with one of 18 cancers from 322 population-based registries in 71 countries. Lancet 391:1023–1075. https://doi.org/10.1016/S0140-6736(17)33326-3

Andreucci E, Pietrobono S, Peppicelli S et al (2018) SOX2 as a novel contributor of oxidative metabolism in melanoma cells. Cell Commun Signal 16:87. https://doi.org/10.1186/s12964-018-0297-z

Bandarchi B, Ma L, Navab R et al (2010) From melanocyte to metastatic malignant melanoma. Dermatol Res Pract 2010:1–8. https://doi.org/10.1155/2010/583748

Bartolomé RA, Torres S, de Val SI et al (2017) VE-cadherin RGD motifs promote metastasis and constitute a potential therapeutic target in melanoma and breast cancers. Oncotarget 8:215–227. https://doi.org/10.18632/oncotarget.13832

Bastian BC (2014) The molecular pathology of melanoma: an integrated taxonomy of melanocytic neoplasia. Annu Rev Pathol 9:239–271. https://doi.org/10.1146/annurev-pathol-012513-104658

Batus M, Waheed S, Ruby C et al (2013) Optimal management of metastatic melanoma: current strategies and future directions. Am J Clin Dermatol 14:179–194. https://doi.org/10.1007/s40257-013-0025-9

Brakebusch C (2003) The integrin-actin connection, an eternal love affair. EMBO J 22:2324–2333. https://doi.org/10.1093/emboj/cdg245

Bray F, Ferlay J, Soerjomataram I et al (2018) Global cancer statistics 2018: GLOBOCAN estimates of incidence and mortality worldwide for 36 cancers in 185 countries. CA Cancer J Clin 68:394–424. https://doi.org/10.3322/caac.21492

Brody JR, Costantino CL, Berger AC et al (2009) Expression of indoleamine 2,3-dioxygenase in metastatic malignant melanoma recruits regulatory T cells to avoid immune detection and affects survival. Cell Cycle 8:1930–1934. https://doi.org/10.4161/cc.8.12.8745

Burián Z, Ladányi A, Barbai T et al (2019) Selective inhibition of HIF1α expression by ZnSO4 has antitumoral effects in human melanoma. Pathol Oncol Res. https://doi.org/10.1007/s12253-018-00573-1

Cesi G, Walbrecq G, Zimmer A et al (2017) ROS production induced by BRAF inhibitor treatment rewires metabolic processes affecting cell growth of melanoma cells. Mol Cancer 16:1–16. https://doi.org/10.1186/s12943-017-0667-y

Chae YC, Vaira V, Caino MC et al (2016) Mitochondrial Akt regulation of hypoxic tumor reprogramming. Cancer Cell 30:257–272. https://doi.org/10.1016/j.ccell.2016.07.004

Chang C, Qiu J, Sullivan DO et al (2015) Metabolic competition in the tumor microenvironment is a driver of cancer progression. Cell 162:1229–1241. https://doi.org/10.1016/j.cell.2015.08.016. Metabolic

Chodon T, Comin-Anduix B, Chmielowski B et al (2014) Adoptive transfer of MART-1 T cell receptor transgenic lymphocytes and dendritic cell vaccination in patients with metastatic melanoma. Clin Cancer Res 20:2457–2465. https://doi.org/10.1158/1078-0432.CCR-13-3017

Cichorek M, Wachulska M, Stasiewicz A, Tymińska A (2013) Skin melanocytes: biology and development. Postep Dermatologii i Alergol:30:30–30:41. https://doi.org/10.5114/pdia.2013.33376

Corazao-Rozas P, Guerreschi P, Jendoubi M et al (2013) Mitochondrial oxidative stress is the achille's heel of melanoma cells resistant to Braf-mutant inhibitor. Oncotarget 4:1986–1998. https://doi.org/10.18632/oncotarget.1420

Couto N, Caja S, Maia J et al (2018) Exosomes as emerging players in cancer biology. Biochimie 155:2–10. https://doi.org/10.1016/j.biochi.2018.03.006

Csóka B, Selmeczy Z, Koscsó B et al (2012) Adenosine promotes alternative macrophage activation via A2A and A2B receptors. FASEB J 26:376–386. https://doi.org/10.1096/fj.11-190934

Dai DL, Martinka M, Li G (2005) Prognostic significance of activated Akt expression in melanoma: a clinicopathologic study of 292 cases. J Clin Oncol 23:1473–1482. https://doi.org/10.1200/JCO.2005.07.168

Dang CV, Le A, Gao P (2009) MYC-induced cancer cell energy metabolism and therapeutic opportunities. Clin Cancer Res 15:6479–6483. https://doi.org/10.1158/1078-0432.CCR-09-0889

Davies H, Bignell GR, Cox C et al (2002) Mutations of the BRAF gene in human cancer. Nature 417:949–954. https://doi.org/10.1038/nature00766

Denker SP, Barber DL (2002) Cell migration requires both ion translocation and cytoskeletal anchoring by the Na-H exchanger NHE1. J Cell Biol 159:1087–1096. https://doi.org/10.1083/jcb.200208050

Dhomen N, Baenke F, Galbraith L et al (2015) Resistance to BRAF inhibitors induces glutamine dependency in melanoma cells. Mol Oncol 10:73–84. https://doi.org/10.1016/j.molonc.2015.08.003

Domingues B, Lopes J, Soares P, Populo H (2018) Melanoma treatment in review. ImmunoTargets Ther 7:35–49. https://doi.org/10.2147/itt.s134842

Durgeau A, Virk Y, Corgnac S, Mami-Chouaib F (2018) Recent advances in targeting CD8 T-cell immunity for more effective cancer immunotherapy. Front Immunol 9:14. https://doi.org/10.3389/fimmu.2018.00014

Emmanouilidi A, Falasca M (2017) Targeting PDK1 for chemosensitization of cancer cells. Cancers (Basel) 9:1–25. https://doi.org/10.3390/cancers9100140

Eroglu Z, Ribas A (2016) Combination therapy with BRAF and MEK inhibitors for melanoma: latest evidence and place in therapy. Ther Adv Med Oncol 8:48–56. https://doi.org/10.1177/1758834015616934

Fischer GM, Gopal YNV, Mcquade JL et al (2019) Metabolic strategies of melanoma cells: mechanisms, interactions with the tumor microenvironment, and therapeutic implications. Pigment Cell Melanoma Res 31:713–745. https://doi.org/10.1111/pcmr.12661. Metabolic

Fisel P, Schaeffeler E, Schwab M (2018) Clinical and functional relevance of the Monocarboxylate trans-

porter family in disease pathophysiology and drug therapy. Clin Transl Sci 11:352–364. https://doi.org/10.1111/cts.12551
Fredholm BB, IJzerman AP, Jacobson KA et al (2001) International Union of pharmacology. XXV. Nomenclature and classification of adenosine receptors. Pharmacol Rev 53:527–552
Friedl P, Wolf K (2003) Tumour-cell invasion and migration: diversity and escape mechanisms. Nat Rev Cancer 3:362–374. https://doi.org/10.1038/nrc1075
Fuchs D, Margreiter R, Brandacher G et al (2011) IDO-mediated tryptophan degradation in the pathogenesis of malignant tumor disease. Int J Tryptophan Res 3:113–120. https://doi.org/10.4137/ijtr.s4157
Fujimura T, Kakizaki A, Kambayashi Y et al (2018) Cytotoxic anti-melanoma drugs suppress the activation of M2 macrophages. Exp Dermatol 27:64–70. https://doi.org/10.1111/exd.13417
Gabrilovich DI, Hurwitz AA (eds) (2014) Tumor-induced immune suppresion – mechanisms and therapeutic reversal, 2nd edn. Springer, New York
Gao J, Shi LZ, Zhao H et al (2017) Loss of IFNγ pathway genes in tumor cells as a mechanism of resistance to anti-CTLA-4 therapy. Cell 167:397–404. https://doi.org/10.1016/j.cell.2016.08.069.Loss
Garbe C, Leiter U (2009) Melanoma epidemiology and trends. Clin Dermatol 27:3–9. https://doi.org/10.1016/j.clindermatol.2008.09.001
Garmy-Susini B, Avraamides CJ, Desgrosellier JS et al (2013) PI3Kα activates integrin α4β1 to establish a metastatic niche in lymph nodes. Proc Natl Acad Sci 110:9042–9047. https://doi.org/10.1073/pnas.1219603110
Gerriets VA, Kishton RJ, Nichols AG et al (2015) Metabolic programming and PDHK1 control CD4+ T cell subsets and inflammation. J Clin Invest 125:194–207. https://doi.org/10.1172/JCI76012
Gerriets VA, Kishton RJ, Johnson MO et al (2016) Foxp3 and toll-like receptor signaling balance T reg cell anabolic metabolism for suppression. Nat Immunol 17:1459–1466. https://doi.org/10.1038/ni.3577
Gogas HJ, Kirkwood JM, Sondak VK (2007) Chemotherapy for metastatic melanoma: time for a change? Cancer 109:455–464. https://doi.org/10.1002/cncr.22427
Goldstein AM, Tucker MA (2013) Dysplastic nevi and melanoma. Cancer Epidemiol Biomarkers 22:528–532. https://doi.org/10.1158/1055-9965.EPI-12-1346
Griffin M, Scotto D, Josephs DH et al (2017) BRAF inhibitors: resistance and the promise of combination treatments for melanoma. Oncotarget 8:78174–78192. https://doi.org/10.18632/oncotarget.19836
Guillot B, Dalac S, Delaunay M et al (2001) Cutaneous malignant melanoma and neurofibromatosis type 1. Melanoma Res 14:159–163. https://doi.org/10.1097/01.cmr.0000124207.72344.38
Guy GP, Thomas CC, Thompson T et al (2015) Vital signs: melanoma incidence and mortality trends and projections – United States, 1982-2030. MMWR Morb Mortal Wkly Rep 64:591–596
Haq R, Fisher DE (2011) Biology and clinical relevance of the micropthalmia family of transcription factors in human cancer. J Clin Oncol 29:3474–3482. https://doi.org/10.1200/JCO.2010.32.6223
Haq R, Shoag J, Andreu-Perez P et al (2013) Oncogenic BRAF regulates oxidative metabolism via PGC1α and MITF. Cancer Cell 23:302–315. https://doi.org/10.1016/j.ccr.2013.02.003
Hauschild A, Grob JJ, Demidov LV et al (2012) Dabrafenib in BRAF-mutated metastatic melanoma: a multicentre, open-label, phase 3 randomised controlled trial. Lancet 380:358–365. https://doi.org/10.1016/S0140-6736(12)60868-X
Heppt MV, Siepmann T, Engel J et al (2017) Prognostic significance of BRAF and NRAS mutations in melanoma: a German study from routine care. BMC Cancer 17:536. https://doi.org/10.1186/s12885-017-3529-5
Hodis E, Watson IR, Kryukov GV et al (2012) A landscape of driver mutations in melanoma. Cell 150:251–263. https://doi.org/10.1016/j.cell.2012.06.024
Hofmann UB, Westphal JR, Van Muijen GNP, Ruiter DJ (2000) Matrix metalloproteinases in human melanoma. J Invest Dermatol 115:337–344. https://doi.org/10.1046/j.1523-1747.2000.00068.x
Holness MJ, Sugden MC (2003) Regulation of pyruvate dehydrogenase complex activity by reversible phosphorylation. Biochem Soc Trans:1143–1151. https://doi.org/10.1042/bst0311143
Hsiao JJ, Fisher DE (2015) The roles of Microphthalmia transcription factor and pigmentation in melanoma. Arch Biochem Biophys 563:28–34. https://doi.org/10.1016/j.abb.2014.07.019.The
Huang R, Rofstad EK (2018) Integrins as therapeutic targets in the organ-specific metastasis of human malignant melanoma. J Exp Clin Cancer Res 37:1–14. https://doi.org/10.1186/s13046-018-0763-x
Itakura E, Huang R, Wen D et al (2011) IL-10 expression by primary tumor cells correlates with melanoma progression from radial to vertical growth phase and development of metastatic competence. Mod Pathol 24:801–809. https://doi.org/10.1038/modpathol.2011.5.IL-10
Johnson LA, Morgan RA, Dudley ME et al (2009) Gene therapy with human and mouse T-cell receptors mediates cancer regression and targets normal tissues expressing cognate antigen. Blood 114:535–546. https://doi.org/10.1182/blood-2009-03-211714
Juneja VR, McGuire KA, Manguso RT et al (2017) PD-L1 on tumor cells is sufficient for immune evasion in immunogenic tumors and inhibits CD8 T cell cytotoxicity. J Exp Med 214:895–904. https://doi.org/10.1084/jem.20160801
Kakizaki A, Fujimura T, Furudate S et al (2015) Immunomodulatory effect of peritumorally administered interferon-beta on melanoma through tumor-associated macrophages. Oncoimmunology 4:1–9. https://doi.org/10.1080/2162402X.2015.1047584
Kalal BS, Upadhya D, Pai VR (2017) Chemotherapy resistance mechanisms in advanced skin cancer. Oncol Rev 11:19–25. https://doi.org/10.4081/oncol.2017.326

Karbowniczek M, Spittle CS, Morrison T et al (2008) mTOR is activated in the majority of malignant melanomas. J Invest Dermatol 128:980–987. https://doi.org/10.1038/sj.jid.5701074

Kim JW, Tchernyshyov I, Semenza GL, Dang CV (2006) HIF-1-mediated expression of pyruvate dehydrogenase kinase: a metabolic switch required for cellular adaptation to hypoxia. Cell Metab 3:177–185. https://doi.org/10.1016/j.cmet.2006.02.002

Kiuru M, Busam KJ (2017) The NF1 gene in tumor syndromes and melanoma. Pathobiol Focus 97:146–157. https://doi.org/10.1038/labinvest.2016.142

Koch A, Lang SA, Wild PJ et al (2015) Glucose transporter isoform 1 expression enhances metastasis of malignant melanoma cells. Oncotarget 6:32748–32760. https://doi.org/10.18632/oncotarget.4977

Kuk D, Shoushtari AN, Barker CA et al (2016) Prognosis of mucosal, uveal, acral, nonacral cutaneous, and unknown primary melanoma from the time of first metastasis. Oncologist 21:848–854. https://doi.org/10.1634/theoncologist.2015-0522

Kumar SM, Yu H, Edwards R et al (2007) Mutant V600E BRAF increases hypoxia inducible factor-1 a expression in melanoma. Cancer Res 67:3177–3185. https://doi.org/10.1158/0008-5472.CAN-06-3312

Kuphal S, Winklmeier A, Warnecke C, Bosserhoff AK (2010) Constitutive HIF-1 activity in malignant melanoma. Eur J Cancer 46:1159–1169. https://doi.org/10.1016/j.ejca.2010.01.031

Kvist A, Lao H, Cirenajwis H et al (2017) NF1 -mutated melanoma tumors harbor distinct clinical and biological characteristics. Mol Oncol 11:438–451. https://doi.org/10.1002/1878-0261.12050

Kwong LN, Davies MA (2013) Navigating the therapeutic complexity of PI3K pathway inhibition in melanoma. Clin Cancer Res 19:5310–5319. https://doi.org/10.1158/1078-0432.CCR-13-0142

Lai K, Killingsworth MC, Lee CS (2015) Gene of the month: PIK3CA. J Clin Pathol 68:253–257. https://doi.org/10.1136/jclinpath-2015-202885

Land SC, Tee AR (2007) Hypoxia-inducible factor 1α is regulated by the mammalian target of rapamycin (mTOR) via an mTOR signaling motif. J Biol Chem 282:20534–20543. https://doi.org/10.1074/jbc.M611782200

Lawrence MS, Stojanov P, Polak P et al (2013) Mutational heterogeneity in cancer and the search for new cancer-associated genes. Nature 499:214–218. https://doi.org/10.1038/nature12213

Lee J, Kefford R, Carlino M (2016) PD-1 and PD-L1 inhibitors in melanoma treatment: past success, present application and future challenges. Immunotherapy 8:733–746. https://doi.org/10.2217/imt-2016-0022

Leonardi GC, Falzone L, Salemi R et al (2018) Cutaneous melanoma: from pathogenesis to therapy. Int J Oncol 52:1071–1080. https://doi.org/10.3892/ijo.2018.4287

Li J, Okino ST, Whitlock JP et al (2002) Regulation of GLUT1 gene transcription by the serine/threonine kinase Akt1. J Biol Chem 274:20281–20286. https://doi.org/10.1074/jbc.274.29.20281

Li XX, Wang ZJ, Zheng Y et al (2018) Nuclear receptor Nur77 facilitates melanoma cell survival under metabolic stress by protecting fatty acid oxidation. Mol Cell 69:480–492.e7. https://doi.org/10.1016/j.molcel.2018.01.001

Lin H, Wei S, Hurt EM et al (2018) Host expression of PD-L1 determines efficacy of PD-L1 pathway blockade–mediated tumor regression. J Clin Invest 128:805–815. https://doi.org/10.1172/jci96113

Liu T, Jin L, Chen M et al (2019) Ku80 promotes melanoma growth and regulates antitumor effect of melatonin by targeting HIF1-α dependent PDK-1 signaling pathway. Redox Biol:101197. https://doi.org/10.1016/j.redox.2019.101197

Lo JA, Fisher DE (2014) The melanoma revolution: from UV carcinogenesis to a new era in therapeutics. Science 346:945–949. https://doi.org/10.1126/science.1253735.The

Mantovani A, Sica A (2010) Macrophages, innate immunity and cancer: balance, tolerance, and diversity. Curr Opin Immunol 22:231–237. https://doi.org/10.1016/j.coi.2010.01.009

Martin GA, Viskoohil D, Bollag G et al (1990) The GAP-related domain of the neurofibromatosis type 1 gene product interacts with ras p21. Cell 63:843–849. https://doi.org/10.1016/0092-8674(90)90150-D

Mcarthur GA, Ribas A (2012) Targeting oncogenic drivers and the immune system in melanoma. J Clin Oncol 31:499–506. https://doi.org/10.1200/JCO.2012.45.5568

McCubrey JA, Steelman LS, Chappell WH et al (2015) Ras/Raf/MEK/ERK and PI3K/PTEN/Akt/mTOR cascade inhibitors: how mutations can result in therapy resistance and how to overcome resistance. Oncotarget 3:1068–1111. https://doi.org/10.18632/oncotarget.659

McQuade JL, Vashisht Gopal Y (2015) Counteracting oxidative phosphorylation-mediated resistance of melanomas to MAPK pathway inhibition. Mol Cell Oncol 2:2–4. https://doi.org/10.4161/23723556.2014.991610

Mendoza MC, Er EE, Blenis J (2011) The Ras-ERK and PI3K-mTOR pathways: cross-talk and compensation. Trends Biochem Sci 36:320–328. https://doi.org/10.1016/j.tibs.2011.03.006

Mezrich JD, Fechner JH, Zhang X et al (2010) Hydrocarbon receptor can generate regulatory T. J Immunol 185:3190–3198. https://doi.org/10.4049/jimmunol.0903670.AN

Morgan RA, Dudley ME, Wunderlich JR et al (2006) Regression in patients after transfer of gentically engineered lymphocytes. Science 314:126–129. https://doi.org/10.1016/j.jsbmb.2011.07.002.Identification

Munn DH, Mellor AL (2007) Indoleamine 2,3-dioxygenase and tumor-induced tolerance. J Clin Invest 117:1147–1154. https://doi.org/10.1172/JCI31178

Munn DH, Mellor AL, Regents G, Place F (2016) IDO in the tumor microenvironment: inflammation, counter-regulation and tolerance. Trends Immunol 37:193–207. https://doi.org/10.1016/j.it.2016.01.002

Muñoz-Couselo E, Adelantado EZ, Ortiz C et al (2017) NRAS-mutant melanoma: current challenges and future prospect. Onco Targets Ther 10:3941–3947. https://doi.org/10.2147/OTT.S117121

Nasi A, Fekete T, Krishnamurthy A et al (2013) Dendritic cell reprogramming by endogenously produced lactic acid. J Immunol 191:3090–3099. https://doi.org/10.4049/jimmunol.1300772

Nasti TH, Cochran JB, Tsuruta Y et al (2016) A murine model for the development of melanocytic nevi and their progression to melanoma. Mol Carcinog 55:646–658. https://doi.org/10.1002/mc.22310

Naves LB, Almeida L, Ramakrishna S (2017) Understanding the microenvironment of melanoma cells for the development of target drug delivery systems. Eur Med J 5:85–92

Nemoz C, Ropars V, Frit P et al (2018) XLF and APLF bind Ku80 at two remote sites to ensure DNA repair by non homologous end joining. Nat Struct Mol Biol 25:971–980. https://doi.org/10.1038/s41594-018-0133-6

Nissan MH, Pratilas CA, Jones AM et al (2014) Loss of NF1 in Cutaneous Melanoma is associated with RAS activation and MEK dependence. Cancer Res 74:2340–2351. https://doi.org/10.1158/0008-5472.CAN-13-2625

Nussenzweig MC (2010) Origin and development of dendritic cells. Immunol Rev 234:45–54. https://doi.org/10.1111/j.0105-2896.2009.00879.x

Ohta A (2016) A metabolic immune checkpoint: adenosine in tumor microenvironment. Front Immunol 7:1–11. https://doi.org/10.3389/fimmu.2016.00109

Ohta A, Sitkovsky M (2014) Extracellular adenosine-mediated modulation of regulatory T cells. Front Immunol 5:1–9. https://doi.org/10.3389/fimmu.2014.00304

Ott PA, Henry T, Baranda SJ et al (2013) Inhibition of both BRAF and MEK in BRAFV600Emutant melanoma restores compromised dendritic cell (DC) function while having differential direct effects on DC properties. Cancer Immunol Immunother 62:811–822. https://doi.org/10.1007/s00262-012-1389-z

Palmer SR, Erickson LA, Ichetovkin I et al (2011) Circulating serologic and molecular biomarkers in malignant melanoma. Mayo Clin Proc 86:981–990. https://doi.org/10.4065/mcp.2011.0287

Papandreou I, Cairns RA, Fontana L et al (2006) HIF-1 mediates adaptation to hypoxia by actively down-regulating mitochondrial oxygen consumption. Cell Metab 3:187–197. https://doi.org/10.1016/j.cmet.2006.01.012

Pardoll DM (2012) The blockade of immune checkpoints in cancer immunotherapy. Nat Rev Cancer 12:252–264. https://doi.org/10.1038/nrc3239

Parmenter TJ, Kleinschmidt M, Kinross KM et al (2014) Response of BRAF mutant melanoma to BRAF inhibition is mediated by a network of transcriptional regulators of glycolysis. Cancer Discov 43:1983–1987. https://doi.org/10.1158/2159-8290.CD-13-0440

Passarelli A, Mannavola F, Stucci LS et al (2017) Immune system and melanoma biology: a balance between immunosurveillance and immune escape. Oncotarget 8:106132–106142. https://doi.org/10.18632/oncotarget.22190

Pathria P, Louis TL, Varner JA (2019) Targeting tumor-associated macrophages in Cancer. Trends Immunol 40:310–327. https://doi.org/10.1016/j.it.2019.02.003

Patsoukis N, Bardhan K, Chatterjee P et al (2015) PD-1 alters T-cell metabolic reprogramming by inhibiting glycolysis and promoting lipolysis and fatty acid oxidation. Nat Commun 6:6692. https://doi.org/10.1038/ncomms7692

Payen VL, Porporato PE, Baselet B, Sonveaux P (2016) Metabolic changes associated with tumor metastasis, part 1: tumor pH, glycolysis and the pentose phosphate pathway. Cell Mol Life Sci 73:1333–1348. https://doi.org/10.1007/s00018-015-2098-5

Picco ME, Castro MV, Quezada MJ et al (2019) STAT3 enhances the constitutive activity of AGC kinases in melanoma by transactivating PDK1. Cell Biosci 9:1–14. https://doi.org/10.1186/s13578-018-0265-8

Pickarski M, Gleason A, Bednar B, Duong LT (2015) Orally active αvβ3 integrin inhibitor MK-0429 reduces melanoma metastasis. Oncol Rep 33:2737–2745. https://doi.org/10.3892/or.2015.3910

Pinheiro C, Miranda-Gonçalves V, Longatto-Filho A et al (2016) The metabolic microenvironment of melanomas : prognostic value of MCT1 and MCT4. Cell Cycle 15:1462–1470. https://doi.org/10.1080/15384101.2016.1175258

Pollock PMÃ, Walker GJ, Glendening JM (2002) PTEN inactivation is rare in melanoma tumours but occurs frequently in melanoma cell lines. Melanoma Res 12:565–575

Pollock PM, Harper UL, Hansen KS et al (2003) High frequency of BRAF mutations in nevi. Nat Genet 33:19–20. https://doi.org/10.1038/ng1054

Porta C, Paglino C, Mosca A (2014) Targeting PI3K/Akt/mTOR signaling in Cancer. Front Oncol 4:1–11. https://doi.org/10.3389/fonc.2014.00064

Potrony M, Badenas C, Aguilera P et al (2015) Update in genetic susceptibility in melanoma. Ann Transl Med 3:210. https://doi.org/10.3978/j.issn.2305-5839.2015.08.11

Ratner N, Miller SJ (2015) A RASopathy gene commonly mutated in cancer: the neurofibromatosis type 1 tumour suppressor. Nat Rev Cancer 15:290–301. https://doi.org/10.1038/nrc3911

Richmond A, Yang J, Su Y (2009) The good and the bad of chemokines/chemokine receptors in melanoma. Pigment Cell Melanoma Res 22:175–186. https://doi.org/10.1111/j.1755-148X.2009.00554.x

Robbins PF, Kassim SH, Tran TLN et al (2015) A pilot trial using lymphocytes genetically engineered with an NY-ESO-1-reactive T-cell receptor: long-term follow-up and correlates with response. Clin Cancer Res 21:1019–1027. https://doi.org/10.1158/1078-0432.CCR-14-2708

Rofstad EK, Mathiesen B, Kindem K, Galappathi K (2006) Acidic extracellular pH promotes experimental metastasis of human melanoma cells in athymic nude mice. Cancer Res 66:6699–6707. https://doi.org/10.1158/0008-5472.CAN-06-0983

Romero-Garcia S, Moreno-Altamirano MMB, Prado-Garcia H, Sánchez-García FJ (2016) Lactate contribution to the tumor microenvironment: mechanisms, effects on immune cells and therapeutic relevance. Front Immunol 7:52. https://doi.org/10.3389/fimmu.2016.00052

Scortegagna M, Lau E, Zhang T et al (2015) PDK1 and SGK3 contribute to the growth of BRAF-mutant melanomas and are potential therapeutic targets. Cancer Res 75:1399–1412. https://doi.org/10.1158/0008-5472.CAN-14-2785

Scott KEN, Cleveland JL (2016) Lactate wreaks havoc on tumor-infiltrating T and NK cells. Cell Metab 24:649–650. https://doi.org/10.1016/j.cmet.2016.10.015

Scott DA, Richardson AD, Filipp FV et al (2011) Comparative metabolic flux profiling of melanoma cell lines. J Biol Chem 286:42626–42634. https://doi.org/10.1074/jbc.m111.282046

Seidel JA, Otsuka A, Kabashima K (2018) Anti-PD-1 and anti-CTLA-4 therapies in Cancer: mechanisms of action, efficacy, and limitations. Front Oncol 8:1–14. https://doi.org/10.3389/fonc.2018.00086

Sensi M, Nicolini G, Petti C et al (2006) Mutually exclusive NRASQ61R and BRAFV600E mutations at the single-cell level in the same human melanoma. Oncogene 25:3357–3364. https://doi.org/10.1038/sj.onc.1209379

Shah DJ, Dronca RS (2015) Latest advances in chemotherapeutic, targeted and immune approaches in the treatment of metastatic melanoma. Mayo Clin Proc 89:504–519. https://doi.org/10.1016/j.mayocp.2014.02.002.Latest

Siegel RL, Miller KD, Jemal A (2019) Cancer statistics, 2019. CA Cancer J Clin 69:7–34. https://doi.org/10.3322/caac.21551

Sitkovsky MV, Lukashev D, Apasov S et al (2004) Physiological control of immune response and inflammatory tissue damage by hypoxia -inducible factors and adenosine A2A receptors. Annu Rev Immunol 22:657–682. https://doi.org/10.1146/annurev.immunol.22.012703.104731

Slominski A, Kim T-K, Brożyna AA et al (2014) The role of melanogenesis in regulation of melanoma behavior: melanogenesis leads to stimulation of HIF-1α expression and HIF-dependent attendant pathways. Arch Biochem Biophys 563:79–93. https://doi.org/10.1016/j.abb.2014.06.030

Smith JW, Ratnikov BI, Ronai ZA et al (2016) Metabolic rewiring in melanoma. Oncogene 36:147–157. https://doi.org/10.1038/onc.2016.198

Sonveaux P, Copetti T, de Saedeleer CJ et al (2012) Targeting the lactate transporter MCT1 in endothelial cells inhibits lactate-induced HIF-1 activation and tumor angiogenesis. PLoS One 7:e33418. https://doi.org/10.1371/journal.pone.0033418

Sosman JA, Kim KB, Schuchter L et al (2012) Survival in BRAF V600–mutant advanced melanoma treated with Vemurafenib. N Engl J Med 366:707–714. https://doi.org/10.1056/NEJMoa1112302

Stewart AA, Dudley ME, Nath A et al (2013) Cancer regression and neurological toxicity following anti-MAGE-A3 TCR gene therapy. J Immunother 36:133–151. https://doi.org/10.1097/cji.0b013e3182829903

Stine ZE, Walton ZE, Altman BJ et al (2016) MYC, metabolism, and Cancer. Cancer Discov 5:1024–1039. https://doi.org/10.1158/2159-8290.CD-15-0507.MYC

Stock C, Gassner B, Hauck CR et al (2005) Migration of human melanoma cells depends on extracellular pH and Na+/H+ exchange. J Physiol 1:225–238. https://doi.org/10.1113/jphysiol.2005.088344

Tau G, Rothman P (1999) Biologic functions of the IFN-γ receptors. Allergy 54:1233–1251. https://doi.org/10.1034/j.1398-9995.1999.00099.x

The Cancer Genome Atlas Network (2015) Genomic classification of cutaneous melanoma. Cell 161:1681–1696. https://doi.org/10.1007/s11065-015-9294-9.Functional

Tobin DJ (2017) Introduction to skin aging. J Tissue Viability 26:37–46. https://doi.org/10.1016/j.jtv.2016.03.002

Tucci M, Stucci S, Passarelli A et al (2014) The immune escape in melanoma: role of the impaired dendritic cell function. Expert Rev Clin Immunol 10:1395–1404. https://doi.org/10.1586/1744666X.2014.955851

Umansky V, Shevchenko I, Bazhin AV, Utikal J (2014) Extracellular adenosine metabolism in immune cells in melanoma. Cancer Immunol Immunother 63:1073–1080. https://doi.org/10.1007/s00262-014-1553-8

Van Allen EM, Wagle N, Sucker A et al (2014) The genetic landscape of clinical resistance to RAF inhibition in metastatic melanoma. Cancer Discov 4:94–109. https://doi.org/10.1158/2159-8290.CD-13-0617

Vara JÁF, Casado E, de Castro J et al (2004) PI3K/Akt signalling pathway and cancer. Cancer Treat Rev 30:193–204. https://doi.org/10.1016/j.ctrv.2003.07.007

Vazquez F, Lim JH, Chim H et al (2013) PGC1α expression defines a subset of human melanoma tumors with increased mitochondrial capacity and resistance to oxidative stress. Cancer Cell 23:287–301. https://doi.org/10.1016/j.ccr.2012.11.020

Vredeveld LCW, Possik PA, Smit MA et al (2012) Abrogation of BRAFV600E-induced senescence by PI3K pathway activation contributes to melanomagenesis. Genes Dev 26:1055–1069. https://doi.org/10.1101/gad.187252.112

Walunas TL, Christina YB, Bluestone JA (1996) CTLA-4 ligation blocks CD28-dependent T cell activation. J Exp Med 183:2541–2550. https://doi.org/10.1084/jem.193.6.2541

Wang L, Leite de Oliveira R, Huijberts S et al (2018) An acquired vulnerability of drug-resistant melanoma with therapeutic potential. Cell 173:1413–1425. https://doi.org/10.1016/j.cell.2018.04.012

Weinstein D, Leininger J, Hamby C, Safai B (2014) Diagnostic and prognostic biomarkers in melanoma. J Clin Aesthet Dermatol 7:13–24

Weir HK, Marrett LD, Cokkinides V et al (2011) Melanoma in adolescents and young adults (ages 15–39 years): United States, 1999–2006. J Am Acad Dermatol 65:S38.e1–S38.e13. https://doi.org/10.1016/j.jaad.2011.04.038

Weller RB, Castellsague X (2017) Skin Cancer: epidemiology, disease burden, pathophysiology, diagnosis, and therapeutic approaches. Dermatol Ther 7:5–19. https://doi.org/10.1007/s13555-016-0165-y

Whiteman DC, Green AC, Olsen CM (2016) The growing burden of invasive melanoma : projections of incidence rates and numbers of new cases in six susceptible populations through 2031. J Invest Dermatol 136:1161–1171. https://doi.org/10.1016/j.jid.2016.01.035

Wolchok JD, Neyns B, Linette G et al (2010) Ipilimumab monotherapy in patients with pretreated advanced melanoma: a randomised, double-blind, multicentre, phase 2, dose-ranging study. Lancet Oncol 11:155–164. https://doi.org/10.1016/S1470-2045(09)70334-1

Yan S, Coffing BN, Li Z et al (2016) Diagnostic and prognostic value of ProEx C and GLUT1 in melanocytic lesions. Anticancer Res 36:2871–2880

Zaal EA, Berkers CR (2018) The influence of metabolism on drug response in Cancer. Front Oncol 8:1–15. https://doi.org/10.3389/fonc.2018.00500

Zeller KI, Jegga AG, Aronow BJ et al (2003) An integrated database of genes responsive to the Myc oncogenic transcription factor: identification of direct genomic targets. Genome Biol 4:R69. https://doi.org/10.1186/gb-2003-4-10-r69

Zhang W, Liu HT (2006) MAPK signal pathways in the regulation of cell proliferation in mammalian cells. Cell Res 12:9–18. https://doi.org/10.1038/sj.cr.7290105

Zhang M, Di Martino JS, Bowman RL et al (2018) Adipocyte-derived lipids mediate melanoma progression via FATP proteins. Cancer Discov 8:1006–1025. https://doi.org/10.1158/2159-8290.CD-17-1371

Zhou X, Gimm O, Hampel H et al (2000) Epigenetic PTEN silencing in malignant melanomas without PTEN mutation. Am J Pathol 157:1123–1128. https://doi.org/10.1016/S0002-9440(10)64627-5

Metabolic Reprogramming and Signaling to Chromatin Modifications in Tumorigenesis

12

Zyanya Díaz-Hirashi, Tian Gao, and Francisco Verdeguer

Abstract

Cellular proliferation relies on a high energetic status, replenished through nutrient intake, that leads to the production of biosynthetic material. A communication between the energetic levels and the control of gene expression is essential to engage in cell division. Multiple nutrient and metabolic sensing mechanisms in cells control transcriptional responses through cell signaling cascades that activate specific transcription factors associated with a concomitant regulation of the chromatin state. In addition to this canonical axis, gene expression could be directly influenced by the fluctuation of specific key intermediary metabolites of central metabolic pathways which are also donors or cofactors of histone and DNA modifications. This alternative axis represents a more direct connection between nutrients and the epigenome function. Cancer cells are highly energetically demanding to sustain proliferation. To reach their energetic demands, cancer cells rewire metabolic pathways. Recent discoveries show that perturbations of metabolic pathways in cancer cells have a direct impact on the epigenome. In this chapter, the interaction between metabolic driven changes of transcriptional programs in the context of tumorigenesis will be discussed.

Z. Díaz-Hirashi · T. Gao · F. Verdeguer (✉)
Department of Molecular Mechanisms of Disease, University of Zürich, Zürich, Switzerland
e-mail: francisco.verdeguer@dmmd.uzh.ch

Keywords

Epigenetics · Intermediary metabolism · Chromatin.

12.1 Influence of Metabolites on Chromatin Modifications and Transcriptional States in Cellular Physiology

The establishment of developmental states or responses to stimuli largely depends on a dynamic regulation of gene transcriptional outputs. The genetic information is stored in the DNA sequence but highly organized through structural proteins named histones to form together with DNA, the chromatin, which has different ordered levels of compaction. Chromatin compaction influences transcriptional output, whereby a loose state is transcriptionally permissive whereas a highly compacted chromatin is associated with transcription repression. The transcriptional status is highly modulated by chromatin modifications which allow changes in chromatin conformation and also signals to the recruitment of specific cofactors. Chromatin modifications are covalent post-translational modifications (PTMs) of histones including the most common, acetylation, methylation, or phosphorylation but also several less abundant PTMs such as ubiquitination, sumoylation, ADP-ribosylation etc. The

J. Serpa (ed.), *Tumor Microenvironment*, Advances in Experimental Medicine and Biology 1219,
https://doi.org/10.1007/978-3-030-34025-4_12

formation of histone modifications depends mainly on the activity of enzymatic writers and erasers and the concentration of the cofactors and donors for each specific catalytic reaction. However, in certain biological conditions or upon some stimuli the cellular concentration of some donors/cofactors could be limiting. This implies a critical regulatory function of these donors/cofactors in the formation of chromatin modifications and thus transcriptional states. In addition, several limiting key cofactors/donors of histone modifications play multiple functions in diverse metabolic pathways. A growing body of knowledge points at histone modifications and chromatin function as direct sensors of the cellular metabolism and effectors of key cellular responses to the environment. This concept has obvious implications in understanding multiple pathological conditions including tumorigenesis and particularly the role of microenvironment in cancer formation. The link between metabolism and chromatin modifications has several evolutionary advantages as sensing the energetic status is key to make crucial cellular decisions such as the timing for the commitment to grow and proliferate. A hypothesis is that these mechanisms could have evolutionary predated more complex endocrine/cell signaling transductions mechanisms in the control of gene expression in mammalian organisms (Fig. 12.1). Identifying the molecular players and understanding the relevance of the metabolic signaling into chromatin modifications may unlock novel therapeutic opportunities caused by dysregulated metabolic and epigenomic processes in cancer.

The canonical cascade of gene regulation is initiated by the binding of growth factors to a receptor tyrosine kinase. This triggers the activation of numerous downstream pathways resulting in the activation of transcription factors which in turn causes an increase in transcription and translation of metabolic enzymes. This leads to an enhanced nutrient uptake and metabolic pathways regulation. However, recent advances suggest that metabolic pathways could directly signal into chromatin modifications and impact on specific gene programs. This metabolic dependent signaling is controlled by nutrient availability which affects nutrient uptake and flux to metabolic pathways. This in turn modifies the concentration of intermediary metabolites (e.g., α-KG, SAM) which impact on chromatin modifications leading to transcription regulation. TF: Transcription factor; TCA: tricarboxylic acid cycle; CME: Chromatin modifier enzymes.

12.1.1 Regulation of Histone Acetylation

12.1.1.1 Acetyl-CoA Metabolism and Acetylation

A key confluent hub of cellular metabolism is the metabolite acetyl-CoA, a two-carbon carrier derived from different sources including carbohydrate, lipid and protein metabolism. For example, acetyl-CoA is produced following glycolysis to feed the tricarboxylic acid cycle (TCA) and from the breakdown of fatty acids, a process called β-oxidation. Conversely, acetyl-CoA is also utilized for the synthesis of fatty acids. Acetyl-CoA is a key metabolic hub for the confluence of multiple metabolic pathways, such as amino acid metabolism or ketoacid formation among others. Acetyl-CoA is the only cellular donor for histone acetylation, and it can essentially link metabolic status with chromatin modifications (Fig. 12.2). Previous studies have shown that acetyl-CoA levels correlate with elevated gene expression of genes involved in cell growth (Cai et al. 2011; Shi and Tu 2015). In fact, when levels of acetyl-CoA are limiting, its concentration determines the activity of histone acetyltransferases. In addition, the cellular pools of acetyl-CoA are regulated through compartmentalization. The mitochondrion is an organelle with high metabolic traffic of acetyl-CoA that is mainly produced by the pyruvate dehydrogenase complex or by fatty acid β-oxidation. Acetyl-CoA is impermeable to the inner mitochondrial membrane and is indirectly transported to the cytosol through a shuttle system. The first step of the TCA cycle is the condensation of oxaloacetate with acetyl-CoA to form citrate taking place in mitochondria. Citrate is exported into the cytosol by the mitochondrial transporter SLC25A1,

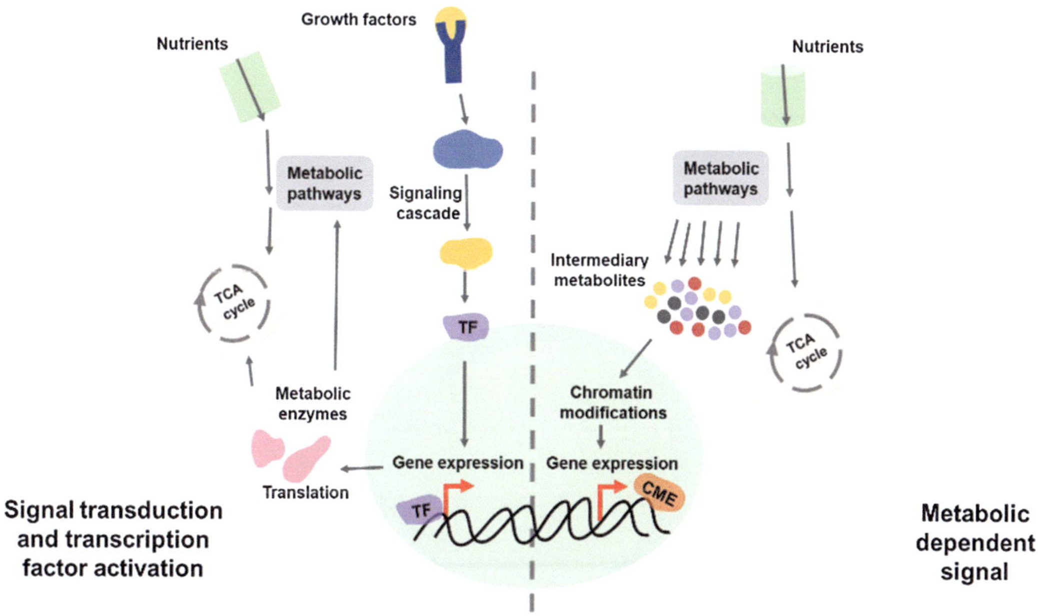

Fig. 12.1 Metabolic dependent transcriptional regulation versus canonical signal to transcription factor control

where a key step catalyzed by ATP citrate lyase (ACLY) leads to the formation of acetyl-CoA and oxaloacetate. Oxaloacetate is recycled back through its conversion into malate which enters the mitochondria. This pathway is required to increase the acetyl-CoA pool in extra-mitochondrial compartments, which is also necessary for fatty acid synthesis as it takes place in the cytoplasm. Acetyl-CoA freely diffuses to the nucleus where it serves as donor for histone acetyltransferases. The expression of ACLY is therefore crucial to the regulation of histone acetylation (Wellen et al. 2009). In agreement, it has been shown that an increase in glucose levels leads to an increase in histone acetylation through ACLY (Zhao et al. 2016). The role of ACLY in cancer will be discussed in following sections.

There are different histone modifications that are linked with energy status and availability of metabolites. Acetylation: acetyl-CoA is the sole donor for histone acetylation. It is mainly produced by ACLY from citrate which is made in mitochondria via TCA cycle and transported out into the cytoplasm. It can also be derived from acetate re-uptake mediated by ACSS2. Once acetyl-CoA is in the nucleus, it is then used as substrate for histone acetylation via HATs. Histone deacetylation is catalyzed by SIRTs and HDACs. Acylation: short chain fatty acids and β-hydroxybutyrate supply the acyl group for histone acylation at lysine residues. Similar to histone acetylation, histone acylation is catalyzed by HATs and deacylation is mediated by SIRTs and HDACs. Acetylation and acylation compete with each other. Methylation: methionine absorbed into the cell is firstly converted to SAM by MATs. SAM then provides the methyl group for histone methylation by HMTs and DNMTs with SAH being produced. αKG which is the TCA cycle intermediate, is used as the substrate by JHDM and TETs for histone demethylation with succinate and CO_2 being produced. ADP-ribosylation: ARTDs catalyze the transfer of ADP-ribose to histone with NAM being produced and they can also promote poly-ADP ribosylation. The removal of ADP ribose monomers from the poly-ADP ribose tail is catalyzed by ARH3 and PARG. The enzymes responsible for the removal of the first ADP ribose group attached to the histone are still unknown, ARH3, PARG and lyase are the main candidates. O-GlcNAcylation: glutamine and glycolytic

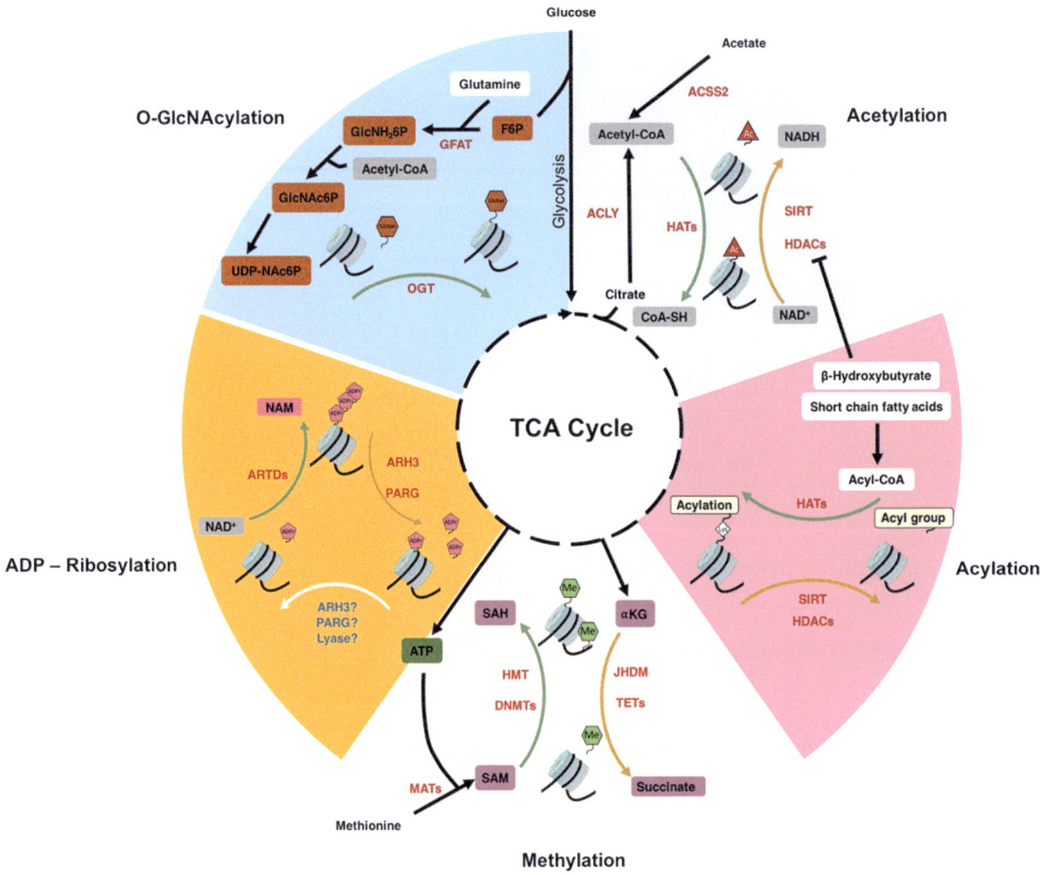

Fig. 12.2 Metabolite induced histone modifications

derived F6P are converted into $GlcNH_2 6P$ by GFAT. $GlcNH_2 6P$ is acetylated to GlcNAc6P with acetyl-CoA supplying the acetyl group. It then receives an UDP transfer to form UDP-GlcNac which is later used as the substrate by OGT for histone O-GlcNAcylation. ACLY: ATP-dependent Citrate Lyase; TCA: tricarboxylic acid cycle; ACSS2: Acyl-CoA Synthetase Short Chain Family Member 2; HATs: Histone acetyl-transferases; SIRTs: Sirtuins; HDACs: Histone deacetylases; SAM: S-adenosylmethionine; MATs: methionine adenosyltransferase; SAH: S-adenosylhomocysteine; HMTs: Histone methyltransferases; DNMTs: DNA methyltransferases; αKG: α-ketoglutarate; JHDM: Jumonji C-domain containing histone demethylase; TETs: ten eleven translocation; ARTDs: ADP-ribosyl-transferases; NAM: Nicotinamide; ARH3: ADP-ribosylhydrolase 3; PARG: Poly(ADP-ribose) glycohydrolase; F6P: fructose-6-phosphate; GFAT: glutamine:fructose-6-phosphate amidotransferase; $GlcNH_2 6P$: glucosamine-6-phosphate; OGT: O-GlcNac Transferase.

12.1.1.2 NAD^+ and Deacetylation

The level of histone acetylation is a dynamic process controlled by histone acetyl-transferases and by histone deacetylases. Histone deacetylases are divided in two large families: the zinc-dependent deacetylases or HDACs and the nicotinamide dinucleotide (NAD^+) dependent deacetylases known as sirtuins. HDAC and sirtuins are involved in deacetylation of also non-histone protein substrates and play a wide variety of cellular functions including metabolic regulation. The role of sirtuins, a family composed of 7 members- SIRT1-SIRT7, is however particularly linked to metabolism. The yeast sirtuin Sir2 was identified as a key regulator of lifespan extension during caloric restriction conditions (Kaeberlein et al. 2004). Since then, sirtuins have been shown to play direct signaling functions in metabolism by deacetylating essential metabolic transcription factors such as the peroxisome proliferator-activated receptor γ co-activator 1α (PGC-1α) (Rodgers et al. 2005; Lagouge et al. 2006), fork-head box protein O1 (FOXO1) (Motta et al. 2004; Cantó et al. 2010), liver X receptors α and β (LXR-α/β) (Li et al. 2007; Rodgers and Puigserver 2007), CREB regulated transcription co-activator (Liu et al. 2008) peroxisome proliferator-activated receptor γ (PPARγ) (Qiang et al. 2012), and acetyl-CoA synthetase (AceCS) (Hallows et al. 2006). Studies in different mouse models showed that increased activity of SIRT1 mediated by pharmacological intervention (resveratrol) or genetic overexpression promotes oxidative metabolism, protection from oxidative stress, life extension and reduced cancer formation (Bordone et al. 2007; Pfluger et al. 2008; Herranz et al. 2010)·. Interestingly, The first evidence of cancer and sirtuins function derived from the observation that the tumour-supressor gene p53 is deacetylated and repressed by SIRT1 in human cells (Luo et al. 2001; Vaziri et al. 2001). Similarly to histone acetylation and acetyl-CoA, the link of sirtuins and metabolic pathways relies on the dependence of NAD^+ for their catalytic activity. Sirtuins consume NAD^+ during deacetylation whereby NAD^+ accepts the acetyl group giving rise to nicotinamide (NAM) and 2-O-acetyl-ADP-ribose (Fig. 12.2). NAD^+ plays a fundamental role in multiple essential energetic pathways. NAD^+ is a coenzyme of dehydrogenases which transfers hydrogen atoms between different substrates through oxido-reductase enzymatic reactions. The oxidized form NAD^+ is reduced to $NADH+H^+$ upon acceptance of hydrogen atoms. The maintenance of the $NAD^+/NADH$ ratio and a sufficient absolute amount of NAD^+ determines proper mitochondrial function and multiple cellular functions. In this way, through the NAD^+ cellular concentration, the activity of sirtuins and histone deacetylation is directly linked to the cellular metabolic status.

Another family of enzymes that utilize NAD^+ is the Poly-(ADP-ribose)-polymerases (PARPs). PARPs lead to a post-translational modification of histone and non-histone proteins through addition of ADP-ribose derived from NAD^+ and will be discussed later (Messner and Hottiger 2011).

12.1.2 Regulation of Histone and DNA Methylation

12.1.2.1 S-Adenosyl-Methionine (SAM) and Regulation of Methylation

Histone and DNA methylation are key covalent modifications that modulate gene transcription. DNA methylation of target genes takes place in the promoter, particularly at regulatory sequences highly enriched in CG dinucleotides, called CpG islands, which correlate with transcriptional repression. Histone lysine methylation, on the other hand, can lead to activation or repression of transcription depending on the modified residue (Kouzarides 2007). The methylation of substrates catalyzed by methyltransferases depend on the availability of the essential amino acid methionine. More specifically, the catalytic activity of methyltransferase uses the intermediate S-adenosyl-methionine (SAM) as a donor of the methyl group (Fig. 12.2). SAM is formed in the cytosol by methionine methyltransferase which uses ATP to add an adenosyl group to methionine. Methyltransferases then use SAM which is converted into S-adenosylhomocysteine (SAH). SAH is next converted into homocysteine after the hydrolysis of the adenosyl group. Homocysteine can be recycled back to form methionine by the methionine synthase. This is a small cycle which depends on the uptake of dietary methionine. The SAM cycle is also interconnected to the folate cycle through tetrahydrofolate (a derivative of the vitamin folic acid) which is an essential intermediate for the synthesis of purines and therefore, DNA replication. Several anticancer drugs target the folate cycle to prevent cell proliferation (Farber et al. 1948; Pascual et al. 2017a).

The dependence of histone and DNA methylation on SAM and its connection with fundamental metabolic pathways places the regulation of methylated histones and DNA at the crossroads between metabolic control and gene regulation.

12.1.2.2 Regulation of Demethylation: TCA Cycle Intermediates and FAD

The counterpart of methylation is the removal of methyl groups by histone or DNA demethylases. Histone lysine demethylases are grouped in two families according to their catalytic domain; a FAD-dependent amine oxidase or a Jumonji (JmjC) domain (Fig. 12.2). Although both families perform oxidative reactions that lead to lysine demethylation, they use different cofactors. FAD-dependent demethylases, such as LSD1, use the redox coenzyme flavin adenine dinucleotide (FAD) as a cofactor to oxidize the amine. This proceeds via reduction of FAD to FADH2 which is recycled to FAD via generation of hydrogen peroxide. The methyl group is released spontaneously as formaldehyde, following the formation of the unstable carbonilamine intermediate. On the other hand, Jumonji demethylases (JHMDs) use the TCA cycle intermediate, α-ketoglutarate (α-KG) and iron (Fe^{2+}) as cofactors to hydroxylate the methylated substrate. The reaction leads to the conversion of α-KG into succinate and production of CO_2 and concludes equally as in the FAD-dependent demethylase reaction, with the release of formaldehyde. Histone lysine demethylation depends on two very important metabolic intermediates FAD and α-KG. FAD is used during TCA cycle from the catalysis of succinate to fumarate which leads to the reduction of FAD into $FADH_2$. This reduced form of FAD is used by the complex II of the respiratory chain to donate 2 electrons necessary for ATP synthesis. FAD is also used in other important metabolic pathways such as fatty acid oxidation which also produces $FADH_2$. The use of α-KG by JmjC demethylases also links their activity to key important metabolic pathways. α-KG is a TCA cycle intermediate derived from the oxidative decarboxylation of isocitrate. In addition, α-KG is a link between TCA cycle and aminoacid metabolism through the reversible reaction catalyzed by glutamate dehydrogenase which deaminates glutamate to form α-KG.

In addition to histone lysine demethylation, DNA demethylation is an essential process for cellular differentiation and development (Wu and Zhang 2010). DNA demethylation is catalyzed by Ten eleven translocase (TET) demethylases which are also α-KG dependent.

12.1.3 Other Histone Modifications at the Crossroads with Metabolism: Acylations, ADP-Ribosylation, Serotonylation

The most abundant post-translational modification in eukaryotic cells is phosphorylation (Cohen 2002). Multiple signaling pathways are regulated by kinases, leading to the phosphorylation of targets including histones. Although ATP is an essential energetic mediator, kinases' activity works with saturating concentrations of ATP and are therefore not limited by ATP abundance (Su et al. 2016). Histone phosphorylation is therefore not linked to the cellular energetic state. On the other hand, as discussed above, histone acetylation and methylation (the most abundant PTMs after phosphorylation) depend on substrate or cofactor availability. Other histone modifications, although less abundant, can also have a relevant function in different contexts. ADP-ribosylation is the transfer of one ADP-ribose moiety from NAD^+ to specific amino acid residues of substrate proteins by releasing nicotinamide (Messner and Hottiger 2011). Therefore, the consumption of NAD^+ by this process links the chromatin modification ADP-ribosylation to the cellular energy status of the cell (Fig. 12.2). ADP-ribose can form longer oligomers of up to 15 chained monomers covalently linked to a histone amino acid residue. The mono or poly-ADP ribosylation, catalyzed by ADP-ribosyl-transferases (ARTs) has a determinant role in DNA repair following DNA damage (Messner and Hottiger 2011), and PARP1 inhibitors are currently used for antitumor therapy by preventing the enzyme DNA repair function (Bryant et al. 2005; Farmer et al. 2005; Fong et al. 2009).

O-linked β-D-N-acetylglucosaminylation (O-GlcNAcylation) is a reversible PTM of cytosolic, nuclear, and mitochondrial proteins that consists in the covalent linkage of a unique residue of N-acetylglucosamine (GlcNAc) to serines and threonines of target proteins including histones (Su et al. 2016). This modification of histones favors transcription and chromatin relaxation or transcription repression and chromatin compaction, depending on the modified histone residue. N-acetylglucosamine is synthesized through the hexosamine pathway which derives from glycolysis and also has links with protein, lipid and nucleotide metabolic pathways. Glycolytic derived fructose-6-phosphate (F6P) is used by glutamine:fructose-6-phosphate amido transferase (GFAT) to produce glucosamine-6-phosphate ($GlcNH_2 6P$) using glutamine. $GlcNH_2 6P$ is consecutively acetylated to GlcNAc6P using acetyl-CoA as substrate and followed by an UDP transfer leading to UDP-GlcNac. O-GlcNAcylation is then catalyzed by O-GlcNac Transferase (OGT) using UDP-GlcNac as a substrate to modify histones and other target proteins (Fig. 12.2). The removal of O-GlcNac is then hydrolyzed by O-GlcNAcase (OGA). Since O-GlcNacylation is indirectly and simultaneously connected to the metabolism of carbohydrates, proteins and lipids through the hexosamine biosynthetic pathway, this histone modification has been proposed to be a central nutrient sensor controlling chromatin dynamics (Dehennaut et al. 2014b). Elevated protein O-GlcNAcylation and changes in OGT and/or OGA expression have been associated to different cancers including breast, lung, colon, liver, bladder, endometrial and chronic lymphocytic leukemia (CCL) (Dehennaut et al. 2014a) (Fardini et al. 2013). In addition, OGT was identified in the complex of TET enzymes through direct interaction, suggesting a regulatory role in DNA demethylation (Chen et al. 2013; Mariappa et al. 2013).

Other less known histone modifications have been recently discovered with a potential effect on transcriptional regulation. A subset of these novel histone modifications are lysine acylations which include Lys propionylation (Kpr), Lys butyrylation (Kbu), Lys 2-hydroxyisobutyrylation (Khib), Lys succinylation (Ksucc), Lys malonylation (Kma), Lys glutarylation (Kglu), Lys crotonylation (Kcr) and Lys β-hydroxybutyrylation (Kbhb) (Sabari et al. 2017a). It has been proposed that lysine acylations mark active regulatory elements in the genome. These modifications compete with acetyl-CoA for histone acetylation, in which under conditions of low acetyl-CoA, lysine acylation modifications, themselves also catalyzed by classical histone acetyl-transferases, increase (Fig. 12.2). Deacylation is likewise catalyzed by HDACs (Sabari et al. 2017b). Short chain fatty acids and the ketone body β-hydroxybutyrate are sources of histone acylation. Interestingly, during ketogenic conditions, such as a low carbohydrate environment, histone acetylation switches to histone acylation, thus connecting this histone modification to the nutritional status (Sabari et al. 2016).

Histone H3 serotonylation at glutamine 5 (Q5ser) has emerged as a very recent histone modification identified in combination with H3 tri-methylated lysine 4 (H3K4me3)-marked nucleosomes present in enriched euchromatin. H3Q5ser was identified in brain and gut, tissues which produce large amounts of serotonin (also known as 5-hydroxytryptamine (5-HT)) (Farrelly et al. 2019).

12.2 Metabolic Reprogramming and Signaling to the Epigenome in Tumorigenesis

Cancer cells undergo a profound metabolic reprogramming driven by the proliferative and survival adaptation in the tumor microenvironment and this constitutes a core hallmark of cancer (Hanahan and Weinberg 2011a). Otto Warburg already observed that proliferative ascites cancer cells underwent fermentation and lactate production even in the presence of oxygen (Warburg et al. 1927). "Warburg" aerobic glycolysis is a profound metabolic adaptation which sustains the needs of proliferative cells and macromolecular synthesis. The scope of all metabolic adaptations is not yet fully understood. Here, the metabolic changes that are intricately connected to the epigenome regulation will be discussed, by mainly covering three different points aspects. First, it has been hypothesized that tumorigenic metabolic reprogramming impacts the epigenome and therefore modulate transcriptional programs. Second, certain cancer mutations have shown to lead to aberrant metabolites named oncometabolites which have a specific impact on chromatin regulators. Finally, some metabolic enzymes paradoxically localize in the nucleus, where they can present enhanced local enzymatic activities or non-canonical functions.

12.2.1 Glucose, Glutamine and One-Carbon Metabolism in Cancer-Dependent Metabolic Rewiring

A phenotypic trait of cancer cells is the increase in fuel demand reflected mainly by the elevation of glucose or other nutrients consumption (Hanahan and Weinberg 2011a) (Fig. 12.3). An overall supply of nutrients could lead to a specific metabolite-dependent perturbation of the metabolism-epigenome axis, causing changes at the chromatin modifications and transcriptional level. Cellular acetyl-CoA levels correlate with cell growth, proliferation and histone acetylation in yeast and mammalian cells (Denisov and Sligar 2012; Shi and Tu 2013; Lee et al. 2014; Henry et al. 2015), suggesting that acetyl-CoA could play a sensor role for basal cellular functions which are energetically costly. Several studies support the regulation of histone acetylation in a nutrient-dependent manner by the acetyl-CoA producing enzyme ACLY (Wellen et al. 2009; Byles et al. 2013; Zhao et al. 2016; Wong et al. 2017; Sivanand et al. 2018). The acetyl-CoA homeostasis is dysregulated by driver mutations in cancer, thus suggesting that acetyl-CoA could also impact the epigenome in tumorigenesis (Kinnaird et al. 2016). For example, MYC and AKT gain of function mutations have been shown to promote acetyl-CoA production through ACLY. MYC has been shown to regulate fatty acid metabolism by controlling acetyl-CoA abundance and in fact, MYC inactivation leads to a reduction in acetyl-CoA levels (Morrish et al. 2009). AKT activates ACLY through its phosphorylation leading to increased acetyl-CoA (Potapova et al. 2000; Berwick et al. 2002), conversely AKT inhibition leads to decreased acetyl-CoA levels and histone acetylation (Lee et al. 2014). Despite that acetyl-CoA levels impact histone acetylation, it is important to address whether acetyl-CoA fluctuations can influence histone acetylation of specific genomic loci. Recent discoveries using glioblastoma multiforme (GBM) cells showed the specific regulation of a set of genes upon high acetyl-CoA levels (Lee et al. 2014). Importantly, this specific regulation is linked to acetyl-CoA dependent changes on H3K27ac on site-specific regulatory regions (Lee et al. 2018). Mechanistically, the transcription factor NFAT1 (nuclear factor of activated T cells 1) mediated the acetyl-CoA dependent transcriptional changes on specific NFAT1 targets (Lee et al. 2018) in GBM cells. This suggests that acetyl-CoA elevation does not trigger a global unspecific chromatin acetylation, but it is rather a regulated process. Similarly, pancreatic adenocarcinoma (PDA) display elevated acetyl-CoA levels due to a mutation in KRAS in acinar cells. This leads to an acinar to ductal metaplasia due to a dysregulation of the mevalonate pathway (cholesterol synthesis) and moreover, the levels of acetyl-CoA could correlate with high stromal content and poor prognosis. Recent data support the role of ACLY in PDA given the increased H3K27 acetylation mark at PDA enhancers of genes involved in the mevalonate pathway (Carrer et al. 2019).

Tumors require large amounts of energy uptake, besides glucose one of the most common nutrient sources of cancer cells is glutamine. *MYC* mutations lead to enhanced utilization of glutamine through metabolic rewiring including increased expression of glutaminase (GLS). GLS deaminates glutamine into glutamate, which is converted into α-KG by the glutamate dehydrogenase (GLDH) and enters TCA cycle to follow

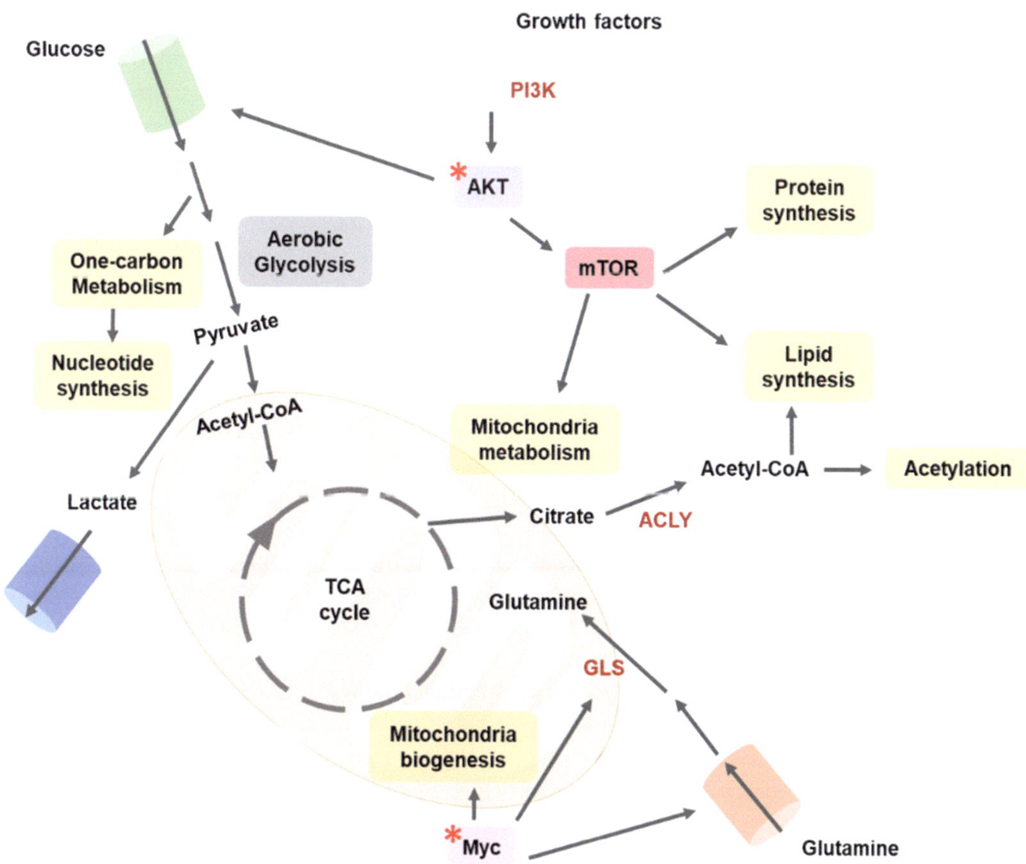

Fig. 12.3 Cancer dependent metabolic rewiring

its oxidation. However, glutamine can also undergo a reductive carboxylation pathway, which is mediated by the reverse direction of the TCA cycle through the catalysis of α-KG to isocitrate by IDH2, in the mitochondria or by IDH1, in the cytoplasm. Importantly, both isoforms use $NADP^+$ as cofactor to form NADPH which is used for fatty acid synthesis. This is in fact an anabolic usage of glutamine through the generation of citrate and then acetyl-CoA by ACLY which can be used for fatty acid synthesis. Cancer cells under hypoxia rely almost exclusively on the reductive carboxylation of glutamine for fatty acid synthesis (Metallo et al. 2012; Mullen et al. 2012). Future studies will possibly investigate how tumorigenic dependent reductive carboxylation of glutamine can impact histone and DNA methylation. One can speculate that it most likely occurs via perturbation of α-KG levels.

The so called, "one carbon metabolism" is a metabolic pathway integrating multiple inputs through the donation of one carbon unit from specific amino acids. The carbon unit is transferred through the folate and methionine cycle into different outputs including cellular biosynthesis, regulation of redox status, histone and DNA methylation, and genome maintenance through the regulation of a nucleotide pool. Due to its relevant implication in nucleotide biosynthesis, one carbon metabolism has been exploited for oncogenic therapy since the 1950's.

One carbon metabolism generates SAM through the methionine cycle. As discussed above, SAM is the substrate used for protein methylation including histone methylation and DNA methylation. Due to the diet dependence on the essential amino acid methionine and the vitamin folate, the nutrient status has been linked with DNA methylation (Balaghi et al. 1993) and could have an effect in cancer. In fact, the serum levels of SAM and SAH in cancer patients and the extent of methylation in tumors correlates with their diet (Poirier et al. 2001; Lim and Song 2012). Specific diet interventions show that colon cancer patients consuming 400 μg/day of folate had more global DNA methylation than patients taking half the dose (Schernhammer et al. 2010). Recent discoveries support these observations showing that modulation of methionine in the diet leads to changes in H3K4me3 in the liver (Mentch et al. 2015).

Cancer cells undergo profound changes in metabolism to promote proliferation and cell growth for survival. Reprograming of tumor cells involves constitutive activation of PI3K/Akt signaling and mTOR activation trough gain of function mutations that confers alterations in AKT signaling and in oncogenes such as *Myc*. AKT signaling promotes glucose uptake, glycolysis and mTOR activation which facilitates anabolic growth and macromolecular synthesis. AKT also activates ACLY consequently increasing acetyl-CoA levels. Moreover, Myc regulates glutamine uptake and conversion into a carbon source by promoting the expression of the enzyme GLS. Myc additionally prompt mitochondria biogenesis by directly regulating transcription of mitochondrial genes and promoting mitochondria metabolite precursors. TCA intermediates are redirected from ATP production towards the synthesis of lipid, proteins and nucleic acids that serve the increased demand of the proliferating transformed cell. PI3K: phosphatidylinositol 3-kinase; AKT: Protein Kinase B; mTOR: mammalian Target of Rapamycin; ACLY: ATP-dependent Citrate Lyase; GLS: glutaminase. Asterisk: mutations.

12.2.2 Oncometabolites Impacting Gene Chromatin Modifying Enzymes

The modulation of certain metabolic pathways through changes in cancer-induced metabolic fluxes and reprogramming are linked to perturbed chromatin modification states. Mutations in specific metabolic enzymes can lead to the accumulation of modified metabolites which impact in tumorigenesis. A paradigmatic example is the identification of mutations of isocitrate dehydrogenase 1 and 2 (IDH1 and IDH2) in several myeloid malignancies, gliomas, chondrosarcomas and several solid tumors (Mardis et al. 2009). In normal catabolic conditions, IDH3 catalyzes the conversion of isocitrate into α-KG in the TCA cycle. α-KG plays a key role in histone and DNA demethylation because it is used as a cofactor of JHDMs and TET enzymes, as mentioned above. The reductive carboxylation of glutamine uses IDH1 and IDH2 which leads to the production of isocitrate from α-KG. However, mutated IDH1 or IDH2 leads to the production of the R enantiomer 2-hydroxyglutarate (2-HG) from α-KG. 2-HG is an oncometabolite and has been shown to competitively inhibit α-KG dependent dioxygenases including JHDMs (Chowdhury et al. 2011; Xu et al. 2011). The inhibition of JHDMs by 2-HG has a direct impact on histone methylation levels (Fig. 12.4). Particularly, the expression of IDH mutants inhibits cell differentiation through hypermethylation of H3K9 and H3K27 in gliomas (Lu et al. 2012). These observations suggest that tumorigenesis can be induced by preventing differentiation through a diminished active histone demethylation of cell growth genes. Additional observations have shown that the differentiation of adipocytes is blocked by 2-HG through the inhibition of KDM4C which leads to increased H3K9me, demonstrating that this mechanism could also operate in non-transformed cells (Lu et al. 2012). Overall, increased intracellular accumulation of 2-HG due to IDH mutations caused increased apoptosis, reduced proliferation, impacted flux through glutaminolytic and reductive carboxylation pathways, impaired mitochondrial respiration, and reduced redox control capacity (Parker and Metallo 2015; Badur et al. 2018; Molenaar et al. 2018). Recent advances have shown a non-cell autonomous effect of 2-HG in the context of T cell immunity and gliomas. T cells uptake the tumor secreted 2-HG leading to the impairment of their anti-tumor activity (Bunse et al. 2018). The mecha-

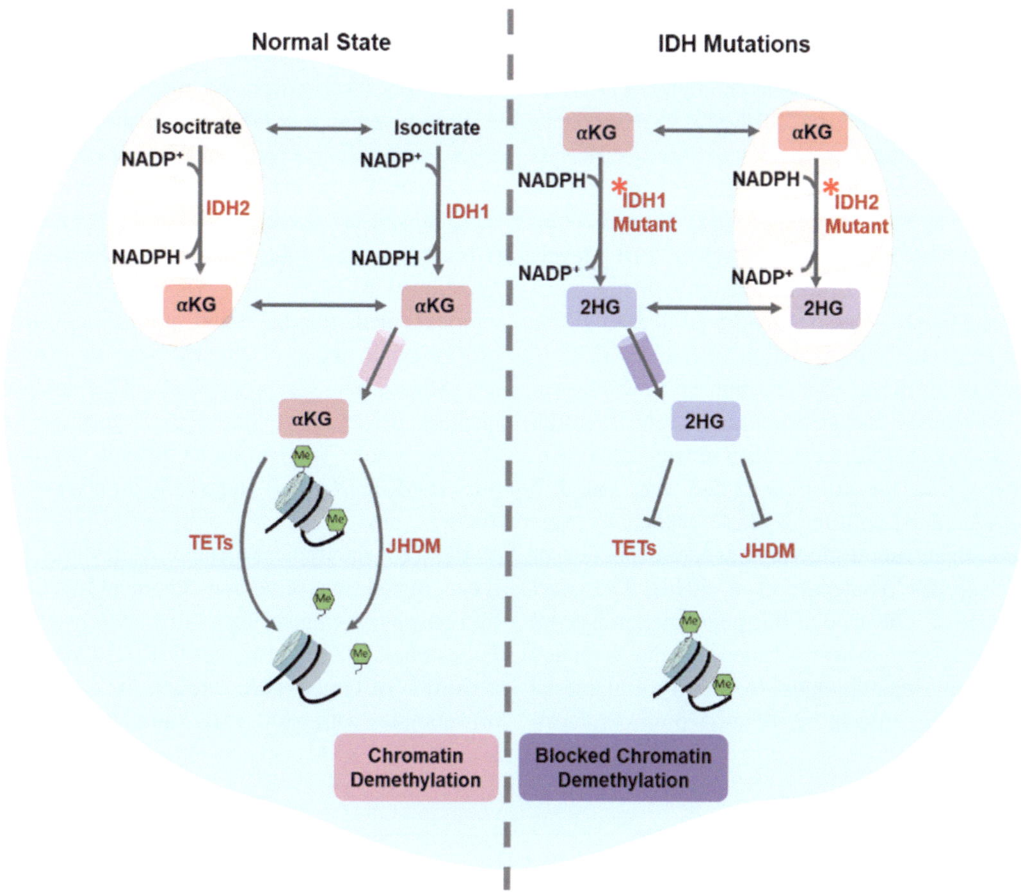

Fig. 12.4 Oncometabolites (ketoglutarate) impacting chromatin function

nism is mediated by an alteration of the calcium-dependent transcriptional activity of the nuclear factor of activated T cells (NFAT) and polyamine synthesis (Bunse et al. 2018).

In normal physiological conditions, isocitrate and $NADP^+$ are used as substrates to be converted to αKG by wild-type IDH with the release of NADPH. Once αKG is shuttle into the nucleus, it is used as the substrate for chromatin demethylation by JHDM and TETs. Cancer-associated mutations in cytosolic IDH1 and mitochondrial IDH2 will lead to the conversion of αKG and NADPH into 2HG and $NADP^+$ instead. 2HG can competitively inhibit αKG-dependent enzymes like JHDM and TETs, resulting in the failure of chromatin demethylation and therefore potentially disrupting normal gene expression and cellular functions. $NADP^+$: Nicotinamide adenine dinucleotide phosphate; αKG: α-ketoglutarate; IDH: isocitrate dehydrogenase; JHDM: Jumonji C-domain containing histone demethylase; TETs: ten eleven translocation.

12.2.3 Nuclear Localization and Function of Metabolic Enzymes

The cellular compartmentalization within organelles allow to physically separate cellular processes that could interfere with other simultaneous cellular reactions. For example the functionally opposite metabolic pathways fatty acid oxidation and fatty acid synthesis are distinctly compartmentalized taking place in mitochondria and cytosol respectively. Recent discoveries have identified non-canonical localization of metabolic enzymes in atypical organelles (Fig. 12.5). The pyruvate dehydrogenase complex (PDC) and pyruvate kinase muscle 2 (PKM2) were shown to translocate into the nucleus to increase acetyl-CoA production locally which facilitates histone acetylation (Sutendra et al. 2014; Matsuda et al. 2016). The elevation of nuclear PDC levels respond to serum, epidermal growth factor, or mitochondrial stress (Sutendra et al. 2014). The authors linked the elevation of nuclear PDC to increased nuclear translocation as a whole protein complex and at expenses of mitochondrial PDC, however the mechanism remains unknown. The nuclear translocation of PDC was found to be associated with the S-phase of the cell cycle thus suggesting a growth dependent function of nuclear PDC (Sutendra et al. 2014). The exact functional relevance of this phenomenon has not yet unveiled; however, it is possibly that is related to the physiopathological conditions of increased demand for histone acetylation during tumorigenic states.

The pyruvate kinase is a rate limiting step of glycolytic ATP production mediated by the phosphoryl transfer from phosphoenolpyruvate into ADP to give rise to ATP and pyruvate. Pyruvate kinase can also be expressed through alternative splicing leading to the isoform PKM2 which has been linked to aerobic glycolysis in tumors (Christofk et al. 2008). Interestingly, PKM2 has been found associated with PDC in the nucleus in a process mediating the transcriptional activation of *Cyp1a*, a target of the transcription factor arylhydrocarbon receptor (AhR). This mechanism is driven by a direct local increase of H3K9ac at the *Cyp1a* promoter through the complex PKM2, PDC, AhR and p300 (Christofk et al. 2008).

PKM2 is an atypical metabolic enzyme since it can, in addition, serve non-canonical functions by phosphorylating or interacting with proteins. For example, PKM2 promotes HIF-1α dependent activation of target genes through the formation of a complex with p300, PHD3 and HIF-1α (Luo et al. 2011). PKM2 also phosphorylates STAT3 in

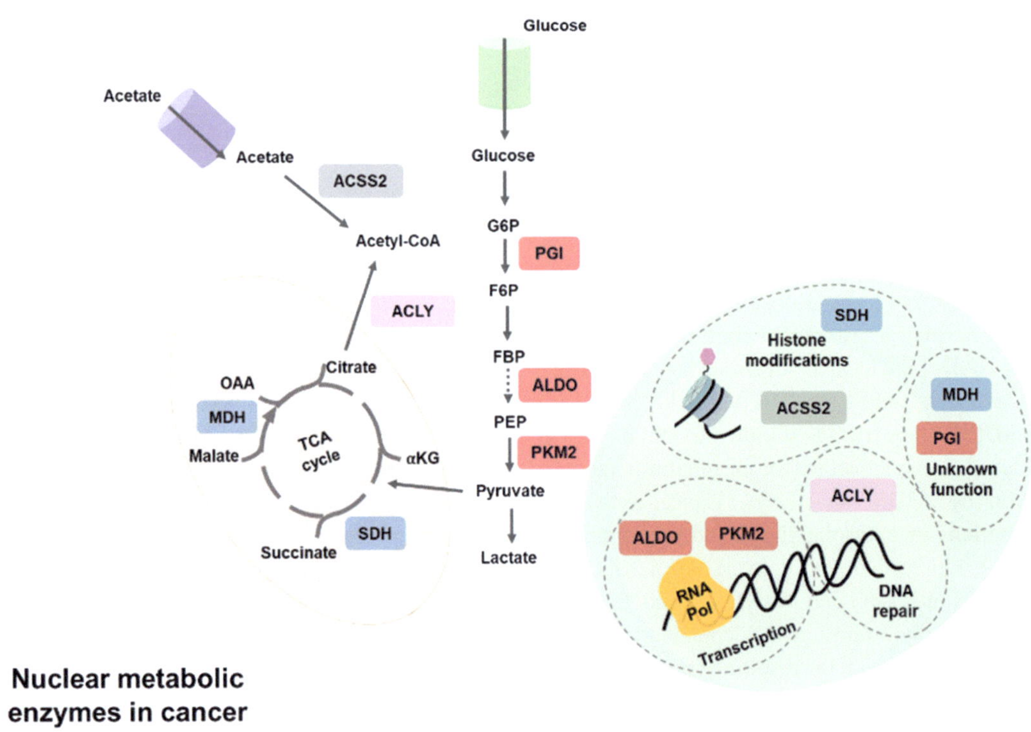

Fig. 12.5 Nuclear metabolic enzymes in cancer

the nucleus leading to increased transcriptional activation of MEK5 and promoting cell proliferation (Gao et al. 2012) PKM2 also undergoes nuclear translocation upon EGFR signaling and leads to beta-catenin transactivation through direct interaction with PKM2 at the Cyclin D1 gene promoter, leading to HDAC3 displacement and increased histone acetylation (Yang et al. 2011). Similarly, PKM2 also phosphorylates H3 upon EGFR activation promoting transcription of Cyclin D1 and c-Myc through HDAC3 displacement of their promoters (Yang et al. 2012). This mechanism leads to tumor cell proliferation, cell-cycle progression, and brain tumorigenesis.

ACLY is also localized in the nucleus suggesting a compartmentalized and local production of acetyl-CoA in response to specific stimuli, such as glucose stimulation (Wellen et al. 2009; Lee et al. 2014; Sivanand et al. 2017). Moreover, ACLY presents a specific function during DNA repair by homologous recombination. ACLY is phosphorylated at S455 by the canonical DNA damage response kinase ATM (Sivanand et al. 2017). This activation is necessary for local histone acetylation preceding BRCA1 recruitment and homologous recombination-mediated DNA repair (Sivanand et al. 2017).

Several other metabolic enzymes have been found in the nucleus where they play a canonical or an atypical function, the latter is also known as a moonlight role. A comprehensive list has been recently reviewed (Boukouris et al. 2016) and includes several glycolytic and TCA cycle enzymes. Interestingly, the mechanism of the nuclear translocation is not understood and remains paradoxical since those metabolic enzymes do not contain a nuclear localization signal. The functional relevance of the nuclear localization of metabolic enzymes has not yet been extensively explored. However, it seems tempting to speculate that metabolic signals are directly transmitted to the transcriptional outputs to adapt to environmental changes. The nuclear localization of metabolic enzymes is particularly utilized by tumorigenic cells whereby the cancer-dependent metabolic rewiring favors proliferation and biosynthetic pathways for tumor growth. Additional molecular mechanisms of how precisely multiple metabolic enzymes impact chromatin modifications will be revealed in the near future and may present novel therapeutic approaches.

The link between metabolism and epigenetic regulation has important roles in pathological processes such as cancer, facilitating the rewiring of transformed cells. Metabolic enzymes typically located in the cytoplasm or mitochondria can be recruited to the nucleus to locally generate metabolites or to serve other functions. These enzymes include the glycolytic enzymes: PGI, ALDO and PKM2; tricarboxylic acid cycle: MDH and SDH; and other metabolic enzymes such as ACSS2 and ACLY. In the nucleus these enzymes exert the role of supply metabolites used for epigenetic regulation. However, they can also have different functions in the regulation of chromatin structure and gene expression that are independent of the production of metabolites. These include direct modification of histones, interaction with transcription regulators and DNA repair. G6P: Glucose 6-phosphate; F6P: Fructose6-phosphate;PEP:Phosphoenolpyruvate; PGI: Phosphoglucose isomerase; ALDO: fructose-bisphosphate aldolase; PKM2: Pyruvate kinase muscle isozyme M2; ACSS2: Acyl-coenzyme A synthetase short-chain family member 2; ACLY: ATP-dependent Citrate Lyase; α -KG: α-ketoglutarate; OAA: oxaloacetate; SDH: Succinate dehydrogenase; MDH: Malate dehydrogenase.

12.3 Tumor Microenvironment Effect on Metabolic Rewiring and Chromatin Modifications in Tumor Cells

Tumors are highly plastic in their metabolic adaptations which lead to their own growth and survival (Hanahan and Weinberg 2011a). Cancerous cells moreover interact with the non-tumor neighboring cells or with components of the extracellular milieu, collectively referred as the tumor microenvironment (TME). A recently identified ability of cancerogenic cells is their capacity to extract energetic substrates from the

TME, particularly from cancer associated fibroblasts (CAFs). A more drastically observed effect in acute myeloid leukemia is the direct transfer of mitochondria from bone marrow mesenchymal stromal cells in the tumor microenvironment to AML malignant cells (Kumar et al. 2018). In another context, recent studies show that cancer cells also take metabolic advantage of adipocytes when existent in the TME. Tumorigenesis could be affected by the proximity and amount of adipose tissue to cancer cells. As a matter of fact, recent epidemiological studies suggest a strong correlation between obesity and the incidence of several tumors including liver, colon pancreas, prostate, breast cancer and others (O'Sullivan et al. 2018). One of the reasons behind this link is the obesity-induced inflammation through secretion of several cytokines such as IL-6 or TNF-α, adipokines (leptin, adiponectin), insulin/insulin-like growth factors (IGFs) or sex hormones which could disrupt tissue homeostasis (Lengyel et al. 2018). Moreover, several studies have shown that tumor cells secret signaling molecules leading to adipose tissue lipolysis, which releases free fatty acids and glycerol utilized by cancer cells (Balaban et al. 2017; Hoy et al. 2017). Likewise, adipocytes drive metabolic reprogramming of cancer cells (Hoy et al. 2017). For example, experiments of co-culture of adipocytes with breast cancer cells have shown that cancer cells increase their lipid accumulation, which is mediated by the activation of the hormone sensitive lipase (HSL) in adipocytes (Balaban et al. 2017). The exact mechanism and signaling pathways are however not yet fully elucidated. Interestingly, CD36, a fatty acid receptor, is required for metastasis in breast cancer and melanoma (Pascual et al. 2017b). How nutrient uptake from neighboring adipocytes impacts chromatin modifications of tumor cells is not known. A recent finding has showed that exogenous lipids are also a source of histone acetylation (Eoin et al. 2016) and this mechanism could also operate in cancer cells which stimulate lipolysis of neighboring adipocytes.

12.4 Conclusion

Nutrient sensing and metabolic pathways have a direct effect on transcription regulation particularly through modulation of key metabolites which impact in chromatin modifications. Tumorigenesis is particularly influenced by different metabolic adaptations which rewire metabolic pathways and therefore the alteration of key metabolites and oncometabolites has an impact in establishing specific chromatin states. Moreover, recent advances are showing how the malignant perturbations of metabolic states could modulate chromatin modifications at specific loci and sustain tumorigenesis. The impact of metabolism in cancer is therefore more than an adaptive response to nutrient preference but an integral part of the signaling events leading to tumorigenic transcriptional states. The challenge remains to further identify regulatory mechanism in the metabolic to chromatin axis, that could be therapeutically targeted.

References

Badur MG, Muthusamy T, Parker SJ et al (2018) Oncogenic R132 IDH1 mutations limit NADPH for De Novo lipogenesis through (D)2-hydroxyglutarate production in fibrosarcoma cells graphical abstract highlights d D2HG production competes with reductive biosynthesis for NADPH in IDH1-mutant cells d 2 H trac. Cell Rep 25. https://doi.org/10.1016/j.celrep.2018.09.074

Balaban S, Shearer RF, Lee LS et al (2017) Adipocyte lipolysis links obesity to breast cancer growth: adipocyte-derived fatty acids drive breast cancer cell proliferation and migration. Cancer Metab 5:1. https://doi.org/10.1186/s40170-016-0163-7

Balaghi M, Wagner C, Wagner C (1993) DNA methylation in folate deficiency use of CpG methylase. Biochem Biophys Res Commun. https://doi.org/10.1006/bbrc.1993.1750

Berwick DC, Hers I, Heesom KJ et al (2002) The identification of ATP-citrate lyase as a protein kinase B (Akt) substrate in primary adipocytes*. J Biol Chem. https://doi.org/10.1074/jbc.M204681200

Bordone L, Cohen D, Robinson A et al (2007) SIRT1 transgenic mice show phenotypes resembling calorie restriction. Aging Cell 6:759–767. https://doi.org/10.1111/j.1474-9726.2007.00335.x

Boukouris AE, Zervopoulos SD, Michelakis ED (2016) Metabolic enzymes moonlighting in the nucleus:

metabolic regulation of gene transcription. Trends Biochem Sci 41:712–730. https://doi.org/10.1016/j.tibs.2016.05.013
Bryant HE, Schultz N, Thomas HD et al (2005) Specific killing of BRCA2-deficient tumours with inhibitors of poly(ADP-ribose) polymerase. Nature. https://doi.org/10.1038/nature03443
Bunse L, Pusch S, Bunse T, et al (2018) Suppression of antitumor T cell immunity by the oncometabolite (R)-2-hydroxyglutarate. Niklas Thon 23, Michael C Burg 6:27. https://doi.org/10.1038/s41591-018-0095-6
Byles V, Covarrubias AJ, Issam B-S et al (2013) The {TSC-mTOR} pathway regulates macrophage polarization. Nat Commun 4:2834. https://doi.org/10.1038/ncomms3834
Cai L, Sutter BM, Li B, Tu BP (2011) {Acetyl-CoA} induces cell growth and proliferation by promoting the acetylation of histones at growth genes. Mol Cell 42:426–437. https://doi.org/10.1016/j.molcel.2011.05.004
Cantó C, Jiang LQ, Deshmukh AS et al (2010) Interdependence of {AMPK} and {SIRT1} for metabolic adaptation to fasting and exercise in skeletal muscle. Cell Metab 11:213–219. https://doi.org/10.1016/j.cmet.2010.02.006
Carrer A, Trefely S, Zhao S et al (2019) Acetyl-CoA metabolism supports multistep pancreatic tumorigenesis. Cancer Discov. https://doi.org/10.1158/2159-8290.CD-18-0567
Chen Q, Chen Y, Bian C et al (2013) TET2 promotes histone O-GlcNAcylation during gene transcription. Nature. https://doi.org/10.1038/nature11742
Chowdhury R, Yeoh KK, Tian YM et al (2011) The oncometabolite 2-hydroxyglutarate inhibits histone lysine demethylases. EMBO Rep. https://doi.org/10.1038/embor.2011.43
Christofk HR, Vander Heiden MG, Harris MH et al (2008) The M2 splice isoform of pyruvate kinase is important for cancer metabolism and tumour growth. Nature 452. https://doi.org/10.1038/nature06734
Cohen P (2002) The origins of protein phosphorylation. Nat Cell Biol 4:E127–E130. https://doi.org/10.1038/ncb0502-e127
Dehennaut V, Leprince D, Lefebvre T (2014a) O-GlcNAcylation, an epigenetic mark. Focus on the histone code, TET family proteins, and polycomb group proteins. Front Endocrinol (Lausanne) 5:155. https://doi.org/10.3389/fendo.2014.00155
Dehennaut V, Leprince D, Lefebvre T, Hansen U (2014b) O-GlcNAcylation, an epigenetic mark. Focus on the histone code, TET family proteins, and polycomb group proteins. Front Endocrinol. https://doi.org/10.3389/fendo.2014.00155
Denisov IG, Sligar SG (2012) A novel type of allosteric regulation: functional cooperativity in monomeric proteins. Arch Biochem Biophys 519:91–102
Eoin M, Crown SB, Fox DB et al (2016) Lipids reprogram metabolism to become a major carbon source for histone acetylation. Cell Rep 17:1463–1472. https://doi.org/10.1016/j.celrep.2016.10.012
Farber S, Diamond LK, Mercer RD et al (1948) Temporary remissions in acute leukemia in children produced by folic acid antagonist, 4-Aminopteroyl-glutamic acid (Aminopterin). N Engl J Med 238:787–793. https://doi.org/10.1056/NEJM194806032382301
Fardini Y, Dehennaut V, Lefebvre T, Issad T (2013) O-GlcNAcylation: a new cancer hallmark? Front Endocrinol (Lausanne) 4:99. https://doi.org/10.3389/fendo.2013.00099
Farmer H, McCabe H, Lord CJ et al (2005) Targeting the DNA repair defect in BRCA mutant cells as a therapeutic strategy. Nature. https://doi.org/10.1038/nature03445
Farrelly LA, Robert T, Zhao S et al (2019) Histone serotonylation is a permissive modification that enhances TFIID binding to H3K4me3. Nature. https://doi.org/10.1038/s41586-019-1024-7
Fong PC, Boss DS, Yap TA et al (2009) Inhibition of Poly(ADP-Ribose) polymerase in tumors from BRCA mutation carriers. N Engl J Med. https://doi.org/10.1056/nejmoa0900212
Gao X, Wang H, Yang JJ et al (2012) Pyruvate kinase M2 regulates gene transcription by acting as a protein kinase. Mol Cell. https://doi.org/10.1016/j.molcel.2012.01.001
Hallows WC, Lee S, Denu JM (2006) Sirtuins deacetylate and activate mammalian acetyl-CoA synthetases. Proc Natl Acad Sci. https://doi.org/10.1073/pnas.0604392103
Hanahan D, Weinberg RA (2011a) Leading edge review hallmarks of cancer: the next generation. Cell 144:646–674. https://doi.org/10.1016/j.cell.2011.02.013
Henry RA, Kuo YM, Bhattacharjee V et al (2015) Changing the selectivity of p300 by acetyl-coa modulation of histone acetylation. ACS Chem Biol. https://doi.org/10.1021/cb500726b
Herranz D, Muñoz-Martin M, Cañamero M et al (2010) Sirt1 improves healthy ageing and protects from metabolic syndrome-associated cancer. Nat Commun 1:1–8. https://doi.org/10.1038/ncomms1001
Hoy AJ, Balaban S, Saunders DN et al (2017) Adipocyte-tumor cell metabolic crosstalk in breast cancer. Trends Mol Med 23. https://doi.org/10.1016/j.molmed.2017.02.009
Kaeberlein M, Kirkland KT, Fields S, Kennedy BK (2004) Sir2-independent life span extension by calorie restriction in yeast. PLoS Biol 2:E296. https://doi.org/10.1371/JOURNAL.PBIO.0020296
Kinnaird A, Zhao S, Wellen KE, Michelakis ED (2016) Metabolic control of epigenetics in cancer. Nat Rev Cancer 16:694–707. https://doi.org/10.1038/nrc.2016.82
Kouzarides T (2007) Chromatin modifications and their function. Cell 128:693–705. https://doi.org/10.1016/j.cell.2007.02.005
Kumar B, Garcia M, Weng L et al (2018) Acute myeloid leukemia transforms the bone marrow niche into a leukemia-permissive microenvironment through exosome secretion. Leukemia. https://doi.org/10.1038/leu.2017.259

Lagouge M, Argmann C, Gerhart-Hines Z et al (2006) Resveratrol improves mitochondrial function and protects against metabolic disease by activating SIRT1 and PGC-1α. Cell. https://doi.org/10.1016/j.cell.2006.11.013

Lee JV, Carrer A, Shah S et al (2014) Article Akt-dependent metabolic reprogramming regulates tumor cell histone acetylation. Cell Metab. https://doi.org/10.1016/j.cmet.2014.06.004

Lee JV, Berry CT, Kim K et al (2018) Acetyl-CoA promotes glioblastoma cell adhesion and migration through Ca2+-NFAT signaling. Genes Dev. https://doi.org/10.1101/gad.311027.117

Lengyel E, Makowski L, DiGiovanni J, Kolonin MG (2018) Cancer as a matter of fat: the crosstalk between adipose tissue and tumors. Trends Cancer 4:374–384. https://doi.org/10.1016/j.trecan.2018.03.004

Li X, Zhang S, Blander G et al (2007) SIRT1 Deacetylates and positively regulates the nuclear receptor LXR. Mol Cell. https://doi.org/10.1016/j.molcel.2007.07.032

Lim U, Song M-A (2012) Dietary and lifestyle factors of DNA methylation. In: Cancer epigenetics. Humana Press, Totowa, pp 359–376

Liu Y, Dentin R, Chen D et al (2008) A fasting inducible switch modulates gluconeogenesis via activator/coactivator exchange. Nature. https://doi.org/10.1038/nature07349

Lu C, Ward PS, Kapoor GS et al (2012) IDH mutation impairs histone demethylation and results in a block to cell differentiation. Nature. https://doi.org/10.1038/nature10860

Luo J, Nikolaev AY, Imai S, Ichiro et al (2001) Negative control of p53 by Sir2α promotes cell survival under stress. Cell. https://doi.org/10.1016/S0092-8674(01)00524-4

Luo W, Hu H, Chang R et al (2011) Pyruvate kinase M2 is a PHD3-stimulated coactivator for hypoxia-inducible factor 1. Cell. https://doi.org/10.1016/j.cell.2011.03.054

Mardis ER, Ding L, Dooling DJ et al (2009) Recurring mutations found by sequencing an acute myeloid leukemia genome. N Engl J Med. https://doi.org/10.1056/NEJMoa0903840

Mariappa D, Pathak S, Van Aalten DMF (2013) A sweet TET-à-tête-synergy of TET proteins and O-GlcNAc transferase in transcription. EMBO J. https://doi.org/10.1038/emboj.2013.26

Matsuda S, Adachi J, Ihara M et al (2016) Nuclear pyruvate kinase M2 complex serves as a transcriptional coactivator of arylhydrocarbon receptor. Nucleic Acids Res 44:636–647. https://doi.org/10.1093/nar/gkv967

Mentch SJ, Mehrmohamadi M, Thalacker-Mercer AE et al (2015) Histone methylation dynamics and gene regulation occur through the sensing of one-carbon metabolism. Cell Metab 22:861–873. https://doi.org/10.1016/j.cmet.2015.08.024

Messner S, Hottiger MO (2011) Histone {ADP-ribosylation} in {DNA} repair, replication and transcription. Trends Cell Biol 21:534–542. https://doi.org/10.1016/j.tcb.2011.06.001

Metallo CM, Gameiro PA, Bell EL et al (2012) Reductive glutamine metabolism by IDH1 mediates lipogenesis under hypoxia. Nature. https://doi.org/10.1038/nature10602

Molenaar RJ, Maciejewski JP, Wilmink JW, van Noorden CJF (2018) Wild-type and mutated IDH1/2 enzymes and therapy responses. Oncogene 37:1949–1960. https://doi.org/10.1038/s41388-017-0077-z

Morrish F, Isern N, Sadilek M et al (2009) C-Myc activates multiple metabolic networks to generate substrates for cell-cycle entry. Oncogene 28(27):2485

Motta MC, Divecha N, Lemieux M et al (2004) Mammalian SIRT1 represses forkhead transcription factors. Cell. https://doi.org/10.1016/S0092-8674(04)00126-6

Mullen AR, Wheaton WW, Jin ES et al (2012) Reductive carboxylation supports growth in tumour cells with defective mitochondria. Nature. https://doi.org/10.1038/nature10642

O'Sullivan J, Lysaght J, Donohoe CL, Reynolds JV (2018) Obesity and gastrointestinal cancer: the interrelationship of adipose and tumour microenvironments. Nat Rev Gastroenterol Hepatol 15:699–714. https://doi.org/10.1038/s41575-018-0069-7

Parker SJ, Metallo CM (2015) Metabolic consequences of oncogenic IDH mutations. Pharmacol Ther. https://doi.org/10.1016/j.pharmthera.2015.05.003

Pascual G, Avgustinova A, Mejetta S et al (2017a) Targeting metastasis-initiating cells through the fatty acid receptor CD36. Nature 541. https://doi.org/10.1038/nature20791

Pascual G, Avgustinova A, Mejetta S et al (2017b) Targeting metastasis-initiating cells through the fatty acid receptorr CD36. Nature 541. https://doi.org/10.1038/nature20791

Pfluger PT, Herranz D, Velasco-Miguel S et al (2008) Sirt1 protects against high-fat diet-induced metabolic damage. Proc Natl Acad Sci 105:9793–9798. https://doi.org/10.1073/pnas.0802917105

Poirier LA, Wise CK, Delongchamp RR, Sinha R (2001) Blood determinations of S-adenosylmethionine, S-adenosylhomocysteine, and homocysteine: correlations with diet. Cancer Epidemiol Biomark Prev 10:649–655

Potapova IA, El-Maghrabi MR, Doronin SV, Benjamin WB (2000) Phosphorylation of recombinant human ATP:citrate lyase by cAMP-dependent protein kinase abolishes homotropic allosteric regulation of the enzyme by citrate and increases the enzyme activity. Allosteric activation of atp:citrate lyase by phosphorylated sugars. Biochemistry. https://doi.org/10.1021/bi992159y

Qiang L, Wang L, Kon N et al (2012) Brown remodeling of white adipose tissue by SirT1-dependent deacetylation of Pparγ. Cell. https://doi.org/10.1016/j.cell.2012.06.027

Rodgers JT, Puigserver P (2007) Fasting-dependent glucose and lipid metabolic response through hepatic sirtuin 1. Proc Natl Acad Sci. https://doi.org/10.1073/pnas.0702509104

Rodgers JT, Lerin C, Haas W et al (2005) Nutrient control of glucose homeostasis through a complex of {PGC-1α} and {SIRT1}. Nature 434:113. https://doi.org/10.1038/nature03354

Sabari BR, Zhang D, Allis DC, Zhao Y (2016) Metabolic regulation of gene expression through histone acylations. Nat Rev Mol Cell Biol 18:nrm.2016.140. https://doi.org/10.1038/nrm.2016.140

Sabari BR, Zhang D, Allis CD, Zhao Y (2017a) Metabolic regulation of gene expression through histone acylations. Nat Rev Mol Cell Biol 18:90–101. https://doi.org/10.1038/nrm.2016.140

Sabari BR, Zhang D, Allis CD, Zhao Y (2017b) Metabolic regulation of gene expression through histone acylations. Nat Rev Mol Cell Biol 18:90–101. https://doi.org/10.1038/nrm.2016.140

Schernhammer ES, Giovannucci E, Kawasaki T et al (2010) Dietary folate, alcohol and B vitamins in relation to LINE-1 hypomethylation in colon cancer. Gut 59:794–799. https://doi.org/10.1136/gut.2009.183707

Shi L, Tu BP (2013) Acetyl-CoA induces transcription of the key G1 cyclin CLN3 to promote entry into the cell division cycle in Saccharomyces cerevisiae. Proc Natl Acad Sci. https://doi.org/10.1073/pnas.1302490110

Shi L, Tu BP (2015) {Acetyl-CoA} and the regulation of metabolism: mechanisms and consequences. Curr Opin Cell Biol 33:125–131. https://doi.org/10.1016/j.ceb.2015.02.003

Sivanand S, Rhoades S, Jiang Q et al (2017) Nuclear {Acetyl-CoA} production by {ACLY} promotes homologous recombination. Mol Cell 67:252–265.e6. https://doi.org/10.1016/j.molcel.2017.06.008

Sivanand S, Viney I, Wellen KE (2018) Spatiotemporal control of {Acetyl-CoA} metabolism in chromatin regulation. Trends Biochem Sci 43:61–74. https://doi.org/10.1016/j.tibs.2017.11.004

Su X, Wellen KE, Rabinowitz JD (2016) Metabolic control of methylation and acetylation. Curr Opin Chem Biol 30:52–60. https://doi.org/10.1016/j.cbpa.2015.10.030

Sutendra G, Kinnaird A, Dromparis P et al (2014) A nuclear pyruvate dehydrogenase complex is important for the generation of Acetyl-CoA and histone acetylation. Cell 158:84–97. https://doi.org/10.1016/j.cell.2014.04.046

Vaziri H, Dessain SK, Eaton EN et al (2001) hSIR2SIRT1functions as an NAD-dependent p53 deacetylase. Cell. https://doi.org/10.1016/S0092-8674(01)00527-X

Warburg O, Wind F, Negelein E (1927) The metabolism of tumors in the body. J Gen Physiol 8:–519

Wellen KE, Hatzivassiliou G, Sachdeva UM et al (2009) {ATP-citrate} lyase links cellular metabolism to histone acetylation. Science 324:1076–1080

Wong CC, Qian Y, Yu J (2017) Interplay between epigenetics and metabolism in oncogenesis: mechanisms and therapeutic approaches. Oncogene. https://doi.org/10.1038/onc.2016.485

Wu SC, Zhang Y (2010) Active DNA demethylation: many roads lead to Rome. Nat Rev Mol Cell Biol 11:607–620. https://doi.org/10.1038/nrm2950

Xu W, Yang H, Liu Y et al (2011) Oncometabolite 2-hydroxyglutarate is a competitive inhibitor of α-ketoglutarate-dependent dioxygenases. Cancer Cell. https://doi.org/10.1016/j.ccr.2010.12.014

Yang W, Xia Y, Ji H et al (2011) Nuclear PKM2 regulates β-catenin transactivation upon EGFR activation. Nature. https://doi.org/10.1038/nature10598

Yang W, Xia Y, Hawke D et al (2012) PKM2 phosphorylates histone H3 and promotes gene transcription and tumorigenesis. Cell. https://doi.org/10.1016/j.cell.2012.07.018

Zhao S, Torres A, Henry RA et al (2016) {ATP-Citrate} Lyase controls a {Glucose-to-Acetate} metabolic switch. Cell Rep 17:1037–1052. https://doi.org/10.1016/j.celrep.2016.09.069

Inflammatory Microenvironment Modulation of Alternative Splicing in Cancer: A Way to Adapt

13

Ana Luísa Silva, Márcia Faria, and Paulo Matos

Abstract

The relationship between inflammation and cancer has been long recognized by the medical and scientific community. In the last decades, it has returned to the forefront of clinical oncology since a wealth of knowledge has been gathered about the cells, cytokines and physiological processes that are central to both inflammation and cancer. It is now robustly established that chronic inflammation can induce certain cancers but also that solid tumors, in turn, can initiate and perpetuate local inflammatory processes that foster tumor growth and dissemination. Inflammation is the hallmark of the innate immune response to tissue damage or infection, but also mediates the activation, expansion and recruitment to the tissues of cells and antibodies of the adaptive immune system. The functional integration of both components of the immune response is crucial to identify and subdue tumor development, progression and dissemination. When this tight control goes awry, altered cells can avoid the immune surveillance and even subvert the innate immunity to promote their full oncogenic transformation. In this chapter, we make a general overview of the most recent data linking the inflammatory process to cancer. We start with the overall inflammatory cues and processes that influence the relationship between tumor and the microenvironment that surrounds it and follow the ever-increasing complexity of processes that end up producing subtle changes in the splicing of certain genes to ascertain survival advantage to cancer cells.

Keywords

Cancer · Inflammation · Tumor microenvironment · Splicing · Signaling

A. L. Silva (✉)
Serviço de Endocrinologia, Diabetes e Metabolismo do CHLN-Hospital Santa Maria, Lisbon, Portugal

ISAMB-Instituto de Saúde Ambiental, Faculdade de Medicina da Universidade de Lisboa, Lisbon, Portugal

M. Faria
Serviço de Endocrinologia, Diabetes e Metabolismo do CHLN-Hospital Santa Maria, Lisbon, Portugal

Faculdade de Ciências, BioISI-Biosystems and Integrative Sciences Institute, Universidade de Lisboa, Lisbon, Portugal

P. Matos
Faculdade de Ciências, BioISI-Biosystems and Integrative Sciences Institute, Universidade de Lisboa, Lisbon, Portugal

Departamento de Genética Humana, Instituto Nacional de Saúde Doutor Ricardo Jorge, Lisbon, Portugal

J. Serpa (ed.), *Tumor Microenvironment*, Advances in Experimental Medicine and Biology 1219,
https://doi.org/10.1007/978-3-030-34025-4_13

13.1 The Tumor Microenvironment

The interaction between the tumor and the tumor microenvironment is nowadays recognized as of vital importance in the tumorigenic process. The tumor microenvironment is composed of extracellular matrix (ECM), soluble factors, and a plethora of cancer-associated cell populations that act in concert to potentiate cancer progression and metastasis (Li et al. 2007). Components of its complex stromal system include immune cell populations such as myeloid-derived suppressor cells (MDSCs), mast cells, monocytes, neutrophils, CD8 and CD4 T-cells, dendritic cells (DCs), natural killer (NK) cells and tumor-associated macrophages (TAMs) as well as endothelial cells, endothelial progenitor cells (EPCs), mesenchymal stem cells (MSCs) and cancer-associated fibroblasts (CAFs). Tumor cells interact with the stromal microenvironment by producing cytokines to which stromal cells respond with the secretion of proteases that remodel the extracellular matrix, of pro-angiogenic factors that attract blood vessels, and of mitogenic factors that feedback on tumor cells, promoting tumor cell growth and survival (Joyce and Pollard 2009; Grivennikov et al. 2010). Colorectal cancer (CRC) cells and their stromal CAFs are a paradigm of this crosstalk in which the transforming growth factor- β (TGF-β), a master cytokine overactivated in both tumor cells and stromal CAFs, serves as a major conduit for communication between both cell populations. TGF-β plays a central role in cell-cell communication that occurs in the tumor microenvironment, which is behind the stromal program that drives CRC metastasis. In fact, TGF-β overexpression has been associated with a subset of CRCs characterized by a marked mesenchymal phenotype with stromal invasion, angiogenesis, refractoriness to treatment, advanced disease and poor prognosis (Picon et al. 1998). In addition, an increasing number of studies show that the metastatic potential of CRC cells is reduced by the targeting of TGF-β signaling (Wakefield and Hill 2013; Gonzalez-Zubeldia et al. 2015; Villalba et al. 2017).

13.2 Inflammatory Microenvironment in Cancer

The growing understanding of the complex interplay between the tumor and the tumor microenvironment has emphasized the impact of an inflammatory microenvironment in cancer development. Normal inflammation, usually associated with tissue healing processes, is usually self-limiting because the production of pro-inflammatory cytokines is followed by the production of anti-inflammatory cytokines. The dysregulation of this controlled process can lead to pathogenesis, as it is the case with neoplastic transformation. The role of inflammation in the development of cancer was described as early as 1863, by Rudolf Virchow, who hypothesized that cancer arises from inflammatory sites (Mantovani et al. 2008). Several factors such as persistent bacterial and viral infections, exposure to environmental pollutants, and inflammatory autoimmune diseases can result in chronic inflammation conditions (Coussens and Werb 2002). Chronic inflammation is presently a well-defined risk factor for tumor development, and several cancers are known to be triggered by this condition (Coussens and Werb 2002; Mantovani et al. 2008). CRC is a distinct example of a malignancy arising in the setting of chronic inflammation, having been established a clear correlation between inflammatory bowel disease and an increased risk for the development of CRC (Romano et al. 2016). Likewise, the associations between Hepatitis C infection and hepatocellular carcinoma, as well as *Helicobacter pylori* infection and gastric cancer (Correa 1995; Kanda et al. 2019) are examples of how chronic inflammation, resulting from persistent infections, may predispose to cancer development. Also, for virus-induced tumors (such as papillomavirus-induced squamous cell carcinomas), which may arise from direct transformation of cells due to active viral oncogene insertion into the host genome, the inflammatory microenvironment resulting from innate antiviral responses has been shown to be essential for tumor development (Coussens and Werb 2002; Giorgio et al. 2013).

There is clear evidence that an inflammatory microenvironment may be a trigger for malignant transformation (Fig. 13.1): (I) being highly oxidative it is prone to induce DNA damage; and (II) being highly vascularized and rich in growth factors, it generates a tumor-promoting stroma (Coussens and Werb 2002; Grivennikov et al. 2010). The chronic exposure to cytokines and chemokines within the microenvironment, such as ROS, TNFũFC;, IL-1, IL-6 and IL-8, can activate intracellular pro-proliferative and pro-survival signaling pathways in the neighboring epithelial cells (Grivennikov 2013; Landskron et al. 2014; Romano et al. 2016).

Again, CRC is paradigmatic of the functional relationship between inflammation and cancer. Despite the Wnt/β-catenin signaling being instrumental for normal cell growth and renewal of intestinal epithelium, it is also a central participant in malignant cell proliferation (Romano et al. 2016). In fact, mutations leading to the activation of Wnt/β-catenin pathway occur in over 90% of CRC cancers, of which inactivating mutations in the adenomatous polyposis coli gene (*APC*) are considered the most frequent (Fodde et al. 2001). However, in CRC that are preceded by clinically detectable inflammatory diseases, rather than driving tumor initiation, alterations in Wnt/β-catenin pathway seem to happen late during cancer development (Grivennikov 2013). Several lines of evidence suggest that inflammatory pathways can promote β-catenin signaling even in the absence of genetic alterations, indicating that the main driver mutations for CRC tumorigenesis can be bypassed by inflammatory signals (Castellone et al. 2005; Grivennikov 2013). Of these are examples inflammation-related pathways such as NF-κB and PI3K-AKT, which can lead to

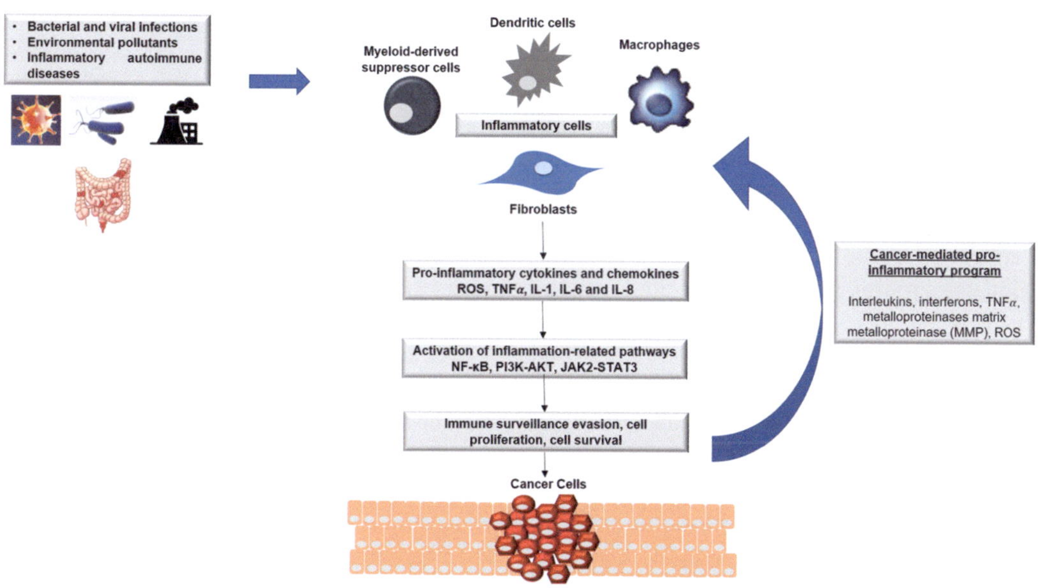

Fig. 13.1 Functional relationship between inflammation and cancer. Bacterial and viral infections, exposure to environmental pollutants, and inflammatory autoimmune diseases can result in chronic inflammation conditions. Cytokines and chemokines within the inflammatory microenvironment, such as ROS, TNFũFC;, IL-1, IL-6 and IL-8, can activate inflammation-related pathways that induce intracellular pro-proliferative and pro-survival signals in the neighboring epithelial cells, triggering malignant transformation. Conversely, a tumor-mediated positive feedback on the inflammatory microenvironment, driven by secreted mediators such as Interleukins, interferons, TNFũFC;, MMPs and ROS, can potentiate the maintenance and progression of the malignant phenotype

β-catenin nuclear accumulation in the absence of APC mutations (Kaler et al. 2009; Lee et al. 2010). Also, cytokines like TNF-α or soluble mediators such as prostaglandin E2 (PGE2) are overexpressed during inflammation and can be particularly responsible for activation of ERK, NF-κB and PI3K-AKT pathways in epithelial cells, thereby increasing β-catenin signaling and thus having a relevant role in initiation of malignancy (Pozzi et al. 2004; Tessner et al. 2004). Another player shown to be of major importance in the tumorigenic process of colitis-associated cancer is IL-6, a pro-inflammatory cytokine related to carcinogenesis in various tissues. Besides NF-kB activation, TNF-α may also stimulate IL-6 expression during chronic inflammation, triggering the canonical IL-6 receptor pathway that activates JAK2-STAT3 signaling, which enables tumor cells to evade immune surveillance allowing tumor growth (Grivennikov et al. 2009).

On the other hand, the induction by the tumor itself of a pro-inflammatory program in the microenvironment seems to act as a strategy to potentiate the maintenance and progression of the malignant phenotype (Fig. 13.1). In fact, tumor cells produce various cytokines and chemokines to which stromal cells respond with the secretion of proteases that remodel the extracellular matrix, or pro-angiogenic factors that attract blood vessels and diverse leukocyte populations. Leukocytes secrete themselves an array of factors with pro-survival and pro-proliferative actions that can contribute in a relevant way to the neoplasm development (Landskron et al. 2014). These also include interleukins, interferons, TNFũFC;, but also matrix metalloproteinase (MMP) and reactive oxygen species (ROS).

Indeed, several members of the MMP family have been identified as poor prognosis markers for epithelial cancer patients and as drivers of many facets of the tumor phenotype in experimental models of breast, colon and pancreatic cancer (Kessenbrock et al. 2010; Radisky and Radisky 2015). Moreover, increased expression of inflammatory cytokines such as IL-1β, IL-8, PGE2 and TNF-α by both stroma and tumor cells has been shown to favor ROS-induced oxidative stress, leading to progression of many types of cancer, including breast, pancreatic and lung carcinomas (Roque et al. 2015; Kumari et al. 2018). In addition, inflammation-associated upregulation of ROS production via extracellular superoxide dismutase (SOD3) has been recently linked to the development of papillary thyroid carcinomas (PTC) from the early stages to the metastasis phase (Parascandolo et al. 2017).

Of all the stromal participants in cancer-mediated inflammation, the tumor-associated macrophages (TAMs) stand out for their role in potentiating the neoplastic process, particularly by driving tumor-associated angiogenesis, thus favoring tumor spreading (Riabov et al. 2014). In addition, by secreting the anti-inflammatory cytokine IL-10, TAMs can also play an important role in blunting the anti-tumor action of type 1 helper T cells, contributing to immune surveillance evasion (Schoppmann et al. 2002; Hassuneh et al. 2013). In CRC, TAMs-producing tumor necrosis factor (TNF)-α, IL-1β, and IL-6 were shown to promote proliferation and migration of tumor cells, which, in turn, have been also shown to stimulate TAMs to produce IL-6, closing a cross-stimulatory loop (Jedinak et al. 2010). TAMs-derived IL-6 also promotes STAT4-induced IL-10 production in CRC, a well-established marker of poor prognosis in these tumors (Galizia et al. 2002; Herbeuval et al. 2004). TAMs in the stroma also strongly express COX-2, and the relationship between COX-2 and colonic adenoma formation is well defined (Adegboyega et al. 2004).

Although TAMs frequently infiltrate thyroid cancer tissue, their role in cancer progression is still unclear. Advanced metastatic thyroid cancers harboring a high density of TAMs were shown to be associated with invasion and decreased cancer-related survival, which suggests that TAMs may also facilitate thyroid cancer progression (Ryder et al. 2008).

13.3 Alternative Splicing in Tumor Cell Adaptation

Cancer adaptive responses rely on the plasticity of gene expression in which the crosstalk between the developing neoplasm and the cancer microenvironment have a pivotal impact. An important means, by which alterations in gene expression resulting from this crosstalk occur, is the modulation of alternative splicing (AS). Generation of alternatively spliced isoforms with oncogenic potential contributes to several cancer hallmarks, such as the cancer cell's ability to sustain proliferation, avoid death, invade and metastasize, as well as to manipulate cellular energetics and evade the immune system (Hanahan and Weinberg 2011; Fouad and Aanei 2017).

AS is the main orchestrator of mRNA diversity, which is essential to the fine tuning of protein expression. Most protein-coding mammalian genes are interrupted by intervening non-coding sequences known as introns. Thus, messenger RNAs are transcribed as intron-containing precursors termed pre-mRNAs. These are subjected to a processing step, termed splicing, through which introns are excised and exons are joint together by means of a macromolecular machine designated the spliceosome. This process produces the mature, translatable form of mRNA (Sanford 2004). AS is a highly regulated and cell-type-specific process whose regulators are themselves tightly regulated. It is thus not unforeseen that defects in AS can lead to several pathological conditions. The neoplastic pathology, in particular, is strongly associated with alterations in AS (Klinck et al. 2008; Venables et al. 2009; Brosseau et al. 2014; Agrawal et al. 2018). These alterations can result from mutations, affecting either the splicing factors or the mRNA cis-acting splicing sequences, or may be as well a consequence of an abnormal expression of factors involved in splicing regulation (Climente-González et al. 2017; Urbanski et al. 2018). The family of heterogeneous nuclear ribonucleoproteins (hnRNP) and the family of serine/arginine-rich proteins (SR), are RNA-binding proteins that represent a plentiful and varied group of splicing modulators (Urbanski et al. 2018; Dvinge 2018). hnRNP A1 and hnRNP A2 are the two most well characterized elements of a family of about twenty members which bind to splicing silencers, promoting exon exclusion. The SR protein family, most of which act as antagonists of hnRNP proteins also comprise more than twenty members, with SFRS1 being the most studied (Dvinge 2018). The polypyrimidine tract binding protein 1 (PTBP1) is another important regulator of AS that acts as a splicing repressor (Xue et al. 2009). The expression of these splicing modulators was shown to be upregulated in several cancers and to play important roles in the establishment and maintenance of cell transformation (Xue et al. 2009; Song et al. 2018b; Urbanski et al. 2018). In colorectal tumors for instance, the comprehensive AS profile retrieved by RNA-seq data analysis allowed the identification of a series of cancer-specific AS events with prognostic value (Liu et al. 2018). Several of the cancer-associated alterations in AS are governed by the oncogenic signaling that occurs in the neoplastic cell. In gliomas, the oncogenic transcription factor c-Myc was shown to upregulate the expression of hnRNPA1/A2 and PTBP1 (David et al. 2010), which modulate the AS of the glycolytic enzyme pyruvate kinase (PK). *PKM*, one of the two paralogous genes encoding PK, undergoes AS of two mutually exclusive exons, generating two isoforms (PKM1 and PKM2) with different substrate affinities (David et al. 2010; Chen et al. 2010). Overexpression of hnRNPA1/A2 and PTBP1 favor the switching from PKM1 to PKM2, which in turn promotes the metabolic shift from oxidative phosphorylation to aerobic glycolysis and provides a selective advantage for tumor progression (Clower et al. 2010; Chen et al. 2010).

A similar mechanism, but driven by c-Myc/PTBP1, has been described to influence PKM AS in CRC (Takahashi et al. 2015; Taniguchi et al. 2015). Also, increased levels of PTBP1, resulting from the oncogene-driven expression of the transcription factors c-Myc and ELK1, were shown to favor the generation of specific NUMB and RAC1 splicing isoforms (the increased inclusion of NUMB exon 9 and the inclusion of RAC1 exon 3b, see below), which have been described as important drivers of CRC tumorigenesis (Hollander et al. 2016).

HnRNPA2 upregulation was also shown to induce an AS switch that downregulates a dominant-negative isoform of A-RAF, which in turn leads to activation of the RAF-MEK-ERK pathway and cellular transformation in hepatocellular carcinoma (Shilo et al. 2014).

13.4 Modulation of Alternative Splicing in Tumor Cell-Microenvironment Communication

Specific AS variants induced in the tumors may be central in the interactions between tumor cells and their microenvironment. CD44 is a cell surface adhesion receptor that is highly expressed in many cancers of epithelial origin. Its interaction with appropriate extracellular matrix ligands promotes the migration and invasion processes involved in metastases (Senbanjo and Chellaiah 2017). CD44 gene undergoes extensive AS generating multiple protein isoforms. Expression of certain CD44 isoforms was linked with progression and metastasis of cancer cells as well as worse patient prognoses (Todaro et al. 2014; Prochazka et al. 2014). A critical mechanism resulting from AS modulation of the CD44 gene is a feed-forward loop regulation that sustains Ras/MAPK activation: activation of Ras/MAPK pathway promotes alternative splicing of CD44, generating a specific isoform (variable exon 6 containing CD44v6 isoform), which in turn, promotes the activation of RTKs, further enhancing Ras/MAPK signaling (Cheng et al. 2006). In the tumoral context this CD44 AS-mediated positive feedback loop represents a pro-oncogenic mechanism, associated with increased malignancy and invasiveness in some tumors (Todaro et al. 2014; Prochazka et al. 2014). Also, the generation of particular splice isoforms of the Focal Adhesion Kinase (FAK), a cytoplasmic tyrosine kinase activated by growth factors and integrins, which plays a critical role in the colon regeneration following tissue damaging, has been shown to be a major player in CRC cell migration and invasion (Devaud et al. 2019).

Another example of AS modulation induced by cell-cell interplay in the tumor microenvironment is the generation of Osteopontin splice isoforms (OPN), which pattern and relative levels within the tumor-microenvironment seem to impact on OPN functions, and to modulate the tumor microenvironment itself (Kazanecki et al. 2007; Castello et al. 2017; Briones-Orta et al. 2017). OPN is a glycoproteinoprotein, overexpressed in many cancer types (Weber 2001; Bellahcène et al. 2008; Wai and Kuo 2008), and has been said to promote several pro-tumorigenic events, such as increased proliferation, survival and cell invasion, due to its involvement with several types of integrins and CD44 receptors (Rangaswami et al. 2006). OPN upregulation has been associated with worse prognosis and poor survival outcome in both CRC and thyroid cancer patients (Likui et al. 2010; Gomaa et al. 2013; Ferreira et al. 2016). Besides full-length OPN (OPN-a), there are two known splice variants (OPN-b and OPN-c). Notably, OPN AS patterns are tissue-specific and distinct malignancies are associated with changes in the specific patterns of OPN splice variants' expression (Gimba and Tilli 2013; Briones-Orta et al. 2017). While, in ovarian cancer, OPNc seems to be the most relevant isoform associated with increased proliferation, migration and invasion (Tilli et al. 2011), in papillary thyroid carcinoma, OPNa was shown to be the predominant isoform, associating with invasiveness (Ferreira et al. 2016).

The regulation of AS can thus be subverted by malignant transformation to favor particular splicing isoforms with oncogenic potential that are central to the progression and maintenance of the malignant phenotype.

13.5 Inflammatory Cues Can Modulate Cancer-Related Alternative Splicing

The crosstalk between the inflammation orchestrated by the developing neoplasm and the inflammatory cell components of the tumor microenvironment induces an adaptive response in the tumor itself, increasing the proliferative and metastatic potential of neoplastic cells, thereby promoting tumorigenesis. Again, modulation of AS provides a plasticity in gene expression that is essential in this scenario. As mentioned above, cancer cells are able to alter the expression of a subset of splicing factors, thus reprogramming the AS of certain groups of transcripts implicated in the modulation of the tumor microenvironment. The above described splicing-regulating factors hnRNP A1/A2, SFRS1 and PTB1 have been also implicated in cancer-related inflammatory processes by regulating mRNA stability and splicing of inflammation-related genes in several tumor types (Venables et al. 2009; Tauler and Mulshine 2009). In cervical cancer, SFRS1 was shown to mediate the IL-17-induced increase in mRNA stability of chemokine CXCL1, widely implicated in the regulation of cancer-associated angiogenesis and metastases (Sun et al. 2011). PTBP1 was also shown to have a strong impact on the pro-inflammatory senescence-associated secretory phenotype (SASP), resulting from oncogene-induced senescence (OIS), by regulating the alternative splicing of genes involved in intracellular trafficking, such as EXOC7 (Georgilis et al. 2018). In fact, PTBP1 overexpression induces a switch in the AS of EXOC7 generating a "shorter" EXOC7 isoform (that lacks exon 7), which was shown to be of central importance for IL-8 and IL-6 expression (representative of the SASP) during OIS (Georgilis et al. 2018).

AS also seems to play an important role in the cancer cell's ability to evade immune destruction. CD45 is a transmembrane tyrosine phosphatase that mediates T Cell Receptor (TCR) signaling, as well as integrin- and cytokine-mediated signaling (Hermiston et al. 2003). Regulation of CD45 AS is one of the mechanisms involved in T cell activation: CD45 transcripts undergo extensive AS in which three variable exons are preferentially skipped in response to immune challenge, being hnRNPLL a critical mediator of this process (Oberdoerffer et al. 2008). The transition from naïve to activated T cells is marked predominantly by increased expression of a smaller CD45 isoform, CD45RO, that excludes all the variable exons. CD45RO isoform diminishes T-cell signaling in response to external stimuli and, therefore, the modulation of CD45 AS favoring the expression of the short isoform may promote the termination of T-cell response in activated T-cells, hampering immune surveillance (Xu and Weiss 2002; Liu and Cheng 2013).

Conversely, signals derived from the tumor microenvironment can induce changes in the generation or stability of the alternative spliced isoforms in cancer cells. Stroma-derived signals such as growth factors were shown to lead to changes in AS patterns of several transcripts in a variety of cancer. Of this are examples the hepatocyte growth factor (HGF), shown to promote the generation of a tumor-promoting AS variant of the tumor suppressor KLF6, by downregulating SRSF3 in hepatocellular carcinomas (Muñoz et al. 2012). Epithelial growth factor (EGF) was also shown to impact on SRPK1 and SRPK2 mediated modulation of AS via AKT signaling activation (Zhou et al. 2012b). AKT-driven activation of SRSF1 induced by growth factors stimulation was also shown to promote the generation of a fibronectin splicing variant with increased pro-proliferative and migration abilities in breast cancer tissue. Likewise, activated SRSF1 was shown to prevent the synthesis of an anti-apoptotic AS isoform of Casp-9 in lung cancer (Shultz et al. 2010). AS patterns in tumors may be as well modulated by signals derived from the inflammatory cell components of its microenvironment. Examples of this include the release of interferon (IFN) by immune cells, which activates the JAK/STAT pathway in breast cancer cells inducing the expression of Interferon regulatory factor-1 (IRF-1). This latter was shown to impact

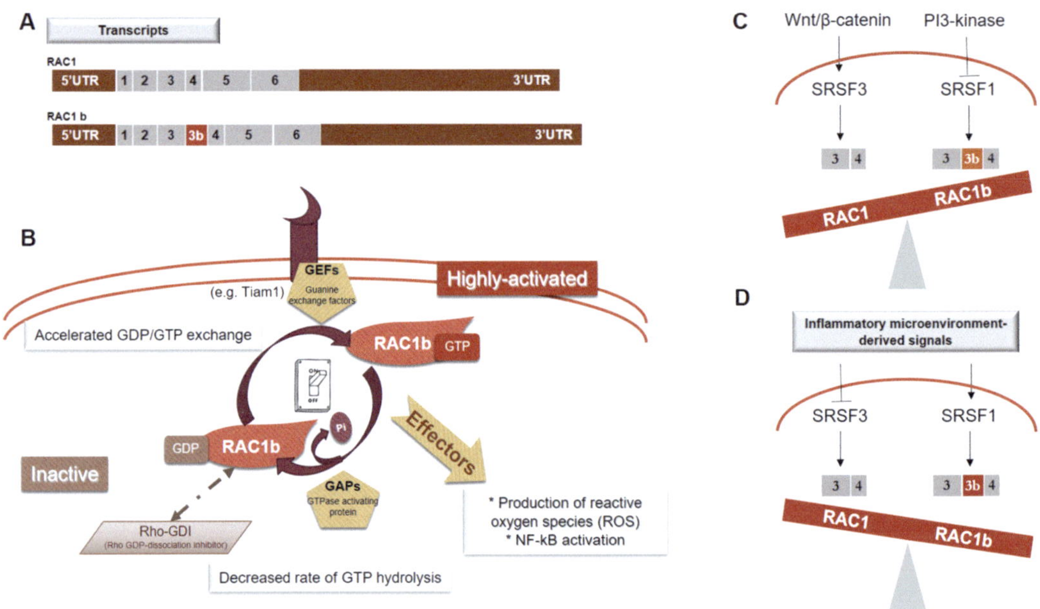

Fig. 13.2 RAC1 alternative splicing in colorectal tumors. (**a**) RAC1b isoform results from an AS event that leads to the inclusion of an additional exon (exon 3b). (**b**) RAC1b, compared to RAC1, shows important pro-tumorigenic properties: it exists predominantly in the GTP-bound active conformation and shows a selective downstream signaling that favors tumor cell survival through the production of ROS and activation of the NF-kB pathway. In CRC, AS of RAC1 is regulated by two SR proteins with antagonistic roles, SRSF1 and SRSF3, which favor exon3b inclusion and skipping, respectively. (**c**) Some cell signaling contexts may impair the generation of the RAC1b variant: Wnt/β-catenin signaling increases the expression of SRSF3 and PI3K/AKT signaling downregulates the expression of SRSF1, both favoring exon 3b skipping. (**d**) Other signaling cues may favor SRSF1 expression relative to SRSF3, thus promoting exon 3b inclusion. In the context of inflammation, inflammatory microenvironment-derived signals can favor the AS switch that generates RAC1b

on AS of several genes involved in regulation of growth and differentiation (Dery et al. 2014). Also, cytokines such as GM-CSF and IL-6 were shown to promote the generation of BCL-x(L), an anti-apoptotic splice variant of BCL2L1, in leukemia cells (Li et al. 2004).

13.6 The Paradigm of RAC1 Alternative Splicing in Serrated Colorectal Tumors

An example of an AS event triggered by inflammation is illustrated by the generation in colonocytes of tumor-related RAC1b, a highly activated splice variant of the GTPase RAC1 (Matos et al. 2003; Fiegen et al. 2004). The RAC1b isoform (Fig. 13.2a) results from an AS event that leads to the inclusion of an additional exon (exon 3b), which confer to RAC1b important pro-tumorigenic properties (Jordan et al. 1999; Melzer et al. 2019). RAC1b lacks down-regulation by the regulatory factor Rho-GDI, existing predominantly in the GTP-bound active conformation and, compared to RAC1, shows a selective downstream signaling favoring specific pathways conducting to the production of reactive oxygen species (ROS) and NF-kB activation ((Fig. 13.2b; Matos et al. 2003; Matos and Jordan 2005, 2006; Faria et al. 2017). RAC1 exon 3b inclusion is promoted by SRSF1, which is activated by phosphorylation that can be accomplished by the SRPK1 kinase (Gonçalves et al. 2014). The overexpression of this kinase has been described in multiple cancers including colon, breast, pancreatic and gastric carcinomas (Hayes et al. 2007; Xu et al. 2017). Yet, the mechanism leading to SRPK1 increased levels remain unclear. Conversely, the splicing factor SRSF3 was shown to promote the skipping of *RAC1* exon 3b, leading to decreased RAC1b expression (Gonçalves et al. 2009). Thus, the expression of these two SR proteins with antagonistic roles may be regulated by different signaling pathways in a concerted action to regulate inclusion or skipping of RAC1 alternative exon 3b. The Wnt/β-catenin signaling, which is frequently dysregulated in CRC through mutations in APC, was shown to increase the expression of SRSF3. The PI3K/AKT signaling, on the other hand, was found to downregulate the expression of SRSF1 (Gonçalves et al. 2014). The activation of both pathways thus promotes the skipping of exon 3b, disfavoring the generation of RAC1b isoform (Fig. 13.2c). Several genetic alterations have been described in CRC, the most prevalent including the oncogenic alterations in APC, TP53, KRAS, PI3K and BRAF (Mármol et al. 2017). Notably, subsets of tumors with oncogenic APC, KRAS (which strongly activates PI3K) or PI3K, do not express the RAC1b isoform. On the other hand, tumors with wild-type KRAS or PI3K but harboring the alternative oncogenic BRAF mutation (which cannot directly activate PI3K) were shown to express increased levels of RAC1b (Matos et al. 2008). In fact, RAC1b was found to be overexpressed in a subset of, so called, serrated colorectal tumors characterized by a high incidence of BRAFV600E mutations and microsatellite instability (MSI) (Matos et al. 2008, 2016). Moreover, survival of these tumor cells was shown to be dependent on the functional cooperation between the overexpression of RAC1b and the constitutive mitogenic signaling of mutant BRAFV600E (Matos and Jordan 2008; Matos et al. 2008). Although the molecular details of the mechanism behind RAC1b overexpression remain unclear, in the context of CRC it may be related to signaling cues derived from driver alterations that initiate colorectal tumorigenesis without, however, enhancing PI3-kinase and β-catenin signaling. These signals may favor SRSF1 expression relative to SRSF3, thus promoting RAC1b overexpression and ensuing increased tumor cell survival (Fig. 13.2d). It is thus proposed that CRC initiated by a BRAFV600E mutation later will require overexpression of hyperactive RAC1b to allow further tumor progression (Matos et al. 2016).

Indeed, it has been shown that inflammatory microenvironment-derived signals can favor the AS switch that generates RAC1b. In fact, the expression of RAC1b was found to be increased in patients with colon inflammatory disorders and following acute colitis induction in mice (Matos et al. 2013). This inflammation-induced increase was counteracted by ibuprofen treatment (Matos et al. 2013), and this involved reduced phosphorylation of SRSF1, which is required for inclusion of the alternative exon 3b (Matos and Jordan 2015; Gonçalves et al. 2017).

In the breast cancer context, stroma-derived signals were also shown to affected RAC1/1b switching, being the activity of extracellular matrix metalloproteinase 3 (MMP3) able to induce an increase in RAC1b expression (Radisky et al. 2005). The mechanism, of how RAC1b expression is induced in this tumor type is also unclear. Moreover, it may likely differ among different tissue types: while in CRC cells, RAC1b AS occurs through the regulation of the two antagonistic splicing factors, SRSF1 and SRSF3 (Gonçalves et al. 2009), in mouse mammary epithelial cells, however, the factor hnRNP A1 seems to mediate exon 3b exclusion, repressing the formation of the RAC1b (Pelisch et al. 2012).

Besides CRC and breast carcinoma, RAC1b was also found to be overexpressed in lung, pancreatic and thyroid cancer (Zhou et al. 2012a; Stallings-Mann et al. 2012; Silva et al. 2013; Mehner et al. 2015; Faria et al. 2016). Notably, in the subset of differentiated thyroid carcinomas (TC) of follicular origin, particularly in the papillary subtype (PTC) in which the oncogenic BRAFV600E mutation is highly prevalent, overexpression of RAC1b was significantly associated with BRAFV600E and unfavorable clinical outcomes (Silva et al. 2013). Similar to that described in CRC, RAC1b overexpression in PTC was shown to induce NF-kB activation, promoting cell proliferation and resistance to apoptosis (Faria et al. 2017).

13.7 Inflammation and RAC1b Splicing in Thyroid Carcinomas

Even though the cellular signaling contexts surrounding RAC1b expression appear to be shared by colorectal and thyroid cancers (association of BRAF V600E with RAC1b overexpression and activation of canonical NF-kB pathway), it remains to be determined whether the tumor's inflammatory microenvironment also modulates the RAC1/1b switch in thyroid cancer.

As it has been said throughout this chapter, in several aspects, CRC represents a paradigm of how the interrelationship between the tumor and the tumor-associated inflammatory microenvironment play a pivotal role in the multiple stages of tumorigenesis, from initiation and maintenance of the malignant phenotype to tumor progression, invasion and metastasis. While chronic inflammation due to long-standing inflammatory bowel disease is a major driving mechanism for the development of CRC, in the thyroid cancer context, this association is not so well defined.

Notably, the thyroid is the organ most affected by chronic inflammation due to autoimmune conditions, with Hashimoto's thyroiditis (HT) being the most common autoimmune disease (Dong and Fu 2014). Accordingly, the relation between autoimmune thyroiditis (AIT) and PTC have long been debated (Boi et al. 2017; Nagayama 2018). AIT was considered for many years a premalignant condition that predisposes patients to thyroid cancer mainly because a high prevalence of thyroiditis surrounding cancerous lesions at the time of thyroidectomy has been observed (Dailey et al. 1955; Muzza et al. 2010; Oh et al. 2014; Lai et al. 2017). Yet, there are additional evidence suggesting otherwise, namely by several studies failing to find an association between AIT and an increased risk of malignancy (Anil et al. 2010; Jankovic et al. 2013; Castagna et al. 2014; Selek et al. 2017). Regardless of whether or not an association between AIT and thyroid cancer exists, the presence of lymphocytic infiltrate is frequently observed in thyroid glands harboring thyroid cancer (Boi et al. 2017; Nagayama 2018). Adding complexity to this subject, a body of evidence suggests that these chronic inflammatory infiltrates may have a protective effect, preventing metastasis and tumor recurrence, and thus their presence being associated with a favorable prognosis (Fugazzola et al. 2011; Liang et al. 2017; Song et al. 2018a).

Considering the molecular context, while CRC with BRAFV600E mutation are characterized by abundant lymphocytic infiltration, PTC-associated autoimmunity is more frequently linked to RET/PTC rearrangement (Muzza et al. 2010; Fugazzola et al. 2011), one of the oncogenic alterations responsible for neoplastic transformation of thyroid cells. Notwithstanding the lack of association between autoimmunity and BRAFV600E mutation in thyroid cancer (Kim et al. 2016, 2018), lymphocytic infiltrates not linked to AIT have been also found in association with BRAF-mutated PTCs (Liang et al. 2017). This is consistent with BRAFV600E mutation in PTCs being associated with tumor-related inflammation but not with the autoimmune process. Indeed, the expression of the inflammation-related genes, CCL20 and CXCL8, was shown to be increased in both BRAFV600E and RET/PTC tumors, when compared to non-neoplastic tissues with thyroiditis, which displayed CCL20 and CXCL8 levels similar to those of normal thyroid (Muzza et al. 2010).

13.8 Concluding Remarks

Although the relevance of both AS and the inflammatory microenvironment to the etiology of many cancer types is no longer disputed, we are still far from fully understand the complex mechanisms regulating the interaction between these events and how they can be potentially explored towards the development of better diagnosis and treatment strategies. In the last few years, the growing awareness of the role played by inflammation and deregulated AS programs in human cancers has fostered the development of promising new therapeutic strategies selectively targeting either process (Lee and Abdel-Wahab 2016; Nakamura and Smyth 2017). These therapeutic strategies range from the inhibition of inflammatory mediators and immune cells, to "re-educate" the inflammatory microenvironment (Nakamura and Smyth 2017), to the identification of small molecules to target components of the splicing machinery and the exploitation of antisense oligonucleotides to manipulate splicing decisions (Lee and Abdel-Wahab 2016). The challenge for the next future is to decipher the connections between the two molecular processes so as to conceive strategies synergistically target their interplay and potentiate their therapeutic efficacy.

References

Adegboyega PA, Ololade O, Saada J, Mifflin R, Di Mari JF, Powell DW (2004) Subepithelial myofibroblasts express cyclooxygenase-2 in colorectal tubular adenomas. Clin Cancer Res Off J Am Assoc Cancer Res 10:5870–5879. https://doi.org/10.1158/1078-0432.CCR-0431-03

Agrawal AA, Yu L, Smith PG, Buonamici S (2018) Targeting splicing abnormalities in cancer. Curr Opin Genet Dev 48:67–74. https://doi.org/10.1016/j.gde.2017.10.010

Anil C, Goksel S, Gursoy A (2010) Hashimoto's thyroiditis is not associated with increased risk of thyroid cancer in patients with thyroid nodules: a single-center prospective study. Thyroid Off J Am Thyroid Assoc 20:601–606. https://doi.org/10.1089/thy.2009.0450

Bellahcène A, Castronovo V, Ogbureke KUE, Fisher LW, Fedarko NS (2008) Small integrin-binding ligand N-linked glycoproteins (SIBLINGs): multifunctional proteins in cancer. Nat Rev Cancer 8:212–226. https://doi.org/10.1038/nrc2345

Boi F, Pani F, Mariotti S (2017) Thyroid autoimmunity and thyroid cancer: review focused on cytological studies. Eur Thyroid J 6:178–186. https://doi.org/10.1159/000468928

Briones-Orta MA, Avendaño-Vázquez SE, Aparicio-Bautista DI, Coombes JD, Weber GF, Syn W-K (2017) Osteopontin splice variants and polymorphisms in cancer progression and prognosis. Biochim Biophys Acta BBA – Rev Cancer 1868:93–108. https://doi.org/10.1016/j.bbcan.2017.02.005

Brosseau J-P, Lucier J-F, Nwilati H, Thibault P, Garneau D, Gendron D, Durand M, Couture S, Lapointe E, Prinos P, Klinck R, Perreault J-P, Chabot B, Abou-Elela S (2014) Tumor microenvironment-associated modifications of alternative splicing. RNA 20:189–201. https://doi.org/10.1261/rna.042168.113

Castagna MG, Belardini V, Memmo S, Maino F, Di Santo A, Toti P, Carli AF, Caruso G, Pacini F (2014) Nodules in autoimmune thyroiditis are associated with increased risk of thyroid cancer in surgical series but not in cytological series: evidence for selection bias. J Clin Endocrinol Metab 99:3193–3198. https://doi.org/10.1210/jc.2014-1302

Castello LM, Raineri D, Salmi L, Clemente N, Vaschetto R, Quaglia M, Garzaro M, Gentilli S, Navalesi P, Cantaluppi V, Dianzani U, Aspesi A, Chiocchetti A (2017) Osteopontin at the crossroads of inflammation and tumor progression. Mediat Inflamm 2017:1–22. https://doi.org/10.1155/2017/4049098

Castellone MD, Teramoto H, Williams BO, Druey KM, Gutkind JS (2005) Prostaglandin E2 promotes colon cancer cell growth through a Gs-axin-beta-catenin signaling axis. Science 310:1504–1510. https://doi.org/10.1126/science.1116221

Chen M, Zhang J, Manley JL (2010) Turning on a fuel switch of cancer: hnRNP proteins regulate alternative splicing of pyruvate kinase mRNA. Cancer Res 70:8977–8980. https://doi.org/10.1158/0008-5472.CAN-10-2513

Cheng C, Yaffe MB, Sharp PA (2006) A positive feedback loop couples Ras activation and CD44 alternative splicing. Genes Dev 20:1715–1720. https://doi.org/10.1101/gad.1430906

Climente-González H, Porta-Pardo E, Godzik A, Eyras E (2017) The functional impact of alternative splicing in cancer. Cell Rep 20:2215–2226. https://doi.org/10.1016/j.celrep.2017.08.012

Clower CV, Chatterjee D, Wang Z, Cantley LC, Vander Heiden MG, Krainer AR (2010) The alternative splicing repressors hnRNP A1/A2 and PTB influence pyruvate kinase isoform expression and cell metabolism. Proc Natl Acad Sci U S A 107:1894–1899. https://doi.org/10.1073/pnas.0914845107

Correa P (1995) Helicobacter pylori and gastric carcinogenesis. Am J Surg Pathol 19(Suppl 1):S37–S43

Coussens LM, Werb Z (2002) Inflammation and cancer. Nature 420:860–867. https://doi.org/10.1038/nature01322

Dailey ME, Lindsay S, Skahen R (1955) Relation of thyroid neoplasms to Hashimoto disease of the thyroid gland. AMA Arch Surg 70:291–297

David CJ, Chen M, Assanah M, Canoll P, Manley JL (2010) HnRNP proteins controlled by c-Myc deregulate pyruvate kinase mRNA splicing in cancer. Nature 463:364–368. https://doi.org/10.1038/nature08697

Dery KJ, Kujawski M, Grunert D, Wu X, Ngyuen T, Cheung C, Yim JH, Shively JE (2014) IRF-1 regulates alternative mRNA splicing of carcinoembryonic antigen-related cell adhesion molecule 1 (CEACAM1) in breast epithelial cells generating an immunore-

ceptor tyrosine-based inhibition motif (ITIM) containing isoform. Mol Cancer 13:64. https://doi.org/10.1186/1476-4598-13-64
Devaud C, Tilkin-Mariamé A-F, Vignolle-Vidoni A, Souleres P, Denadai-Souza A, Rolland C, Duthoit C, Blanpied C, Chabot S, Bouillé P, Lluel P, Vergnolle N, Racaud-Sultan C, Ferrand A (2019) FAK alternative splice mRNA variants expression pattern in colorectal cancer: FAK alternative splice mRNA variants expression pattern in colorectal cancer. Int J Cancer. https://doi.org/10.1002/ijc.32120
Dong YH, Fu DG (2014) Autoimmune thyroid disease: mechanism, genetics and current knowledge. Eur Rev Med Pharmacol Sci 18:3611–3618
Dvinge H (2018) Regulation of alternative mRNA splicing: old players and new perspectives. FEBS Lett 592:2987–3006. https://doi.org/10.1002/1873-3468.13119
Faria M, Capinha L, Simões-Pereira J, Bugalho MJ, Silva AL (2016) Extending the impact of RAC1b overexpression to follicular thyroid carcinomas. Int J Endocrinol 2016:1972367. https://doi.org/10.1155/2016/1972367
Faria M, Matos P, Pereira T, Cabrera R, Cardoso BA, Bugalho MJ, Silva AL (2017) RAC1b overexpression stimulates proliferation and NF-kB-mediated anti-apoptotic signaling in thyroid cancer cells. PLoS One 12:e0172689. https://doi.org/10.1371/journal.pone.0172689
Ferreira LB, Tavares C, Pestana A, Pereira CL, Eloy C, Pinto MT, Castro P, Batista R, Rios E, Sobrinho-Simões M, Gimba ERP, Soares P (2016) Osteopontin-a splice variant is overexpressed in papillary thyroid carcinoma and modulates invasive behavior. Oncotarget 7. https://doi.org/10.18632/oncotarget.10468
Fiegen D, Haeusler L-C, Blumenstein L, Herbrand U, Dvorsky R, Vetter IR, Ahmadian MR (2004) Alternative splicing of Rac1 generates Rac1b, a self-activating GTPase. J Biol Chem 279:4743–4749. https://doi.org/10.1074/jbc.M310281200
Fodde R, Smits R, Clevers H (2001) APC, signal transduction and genetic instability in colorectal cancer. Nat Rev Cancer 1:55–67. https://doi.org/10.1038/35094067
Fouad YA, Aanei C (2017) Revisiting the hallmarks of cancer. Am J Cancer Res 7:1016–1036
Fugazzola L, Colombo C, Perrino M, Muzza M (2011) Papillary thyroid carcinoma and inflammation. Front Endocrinol 2. https://doi.org/10.3389/fendo.2011.00088
Galizia G, Orditura M, Romano C, Lieto E, Castellano P, Pelosio L, Imperatore V, Catalano G, Pignatelli C, De Vita F (2002) Prognostic significance of circulating IL-10 and IL-6 serum levels in colon cancer patients undergoing surgery. Clin Immunol Orlando Fla 102:169–178. https://doi.org/10.1006/clim.2001.5163
Georgilis A, Klotz S, Hanley CJ, Herranz N, Weirich B, Morancho B, Leote AC, D'Artista L, Gallage S, Seehawer M, Carroll T, Dharmalingam G, Wee KB, Mellone M, Pombo J, Heide D, Guccione E, Arribas J, Barbosa-Morais NL, Heikenwalder M, Thomas GJ, Zender L, Gil J (2018) PTBP1-mediated alternative splicing regulates the inflammatory Secretome and the pro-tumorigenic effects of senescent cells. Cancer Cell 34:85–102.e9. https://doi.org/10.1016/j.ccell.2018.06.007
Gimba ER, Tilli TM (2013) Human osteopontin splicing isoforms: known roles, potential clinical applications and activated signaling pathways. Cancer Lett 331:11–17. https://doi.org/10.1016/j.canlet.2012.12.003
Giorgio M, Maria Z, Maria C, Marco I, Martina S, Gianna F, Eliana C, Giovanna R (2013) Pro-inflammatory cytokines analysis in HPV-positive cancer cells. Front Immunol 4. https://doi.org/10.3389/conf.fimmu.2013.02.00306
Gomaa W, Al-Ahwal M, Hamour O, Al-Maghrabi J (2013) Osteopontin cytoplasmic immunoexpression is a predictor of poor disease-free survival in thyroid cancer. J Microsc Ultrastruct 1:8. https://doi.org/10.1016/j.jmau.2013.07.001
Gonçalves V, Matos P, Jordan P (2009) Antagonistic SR proteins regulate alternative splicing of tumor-related Rac1b downstream of the PI3-kinase and Wnt pathways. Hum Mol Genet 18:3696–3707. https://doi.org/10.1093/hmg/ddp317
Gonçalves V, Henriques AFA, Henriques A, Pereira JFS, Pereira J, Neves Costa A, Moyer MP, Moita LF, Gama-Carvalho M, Matos P, Jordan P (2014) Phosphorylation of SRSF1 by SRPK1 regulates alternative splicing of tumor-related Rac1b in colorectal cells. RNA 20:474–482. https://doi.org/10.1261/rna.041376.113
Gonçalves V, Pereira J, Jordan P (2017) Signaling pathways driving aberrant splicing in cancer cells. Genes 9:9. https://doi.org/10.3390/genes9010009
Gonzalez-Zubeldia I, Dotor J, Redrado M, Bleau A-M, Manrique I, de Aberasturi AL, Villalba M, Calvo A (2015) Co-migration of colon cancer cells and CAFs induced by TGFβ_1 enhances liver metastasis. Cell Tissue Res 359:829–839. https://doi.org/10.1007/s00441-014-2075-6
Grivennikov SI (2013) Inflammation and colorectal cancer: colitis-associated neoplasia. Semin Immunopathol 35:229–244. https://doi.org/10.1007/s00281-012-0352-6
Grivennikov S, Karin E, Terzic J, Mucida D, Yu G-Y, Vallabhapurapu S, Scheller J, Rose-John S, Cheroutre H, Eckmann L, Karin M (2009) IL-6 and Stat3 are required for survival of intestinal epithelial cells and development of colitis-associated cancer. Cancer Cell 15:103–113. https://doi.org/10.1016/j.ccr.2009.01.001
Grivennikov SI, Greten FR, Karin M (2010) Immunity, inflammation, and cancer. Cell 140:883–899. https://doi.org/10.1016/j.cell.2010.01.025
Hanahan D, Weinberg RA (2011) Hallmarks of cancer: the next generation. Cell 144:646–674. https://doi.org/10.1016/j.cell.2011.02.013
Hassuneh MR, Nagarkatti M, Nagarkatti PS (2013) Role of interleukin-10 in the regulation of tumorigenicity of a T cell lymphoma. Leuk Lymphoma 54:827–834. https://doi.org/10.3109/10428194.2012.726721
Hayes GM, Carrigan PE, Miller LJ (2007) Serine-arginine protein kinase 1 overexpression is associated with tumorigenic imbalance in mitogen-activated protein

kinase pathways in breast, colonic, and pancreatic carcinomas. Cancer Res 67:2072–2080. https://doi.org/10.1158/0008-5472.CAN-06-2969

Herbeuval J-P, Lelievre E, Lambert C, Dy M, Genin C (2004) Recruitment of STAT3 for production of IL-10 by colon carcinoma cells induced by macrophage-derived IL-6. J Immunol Baltim Md 172:4630–4636

Hermiston ML, Xu Z, Weiss A (2003) CD45: a critical regulator of signaling thresholds in immune cells. Annu Rev Immunol 21:107–137. https://doi.org/10.1146/annurev.immunol.21.120601.140946

Hollander D, Donyo M, Atias N, Mekahel K, Melamed Z, Yannai S, Lev-Maor G, Shilo A, Schwartz S, Barshack I, Sharan R, Ast G (2016) A network-based analysis of colon cancer splicing changes reveals a tumorigenesis-favoring regulatory pathway emanating from ELK1. Genome Res 26:541–553. https://doi.org/10.1101/gr.193169.115

Jankovic B, Le KT, Hershman JM (2013) Hashimoto's thyroiditis and papillary thyroid carcinoma: is there a correlation? J Clin Endocrinol Metab 98:474–482. https://doi.org/10.1210/jc.2012-2978

Jedinak A, Dudhgaonkar S, Sliva D (2010) Activated macrophages induce metastatic behavior of colon cancer cells. Immunobiology 215:242–249. https://doi.org/10.1016/j.imbio.2009.03.004

Jordan P, Brazåo R, Boavida MG, Gespach C, Chastre E (1999) Cloning of a novel human Rac1b splice variant with increased expression in colorectal tumors. Oncogene 18:6835–6839. https://doi.org/10.1038/sj.onc.1203233

Joyce JA, Pollard JW (2009) Microenvironmental regulation of metastasis. Nat Rev Cancer 9:239–252. https://doi.org/10.1038/nrc2618

Kaler P, Godasi BN, Augenlicht L, Klampfer L (2009) The NF-κB/AKT-dependent induction of Wnt signaling in colon cancer cells by macrophages and IL-1β. Cancer Microenviron Off J Int Cancer Microenviron Soc 2:69–80. https://doi.org/10.1007/s12307-009-0030-y

Kanda T, Goto T, Hirotsu Y, Moriyama M, Omata M (2019) Molecular mechanisms driving progression of liver cirrhosis towards hepatocellular carcinoma in chronic hepatitis B and C infections: a review. Int J Mol Sci 20. https://doi.org/10.3390/ijms20061358

Kazanecki CC, Uzwiak DJ, Denhardt DT (2007) Control of osteopontin signaling and function by post-translational phosphorylation and protein folding. J Cell Biochem 102:912–924. https://doi.org/10.1002/jcb.21558

Kessenbrock K, Plaks V, Werb Z (2010) Matrix metalloproteinases: regulators of the tumor microenvironment. Cell 141:52–67. https://doi.org/10.1016/j.cell.2010.03.015

Kim S, Myong JP, Jee H-G, Chai YJ, Choi JY, Min HS, Lee KE, Youn Y-K (2016) Combined effect of Hashimoto's thyroiditis and BRAF(V600E) mutation status on aggressiveness in papillary thyroid cancer. Head Neck 38:95–101. https://doi.org/10.1002/hed.23854

Kim WW, Ha TK, Bae SK (2018) Clinical implications of the BRAF mutation in papillary thyroid carcinoma and chronic lymphocytic thyroiditis. J Otolaryngol – Head Neck Surg 47. https://doi.org/10.1186/s40463-017-0247-6

Klinck R, Bramard A, Inkel L, Dufresne-Martin G, Gervais-Bird J, Madden R, Paquet ER, Koh C, Venables JP, Prinos P, Jilaveanu-Pelmus M, Wellinger R, Rancourt C, Chabot B, Abou Elela S (2008) Multiple alternative splicing markers for ovarian cancer. Cancer Res 68:657–663. https://doi.org/10.1158/0008-5472.CAN-07-2580

Kumari S, Badana AK, MM G, S G, Malla R (2018) Reactive oxygen species: a key constituent in cancer survival. Biomark Insights 13:1177271918755391. https://doi.org/10.1177/1177271918755391

Lai X, Xia Y, Zhang B, Li J, Jiang Y (2017) A meta-analysis of Hashimoto's thyroiditis and papillary thyroid carcinoma risk. Oncotarget 8:62414–62424. https://doi.org/10.18632/oncotarget.18620

Landskron G, De la Fuente M, Thuwajit P, Thuwajit C, Hermoso MA (2014) Chronic inflammation and cytokines in the tumor microenvironment. J Immunol Res 2014:149185. https://doi.org/10.1155/2014/149185

Lee SC-W, Abdel-Wahab O (2016) Therapeutic targeting of splicing in cancer. Nat Med 22:976–986. https://doi.org/10.1038/nm.4165

Lee G, Goretsky T, Managlia E, Dirisina R, Singh AP, Brown JB, May R, Yang G-Y, Ragheb JW, Evers BM, Weber CR, Turner JR, He XC, Katzman RB, Li L, Barrett TA (2010) Phosphoinositide 3-kinase signaling mediates beta-catenin activation in intestinal epithelial stem and progenitor cells in colitis. Gastroenterology 139(869–881):881–889. https://doi.org/10.1053/j.gastro.2010.05.037

Li CY, Chu JY, Yu JK, Huang XQ, Liu XJ, Shi L, Che YC, Xie JY (2004) Regulation of alternative splicing of Bcl-x by IL-6, GM-CSF and TPA. Cell Res 14:473–479. https://doi.org/10.1038/sj.cr.7290250

Li H, Fan X, Houghton J (2007) Tumor microenvironment: the role of the tumor stroma in cancer. J Cell Biochem 101:805–815. https://doi.org/10.1002/jcb.21159

Liang J, Zeng W, Fang F, Yu T, Zhao Y, Fan X, Guo N, Gao X (2017) Clinical analysis of Hashimoto thyroiditis coexistent with papillary thyroid cancer in 1392 patients. Acta Otorhinolaryngol Ital Organo Uff Della Soc Ital Otorinolaringol E Chir Cerv-facc 37:393–400. https://doi.org/10.14639/0392-100X-1709

Likui W, Hong W, Shuwen Z (2010) Clinical significance of the upregulated Osteopontin mRNA expression in human colorectal Cancer. J Gastrointest Surg 14:74–81. https://doi.org/10.1007/s11605-009-1035-z

Liu S, Cheng C (2013) Alternative RNA splicing and cancer: alternative RNA splicing and cancer. Wiley Interdiscip Rev RNA 4:547–566. https://doi.org/10.1002/wrna.1178

Liu J, Li H, Shen S, Sun L, Yuan Y, Xing C (2018) Alternative splicing events implicated in carcinogenesis and prognosis of colorectal cancer. J Cancer 9:1754–1764. https://doi.org/10.7150/jca.24569

Mantovani A, Allavena P, Sica A, Balkwill F (2008) Cancer-related inflammation. Nature 454:436–444. https://doi.org/10.1038/nature07205

Mármol I, Sánchez-de-Diego C, Pradilla Dieste A, Cerrada E, Rodriguez Yoldi MJ (2017) Colorectal carcinoma: a general overview and future perspectives in colorectal cancer. Int J Mol Sci 18. https://doi.org/10.3390/ijms18010197

Matos P, Jordan P (2005) Expression of Rac1b stimulates NF-kappaB-mediated cell survival and G1/S progression. Exp Cell Res 305:292–299. https://doi.org/10.1016/j.yexcr.2004.12.029

Matos P, Jordan P (2006) Rac1, but not Rac1B, stimulates RelB-mediated gene transcription in colorectal cancer cells. J Biol Chem 281:13724–13732. https://doi.org/10.1074/jbc.M513243200

Matos P, Jordan P (2008) Increased Rac1b expression sustains colorectal tumor cell survival. Mol Cancer Res MCR 6:1178–1184. https://doi.org/10.1158/1541-7786.MCR-08-0008

Matos P, Jordan P (2015) Beyond COX-inhibition: "side-effects" of ibuprofen on neoplastic development and progression. Curr Pharm Des 21:2978–2982

Matos P, Collard JG, Jordan P (2003) Tumor-related alternatively spliced Rac1b is not regulated by Rho-GDP dissociation inhibitors and exhibits selective downstream signaling. J Biol Chem 278:50442–50448. https://doi.org/10.1074/jbc.M308215200

Matos P, Oliveira C, Velho S, Gonçalves V, da Costa LT, Moyer MP, Seruca R, Jordan P (2008) B-Raf(V600E) cooperates with alternative spliced Rac1b to sustain colorectal cancer cell survival. Gastroenterology 135:899–906. https://doi.org/10.1053/j.gastro.2008.05.052

Matos P, Kotelevets L, Jordan P, Gonçalves V, Henriques A, Zerbib P, Moyer MP, Chastre E (2013) Ibuprofen inhibits colitis-induced overexpression of tumor related Rac1b. Neoplasia 15:102–111. https://doi.org/10.1593/neo.121890

Matos P, Gonçalves V, Jordan P (2016) Targeting the serrated pathway of colorectal cancer with mutation in BRAF. Biochim Biophys Acta 1866:51–63. https://doi.org/10.1016/j.bbcan.2016.06.003

Mehner C, Miller E, Nassar A, Bamlet WR, Radisky ES, Radisky DC (2015) Tumor cell expression of MMP3 as a prognostic factor for poor survival in pancreatic, pulmonary, and mammary carcinoma. Genes Cancer 6:480–489. https://doi.org/10.18632/genesandcancer.90

Melzer C, Hass R, Lehnert H, Ungefroren H (2019) RAC1B: a Rho GTPase with versatile functions in malignant transformation and tumor progression. Cell 8. https://doi.org/10.3390/cells8010021

Muñoz Ú, Puche JE, Hannivoort R, Lang UE, Cohen-Naftaly M, Friedman SL (2012) Hepatocyte growth factor enhances alternative splicing of the Kruppel-like factor 6 (KLF6) tumor suppressor to promote growth through SRSF1. Mol Cancer Res MCR 10:1216–1227. https://doi.org/10.1158/1541-7786.MCR-12-0213

Muzza M, Degl'Innocenti D, Colombo C, Perrino M, Ravasi E, Rossi S, Cirello V, Beck-Peccoz P, Borrello MG, Fugazzola L (2010) The tight relationship between papillary thyroid cancer, autoimmunity and inflammation: clinical and molecular studies. Clin Endocrinol 72:702–708. https://doi.org/10.1111/j.1365-2265.2009.03699.x

Nagayama Y (2018) Thyroid autoimmunity and thyroid cancer – the pathogenic connection: a 2018 update. Horm Metab Res Horm Stoffwechselforschung Horm Metab 50:922–931. https://doi.org/10.1055/a-0648-4593

Nakamura K, Smyth MJ (2017) Targeting cancer-related inflammation in the era of immunotherapy. Immunol Cell Biol 95:325–332. https://doi.org/10.1038/icb.2016.126

Oberdoerffer S, Moita LF, Neems D, Freitas RP, Hacohen N, Rao A (2008) Regulation of CD45 alternative splicing by heterogeneous ribonucleoprotein, hnRNPLL. Science 321:686–691. https://doi.org/10.1126/science.1157610

Oh C-M, Park S, Lee JY, Won Y-J, Shin A, Kong H-J, Choi K-S, Lee YJ, Chung K-W, Jung K-W (2014) Increased prevalence of chronic lymphocytic thyroiditis in Korean patients with papillary thyroid cancer. PLoS One 9:e99054. https://doi.org/10.1371/journal.pone.0099054

Parascandolo A, Rappa F, Cappello F, Kim J, Cantu DA, Chen H, Mazzoccoli G, Hematti P, Castellone MD, Salvatore M, Laukkanen MO (2017) Extracellular superoxide dismutase expression in papillary thyroid cancer mesenchymal stem/stromal cells modulates cancer cell growth and migration. Sci Rep 7:41416. https://doi.org/10.1038/srep41416

Pelisch F, Khauv D, Risso G, Stallings-Mann M, Blaustein M, Quadrana L, Radisky DC, Srebrow A (2012) Involvement of hnRNP A1 in the matrix metalloprotease-3-dependent regulation of Rac1 pre-mRNA splicing. J Cell Biochem 113:2319–2329. https://doi.org/10.1002/jcb.24103

Picon A, Gold LI, Wang J, Cohen A, Friedman E (1998) A subset of metastatic human colon cancers expresses elevated levels of transforming growth factor beta1. Cancer Epidemiol Biomark Prev Publ Am Assoc Cancer Res Cosponsored Am Soc Prev Oncol 7:497–504

Pozzi A, Yan X, Macias-Perez I, Wei S, Hata AN, Breyer RM, Morrow JD, Capdevila JH (2004) Colon carcinoma cell growth is associated with prostaglandin E2/EP4 receptor-evoked ERK activation. J Biol Chem 279:29797–29804. https://doi.org/10.1074/jbc.M313989200

Prochazka L, Tesarik R, Turanek J (2014) Regulation of alternative splicing of CD44 in cancer. Cell Signal 26:2234–2239. https://doi.org/10.1016/j.cellsig.2014.07.011

Radisky ES, Radisky DC (2015) Matrix metalloproteinases as breast cancer drivers and therapeutic targets. Front Biosci Landmark Ed 20:1144–1163

Radisky DC, Levy DD, Littlepage LE, Liu H, Nelson CM, Fata JE, Leake D, Godden EL, Albertson DG, Nieto MA, Werb Z, Bissell MJ (2005) Rac1b and reactive oxygen species mediate MMP-3-induced EMT and genomic instability. Nature 436:123–127. https://doi.org/10.1038/nature03688

Rangaswami H, Bulbule A, Kundu GC (2006) Osteopontin: role in cell signaling and cancer progression. Trends Cell Biol 16:79–87. https://doi.org/10.1016/j.tcb.2005.12.005

Riabov V, Gudima A, Wang N, Mickley A, Orekhov A, Kzhyshkowska J (2014) Role of tumor associated macrophages in tumor angiogenesis and lymphangiogenesis. Front Physiol 5:75. https://doi.org/10.3389/fphys.2014.00075

Romano M, DE Francesco F, Zarantonello L, Ruffolo C, Ferraro GA, Zanus G, Giordano A, Bassi N, Cillo U (2016) From inflammation to Cancer in inflammatory bowel disease: molecular perspectives. Anticancer Res 36:1447–1460

Roque AT, Gambeloni RZ, Felitti S, Ribeiro ML, Santos JC (2015) Inflammation-induced oxidative stress in breast cancer patients. Med Oncol Northwood Lond Engl 32:263. https://doi.org/10.1007/s12032-015-0709-5

Ryder M, Ghossein RA, Ricarte-Filho JCM, Knauf JA, Fagin JA (2008) Increased density of tumor-associated macrophages is associated with decreased survival in advanced thyroid cancer. Endocr Relat Cancer 15:1069–1074. https://doi.org/10.1677/ERC-08-0036

Sanford JR (2004) Pre-mRNA splicing: life at the centre of the central dogma. J Cell Sci 117:6261–6263. https://doi.org/10.1242/jcs.01513

Schoppmann SF, Birner P, Stöckl J, Kalt R, Ullrich R, Caucig C, Kriehuber E, Nagy K, Alitalo K, Kerjaschki D (2002) Tumor-associated macrophages express lymphatic endothelial growth factors and are related to peritumoral lymphangiogenesis. Am J Pathol 161:947–956. https://doi.org/10.1016/S0002-9440(10)64255-1

Selek A, Cetinarslan B, Tarkun I, Canturk Z, Ustuner B, Akyay Z (2017) Thyroid autoimmunity: is really associated with papillary thyroid carcinoma? Eur Arch Oto-Rhino-Laryngol Off J Eur Fed Oto-Rhino-Laryngol Soc EUFOS Affil Ger Soc Oto-Rhino-Laryngol – Head Neck Surg 274:1677–1681. https://doi.org/10.1007/s00405-016-4414-6

Senbanjo LT, Chellaiah MA (2017) CD44: a multifunctional cell surface adhesion receptor is a regulator of progression and metastasis of Cancer cells. Front Cell Dev Biol 5. https://doi.org/10.3389/fcell.2017.00018

Shilo A, Ben Hur V, Denichenko P, Stein I, Pikarsky E, Rauch J, Kolch W, Zender L, Karni R (2014) Splicing factor hnRNP A2 activates the Ras-MAPK-ERK pathway by controlling a-Raf splicing in hepatocellular carcinoma development. RNA 20:505–515. https://doi.org/10.1261/rna.042259.113

Shultz JC, Goehe RW, Wijesinghe DS, Murudkar C, Hawkins AJ, Shay JW, Minna JD, Chalfant CE (2010) Alternative splicing of caspase 9 is modulated by the phosphoinositide 3-kinase/Akt pathway via phosphorylation of SRp30a. Cancer Res 70:9185–9196. https://doi.org/10.1158/0008-5472.CAN-10-1545

Silva AL, Carmo F, Bugalho MJ (2013) RAC1b overexpression in papillary thyroid carcinoma: a role to unravel. Eur J Endocrinol Eur Fed Endocr Soc 168:795–804. https://doi.org/10.1530/EJE-12-0960

Song E, Jeon MJ, Park S, Kim M, Oh H-S, Song DE, Kim WG, Kim WB, Shong YK, Kim TY (2018a) Influence of coexistent Hashimoto's thyroiditis on the extent of cervical lymph node dissection and prognosis in papillary thyroid carcinoma. Clin Endocrinol 88:123–128. https://doi.org/10.1111/cen.13475

Song X, Zeng Z, Wei H, Wang Z (2018b) Alternative splicing in cancers: from aberrant regulation to new therapeutics. Semin Cell Dev Biol 75:13–22. https://doi.org/10.1016/j.semcdb.2017.09.018

Stallings-Mann ML, Waldmann J, Zhang Y, Miller E, Gauthier ML, Visscher DW, Downey GP, Radisky ES, Fields AP, Radisky DC (2012) Matrix metalloproteinase induction of Rac1b, a key effector of lung cancer progression. Sci Transl med 4:142ra95. https://doi.org/10.1126/scitranslmed.3004062

Sun D, Novotny M, Bulek K, Liu C, Li X, Hamilton T (2011) Treatment with IL-17 prolongs the half-life of chemokine CXCL1 mRNA via the adaptor TRAF5 and the splicing-regulatory factor SF2 (ASF). Nat Immunol 12:853–860. https://doi.org/10.1038/ni.2081

Takahashi H, Nishimura J, Kagawa Y, Kano Y, Takahashi Y, Wu X, Hiraki M, Hamabe A, Konno M, Haraguchi N, Takemasa I, Mizushima T, Ishii M, Mimori K, Ishii H, Doki Y, Mori M, Yamamoto H (2015) Significance of polypyrimidine tract-binding protein 1 expression in colorectal cancer. Mol Cancer Ther 14:1705–1716. https://doi.org/10.1158/1535-7163.MCT-14-0142

Taniguchi K, Sugito N, Kumazaki M, Shinohara H, Yamada N, Matsuhashi N, Futamura M, Ito Y, Otsuki Y, Yoshida K, Uchiyama K, Akao Y (2015) Positive feedback of DDX6/c-Myc/PTB1 regulated by miR-124 contributes to maintenance of the Warburg effect in colon cancer cells. Biochim Biophys Acta 1852:1971–1980. https://doi.org/10.1016/j.bbadis.2015.06.022

Tauler J, Mulshine JL (2009) Lung cancer and inflammation: interaction of chemokines and hnRNPs. Curr Opin Pharmacol 9:384–388. https://doi.org/10.1016/j.coph.2009.06.004

Tessner TG, Muhale F, Riehl TE, Anant S, Stenson WF (2004) Prostaglandin E2 reduces radiation-induced epithelial apoptosis through a mechanism involving AKT activation and bax translocation. J Clin Invest 114:1676–1685. https://doi.org/10.1172/JCI22218

Tilli TM, Franco VF, Robbs BK, Wanderley JLM, de Azevedo da Silva FR, de Mello KD, Viola JPB, Weber GF, Gimba ER (2011) Osteopontin-c splicing isoform contributes to ovarian Cancer progression. Mol Cancer Res 9:280–293. https://doi.org/10.1158/1541-7786.MCR-10-0463

Todaro M, Gaggianesi M, Catalano V, Benfante A, Iovino F, Biffoni M, Apuzzo T, Sperduti I, Volpe S, Cocorullo

G, Gulotta G, Dieli F, De Maria R, Stassi G (2014) CD44v6 is a marker of constitutive and reprogrammed Cancer stem cells driving colon cancer metastasis. Cell Stem Cell 14:342–356. https://doi.org/10.1016/j.stem.2014.01.009

Urbanski LM, Leclair N, Anczuków O (2018) Alternative-splicing defects in cancer: splicing regulators and their downstream targets, guiding the way to novel cancer therapeutics. Wiley Interdiscip Rev RNA 9:e1476. https://doi.org/10.1002/wrna.1476

Venables JP, Klinck R, Koh C, Gervais-Bird J, Bramard A, Inkel L, Durand M, Couture S, Froehlich U, Lapointe E, Lucier J-F, Thibault P, Rancourt C, Tremblay K, Prinos P, Chabot B, Elela SA (2009) Cancer-associated regulation of alternative splicing. Nat Struct Amp Mol Biol 16:670

Villalba M, Evans SR, Vidal-Vanaclocha F, Calvo A (2017) Role of TGF-β in metastatic colon cancer: it is finally time for targeted therapy. Cell Tissue Res 370:29–39. https://doi.org/10.1007/s00441-017-2633-9

Wai PY, Kuo PC (2008) Osteopontin: regulation in tumor metastasis. Cancer Metastasis Rev 27:103–118. https://doi.org/10.1007/s10555-007-9104-9

Wakefield LM, Hill CS (2013) Beyond TGFβ: roles of other TGFβ superfamily members in cancer. Nat Rev Cancer 13:328–341. https://doi.org/10.1038/nrc3500

Weber GF (2001) The metastasis gene osteopontin: a candidate target for cancer therapy. Biochim Biophys Acta BBA – Rev Cancer 1552:61–85. https://doi.org/10.1016/S0304-419X(01)00037-3

Xu Z, Weiss A (2002) Negative regulation of CD45 by differential homodimerization of the alternatively spliced isoforms. Nat Immunol 3:764–771. https://doi.org/10.1038/ni822

Xu X, Wei Y, Wang S, Luo M, Zeng H (2017) Serine-arginine protein kinase 1 (SRPK1) is elevated in gastric cancer and plays oncogenic functions. Oncotarget 8:61944–61957. https://doi.org/10.18632/oncotarget.18734

Xue Y, Zhou Y, Wu T, Zhu T, Ji X, Kwon Y-S, Zhang C, Yeo G, Black DL, Sun H, Fu X-D, Zhang Y (2009) Genome-wide analysis of PTB-RNA interactions reveals a strategy used by the general splicing repressor to modulate exon inclusion or skipping. Mol Cell 36:996–1006. https://doi.org/10.1016/j.molcel.2009.12.003

Zhou C, Licciulli S, Avila JL, Cho M, Troutman S, Jiang P, Kossenkov AV, Showe LC, Liu Q, Vachani A, Albelda SM, Kissil JL (2012a) The Rac1 splice form Rac1b promotes K-Ras-induced lung tumorigenesis. Oncogene. https://doi.org/10.1038/onc.2012.99

Zhou Z, Qiu J, Liu W, Zhou Y, Plocinik RM, Li H, Hu Q, Ghosh G, Adams JA, Rosenfeld MG, Fu X-D (2012b) The Akt-SRPK-SR axis constitutes a major pathway in transducing EGF signaling to regulate alternative splicing in the nucleus. Mol Cell 47:422–433. https://doi.org/10.1016/j.molcel.2012.05.014

The Bone Marrow Niche – The Tumor Microenvironment That Ensures Leukemia Progression

14

Bruno António Cardoso

Abstract

The human body requires a constant delivery of fresh blood cells that are needed to maintain body homeostasis. Hematopoiesis is the process that drives the formation of new blood cells from a single stem cell. This is a complex, orchestrated and tightly regulated process that occurs within the bone marrow. When such process is faulty or deregulated, leukemia arises, develops and thrives by subverting normal hematopoiesis and availing the supplies of this rich milieu.

In this book chapter we will describe and characterize the bone marrow microenvironment and its key importance for leukemia expansion. The several components of the bone marrow niche, their interaction with the leukemic cells and the cellular pathways activated within the malignant cells will be emphasized. Finally, novel therapeutic strategies to target this sibling interaction will also be discussed.

Keywords

Hematopoiesis · Leukemia · Bone marrow · Microenvironment · Signaling pathways · Dual-targeting

B. A. Cardoso (✉)
Instituto de Medicina Molecular João Lobo Antunes, Faculdade de Medicina da Universidade de Lisboa, Lisbon, Portugal
e-mail: bcardoso@medicina.ulisboa.pt

Abbreviations

PDK	3-Phosphinositide-dependent Protein Kinase
Ang	Angiopoietin
AGM	Aorta-gonad-mesonephros
ALL	Acute Lymphoblastic Leukemia
AML	Acute Myeloid Leukemia
B-ALL	B-cell Acute Lymphoblastic Leukemia
BM	Bone Marrow
BMP	Bone Morphogenetic Protein
CNS	Central Nervous System
CML	Chronic Myeloid Leukemia
CSF	Colony-stimulating Factor
CLP	Common Lymphoid Progenitor
CMP	Common Myeloid Progenitor
CAR	CXCL12 Abundant Reticular cells
CXCR4	C-X-C chemokine receptor 4
CXCL12	C-X-C motif chemokine ligand 12
Ara-C	Cytarabine
Dll-1	Delta-like-1
DHH	Desert Hedgehog
ETP-ALL	Early T-cell Precursor Acute Lymphoblastic Leukemia
EC	Endothelial Cells
ECM	Extracellular Matrix
FABP4	Fatty Acid Binding Protein 4
FAO	Fatty Acid Oxidation
FGF	Fibroblast Growht Factor
FL	Fetal Liver
GAL	Galectin

J. Serpa (ed.), *Tumor Microenvironment*, Advances in Experimental Medicine and Biology 1219,
https://doi.org/10.1007/978-3-030-34025-4_14

GMP	Granulocyte-monocyte Progenitors
G-CSF	Granulocyte-stimulating factor
GLI	Glioma Zinc Finger Transcription Factor
HSC	Hematopoietic Stem Cell
HIF	Hypoxia Inducible Factor
IHH	Indian Hedgehog
IFN	Interferon
IGFBP	Insulin-like Growth Factor Binding Protein
IL	Interleukin
ICAM-1	Intracellular Adhesion Molecule-1
ICN	Intracellular Notch
JAK	Janus kinase
LepR	Leptin receptor
LIC	Leukemia Initiating Cell
LSC	Leukemic Stem Cell
LSK	Lin^{-} Sca-1^{+} c-Kit^{+}
LT-HSC	Long-Term Hematopoietic Stem Cell
LFA-1	Lymphocyte Function Associated Antigen-1
mTOR	Mammalian Target of Rapamycin
MEP	Megakaryocyte-erythrocyte Progenitors
MSC	Mesenchymal Stem Cell
OB	Osteoblast
OC	Osteoclast
OPN	Osteopontin
PDX	Patient-derived Xenograft
PTEN	Phosphatase and Tensin Homologue
PIP2	Phosphatidyl-Inositol 4,5-Bisphosphate
PIP3	Phosphatidyl-Inositol 3,4,5-Trisphosphate
PI3K	Phospho-Inositol-3-Kinase
PKB/Akt	Protein Kinase B
RBC	Red Blood Cell
ST-HSC	Short-Term Hematopoietic Stem Cell
STAT	Signal Transducer and Activator of Transcription
SMO	Smoothened
SHH	Sonic Hedgehog
SCF	Stem Cell Factor
SC	Stromal Cell
SDF-1	Stromal Derived Factor -1
SNC	Sympathetic neural cells
T-ALL	T-cell Acute Lymphoblastic Leukemia
TCF/LEF	T-cell factor/Lymphoid enhancer binding factor
TGF	Transforming Growth Factor
TPO	Thrombopoietin
TNF	Tumor Necrosis Factor
TKI	Tyrosine Kinase Inhibitor
VCAM-1	Vascular Cell Adhesion Molecule-1
VE	Vascular Endothelial
VEGF	Vascular Endothelial Growth Factor
VEGFR	Vascular Endothelial Growth Factor Receptor
VLA-4	Very Late Antigen-4
VHL	von Hippel-Lindau
WNT	Wingless and INT-1
WHO	World Health Organization
YS	Yolk-Sac

14.1 Introduction

The human body needs a constant and steady supply of blood cells, which includes red and white blood cells, in order to mount a response against infections and also to account for some blood loss upon injury (Ogawa 1993; Rossi et al. 2007; Reya et al. 2001).

Hematopoiesis is the process by which a single hematopoietic stem cell (HSC) is capable of differentiating in the different mature blood lineages. As shown in Fig. 14.1, this is a very complex, orchestrated and tightly regulated process, as multiple factors play a crucial role in it. These factors can be divided in: (i) intrinsic factors like transcription factors; and also (ii) extrinsic factors like cytokines and other soluble growth factors that lie within the vicinity and in the specific microenvironment of the HSC. In the adult organism (higher mammals) hematopoiesis occurs in a very specialized location that provides all the

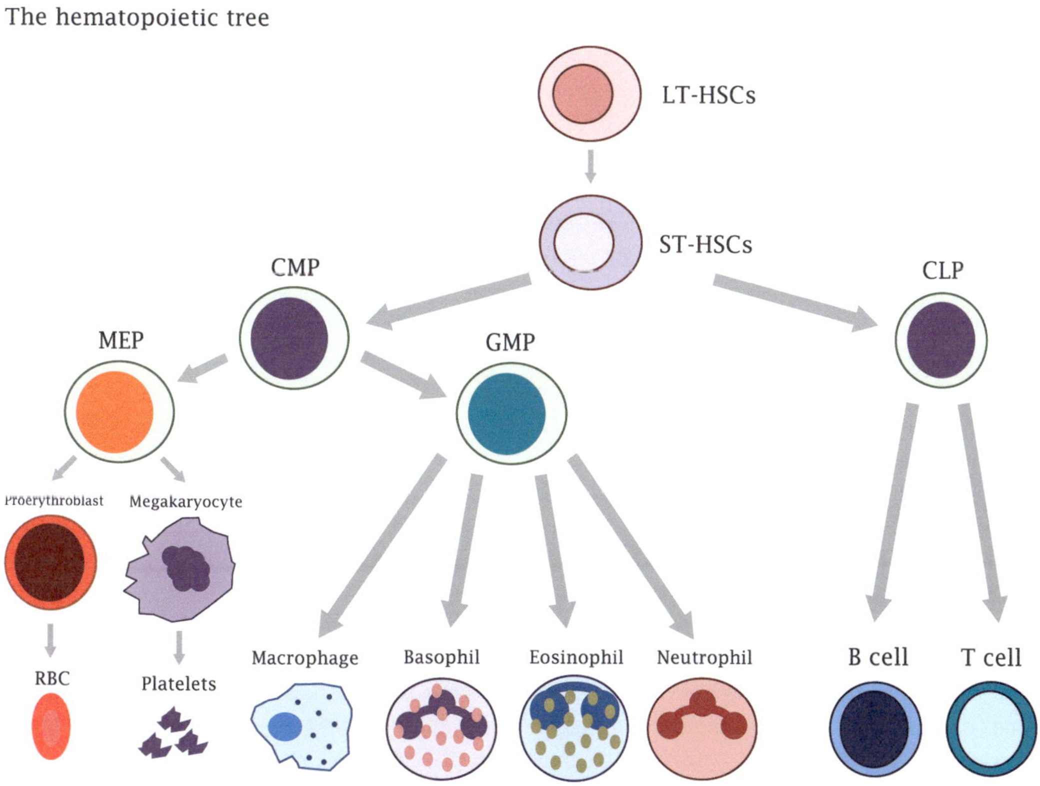

Fig. 14.1 The Hematopoietic tree. Simple representation of the crucial lineage commitment steps that occur in hematopoiesis. The hematopoietic stem cell (HSC) is the basis of the hematopoietic hierarchy and is composed of two main cellular populations, the long-term hematopoietic stem cells (LT-HSCs) and short-term hematopoietic stem cells (ST-HSCs). Subsequent differentiation branches the committed progenitors into two main lineages: (i) common lymphoid progenitor (CLP) that gives rise to B and T cells and ii) the common myeloid progenitor (CMP) that produces all the cells from the myeloid branch. Further differentiation from CMPs gives rise to megakaryocyte-erythrocyte progenitors (MEP) and granulocyte-monocyte progenitors (GMP). MEP progenitors produce red blood cells (RBC) and platelets from proerythroblasts and megakaryocytes, respectively. The GMP progenitors produce all the granulocyte lineage of mature blood cells, which include macrophages, basophils, eosinophils and neutrophils among others that for simplicity reasons were not include in this figure. See text for further details

necessary components for an efficient differentiation and maturation of blood cells – the bone marrow (BM) (Tavian and Peault 2005; Seita and Weissman 2010).

Although being a tightly regulated process, hematopoiesis is not error safe. The current paradigm of leukemogenesis – the process involving the initiation and development of leukemia – suggests that in a developing HSC a genetic alteration (single nucleotide alterations, deletions, insertions gene fusions or even large chromosomal alterations – first hit) occurs in a gene that controls hematopoietic differentiation. Notwithstanding, it should be also highlighted that in the recent years growing amount of evidence demonstrated these alterations can also occur in cells that support hematopoiesis, providing a pre-leukemic niche that would favor a genetic alteration in the developing HSC (Walkley et al. 2007a; Kim et al. 2008; Dong et al. 2016; Raaijmakers et al. 2010, Kode et al. 2014), these alterations will be enlisted and further described in this book chapter. Nevertheless, these modifications block and stall HSC differentiation, allowing for a second genetic alteration (second hit) to occur, usually in genes that control cell cycle and apoptosis. These malignant alterations result in the accumulation of immature blasts that display uncontrolled growth, disabled differentiation potential, abnormal self-renewal capacities and also decreased sensitivity to apoptosis within the bone marrow (BM) milieu but also in circulation. This accumulation of immature blasts in turn will impair the normal hematopoietic development leading to BM failure and consequently aggressive cytopenias as a result from a disrupted hematopoiesis (Askmyr et al. 2011).

This book chapter does not intend to provide an in depth description of hematopoiesis nor the leukemogenic process, neither exhaustively describe in detail the genetic alterations associated with all types of leukemia. We will rather focus our attention on how leukemic cells might interact with the surrounding environment, particularly the BM, and how this interaction might influence leukemia progression, expansion and importantly, drug resistance.

The World Health Organization (WHO) recently revised the classification of the different types of leukemia and this information is summarized in two different review articles (Arber et al. 2016; Swerdlow et al. 2016). This last update in the WHO classification includes many types of leukemia that can be divided in four major groups: (i) myeloid versus (ii) lymphoid leukemia, which reflects the lineage where the differentiation block occurred during hematopoiesis; and (iii) acute versus (iv) chronic which reflects the proliferation state of the leukemic cells. We will focus our attention on Acute Lymphoblastic Leukemia (ALL, B- and T-), Acute Myeloid Leukemia (AML) and Chronic Myeloid Leukemia (CML) because the vast majority of studies regarding the interaction of leukemic cells and the BM milieu are performed under the context of these pathologies.

14.2 Supporting Hematopoiesis – The Bone Marrow Niche

14.2.1 Hematopoiesis and the BM Niche

As discussed above, hematopoiesis is a complex process regarding its regulation and also its location that shifts throughout development. The primitive, or embryonic, hematopoiesis occurs very early during embryogenesis in specialized locations within the embryo, first within the yolk-sac (YS), secondly in the aorta-gonad-mesonephros (AGM) complex and finally the fetal liver (FL). In the adult organism the location of hematopoietic development shifts definitively to the BM (Medvinsky et al. 2011; Mikkola and Orkin 2006).

The BM is a soft and viscous tissue that occupies the central cavity of several bones. This specialized tissue is composed of several types of cells, these include – (i) hematopoietic cells, like the HSC and all of its progeny (mature and non-mature blood cells); (ii) mature hematopoietic cells that support hematopoiesis and also (iii) non-hematopoietic cells whose main function is to support hematopoiesis (Yin and Li 2006).

In 1978 the prominent researcher Raymond Schofield proposed the term functional "niche", specific locations within the BM that protect HSCs from environmental stress and provide them with adequate support and regulation for their self-renewal, growth, viability, distribution and long-term differentiation (Schofield 1978). The non-hematopoietic component of the niche provides this support to HSCs through the secretion of soluble factors, like cytokine and growth factors, but also through direct physical interaction between the HSCs and the supportive cells. In fact, pivotal research in the last 30 years have identified several non-hematopoietic and also hematopoietic components that orchestrate and regulate HSC differentiation, these include osteoblasts (OB), osteoclasts (OC), adipocytes, endothelial cells (EC), mesenchymal stem cells

(MSC) and sympathetic neuronal cells (SNC) (Seita and Weissman 2010). The description of this supportive microenvironment will be further discussed below this chapter by highlighting their effect on the leukemic cells.

The HSCs reside within the BM and may stay there until they fully mature, depending on the cell types. As shown in Fig. 14.1, hematopoiesis is an hierarchical process and HSCs are at the top of this hierarchy with high self-renewal and differentiation capacity; once these cells become more differentiated and restricted to a particular lineage or cell type, those capabilities become more sparse. Importantly, pivotal research in HSC field demonstrated that these cells can be isolated using cell surface markers: Lin- (absence of any mature linage markers) Sca-1^+ c-Kit^+ (in the adult mouse), commonly referred to as LSK population, and Lin- $CD34^+$ $CD38^-$ $CD90^+$ (in human bone marrow) (Seita and Weissman 2010). The current view of hematopoietic differentiation propose that HSCs can be functionally divided into: (i) long-term HSCs (LT-HSCs) which can provide long term hematopoietic reconstitution in recipient hosts and display significant self-renewal activity; and the (ii) short-term HSCs (ST-HSCs) which display short term hematopoietic reconstitution in recipient hosts and reduced self-renewal capabilities (Seita and Weissman 2010; Mikkola and Orkin 2006; Martinez-Agosto et al. 2007). As shown in Fig. 14.1, the ST-HSCs can branch in two different committed progenitors that give rise to two different and important lineages: (i) the common lymphoid progenitors (CLPs) which, in turn, further differentiate into the adaptive immune cells like B and T lymphocytes; (ii) the common myeloid progenitors (CMPs) that further differentiate into megakaryocytes, erythrocytes and macrophages/granulocytes, the latter being innate immune cells (Yang et al. 2005; Seita and Weissman 2010).

The description of hematopoiesis in the paragraph above does not intent to be an exhaustive narrative of the process, which, in fact, as research has demonstrated, is highly complex and constantly being refined. Our objective is to introduce the building blocks on this hematopoiesis so that the reader can fully grasp the important interactions of leukemic cells with this very special microenvironment – the BM.

14.2.2 Disrupting Hematopoiesis – Setting the Ground for Leukemia

As discussed before, hematopoiesis is a tightly regulated process that ensures a constant, steady and structured production of blood cells (Seita and Weissman 2010), however, the deregulation of such process might set the pace for the development of hematological malignancies.

Over the years, research has shown that mutations occurring in HSCs would result in leukemia development. There are many examples of such mutations, the BCR-ABL translocation in CML (Goldman 2010), the *MLL-AF9* fusion (Milne 2017), and the *NOTCH1* activating mutations in T-cell Acute Lymphoblastic Leukemia (T-ALL) (Pear et al. 1996; Weng et al. 2004; O'Neil et al. 2007). However, not all mutations are capable of driving leukemia development *per se*, in fact, for most of the mutations or genetic alterations, often, a second alteration is required (second hit). Importantly, according to the two-hit hypothesis discussed above, these mutations, or translocations, occur in two types of genes, genes that encode for kinases or other genes associated with cellular proliferation and survival that render these cells highly proliferative potential allowing the malignant clone to proliferate and thrive – class I mutations – and in genes that encode transcription regulators of hematopoiesis; such mutations are able to block hematopoiesis – class II mutations. An example of class I mutations is the *JAK2V617F* mutation (Baxter et al. 2005; James et al. 2005; Kralovics et al. 2005; Levine et al. 2005) that occurs in a subset of patients with myeloproliferation (a pre-leukemic condition characterized by the abnormal and deregulated expansion of myeloid cells often leading to AML) (Vannucchi et al. 2009), this mutation renders the developing hematopoietic cells growth factor independency and hypersensitivity but it is not sufficient to develop leukemia. In addition to this, the t(8;21) chromosomal translocation *AML1-ETO* is a class II mutation and act as a dominant-negative of regular AML1 function by blocking hematopoiesis and promoting self-renewal of the developing hematopoietic progenitors, as for the class I mutations, this translocation is insufficient to drive leukemia development (Mulloy et al. 2002; Yuan et al. 2001; Schneider et al. 2007).

The leukemia-initiating cells (LICs) or leukemic stem cells (LSCs) are the cells thought to be at the origin of leukemogenesis, they are organized hierarchically as the HSCs and are able to sustain leukemia development in recipient hosts, and similarly to the HSCs, LSCs also possess the capacity of quiescence, self-renewal, and differentiation (Warner et al. 2004; Bonnet and Dick 1997). Furthermore, increasing amount of evidence demonstrates that LSCs can resist elimination by chemotherapy and are responsible for patient refraction to therapy and disease relapse (Essers and Trumpp 2010).

Importantly, LSCs or LICs will alter the dynamic of the BM microenvironment, since these cells will outgrow the HSCs that are developing within this niche by hijacking the niches that support normal hematopoiesis, either by uptaking cytokine signaling and other soluble growth factors and also through the interaction with the neighboring cells (this will be described later in this chapter). Interestingly, not only the LSCs hijack the normal HSC-supportive niche, but also are able to shape this supportive microenvironment and shift them from HSC-supportive to LSC-supportive leading to leukemic growth and decreased/disrupted hematopoiesis, which in fact accounts for the normal feature of disrupted hematopoiesis and decreased number of mature leukocytes observed in the vast majority of leukemia patients. The signals that are secreted from the LSCs and shape the surrounding microenvironment are mainly cytokines and chemokines with pro-inflammatory and angiogenic properties such as Stromal Derived Factor – 1 (SDF-1) (C-X-C Motif Chemokine ligand – 12 (CXCL12)), Vascular Endothelial Growth Factor (VEGF), Tumor Necrosis Factor (TNF)-α, Interleukin (IL)-1, IL-6, IL-8 and also Stem Cell Factor (SCF) (Medyouf et al. 2014; Zambetti et al. 2016; Zhang et al. 2016).

Interestingly, several researchers have reported over the last decade several cases of how leukemic cells can shape their own microenvironment and create a leukemia sustainable niche. One of such examples are the elegant experiments conducted by Dorothy Sipkins group where they show that, in the context of ALL and AML, leukemic cells disrupted normal hematopoiesis by displacing HSCs from their supporting niche to abnormal “leukemic” niches through the secretion of SCF; this displacement reduced the numbers of HSCs and could be restored to normal levels upon treatment with a neutralizing antibody to SCF (Colmone et al. 2008). More recently, in the context of CML, Schepers and colleagues demonstrated that leukemic cells directly interact with MSCs “instructing” them to increase the production of OBs and to secrete inflammatory mediators like the CCL3 chemokine and Thrombopoietin (TPO) that, in turn, will remodel and render their niche more pro-inflammatory, which will favor their abnormal growth (Schepers et al. 2013). Also in line with this concept, Zhang and colleagues demonstrated that in CML mouse model LSCs modulate the BM niche for their own benefit. The mechanism behind such effect relies on sustained and increased secretion of G-CSF by LSCs, which in turn, lead to the decreased levels of SDF-1 which promoted LSC expansion at the expenses of impaired HSC growth. Interestingly, imatinib treatment reverted, to a certain extent, such LSC-mediated effect (Zhang et al. 2012a). In addition to this, loss of OB cells has been reported in the context of AML and also T-ALL and is associated with worse overall survival of patients (Wang et al. 2016; Frisch et al. 2012).

Interestingly, the relationship between the leukemic cells and their microenvironment is quite specific, in relation to both disease stage and subtype. As an example, LSCs in AML and CML malignancies display altered dependency of OBs on the endosteal niche. In AML the OB cells promote the expansion of MLL-AF9 driven AML, while in BCR-ABL-driven CML the same OB cells impair their propagation (Krause et al. 2013). Moreover, response to therapy also influences this dual interaction of leukemic cells and the BM microenvironment; in CML disease it was demonstrated in a very elegant manner that in response to imatinib, a well-known tyrosine kinase inhibitor (TKI) that specifically inhibits the BCR-ABL fusion protein, LSCs change their localization in the BM by moving closer to the endosteal niche and, therefore, they become more protected from the cytotoxic actions of this drug (Jin et al. 2008; Meister et al. 2014).

All of the reports described provide the clear evidence that not only leukemic cells are dependent on the BM niche for their support, survival and drug resistance but are also able to transform this niche into a leukemic niche that completely disrupts normal hematopoiesis either by directly influencing HSCs or by affecting the supportive cells, like the OB lineage.

14.2.2.1 Disrupting Hematopoiesis – From the Niche Standpoint

Above we described several examples of how leukemic cells disrupt normal hematopoiesis by controlling and modulating the BM microenvironment at their own will and shift it to a leukemic microenvironment. Here, we will describe and exemplify a novel hypothesis that has started to emerge within the scientific community; alterations within the BM microenvironment, and not in the developing HSCs, might lead to malignant transformation and leukemia development by disrupting normal hematopoiesis.

One of the first reports that originate this concept of niche-induced leukemogenesis was the elegant work of Walkley and colleagues where it was demonstrated that the conditional deletion of the retinoblastoma protein (Rb) in both the myeloid lineage and in the BM microenvironment was required to drive myeloproliferation, suggesting a key role of the BM niche in myeloproliferation (Walkley et al. 2007b). Moreover, the same research group reported that myeloproliferative disease was induced in RARγ knock-out mice. Interestingly the effect was only observed when wild-type hematopoietic cells were transplanted in RARγ knock-out host, again highlighting the importance of the BM microenvironment in the development of myeloproliferative disease (Walkley et al. 2007a). Other research studies demonstrated that defective Notch signaling in MSCs and activating mutations on the *Ptpn11* gene in osteoblastic progenitors are also implicated in leukemia development (Dong et al. 2016).

Remarkably, it seems that the osteoblastic lineage plays a pivoltal role in the regulation of myeloid differentiation since alterations in these cells often lead to AML development. As an example, the Scadden group also reported that deletion of *Dicer 1* (microRNA processing enzyme) in immature osteoblastic progenitors resulted in myelodisplasia (a pre-leukemic condition characterized by abnormal differentiation and reduction of myeloid cells often leading to AML) with further transformation into AML (Raaijmakers et al. 2010). Furthermore, ectopic activation of β-Catenin signaling and FoxO1 in mouse osteoblasts resulted in the development of AML (Kode et al. 2014, 2016).

It is interesting to note that this effect of altered BM microenvironment was observed experimentally in mouse models and clearly demonstrate that perturbations in the BM microenvironment may lead to leukemia development. It is still unclear whether such alterations may cause human leukemia, nevertheless, the concept itself is supported by the observations that leukemia is originated upon allogeneic BM transplantation from healthy donor cells (Wiseman 2011; Sanchez-Aguilera and Mendez-Ferrer 2017).

14.2.2.2 The Leukemic Niche – BM Support to Leukemia Progression

14.2.2.2.1 Signaling Pathways Activated by the BM Microenvironment in Leukemia Cells

As described above, the BM microenvironment supports HSC survival and differentiation and is able to initiate and sustain leukemia development, progression and expansion. Here we will describe in more detail the signaling pathways activated in the leukemic cells, either by direct involvement of the BM microenvironment (soluble factor secretion or direct cell-to-cell contact) but also by aberrant activation as a result of genetic alteration.

SDF-1 Signaling Chemokines are small chemotactic proteins that regulate the homing of leukocytes to the sites of immune response and also to the locations where hematopoiesis occurs (Olson and Ley 2002). Among these chemotactic proteins, the SDF-1 chemokine is of outmost importance. Upon binding of SDF-1 to its receptor, C-X-C chemokine receptor type – 4 (CXCR4), this activates a signaling cascade with a multitude of outputs, including PI3K, protein kinase C, Janus kinase/signal transducer and activator of transcription (JAK/STAT), MAPK and also NF-κB activation which in turn modulate several cellular processes like cell growth, survival and migration (de Lourdes Perim et al. 2015). Several cellular components of the BM microenvironment secrete SDF-1: OB progenitors, ECs, CXCL12 abundant reticular cells (CAR) and also MSCs Nestin^{+} (Christopher et al. 2009; Sugiyama et al. 2006). Importantly, HSCs and several of its downstream progeny express the CXCR4 receptor, and are sensitive to SDF-1 gradients produced by the BM niches (Aiuti et al. 1997).

In the leukemic context the SDF-1/CXCR4 signaling axis has been extensively studied. CXCR4 expression increased leukemic blasts, which correlated with decreased outcome (Schneider et al. 2002; van den Berk et al. 2014; Spoo et al. 2007) Importantly, SDF-1/CXCR4 was demonstrated to be crucial for the retention of leukemic cells within the BM niche (Becker 2012; Tavor et al. 2004). Elegant research studies demonstrated that this signaling axis is responsible for BM-derived MSCs mediated protection to chemotherapeutic and TKI drugs (Parameswaran et al. 2011). Importantly, abrogation of this signaling axis decreased leukemia cellular survival *in vitro* and leukemia burden *in vivo* (Juarez et al. 2007b; Li et al. 2015).

Wingless and INT-1 (WNT)/β-Catenin signaling As many other signaling pathways, the canonical WNT/β-Catenin signaling pathway is fundamental for several processes during embryogenesis, these include cell differentiation, proliferation, survival and polarity (van Amerongen and Nusse 2009). This signaling pathway includes 19 WNT ligands, 10 frizzled (FZD) receptors and several intermediate mediators of the pathway. Several factors including developmental stage of the cells, the WNT ligand dosage, and BM microenvironmental factors modulate the strength and the downstream effect of WNT signaling. Activation of WNT/β-Catenin signaling occurs upon binding of a WNT ligand to a FZD receptor, this interaction relieves β-Catenin from its proteosomal degradation allowing its translocation to the nucleus where it associates with a complex of transcription factors, the T-cell factor/Lymphoid enhancer binding factor (TCF/LEF) complex, to activate WNT signaling downstream targets (van Amerongen and Nusse 2009; Polakis 2012). Extensive research demonstrated the role of WNT signaling during normal hematopoiesis. It was demonstrated to have a crucial role during embryonic hematopoiesis and to a lesser extend in adult hematopoiesis. Elegant research by Prof. Staal showed that the levels WNT signaling dictate the balance between HSCs self-renewal and differentiation (Luis et al. 2011). Moreover, other studies highlighted the importance of WNT signaling in the BM microenvironment that supports normal hematopoiesis (Ling et al. 2009).

In the leukemic perspective the WNT signaling pathway has been highlighted in the several types of leukemia. In CML, increased β-Catenin activity was demonstrated to promote the blast crisis CML (a condition in CML disease characterized by increased expansion of leukemic blasts resembling AML disease) (Jamieson et al. 2004; Minami et al. 2008). Importantly, the integrity of the pathway is crucial in CML, as deletion of β-Catenin reduced leukemia burden and cooperated with TKI treatment in order to eliminate LSC (Hu et al. 2009; Zhao et al. 2007). Also in line with this, enhanced WNT/β-Catenin activity has been shown to promote resistance to TKI treatment (Heidel et al. 2012). Wang and colleagues reported in an elegant study that the WNT/β-Catenin is active and required for the development of LSC in AML (Wang et al. 2010). More recent studies implicated Galectin (GAL) -3 and − 9 in AML by linking the BM-MSCs to LSCs increased self-renewal and drug resistance through the stabilization β-Catenin (Hu et al. 2015; Yamamoto-Sugitani et al. 2011; Kikushige et al. 2015).

In the B-ALL the involvement of WNT/β-Catenin is scarce, nevertheless, BM-derived MSCs protected B-ALL cell lines and patient samples from the cytotoxic effects of cytarabine (Ara-C) and this protection paralleled with the induction of WNT signaling transcriptional program (Khan et al. 2007). Importantly, the inhibition of WNT signaling sensitized B-ALL cells *in vitro* and *in vivo* to Ara-C-induced apoptosis (Yang et al. 2013). In T-ALL, the involvement of WNT/β-Catenin signaling was only recently demonstrated. Firstly, Guo and colleagues showed that stabilization of β-Catenin in developing T-cells resulted in the accumulation immature T-cells with consequent development of leukemia with MYC activation (Guo et al. 2007). Inactivating mutations in the *LEF1* gene were also reported in a significant number of T-ALL patients and are concomitant with *NOTCH1* and PI3K pathway activating mutations (Gutierrez et al. 2010). Furthermore, in an elegant study performed in Prof. Weng's laboratory, the WNT/β-Catenin signaling and hypoxia-inducible factor (HIF)-1 was shown to be active in LSCs in a Notch1-induced mouse model. Upon genetic ablation of β-Catenin or HIF-1 the LSC population is diminished. Pharmacological targeting of the WNT/β-Catenin signaling in human cell lines *in vitro* and patient derived-xenografts (PDX) cultured *ex vivo* also impacted on the viability of these cells further demonstrating the importance of this signaling pathway in T-ALL (Giambra et al. 2015).

Hedgehog Signaling The Hedgehog signaling pathway was discovered during a genetic screen in Drosophila in early 80's and plays a critical role in embryonic development and body segmentation (Nusslein-Volhard and Wieschaus 1980). The activation of this pathway is not the typical canonical engagement of ligand-receptor-transducer of signal associated with several signaling pathways (Briscoe and Therond 2013, Crompton et al. 2007). There are three hedgehog ligands described so far, the sonic hedgehog (SHH), the indian hedgehog (IHH) and the desert hedgehog (DHH); they are secreted proteins that bind and block receptors within the target cells. Upon engagement of the hedgehog ligands, the PTCH1 and PTCH2 transmembrane proteins receptors are displaced from the smothened (SMO), another transmembrane receptor, allowing its activation and consequent translocation of the glioma zinc finger transcription factors (GLI) −1, - 2 and -3 to the nucleus to activate Hedgehog target genes (Briscoe and Therond 2013; Crompton et al. 2007).

In the CML context, the integrity of this signaling pathway is crucial for leukemia progression and expansion, since genetic and pharmacological inhibition reduces leukemia burden and prolongs mice survival (Zhao et al. 2009; Dierks et al. 2008; Katagiri et al. 2013). Moreover, in the T-ALL field two recent studies highlighted the importance of this pathway in this malignancy. Firstly, somatic mutations were identified in the SMO receptor (*SMOR726* and *SMOR763*) and in the GLI transcription factors (*GLI1S538F* and *GLI3G727R*) in T-ALL patients. Importantly, SMO mutations render it insensitive to PTCH inhibition *in vitro* and *in vivo* and this resulted in lymph node infiltration with immature T-cells (Dagklis et al. 2015). Another report demonstrated that this pathway is aberrantly hyperactived in 20% of T-ALL patients by ectopic expression of the SHH and IHH ligands as well as up-regulation of GLI-1. Importantly, human T-ALL cell lines and patient samples with activated hedgehog signaling decrease their viability *in vitro* upon treatment with SMO and GLI inhibitors, in xenotransplantation models these inhibitors also reduce leukemia burden and aggressiveness (Dagklis et al. 2016).

Notch Signaling The Notch signaling pathway is a highly conserved signaling pathway critical for the regulation of cell fate decisions in stem cell maintenance, neurogenesis and also T-cell differentiation. The Notch receptors are heterodimeric proteins composed of an extracellular subunit and a transmembrane subunit that are non-covalently bound through a heterodimerization domain. The binding of the ligand to the extracellular subunit of the Notch receptor activates several proteolytic cleavages. The final cleavage is catalyzed by the γ-secretase complex that releases the intracellular Notch (ICN) receptor, which in turn activates the transcription of Notch target genes. The members of this signaling pathway are the four Notch receptors (1–4) and the five cognate ligands Delta-like (Dll) -1, -3 and -4 and Jagged-1 and Jagged-2 (Penton et al. 2012; Andersson et al. 2011; Artavanis-Tsakonas et al. 1999).

Several cellular components of the BM microenvironment express Notch ligands which upon interaction with Notch receptors present in the HSCs and their progeny are able to regulate hematopoiesis. For example, OBs in the endosteal niche express high levels of Jagged-1, which interact with Notch receptors on HSCs increasing their repopulating activity (Weber et al. 2006). Moreover, it was also demonstrated that engagement of Jagged-1 ligand on OBs inhibits HSC cell-cycle contributing to HSCs quiescence (Calvi et al. 2003). In addition to this, Notch signaling has been shown to play a critical role in T-cell differentiation (Jaleco et al. 2001; Schmitt and Zuniga-Pflucker 2002).

In the leukemic context, Notch signaling has been extensively studied in T-ALL, but scarcely in B-cell acute lymphoblastic leukemia (B-ALL) and AML. NOTCH1 activating mutations are the hallmark of T-ALL, as more than 60% of T-ALL patients display such mutations. The mutations were found in two distinct regions of the *NOTCH1* gene, in the heterodimerization domain (44%) and in the PEST domain located in the C-terminus (30%), with a significant percentage of the patients (17%) displaying mutations in the both regions (Weng et al. 2004). In fact, not only T-ALL patients display *NOTCH1* activating mutations, also *ICN* is a very powerful T-cell oncogene, as BM transplantation of hematopoietic progenitors transduced with ICN resulted in increased aggressiveness and decreased overall survival of the transplanted mice (Pear et al. 1996; Fragoso et al. 2012). Moreover, Dll-4 ligand was demonstrated to increase Notch1 and Notch3 signaling of human T-ALL primary samples and pharmacological blockade of the Dll-4 ligand with a therapeutic antibody dampened Notch signaling and delayed tumor growth in xenotransplanted mice (Minuzzo et al. 2015). In B-ALL, Notch signaling has been also implicated, more specifically, the NOTCH3 and NOTCH4 receptors and the Dll-1, Jagged-1 and the Jagged-2 ligands. The engagement between these proteins sustained cellular proliferation and survival upon *in-vitro* co-culture with BM-derived MSCs. Moreover, protection from corticoid induced cytotoxicity was also dependent on the NOTCH3 and NOTCH4 receptors in B-ALL (Nwabo Kamdje et al. 2011). In AML Notch signaling activation resulted in decreased cellular survival *in vitro* and induced leukemic growth arrest and apoptosis *in vivo* demonstrating the role of Notch signaling as a myeloid tumor-suppressor (Lobry et al. 2013). Accordingly, Notch signaling is silenced in AML patient samples and in mouse model of AML (Kannan et al. 2013; Lobry et al. 2013).

Phospho-Inositol-3-Kinase (PI3K) Signaling The PI3K signaling pathway is a pleiotropic pathway involved in several processes within a living cell, these include survival, proliferation, differentiation, motility and also metabolism (Polak and Buitenhuis 2012; Song et al. 2005; Cantley 2002). Receptor tyrosine kinases, cytokines signaling and other soluble growth factors and direct cell-to-cell contact through adhesion molecules activate the PI3K pathway. The PI3K complex consists of a catalytic subunit and a regulatory subunit. The activation of this complex leads to the specific phosphorylation of the Phosphatidyl-Inositol 4,5-Bisphosphate (PIP2) lipid in the position 3 leading to the formation Phosphatidyl-Inositol 3,4,5-Trisphosphate (PIP3). The Phosphatase and Tensin Homologue (PTEN) tumor suppressor catalyzes the reverse reaction. Upon formation of PIP3, the PI3K downstream targets Protein Kinase B (PKB/Akt) and 3-phosphinositide-dependent protein kinase (PDK) -1 bind to the plasma membrane through their plecstrin homology domains that anchor them to the PIP3. Consequently, the PDK-1 kinase phosphorylates PKB/Akt in the Threonine 308 residue and the full activation of PKB/Akt is achieved upon phosphorylation in the Serine 473 residue by PDK-2. Activated PKB/Akt is known to phosphorylate a wide variety of cellular substrates, including FOXO family of transcription factors, the GSK-3α/β kinase, the pro-apoptotic protein BAD, the negative regulator of mammalian target of rapamycin (mTOR) complex 1 – TSC2 and mTOR protein, amongst others. These activated substrates ensure the pleiotropic effects of the PI3K pathway by controlling cell cycle, proliferation, apoptosis and protein synthesis (Polak and Buitenhuis 2012; Song et al. 2005; Cantley 2002).

The role of PI3K signaling has been extensively studied and it has been implicated in HSC maintenance and lineage commitment upon differentiation (Polak and Buitenhuis 2012). Regarding malignant hematopoiesis, the PI3K pathway has been also widely associated. Firstly, mutations in members of the signaling pathway have been identified in myeloid and lymphoid leukemias and include: (i) activating mutations in the catalytic subunit of PI3K (Horn et al. 2008); (ii) inactivating mutations in the regulatory subunit of PI3K (Gutierrez et al. 2009); (iii) deletions and sequence alterations in the PTEN phosphatase (Liu et al. 2000); (iv) other mutations include inactivating mutations in SHIP1 (another negative regulator of PI3K) (Lo et al. 2009). Moreover, this signaling pathway can be activated through cross-talk and due to upstream activation of unrelated oncogenes. Such examples include PI3K activation by the BCR-ABL fusion protein in CML (Kim et al. 2005), by mutations in the Fms-like tyrosine kinase (FLT) -3 in AML (Lindblad et al. 2016) and also the *JAK2V617F* mutation in myeloproliferative disease (Fiskus et al. 2013). In line with this, Prof. Barata laboratory reported, in a very elegant study, that the vast majority of T-ALL patients (up to 85%) display constitutive activation of PI3K signaling pathway. The PI3K constitutive activation results from the phosphorylation and consequent inhibition of PTEN activity by the CK2 kinase (Silva et al. 2008).

Importantly, and regarding the BM microenvironment, PI3K signaling can be activated in leukemic cells by growth factor engagement (cytokines – it will be addressed below) but also by direct contact with the BM cellular components. In fact, Jacamo and colleagues reported that the activation of PI3K signaling dampens the response of AML to chemotherapy upon co-culture with BM-derived MSCs (Jacamo et al. 2014). In addition to this, PI3K mediated resistance to TKI treatment has also been reported. In AML with FLT3 mutations, PI3K activation mediated BM-derived MSC resistance to Sorafenib (a TKI drug – it inhibits downstream signaling from the FLT3 kinase), since co-treatment of AML cells with this TKI and a selective PI3K inhibitor reverted MSC mediated resistance (Jin et al. 2013). Similarly, in myeloproliferative disease, activation of PI3K signaling also mediated BM stromal resistance to Ruxolitinib (a TKI – it inhibits JAK2 kinase) (Cardoso et al. 2015).

Hypoxia-Inducible Factor (HIF) Signaling Oxygen levels play a critical role in regulating HSCs differentiation, self-renewal and also mobilization in the BM microenvironment (Majmundar et al. 2010; Parmar et al. 2007). Several lines of evidence suggest that a gradient of oxygen occurs within the BM niche, with high oxygen levels in the perivascular niche (discussed below) close to ECs and lower oxygen levels in the endosteal niche (discussed below) (Nombela-Arrieta et al. 2013; Parmar et al. 2007). However, real-time measurements within the BM of an adult mouse show that both niches are hypoxic (Spencer et al. 2014). Importantly the most immature HSCs (LT-HSCs) are attracted to these niches with low oxygen levels – hypoxic niches (typically around 1.5% of oxygen tension) (Parmar et al. 2007). Hypoxia is a well regulated cellular process and in a simplified manner relies on the activity of the HIF transcription factor. Under normal oxygen levels – normoxia – (close to 5% of oxygen tension), HIF-1/2 subunits are hydroxylated and constutively degraded by the von Hippel-Lindau (VHL) E3-ubiquitin ligase, when oxygen tension decrease below 2%, HIF hydroxylation is diminished allowing for its stabilization and nuclear translocation in order to activate its transcriptional program which includes target genes known to control cellular metabolism, angiogenesis, apoptosis and cell cycle progression (Jiang et al. 1996; Majmundar et al. 2010). Interestingly, SDF-1 is a HIF target gene suggesting that HSCs might home to hypoxic niches through SDF-1/CXCR4 signaling that we described below (Ceradini et al. 2004).

In myeloid leukemias, HIF-2 activity is fundamental: (i) it promotes survival of primary AML cells; (ii) down-regulation of HIF-2 decreases engraftment of human AML cells; (iii) and HIF-2 ectopic expression protects AML cells from stress induced apoptosis (He et al. 2013; Forristal et al. 2015). Interestingly in the case CML, HIF-1 was sufficient to maintain LSC activity in a mouse model of this disease even upon inhibition of BCR-ABL with TKI treatment (Ng et al. 2014). In lymphoid leukemia, HIF-1 appears to be the most prominent HIF protein. In fact, HIF-1 is overexpressed in ALL patient samples and is associated with poor prognosis (Benito et al. 2011). Notably, HIF-1 is induced in B-ALL blasts co-cultured with BM-derived MSCs and this induction protected leukemic cells from cytotoxic therapy and increased glycolytic rate, both effects were reverted upon inhibition of mTOR signaling (with RAD001 – Everolimus) suggesting a crosstalk between the HIF1 transcription factor and mTOR signaling (Frolova et al. 2012). Moreover, in T-ALL the role of HIF-1 has also been depicted. As described above, in a Notch1 mouse model of T-ALL, Prof. Weng laboratory demonstrated that HIF-1 increased WNT/β-Catenin signaling under hypoxic conditions in LSCs of this mouse model. Moreover inactivation of both HIF-1 and β-Catenin diminished LSCs without impacting the viability and proliferation of the majority of T-ALL blasts (Giambra et al. 2015).

Adhesion Molecules Adhesion molecules play a pivotal role in BM homeostasis and in hematopoietic differentiation. These include osteopontin (OPN) – which lies within the extracellular matrix (ECM), Very late antigen-4 (VLA-4)/Vascular cell adhesion molecule-1 (VCAM-1), Lymphocyte function associated antigen-1 (LFA-1)/Intracellular adhesion molecule-1 (ICAM-1), N-cadherin, E-selectin and also CD44 (Seita and Weissman 2010). Importantly, depending on the cellular components (see description below) adhesion molecules might deliver signals for HSCs to proliferate or to maintain quiescence (Seita and Weissman 2010). As an example, E-selectin ligand-1 is expressed in both HSCs and AML blasts, and mediates cell proliferation and exit from quiescence (Winkler et al. 2012). In addition to this, CD44 transmembrane glycoprotein is expressed within the BM microenvironment and was shown to mediate myeloid progenitor migration and also BM colonization (Dimitroff et al. 2001; Schmits et al. 1997).

In AML, adhesion molecules, in particular the VLA-4/VCAM-1 interaction, regulate leukemic blast adherence to endothelial cells and are pivotal in maintaining cell survival. Moreover, the VLA-4 molecule provided resistance mediated by BM-derived MSCs to apoptosis induced by chemotherapeutic agents in AML blasts (Jacamo et al. 2014). E-selectin has been implicated in refractory AML disease, as relapsed/refractory patient samples have higher expression of E-selectin than in those with *de novo* disease. Moreover, this molecule has been shown to enhance the survival of AML blasts upon binding to the perivascular niche via E-selectin ligands and activation of WNT/β-Catenin signaling (Chien et al. 2013). With regards to CD44, this adhesion molecule is highly expressed in AML patients samples and the expression of some spliced variants correlates with poor prognosis. Moreover, CD44 is also a key regulator LSCs homing to BM niches and mediates resistance to drug-induced apoptosis in the AML context (Allouche et al. 2000; Dimitroff et al. 2001). Inhibition of CD44 engagement with a monoclonal antibody resulted in a marked decrease in leukemia burden in PDX experiments, thus implying CD44 as a key regulator in AML disease biology (Jin et al. 2006). Moreover in the context of CML, the CD44 adhesion molecule is absolutely required for the homing of LSCs to the BM niche (Krause et al. 2006) and the treatment of a CML mouse model with a human CD44 antibody reduces LSC survival highlighting the therapeutic potential of CD44 inhibition, which demonstrates its importance (Hellqvist et al. 2013).

Regarding lymphoid malignancies, one of the first reports implicating adhesion molecules demonstrated that the LFA-1/ICAM-1 interaction promoted cell survival of T-ALL blasts (cell lines and primary cells) co-cultured in BM stroma (Winter et al. 2001). Other studies in B-ALL showed that elevated levels of VLA-4 correlated with poor prognosis in these leukemic patients (Shalapour et al. 2011), and similarly to the AML context, VLA-4 molecule also act as a gatekeeper of chemo-protection mediated by the BM-derived MSCs upon activation of several downstream signaling pathways like PI3K (Jacamo et al. 2014). Moreover, recent research studies also associate CD44 molecule to chemoresistance in T-ALL. Hoofd and colleagues reported that cytotoxic chemotherapy selects for cells with high CD44 expression both in mouse models and in the clinical setting and enforced expression of CD44 in human T-ALL further enhance this chemoresistance. Such effect was dampened upon genetic ablation of CD44 or therapeutic targeting with a blocking antibody (Hoofd et al. 2016).

Cytokine Signaling Cytokines are soluble proteins with numerous functions like regulating immune response, ensuring immune homeostasis and also regulating hematopoiesis. These proteins are produced in the BM microenvironment, in other tissues where hematopoiesis might occur (like the thymus and spleen) and also in peripheral tissues (Bociek and Armitage 1996; Miyajima et al. 1999). Cytokines include colony-stimulating factors (CSFs), interleukins (ILs), interferons (IFNs) and other soluble growth factors (Jatiani et al. 2010). The binding of cytokines to their receptors engages a series of signaling events that can result in increased proliferation, cell survival, differentiation and also apoptosis of the target cells. The canonical signaling driven upon cytokine engagement with its receptor is the induction of JAK/STAT signaling which results in increased phosphorylation of STAT proteins and subsequent translocation into the nucleus to activate the transcription of target genes, in addition to this other signaling pathways might be activated, like PI3K and MEK-ERK (Barata et al. 2005). In the context of hematopoiesis, cytokine signaling provides the fine-tuning between the different cellular components of the BM microenvironment ensuring proper HSCs differentiation, survival, self-renewal and also proliferation (Seita and Weissman 2010).

As we have been discussing, leukemic cells take advantage of this growth factors to thrive and expand. An important example of this feature is the role of IL-7 in T-ALL biology. BM-derived MSCs increased T-ALL blast survival and proliferation upon engagement of IL-7 signaling (Scupoli et al. 2007). Moreover, extensive research carried out by Prof. Barata group demonstrated that IL-7 is a key cytokine in supporting T-ALL survival, proliferation and also in the regulation of glucose metabolism through the activation of the PI3K signaling pathway *in vitro* (Barata et al. 2001, 2004a, b, c; Silva et al. 2011a), and *in vivo* IL-7 is able to sustain T-ALL malignancy by modulating BCL-2 and $p27^{Kip1}$ protein levels (Silva et al. 2011b). Moreover, the same group also identified activating mutations in IL-7Rα chain that sustain constitutive JAK-STAT signaling in T-ALL cells (Zenatti et al. 2011). Altogether, these results highlight the importance of IL-7/IL-7Rα signaling in T-ALL being supportive on one side (through IL-7) and initiator on the other (through constitutive IL-7Rα signaling) (Oliveira et al. 2019; Ribeiro et al. 2013).

Interestingly, in the B-ALL context, very early reports indicated that leukemic blasts expressed higher levels of several cytokines (IL-7, IL-10, IL-15 and IFN-γ) and their cognate receptors which were suggestive of a autocrine loop in leukemic cells (Kebelmann-Betzing et al. 2001). Moreover, IL-7 and IL-3 increase proliferation of B-ALL cells in combination with the SDF-1α chemokine through the activation of several signaling pathways, like the PI3K pathway among others (Juarez et al. 2007a).

14.2.2.2.2 Cellular Components of the BM Niche

Here we will describe more deeply the cellular components that are responsible for this support, both normal hematopoiesis and more importantly, the malignant hematopoiesis. The BM niche is divided into different anatomic regions: the endosteal (or osteoblastic) – closer to the endosteal bone – and the perivascular niche – closer to the BM center. These important anatomic regions play a critical role in HSC and LSC maintenance, as an example, in the adult mouse the HSCs reside closer to the endosteal bone surface and upon transplantation HSCs migrate preferentially to this region. Moreover, HSCs isolated from this anatomic region are enriched in LT-HSCs, whereas more differentiated progenitors are found in the central BM region near the perivascular niche. The different niches and their cellular components are illustrated in Fig. 14.2.

The endosteal niche

Osteoblasts (OBs) OBs are the bone forming cells that are present along the bone within the endosteal niche and are required for HSCs differentiation as they deliver that regulate HSC homing, self-renewal and quiescence (Calvi et al. 2003; Zhang et al. 2003). This notion stems from studies where OB ablation resulted in reduced BM cellularity, decreased HSCs and progenitor cells and increased extra-medullary hematopoiesis (Visnjic et al. 2004; Bowers et al. 2015). More recent studies suggest that OB regulation of HSC differentiation may rely on the OB-differentiation state or even on other cellular components of the endosteal niche (Cordeiro-Spinetti et al. 2015). Importantly, as described above, transplantation studies in mouse models show that HSCs preferentially localize in the endosteal niche, a region that is highly enriched in OBs (Lo Celso et al. 2009; Nilsson et al. 2001). Several soluble growth factors and cytokines are produce by OBs that might regulate HSCs function, these include: SDF-1, SCF, OPN, Granulocyte-stimulating factor (G-CSF), Annexin-2, Angiopoeitin-1 (Ang-1) and Thrombopoietin (TPO) (Calvi et al. 2003; Arai et al. 2004; Jung et al. 2007; Ponomaryov et al. 2000; Stier et al. 2005; Taichman and Emerson 1994; Yoshihara et al. 2007). SDF-1 and SCF are also produced by other cellular components of the BM, other than the OBs (as described below), and play fundamental roles in HSCs function (Christopher et al. 2009; Sugiyama et al. 2006; Aiuti et al. 1997). Another example of growth factor is OPN that is secreted by OBs in the endosteal niche and is critical in the regulation of HSC migration, proliferation and differentiation (Stier et al. 2005; Nilsson et al. 2005; Grassinger et al. 2009).

In addition to this role in supporting hematopoiesis, OBs have a tumor-suppressor role in the context of myeloid leukemia, as their numbers are reduced in MDS/AML patients samples. Moreover, in a mouse model, ablation of OBs resulted in increased engraftment and expansion of blast cells within the BM milieu and decreased survival of the mice. Inversely, when OBs frequency is restored, leukemic burden decreases and mice survival is prolonged (Krevvata et al. 2014). Moreover, another study demonstrated that in a transplanted model of AML, Osteocalcin (a soluble molecule that stimulates OBs proliferation) is reduced, further reinforcing the tumor-suppressor role of these cellular cues in the BM microenvironment (Frisch et al. 2012).

Osteoclasts (OCs) Bone formation is a dynamic process in which two cellular entities are

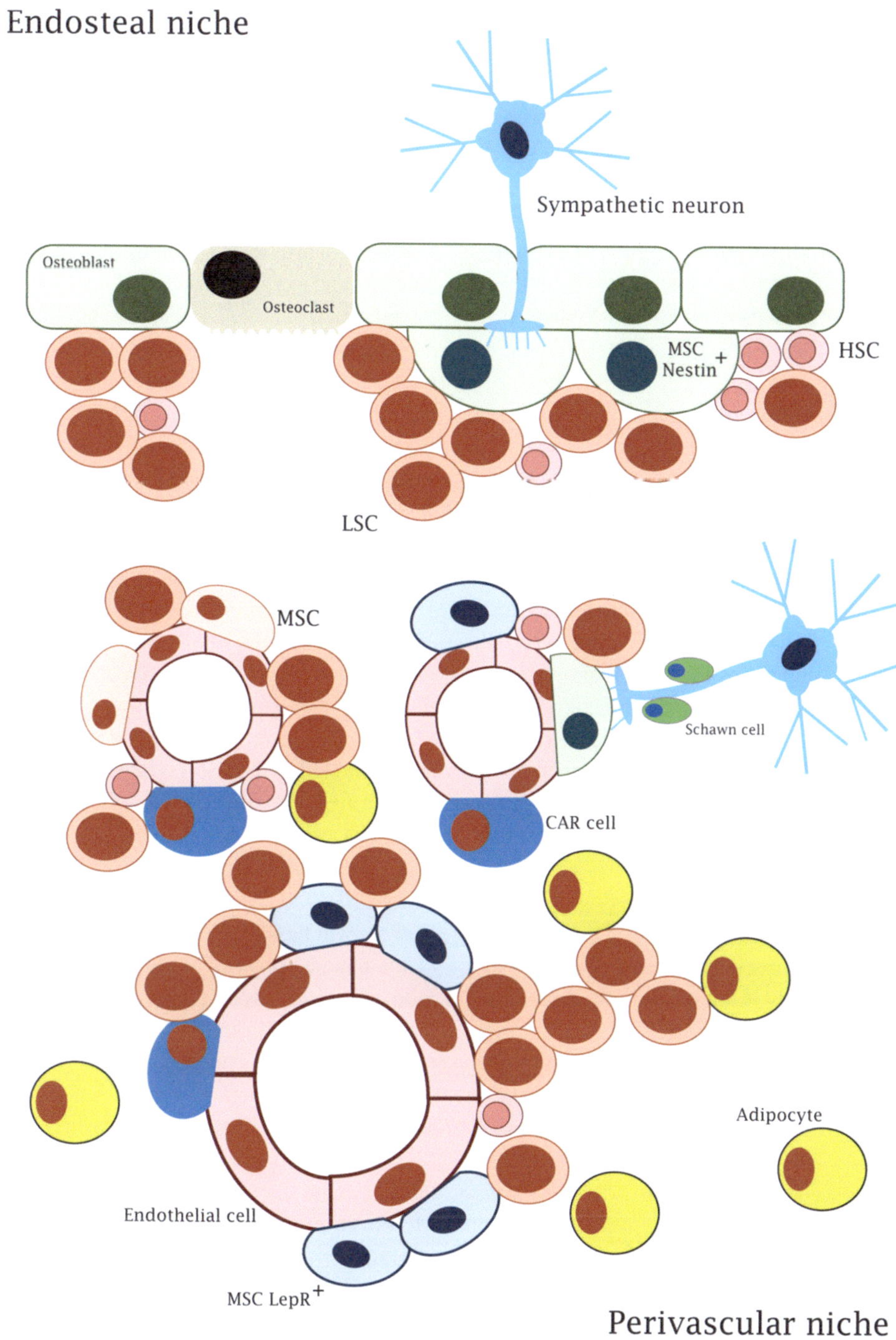

Fig. 14.2 The leukemic bone marrow microenvironment. As described in this chapter, leukemic cells disrupt normal hematopoiesis and create specialized niches that attract and expand leukemic stem cells (LSC) at the expense of hematopoietic stem cells (HSCs) maintenance. Several cellular components partake and regulate leukemia expansion which includes mesenchymal stem cells (MSC) – Nestin⁺, Leptin receptor⁺ and C-X-C motif chemokine ligand 12 (CXCL12) abundant reticular (CAR) cells – osteoblasts, osteoclasts, endothelial cells, adipocytes and also neuronal Schawn cells. See text for further details

involved, OBs which are bone forming cells, and OCs which are bone reabsorbing cells (Suda et al. 1992; Boyle et al. 2003). OCs are derived from the myeloid/macrophage lineage and have been shown to regulate HSC function. Firstly, by physical creation of the endosteal niches and secondly by decreasing the levels of soluble factors implicated in HSCs function like SCF, SDF-1 and also OPN; and as a consequence activated OCs promote HSC mobilization (Lymperi et al. 2011; Kollet et al. 2006).

The perivascular niche

Adipocytes These cells are differentiated from MSCs (described below) and are the core components of the yellow marrow that expands with age; they are located mainly within the perivascular niche closer to differentiating HSCs (Mendez-Ferrer et al. 2010; Seita and Weissman 2010). Importantly, adipocytes have been shown to act as a reservoir of HSCs and progenitors cells, highlighting their role in supporting hematopoiesis (Han et al. 2010). However, emerging data suggests that adipocytes may act as negative regulators of hematopoiesis (Naveiras et al. 2009) and obesity (increased fat tissue) has been linked to a poor outcome in leukemia patients (Meloni et al. 2001).

One of the revised hallmarks of cancer that were postulated by Hanahan and Weinberg is the deregulated cellular energetics (Hanahan and Weinberg 2000, 2011). In the context of leukemia, adipocytes may play a crucial role in this process by providing fatty acids that may serve as substrate for fatty acid oxidation (FAO) in leukemic cells for energetic purposes – this phenomenon is described as metabolic symbiosis (Herroon et al. 2013). In fact, several studies demonstrated that BM-derived adipocytes can increase cell survival, migration, expression of adhesion molecules and repression of oxidative phosphorylation in AML cells in a FAO-dependent manner, as pharmacological inhibition of FAO abrogates such effects (Ye et al. 2016; Tabe et al. 2017). Importantly, BM-derived adipocytes can also protect AML cells from the cytotoxic effects of chemotherapy (Shafat et al. 2017b) and the chaperone protein fatty acid binding protein 4 (FABP4) has been demonstrated to be pivotal in maintaining AML cellular survival even under the chemotherapeutic stress (Tabe et al. 2017).

Endothelial Cells (EC) Recent research studies demonstrated that BM vasculature is composed of a network of arterioles and sinusoids which deliver nutrients and oxygen to the BM microenvironment (Nombela-Arrieta et al. 2013). The building blocks that composed these small vasculature elements are the ECs, which are localized at the interface between these blood vessels and the BM microenvironment (Rafii et al. 1994). These cells can be easily detected using specific endothelial cell surface markers like CD31, MECA-32 (pan-endothelial cell antigen), Vascular Endothelial (VE) –Cadherin, VCAM-1 and Vascular Endothelial growth factor receptor (VEGFR) -2 (Winkler et al. 2012; Butler et al. 2010). Importantly, ECs express several growth factors that regulate HSC homeostasis (expansion and self-renewal) like SDF-1 and SCF but also several Notch ligands like Jagged-1 and -2 and Dll-1 and Dll-4 (Ding et al. 2012; Ding and Morrison 2013; Greenbaum et al. 2013). Moreover, other soluble factors might by produced by ECs depending on the signaling pathway activation: (i) PI3K-Akt pathway up-regulate fibroblast growth factor-2 (FGF-2), insulin-like growth factor binding protein-2 (IGFBP2), Ang-1, bone morphogenetic protein-4 (BMP-4) and DHH secretion which promotes HSC self-renewal and expansion; (ii) by contrast, MEK-Erk activation shifts EC production to Ang-2 and IL-6 which favors HSC differentiation (Kobayashi et al. 2010).

As discussed above, leukemic cells also use these growth factors to thrive and proliferate. Moreover, in the AML context, leukemic blasts migrate close to ECs and directly interact with them using cell adhesion molecules like integrins which confer protection against the cytotoxic effects of chemotherapy (Becker 2012; Zhang et al. 2013; Tran et al. 2002).

Mesenchymal Stem Cells (MSCs) Within the BM microenvironment MSCs are capable of developing into bone cells, adipocytes and also fibroblast-like stromal cells (Mendez-Ferrer et al.

2010). These are located in close proximity to ECs, actually, these remain entangled within the EC network in the BM. Several surface markers are used to detect MSCs, which include Nestin, Leptin receptor (LepR), CD51, CD140a and also Sca-1. MSCs are an heterogeneous population with regards to its self-renewal and multipotency capacities, distribution within the BM microenvironment and specific association with different blood vessels (Pinho et al. 2013; Houlihan et al. 2012). These cells express several HSC supportive factors, and are the major source of SCF and SDF-1 within the BM. SCF, which is secreted by endothelial cells, is also produced by a cell population derived from MSCs, the **perivascular stromal cells LepR⁺** (Ding and Morrison 2013; Ding et al. 2012). On the other hand, the CAR cells (**MSC Nestin⁺ LepR⁺**) cells are a specific subset of MSCs that are scattered along the BM forming a network (Sugiyama et al. 2006) and they secrete very high levels of SDF-1 contributing to the maintenance of HSCs within the BM niche (Omatsu et al. 2010; Sugiyama et al. 2006). Interestingly, CAR cells can also be found within the endosteal niche where they differentiate into osteoblasts and adipocytes (Omatsu et al. 2010). Notably, given that leukemic cells (from both lineages, lymphoid and myeloid) commonly express CXCR4 (the receptor for SDF-1), the CAR cells are also the main responsible for the homing and the retention of leukemic blasts within the BM microenvironment allowing their protection from cytotoxic therapies (Tavor et al. 2004).

Sympathetic Neural Cells (SNC) A considerable amount of data has demonstrated that the central nervous system (CNS) is able to control and regulate HSCs homeostasis and differentiation (Seita and Weissman 2010; Cosentino et al. 2015) and this process follows a circadian pattern under steady-state conditions (Mendez-Ferrer et al. 2008). Sympathetic neuronal fibers innervate Nestin⁺ MSCs, and through the release of catecholamines (like noraepinephrine), that bind to the β3-adrenergic receptors of several BM stromal cellular components like MSCs and OBs are the major sources of the SDF-1 and SCF, the major regulators of HSC homeostasis (Asada et al. 2013; Kalinkovich et al. 2009). **Non-myelinating Schwann cells** are another neuronal type of cells that are able to regulate HSC quiescence through the activation of transforming growth factor (TGF) -β1 signaling (Yamazaki et al. 2011).

The impact of SNCs in leukemia was recently demonstrated in two recent reports. Firstly, in a mouse model of Myeloproliferative disease, a pre-leukemic condition described above, the mutant HSCs (with constitutive activation of JAK2 kinase by expressing the JAK2V617F mutation in their HSCs) secrete IL-1β that, in turn, will remodel the BM milieu in a way that decreased the population of MSCs Nestin⁺, sympathetic neurons and also non-myelinating Schwann cells. The consequence of such remodeling is the expansion of these mutant HSCs with constitutive JAK2 kinase. Importantly this effect is abrogated when the mice are treated with a neuroprotective drug, BRL37344 – a β_3-adrenergic agonist (Arranz et al. 2014). Moreover, also in the context of myeloid leukemia, the chemical removal of adrenergic nerve fibers also resulted in the expansion of leukemic blasts within the BM niche. This leukemic growth induced a BM remodeling that not only amplified malignant expansion, but also induced a shift from a HSC supporting activity to leukemia supportive activity of MSCs population (Hanoun et al. 2014).

14.3 Targeting the BM Niche – New Toys in the Anti-leukemic Armentarium?

14.3.1 Standard Chemotherapy

The treatment of leukemia patients, the treatment response and the prognosis associated to each patient naturally depends on the type of leukemia the patient is diagnosed with. Childhood lymphoid leukemia, as B- and T-ALL are successful stories of modern medicine with survival rate at 5 years post-diagnosis reaching up to 80%. This was achieved with the use of risk-adjusted and multi-agent chemotherapy (Pui and Evans 2006). However, the outcome is far more dismal in adult B- and T-ALL than in the childhood patients where treatment fails in patients with relapsed disease (Nguyen et al. 2008). The standard treatment of ALL patients is a multi-agent chemotherapy protocol that includes glucocorticoids, anthracyclins, vincristine, L-asparagine, methotrexate, amongst others (Pui and Evans 2006; Pui et al. 2008).

In the case of myeloid leukemia – and here we will discuss only AML (CML is a very special case in hematology that we will address below) – the treatment options and outcome of patients is poor. This is a very aggressive disease characterized by the uncontrolled expansion of undifferentiated myeloid progenitors within the BM. The treatment protocol for this disease is simple and involves two drugs, 3 days of an anthraycin followed by 7 days of treatment with cytarabine. One of the major drawbacks of dealing with AML is the fact these patients often display disease relapse. Because of this, under certain conditions, AML patients undergo bone marrow transplantation but also hold some severe side effects associated with it, like graft versus host disease (Dohner et al. 2015).

The clinical use of these chemotherapeutic drugs relies on the fact that leukemic cells are more proliferative than their normal counterparts and are not specific. In fact, these compounds were introduced in the clinical almost 60 years ago and their mechanism of action is due to the inhibition of important cellular functions: DNA replication (cytarabine, vincristine and anthracyclins) (Betcher and Burnham 1990; Jordan 2002; Gewirtz 1999), protein synthesis (L-asparaginase) (Broome 1981), apoptosis (glucocorticoids) (Ramamoorthy and Cidlowski 2016) and also metabolism (methotrexate) (Goodsell 1999). Unfortunately, as we discussed before, this lack of specificity the chemotherapeutic regiments are associated with severe side effects.

14.3.2 The First Treatment Revolution in CML: Target-Therapies with TKIs

In the context of myeloid leukemia, CML is a separate case, regarding its pathophysiology and the outcome in patients with such condition (Goldman 2010). This condition results from the clonal expansion of myeloid progenitors carrying the Philadelphia chromosome t(9;22)(q34;q11), this reciprocal translocation fuses the ABL1 kinase to the *BCR* gene creating the BCR-ABL fusion protein with constitutive ABL kinase activity, which, in turn, activates several downstream signaling pathways like JAK-STAT and PI3K (Chai et al. 1997; Kim et al. 2005). In the end of the twentieth century, CML patients were still facing adverse outcomes, these were treated with cytoreductive agents, transplantation and also IFN-α which had some adverse effects associated (Goldman 2010; Hehlmann 2015).

However, in the late 90's and in the beginning of this century pioneered research directed by Prof. Brian Druker led to the development of a compound that specifically inhibit the BCR-ABL fusion protein (Druker et al. 2006; Deininger et al. 2005). This compound, imatinib-mesylate – a TKI was introduced in the clinical practice as Gleevec and marked a whole revolution in the medical field. For the first time medical community had in their armamentarium a "magic bullet" that specifically targeted malignant cells with a specific gene alteration. Importantly, not only the compound was capable of inhibiting BCR-ABL activity but in clinical terms eradicated leukemic cells and prolonged survival of CML patients (Goldman 2010). In fact, recent studies indicated that the survival of CML patients treated with imatinib at 5 years is around 90–95% demonstrating the fantastic potency of this compound (Hehlmann 2015).

The development of imatinib for CML and the extraordinary results it produced led the research community to invest on other mutations or intracellular deregulated pathways to target with TKIs in solid tumors and in hematological malignancies (Vergoulidou 2015; Kosior et al. 2011). In the case of the hematological malignancies these include the use of FLT3 inhibitors in AML, JAK2 inhibitors in Myeloproliferative disease and more recently in T-ALL.

Intensive research in the AML field identified several recurrent genetic alterations. Among these alterations are the mutations in the *FLT3* gene that encodes the FLT3 kinase. *FLT3* mutations occur is roughly around 30% of AML patients, are associated with poor prognosis and were shown to act as drivers of AML disease (Daver et al. 2019). The importance and high frequency of *FLT3* mutations in AML patients lead to the development of specific inhibitor that would target exclusively the leukemic blasts with the FLT3 kinase. However, FLT3 inhibitors only showed modest anti-leukemic activity when used as single-agent in clinical trials (Rollig et al. 2015) (Fiedler et al. 2015). Importantly, novel FLT3 inhibitors are being developed with reduced side effects and more potent TKI activity to be tested in clinical trials in combination with standard chemotherapy (Stone et al. 2017). Remarkably, there are more than 150 active clinical trials investigating FLT3 inhibitors in AML in the clinical trial registry website.

Myeloproliferative diseases are a group of heterogeneous conditions that are characterized by the clonal expansion of myeloid progenitors without the BCR-ABL fusion gene, also called BCR-ABL-negative Myeloproliferative neoplasms (Vannucchi et al. 2009). In 2005 a mutation in the gene that encodes the JAK2 kinase (the *JAK2V617F* mutation) was identified as a driver of myeloproliferative disease which results in constitutive signaling driven by the JAK-STAT signaling pathway (Baxter et al. 2005; James et al. 2005; Levine et al. 2005; Kralovics et al. 2005). This discovery led to the development of JAK inhibitors for the treatment of these diseases (Mascarenhas and Hoffman 2013). The introduction of JAK2 inhibitors in the clinical setting for the treatment of myeloproliferative disease resulted in a amelioration of symptoms and slight improvement in the overall survival (Verstovsek et al. 2012; Harrison et al. 2012; Mascarenhas and Hoffman 2012).

Another example of TKI introduction into the clinical is the testing of ruxolitinib in early T-cell precursor acute lymphoblastic leukemia (ETP-ALL). As described above, activating mutations in the *IL7Rα* gene in T-ALL were described (Zenatti et al. 2011; Shochat et al. 2011). ETP-ALL is an aggressive malignancy associated with high risk of treatment failure that accounts for about 15% of T-ALL cases (Haydu and Ferrando 2013). In a recent genomic study in ETP-ALL a high prevalence of mutations affecting cytokine signaling where identified, including activating mutations in the *IL7RA* gene (Zhang et al. 2012b) which led to the development of a clinical trial to test the efficacy of ruxolitinib (a JAK1/2 TKI) in combination with standard therapy for ETP-ALL patients (NCT03613428).

In summary, the results described highlight the importance and the revolution that was the introduction of TKIs in the clinical practice. The TKIs not only are more specific, which reduces the severe side effects of chemotherapy, but also, are more effective in the decrease of leukemia burden.

14.3.3 The Dual Targeting Concept

Throughout the chapter we have been arguing and discussing the importance of the BM microenvironment in the modulation of leukemia survival, expansion and response to chemotherapy.

As discussed, the standard chemotherapy is able to eliminate fast cycling cells but produces severe side effects. Targeted therapies using TKIs are more precise by targeting cells with specific gene alterations or signaling pathway, and for this reason the side effects are less severe. However, leukemic patients are displaying refraction to targeted therapies and this can be attributed to at least two different factors: (i) cell-intrinsic development of resistance in leukemic cells to a certain compound (Mahadevan and List 2004); and (ii) as we have been discussing in the previous pages, the BM microenvironment act as a sanctuary for providing support and protection to leukemic cells. Importantly, the protective action of the BM microenvironment on leukemic cells led the research community to develop novel strategies to overcome this hurdle. In this way, the concept of targeting "seed (leukemic cells) and soil (the supportive BM microenvironment)" or as we like to address it – dual targeting (Fig. 14.3) – has emerged over the last years to overcome BM-induced resistance in leukemia treatment protocols (Agarwal and Bhatia 2015; Shafat et al. 2017a). The concept stems from the fact that pre-

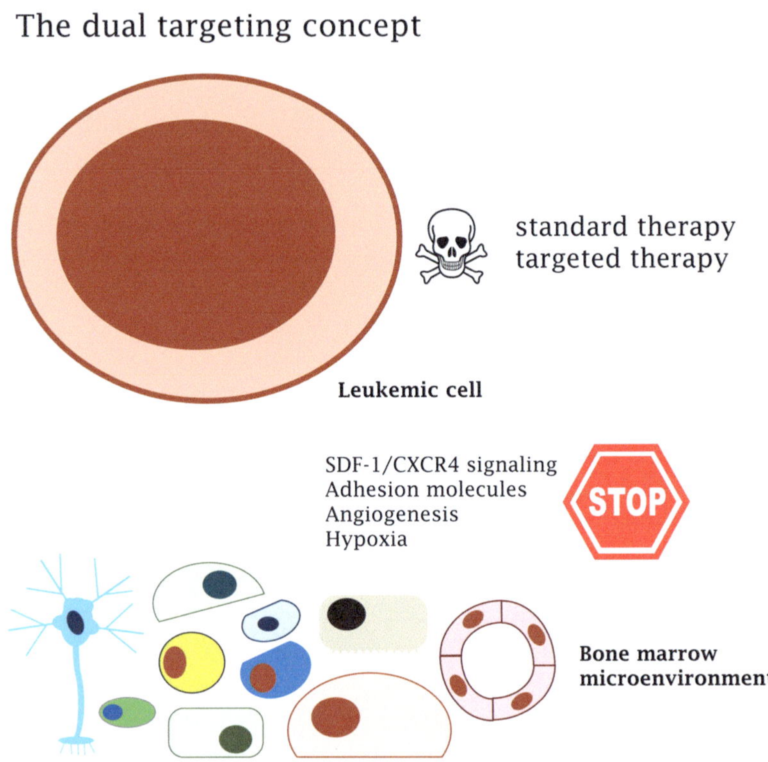

Fig. 14.3 The dual targeting concept. This concept stems from the fact that targeting leukemic cells (with standard chemotherapy or targeted therapy) in combination with the inhibition of the bone marrow microenvironmental factors that shown to protect leukemic cells (Stromal derived factor-1 (SDF-1)/C-X-C chemokine receptor type -4 (CXCR4) signaling, adhesion molecules, angiogenesis and hypoxia) will be more effective in the clinical setting than used as single agents. The details in the text describe how such agents are being tested

clinical studies show that employing a treatment protocol that targets the both leukemic cell and BM-leukemia interaction increases the efficacy of the protocol when compared with the action of both drugs acting separately.

Next, we will discuss the approaches that the medical research community has been investigating that could play a role in the dual-targeting concept for leukemia.

SDF-1 Signaling Leukemic cells home to specific niches in the BM through SDF-1 signaling and this interaction has been widely investigated from the clinical standpoint. Inhibition of SDF-1/CXCR4 signaling axis with several CXCR4 inhibitors (AMD3100) (Nervi et al. 2009; Uy et al. 2012; Roboz et al. 2018; Andreeff et al. 2012), BMS936564 (Becker et al. 2014) and BL-8040 (Borthakur et al. 2014; Borthakur et al. 2015) led to mobilization of leukemic cells to the periphery and increased efficacy of chemotherapy and targeted therapies, mainly in AML context, but also in the ALL as well (Cooper et al. 2017). As shown, various clinical trials were developed and others are still ongoing to test novel compounds and novel treatment protocols to inhibit this signaling axis in combination with other therapies (NCT01236144; NCT00512252; NCT01120457; NCT02763384). Altogether, these facts highlight the importance that such pathway in the context of BM-mediated protection of leukemic cells.

Adhesion Molecules Physical interaction between leukemic cells and the BM cellular components has been shown to be of absolute importance for the survival of leukemic cells. Currently, several adhesion molecules are under ongoing investigation.

Natalizumab is a humanized VLA-4 monoclonal antibody used for the treatment of autoimmune diseases that produced interesting pre-clinical results (Goodman and Picard 2012; Hsieh et al. 2013), however patients with hematological malignancies are not candidates to be treated with this antibody (Bloomgren et al. 2012). Another VLA-4 Inhibitor is AS101 which in pre-clinical studies induced chemosensitivity of AML cells and prolonged mice survival (Layani-Bazar et al. 2014). A clinical trial was registered (NCT01010373) to test the impact of AS101 in AML patients but was suspended by the sponsor.

Another important molecule that is currently being actively investigated is E-selectin – GMI-1271 is a small molecule inhibitor of E-selectin (Laird et al. 2018) – and in pre-clinical studies it was demonstrated to enhance chemotherapy effects and decreased leukemia burden in xenotransplant model of AML disease (Chien et al. 2013). Currently, this molecule is being studied in three different clinical trials (NCT02306291, NCT03701308, NCT03616470). The last two are still recruiting patients, but preliminary data from the first trial was highly encouraging since GMI-1271 demonstrated efficacy when combined with standard chemotherapy in refractory/relapsed AML (DeAngelo et al. 2017, 2018). CD44 is an additional molecule that could be tested in the leukemic context. Pre-clinical data from AML PDX treated with an anti-CD44 monoclonal antibody demonstrated a decrease in the LSC population of the treated mice (Jin et al. 2006). It is our view that such result warrants further validation and testing with Bivatuzumab, a humanized monoclonal antibody against CD44 that is currently under investigation for other malignant conditions.

Angiogenesis Another BM microenvironmental factor that might be a potential target in the leukemic context is angiogenesis. Angiogenesis (growth and expansion of blood vessels) facilitates the delivery of oxygen, nutrients and growth factors to the BM microenvironment (Sullivan and Brekken 2010). Bevacizumab is a monoclonal antibody that blocks VEGF action by preventing it to bind to its receptor and exert its angiogenic actions (Presta et al. 1997). This antibody has been approved for use in solid cancers (Presta et al. 1997) and was tested in clinical trials in the AML context. However, a randomised trial of Bevacizumab in AML patients as a single agent and in combination with standard chemotherapy did not show any improvement in the therapeutic outcome of these patients (Ossenkoppele et al. 2012). Other promising anti-angiogenic compound is Combretastatin, an endothelial disruptor that acts by inducing cell-cycle arrest in endothelial cells (Tozer et al. 2002). In a phase I clinical trial (NCT 01085656) in AML patients Combretastatin showed promising results in reducing leukemia burden in some patients and is well tolerated (Stockton et al. 2015; Turner et al. 2013). Currently, a clinical trial is ongoing (NCT02576301) for evaluation of the response in AML patients to Combretastatin treatment in combination with standard chemotherapy. Inhibition of Ang-1/2 with a neutralizing antibody is also a strategy that is being pursued in an ongoing clinical trial (NCT01555268) with no clear results thus far (Wang et al. 2013).

Hypoxia The BM microenvironment is mainly hypoxic, which activates HIF transcription factor, known to drive chemoresistance in leukemic cells (Frolova et al. 2012). Hypoxia-activated drugs are a class of drugs that, upon activation under hypoxic conditions, interfere with DNA synthesis leading to cell death (Mistry et al. 2017). Two of such drugs (TH-302 and PR-104) have demonstrated efficacy in pre-clinical models of leukemia (Portwood et al. 2013; Benito et al. 2011) and were tested in the clinical setting (NCT01037556 and NCT01149915). Unfortunately both compounds showed limited efficacy, probably because they were used as single dose therapies (Konopleva et al. 2015). Nevertheless some response was observed which demonstrates the need for other clinical investigations to test the combination of these compounds with standard or targeted-therapies.

14.4 Concluding Remarks

As we have discussed throughout this chapter, hematopoiesis is a very complex process that relies on specific and controlled interaction between the hematopoietic cells and the BM microenvironment. Leukemia arises when such process becomes deregulated and takes advantage of the BM supportive microenvironment to thrive and expand. The interactions between the leukemic cells and this supportive microenvironment are complex and rely on many cues. Importantly, the paradigm of the treatment protocols is shifting and the scientific community has invested a lot of time in understanding the complex interactions between the BM microenvironment and leukemic cells with a therapeutical purpose. It is our particular belief, and from several other prominent researchers as well, that the standard treatment protocols for leukemia patients should target not only the leukemic cell, by using standard cytotoxic drugs or targeted therapies for specific oncogene activation, but also to target the interaction of the leukemic cells with the BM microenvironment. We call this treatment protocol dual targeting because it combines both strategies to eradicate the leukemic clone is currently under important investigation in the leukemic field.

References

Agarwal P, Bhatia R (2015) Influence of bone marrow microenvironment on leukemic stem cells: breaking up an intimate relationship. Adv Cancer Res 127:227–252. https://doi.org/10.1016/bs.acr.2015.04.007. S0065-230X(15)00035-4 [pii]

Aiuti A, Webb IJ, Bleul C, Springer T, Gutierrez-Ramos JC (1997) The chemokine SDF-1 is a chemoattractant for human CD34+ hematopoietic progenitor cells and provides a new mechanism to explain the mobilization of CD34+ progenitors to peripheral blood. J Exp Med 185(1):111–120. https://doi.org/10.1084/jem.185.1.111

Allouche M, Charrad RS, Bettaieb A, Greenland C, Grignon C, Smadja-Joffe F (2000) Ligation of the CD44 adhesion molecule inhibits drug-induced apoptosis in human myeloid leukemia cells. Blood 96(3):1187–1190

Andersson ER, Sandberg R, Lendahl U (2011) Notch signaling: simplicity in design, versatility in function. Development 138(17):3593–3612. https://doi.org/10.1242/dev.063610. 138/17/3593 [pii]

Andreeff M, Zeng Z, Kelly MA, Wang R-y, McQueen T, Duvvuri S, Nowshah G, Borthakur G, Burger JA, Kadia TM, Jabbour E, Cortes JE, Kantarjian HM, Konopleva M (2012) Mobilization and elimination of FLT3-ITD+ Acute Myelogenous Leukemia (AML) stem/progenitor cells by Plerixafor/G-CSF/Sorafenib: results from a phase I trial in relapsed/refractory AML patients. Blood 120:142

Arai F, Hirao A, Ohmura M, Sato H, Matsuoka S, Takubo K, Ito K, Koh GY, Suda T (2004) Tie2/angiopoietin-1 signaling regulates hematopoietic stem cell quiescence in the bone marrow niche. Cell 118(2):149–161. https://doi.org/10.1016/j.cell.2004.07.004. S0092867404006622 [pii]

Arber DA, Orazi A, Hasserjian R, Thiele J, Borowitz MJ, Le Beau MM, Bloomfield CD, Cazzola M, Vardiman JW (2016) The 2016 revision to the World Health Organization classification of myeloid neoplasms and acute leukemia. Blood 127(20):2391–2405. https://doi.org/10.1182/blood-2016-03-643544. blood-2016-03-643544 [pii]

Arranz L, Sanchez-Aguilera A, Martin-Perez D, Isern J, Langa X, Tzankov A, Lundberg P, Muntion S, Tzeng YS, Lai DM, Schwaller J, Skoda RC, Mendez-Ferrer S (2014) Neuropathy of haematopoietic stem cell niche is essential for myeloproliferative neoplasms. Nature 512(7512):78–81. https://doi.org/10.1038/nature13383. nature13383 [pii]

Artavanis-Tsakonas S, Rand MD, Lake RJ (1999) Notch signaling: cell fate control and signal integration in development. Science 284(5415):770–776. https://doi.org/10.1126/science.284.5415.770

Asada N, Katayama Y, Sato M, Minagawa K, Wakahashi K, Kawano H, Kawano Y, Sada A, Ikeda K, Matsui T, Tanimoto M (2013) Matrix-embedded osteocytes regulate mobilization of hematopoietic stem/progenitor cells. Cell Stem Cell 12(6):737–747. https://doi.org/10.1016/j.stem.2013.05.001. S1934-5909(13)00194-X [pii]

Askmyr M, Quach J, Purton LE (2011) Effects of the bone marrow microenvironment on hematopoietic malignancy. Bone 48(1):115–120. https://doi.org/10.1016/j.bone.2010.06.003. S8756-3282(10)01296-2 [pii]

Barata JT, Cardoso AA, Nadler LM, Boussiotis VA (2001) Interleukin-7 promotes survival and cell cycle progression of T-cell acute lymphoblastic leukemia cells by down-regulating the cyclin-dependent kinase inhibitor p27(kip1). Blood 98(5):1524–1531. https://doi.org/10.1182/blood.v98.5.1524

Barata JT, Boussiotis VA, Yunes JA, Ferrando AA, Moreau LA, Veiga JP, Sallan SE, Look AT, Nadler LM, Cardoso AA (2004a) IL-7-dependent human leukemia T-cell line as a valuable tool for drug discovery in T-ALL. Blood 103(5):1891–1900. https://doi.org/10.1182/blood-2002-12-3861. 2002-12-3861 [pii]

Barata JT, Keenan TD, Silva A, Nadler LM, Boussiotis VA, Cardoso AA (2004b) Common gamma chain-signaling cytokines promote proliferation of T-cell acute lymphoblastic leukemia. Haematologica 89(12):1459–1467

Barata JT, Silva A, Brandao JG, Nadler LM, Cardoso AA, Boussiotis VA (2004c) Activation of PI3K is indispensable for interleukin 7-mediated viability, proliferation, glucose use, and growth of T cell acute lymphoblastic leukemia cells. J Exp Med 200(5):659–669. https://doi.org/10.1084/jem.20040789. jem.20040789 [pii]

Barata JT, Cardoso AA, Boussiotis VA (2005) Interleukin-7 in T-cell acute lymphoblastic leukemia: an extrinsic factor supporting leukemogenesis? Leuk Lymphoma 46(4):483–495. https://doi.org/10.1080/10428190400027852.VNAAJ4KD7X23MEBE [pii]

Baxter EJ, Scott LM, Campbell PJ, East C, Fourouclas N, Swanton S, Vassiliou GS, Bench AJ, Boyd EM, Curtin N, Scott MA, Erber WN, Green AR (2005) Acquired mutation of the tyrosine kinase JAK2 in human myeloproliferative disorders. Lancet 365(9464):1054–1061. doi: S0140-6736(05)71142-9 [pii]. https://doi.org/10.1016/S0140-6736(05)71142-9

Becker PS (2012) Dependence of acute myeloid leukemia on adhesion within the bone marrow microenvironment. Sci World J 2012:856467. https://doi.org/10.1100/2012/856467

Becker PS, Foran JM, Altman JK, Yacoub A, Castro JE, Sabbatini P, Dilea C, Wade M, Xing G, Gutierrez A, Cohen L, Smith BD (2014) Targeting the CXCR4 pathway: safety, tolerability and clinical activity of ulocuplumab (BMS-936564), an anti-CXCR4 antibody, in relapsed/refractory acute myeloid Leukemia. Blood 124:386

Benito J, Shi Y, Szymanska B, Carol H, Boehm I, Lu H, Konoplev S, Fang W, Zweidler-McKay PA, Campana D, Borthakur G, Bueso-Ramos C, Shpall E, Thomas DA, Jordan CT, Kantarjian H, Wilson WR, Lock R, Andreeff M, Konopleva M (2011) Pronounced hypoxia in models of murine and human leukemia: high efficacy of hypoxia-activated prodrug PR-104. PLoS One 6(8):e23108. https://doi.org/10.1371/journal.pone.0023108. PONE-D-11-04798 [pii]

Betcher DL, Burnham N (1990) Cytarabine. J Pediatr Oncol Nurs 7(4):154–157. https://doi.org/10.1177/104345429000700406

Bloomgren G, Richman S, Hotermans C, Subramanyam M, Goelz S, Natarajan A, Lee S, Plavina T, Scanlon JV, Sandrock A, Bozic C (2012) Risk of natalizumab-associated progressive multifocal leukoencephalopathy. N Engl J Med 366(20):1870–1880. https://doi.org/10.1056/NEJMoa1107829

Bociek RG, Armitage JO (1996) Hematopoietic growth factors. CA Cancer J Clin 46(3):165–184

Bonnet D, Dick JE (1997) Human acute myeloid leukemia is organized as a hierarchy that originates from a primitive hematopoietic cell. Nat Med 3(7):730–737

Borthakur G, Nagler A, Ofran Y, Rowe JM, Altman JK, Frankfurt O, Tallman MS, Avivi I, Peled A, Pereg Y, Foley-Comer A, Russovsky L, Aharon A, McQueen T, Pemmaraju N, Bueso-Ramos CE, Cortes JE, Andreeff M (2014) BL-8040, a Peptidic CXCR4 antagonist, induces leukemia cell death and specific leukemia cell mobilization in relapsed/refractory acute myeloid leukemia patients in an ongoing phase IIa clinical trial. Blood 124:950

Borthakur G, Ofran Y, Nagler A, Rowe JM, Foran JM, Uy GL, DiPersio JF, Altman JK, Frankfurt O, Tallman MS, Peled A, Pereg Y, Vainstein A, Aharon A, AlRawi A, McQueen T, Pemmaraju N, Bueso-Ramos CE, Cortes JE, Andreeff M (2015) The peptidic CXCR4 antagonist, BL-8040, significantly reduces bone marrow immature leukemia progenitors by inducing differentiation, apoptosis and mobilization: results of the dose escalation clinical trial in acute myeloid leukemia. Blood 126:2546

Bowers M, Zhang B, Ho Y, Agarwal P, Chen CC, Bhatia R (2015) Osteoblast ablation reduces normal long-term hematopoietic stem cell self-renewal but accelerates leukemia development. Blood 125(17):2678–2688. https://doi.org/10.1182/blood-2014-06-582924. blood-2014-06-582924 [pii]

Boyle WJ, Simonet WS, Lacey DL (2003) Osteoclast differentiation and activation. Nature 423(6937):337–342. https://doi.org/10.1038/nature01658. nature01658 [pii]

Briscoe J, Therond PP (2013) The mechanisms of Hedgehog signalling and its roles in development and disease. Nat Rev Mol Cell Biol 14(7):416–429. https://doi.org/10.1038/nrm3598. nrm3598 [pii]

Broome JD (1981) L-Asparaginase: discovery and development as a tumor-inhibitory agent. Cancer Treat Rep 65(Suppl 4):111–114

Butler JM, Kobayashi H, Rafii S (2010) Instructive role of the vascular niche in promoting tumour growth and tissue repair by angiocrine factors. Nat Rev Cancer 10(2):138–146. https://doi.org/10.1038/nrc2791. nrc2791 [pii]

Calvi LM, Adams GB, Weibrecht KW, Weber JM, Olson DP, Knight MC, Martin RP, Schipani E, Divieti P, Bringhurst FR, Milner LA, Kronenberg HM, Scadden DT (2003) Osteoblastic cells regulate the haematopoietic stem cell niche. Nature 425(6960):841–846. https://doi.org/10.1038/nature02040. nature02040 [pii]

Cantley LC (2002) The phosphoinositide 3-kinase pathway. Science 296(5573):1655–1657. https://doi.org/10.1126/science.296.5573.1655. 296/5573/1655 [pii]

Cardoso BA, Belo H, Barata JT, Almeida AM (2015) The bone marrow-mediated protection of myeloproliferative neoplastic cells to Vorinostat and Ruxolitinib relies on the activation of JNK and PI3K signalling pathways. PLoS One 10(12):e0143897. https://doi.org/10.1371/journal.pone.0143897. PONE-D-15-39114 [pii]

Ceradini DJ, Kulkarni AR, Callaghan MJ, Tepper OM, Bastidas N, Kleinman ME, Capla JM, Galiano RD, Levine JP, Gurtner GC (2004) Progenitor cell trafficking is regulated by hypoxic gradients through HIF-1 induction of SDF-1. Nat Med 10(8):858–864. https://doi.org/10.1038/nm1075. nm1075 [pii]

Chai SK, Nichols GL, Rothman P (1997) Constitutive activation of JAKs and STATs in BCR-Abl-expressing cell lines and peripheral blood cells derived from leukemic patients. J Immunol 159(10):4720–4728

Chien S, Haq S, Pawlus M, Moon RT, Estey EH, Appelbaum FR, Othus M, Magnani JL, Becker PS (2013) Adhesion of acute myeloid leukemia blasts to

E-selectin in the vascular niche enhances their survival by mechanisms such as Wnt activation. Blood 122:61

Christopher MJ, Liu F, Hilton MJ, Long F, Link DC (2009) Suppression of CXCL12 production by bone marrow osteoblasts is a common and critical pathway for cytokine-induced mobilization. Blood 114(7):1331–1339. https://doi.org/10.1182/blood-2008-10-184754. blood-2008-10-184754 [pii]

Colmone A, Amorim M, Pontier AL, Wang S, Jablonski E, Sipkins DA (2008) Leukemic cells create bone marrow niches that disrupt the behavior of normal hematopoietic progenitor cells. Science 322(5909):1861–1865. https://doi.org/10.1126/science.1164390. 322/5909/1861 [pii]

Cooper TM, Sison EAR, Baker SD, Li L, Ahmed A, Trippett T, Gore L, Macy ME, Narendran A, August K, Absalon MJ, Boklan J, Pollard J, Magoon D, Brown PA (2017) A phase 1 study of the CXCR4 antagonist plerixafor in combination with high-dose cytarabine and etoposide in children with relapsed or refractory acute leukemias or myelodysplastic syndrome: a pediatric oncology experimental therapeutics investigators' consortium study (POE 10-03). Pediatr Blood Cancer 64(8). https://doi.org/10.1002/pbc.26414

Cordeiro-Spinetti E, Taichman RS, Balduino A (2015) The bone marrow endosteal niche: how far from the surface? J Cell Biochem 116(1):6–11. https://doi.org/10.1002/jcb.24952

Cosentino M, Marino F, Maestroni GJ (2015) Sympathoadrenergic modulation of hematopoiesis: a review of available evidence and of therapeutic perspectives. Front Cell Neurosci 9:302. https://doi.org/10.3389/fncel.2015.00302

Crompton T, Outram SV, Hager-Theodorides AL (2007) Sonic hedgehog signalling in T-cell development and activation. Nat Rev Immunol 7(9):726–735. doi: nri2151 [pii]. https://doi.org/10.1038/nri2151

Dagklis A, Pauwels D, Lahortiga I, Geerdens E, Bittoun E, Cauwelier B, Tousseyn T, Uyttebroeck A, Maertens J, Verhoef G, Vandenberghe P, Cools J (2015) Hedgehog pathway mutations in T-cell acute lymphoblastic leukemia. Haematologica 100(3):e102–e105. https://doi.org/10.3324/haematol.2014.119248. haematol.2014.119248 [pii]

Dagklis A, Demeyer S, De Bie J, Radaelli E, Pauwels D, Degryse S, Gielen O, Vicente C, Vandepoel R, Geerdens E, Uyttebroeck A, Boeckx N, de Bock CE, Cools J (2016) Hedgehog pathway activation in T-cell acute lymphoblastic leukemia predicts response to SMO and GLI1 inhibitors. Blood 128(23):2642–2654. doi: blood-2016-03-703454 [pii]. https://doi.org/10.1182/blood-2016-03-703454

Daver N, Schlenk RF, Russell NH, Levis MJ (2019) Targeting FLT3 mutations in AML: review of current knowledge and evidence. Leukemia 33(2):299–312. https://doi.org/10.1038/s41375-018-0357-9. [pii]

de Lourdes Perim A, Amarante MK, Guembarovski RL, de Oliveira CE, Watanabe MA (2015) CXCL12/CXCR4 axis in the pathogenesis of acute lymphoblastic leukemia (ALL): a possible therapeutic target. Cell Mol Life Sci 72(9):1715–1723. https://doi.org/10.1007/s00018-014-1830-x

DeAngelo DJ, Jonas BA, Liesveld JL, Bixby DL, Advani AS, Marlton P, O'Dwyer M, Magnani JL, Thackray HM, Becker PS (2017) GMI-1271 improves efficacy and safety of chemotherapy in R/R and newly diagnosed older patients with AML: results of a phase 1/2 study. Blood 130:894

DeAngelo DJ, Jonas BA, Liesveld JL, Bixby DL, Advani AS, Marlton P, O'Dwyer ME, Fogler WE, Wolfgang CD, Magnani JL, Thackray HM, Becker PS (2018) Uproleselan (GMI-1271), an E-selectin antagonist, improves the efficacy and safety of chemotherapy in relapsed/refractory (R/R) and newly diagnosed older patients with acute myeloid Leukemia: final, correlative, and subgroup analyses. Blood 132:331. https://doi.org/10.1182/blood-2018-99-114286

Deininger M, Buchdunger E, Druker BJ (2005) The development of imatinib as a therapeutic agent for chronic myeloid leukemia. Blood 105(7):2640–2653. doi: 2004-08-3097 [pii]. https://doi.org/10.1182/blood-2004-08-3097

Dierks C, Beigi R, Guo GR, Zirlik K, Stegert MR, Manley P, Trussell C, Schmitt-Graeff A, Landwerlin K, Veelken H, Warmuth M (2008) Expansion of Bcr-Abl-positive leukemic stem cells is dependent on Hedgehog pathway activation. Cancer Cell 14(3):238–249. https://doi.org/10.1016/j.ccr.2008.08.003. S1535-6108(08)00257-2 [pii]

Dimitroff CJ, Lee JY, Rafii S, Fuhlbrigge RC, Sackstein R (2001) CD44 is a major E-selectin ligand on human hematopoietic progenitor cells. J Cell Biol 153(6):1277–1286. https://doi.org/10.1083/jcb.153.6.1277

Ding L, Morrison SJ (2013) Haematopoietic stem cells and early lymphoid progenitors occupy distinct bone marrow niches. Nature 495(7440):231–235. https://doi.org/10.1038/nature11885. nature11885 [pii]

Ding L, Saunders TL, Enikolopov G, Morrison SJ (2012) Endothelial and perivascular cells maintain haematopoietic stem cells. Nature 481(7382):457–462. https://doi.org/10.1038/nature10783. nature10783 [pii]

Dohner H, Weisdorf DJ, Bloomfield CD (2015) Acute Myeloid leukemia. N Engl J Med 373(12):1136–1152. https://doi.org/10.1056/NEJMra1406184

Dong L, Yu WM, Zheng H, Loh ML, Bunting ST, Pauly M, Huang G, Zhou M, Broxmeyer HE, Scadden DT, Qu CK (2016) Leukaemogenic effects of Ptpn11 activating mutations in the stem cell microenvironment. Nature 539(7628):304–308. https://doi.org/10.1038/nature20131. nature20131 [pii]

Druker BJ, Guilhot F, O'Brien SG, Gathmann I, Kantarjian H, Gattermann N, Deininger MW, Silver RT, Goldman JM, Stone RM, Cervantes F, Hochhaus A, Powell BL, Gabrilove JL, Rousselot P, Reiffers J, Cornelissen JJ, Hughes T, Agis H, Fischer T, Verhoef G, Shepherd J, Saglio G, Gratwohl A, Nielsen JL, Radich JP, Simonsson B, Taylor K, Baccarani M, So C, Letvak L, Larson RA (2006) Five-year follow-up of patients receiving imatinib for chronic myeloid leukemia. N

Engl J Med 355(23):2408–2417. doi: 355/23/2408 [pii]. https://doi.org/10.1056/NEJMoa062867
Essers MA, Trumpp A (2010) Targeting leukemic stem cells by breaking their dormancy. Mol Oncol 4(5):443–450. https://doi.org/10.1016/j.molonc.2010.06.001. S1574-7891(10)00047-5 [pii]
Fiedler W, Kayser S, Kebenko M, Janning M, Krauter J, Schittenhelm M, Gotze K, Weber D, Gohring G, Teleanu V, Thol F, Heuser M, Dohner K, Ganser A, Dohner H, Schlenk RF (2015) A phase I/II study of sunitinib and intensive chemotherapy in patients over 60 years of age with acute myeloid leukaemia and activating FLT3 mutations. Br J Haematol 169(5):694–700. https://doi.org/10.1111/bjh.13353
Fiskus W, Verstovsek S, Manshouri T, Smith JE, Peth K, Abhyankar S, McGuirk J, Bhalla KN (2013) Dual PI3K/AKT/mTOR inhibitor BEZ235 synergistically enhances the activity of JAK2 inhibitor against cultured and primary human myeloproliferative neoplasm cells. Mol Cancer Ther 12(5):577–588. https://doi.org/10.1158/1535-7163.MCT-12-0862. 1535-7163.MCT-12-0862 [pii]
Forristal CE, Brown AL, Helwani FM, Winkler IG, Nowlan B, Barbier V, Powell RJ, Engler GA, Diakiw SM, Zannettino AC, Martin S, Pattabiraman D, D'Andrea RJ, Lewis ID, Levesque JP (2015) Hypoxia inducible factor (HIF)-2alpha accelerates disease progression in mouse models of leukemia and lymphoma but is not a poor prognosis factor in human AML. Leukemia 29(10):2075–2085. https://doi.org/10.1038/leu.2015.102. leu2015102 [pii]
Fragoso R, Mao T, Wang S, Schaffert S, Gong X, Yue S, Luong R, Min H, Yashiro-Ohtani Y, Davis M, Pear W, Chen CZ (2012) Modulating the strength and threshold of NOTCH oncogenic signals by mir-181a-1/b-1. PLoS Genet 8(8):e1002855. https://doi.org/10.1371/journal.pgen.1002855. PGENETICS-D-12-00550 [pii]
Frisch BJ, Ashton JM, Xing L, Becker MW, Jordan CT, Calvi LM (2012) Functional inhibition of osteoblastic cells in an in vivo mouse model of myeloid leukemia. Blood 119(2):540–550. https://doi.org/10.1182/blood-2011-04-348151. blood-2011-04-348151 [pii]
Frolova O, Samudio I, Benito JM, Jacamo R, Kornblau SM, Markovic A, Schober W, Lu H, Qiu YH, Buglio D, McQueen T, Pierce S, Shpall E, Konoplev S, Thomas D, Kantarjian H, Lock R, Andreeff M, Konopleva M (2012) Regulation of HIF-1alpha signaling and chemoresistance in acute lymphocytic leukemia under hypoxic conditions of the bone marrow microenvironment. Cancer Biol Ther 13(10):858–870. https://doi.org/10.4161/cbt.20838. 20838 [pii]
Gewirtz DA (1999) A critical evaluation of the mechanisms of action proposed for the antitumor effects of the anthracycline antibiotics adriamycin and daunorubicin. Biochem Pharmacol 57(7):727–741. https://doi.org/10.1016/s0006-2952(98)00307-4. S0006295298003074 [pii]
Giambra V, Jenkins CE, Lam SH, Hoofd C, Belmonte M, Wang X, Gusscott S, Gracias D, Weng AP (2015) Leukemia stem cells in T-ALL require active Hif1alpha and Wnt signaling. Blood 125(25):3917–3927. https://doi.org/10.1182/blood-2014-10-609370. blood-2014-10-609370 [pii]
Goldman JM (2010) Chronic myeloid leukemia: a historical perspective. Semin Hematol 47(4):302–311. https://doi.org/10.1053/j.seminhematol.2010.07.001. S0037-1963(10)00083-1 [pii]
Goodman SL, Picard M (2012) Integrins as therapeutic targets. Trends Pharmacol Sci 33(7):405–412. https://doi.org/10.1016/j.tips.2012.04.002. S0165-6147(12)00057-0 [pii]
Goodsell DS (1999) The molecular perspective: methotrexate. Oncologist 4(4):340–341
Grassinger J, Haylock DN, Storan MJ, Haines GO, Williams B, Whitty GA, Vinson AR, Be CL, Li S, Sorensen ES, Tam PP, Denhardt DT, Sheppard D, Choong PF, Nilsson SK (2009) Thrombin-cleaved osteopontin regulates hemopoietic stem and progenitor cell functions through interactions with alpha9beta1 and alpha4beta1 integrins. Blood 114(1):49–59. https://doi.org/10.1182/blood-2009-01-197988. blood-2009-01-197988 [pii]
Greenbaum A, Hsu YM, Day RB, Schuettpelz LG, Christopher MJ, Borgerding JN, Nagasawa T, Link DC (2013) CXCL12 in early mesenchymal progenitors is required for haematopoietic stem-cell maintenance. Nature 495(7440):227–230. https://doi.org/10.1038/nature11926. nature11926 [pii]
Guo Z, Dose M, Kovalovsky D, Chang R, O'Neil J, Look AT, von Boehmer H, Khazaie K, Gounari F (2007) Beta-catenin stabilization stalls the transition from double-positive to single-positive stage and predisposes thymocytes to malignant transformation. Blood 109(12):5463–5472. doi: blood-2006-11-059071 [pii]. https://doi.org/10.1182/blood-2006-11-059071
Gutierrez A, Sanda T, Grebliunaite R, Carracedo A, Salmena L, Ahn Y, Dahlberg S, Neuberg D, Moreau LA, Winter SS, Larson R, Zhang J, Protopopov A, Chin L, Pandolfi PP, Silverman LB, Hunger SP, Sallan SE, Look AT (2009) High frequency of PTEN, PI3K, and AKT abnormalities in T-cell acute lymphoblastic leukemia. Blood 114(3):647–650. https://doi.org/10.1182/blood-2009-02-206722. blood-2009-02-206722 [pii]
Gutierrez A, Sanda T, Ma W, Zhang J, Grebliunaite R, Dahlberg S, Neuberg D, Protopopov A, Winter SS, Larson RS, Borowitz MJ, Silverman LB, Chin L, Hunger SP, Jamieson C, Sallan SE, Look AT (2010) Inactivation of LEF1 in T-cell acute lymphoblastic leukemia. Blood 115(14):2845–2851. https://doi.org/10.1182/blood-2009-07-234377. blood-2009-07-234377 [pii]
Han J, Koh YJ, Moon HR, Ryoo HG, Cho CH, Kim I, Koh GY (2010) Adipose tissue is an extramedullary reservoir for functional hematopoietic stem and progenitor cells. Blood 115(5):957–964. https://doi.org/10.1182/blood-2009-05-219923. blood-2009-05-219923 [pii]
Hanahan D, Weinberg RA (2000) The hallmarks of cancer. Cell 100(1):57–70. https://doi.org/10.1016/s0092-8674(00)81683-9. S0092-8674(00)81683-9 [pii]

Hanahan D, Weinberg RA (2011) Hallmarks of cancer: the next generation. Cell 144(5):646–674. https://doi.org/10.1016/j.cell.2011.02.013. S0092-8674(11)00127-9 [pii]
Hanoun M, Zhang D, Mizoguchi T, Pinho S, Pierce H, Kunisaki Y, Lacombe J, Armstrong SA, Duhrsen U, Frenette PS (2014) Acute myelogenous leukemia-induced sympathetic neuropathy promotes malignancy in an altered hematopoietic stem cell niche. Cell Stem Cell 15(3):365–375. doi: S1934-5909(14)00296-3 [pii]. https://doi.org/10.1016/j.stem.2014.06.020
Harrison C, Kiladjian JJ, Al-Ali HK, Gisslinger H, Waltzman R, Stalbovskaya V, McQuitty M, Hunter DS, Levy R, Knoops L, Cervantes F, Vannucchi AM, Barbui T, Barosi G (2012) JAK inhibition with ruxolitinib versus best available therapy for myelofibrosis. N Engl J Med 366(9):787–798. https://doi.org/10.1056/NEJMoa1110556
Haydu JE, Ferrando AA (2013) Early T-cell precursor acute lymphoblastic leukaemia. Curr Opin Hematol 20(4):369–373. https://doi.org/10.1097/MOH.0b013e3283623c61
He M, Wang QY, Yin QQ, Tang J, Lu Y, Zhou CX, Duan CW, Hong DL, Tanaka T, Chen GQ, Zhao Q (2013) HIF-1alpha downregulates miR-17/20a directly targeting p21 and STAT3: a role in myeloid leukemic cell differentiation. Cell Death Differ 20(3):408–418. https://doi.org/10.1038/cdd.2012.130. cdd2012130 [pii]
Hehlmann R (2015) CML--where do we stand in 2015? Ann Hematol 94(Suppl 2):S103–S105. https://doi.org/10.1007/s00277-015-2331-1
Heidel FH, Bullinger L, Feng Z, Wang Z, Neff TA, Stein L, Kalaitzidis D, Lane SW, Armstrong SA (2012) Genetic and pharmacologic inhibition of beta-catenin targets imatinib-resistant leukemia stem cells in CML. Cell Stem Cell 10(4):412–424. https://doi.org/10.1016/j.stem.2012.02.017. S1934-5909(12)00076-8 [pii]
Hellqvist E, Holm F, Mason CN, Runza V, Weigand S, Sadarangani A, Jamieson CHM (2013) CD44 monoclonal antibody-enhanced clearance of chronic myeloid leukemia stem cells from the malignant niche. Blood 122:858
Herroon MK, Rajagurubandara E, Hardaway AL, Powell K, Turchick A, Feldmann D, Podgorski I (2013) Bone marrow adipocytes promote tumor growth in bone via FABP4-dependent mechanisms. Oncotarget 4(11):2108–2123. doi: 1482 [pii]. https://doi.org/10.18632/oncotarget.1482
Hoofd C, Wang X, Lam S, Jenkins C, Wood B, Giambra V, Weng AP (2016) CD44 promotes chemoresistance in T-ALL by increased drug efflux. Exp Hematol 44(3):166–171 e117. https://doi.org/10.1016/j.exphem.2015.12.001 S0301-472X(15)00798-5 [pii]
Horn S, Bergholz U, Jucker M, McCubrey JA, Trumper L, Stocking C, Basecke J (2008) Mutations in the catalytic subunit of class IA PI3K confer leukemogenic potential to hematopoietic cells. Oncogene 27(29):4096–4106. https://doi.org/10.1038/onc.2008.40. onc200840 [pii]
Houlihan DD, Mabuchi Y, Morikawa S, Niibe K, Araki D, Suzuki S, Okano H, Matsuzaki Y (2012) Isolation of mouse mesenchymal stem cells on the basis of expression of Sca-1 and PDGFR-alpha. Nat Protoc 7(12):2103–2111. https://doi.org/10.1038/nprot.2012.125. nprot.2012.125 [pii]
Hsieh Y-T, Jiang E, Pham J, Kim H-N, Abdel-Azim H, Khazal S, Bug G, Spohn G, Bonig H, Kim Y-M (2013) VLA4 blockade in acute myeloid Leukemia. Blood 122:3944
Hu Y, Chen Y, Douglas L, Li S (2009) Beta-catenin is essential for survival of leukemic stem cells insensitive to kinase inhibition in mice with BCR-ABL-induced chronic myeloid leukemia. Leukemia 23(1):109–116. https://doi.org/10.1038/leu.2008.262. leu2008262 [pii]
Hu K, Gu Y, Lou L, Liu L, Hu Y, Wang B, Luo Y, Shi J, Yu X, Huang H (2015) Galectin-3 mediates bone marrow microenvironment-induced drug resistance in acute leukemia cells via Wnt/beta-catenin signaling pathway. J Hematol Oncol 8(1). https://doi.org/10.1186/s13045-014-0099-8. s13045-014-0099-8 [pii]
Jacamo R, Chen Y, Wang Z, Ma W, Zhang M, Spaeth EL, Wang Y, Battula VL, Mak PY, Schallmoser K, Ruvolo P, Schober WD, Shpall EJ, Nguyen MH, Strunk D, Bueso-Ramos CE, Konoplev S, Davis RE, Konopleva M, Andreeff M (2014) Reciprocal leukemia-stroma VCAM-1/VLA-4-dependent activation of NF-kappaB mediates chemoresistance. Blood 123(17):2691–2702. https://doi.org/10.1182/blood-2013-06-511527. blood-2013-06-511527 [pii]
Jaleco AC, Neves H, Hooijberg E, Gameiro P, Clode N, Haury M, Henrique D, Parreira L (2001) Differential effects of Notch ligands Delta-1 and Jagged-1 in human lymphoid differentiation. J Exp Med 194(7):991–1002. https://doi.org/10.1084/jem.194.7.991
James C, Ugo V, Le Couedic JP, Staerk J, Delhommeau F, Lacout C, Garcon L, Raslova H, Berger R, Bennaceur-Griscelli A, Villeval JL, Constantinescu SN, Casadevall N, Vainchenker W (2005) A unique clonal JAK2 mutation leading to constitutive signalling causes polycythaemia vera. Nature 434(7037):1144–1148. doi: nature03546 [pii]. https://doi.org/10.1038/nature03546
Jamieson CH, Ailles LE, Dylla SJ, Muijtjens M, Jones C, Zehnder JL, Gotlib J, Li K, Manz MG, Keating A, Sawyers CL, Weissman IL (2004) Granulocyte-macrophage progenitors as candidate leukemic stem cells in blast-crisis CML. N Engl J Med 351(7):657–667. https://doi.org/10.1056/NEJMoa040258. 351/7/657 [pii]
Jatiani SS, Baker SJ, Silverman LR, Reddy EP (2010) Jak/STAT pathways in cytokine signaling and myeloproliferative disorders: approaches for targeted therapies. Genes Cancer 1(10):979–993. https://doi.org/10.1177/1947601910397187. [pii]
Jiang BH, Semenza GL, Bauer C, Marti HH (1996) Hypoxia-inducible factor 1 levels vary exponentially over a physiologically relevant range of O_2 tension.

Am J Phys 271(4 Pt 1):C1172–C1180. https://doi.org/10.1152/ajpcell.1996.271.4.C1172

Jin L, Hope KJ, Zhai Q, Smadja-Joffe F, Dick JE (2006) Targeting of CD44 eradicates human acute myeloid leukemic stem cells. Nat Med 12(10):1167–1174. doi: nm1483 [pii]. https://doi.org/10.1038/nm1483

Jin L, Tabe Y, Konoplev S, Xu Y, Leysath CE, Lu H, Kimura S, Ohsaka A, Rios MB, Calvert L, Kantarjian H, Andreeff M, Konopleva M (2008) CXCR4 up-regulation by imatinib induces chronic myelogenous leukemia (CML) cell migration to bone marrow stroma and promotes survival of quiescent CML cells. Mol Cancer Ther 7(1):48–58. https://doi.org/10.1158/1535-7163.MCT-07-0042. 7/1/48 [pii]

Jin L, Tabe Y, Lu H, Borthakur G, Miida T, Kantarjian H, Andreeff M, Konopleva M (2013) Mechanisms of apoptosis induction by simultaneous inhibition of PI3K and FLT3-ITD in AML cells in the hypoxic bone marrow microenvironment. Cancer Lett 329(1):45–58. https://doi.org/10.1016/j.canlet.2012.09.020. S0304-3835(12)00578-2 [pii]

Jordan MA (2002) Mechanism of action of antitumor drugs that interact with microtubules and tubulin. Curr Med Chem Anticancer Agents 2(1):1–17

Juarez J, Baraz R, Gaundar S, Bradstock K, Bendall L (2007a) Interaction of interleukin-7 and interleukin-3 with the CXCL12-induced proliferation of B-cell progenitor acute lymphoblastic leukemia. Haematologica 92(4):450–459. https://doi.org/10.3324/haematol.10621

Juarez J, Dela Pena A, Baraz R, Hewson J, Khoo M, Cisterne A, Fricker S, Fujii N, Bradstock KF, Bendall LJ (2007b) CXCR4 antagonists mobilize childhood acute lymphoblastic leukemia cells into the peripheral blood and inhibit engraftment. Leukemia 21(6):1249–1257. doi: 2404684 [pii]. https://doi.org/10.1038/sj.leu.2404684

Jung Y, Wang J, Song J, Shiozawa Y, Havens A, Wang Z, Sun YX, Emerson SG, Krebsbach PH, Taichman RS (2007) Annexin II expressed by osteoblasts and endothelial cells regulates stem cell adhesion, homing, and engraftment following transplantation. Blood 110(1):82–90. doi: blood-2006-05-021352 [pii]. https://doi.org/10.1182/blood-2006-05-021352

Kalinkovich A, Spiegel A, Shivtiel S, Kollet O, Jordaney N, Piacibello W, Lapidot T (2009) Blood-forming stem cells are nervous: direct and indirect regulation of immature human CD34+ cells by the nervous system. Brain Behav Immun 23(8):1059–1065. https://doi.org/10.1016/j.bbi.2009.03.008. S0889-1591(09)00100-7 [pii]

Kannan S, Sutphin RM, Hall MG, Golfman LS, Fang W, Nolo RM, Akers LJ, Hammitt RA, McMurray JS, Kornblau SM, Melnick AM, Figueroa ME, Zweidler-McKay PA (2013) Notch activation inhibits AML growth and survival: a potential therapeutic approach. J Exp Med 210(2):321–337. https://doi.org/10.1084/jem.20121527. jem.20121527 [pii]

Katagiri S, Tauchi T, Okabe S, Minami Y, Kimura S, Maekawa T, Naoe T, Ohyashiki K (2013) Combination of ponatinib with hedgehog antagonist vismodegib for therapy-resistant BCR-ABL1-positive leukemia. Clin Cancer Res 19(6):1422–1432. https://doi.org/10.1158/1078-0432.CCR-12-1777. 1078-0432.CCR-12-1777 [pii]

Kebelmann-Betzing C, Korner G, Badiali L, Buchwald D, Moricke A, Korte A, Kochling J, Wu S, Kappelmeier D, Oettel K, Henze G, Seeger K (2001) Characterization of cytokine, growth factor receptor, costimulatory and adhesion molecule expression patterns of bone marrow blasts in relapsed childhood B cell precursor all. Cytokine 13(1):39–50. https://doi.org/10.1006/cyto.2000.0794. S1043-4666(00)90794-4 [pii]

Khan NI, Bradstock KF, Bendall LJ (2007) Activation of Wnt/beta-catenin pathway mediates growth and survival in B-cell progenitor acute lymphoblastic leukaemia. Br J Haematol 138(3):338–348. doi: BJH6667 [pii]. https://doi.org/10.1111/j.1365-2141.2007.06667.x

Kikushige Y, Miyamoto T, Yuda J, Jabbarzadeh-Tabrizi S, Shima T, Takayanagi S, Niiro H, Yurino A, Miyawaki K, Takenaka K, Iwasaki H, Akashi K (2015) A TIM-3/gal-9 autocrine stimulatory loop drives self-renewal of human myeloid leukemia stem cells and leukemic progression. Cell Stem Cell 17(3):341–352. https://doi.org/10.1016/j.stem.2015.07.011. S1934-5909(15)00310-0 [pii]

Kim JH, Chu SC, Gramlich JL, Pride YB, Babendreier E, Chauhan D, Salgia R, Podar K, Griffin JD, Sattler M (2005) Activation of the PI3K/mTOR pathway by BCR-ABL contributes to increased production of reactive oxygen species. Blood 105(4):1717–1723. doi: 2004-03-0849 [pii]. https://doi.org/10.1182/blood-2004-03-0849

Kim YW, Koo BK, Jeong HW, Yoon MJ, Song R, Shin J, Jeong DC, Kim SH, Kong YY (2008) Defective Notch activation in microenvironment leads to myeloproliferative disease. Blood 112(12):4628–4638. https://doi.org/10.1182/blood-2008-03-148999. blood-2008-03-148999 [pii]

Kobayashi H, Butler JM, O'Donnell R, Kobayashi M, Ding BS, Bonner B, Chiu VK, Nolan DJ, Shido K, Benjamin L, Rafii S (2010) Angiocrine factors from Akt-activated endothelial cells balance self-renewal and differentiation of haematopoietic stem cells. Nat Cell Biol 12(11):1046–1056. https://doi.org/10.1038/ncb2108. ncb2108 [pii]

Kode A, Manavalan JS, Mosialou I, Bhagat G, Rathinam CV, Luo N, Khiabanian H, Lee A, Murty VV, Friedman R, Brum A, Park D, Galili N, Mukherjee S, Teruya-Feldstein J, Raza A, Rabadan R, Berman E, Kousteni S (2014) Leukaemogenesis induced by an activating beta-catenin mutation in osteoblasts. Nature 506(7487):240–244. https://doi.org/10.1038/nature12883. nature12883 [pii]

Kode A, Mosialou I, Manavalan SJ, Rathinam CV, Friedman RA, Teruya-Feldstein J, Bhagat G, Berman E, Kousteni S (2016) FoxO1-dependent induction of acute myeloid leukemia by osteoblasts in mice.

Leukemia 30(1):1–13. https://doi.org/10.1038/leu.2015.161. leu2015161 [pii]

Kollet O, Dar A, Shivtiel S, Kalinkovich A, Lapid K, Sztainberg Y, Tesio M, Samstein RM, Goichberg P, Spiegel A, Elson A, Lapidot T (2006) Osteoclasts degrade endosteal components and promote mobilization of hematopoietic progenitor cells. Nat Med 12(6):657–664. doi: nm1417 [pii]. https://doi.org/10.1038/nm1417

Konopleva M, Thall PF, Yi CA, Borthakur G, Coveler A, Bueso-Ramos C, Benito J, Konoplev S, Gu Y, Ravandi F, Jabbour E, Faderl S, Thomas D, Cortes J, Kadia T, Kornblau S, Daver N, Pemmaraju N, Nguyen HQ, Feliu J, Lu H, Wei C, Wilson WR, Melink TJ, Gutheil JC, Andreeff M, Estey EH, Kantarjian H (2015) Phase I/II study of the hypoxia-activated prodrug PR104 in refractory/relapsed acute myeloid leukemia and acute lymphoblastic leukemia. Haematologica 100(7):927–934. https://doi.org/10.3324/haematol.2014.118455. haematol.2014.118455 [pii]

Kosior K, Lewandowska-Grygiel M, Giannopoulos K (2011) Tyrosine kinase inhibitors in hematological malignancies. Postepy Hig Med Dosw (Online) 65:819–828. https://doi.org/10.5604/17322693.968778. [pii]

Kralovics R, Passamonti F, Buser AS, Teo SS, Tiedt R, Passweg JR, Tichelli A, Cazzola M, Skoda RC (2005) A gain-of-function mutation of JAK2 in myeloproliferative disorders. N Engl J Med 352(17):1779–1790. doi: 352/17/1779 [pii]. https://doi.org/10.1056/NEJMoa051113

Krause DS, Lazarides K, von Andrian UH, Van Etten RA (2006) Requirement for CD44 in homing and engraftment of BCR-ABL-expressing leukemic stem cells. Nat Med 12(10):1175–1180. https://doi.org/10.1038/nm1489. nm1489 [pii]

Krause DS, Fulzele K, Catic A, Sun CC, Dombkowski D, Hurley MP, Lezeau S, Attar E, Wu JY, Lin HY, Divieti-Pajevic P, Hasserjian RP, Schipani E, Van Etten RA, Scadden DT (2013) Differential regulation of myeloid leukemias by the bone marrow microenvironment. Nat Med 19(11):1513–1517. https://doi.org/10.1038/nm.3364. nm.3364 [pii]

Krevvata M, Silva BC, Manavalan JS, Galan-Diez M, Kode A, Matthews BG, Park D, Zhang CA, Galili N, Nickolas TL, Dempster DW, Dougall W, Teruya-Feldstein J, Economides AN, Kalajzic I, Raza A, Berman E, Mukherjee S, Bhagat G, Kousteni S (2014) Inhibition of leukemia cell engraftment and disease progression in mice by osteoblasts. Blood 124(18):2834–2846. https://doi.org/10.1182/blood-2013-07-517219. blood-2013-07-517219 [pii]

Laird CT, Hassanein W, O'Neill NA, French BM, Cheng X, Fogler WE, Magnani JL, Parsell D, Cimeno A, Phelps CJ, Ayares D, Burdorf L, Azimzadeh AM, Pierson RN 3rd (2018) P- and E-selectin receptor antagonism prevents human leukocyte adhesion to activated porcine endothelial monolayers and attenuates porcine endothelial damage. Xenotransplantation 25(2):e12381. https://doi.org/10.1111/xen.12381

Layani-Bazar A, Skornick I, Berrebi A, Pauker MH, Noy E, Silberman A, Albeck M, Longo DL, Kalechman Y, Sredni B (2014) Redox modulation of adjacent thiols in VLA-4 by AS101 converts myeloid leukemia cells from a drug-resistant to drug-sensitive state. Cancer Res 74(11):3092–3103. https://doi.org/10.1158/0008-5472. CAN-13-2159 0008-5472.CAN-13-2159 [pii]

Levine RL, Wadleigh M, Cools J, Ebert BL, Wernig G, Huntly BJ, Boggon TJ, Wlodarska I, Clark JJ, Moore S, Adelsperger J, Koo S, Lee JC, Gabriel S, Mercher T, D'Andrea A, Frohling S, Dohner K, Marynen P, Vandenberghe P, Mesa RA, Tefferi A, Griffin JD, Eck MJ, Sellers WR, Meyerson M, Golub TR, Lee SJ, Gilliland DG (2005) Activating mutation in the tyrosine kinase JAK2 in polycythemia vera, essential thrombocythemia, and myeloid metaplasia with myelofibrosis. Cancer Cell 7(4):387–397. doi: S1535-6108(05)00094-2 [pii]. https://doi.org/10.1016/j.ccr.2005.03.023

Li X, Guo H, Duan H, Yang Y, Meng J, Liu J, Wang C, Xu H (2015) Improving chemotherapeutic efficiency in acute myeloid leukemia treatments by chemically synthesized peptide interfering with CXCR4/CXCL12 axis. Sci Rep 5:16228. https://doi.org/10.1038/srep16228. srep16228 [pii]

Lindblad O, Cordero E, Puissant A, Macaulay L, Ramos A, Kabir NN, Sun J, Vallon-Christersson J, Haraldsson K, Hemann MT, Borg A, Levander F, Stegmaier K, Pietras K, Ronnstrand L, Kazi JU (2016) Aberrant activation of the PI3K/mTOR pathway promotes resistance to sorafenib in AML. Oncogene 35(39):5119–5131. https://doi.org/10.1038/onc.2016.41. onc201641 [pii]

Ling L, Nurcombe V, Cool SM (2009) Wnt signaling controls the fate of mesenchymal stem cells. Gene 433(1–2):1–7. https://doi.org/10.1016/j.gene.2008.12.008. S0378-1119(08)00636-7 [pii]

Liu TC, Lin PM, Chang JG, Lee JP, Chen TP, Lin SF (2000) Mutation analysis of PTEN/MMAC1 in acute myeloid leukemia. Am J Hematol 63(4):170–175. https://doi.org/10.1002/(SICI)1096-8652(200004)63:4<170::AID-AJH2>3.0.CO;2-0. [pii]

Lo Celso C, Fleming HE, Wu JW, Zhao CX, Miake-Lye S, Fujisaki J, Cote D, Rowe DW, Lin CP, Scadden DT (2009) Live-animal tracking of individual haematopoietic stem/progenitor cells in their niche. Nature 457(7225):92–96. https://doi.org/10.1038/nature07434. nature07434 [pii]

Lo TC, Barnhill LM, Kim Y, Nakae EA, Yu AL, Diccianni MB (2009) Inactivation of SHIP1 in T-cell acute lymphoblastic leukemia due to mutation and extensive alternative splicing. Leuk Res 33(11):1562–1566. https://doi.org/10.1016/j.leukres.2009.04.032. S0145-2126(09)00224-0 [pii]

Lobry C, Ntziachristos P, Ndiaye-Lobry D, Oh P, Cimmino L, Zhu N, Araldi E, Hu W, Freund J, Abdel-Wahab O, Ibrahim S, Skokos D, Armstrong SA, Levine RL, Park CY, Aifantis I (2013) Notch pathway activation targets AML-initiating cell homeostasis and differentiation.

J Exp Med 210(2):301–319. https://doi.org/10.1084/jem.20121484. jem.20121484 [pii]

Luis TC, Naber BA, Roozen PP, Brugman MH, de Haas EF, Ghazvini M, Fibbe WE, van Dongen JJ, Fodde R, Staal FJ (2011) Canonical wnt signaling regulates hematopoiesis in a dosage-dependent fashion. Cell Stem Cell 9(4):345–356. https://doi.org/10.1016/j.stem.2011.07.017. S1934-5909(11)00380-8 [pii]

Lymperi S, Ersek A, Ferraro F, Dazzi F, Horwood NJ (2011) Inhibition of osteoclast function reduces hematopoietic stem cell numbers in vivo. Blood 117(5):1540–1549. https://doi.org/10.1182/blood-2010-05-282855. blood-2010-05-282855 [pii]

Mahadevan D, List AF (2004) Targeting the multidrug resistance-1 transporter in AML: molecular regulation and therapeutic strategies. Blood 104(7):1940–1951. https://doi.org/10.1182/blood-2003-07-2490. 2003-07-2490 [pii]

Majmundar AJ, Wong WJ, Simon MC (2010) Hypoxia-inducible factors and the response to hypoxic stress. Mol Cell 40(2):294–309. https://doi.org/10.1016/j.molcel.2010.09.022. S1097-2765(10)00750-1 [pii]

Martinez-Agosto JA, Mikkola HK, Hartenstein V, Banerjee U (2007) The hematopoietic stem cell and its niche: a comparative view. Genes Dev 21(23):3044–3060. doi: 21/23/3044 [pii]. https://doi.org/10.1101/gad.1602607

Mascarenhas J, Hoffman R (2012) Ruxolitinib: the first FDA approved therapy for the treatment of myelofibrosis. Clin Cancer Res 18(11):3008–3014. https://doi.org/10.1158/1078-0432.CCR-11-3145. 1078-0432.CCR-11-3145 [pii]

Mascarenhas J, Hoffman R (2013) A comprehensive review and analysis of the effect of ruxolitinib therapy on the survival of patients with myelofibrosis. Blood 121(24):4832–4837. https://doi.org/10.1182/blood-2013-02-482232. blood-2013-02-482232 [pii]

Medvinsky A, Rybtsov S, Taoudi S (2011) Embryonic origin of the adult hematopoietic system: advances and questions. Development 138(6):1017–1031. https://doi.org/10.1242/dev.040998. 138/6/1017 [pii]

Medyouf H, Mossner M, Jann JC, Nolte F, Raffel S, Herrmann C, Lier A, Eisen C, Nowak V, Zens B, Mudder K, Klein C, Oblander J, Fey S, Vogler J, Fabarius A, Riedl E, Roehl H, Kohlmann A, Staller M, Haferlach C, Muller N, John T, Platzbecker U, Metzgeroth G, Hofmann WK, Trumpp A, Nowak D (2014) Myelodysplastic cells in patients reprogram mesenchymal stromal cells to establish a transplantable stem cell niche disease unit. Cell Stem Cell 14(6):824–837. https://doi.org/10.1016/j.stem.2014.02.014. S1934-5909(14)00078-2 [pii]

Meister M, Spencer JA, Wu J, Zhao C, Stefania L, Ferraro F, Lo Celso C, Scadden DT, Van Etten RA, Lin C, Krause DS (2014) The microanatomy of the leukemic stem cell niche in murine chronic myelogenous leukemia. Blood 124:351

Meloni G, Proia A, Capria S, Romano A, Trape G, Trisolini SM, Vignetti M, Mandelli F (2001) Obesity and autologous stem cell transplantation in acute myeloid leukemia. Bone Marrow Transplant 28(4):365–367. https://doi.org/10.1038/sj.bmt.1703145

Mendez-Ferrer S, Lucas D, Battista M, Frenette PS (2008) Haematopoietic stem cell release is regulated by circadian oscillations. Nature 452(7186):442–447. https://doi.org/10.1038/nature06685. nature06685 [pii]

Mendez-Ferrer S, Michurina TV, Ferraro F, Mazloom AR, Macarthur BD, Lira SA, Scadden DT, Ma'ayan A, Enikolopov GN, Frenette PS (2010) Mesenchymal and haematopoietic stem cells form a unique bone marrow niche. Nature 466(7308):829–834. https://doi.org/10.1038/nature09262. nature09262 [pii]

Mikkola HK, Orkin SH (2006) The journey of developing hematopoietic stem cells. Development 133(19):3733–3744. https://doi.org/10.1242/dev.02568. 133/19/3733 [pii]

Milne TA (2017) Mouse models of MLL leukemia: recapitulating the human disease. Blood 129(16):2217–2223. https://doi.org/10.1182/blood-2016-10-691428. blood-2016-10-691428 [pii]

Minami Y, Stuart SA, Ikawa T, Jiang Y, Banno A, Hunton IC, Young DJ, Naoe T, Murre C, Jamieson CH, Wang JY (2008) BCR-ABL-transformed GMP as myeloid leukemic stem cells. Proc Natl Acad Sci U S A 105(46):17967–17972. https://doi.org/10.1073/pnas.0808303105. 0808303105 [pii]

Minuzzo S, Agnusdei V, Pusceddu I, Pinazza M, Moserle L, Masiero M, Rossi E, Crescenzi M, Hoey T, Ponzoni M, Amadori A, Indraccolo S (2015) DLL4 regulates NOTCH signaling and growth of T acute lymphoblastic leukemia cells in NOD/SCID mice. Carcinogenesis 36(1):115–121. https://doi.org/10.1093/carcin/bgu223. bgu223 [pii]

Mistry IN, Thomas M, Calder EDD, Conway SJ, Hammond EM (2017) Clinical advances of hypoxia-activated prodrugs in combination with radiation therapy. Int J Radiat Oncol Biol Phys 98(5):1183–1196. https://doi.org/10.1016/j.ijrobp.2017.03.024. S0360-3016(17)30710-1 [pii]

Miyajima A, Ito Y, Kinoshita T (1999) Cytokine signaling for proliferation, survival, and death in hematopoietic cells. Int J Hematol 69(3):137–146

Mulloy JC, Cammenga J, MacKenzie KL, Berguido FJ, Moore MA, Nimer SD (2002) The AML1-ETO fusion protein promotes the expansion of human hematopoietic stem cells. Blood 99(1):15–23. https://doi.org/10.1182/blood.v99.1.15

Naveiras O, Nardi V, Wenzel PL, Hauschka PV, Fahey F, Daley GQ (2009) Bone-marrow adipocytes as negative regulators of the haematopoietic microenvironment. Nature 460(7252):259–263. https://doi.org/10.1038/nature08099. nature08099 [pii]

Nervi B, Ramirez P, Rettig MP, Uy GL, Holt MS, Ritchey JK, Prior JL, Piwnica-Worms D, Bridger G, Ley TJ, DiPersio JF (2009) Chemosensitization of acute myeloid leukemia (AML) following mobilization by the CXCR4 antagonist AMD3100. Blood 113(24):6206–6214. https://doi.org/10.1182/blood-2008-06-162123. blood-2008-06-162123 [pii]

Ng KP, Manjeri A, Lee KL, Huang W, Tan SY, Chuah CT, Poellinger L, Ong ST (2014) Physiologic hypoxia promotes maintenance of CML stem cells despite effective BCR-ABL1 inhibition. Blood 123(21):3316–3326. https://doi.org/10.1182/blood-2013-07-511907. blood-2013-07-511907 [pii]

Nguyen K, Devidas M, Cheng SC, La M, Raetz EA, Carroll WL, Winick NJ, Hunger SP, Gaynon PS, Loh ML (2008) Factors influencing survival after relapse from acute lymphoblastic leukemia: a Children's oncology group study. Leukemia 22(12):2142–2150. https://doi.org/10.1038/leu.2008.251. leu2008251 [pii]

Nilsson SK, Johnston HM, Coverdale JA (2001) Spatial localization of transplanted hemopoietic stem cells: inferences for the localization of stem cell niches. Blood 97(8):2293–2299. https://doi.org/10.1182/blood.v97.8.2293

Nilsson SK, Johnston HM, Whitty GA, Williams B, Webb RJ, Denhardt DT, Bertoncello I, Bendall LJ, Simmons PJ, Haylock DN (2005) Osteopontin, a key component of the hematopoietic stem cell niche and regulator of primitive hematopoietic progenitor cells. Blood 106(4):1232–1239. doi: 2004-11-4422 [pii]. https://doi.org/10.1182/blood-2004-11-4422

Nombela-Arrieta C, Pivarnik G, Winkel B, Canty KJ, Harley B, Mahoney JE, Park SY, Lu J, Protopopov A, Silberstein LE (2013) Quantitative imaging of haematopoietic stem and progenitor cell localization and hypoxic status in the bone marrow microenvironment. Nat Cell Biol 15(5):533–543. https://doi.org/10.1038/ncb2730. ncb2730 [pii]

Nusslein-Volhard C, Wieschaus E (1980) Mutations affecting segment number and polarity in Drosophila. Nature 287(5785):795–801. https://doi.org/10.1038/287795a0

Nwabo Kamdje AH, Mosna F, Bifari F, Lisi V, Bassi G, Malpeli G, Ricciardi M, Perbellini O, Scupoli MT, Pizzolo G, Krampera M (2011) Notch-3 and Notch-4 signaling rescue from apoptosis human B-ALL cells in contact with human bone marrow-derived mesenchymal stromal cells. Blood 118(2):380–389. https://doi.org/10.1182/blood-2010-12-326694. blood-2010-12-326694 [pii]

O'Neil J, Grim J, Strack P, Rao S, Tibbitts D, Winter C, Hardwick J, Welcker M, Meijerink JP, Pieters R, Draetta G, Sears R, Clurman BE, Look AT (2007) FBW7 mutations in leukemic cells mediate NOTCH pathway activation and resistance to gamma-secretase inhibitors. J Exp Med 204(8):1813–1824. doi: jem.20070876 [pii]. https://doi.org/10.1084/jem.20070876

Ogawa M (1993) Differentiation and proliferation of hematopoietic stem cells. Blood 81(11):2844–2853

Oliveira ML, Akkapeddi P, Ribeiro D, Melao A, Barata JT (2019) IL-7R-mediated signaling in T-cell acute lymphoblastic leukemia: an update. Adv Biol Regul 71:88–96. doi: S2212-4926(18)30138-6 [pii]. https://doi.org/10.1016/j.jbior.2018.09.012

Olson TS, Ley K (2002) Chemokines and chemokine receptors in leukocyte trafficking. Am J Physiol Regul Integr Comp Physiol 283(1):R7–R28. https://doi.org/10.1152/ajpregu.00738.2001

Omatsu Y, Sugiyama T, Kohara H, Kondoh G, Fujii N, Kohno K, Nagasawa T (2010) The essential functions of adipo-osteogenic progenitors as the hematopoietic stem and progenitor cell niche. Immunity 33(3):387–399. https://doi.org/10.1016/j.immuni.2010.08.017. S1074-7613(10)00322-5 [pii]

Ossenkoppele GJ, Stussi G, Maertens J, van Montfort K, Biemond BJ, Breems D, Ferrant A, Graux C, de Greef GE, Halkes CJ, Hoogendoorn M, Hollestein RM, Jongen-Lavrencic M, Levin MD, van de Loosdrecht AA, van Marwijk Kooij M, van Norden Y, Pabst T, Schouten HC, Vellenga E, Verhoef GE, de Weerdt O, Wijermans P, Passweg JR, Lowenberg B (2012) Addition of bevacizumab to chemotherapy in acute myeloid leukemia at older age: a randomized phase 2 trial of the Dutch-Belgian cooperative trial group for Hemato-Oncology (HOVON) and the Swiss Group for Clinical Cancer Research (SAKK). Blood 120(24):4706–4711. https://doi.org/10.1182/blood-2012-04-420596. blood-2012-04-420596 [pii]

Parameswaran R, Yu M, Lim M, Groffen J, Heisterkamp N (2011) Combination of drug therapy in acute lymphoblastic leukemia with a CXCR4 antagonist. Leukemia 25(8):1314–1323. https://doi.org/10.1038/leu.2011.76. leu201176 [pii]

Parmar K, Mauch P, Vergilio JA, Sackstein R, Down JD (2007) Distribution of hematopoietic stem cells in the bone marrow according to regional hypoxia. Proc Natl Acad Sci U S A 104(13):5431–5436. doi: 0701152104 [pii]. https://doi.org/10.1073/pnas.0701152104

Pear WS, Aster JC, Scott ML, Hasserjian RP, Soffer B, Sklar J, Baltimore D (1996) Exclusive development of T cell neoplasms in mice transplanted with bone marrow expressing activated Notch alleles. J Exp Med 183(5):2283–2291. https://doi.org/10.1084/jem.183.5.2283

Penton AL, Leonard LD, Spinner NB (2012) Notch signaling in human development and disease. Semin Cell Dev Biol 23(4):450–457. https://doi.org/10.1016/j.semcdb.2012.01.010. S1084-9521(12)00014-6 [pii]

Pinho S, Lacombe J, Hanoun M, Mizoguchi T, Bruns I, Kunisaki Y, Frenette PS (2013) PDGFRalpha and CD51 mark human nestin+ sphere-forming mesenchymal stem cells capable of hematopoietic progenitor cell expansion. J Exp Med 210(7):1351–1367. https://doi.org/10.1084/jem.20122252. jem.20122252 [pii]

Polak R, Buitenhuis M (2012) The PI3K/PKB signaling module as key regulator of hematopoiesis: implications for therapeutic strategies in leukemia. Blood 119(4):911–923. https://doi.org/10.1182/blood-2011-07-366203. blood-2011-07-366203 [pii]

Polakis P (2012) Wnt signaling in cancer. Cold Spring Harb Perspect Biol 4(5). https://doi.org/10.1101/cshperspect.a008052. a008052 [pii] cshperspect.a008052 [pii]

Ponomaryov T, Peled A, Petit I, Taichman RS, Habler L, Sandbank J, Arenzana-Seisdedos F, Magerus A, Caruz A, Fujii N, Nagler A, Lahav M, Szyper-Kravitz M, Zipori D, Lapidot T (2000) Induction of the chemokine stromal-derived factor-1 following DNA damage improves human stem cell function. J Clin Invest 106(11):1331–1339. https://doi.org/10.1172/JCI10329

Portwood S, Lal D, Hsu YC, Vargas R, Johnson MK, Wetzler M, Hart CP, Wang ES (2013) Activity of the hypoxia-activated prodrug, TH-302, in preclinical human acute myeloid leukemia models. Clin Cancer Res 19(23):6506–6519. https://doi.org/10.1158/1078-0432.CCR-13-0674. 1078-0432.CCR-13-0674 [pii]

Presta LG, Chen H, O'Connor SJ, Chisholm V, Meng YG, Krummen L, Winkler M, Ferrara N (1997) Humanization of an anti-vascular endothelial growth factor monoclonal antibody for the therapy of solid tumors and other disorders. Cancer Res 57(20):4593–4599

Pui CH, Evans WE (2006) Treatment of acute lymphoblastic leukemia. N Engl J Med 354(2):166–178. doi: 354/2/166 [pii]. https://doi.org/10.1056/NEJMra052603

Pui CH, Robison LL, Look AT (2008) Acute lymphoblastic leukaemia. Lancet 371(9617):1030–1043. https://doi.org/10.1016/S0140-6736(08)60457-2. S0140-6736(08)60457-2 [pii]

Raaijmakers MH, Mukherjee S, Guo S, Zhang S, Kobayashi T, Schoonmaker JA, Ebert BL, Al-Shahrour F, Hasserjian RP, Scadden EO, Aung Z, Matza M, Merkenschlager M, Lin C, Rommens JM, Scadden DT (2010) Bone progenitor dysfunction induces myelodysplasia and secondary leukaemia. Nature 464(7290):852–857. https://doi.org/10.1038/nature08851

Rafii S, Shapiro F, Rimarachin J, Nachman RL, Ferris B, Weksler B, Moore MA, Asch AS (1994) Isolation and characterization of human bone marrow microvascular endothelial cells: hematopoietic progenitor cell adhesion. Blood 84(1):10–19

Ramamoorthy S, Cidlowski JA (2016) Corticosteroids: mechanisms of action in health and disease. Rheum Dis Clin N Am 42(1):15–31, vii. https://doi.org/10.1016/j.rdc.2015.08.002 S0889-857X(15)00066-6 [pii]

Reya T, Morrison SJ, Clarke MF, Weissman IL (2001) Stem cells, cancer, and cancer stem cells. Nature 414(6859):105–111. https://doi.org/10.1038/35102167. 35102167 [pii]

Ribeiro D, Melao A, Barata JT (2013) IL-7R-mediated signaling in T-cell acute lymphoblastic leukemia. Adv Biol Regul 53(2):211–222. https://doi.org/10.1016/j.jbior.2012.10.005. S2212-4926(12)00100-5 [pii]

Roboz GJ, Ritchie EK, Dault Y, Lam L, Marshall DC, Cruz NM, Hsu HC, Hassane DC, Christos PJ, Ippoliti C, Scandura JM, Guzman ML (2018) Phase I trial of plerixafor combined with decitabine in newly diagnosed older patients with acute myeloid leukemia. Haematologica 103(8):1308–1316. https://doi.org/10.3324/haematol.2017.183418. haematol.2017.183418 [pii]

Rollig C, Serve H, Huttmann A, Noppeney R, Muller-Tidow C, Krug U, Baldus CD, Brandts CH, Kunzmann V, Einsele H, Kramer A, Schafer-Eckart K, Neubauer A, Burchert A, Giagounidis A, Krause SW, Mackensen A, Aulitzky W, Herbst R, Hanel M, Kiani A, Frickhofen N, Kullmer J, Kaiser U, Link H, Geer T, Reichle A, Junghanss C, Repp R, Heits F, Durk H, Hase J, Klut IM, Illmer T, Bornhauser M, Schaich M, Parmentier S, Gorner M, Thiede C, von Bonin M, Schetelig J, Kramer M, Berdel WE, Ehninger G (2015) Addition of sorafenib versus placebo to standard therapy in patients aged 60 years or younger with newly diagnosed acute myeloid leukaemia (SORAML): a multicentre, phase 2, randomised controlled trial. Lancet Oncol 16(16):1691–1699. https://doi.org/10.1016/S1470-2045(15)00362-9. S1470-2045(15)00362-9 [pii]

Rossi DJ, Seita J, Czechowicz A, Bhattacharya D, Bryder D, Weissman IL (2007) Hematopoietic stem cell quiescence attenuates DNA damage response and permits DNA damage accumulation during aging. Cell Cycle 6(19):2371–2376. doi: 4759 [pii]. https://doi.org/10.4161/cc.6.19.4759

Sanchez-Aguilera A, Mendez-Ferrer S (2017) The hematopoietic stem-cell niche in health and leukemia. Cell Mol Life Sci 74(4):579–590. https://doi.org/10.1007/s00018-016-2306-y. 10.1007/s00018-016-2306-y [pii]

Schepers K, Pietras EM, Reynaud D, Flach J, Binnewies M, Garg T, Wagers AJ, Hsiao EC, Passegue E (2013) Myeloproliferative neoplasia remodels the endosteal bone marrow niche into a self-reinforcing leukemic niche. Cell Stem Cell 13(3):285–299. https://doi.org/10.1016/j.stem.2013.06.009. S1934-5909(13)00267-1 [pii]

Schmits R, Filmus J, Gerwin N, Senaldi G, Kiefer F, Kundig T, Wakeham A, Shahinian A, Catzavelos C, Rak J, Furlonger C, Zakarian A, Simard JJ, Ohashi PS, Paige CJ, Gutierrez-Ramos JC, Mak TW (1997) CD44 regulates hematopoietic progenitor distribution, granuloma formation, and tumorigenicity. Blood 90(6):2217–2233

Schmitt TM, Zuniga-Pflucker JC (2002) Induction of T cell development from hematopoietic progenitor cells by delta-like-1 in vitro. Immunity 17(6):749–756. doi: S1074-7613(02)00474-0 [pii]

Schneider P, Vasse M, Al Bayati A, Lenormand B, Vannier JP (2002) Is high expression of the chemokine receptor CXCR-4 of predictive value for early relapse in childhood acute lymphoblastic leukaemia? Br J Haematol 119(2):579–580. doi:3825_6 [pii]. https://doi.org/10.1046/j.1365-2141.2002.03835_6.x

Schneider F, Bohlander SK, Schneider S, Papadaki C, Kakadyia P, Dufour A, Vempati S, Unterhalt M, Feuring-Buske M, Buske C, Braess J, Wandt H, Hiddemann W, Spiekermann K (2007) AML1-ETO meets JAK2: clinical evidence for the two hit model of leukemogenesis from a myeloproliferative syndrome progressing to acute myeloid leukemia. Leukemia 21(10):2199–2201. doi: 2404830 [pii]. https://doi.org/10.1038/sj.leu.2404830

Schofield R (1978) The relationship between the spleen colony-forming cell and the haemopoietic stem cell. Blood Cells 4(1–2):7–25

Scupoli MT, Perbellini O, Krampera M, Vinante F, Cioffi F, Pizzolo G (2007) Interleukin 7 requirement for survival of T-cell acute lymphoblastic leukemia and human thymocytes on bone marrow stroma. Haematologica 92(2):264–266. https://doi.org/10.3324/haematol.10356

Seita J, Weissman IL (2010) Hematopoietic stem cell: self-renewal versus differentiation. Wiley Interdiscip Rev Syst Biol Med 2(6):640–653. https://doi.org/10.1002/wsbm.86

Shafat MS, Gnaneswaran B, Bowles KM, Rushworth SA (2017a) The bone marrow microenvironment – home of the leukemic blasts. Blood Rev 31(5):277–286. doi: S0268-960X(16)30073-X [pii]. https://doi.org/10.1016/j.blre.2017.03.004

Shafat MS, Oellerich T, Mohr S, Robinson SD, Edwards DR, Marlein CR, Piddock RE, Fenech M, Zaitseva L, Abdul-Aziz A, Turner J, Watkins JA, Lawes M, Bowles KM, Rushworth SA (2017b) Leukemic blasts program bone marrow adipocytes to generate a protumoral microenvironment. Blood 129(10):1320–1332. https://doi.org/10.1182/blood-2016-08-734798. blood-2016-08-734798 [pii]

Shalapour S, Hof J, Kirschner-Schwabe R, Bastian L, Eckert C, Prada J, Henze G, von Stackelberg A, Seeger K (2011) High VLA-4 expression is associated with adverse outcome and distinct gene expression changes in childhood B-cell precursor acute lymphoblastic leukemia at first relapse. Haematologica 96(11):1627–1635. https://doi.org/10.3324/haematol.2011.047993. haematol.2011.047993 [pii]

Shochat C, Tal N, Bandapalli OR, Palmi C, Ganmore I, te Kronnie G, Cario G, Cazzaniga G, Kulozik AE, Stanulla M, Schrappe M, Biondi A, Basso G, Bercovich D, Muckenthaler MU, Izraeli S (2011) Gain-of-function mutations in interleukin-7 receptor-alpha (IL7R) in childhood acute lymphoblastic leukemias. J Exp Med 208(5):901–908. https://doi.org/10.1084/jem.20110580. jem.20110580 [pii]

Silva A, Yunes JA, Cardoso BA, Martins LR, Jotta PY, Abecasis M, Nowill AE, Leslie NR, Cardoso AA, Barata JT (2008) PTEN posttranslational inactivation and hyperactivation of the PI3K/Akt pathway sustain primary T cell leukemia viability. J Clin Invest 118(11):3762–3774. https://doi.org/10.1172/JCI34616

Silva A, Girio A, Cebola I, Santos CI, Antunes F, Barata JT (2011a) Intracellular reactive oxygen species are essential for PI3K/Akt/mTOR-dependent IL-7-mediated viability of T-cell acute lymphoblastic leukemia cells. Leukemia 25(6):960–967. https://doi.org/10.1038/leu.2011.56. leu201156 [pii]

Silva A, Laranjeira AB, Martins LR, Cardoso BA, Demengeot J, Yunes JA, Seddon B, Barata JT (2011b) IL-7 contributes to the progression of human T-cell acute lymphoblastic leukemias. Cancer Res 71(14):4780–4789. https://doi.org/10.1158/0008-5472.CAN-10-3606. 0008-5472.CAN-10-3606 [pii]

Song G, Ouyang G, Bao S (2005) The activation of Akt/PKB signaling pathway and cell survival. J Cell Mol Med 9(1):59–71. doi: 009.001.07 [pii]. https://doi.org/10.1111/j.1582-4934.2005.tb00337.x

Spencer JA, Ferraro F, Roussakis E, Klein A, Wu J, Runnels JM, Zaher W, Mortensen LJ, Alt C, Turcotte R, Yusuf R, Cote D, Vinogradov SA, Scadden DT, Lin CP (2014) Direct measurement of local oxygen concentration in the bone marrow of live animals. Nature 508(7495):269–273. https://doi.org/10.1038/nature13034. nature13034 [pii]

Spoo AC, Lubbert M, Wierda WG, Burger JA (2007) CXCR4 is a prognostic marker in acute myelogenous leukemia. Blood 109(2):786–791. doi: blood-2006-05-024844 [pii]. https://doi.org/10.1182/blood-2006-05-024844

Stier S, Ko Y, Forkert R, Lutz C, Neuhaus T, Grunewald E, Cheng T, Dombkowski D, Calvi LM, Rittling SR, Scadden DT (2005) Osteopontin is a hematopoietic stem cell niche component that negatively regulates stem cell pool size. J Exp Med 201(11):1781–1791. doi: jem.20041992 [pii]. https://doi.org/10.1084/jem.20041992

Stockton SS, Pettiford L, Cline C, Chaplin D, Hsu JW, Wingard JR, Cogle CR (2015) The vascular disrupting agent OXi4503 in relapsed and refractory AML and MDS. Blood 126:4936

Stone RM, Mandrekar SJ, Sanford BL, Laumann K, Geyer SM, Bloomfield CD, Dohner K, Thiede C, Marcucci G, Lo Coco F, Klisovic RB, Wei A, Sierra J, Sanz MA, Brandwein JM, de Witte TMM, Niederwieser D, Appelbaum FR, Medeiros BC, Tallman MS, Krauter J, Schlenk RF, Ganser A, Serve H, Ehninger G, Amadori S, Dohner H, Larson RA (2017) The addition of Midostaurin to standard chemotherapy decreases cumulative incidence of relapse (CIR) in the international prospective randomized, placebo-controlled, double-blind trial (CALGB 10603 / RATIFY [Alliance]) for newly diagnosed acute myeloid Leukemia (AML) patients with FLT3 mutations. Blood 130:2580

Suda T, Takahashi N, Martin TJ (1992) Modulation of osteoclast differentiation. Endocr Rev 13(1):66–80. https://doi.org/10.1210/edrv-13-1-66

Sugiyama T, Kohara H, Noda M, Nagasawa T (2006) Maintenance of the hematopoietic stem cell pool by CXCL12-CXCR4 chemokine signaling in bone marrow stromal cell niches. Immunity 25(6):977–988. doi: S1074-7613(06)00515-2 [pii]. https://doi.org/10.1016/j.immuni.2006.10.016

Sullivan LA, Brekken RA (2010) The VEGF family in cancer and antibody-based strategies for their inhibition. MAbs 2(2):165–175. doi: 11360 [pii]. https://doi.org/10.4161/mabs.2.2.11360

Swerdlow SH, Campo E, Pileri SA, Harris NL, Stein H, Siebert R, Advani R, Ghielmini M, Salles GA,

Zelenetz AD, Jaffe ES (2016) The 2016 revision of the World Health Organization classification of lymphoid neoplasms. Blood 127(20):2375–2390. https://doi.org/10.1182/blood-2016-01-643569. blood-2016-01-643569 [pii]

Tabe Y, Yamamoto S, Saitoh K, Sekihara K, Monma N, Ikeo K, Mogushi K, Shikami M, Ruvolo V, Ishizawa J, Hail N Jr, Kazuno S, Igarashi M, Matsushita H, Yamanaka Y, Arai H, Nagaoka I, Miida T, Hayashizaki Y, Konopleva M, Andreeff M (2017) Bone marrow adipocytes facilitate fatty acid oxidation activating AMPK and a transcriptional network supporting survival of acute Monocytic Leukemia cells. Cancer Res 77(6):1453–1464. https://doi.org/10.1158/0008-5472.CAN-16-1645. 0008-5472.CAN-16-1645 [pii]

Taichman RS, Emerson SG (1994) Human osteoblasts support hematopoiesis through the production of granulocyte colony-stimulating factor. J Exp Med 179(5):1677–1682. https://doi.org/10.1084/jem.179.5.1677

Tavian M, Peault B (2005) Embryonic development of the human hematopoietic system. Int J Dev Biol 49(2–3):243–250. doi: 041957mt [pii]. https://doi.org/10.1387/ijdb.041957mt

Tavor S, Petit I, Porozov S, Avigdor A, Dar A, Leider-Trejo L, Shemtov N, Deutsch V, Naparstek E, Nagler A, Lapidot T (2004) CXCR4 regulates migration and development of human acute myelogenous leukemia stem cells in transplanted NOD/SCID mice. Cancer Res 64(8):2817–2824

Tozer GM, Kanthou C, Parkins CS, Hill SA (2002) The biology of the combretastatins as tumour vascular targeting agents. Int J Exp Pathol 83(1):21–38. doi: 211 [pii]. https://doi.org/10.1046/j.1365-2613.2002.00211.x

Tran J, Master Z, Yu JL, Rak J, Dumont DJ, Kerbel RS (2002) A role for survivin in chemoresistance of endothelial cells mediated by VEGF. Proc Natl Acad Sci U S A 99(7):4349–4354. https://doi.org/10.1073/pnas.072586399. 072586399 [pii]

Turner D, Gonzalez A, Pettiford L, Meacham A, Wise E, Bosse RC, Chaplin D, Hsu JW, Brown RA, Hiemenz JW, Norkin M, Wingard JR, Cogle CR (2013) A phase I study of the vascular disrupting Combretastatin, OXi4503, in patients with relapsed and refractory Acute Myeloid Leukemia (AML) and Myelodysplastic Syndromes (MDS). Blood 122:1463

Uy GL, Rettig MP, Motabi IH, McFarland K, Trinkaus KM, Hladnik LM, Kulkarni S, Abboud CN, Cashen AF, Stockerl-Goldstein KE, Vij R, Westervelt P, DiPersio JF (2012) A phase 1/2 study of chemosensitization with the CXCR4 antagonist plerixafor in relapsed or refractory acute myeloid leukemia. Blood 119(17):3917–3924. https://doi.org/10.1182/blood-2011-10-383406. blood-2011-10-383406 [pii]

van Amerongen R, Nusse R (2009) Towards an integrated view of Wnt signaling in development. Development 136(19):3205–3214. https://doi.org/10.1242/dev.033910. 136/19/3205 [pii]

van den Berk LC, van der Veer A, Willemse ME, Theeuwes MJ, Luijendijk MW, Tong WH, van der Sluis IM, Pieters R, den Boer ML (2014) Disturbed CXCR4/CXCL12 axis in paediatric precursor B-cell acute lymphoblastic leukaemia. Br J Haematol 166(2):240–249. https://doi.org/10.1111/bjh.12883

Vannucchi AM, Guglielmelli P, Tefferi A (2009) Advances in understanding and management of myeloproliferative neoplasms. CA Cancer J Clin 59(3):171–191. https://doi.org/10.3322/caac.20009. caac.20009 [pii]

Vergoulidou M (2015) More than a decade of tyrosine kinase inhibitors in the treatment of solid tumors: what we have learned and what the future holds. Biomark Insights 10(Suppl 3):33–40. https://doi.org/10.4137/BMI.S22436. bmi-suppl.3-2015-033 [pii]

Verstovsek S, Mesa RA, Gotlib J, Levy RS, Gupta V, DiPersio JF, Catalano JV, Deininger M, Miller C, Silver RT, Talpaz M, Winton EF, Harvey JH Jr, Arcasoy MO, Hexner E, Lyons RM, Paquette R, Raza A, Vaddi K, Erickson-Viitanen S, Koumenis IL, Sun W, Sandor V, Kantarjian HM (2012) A double-blind, placebo-controlled trial of ruxolitinib for myelofibrosis. N Engl J Med 366(9):799–807. https://doi.org/10.1056/NEJMoa1110557

Visnjic D, Kalajzic Z, Rowe DW, Katavic V, Lorenzo J, Aguila HL (2004) Hematopoiesis is severely altered in mice with an induced osteoblast deficiency. Blood 103(9):3258–3264. https://doi.org/10.1182/blood-2003-11-4011. 2003-11-4011 [pii]

Walkley CR, Olsen GH, Dworkin S, Fabb SA, Swann J, McArthur GA, Westmoreland SV, Chambon P, Scadden DT, Purton LE (2007a) A microenvironment-induced myeloproliferative syndrome caused by retinoic acid receptor gamma deficiency. Cell 129(6):1097–1110. doi: S0092-8674(07)00608-3 [pii]. https://doi.org/10.1016/j.cell.2007.05.014

Walkley CR, Shea JM, Sims NA, Purton LE, Orkin SH (2007b) Rb regulates interactions between hematopoietic stem cells and their bone marrow microenvironment. Cell 129(6):1081–1095. doi: S0092-8674(07)00536-3 [pii]. https://doi.org/10.1016/j.cell.2007.03.055

Wang Y, Krivtsov AV, Sinha AU, North TE, Goessling W, Feng Z, Zon LI, Armstrong SA (2010) The Wnt/beta-catenin pathway is required for the development of leukemia stem cells in AML. Science 327(5973):1650–1653. https://doi.org/10.1126/science.1186624. 327/5973/1650 [pii]

Wang ES, Fetterly G, Brady W, Tan W, Greene J, Gaudy A, Vigil CE, Mendler JH, Becker MW, O'Dwyer K, Liesveld JL, Wetzler M (2013) Clinical and biologic effects of the angiopoietin 1/2 neutralizing Peptibody, Trebananib (AMG 386), in acute myeloid Leukemia patients. Blood 122:2701

Wang W, Zimmerman G, Huang X, Yu S, Myers J, Wang Y, Moreton S, Nthale J, Awadallah A, Beck R, Xin W, Wald D, Huang AY, Zhou L (2016) Aberrant notch signaling in the bone marrow microenvironment of acute lymphoid leukemia suppresses osteoblast-mediated support of hematopoietic niche function. Cancer Res 76(6):1641–1652. https://doi.org/10.1158/0008-5472.CAN-15-2092. 0008-5472.CAN-15-2092 [pii]

Warner JK, Wang JC, Hope KJ, Jin L, Dick JE (2004) Concepts of human leukemic development. Oncogene 23(43):7164–7177. https://doi.org/10.1038/sj.onc.1207933. 1207933 [pii]

Weber JM, Forsythe SR, Christianson CA, Frisch BJ, Gigliotti BJ, Jordan CT, Milner LA, Guzman ML, Calvi LM (2006) Parathyroid hormone stimulates expression of the notch ligand Jagged1 in osteoblastic cells. Bone 39(3):485–493. doi: S8756-3282(06)00370-X [pii]. https://doi.org/10.1016/j.bone.2006.03.002

Weng AP, Ferrando AA, Lee W, Morris JP, Silverman LB, Sanchez-Irizarry C, Blacklow SC, Look AT, Aster JC (2004) Activating mutations of NOTCH1 in human T cell acute lymphoblastic leukemia. Science 306(5694):269–271. doi: 306/5694/269 [pii]. https://doi.org/10.1126/science.1102160

Winkler IG, Barbier V, Nowlan B, Jacobsen RN, Forristal CE, Patton JT, Magnani JL, Levesque JP (2012) Vascular niche E-selectin regulates hematopoietic stem cell dormancy, self renewal and chemoresistance. Nat Med 18(11):1651–1657. https://doi.org/10.1038/nm.2969. nm.2969 [pii]

Winter SS, Sweatman JJ, Lawrence MB, Rhoades TH, Hart AL, Larson RS (2001) Enhanced T-lineage acute lymphoblastic leukaemia cell survival on bone marrow stroma requires involvement of LFA-1 and ICAM-1. Br J Haematol 115(4):862–871. doi: 3182 [pii]. https://doi.org/10.1046/j.1365-2141.2001.03182.x

Wiseman DH (2011) Donor cell leukemia: a review. Biol Blood Marrow Transplant 17(6):771–789. https://doi.org/10.1016/j.bbmt.2010.10.010. S1083-8791(10)00448-9 [pii]

Yamamoto-Sugitani M, Kuroda J, Ashihara E, Nagoshi H, Kobayashi T, Matsumoto Y, Sasaki N, Shimura Y, Kiyota M, Nakayama R, Akaji K, Taki T, Uoshima N, Kobayashi Y, Horiike S, Maekawa T, Taniwaki M (2011) Galectin-3 (Gal-3) induced by leukemia microenvironment promotes drug resistance and bone marrow lodgment in chronic myelogenous leukemia. Proc Natl Acad Sci U S A 108(42):17468–17473. https://doi.org/10.1073/pnas.1111138108. 1111138108 [pii]

Yamazaki S, Ema H, Karlsson G, Yamaguchi T, Miyoshi H, Shioda S, Taketo MM, Karlsson S, Iwama A, Nakauchi H (2011) Nonmyelinating Schwann cells maintain hematopoietic stem cell hibernation in the bone marrow niche. Cell 147(5):1146–1158. https://doi.org/10.1016/j.cell.2011.09.053. S0092-8674(11)01269-4 [pii]

Yang L, Bryder D, Adolfsson J, Nygren J, Mansson R, Sigvardsson M, Jacobsen SE (2005) Identification of Lin(−)Sca1(+)kit(+)CD34(+)Flt3- short-term hematopoietic stem cells capable of rapidly reconstituting and rescuing myeloablated transplant recipients. Blood 105(7):2717–2723. doi: 2004-06-2159 [pii]. https://doi.org/10.1182/blood-2004-06-2159

Yang Y, Mallampati S, Sun B, Zhang J, Kim SB, Lee JS, Gong Y, Cai Z, Sun X (2013) Wnt pathway contributes to the protection by bone marrow stromal cells of acute lymphoblastic leukemia cells and is a potential therapeutic target. Cancer Lett 333(1):9–17. https://doi.org/10.1016/j.canlet.2012.11.056. S0304-3835(13)00037-2 [pii]

Ye H, Adane B, Khan N, Sullivan T, Minhajuddin M, Gasparetto M, Stevens B, Pei S, Balys M, Ashton JM, Klemm DJ, Woolthuis CM, Stranahan AW, Park CY, Jordan CT (2016) Leukemic stem cells evade chemotherapy by metabolic adaptation to an adipose tissue niche. Cell Stem Cell 19(1):23–37. https://doi.org/10.1016/j.stem.2016.06.001. S1934-5909(16)30151-5 [pii]

Yin T, Li L (2006) The stem cell niches in bone. J Clin Invest 116(5):1195–1201. https://doi.org/10.1172/JCI28568

Yoshihara H, Arai F, Hosokawa K, Hagiwara T, Takubo K, Nakamura Y, Gomei Y, Iwasaki H, Matsuoka S, Miyamoto K, Miyazaki H, Takahashi T, Suda T (2007) Thrombopoietin/MPL signaling regulates hematopoietic stem cell quiescence and interaction with the osteoblastic niche. Cell Stem Cell 1(6):685–697. https://doi.org/10.1016/j.stem.2007.10.020. S1934-5909(07)00237-8 [pii]

Yuan Y, Zhou L, Miyamoto T, Iwasaki H, Harakawa N, Hetherington CJ, Burel SA, Lagasse E, Weissman IL, Akashi K, Zhang DE (2001) AML1-ETO expression is directly involved in the development of acute myeloid leukemia in the presence of additional mutations. Proc Natl Acad Sci U S A 98(18):10398–10403. https://doi.org/10.1073/pnas.171321298. 98/18/10398 [pii]

Zambetti NA, Ping Z, Chen S, Kenswil KJG, Mylona MA, Sanders MA, Hoogenboezem RM, Bindels EMJ, Adisty MN, Van Strien PMH, van der Leije CS, Westers TM, Cremers EMP, Milanese C, Mastroberardino PG, van Leeuwen J, van der Eerden BCJ, Touw IP, Kuijpers TW, Kanaar R, van de Loosdrecht AA, Vogl T, Raaijmakers M (2016) Mesenchymal inflammation drives genotoxic stress in hematopoietic stem cells and predicts disease evolution in human preleukemia. Cell Stem Cell 19(5):613–627. doi: S1934-5909(16)30268-5 [pii]. https://doi.org/10.1016/j.stem.2016.08.021

Zenatti PP, Ribeiro D, Li W, Zuurbier L, Silva MC, Paganin M, Tritapoe J, Hixon JA, Silveira AB, Cardoso BA, Sarmento LM, Correia N, Toribio ML, Kobarg J, Horstmann M, Pieters R, Brandalise SR, Ferrando AA, Meijerink JP, Durum SK, Yunes JA, Barata JT (2011) Oncogenic IL7R gain-of-function mutations in childhood T-cell acute lymphoblastic leukemia. Nat Genet 43(10):932–939. https://doi.org/10.1038/ng.924. ng.924 [pii]

Zhang J, Niu C, Ye L, Huang H, He X, Tong WG, Ross J, Haug J, Johnson T, Feng JQ, Harris S, Wiedemann LM, Mishina Y, Li L (2003) Identification of the haematopoietic stem cell niche and control of the niche size. Nature 425(6960):836–841. https://doi.org/10.1038/nature02041. nature02041 [pii]

Zhang B, Ho YW, Huang Q, Maeda T, Lin A, Lee SU, Hair A, Holyoake TL, Huettner C, Bhatia R (2012a) Altered microenvironmental regulation of leukemic and normal stem cells in chronic myelogenous leukemia. Cancer Cell 21(4):577–592. https://doi.org/10.1016/j.ccr.2012.02.018. S1535-6108(12)00080-3 [pii]

Zhang J, Ding L, Holmfeldt L, Wu G, Heatley SL, Payne-Turner D, Easton J, Chen X, Wang J, Rusch M, Lu C, Chen SC, Wei L, Collins-Underwood JR, Ma J, Roberts KG, Pounds SB, Ulyanov A, Becksfort J, Gupta P, Huether R, Kriwacki RW, Parker M, McGoldrick DJ, Zhao D, Alford D, Espy S, Bobba KC, Song G, Pei D, Cheng C, Roberts S, Barbato MI, Campana D, Coustan-Smith E, Shurtleff SA, Raimondi SC, Kleppe M, Cools J, Shimano KA, Hermiston ML, Doulatov S, Eppert K, Laurenti E, Notta F, Dick JE, Basso G, Hunger SP, Loh ML, Devidas M, Wood B, Winter S, Dunsmore KP, Fulton RS, Fulton LL, Hong X, Harris CC, Dooling DJ, Ochoa K, Johnson KJ, Obenauer JC, Evans WE, Pui CH, Naeve CW, Ley TJ, Mardis ER, Wilson RK, Downing JR, Mullighan CG (2012b) The genetic basis of early T-cell precursor acute lymphoblastic leukaemia. Nature 481(7380):157–163. https://doi.org/10.1038/nature10725. nature10725 [pii]

Zhang J, Ye J, Ma D, Liu N, Wu H, Yu S, Sun X, Tse W, Ji C (2013) Cross-talk between leukemic and endothelial cells promotes angiogenesis by VEGF activation of the Notch/Dll4 pathway. Carcinogenesis 34(3):667–677. https://doi.org/10.1093/carcin/bgs386. bgs386 [pii]

Zhang B, Chu S, Agarwal P, Campbell VL, Hopcroft L, Jorgensen HG, Lin A, Gaal K, Holyoake TL, Bhatia R (2016) Inhibition of interleukin-1 signaling enhances elimination of tyrosine kinase inhibitor-treated CML stem cells. Blood 128(23):2671–2682. doi: blood-2015-11-679928 [pii]. https://doi.org/10.1182/blood-2015-11-679928

Zhao C, Blum J, Chen A, Kwon HY, Jung SH, Cook JM, Lagoo A, Reya T (2007) Loss of beta-catenin impairs the renewal of normal and CML stem cells in vivo. Cancer Cell 12(6):528–541. doi: S1535-6108(07)00334-0 [pii]. https://doi.org/10.1016/j.ccr.2007.11.003

Zhao C, Chen A, Jamieson CH, Fereshteh M, Abrahamsson A, Blum J, Kwon HY, Kim J, Chute JP, Rizzieri D, Munchhof M, VanArsdale T, Beachy PA, Reya T (2009) Hedgehog signalling is essential for maintenance of cancer stem cells in myeloid leukaemia. Nature 458(7239):776–779. https://doi.org/10.1038/nature07737. nature07737 [pii]

Part III

Metabolic Fitness and Therapy Response in Cancer

15 Exploiting Cancer Cells Metabolic Adaptability to Enhance Therapy Response in Cancer

Sofia C. Nunes

Abstract

Despite all the progresses developed in prevention and new treatment approaches, cancer is the second leading cause of death worldwide, being chemoresistance a pivotal barrier in cancer management. Cancer cells present several mechanisms of drug resistance/tolerance and recently, growing evidence have been supporting a role of metabolism reprograming per se as a driver of chemoresistance. In fact, cancer cells display several adaptive mechanisms that allow the emergency of chemoresistance, revealing cancer as a disease that adapts and evolve along with the treatment. Therefore, clinical protocols that take into account the adaptive potential of cancer cells should be more effective than the current traditional standard protocols on the fighting against cancer.

In here, some of the recent findings on the role of metabolism reprograming in cancer chemoresistance emergence will be discussed, as the potential evolutionary strategies that could unable these adaptations, hence allowing to prevent the emergency of treatment resistance, changing cancer outcome.

S. C. Nunes (✉)
CEDOC, Chronic Diseases Research Centre, NOVA Medical School | Faculdade de Ciências Médicas, Universidade NOVA de Lisboa, Lisbon, Portugal

Instituto Português de Oncologia de Lisboa Francisco Gentil (IPOLFG), Lisbon, Portugal

Keywords

Adaptation · Cancer · Chemoresistance · Evolution · Metabolism

15.1 Cancer: From Hanahan and Weinberg to Darwin

Despite all the progresses developed in prevention and new treatment approaches, cancer is the second leading cause of death worldwide (Fitzmaurice et al. 2015). In accordance with the International Agency for Research on Cancer, 14.1 million cancer cases (Ferlay et al. 2013a) and 8.2 million cancer deaths (Ferlay et al. 2013b) were estimated worldwide in 2012. For 2020, 17.1 million incidences and 10.05 million cancer deaths (Ferlay et al. 2013a) are estimated. Metastatic disease accounts for over 90% of all cancer-related deaths, where the treatment with surgery, conventional chemotherapy and radiation is ineffective (Rankin and Giaccia 2016). The late diagnosis combined with resistance to the conventional anti-cancer drugs used, are the major causes of cancer poor prognosis.

More than 200 different types of cancer exist (cancerresearchuk.org 2018), however, the physiological alterations that entail the malignant transformation were proposed to be common to the majority or even to all types of human tumours (Hanahan and Weinberg 2000). Therefore, in

J. Serpa (ed.), *Tumor Microenvironment*, Advances in Experimental Medicine and Biology 1219,
https://doi.org/10.1007/978-3-030-34025-4_15

2000, Hanahan and Weinberg proposed the existence of six core hallmarks of cancer cells: self-sufficiency in growth signals, insensitivity to growth-inhibitory signals, evasion of programmed cell death, limitless replicative potential, sustained angiogenesis, and tissue invasion and metastasis (Hanahan and Weinberg 2000). Eleven years later, the authors revisited those original hallmarks, and included energy metabolism reprogramming and evading immune destruction, as emerging hallmarks of cancer (Hanahan and Weinberg 2011). Underlying these hallmarks, the authors suggested two consequential characteristics of neoplastic cells that facilitate the acquisition of both core and emerging hallmarks: genome instability, and inflammation (Hanahan and Weinberg 2011). The acquisition of these hallmarks is an evolutionary process, involving natural selection among the neoplastic cells, allowing cancer initiation, progression and chemoresistance (Crespi and Summers 2005). In fact, cancer cells evolve under the same rules as Darwin's finches on the Galapagos, in which several genetically heterogeneous individual cells share the tumour microenvironment, competing for growth and survival in continuously changing environments (Polyak 2007).

Cairns and Nowell firstly introduced the evolutionary perspective to cancer. In 1975, Cairns had argued cancer as an evolutionary process, driven by mutation and natural selection (Cairns 1975). In 1976, Nowell's proposed that the majority of neoplasms present a unicellular origin, and that the tumour progression results from acquired genetic variability within the original clone, allowing the sequential selection of more aggressive subclones (Nowell 1976). Nowell have then established the clonal evolution theory of tumour progression (Nowell 1976).

Besides being an evolutionary process, cancer is also an ecological process, being cancer cells subject to competition for space and resources, predation by the immune system and cooperation to disperse and colonise new organs (Axelrod et al. 2006; Merlo et al. 2006). Strengthening the relevance of evolution and ecology on cancer, recently, Maley and colleagues have developed an evolutionary and ecological classification system for neoplasms in order to improve the clinical management of cancer. Hence, the authors proposed the classification of neoplasms based on the Evo-index, including the intratumoural heterogeneity and its changes over time, and the Eco-index, including the hazards to neoplastic cell survival and the resources available to these cells (Maley et al. 2017).

Hypoxia and acidosis are common features of the tumour microenvironment, being highly selective and inducing genetic instability, hence promoting somatic evolution (Gillies et al. 2012). Cytotoxic anti-cancer drugs also drive evolution of cancer cells, by imposing strong evolutionary selection pressures on the surviving cells (Gillies et al. 2012).

We have to highlight that besides genetic variation, other non-genetic features as epigenetic mechanisms may also be pivotal for the adaptation of cancer cells to new environments. In fact, Salgia and Kulkarni have recently published a reflexion on this duality of genetic/non-genetic features of chemoresistance (Salgia and Kulkarni 2018) that merits further attention.

In the next section, I will focus on some metabolic adaptive strategies that cancer cells undergo in order to cope with anti-cancer drugs, allowing disease progression and resistance to treatment.

15.2 Metabolism Reprograming in Cancer: A Driving Force of Adaptation to Challenging Environments

The metabolism reprograming is well known to be a key feature of tumorigenesis and recently, evidence have been supporting also a role of altered metabolism in anti-cancer drugs response and adaptation (Morandi and Indraccolo 2017).

The best characterised metabolic phenotype observed in tumour cells is the Warburg effect, proposing that cancer cells present increased rate of glycolysis even under normal oxygen concentrations due to defective mitochondrial oxidative phosphorylation (OXPHOS) (Warburg 1956). However, evidence accumulate showing that mitochondrial OXPHOS function is intact in most tumours (Alam et al. 2016; Guppy et al. 2002; Rodríguez-Enríquez et al. 2000, 2006; Viale et al. 2015). Moreover, evidence also support that the bioenergetics of tumour cells is highly complex, where cancer cells have the ability to use several substrates in order to support energy production, including glucose, glutamine, fatty acids, and acetate (Alam et al. 2016). Also, within a tumour, subpopulations of cells with glycolytic and oxidative metabolisms coexist, enhancing metabolic plasticity and improving tumorigenesis and metastasis (Viale et al. 2015; Yu et al. 2017), hence highlighting the metabolic complexity of cancer cells that allows coping with changing environments. Recent studies have disclosed the Warburg effect as a way of cancer cells to sustain cell proliferation rather than producing energy (Liang et al. 2017; Liu and Yin 2017; Lopes-Coelho et al. 2017), once the glycolytic intermediates are deviated to serve as building blocks needed for replicating DNA and cellular machinery prior to mitosis (Lopes-Coelho et al. 2017). Other hypothesis that explain the advantage of the Warburg effect on cancer cells is that it supports an ideal tumour microenvironment, sustaining cancer cells proliferation (e.g. acid-mediated invasion hypothesis) and that altered glucose metabolism alters cancer cell signalling, promoting tumorigenesis via reactive oxygen species (ROS) and the modulation of chromatin state (reviewed in (Liberti and Locasale 2016)).

In the next section, the mechanisms of drug resistance will be briefly addressed and the role of metabolic reprograming *per se* as a driver of cancer cells adaptation and resistance to anti-cancer drugs will be discussed.

15.3 Metabolism Reprograming as a Driver of Cancer Cells Adaptation and Resistance to Anti-cancer Drugs

Drug resistance can be intrinsic (exists prior to treatment) or acquired during treatment (Holohan et al. 2013) and two general causes of drug resistance/tolerance exist: host factors and specific genetic or epigenetic alterations in the cancer cells (Gottesman 2002). Importantly, tumours present a high molecular heterogeneity (Swanton 2013), allowing therapy-induced selection of a resistant subpopulation of cells, thus leading to drug resistance emergence (Holohan et al. 2013).

As Salgia and Kulkarni emphasized, drug resistance, tolerance and persistence terms have been ambiguously and inadvertently used (Salgia and Kulkarni 2018). Whereas genetics strongly underlies drug resistance, tolerance may be inherited or not and is commonly used to describe the survival capacity upon the transient exposure to high drug concentrations. Persistence refers to the survival capacity of a subpopulation of a clonal population upon the exposure to high drug concentrations (Salgia and Kulkarni 2018).

Several mechanisms were already associated with drug resistance/tolerance, including the increased drug efflux and decreased drug influx, drug inactivation, alterations in drug target, increased DNA damage repair, deregulation of apoptosis, autophagy, activation of prosurvival signalling, oncogenic bypass and pathway redundancy and epithelial–mesenchymal transition (Holohan et al. 2013). The tumour microenvironment has been implicated not only in tumour growth, invasion, and metastasis but also in acquired drug resistance, mediated by myeloid cells, cancer-associated fibroblasts, mesenchymal stem cells and the interaction with the extracellular matrix (Son et al. 2017). Moreover, hypoxia is a common tumour microenvironmental condition that is intimately related to chemoresistance (Semenza 2012; Vaupel and Mayer 2007).

In here, I will focus on some adaptive strategies (inherited or not) that favour drug resistance/tolerance, focusing on metabolic adaptations that allow cancer cells survival upon cytotoxic drugs exposure. It is not my goal to focus on oncogenes, tumour suppressor genes or signalling cascades known to play important roles in the metabolic shifting of cancer cells and on chemoresistance. Instead, it is my goal to explore the role of metabolic reprograming *per se* as a driver of cancer cells adaptation and resistance to anti-cancer drugs.

Albeit the well known role of the Warburg effect on tumorigenesis, its causative effect in chemoresistance is still unclear (Morandi and Indraccolo 2017). Some studies already proposed that targeting glycolysis could be an efficient way to revert both 5-fluorouracil (5-Fu) (Zhao et al. 2014) and doxorubicin (Ma et al. 2015) resistance. Interestingly, Zhao and co-workers have reported that 5-Fu-resistant A549 cells presented an increased glucose metabolism, whereas cisplatin-resistant cells presented a decreased glucose metabolism. In addition, 5-Fu combined with cisplatin contributed to the synergistic anti-cancer effect through the inhibition of glucose metabolism, suggesting that targeting this metabolic pathway should be effective for overcoming 5-Fu resistance (Zhao et al. 2014). Ma et al. reported an enhanced doxorubicin activity in MCF-7 resistant cells treated with a glucose analogous, 2-deoxy-D-glucose, that inhibits glucose metabolism by competitively inhibiting its uptake and utilization (Ma et al. 2015). This effect on doxorubin reversion of resistance by 2-deoxy-D-glucose was reported to be via intracellular ATP depletion, via the inactivation of drug-efflux pump, and by downregulation of transmembrane transporters (Ma et al. 2015). Zhou et al. have reported that intracellular ATP levels are pivotal in the development of oxaliplatin resistance in human colon cancer cells that present distinct genetic backgrounds (Zhou et al. 2012). The increased ATP levels were shown to be driven by an enhanced aerobic glycolysis in the chemoresistant cells albeit these cells consumed more oxygen without increased mitochondrial ATP production (Zhou et al. 2012). Zhang and colleagues reported that aerobic glycolysis mediated by AMPK/mTOR/HIF1α pathways probably plays a role in resistance to carmustine of mitochondrial hydroxylase Clk1 deficient glioma cells (Zhang et al. 2017a). Moreover, an acidic extracellular environment due to lactate accumulation was also reported to have a role in drug resistance both in vivo and in vitro (reviewed in (Morandi and Indraccolo 2017)). Contrarily to these observations, Pastò and colleagues data suggested that ovarian cancer platinum-sensitive cells (both epithelial ovarian cancer cells from patients and in a xenograft model) rely more on glucose metabolism than their resistant counterparts (Pastò et al. 2017). However, it is unclear if platinum modulates the metabolic shift of cancer cells or if it selects a population of cells that rely less on glucose metabolism (Pastò et al. 2017).

Komurov and colleagues have reported that lapatinib resistance (an epidermal growth factor receptor – EGFR/erb-b2 receptor tyrosine kinase 2 – ErbB2 inhibitor), induced the expression of the glucose deprivation response pathway, including glucagon signalling, glucose uptake and gluconeogenesis (Komurov et al. 2012). They also found that the glucose deprivation pathway was significantly correlated with higher rates of clinical relapse in ErbB2-positive breast cancer patients and that glucose deprivation was able to increase lapatinib-sensitive cells resistance (Komurov et al. 2012). Moreover, they also observed higher glycolysis rates in resistant cells and, since the lactate/glucose ratio was significantly decreased in these cells, they have suggested a switch from glycolysis to the pentose phosphate pathway, leading to increased NADPH and, consequently, to an increased capacity of the resistant cells to overcome oxidative stress (Komurov et al. 2012).

Recently, the hexosamine biosynthetic pathway, which is also involved in glucose metabolism, was reported to play an important role in chemoresistance through the regulation of O-GlcNAcylation in the presence of doxorubicin or camptothecin in several cancer cell lines (Liu et al. 2018). Importantly, the suppression of this pathway or O-GlcNAcylation decreased cancer cells chemoresistance (Liu et al. 2018).

Collectively, data supports an active role of glucose metabolism in the ability of cancer cells to survive upon cytotoxic drugs exposure, weather by favouring it or, on the contrary, by avoiding it, hence favouring other metabolic pathways.

Regarding OXPHOS role on the ability of cancer cells to adapt to anti-cancer drugs, interestingly, Qian and co-workers have shown a positive correlation between cellular density of mitochondria and cisplatin sensitivity both in vivo and in vitro (Qian et al. 2005). Contrarily, Denise and colleagues have found a mesenchymal stem-like phenotype and an addicted-OXPHOS phenotype in colon cancer cells treated with 5-Fu (Denise et al. 2015). In ovarian cancer, it was shown that chemotherapy treatment induces metabolic plasticity in ovarian cancer stem cells-like recurrent cells, favouring pathways that rely on OXPHOS-mediated lipid metabolism (Ahmed et al. 2018). Ippolito and co-workers have shown that docetaxel treatment induces a glycolytic phenotype shift to an OXPHOS phenotype in resistant prostate cancer cells (Ippolito et al. 2016). Importantly, reverting the OXPHOS phenotype via miR-205 re-sensitized the resistant cells to docetaxel (Ippolito et al. 2016). These opposite observations strongly supports that the metabolic reprograming causative of drug resistance/tolerance of cancer cells is dependent on the type of chemotherapy agents used (Morandi and Indraccolo 2017). Interestingly, in ovarian cancer context, Dar and colleagues have reported that chemosensitive cancer cell lines presented a glycolytic phenotype whereas the chemoresistant cells exhibited a high metabolically active phenotype, with metabolic switching between OXPHOS and glycolysis (Dar et al. 2017). Importantly, while the chemosensitive cells were glucose-dependent, the chemoresistant ones presented metabolic adaptability (Dar et al. 2017). Moreover, patient derived ovarian cancer cells also presented a similar pattern of chemoresistance, where cells presented a high metabolically active phenotype (Dar et al. 2017). However, the authors could not state if the metabolic adaptation of chemoresistant cells was a driver or an outcome event of chemoresistance (Dar et al. 2017).

It is important to highlight that in cancer, subpopulations of cells with both glycolytic and oxidative metabolisms coexist, providing metabolic plasticity, thus allowing tumour cells survival under different microenvironments, hence possibly supporting tumour metastasis and chemoresistance (Jia et al. 2018). Corroborating this hypothesis, Sancho and co-workers have reported that during metformin exposure, an anti-diabetic drug, the resistant pancreatic cancer stem cells arise with an intermediate glycolytic/respiratory phenotype (Sancho et al. 2015). Moreover, in a very interesting publication, in the context of pancreatic neuroendocrine tumors, Allen and colleagues found that metabolic symbiosis can function as a mechanism of adaptive resistance (Allen et al. 2016). They described this adaptive mechanism in response to anti-angiogenic therapies that lead to hypoxia (Allen et al. 2016). Thus, they have found that hypoxic cancer cells metabolise glucose and secrete lactate, whereas the normoxic cells, which are proximal to the vessels, import and use lactate for energy metabolism, by favouring glutamine metabolism (Allen et al. 2016). Though NMR spectroscopy and using $^{3\text{-}13C}$ lactate in glucose-free media, the authors reported that the normoxic cells catabolised $^{3\text{-}13C}$ lactate to C4-glutamate, C2- and C3-aspartate, and C3-alanine (Allen et al. 2016). Glutamate can be then converted into α-ketoglutarate, replenishing intermediates for the mitochondrial Tricarboxylic acid cycle (TCA) cycle, crucial for energy production and biosynthesis of cellular building blocks (Allen et al. 2016). This publication deeply reflects the enormous complexity involved in the adaptive mechanisms of cancer cells to anti-cancer drugs.

Recently, a role of energy metabolism mediated by miRNAs regulation in chemoresistance was also suggested (reviewed in (Ye et al. 2018)).

Glutamine metabolism was also reported to drive chemoresistance. For instance, Gastel and colleagues have reported the activation of glutamine metabolism as a driver of chemoresistance in in vivo models of acute myeloid leukemia (Gastel et al. 2017). Gallipoli and colleagues have confirmed a role of glutamine metabolism in this disease (Gallipoli et al. 2018). In acute myeloid leukemia, mutations that activate tyrosine kinases (TK) are common and are associated with poor prognosis, including mutations in the type-III receptor TK fms related tyrosine kinase 3 (FLT3), that frequently result from an internal tandem duplication ($FLT3^{ITD}$) (Gallipoli et al. 2018). Importantly, the authors have reported that following FLT3 inhibition in $FLT3^{ITD}$ cells, glutamine metabolism is protective, allowing an adaptive response to FLT3-TK inhibitors (Gallipoli et al. 2018).

Glutamine is pivotal for several functions in cancer cells, including cellular bioenergetics, nucleotide biosynthesis, and redox homeostasis, as a precursor of glutamate that is used in the synthesis of glutathione (GSH) (reviewed in (Nguyen and Durán 2018)). In fact, another important metabolic adaptation of cancer cells that allows resistance to cytotoxic drugs is the increased cellular antioxidant capacity (Ju et al. 2015; Landriscina et al. 2009). The transcription factor nuclear factor-erythroid 2 p45-related factor 2 (Nrf2) is a pivotal player in cellular redox homeostasis regulation, strongly influencing intrinsic resistance to oxidative stress and controlling adaptive responses to several stressful environmental conditions (Hayes and Dinkova-Kostova 2014). Nrf2 is not only involved in the regulation of the GSH-based antioxidant system, but also regulates the expression of cytosolic thioredoxin (TRX1), TrxR1 and sulphiredoxin1 (Hayes and Dinkova-Kostova 2014). Recently, Khamari and colleagues have shown that the acquisition of B-Raf proto-oncogene, serine/threonine kinase (BRAF) inhibitors resistance was linked with both an increased mitochondrial OXPHOS and with glutamine metabolism (Khamari et al. 2018). They also reported a role of the Nrf2 pathway on melanoma with acquired resistance to BRAF inhibitors, where its strong activation was found to be responsible for an increased pentose phosphate pathway, that is involved in the regeneration of reduced GSH (Khamari et al. 2018). The authors also observed an increased expression of the xCT transporter (Khamari et al. 2018). Thus, they have linked chemoresistance with mitochondrial metabolism adaptations that favour glucose-derived glutamate synthesis, cysteine uptake and GSH synthesis (Khamari et al. 2018), hence strengthening the complex adaptive responses of cancer cells to anti-cancer drugs. Kerr and colleagues found similar metabolic reprogramming features during lung cancer malignant progression in vivo (Kerr et al. 2016). They found that in spontaneous advanced murine lung tumours that present a high frequency of $KRAS^{G12D}$ copy gain, the cells presented a glycolytic switch combined with increased glucose-derived metabolites canalized into the TCA cycle and GSH biosynthesis, leading to an enhanced GSH-mediated detoxification (Kerr et al. 2016). However, this metabolic shifting was not present in the corresponding early tumours ($Kras^{G12D}$ heterozygous). Importantly, the authors also found a plausible role of Nrf2-mediated detoxification in this metabolic switch (Kerr et al. 2016).

An increased antioxidant capacity was also found to contribute to paclitaxel resistance. Hence, Datta and colleagues have shown a gradual increase in GSH content and in the activities of catalase and glutathione peroxidase (GPX) along with paclitaxel resistance development in A549 human lung adenocarcinoma cells (Datta et al. 2017). The authors reported that increased rates of extracellular acidification and oxygen consumption were directly correlated with the acquisition of resistance (Datta et al. 2017).

Strikingly, Roh et al. reported that the inhibition of both GSH and Thioredoxin (Trx) systems presented a synergistic effect on head and neck cancer cells death, but the effect was suboptimal due to the activation of Nrf2-antioxidant response element pathway in resistant cells (Roh et al. 2017). However, with the simultaneously blocking of GSH, Trx and the Nrf2-ARE pathways, the authors were able to eliminate the resistant head and neck cancers (Roh et al. 2017).

Collectively, these results strongly support a key role of both cellular bioenergetics pathways and antioxidant defence systems in cancer biology, thus suggesting that their targeting from an evolutionary perspective could be a successful strategy to fight several types of cancer.

Deblois and co-workers have recently reported that taxane-resistant triple-negative breast cancer cells endure metabolic adaptations by impairing methionine metabolism and S-adenosylmethionine availability, leading to a global decrease in DNA methylation that H3K27me3 forming large organized chromatin domains of lysine modification compensate (Deblois et al. 2018). Moreover, this epigenetic reprogramming induced by metabolic adaptations, lead to an epigenetic-targeted opportunity to re-sensitize the taxane-resistant cells with chemical inhibitors of EZH2, the H3K27me3 methyltransferase (Deblois et al. 2018). Hence, this work has shown the vast possible complex consequences of metabolism alterations in epigenetics reprograming and drug resistance.

The goal of this section was to illustrate the complexity involved in the metabolic adaptive strategies that cancer cells undergo allowing their survival upon exposure to anti-cancer drugs. In the next section, the relevance of evolutionary principles in preventing the spread of chemoresistant phenotypes will be explored. These strategies could, therefore, counteract the emergency of these metabolic adaptive strategies in cancer cells, culminating possibly in the overcome of drug-resistance/tolerance.

15.4 Turning Cancer Cells Adaptability Against Themselves: The Power of Evolutionary Strategies in Overcoming Chemoresistance

In the previous section, several examples of active metabolism reprogramming as a causative effect of cancer cells adaptation to anti-cancer drugs were presented. The link with the development of metabolic pathways-targeting drugs is then obvious, but I do not intend to explore the drugs that were already developed following this rational. Instead, given the role of adaptive evolution in cancer cells resistance/tolerance to treatment, it is my objective to address the treatment strategies that exploit the dynamics of cancer cells adaptation to anti-cancer drugs. The ultimate goal of these strategies is, therefore, to prevent the possibility of cancer cells to adapt to anti-cancer drugs, regardless the adaptive mechanism. In the next sections, some of the different evolutionary perspectives that were already explored in cancer research will be addressed, namely the adaptive therapy and the fitness threshold model. Other perspectives will be also discussed.

15.4.1 Exploiting the Cost of Resistance: Playing with the Ecology of Cancer Cells

It is important to highlight that the conventional cancer therapies, which administer cytotoxic drugs at maximum tolerated doses until progression, strongly select for resistant phenotypes and, by eliminating the sensitive cells, eliminate competition, allowing a rapid proliferation of the resistant populations even in the absence of drugs – an evolutionary phenomenon designated "competitive release" (Enriquez-navas et al. 2015; Enriquez-Navas et al. 2016; Zhang et al. 2017a, b). However, as more and more evidence accumulates highlighting cancer as an evolutionary disease, in 2011, Atkipis et al. analysed 6228 publications concerning therapeutic resistance and/or cancer relapse and reported that in abstracts, evolution terms were present in only about 1% since the 1980s (Aktipis et al. 2011). Moreover, Darwinian dynamics are still rarely integrated into anti-cancer protocols in clinical contexts (Zhang et al. 2017b).

In 2009, Gatenby and colleagues have explored the conceptual model of adaptive therapy that defends that, since the tumour populations that are exposed to treatment are dynamic, the treatment should be also dynamic with continuous adjustment of drugs, dose, and timing (Gatenby et al. 2009), thus evolving along with cancer cells. The authors have developed mathematical models that predicted that an optimal treatment strategy adjust therapy in order to maintain a stable population of chemosensitive cells that are more fitted in the absence of therapy, being able to compete and inhibit the growth of resistant populations due to fitness costs of resistance (Fig. 15.1) (Gatenby et al. 2009). The same authors confirmed the benefits of the adaptive therapy in in vivo experiments with OVCAR3

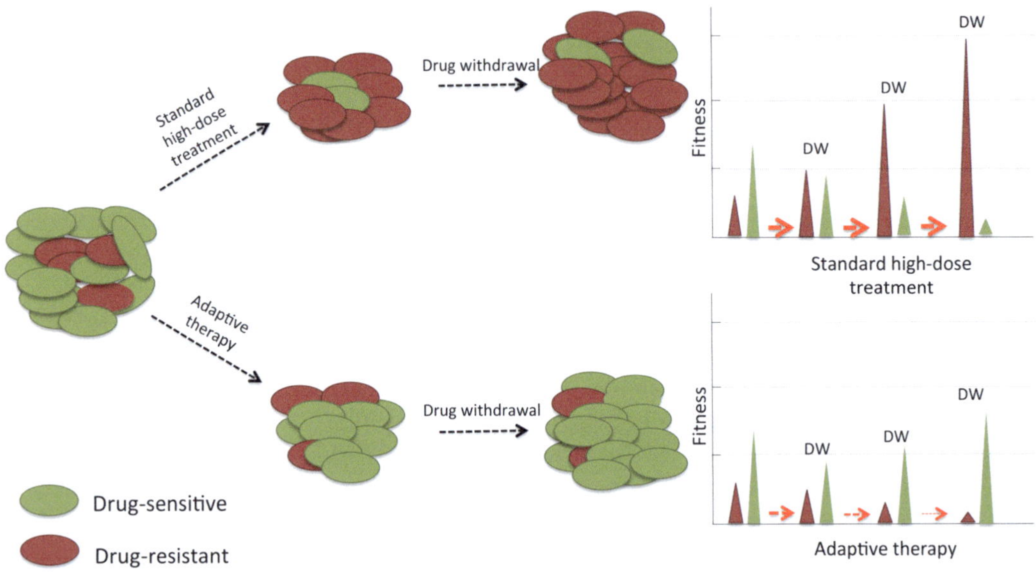

Fig. 15.1 The power of adaptive therapy in maintaining a stable tumour population, by playing with competition among resistant and sensitive cells
The standard high-dose treatment strongly selects resistant phenotypes by eliminating the sensitive cells that competes with the resistant cells, allowing the rapid spread of resistant cells even in the absence of drugs. Hence, albeit an initial tumour shrinkage can be observed, this tumour is mainly composed by resistant cells that gain fitness during treatment, even in the absence of the drug (upper panel). On the contrary, by administering lower doses with continuous adjustments, the resistant cells undergo competition with sensitive cells, allowing the maintenance of a stable tumour population for a prolonged period of time, hence maintaining the resistant cells with a constant lower fitness. The red arrows correspond to the treatment administration, where the width reflects the dose used during treatment and DW to drug withdrawal. (Adapted from Enriquez-Navas and Gatenby 2017 and Salgia and Kulkarni 2018)

xenografts treated with carboplatin, showing that this sstrategy was able to maintain a stable tumour population for a prolonged period of time, allowing a long-term survival (Gatenby et al. 2009). Enriquez-Navas and colleagues reported similar findings in different preclinical models of breast cancer using paclitaxel (Enriquez-Navas et al. 2016). Gallaher et al. went further and identified two different adaptive strategies that are effective in heterogeneous tumours, a dose modulation strategy that is efficient in the majority of tumours with fewer drug, and a more vacation-oriented strategy that is able to control more invasive tumours (Gallaher et al. 2017). Importantly, Silva and colleagues have reported that low doses of verapamil and 2-deoxyglucose, were able to increase the cost of resistance and to decrease energy production, abolishing drug-resistant cells proliferation in vivo (Silva et al. 2012). In breast cancer tumour models, this strategy allowed to increase the time to progression by 2- to ten-fold compared to standard high dose treatments (Silva et al. 2012). Hence, these authors have shown that these evolutionary strategies are also effective when targeting metabolic pathways of cancer cells.

Recently, Zhang and colleagues have integrated evolutionary dynamics into a pilot clinical trial of patients with metastatic castrate-resistant prostate cancer in order to avoid the evolution of resistance to abiraterone (that inhibits CYP17A, an enzyme responsible for testosterone autoproduction). Outstandingly, the authors have reported that the adaptive therapy treatment was able to increase the time to progression and to reduce the cumulative drug dose to less than a half compared to the standard strategy (Zhang et al. 2017b).

The cost of resistance in the absence of drugs was also explored in a different perspective, as

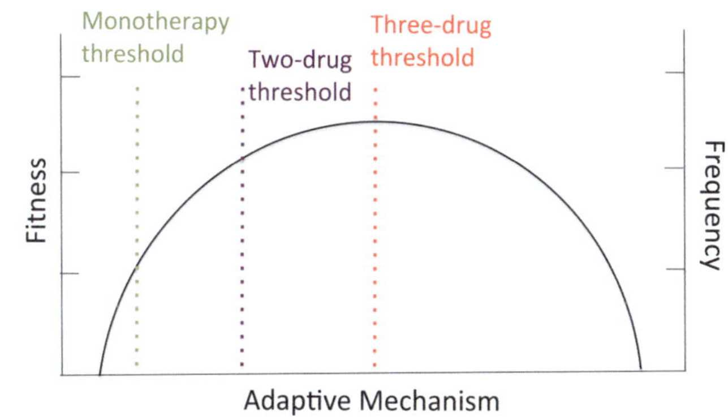

Fig. 15.2 The fitness threshold model as a tool to prevent the emergency of resistance
In accordance with this model, the fitness threshold corresponds to the barrier that subclonal populations need to overcome in order to recover fitness during drug treatment (Xue et al. 2017). The model predicted that, whereas the sequential treatment with RAF inhibitor followed by an ERK inhibitor was not effective, an intermittent three-drug treatment combination was, allowing the increase of the fitness threshold and counteracting the adaptive mechanisms of cancer cells (Xue et al. 2017). This approach could be possibly used to counteract other types of adaptive mechanisms beyond BRAF copy number gain with different anti-cancer drugs, when both the monotherapy and a two-drug combination are not effective. (Adapted from Xue et al. 2017)

chemoresistance may induce drug addiction due to the high fitness costs upon drug withdrawal. Therefore, drug addiction is the dependency of tumour cells on the anti-cancer drugs to which they have developed resistance (Kong et al. 2017), that may allow clinical benefits. In the context of melanoma, Kong and colleagues observed that even after an extended drug withdrawal, resistant clones could arise (Kong et al. 2017), thus, surpassing the drug addicted phenotype. Therefore, in a patient setting, they combined the drug withdrawal of BRAF inhibition with the introduction of dacarbazine, an alkylating agent generally used as a monotherapy in metastatic melanoma, even with poor response rates (Kong et al. 2017). Whereas dacarbazine showed low cytotoxic effects in the presence of BRAF inhibitor on melanoma cell lines, the administration of dacarbazine upon BRAF inhibitor withdrawal presented a strong synergetic effect (Kong et al. 2017). The authors argued that gaining insights into the molecular mechanisms of drug addiction may open the opportunity to develop alternating more efficient treatment strategies in order to fight chemoresistance (Kong et al. 2017).

Together, growing evidence had strengthened the use of evolutionary principles in clinical settings as an efficient and powerful way to prevent the spread of chemoresistant phenotypes. These studies have also shown that from identical evolutionary points of view (e.g. the cost of resistance in the absence of the drug), different evolutionary strategies may be developed. Moreover, evidence also support that these principles are effective for several types of anti-cancer drugs and in several cancer contexts, hence supporting its general use in cancer management.

15.4.2 The Fitness Threshold Model and Beyond

In a different evolutionary perspective, by using single-cell DNA sequencing, Xue and colleagues have found that parallel evolution lead to the selection and spread of different *BRAF*-amplified subclones, allowing the tumours to adapt to ERK inhibitor treatment while maintaining intratumoral heterogeneity (Xue et al. 2017). They proposed the fitness threshold model (Fig. 15.2) to explain their findings, being the fitness threshold the barrier that subclonal populations have to overcome in order to recover fitness during drug treatment. The model predicted that sequential treatment was not effective, prediction that was supported by their results showing that treatment with a RAF inhibitor followed by an ERK inhibitor induced a gradual increase in *BRAF* copy number, allowing a fitness advantage in the presence of the drugs (Xue et al. 2017). Moreover, the same authors reported that an intermittent three-drug treatment combination was able to inhibit tumour growth in $BRAF^{V600E}$ patient-derived tumour xenografts models for lung cancer and melanoma, hence being able to increase the fitness threshold and counteracting the spread of subclones with *BRAF*-amplification (Xue et al. 2017). However, the authors did not address the hypothesis of resistance emergency with the intermittent three-drug treatment combination and, if so, if other alternative treatments would be plausible.

Noticeably, Xue and colleagues have tested different scenarios of drugs administration, including the continuous versus intermittent administration and also different sequences of drug administration (Xue et al. 2017). However, in their model, a lower efficiency of regimens in which the drugs were not given simultaneously was found (Xue et al. 2017).

The idea that therapy response is dependent on the sequence of administration of anti-cancer drugs is gaining prominence (Goldman et al. 2015). Goldman reported that the administration of a chemotherapy drug pair in a specific temporal sequence was able to surpass the adaptive resistance by targeting a vulnerable drug-induced phenotypic transition (Goldman et al. 2015). They found that the treatment of breast cancer cells with Src Family Kinase inhibitors after a taxane-based treatment, but not the co-administration, significantly sensitised the cells to the treatment, resulting in an enhanced anti-cancer outcome (Goldman et al. 2015). This is in accordance with Kent and Green that reported that the order in which genetic mutations arise impacts cancer evolution (Kent and Green 2017). Moreover, the case study reported by Shaw and colleagues truly reflects the power of drug sequence in therapy outcome, by describing the dynamics of response to lorlatinib and crizotinib in a non–small-cell lung cancer patient (Shaw et al. 2016).

In a different evolutionary perspective, Niekerk and colleagues have defended the clinical relevance of synthetic lethality (meaning that the concurrent loss of function in two genes results in lethality, whereas the loss of function in each single gene is tolerated due to compensatory effects) in the context of cancer (van Niekerk et al. 2017). The authors argued that cancer cells are subject to evolutionary trajectories selecting for functional dependencies similar to synthetic lethality, being the auxotrophic induction a way to "turn the evolvability of cancer cells against themselves" (van Niekerk et al. 2017). In fact, evidence suggests that cancer cells display evolution of auxotrophic phenotypes, such as auxotrophy toward arginine or the "oncogene addiction" (van Niekerk et al. 2017).

Noticeably, Russo and colleagues have reported the simultaneously emergence of different acquired resistance mechanisms in separate metastases within the same colorectal cancer patient, leading to diverse responses to the following targeted therapies (Russo et al. 2016). This observation strengthens the pivotal role of evolutionary strategies in the clinical settings, as these could help to trace alternative effective strategies, by "playing" with the different adaptive/resistance mechanisms present in the different metastases within the same individual.

Importantly, Sun and colleagues performed a systematic computational analysis in order to address the effects of different drug-imposed selective pressures on long-term therapeutic outcomes of cancer cells (Sun et al. 2016). They observed that the initial tumour response may not be the best prognosis predictor, since when the initial selective pressure imposed by the drug was identical (meaning an identical cells eradication), different therapeutic outcomes were observed due to differential selective pressure on the subpopulations of cells (Sun et al. 2016). Moreover, their findings were corroborated with a preclinical murine model of Burkitt's lymphoma (Sun et al. 2016). Importantly, they reported the existence of an intrinsic trade-off in maximizing overall tumour cells killing and a higher resistance potential, hence showing that the traditional chemotherapy regimens may lead to tumour shrinkage at the cost of drug sensitivity (Sun et al. 2016).

Taken together, evidence strongly supports the use of evolutionary principles in several and diverse ways in the clinical context of cancer. Clinical protocols that join evolutionary dynamics of cancer cells response to therapy should be of extreme importance as it would possibly allow not only to predict the emergence of resistance, but also to overcome it, hence allowing to change the outcome of this complex group of diseases. These clinical evolutionary strategies could then counteract the evolution of the adaptive strategies of cancer cells, such as metabolic reprograming, hence allowing to overcome drug resistance/tolerance, probably impacting profoundly cancer outcome.

15.5 Final Remarks

More and more evidence supports that cancer cells exhibit metabolic plasticity that enables their survival in changing and challenging environments. Recently, this metabolic plasticity of cancer cells has been found to be itself a driver of chemoresistance. Hence, the knowledge of these metabolic adaptations should be of extreme importance for disease outcome, as more efficient strategy treatments could be developed. More than developing new drugs that target these metabolic adaptations directly, treatments that exploit the evolutionary dynamics of cancer cells response and adaptation to anti-cancer drugs may allow the avoidance of chemoresistance emergency and spread, possibly by preventing these same metabolic adaptations. These evolutionary principles were found to be effective in several cancer types and with several types of drugs, hence opening the opportunity to develop general evolution-guided protocols with drugs that are already used in the clinical setting. This also opens the opportunity to rethink the way anticancer drugs are being administered, the dose used, its schedule and the sequence of the drugs that are used, details that may impact profoundly the disease outcome. Trying to avoid the adaptability and evolvability of cancer cells is only possible if the treatments also evolve along with cancer cells. This would ultimately allow to predict and to overcome chemoresistance, changing cancer prognosis.

Acknowledgments The authors acknowledge iNOVA-4Health – UID/Multi/04462/2013, a program financially supported by Fundação para a Ciência e Tecnologia/Ministério da Educação e Ciência, through national funds and co-funded by FEDER under the PT2020 Partnership Agreement.

References

Ahmed N, Escalona R, Leung D, Chan E, Kannourakis G (2018) Tumour microenvironment and metabolic plasticity in cancer and cancer stem cells: perspectives on metabolic and immune regulatory signatures in chemoresistant ovarian cancer stem cells. Semin Cancer Biol 53:265–281. https://doi.org/10.1016/J.SEMCANCER.2018.10.002

Aktipis CA, Kwan VSY, Johnson KA, Neuberg SL, Maley CC (2011) Overlooking evolution: a systematic analysis of cancer relapse and therapeutic resistance research. PLoS One 6:e26100.1–e26100.9. https://doi.org/10.1371/journal.pone.0026100

Alam MM, Lal S, FitzGerald KE, Zhang L (2016) A holistic view of cancer bioenergetics: mitochondrial function and respiration play fundamental roles in the development and progression of diverse tumors. Clin Transl Med 5(3). https://doi.org/10.1186/s40169-016-0082-9

Allen E, Ville PM, Warren CM, Saghafinia S, Li L, Peng MW, Hanahan D (2016) Metabolic symbiosis enables adaptive resistance to anti-angiogenic therapy that is dependent on mTOR signaling. Cell Rep 15:1144–1160. https://doi.org/10.1016/j.celrep.2016.04.029

Axelrod R, Axelrod DE, Pienta KJ (2006) Evolution of cooperation among tumor cells. Proc Natl Acad Sci U S A 103:13474–13479. https://doi.org/10.1073/pnas.0606053103

Cairns J (1975) Mutation selection and the natural history of cancer. Nature 255:197–200. https://doi.org/10.1038/255197a0. https://www.cancerresearchuk.org/ (2018)

Crespi B, Summers K (2005) Evolutionary biology of cancer. Trends Ecol Evol 20:545–552. https://doi.org/10.1016/j.tree.2005.07.007

Dar S, Chhina J, Mert I, Chitale D, Buekers T, Kaur H et al (2017) Bioenergetic adaptations in chemoresistant ovarian cancer cells. Sci Rep 7:1–17. https://doi.org/10.1038/s41598-017-09206-0

Datta S, Choudhury D, Das A, Das Mukherjee D, Das N, Roy SS, Chakrabarti G (2017) Paclitaxel resistance development is associated with biphasic changes in reactive oxygen species, mitochondrial membrane potential and autophagy with elevated energy production capacity in lung cancer cells: a chronological study. Tumor Biol 39:1–14. https://doi.org/10.1177/1010428317694314

Deblois G, Tonekaboni SAM, Kao YI, Tai F, Liu X, Ettayebi I et al (2018) Metabolic adaptations underlie epigenetic vulnerabilities in chemoresistant breast cancer. bioRxiv:1–51. https://doi.org/10.1101/286054

Denise C, Paoli P, Calvani M, Taddei ML, Giannoni E, Kopetz S et al (2015) 5-fluorouracil resistant colon cancer cells are addicted to OXPHOS to survive and enhance stem-like traits. Oncotarget 6:41706–41721. https://doi.org/10.18632/oncotarget.5991

Enriquez-navas PM, and Gatenby RA (2017) Applying tools from evolutionary biology to cancer research. Ecol Evol Cancer. Chapter 14:193–200. https://doi.org/10.1016/B978-0-12-804310-3/00014-4

Enriquez-navas PM, Wojtkowiak JW, Gatenby RA (2015) Application of evolutionary principles to cancer therapy. Cancer Res 75:4675–4680. https://doi.org/10.1158/0008-5472.CAN-15-1337

Enriquez-Navas PM, Kam Y, Das T, Hassan S, Silva A, Foroutan P et al (2016) Exploiting evolutionary principles to prolong tumor control in preclinical models of breast cancer. Sci Transl Med 8:1–9. https://doi.org/10.1126/scitranslmed.aad7842

Ferlay J, Soerjomataram I, Ervik M, Dikshit R, Eser S, Mathers C et al. (2013a) GLOBOCAN 2012a v1.0, cancer incidence and mortality worldwide: IARC cancerbase No. 11. Retrieved August 24, 2018, from http://globocan.iarc.fr

Ferlay J, Soerjomataram I, Ervik M, Dikshit R, Eser S, Mathers C et al. (2013b) GLOBOCAN 2012b v1.0, cancer incidence and mortality worldwide: IARC cancerbase no. 11. Retrieved August 30, 2018, from http://globocan.iarc.fr

Fitzmaurice C, Dicker D, Pain A, Hamavid H, Moradi-Lakeh M, MacIntyre MF et al (2015) The global burden of cancer 2013. JAMA Oncol 1:505–527. https://doi.org/10.1001/jamaoncol.2015.0735

Gallaher JA, Enriquez-Navas PM, Luddy KA, Gatenby RA, Anderson ARA (2017) Spatial heterogeneity and evolutionary dynamics modulate time to recurrence in continuous and adaptive cancer therapies. bioRxiv:1–21. https://doi.org/10.1101/128959

Gallipoli P, Giotopoulos G, Tzelepis K, Costa ASH, Vohra S, Medina-Perez P et al (2018) Glutaminolysis is a metabolic dependency in FLT3ITDacute myeloid leukemia unmasked by FLT3 tyrosine kinase inhibition. Blood 131:1639–1653. https://doi.org/10.1182/blood-2017-12-820035

Gastel N, van Schajnovitz A, Vidoudez C, Oki T, Sharda A, Trauger SA, Scadden DT (2017) Untargeted metabolomics identifies glutamine metabolism as a driver of chemoresistance in acute myeloid Leukemia. Blood 130:2523

Gatenby RA, Silva AS, Gillies RJ, Frieden BR (2009) Adaptive therapy. Cancer Res 69:4894–4903. https://doi.org/10.1158/0008-5472.CAN-08-3658

Gillies RJ, Verduzco D, Gatenby RA (2012) Evolutionary dynamics of carcinogenesis and why targeted therapy does not work. Nat Rev Cancer 12:487–493. https://doi.org/10.1038/nrc3298

Goldman A, Majumder B, Dhawan A, Ravi S, Goldman D, Kohandel M et al (2015) Temporally sequenced anticancer drugs overcome adaptive resistance by targeting a vulnerable chemotherapy-induced phenotypic transition. Nat Commun 6:1–13. https://doi.org/10.1038/ncomms7139

Gottesman MM (2002) Mechanisms of cancer drug resistance. Annu Rev Med 53:615–627. https://doi.org/10.1146/annurev.med.53.082901.103929

Guppy M, Leedman P, Zu X, Russell V (2002) Contribution by different fuels and metabolic pathways to the total ATP turnover of proliferating MCF-7 breast cancer cells. Biochem J 364:309–315. https://doi.org/10.1042/bj3640309

Hanahan D, Weinberg RA (2000) The hallmarks of cancer. Cell 100:57–70. https://doi.org/10.1007/s00262-010-0968-0

Hanahan D, Weinberg RA (2011) Hallmarks of cancer: the next generation. Cell 144:646–674. https://doi.org/10.1016/j.cell.2011.02.013

Hayes JD, Dinkova-Kostova AT (2014) The Nrf2 regulatory network provides an interface between redox and intermediary metabolism. Trends Biochem Sci 39:199–218. https://doi.org/10.1016/j.tibs.2014.02.002

Holohan C, Van Schaeybroeck S, Longley DB, Johnston PG (2013) Cancer drug resistance: an evolving paradigm. Nat Rev Cancer 13:714–726. https://doi.org/10.1038/nrc3599

Ippolito L, Marini A, Cavallini L, Morandi A, Pietrovito L, Pintus G et al (2016) Metabolic shift toward oxidative phosphorylation in docetaxel resistant prostate cancer cells. Oncotarget 7:61890–61904. https://doi.org/10.18632/oncotarget.11301

Jia D, Park J, Jung K, Levine H, Kaipparettu B (2018) Elucidating the metabolic plasticity of cancer: mitochondrial reprogramming and hybrid metabolic states. Cell 7:21. https://doi.org/10.3390/cells7030021

Ju HQ, Gocho T, Aguilar M, Wu M, Zhuang ZN, Fu J et al (2015) Mechanisms of overcoming intrinsic resistance to gemcitabine in pancreatic ductal adenocarcinoma through the redox modulation. Mol Cancer Ther 14:788–798. https://doi.org/10.1158/1535-7163.MCT-14-0420

Kent DG, Green AR (2017) Order matters: the order of somatic mutations influences cancer evolution. Cold Spring Harb Perspect Med 7:1–16. https://doi.org/10.1101/cshperspect.a027060

Kerr EM, Gaude E, Turrell FK, Frezza C, Martins CP (2016) Mutant Kras copy number defines metabolic reprogramming and therapeutic susceptibilities. Nature 531:110–113. https://doi.org/10.1038/nature16967

Khamari R, Trinh A, Gabert PE, Corazao-Rozas P, Riveros-Cruz S, Balayssac S et al (2018) Glucose metabolism and NRF2 coordinate the antioxidant response in melanoma resistant to MAPK inhibitors.

Cell Death Dis 9:325–338. https://doi.org/10.1038/s41419-018-0340-4
Komurov K, Tseng JT, Muller M, Seviour EG, Moss TJ, Yang L et al (2012) The glucose-deprivation network counteracts lapatinib-induced toxicity in resistant ErbB2-positive breast cancer cells. Mol Syst Biol 8:1–10. https://doi.org/10.1038/msb.2012.25
Kong X, Kuilman T, Shahrabi A, Boshuizen J, Kemper K, Song JY et al (2017) Cancer drug addiction is relayed by an ERK2-dependent phenotype switch. Nature 550:270–274. https://doi.org/10.1038/nature24037
Landriscina M, Maddalena F, Laudiero G, Esposito F (2009) Adaptation to oxidative stress, chemoresistance, and cell survival. Antioxid Redox Signal 11:2701–2716. https://doi.org/10.1089/ars.2009.2692
Liang C, Qin Y, Zhang B, Ji S, Shi S, Xu W et al (2017) ARF6, induced by mutant Kras, promotes proliferation and Warburg effect in pancreatic cancer. Cancer Lett 388:303–311. https://doi.org/10.1016/j.canlet.2016.12.014
Liberti MV, Locasale JW (2016) The Warburg effect: how does it benefit cancer cells? Trends Biochem Sci 41:211–218. https://doi.org/10.1016/j.tibs.2015.12.001
Liu T, Yin H (2017) PDK1 promotes tumor cell proliferation and migration by enhancing the Warburg effect in non-small cell lung cancer. Oncol Rep 37:193–200. https://doi.org/10.3892/or.2016.5253
Liu Y, Cao Y, Pan X, Shi M, Wu Q, Huang T et al (2018) O-GlcNAc elevation through activation of the hexosamine biosynthetic pathway enhances cancer cell chemoresistance. Cell Death Dis 9:485–496. https://doi.org/10.1038/s41419-018-0522-0
Lopes-Coelho F, Nunes C, Gouveia-Fernandes S, Rosas R, Silva F, Gameiro P et al (2017) Monocarboxylate transporter 1 (MCT1), a tool to stratify acute myeloid leukemia (AML) patients and a vehicle to kill cancer cells. Oncotarget 8:82803–82823. https://doi.org/10.18632/oncotarget.20294
Ma S, Jia R, Li D, Shen B (2015) Targeting cellular metabolism chemosensitizes the doxorubicin-resistant human breast adenocarcinoma cells. Biomed Res Int 2015:1–8. https://doi.org/10.1155/2015/453986
Maley CC, Aktipis A, Graham TA, Sottoriva A, Boddy AM, Janiszewska M et al (2017) Classifying the evolutionary and ecological features of neoplasms. Nat Rev Cancer 17:605–619. https://doi.org/10.1038/nrc.2017.69
Merlo LMF, Pepper JW, Reid BJ, Maley CC (2006) Cancer as an evolutionary and ecological process. Nat Rev Cancer 6:924–935. https://doi.org/10.1038/nrc2013
Morandi A, Indraccolo S (2017) Linking metabolic reprogramming to therapy resistance in cancer. Biochimica et Biophysica Acta – Rev Cancer 1868:1–6. https://doi.org/10.1016/j.bbcan.2016.12.004
Nguyen TL, Durán RV (2018) Glutamine metabolism in cancer therapy. Cancer Drug Resist 1:126–138. https://doi.org/10.20517/cdr.2018.08
Nowell PC (1976) The clonal evolution of tumor cell populations. Science 194:23–28. https://doi.org/10.1126/science.191.4224.241-a
Pastò A, Pagotto A, Pilotto G, De Paoli A, De Salvo GL, Baldoni A et al (2017) Resistance to glucose starvation as metabolic trait of platinum- resistant human epithelial ovarian cancer cells. Oncotarget 8:6433–6445. https://doi.org/10.18632/oncotarget.14118
Polyak K (2007) Breast cancer stem cells: a case of mistaken identity? Stem Cell Rev 3:107–109. https://doi.org/10.1007/s12015-007-0020-8
Qian W, Nishikawa M, Haque AM, Hirose M, Mashimo M, Sato E, Inoue M (2005) Mitochondrial density determines the cellular sensitivity to cisplatin-induced cell death. Am J Phys Cell Phys 289:C1466–C1475. https://doi.org/10.1152/ajpcell.00265.2005
Rankin EB, Giaccia AJ (2016) Hypoxic control of metastasis. Science 352:175–180. https://doi.org/10.1126/science.aaf4405
Rodríguez-Enríquez S, Torres-Márquez ME, Moreno-Sánchez R (2000) Substrate oxidation and ATP supply in AS-30D hepatoma cells. Arch Biochem Biophys 375:21–30. https://doi.org/10.1006/abbi.1999.1582
Rodríguez-Enríquez S, Vital-González PA, Flores-Rodríguez FL, Marín-Hernández A, Ruiz-Azuara L, Moreno-Sánchez R (2006) Control of cellular proliferation by modulation of oxidative phosphorylation in human and rodent fast-growing tumor cells. Toxicol Appl Pharmacol 215:208–217. https://doi.org/10.1016/j.taap.2006.02.005
Roh JL, Jang H, Kim EH, Shin D (2017) Targeting of the glutathione, thioredoxin, and Nrf2 antioxidant systems in head and neck cancer. Antioxid Redox Signal 27:106–114. https://doi.org/10.1089/ars.2016.6841
Russo M, Siravegna G, Blaszkowsky LS, Corti G, Crisafulli G, Ahronian LG et al (2016) Tumor heterogeneity and lesion-specific response to targeted therapy in colorectal cancer. Cancer Discov 6:147–153. https://doi.org/10.1158/2159-8290.CD-15-1283
Salgia R, Kulkarni P (2018) The genetic/non-genetic duality of drug "resistance" in cancer. Trends Cancer 4:110–118. https://doi.org/10.1016/j.trecan.2018.01.001
Sancho P, Burgos-Ramos E, Tavera A, Bou Kheir T, Jagust P, Schoenhals M et al (2015) MYC/PGC-1α balance determines the metabolic phenotype and plasticity of pancreatic cancer stem cells. Cell Metab 22:590–605. https://doi.org/10.1016/j.cmet.2015.08.015
Semenza GL (2012) Hypoxia-inducible factors: mediators of cancer progression and targets for cancer therapy. Trends Pharmacol Sci 33:207–214. https://doi.org/10.1016/j.tips.2012.01.005
Shaw AT, Friboulet L, Leshchiner I, Gainor JF, Bergqvist S, Brooun A et al (2016) Resensitization to crizotinib by the lorlatinib ALK resistance mutation L1198F. N Engl J Med 374:54–61. https://doi.org/10.1038/nrg3575.Systems
Silva AS, Kam Y, Khin ZP, Minton SE, Gillies RJ, Gatenby RA (2012) Evolutionary approaches to prolong progression-free survival in breast cancer.

Cancer Res 72:6362–6370. https://doi.org/10.1002/bmb.20244.DNA

Son B, Lee S, Youn H, Kim E, Kim W, Youn B (2017) The role of tumor microenvironment in therapeutic resistance. Oncotarget 8:3933–3945. https://doi.org/10.18632/oncotarget.13907

Sun D, Dalin S, Hemann MT, Lauffenburger DA, Zhao B (2016) Differential selective pressure alters rate of drug resistance acquisition in heterogeneous tumor populations. Sci Rep 6:1–13. https://doi.org/10.1038/srep36198

Swanton C (2013) Intratumour heterogeneity : evolution through space and time an evolutionary perspective on cancer heterogeneity. Cancer Res 72:4875–4882. https://doi.org/10.1158/0008-5472.CAN-12-2217.Intratumour

van Niekerk G, Nell T, Engelbrecht AM (2017) Domesticating cancer: an evolutionary strategy in the war on Cancer. Front Oncol 7:1–8. https://doi.org/10.3389/fonc.2017.00304

Vaupel P, Mayer A (2007) Hypoxia in cancer: significance and impact on clinical outcome. Cancer Metastasis Rev 26:225–239. https://doi.org/10.1007/s10555-007-9055-1

Viale A, Corti D, Draetta GF (2015) Tumors and mitochondrial respiration: a neglected connection. Cancer Res 75:1–5. https://doi.org/10.1158/0008-5472.CAN-15-0491

Warburg O (1956) On the origin of cancer cells on the origin of cancer. Science 123:309–314. https://doi.org/10.1126/science.123.3191.309

Xue Y, Martelotto L, Baslan T, Vides A, Solomon M, Mai TT et al (2017) An approach to suppress the evolution of resistance in BRAF V600E-mutant cancer. Nat Med 23:929–937. https://doi.org/10.1038/nm.4369

Ye J, Zou M, Li P, Liu H (2018) MicroRNA regulation of energy metabolism to induce chemoresistance in cancers. Technol Cancer Res Treat 17:1–6. https://doi.org/10.1177/1533033818805997

Yu L, Lu M, Jia D, Ma J, Ben-Jacob E, Levine H et al (2017) Modeling the genetic regulation of cancer metabolism: interplay between glycolysis and oxidative phosphorylation. Cancer Res 77:1564–1574. https://doi.org/10.1158/0008-5472.CAN-16-2074.Modeling

Zhang L, Yang H, Zhang W, Liang Z, Huang Q, Guoqiang X et al (2017a) Clk1–regulated aerobic glycolysis is involved in gliomas chemoresistance. J Neurochem 142:574–588. https://doi.org/10.1111/ijlh.12426

Zhang J, Cunningham JJ, Brown JS, Gatenby RA (2017b) Integrating evolutionary dynamics into treatment of metastatic castrate-resistant prostate cancer. Nat Commun 8:1–9. https://doi.org/10.1038/s41467-017-01968-5

Zhao JG, Ren KM, Tang J (2014) Overcoming 5-Fu resistance in human non-small cell lung cancer cells by the combination of 5-Fu and cisplatin through the inhibition of glucose metabolism. Tumor Biol 35:12305–12315. https://doi.org/10.11862/CJIC.2015.154

Zhou Y, Tozzi F, Chen J, Fan F, Xia L, Wang J et al (2012) Intracellular ATP levels are a pivotal determinant of chemoresistance in colon cancer cells. Cancer Res 72:304–314. https://doi.org/10.1158/0008-5472.CAN-11-1674

The Metabolic Remodelling in Lung Cancer and Its Putative Consequence in Therapy Response

16

Ana Hipólito, Cindy Mendes, and Jacinta Serpa

Abstract

Lung cancer is the leading cause of cancer-related deaths worldwide in both men and women. Conventional chemotherapy has failed to provide long-term benefits for many patients and in the past decade, important advances were made to understand the underlying molecular/genetic mechanisms of lung cancer, allowing the unfolding of several other pathological entities. Considering these molecular subtypes, and the appearance of promising targeted therapies, an effective personalized control of the disease has emerged, nonetheless benefiting a small proportion of patients. Although immunotherapy has also appeared as a new hope, it is still not accessible to the majority of patients with lung cancer.

The metabolism of energy and biomass is the basis of cellular survival. This is true for normal cells under physiological conditions and it is also true for pathophysiologically altered cells, such as cancer cells. Thus, knowledge of the metabolic remodelling that occurs in cancer cells in the sense of, on one hand, surviving in the microenvironment of the organ in which the tumour develops and, on the other hand, escaping from drugs conditioned microenvironment, is essential to understand the disease and to develop new therapeutic approaches.

Keywords

Metabolic remodelling · Tumor microenvironment · Lung cancer · Targeted therapy · New therapeutic approaches

16.1 Lung Cancer: Understanding Its Molecular Pathology

Lung cancer is a highly mortal disease, having the overall 5-year survival rate increased only 4% (from 12% to 16%) over the past four decades, and late diagnosis is a major obstacle in improving lung cancer prognosis (Inamura 2017; I and Cho 2015).

Lung cancer is categorized in two main histological groups: small cell lung carcinoma (SCLC, 15% of all lung cancers) and non-small cell lung carcinoma (NSCLC, 85% of all lung cancers). NSCLCs are generally subcategorized

Authors Ana Hipólito and Cindy Mendes have equally contributed to this chapter.

A. Hipólito · C. Mendes · J. Serpa (✉)
CEDOC, Chronic Diseases Research Centre, NOVA Medical School | Faculdade de Ciências Médicas, Universidade NOVA de Lisboa, Lisbon, Portugal

Instituto Português de Oncologia de Lisboa Francisco Gentil (IPOLFG), Lisbon, Portugal
e-mail: jacinta.serpa@nms.unl.pt

J. Serpa (ed.), *Tumor Microenvironment*, Advances in Experimental Medicine and Biology 1219, https://doi.org/10.1007/978-3-030-34025-4_16

as adenocarcinoma, squamous cell carcinoma and large cell carcinoma (Inamura 2017; Lemjabbar-Alaoui et al. 2015).

About 90% of lung cancer cases are caused by smoking habits and the use of tobacco products. However, other factors such as polluted air exposure and chronic infections can also contribute to lung carcinogenesis (Lemjabbar-Alaoui et al. 2015). Given the highly carcinogenic compounds present in tobacco smoke, passive smoking is thought to be one of the major causative factors of lung cancer in never smokers (Torok et al. 2011). In addition, multiple inherited and acquired mechanisms of susceptibility to lung cancer have been proposed (Lemjabbar-Alaoui et al. 2015). Studies have identified differences in chromosomal aberrations, genetic polymorphisms and gene mutations and methylation status between lung cancer in never smokers and tobacco-associated lung cancer.

Lung cancer is a highly intricate and heterogeneous disease with genomic diversity in each histological class, presenting different mutated genes. In NSCLC tumours, 40–60% are associated with mutations of the tumour suppressor gene *TP53*, presenting higher prevalence in tobacco-associated lung cancer than in lung cancer in never smokers. Moreover, mutations in *EGFR* (epidermal growth factor receptor gene), encoding a tyrosine kinase receptor, in NSCLC can cause oncogenic transformation and lead to sensitivity to tyrosine kinase inhibitors (Subramanian and Govindan 2008). These are much more common in people who have never smoked than in patients who had. In contrast to *EGFR*, mutations in the *KRAS* (Kirsten rat sarcoma viral oncogene homolog) are rare in patients who have never smoked. The KRAS protein is downstream to the EGFR activation pathway and *KRAS* and *EGFR* mutations seem almost mutually exclusive of each other in both nonsmokers and smokers (Subramanian and Govindan 2008; Shigematsu et al. 2005). Interestingly, the presence of *KRAS* mutations is associated with poor response to tyrosine kinase inhibitors (Subramanian and Govindan 2008). Genetic alterations in the anaplastic lymphoma kinase (*ALK*) gene occur in 2–9% of NSCLCs (Woo et al. 2016). However, driver oncogenes and genetic regulatory mechanisms that result in the initiation and progression of each type of lung cancer are yet far from being totally understood (Zhang et al. 2017a).

Besides radiation and cytostatic therapy, molecular targeted therapies have advanced most for younger patients with adenocarcinoma, who are mostly never-smokers. For patients with advanced NSCLC who do not fit an approved molecular targeted therapy, the standard first-line treatment remains platinum-based doublet therapy with or without bevacizumab (anti-angiogenic monoclonal antibody). Over the years, several inhibitors have been tested for clinical use, targeting specific oncogenic proteins in lung cancer, such as the receptor tyrosine kinase (RTK) inhibitors gefitinib/erlotinib (EGFR inhibitors) and crizotinib (EML4-ALK inhibitor). However, these treatments benefit only a small proportion (15–20%) of patients, harboring these driver mutations, and the acquired resistance to these therapies presents a major impediment to the effective treatment of NSCLC patients with these mutations (Hirsch et al. 2017; Mittal et al. 2016a). Beyond surgery, radiation and chemotherapy, immunotherapy has emerged in recent years as a fourth pillar in the therapeutic approach against lung cancer, taking advantage of the native antitumor immune response (Qin et al. 2016). So far, the clinical results on the application of immune check points inhibition are really enthusiastic, showing benefits on clinical outcome mainly in combined therapeutic protocols, using immune check points inhibition with conventional chemo and/or radiotherapy (Melosky et al. 2019; Vansteenkiste et al. 2019).

16.2 Tumour Microenvironment (TME) in Lung Cancer

Cancer is a complex group of diseases in which several cellular and molecular components of the tumour microenvironment (TME) contribute to the survival of cancer cells. The TME is composed of both cellular (fibroblasts, adipocytes, endothelial and immune cells) and non-cellular components (e.g. growth factors, chemokines, cytokines, proteases, extracellular matrix- ECM) which synergistically play a role in cancer progression (Hanahan and Coussens 2012; Lopes-Coelho et al. 2018). Intercellular communication between cancer cells and non-malignant cells is driven by a complex and dynamic network of cytokines, growth factors and organic molecules that support cellular viability and proliferation (Serpa and Dias 2011). This interplay between cells promotes carcinogenesis by contributing to inflammation, immune suppression, therapeutic resistance and generating premetastatic niches that support the initiation and establishment of distant metastasis (Chen et al. 2015). Therefore, targeting the cellular components of the TME has emerged as a promising therapeutic approach constituting the basis for anti-angiogenic and anti-inflammatory therapies applied to different types of tumour (Hanahan and Coussens 2012; Ebos and Kerbel 2011).

The lung presents a unique milieu in which tumours progress in collusion with the TME, as demonstrated by regions of aberrant angiogenesis, acidosis and hypoxia (Mittal et al. 2016b). The anatomical and cellular characteristics of normal lungs act as a defence barrier against foreign microorganisms. In inflammatory states such as chronic obstructive pulmonary disease, the lung microenvironment shows features that may promote carcinogenesis (Altorki et al. 2019; Houghton 2013). Notably, extensive stage-dependent immune and inflammatory cell infiltration in human lung cancer samples was observed (Banat et al. 2015; Kargl et al. 2017). Lung adenocarcinomas comprise unique lung cancer subtypes with distinct cellular and mutational heterogeneity (Altorki et al. 2019; Chen et al. 2014a). Heterogeneity is also a rule in TME, which includes vasculature, cancer associated fibroblasts (CAFs), ECM and infiltrating immune cells (Fig. 16.1). Depending on the composition of these cells and the local cytokine milieu, the levels of tumour oxygenation, nutrients, interstitial pressure and pH can be extremely variable within the same tumour (Graves et al. 2010). In the lung TME, malignant cells are able to reprogramme the tumour infiltrating stromal cells, which consequently contributes to carcinogenesis (Quail and Joyce 2013). Moreover, the TME has been recognized as a target for the development of novel anticancer agents in both primary and secondary lung tumours (Altorki et al. 2019).

Next, we will describe the contribution of some cancer associated cells and special microenvironmental conditions for lung cancer survival and progression.

16.2.1 Cancer Associated Fibroblasts (CAFs), Metabolic Partners of Lung Cancer Cells

CAFs differ morphologically and functionally from normal fibroblasts and exhibit similar activities with wound-activated fibroblasts, suggesting that the supportive and reparative roles of activated fibroblasts in wound healing contribute to the pro-tumorigenic activities of CAFs (Mittal et al. 2016a; Wang et al. 2017a, 2019a). The origin of CAFs is not clear, yet it is likely that they arise from a reprogramming of tissue resident fibroblasts, as well as differentiating from bone marrow cells recruited to the tumour. CAFs have been reported to support chemotherapy resistance, tumour progression and metastasis through a wide variety of mechanisms. These mechanisms include the paracrine support of cancer cells via the secretion of growth factors, cytokines, and chemokines, accounting for pro-angiogenic effects, remodelling the ECM, epithelial-to-mesenchymal transition (EMT) and expression of metastasis-related genes (Mittal et al. 2016a; Wang et al. 2017a, 2019a).

The role of CAFs at the metabolic level has been described as the Reverse Warburg effect. According to this model, tumour cells within the microenvironment would activate CAFs by different factors and would control them, taking advantage of their metabolism. Proliferating CAFs present an increased glycolytic flux and glutamine metabolism, but they also have a truncated tricarboxylic acid (TCA) cycle (Fig. 16.2). In fact, tumour cells are nutrient supplied by

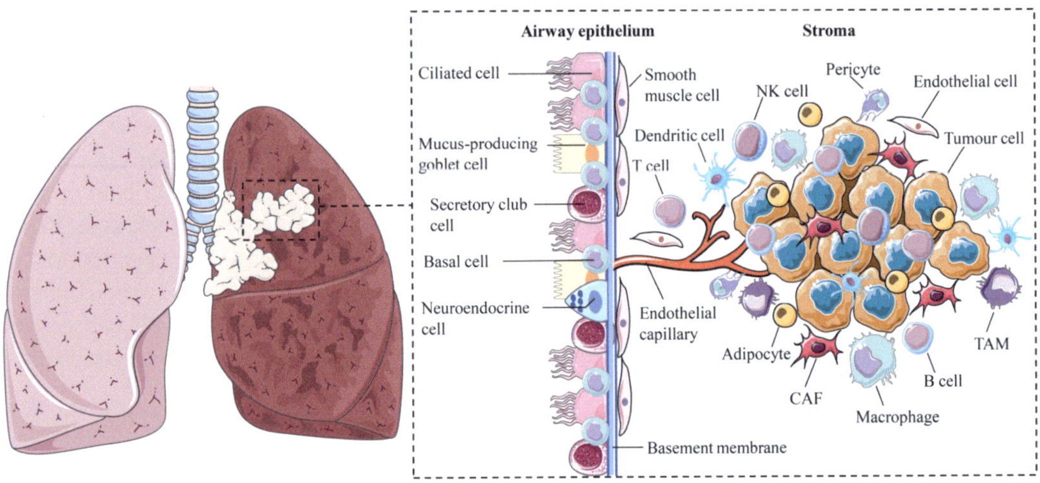

Fig. 16.1 The heterogeneous TME of lung cancer. A schematic of the proximal airway of the lung composed of: ciliated cells, which provide the mechanism for moving the mucus blanket and have also been reported to be involved in epithelial cell trans-differentiation and repair; secretory club cells, crucial for airway repair after injury; undifferentiated basal cells, which are the progenitor cells for the epithelium and differentiate to form the other cells in injury and repair; the goblet cells responsible for the production and secretion of mucus; and neuroendocrine cells which modulate early lung development as well as airway chemoreceptors and have been speculated to constitute the cells of origin of SCLC (Song et al. 2014; Song et al. 2012). Other cell types in the lung microenvironment include smooth muscle cells, CAFs, endothelial cells, pericytes and immune cells, including resident alveolar macrophages and dendritic cells. Recruitment, activation and reprogramming of these cells in the extracellular space are the consequences of reciprocal interaction between TME and cancer cells. Endothelium-derived angiocrine signalling induces and sustains regenerative lung alveolarization (Sen et al. 2011). Resident alveolar macrophages maintain immune homeostasis but can also contribute to inflammation and development of premalignant lung lesions in mice (Morales-Nebreda et al. 2015)

CAFs, through the production and release of lactic acid, amino acids and ketone bodies. Tumour cells in turn produce reactive oxygen species (ROS) that activate and lead to the maintenance of CAFs glycolytic metabolism/cancer-supplying phenotype (Biswas 2015; Cruz-Bermúdez et al. 2019; Koukourakis et al. 2017). All this dynamics unravels an important metabolic interplay between cancer cells and CAFs, based on the exchange of organic compounds to fulfil the tumoral demands.

16.2.2 Endothelial Cells (ECs), an Important Component of Vessels and a Nutrients Supplier for Tumour

In the lung, the histological architecture provides an intimal contact between squamous epithelial cells and sinusoid vessels. Hence, upon lung cancer development the recruitment of new blood vessels can be an early event. ECs are an important component of the vessels, essential in TME to provide nutrients and oxygen to the tumour. ECs actively regulate the inflammatory response in normal and "unhealthy" tissues (Pober and Sessa 2007) and recent findings indicate that ECs directly influence tumour behaviour (Franses et al. 2011). In NSCLC, the extent of tumour associated angiogenesis correlates with disease progression, predicting poor survival outcome (Dundar et al. 2008; Herbst et al. 2005). Recent findings showed that ECs-derived angiocrine signals induced regenerative lung alveolarization. The activation of vascular endothelial growth factor receptor 2 (VEGR2) and fibroblast growth factor receptor 1 (FGFR1) in ECs from pulmonary capillaries induce the expression of MMP14, which unmasked EGF receptor ligands to enhance alveologenesis (Sen et al. 2011). Lung ECs also regulate lung stem cell differentiation as

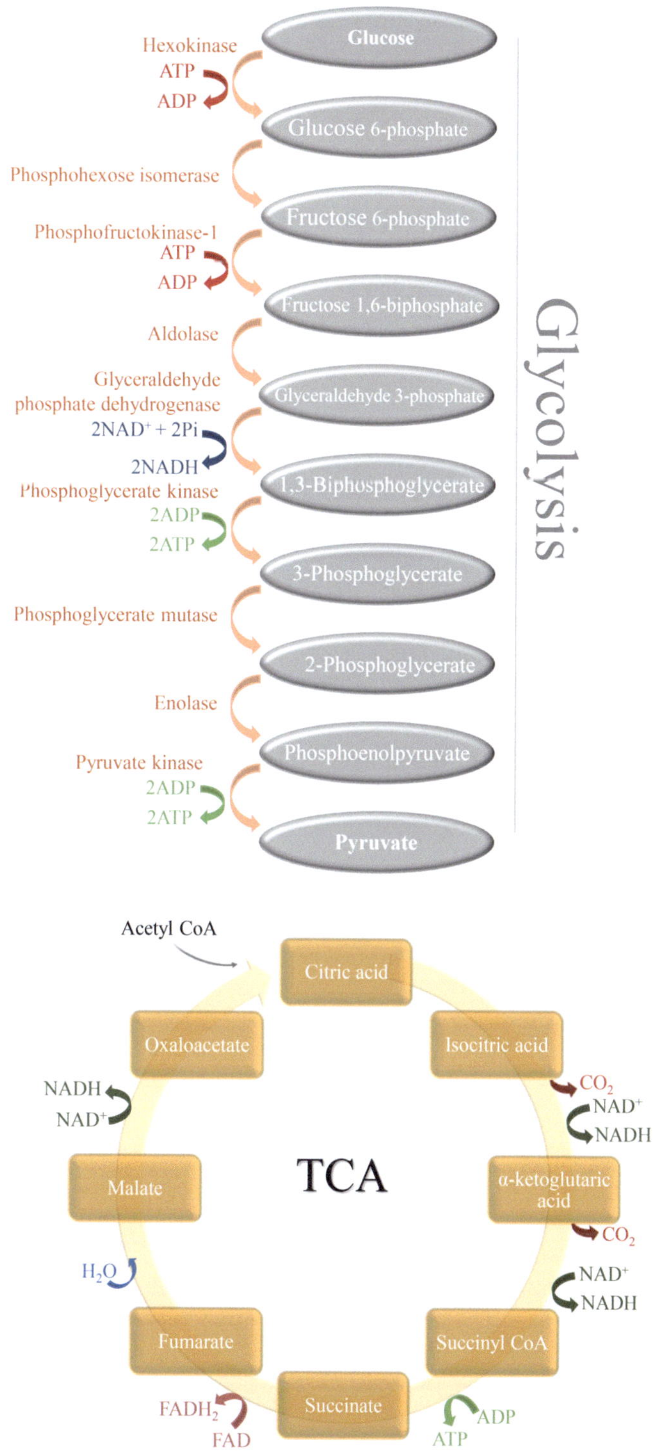

Fig. 16.2 Schematic representation of glycolysis and TCA pathways and the participating compounds and enzymes. Cells obtain their energy mainly through glycolysis, a nine-step catabolic pathway, where glucose that enters the pathway is converted to pyruvate producing an energy yield of two molecules of pyruvate. Pyruvate is then decarboxylated into acetyl-CoA, in cytoplasm, to be further transported into mitochondria to enter the TCA cycle. After the electron transport chain (ETC), the total energy gain under these conditions is roughly 36 ATP *per* each molecule of glucose (Palsson-McDermott and O'Neill 2013)

bone morphogenetic protein 4 (BMP4)-BMPR1A signalling activates calcineurin/NFATc1-dependent expression of thrombospondin-1 (Tsp-1) in lung ECs, promoting alveolar lineage-specific bronchioalveolar stem cell differentiation (Lee et al. 2014). In lung cancer models, controversially, ECs potentiate EMT and invasiveness (Kim et al. 2019), but they also block proliferation and invasiveness by disturbing pro-inflammatory pathways (Franses et al. 2011).

Little is known about the metabolic cooperation between lung cancer cells and ECs. Nevertheless, a recent study shows in lung cancer models that the inflammatory TME not only stimulates angiogenesis but also the ECs metabolic remodelling, inducing an increased degradation of fatty acids (β-oxidation) (Wang et al. 2019b). Other studies showed that ECs have a high plasticity and easily adapt their metabolic course to microenvironment conditions. ECs rely mostly on glycolytic phenotype to migrate and proliferate, but when subjected to glucose scarcity ECs fulfil oxidative phosphorylation (OXPHOS) (Teuwen et al. 2017; Dagher et al. 2001; Schoors et al. 2015; De Bock et al. 2013; Dranka et al. 2010), which can be sustained by β-oxidation. This metabolic plasticity gives us a clue that, also in lung cancer context, ECs can serve as metabolic factories to a certain point as CAFs, contributing to cancer cells survival through nutrients supply.

16.2.3 Immune Cells, from Enemies to Facilitators

Although immune cells should in principle detect and eliminate transformed cells, their interaction with tumour cells, can both antagonize and enhance tumour development and progression by leading to changes in their phenotype. For example, the release of ROS that are actively mutagenic for nearby cancer cells, can accelerate their genetic evolution toward its malignancy state. This results in the establishment of a tumour-supporting environment in various cancer settings, including lung cancer (Banat et al. 2015; Hanahan and Weinberg 2011). Pathologists have long recognized that some tumours are densely infiltrated by cells of both the innate and adaptive immune system and thereby mirror inflammatory conditions arising in non-neoplastic tissues (Hanahan and Weinberg 2011). Macrophages represent one of the major immune infiltrates in solid tumours and are known to influence cancer progression by enhancing survival and proliferation of cancer cells, angiogenesis, metastasis, cancer-related inflammation, and immune suppression. Similarly, other studies have indicated the involvement of almost every other immune cell type including: T cells, B cells, NK cells, NKT cells, basophils, neutrophils, dendritic cells (DCs), and myeloid derived suppressor cells (MDSCs) in the regulation of cancer progression (Banat et al. 2015; Biswas 2015; Bindea et al. 2013). However, generically in cancer context, T-cells and tumour associated macrophages (TAMs) are the more frequently addressed immune cells that can contribute for tumour progression (Candido and Hagemann 2013; Stathopoulos et al. 2008; Zengin 2019) By itself, the differentiation and proliferation of immune cells for sure involves metabolic adjustments. For example, HIF transcription factors play important roles in regulating adaptive and innate immunity (Gnanaprakasam et al. 2017). HIF-1α accumulation favours the differentiation into T helper 17 (Th17) cells, through increased production of IL-17 and IL-6, in a STAT3 dependent manner. It has also been reported that HIF-1α promotes TH17 differentiation by inducing RORγt expression and inhibits Treg differentiation by decreasing the expression of FOXp3, contributing for the balance between TH17 and Treg (Dang et al. 2011; Shi et al. 2011).

Moreover, T cell activation underlies metabolic adaptations that involve changes in the aerobic glycolytic and OXPHOS as well as changes in glutamine and leucine uptake, depending upon their activation or differentiation state (Chapman et al. 2017). T cell activation requires a bio-energetic favourable metabolic reprogramming: Glucose transporter 1 (GLUT1) is shown to be strongly upregulated to increase glucose uptake; mitOXPHOS is suppressed; the production of lactate indicates that pyruvate generated during glycolysis has not entered the oxidative pathway in the mitochondria (Dugnani et al. 2017; Wang et al. 2011; Jacobs et al. 2008; Cammann et al.

2016). Furthermore, pathways required for T cell activation also control metabolic reprogramming, which includes CD28-mediated Akt-dependent and independent pathways (Frauwirth et al. 2002).

The complex relationship between shifts in metabolism and functional reprogramming of the macrophages has been studied in different experimental models. However, less is known about the specific metabolic phenotype of TAMs and how it shapes the functional phenotype of these cells in the TME. As the metabolic program of TAMs can regulate their pro-tumoral functions, there is considerable interest in understanding the cellular pathways that underpin the phenotype of TAMs (Rabold et al. 2017). For instance, arginine is very important for T cells proliferation (Rodriguez et al. 2007), and some leukocytes associated to tumour vasculature deplete T-cells by producing arginase 1 (Bak et al. 2008). A similar mechanism preconized by macrophages that increase the arginine consume, seems to prompt the differentiation of certain subsets of T-cells (Rodriguez et al. 2003). In TAMs an increased glycolytic and fatty acids oxidative phenotype together with glutamate/glutamine, cysteine and arginine reliance has been described as important alterations in TAMs, accounting for cancer progression in a metabolic symbiosis context (Lopes-Coelho et al. 2018).

The metabolic cross talk between lung cancer cells and immune cells is far from being explored. But a very recent and interesting paper by Stoll et al., stated that the expression of relevant enzymes in lipids metabolism, aldehyde dehydrogenase 7 family, member A1 (ALDH7A1) and lipase C (LiPC) in cancer cells control the myeloid and lymphoid cells infiltrate in tumours. The same study indicates that in NSCLC the increased expression of ALDH7A1 and LiPC correlate with immune cells paucity in tumours (Stoll et al. 2019). This study re-enforces the evidence that cancer metabolism comes together with immune metabolism; and cancer and immune cells can press and modulated each other metabolic remodelling towards cancer promotion.

16.2.4 Hypoxia, Enhancing Cancer in Adversity

Hypoxia (low oxygen levels) is a feature of solid tumours that promotes genomic instability, enhanced aggressiveness, and metastasis and it is an important factor in treatment resistance and poor survival (Salem et al. 2018). Although hypoxia is toxic to both cancer cells and normal cells, cancer cells undergo genetic and adaptive changes that allow them to survive and even proliferate in a hypoxic environment, which contributes to the malignant phenotype and aggressiveness of the tumour (Harris 2002). Hypoxia inducible factor–1α (HIF-1 α) is a pivotal transcription factor that mediates hypoxia consequences. In response to decreased oxygen levels, HIF-1α activates the expression of numerous hypoxia-responsive genes that are associated to a number of hallmarks of cancer, such as suppression of apoptosis, motility, invasion, energy metabolism reprogramming and angiogenesis (Mittal et al. 2016a; Hanahan and Weinberg 2011; Salem et al. 2018; Harris 2002). In NSCLC, HIF-1α expression is associated with resistance to cancer therapy, including EGFR inhibitors (Mittal et al. 2016a; Salem et al. 2018).

Hypoxia can be categorized as acute or chronic. Chronic hypoxia is characterized by presenting a necrotic centre containing cancer cells beyond the capillary diffusion distance, while viable cancer cells exist in an environment of decreasing hypoxia away from the centre. Acute hypoxia, which presents an intermittent pattern, occurs in areas adjacent to blood supply due to transient vessel occlusion. This is due to vessel fragility and increased interstitial pressure resulting from tumour cell proliferation outstripping new capillary growth. The two hypoxia types are not exclusive and do coexist, resulting in spatial intra-tumour and inter-tumour heterogeneity (Salem et al. 2018; Harris 2002).

The state of hypoxia triggers molecular changes that facilitate metabolic adaptations in cancer cells aiming to maintain tumour growth (Justus et al. 2015).

HIF-1 positively regulates the transcription of over 100 genes, of which many directly upregulate glycolysis, among them are pyruvate dehydrogenase kinase 1 (*PDK1*) and *LDH-A (*Ke and Costa 2006*;* Papandreou et al. 2006*)*. In order to increase glycolytic flux, glucose transporter type 1 (GLUT1) and 3 (GLUT3) expression is increased by HIF-1, which enhances the availability of glucose within the cytoplasm. Furthermore, HIF-1 facilitates the conversion of glucose to pyruvate by increasing the expression of glycolytic enzymes such as hexokinase 1/2 (HK-1/-2) and pyruvate kinase M2 (PKM2). HIF-1 activation not only increases glycolysis, but also directly inhibits OXPHOS by blocking pyruvate entrance into the TCA cycle (Papandreou et al. 2006).

Giatromanolaki and colleagues, showed that lung cancer cells (along with lung fibroblasts) respond to acute hypoxia towards pyruvate transformation to lactate, which is extruded out of cells through MCT1, leading to a deviation in the normal metabolic flux (Giatromanolaki et al. 2017), and showing the capacity of cancer cells of adapting to stressful metabolic conditions.

16.2.5 Inflammation, a Deleterious Protective Reaction

Chronic lung inflammation has been linked to an increased risk of lung cancer. Carcinogens including cigarette smoke, asbestos and other pollutants induce a chronic inflammatory state which in turn promotes carcinogenesis (Candido and Hagemann 2013). Chronic obstructive pulmonary disease and pulmonary fibrosis, characterized by high inflammatory content, are well known to be associated with greater risk of developing lung cancer (Houghton 2013). Although it remains unclear whether inflammation disturbs the incidence of driver oncogenic mutations, lipopolysaccharide (LPS), an endotoxin capable of promoting lung inflammation, significantly increased the risk of lung tumorigenesis in mice treated with carcinogens through *KRAS* gene activation by point mutations (Keohavong et al. 2011). Inflammation also has implications in the generation of lung metastasis from extrapulmonary neoplasms, since clinical studies suggested a correlation between smoking and a higher risk of lung metastasis in patients suffering from other types of cancer (e.g. oesophageal cancer and breast cancer) (Abrams et al. 2008; Murin and Inciardi 2003). The association with increased lung metastasis was also observed in an arthritic mice model of cancer, which exhibit lung inflammation characterized by neutrophil and mast cell infiltration associated with high levels of circulating pro-inflammatory cytokines (Das Roy et al. 2009). Moreover, TNF-α signalling through NF-κB in resident macrophages creates an inflammatory microenvironment which enhances Lewis lung carcinoma cells to metastasize (Stathopoulos et al. 2008). On the contrary, depletion of alveolar macrophages by intratracheal CLL (clodronate liposome) injection abrogated this enhanced metastasis (Stathopoulos et al. 2008).

It has been recently demonstrated that the expression of adipose triglyceride lipase (ATGL) which catalyses triacylglycerols hydrolysis is down-regulated in lung cancer (Vegliante et al. 2018). ATGL involvement in cancer cell metabolism is connected to the peroxisome proliferator-activated receptor-α (PPAR-α) signalling and to the pathways involved in inflammation, redox homeostasis and autophagy (Haemmerle et al. 2011). Indeed, PPAR-α and PPAR-γ are known to be implicated in the regulation of macrophage and endothelial cell inflammatory responses (Daynes and Jones 2002).

It is well recognised that mitochondria are at the centre of pro-inflammatory signalling and the pro-inflammatory milieu can modify mitochondrial physiology (West 2017). Damage-associated molecular patterns (DAMPs) are formed upon mitochondrial damage and contribute to inflammasome formation and caspase-1 activation (West 2017). Additionally, several metabolic inducers such as ATP and ROS trigger inflammasome (complex of proteins responsible for the activation of inflammation) (Mariathasan et al. 2004) activation. In fact, ATP induces the assembly of the inflammasome and the initiation of IL-1β generation (West 2017), a key mediator of inflammation, emphasizing the interaction between inflammation and metabolism in tumour tissue (Fig. 16.1).

Evidences show that RNA binding proteins (RBPs) are involved in metabolism, indicating a correlation between transcriptomic traits of metabolism and inflammation in cancer (Lujan et al. 2018). Lujan et al. showed that cold-inducible RNA binding protein (CIRP) binds to the serum TLR4-MD2 complex, relevant for anti-LPS response (Achek et al. 2017), and acts as a damage-associated molecular pattern (DAMP), crucial in inflammation (O'Reilly and van Laar 2018). Moreover, CIRP also activates the NF-κB pathway inducing higher levels of pro-inflammatory cytokines thereby acting as a tumour promoter in some cancer types, including lung (Lujan et al. 2018; Lee et al. 2016).

16.3 Metabolic Remodelling in Lung Cancer

Metabolic reprogramming is one of the emerging hallmarks of cancer (Hanahan and Weinberg 2011) and it is well acknowledged that cancer cells have a vast metabolic plasticity in order to support continuous cell growth and proliferation, meeting their energetic and biomass demands (Lopes-Coelho et al. 2017). The metabolic adaptation of tumour cells not only allows the development and establishment of a tumour in a certain microenvironment but also influences the response to therapy (Lopes-Coelho et al. 2018). Metabolic remodelling in the TME is not exclusive to cancer cells, since non-malignant cells share the same TME (Lopes-Coelho et al. 2018). Indeed, a tight metabolic synergy occurs between cancer cells and stromal cells in which the normal cells act as suppliers of energy sources and precursors for macromolecules synthesis (Lopes-Coelho et al. 2017; Martinez-Outschoorn et al. 2014).

Cancer cells become dependent on the activation of particular metabolic pathways during malignant transformation. Warburg verified that cancer cells preferred to fulfil glycolysis (Fig. 16.2) instead of the more energy efficient OXPHOS, occurring independently of oxygen levels (Warburg and Minami 1923). This glycolytic switch known as the Warburg effect (Fig. 16.3) was initially described as a compensatory mechanism for mitochondrial dysfunctions in tumours (Warburg 1956). Some studies associated this cancer cell specialization with mutations in metabolic genes, but afterwards it was noticed that not all cancer cells that fulfil aerobic glycolysis have a deficient mitochondrial metabolism. Although, in some cases, enzymes can have their activity modulated by phosphorylation (Zimmer et al. 2016) or the translation of mitochondrial metabolism related genes can be limited by oncogenes as *EGFR* (Dittmann et al. 2015). Glucose metabolism rewiring is more likely to be driven by the high demand of reducing equivalents and molecular precursors of proteins, nucleotides and lipids, which are the building blocks required to maintain cancer cells growth and proliferation (Pavlova and Thompson 2016). Although most tumours experience metabolic remodelling, multiple lines of evidence suggest that tumour metabolic signatures are context dependent, being influenced by numerous factors as oncogenic signalling, tissue of origin, TME and tumour grade (Kerr and Martins 2018).

Metabolic alterations in glucose, lipids, amino acids and nucleic acids metabolism were found in NSCLC cells in recent studies. Fresh surgical resections of NSCLC with mixed histology showed increased levels of lactate, demonstrating an upregulation in glycolysis relative to normal tissue (Fan et al. 2009). The same authors also verified an increase in glucose-derived TCA cycle intermediates in tumour samples, indicating that TCA cycle activity is enhanced in NSCLC comparatively to normal lung. The activity of PC (pyruvate carboxylase), an enzyme responsible for the conversion of pyruvate to oxaloacetate was elevated in NSCLC tumours (Sellers et al. 2015; Hensley et al. 2016). Silencing PC significantly reduced the proliferative and colony-forming capacity of NSCLC cell lineages and reduced tumour growth in murine xenograft models, suggesting a dependence on PC mediated anaplerosis (Sellers et al. 2015). Moreover, it was found that glycolysis and glucose oxidation via PDH (pyruvate dehydrogenase) and the TCA cycle were enhanced in NSCLC comparing to adjacent benign lung (Hensley et al. 2016). The authors demonstrated

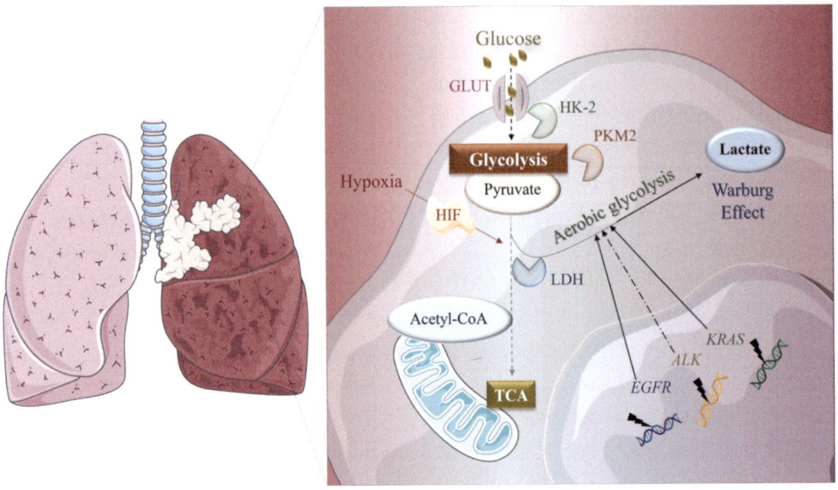

Fig. 16.3 Schematic representation of metabolic flux in lung cancer cells. In lung cancer cells, energy is obtained through the conversion of glucose into pyruvate – glycolysis. Pyruvate is decarboxylated into acetyl-CoA, in cytoplasm, to be further transported into mitochondria to enter the TCA cycle (Palsson-McDermott and O'Neill 2013). In hypoxia, a cell has the ability to convert pyruvate into lactate – Warburg effect (Biswas 2015; Pavlova and Thompson 2016; Palsson-McDermott and O'Neill 2013; Bhattacharya et al. 2016; De Alteriis et al. 2018). The GLUT family of membrane transport proteins mediates the import of glucose by a process of facilitative diffusion and it is known to be deregulated in cancer (Thorens and Mueckler 2010; Adekola et al. 2012). Hexokinase 2 (HK-2) is the first rate-limiting enzyme in the glycolytic pathway and it is known to be up-regulated in many cancers and to induce drug resistance (Bhattacharya et al. 2016; Min et al. 2013; Lis et al. 2016). Likewise, pyruvate kinase M2 (PKM2) plays a role in the regulation of glycolysis and it has been reported as promoting cell survival and preventing apoptosis (Kwon et al. 2012) as well as activating transcription factors (Gatenby and Gillies 2007; Jang et al. 2013; Vander Heiden et al. 2009). Approximately 15–30% of NSCLC patients have mutations in *EGFR*, which constitutively activates the EGFR protein tyrosine kinase domain (Min and Lee 2018; Zhang et al. 2016), leading to metabolic reprogramming in NSCLC, which includes enhanced aerobic glycolysis (Min and Lee 2018; Makinoshima et al. 2014). *ALK* rearrangement is detected in 3–7% of patients with NSCLC (Katayama et al. 2015; Hofman 2017). The impact of *ALK* rearrangements on metabolism in lung adenocarcinoma has not been well characterized, however, a recent study has observed the presence of upregulated glucose metabolism in highly metastatic phenotypes in this subset of lung cancer (Choi et al. 2013). Furthermore, mutations in *KRAS* affect ~30% of lung adenocarcinomas (Kerr and Martins 2018) and several studies demonstrate the involvement of mutant *KRAS* in the metabolic remodelling of different types of cancer (Kerr and Martins 2018; Kimmelman 2015; Kawada et al. 2017) with an upregulation of glucose uptake and aerobic glycolysis (Ying et al. 2012; Son et al. 2013; Onetti et al. 1997)

that tissue perfusion dictated preferential nutrient utilization in a specific region, as well as its metabolic profile suggesting that the metabolic heterogeneity of lung tumours is regulated by the TME. Additionally, the use of lactate as the main carbon source for the TCA cycle in tumours from NSCLC patients and tumour xenografts was demonstrated by Faubert et al. (Faubert et al. 2017). The expression of ATP citrate lyase (ACLY), a key enzyme in fatty acid synthesis involved in the synthesis of acetyl-CoA and oxaloacetate, was found to be upregulated in NSCLC and it was also associated with the disease poor prognosis (Migita et al. 2008). In addition, glycine decarboxylase (GLDC), couples decarboxylation of glycine to the biosynthesis of serine and it takes part in pyrimidine metabolism; is upregulated in NSCLC tumour-initiating cells (Zhang et al. 2012). xCT (SLC7A11), a cystine/glutamate antiporter, was shown to be overexpressed in the plasma membrane in NSCLC, correlating with patients' worse survival (Ji et al. 2018). The authors found that cancer cells expressing high levels of xCT, relied on glutamine dependency for OXPHOS. This glutamine consumer phenotype can also be related to cyst(e) ine dependency, as xCT concomitantly exports glutamate and imports cyst(e)ine. Glutamate is a direct product of glutamine degradation and maintaining the glutamine import sustains the import of cyst(e)ine, which has been related to increased therapy resistance in different cancer models (Nunes et al. 2018), mainly due to its role in glutathione synthesis (Colla et al. 2016; Mallappa et al. 2019; Nunes and Serpa 2018; Ciamporcero et al. 2018).

Despite similarities in metabolic reprogramming, the metabolic alterations in individual NSCLC cells or tumours are highly heterogeneous (Hensley et al. 2016; Min and Lee 2018; Chen et al. 2014b). Thus, understanding the influence of cellular or environmental factors, such as oncogene-induced metabolic switches, on cancer cell metabolism is crucial for the development of better therapeutic approaches targeting metabolic remodelling in cancer cells.

16.3.1 Metabolic Cues in Lung Cancer TME

Non-malignant components of the tumour stroma, such as immune cells and fibroblasts provide structural support as well as immune protection promoting invasion and metastasis (Bremnes et al. 2011). Stromal cells may affect tumour cell metabolism in different ways, including competition for nutrients, provision of alternative metabolic substrates or modulation of tumour cell signalling through cell to cell contacts (Pavlides et al. 2009; Chang et al. 2015). Indeed, lactate, amino acids and fatty acids act as signalling molecules that can be exchanged between tumour and stromal cells, regulating signal transduction, gene expression and neighbouring cells' characteristics (Lyssiotis and Kimmelman 2017). In comparison with normal fibroblasts, basal autophagy was enhanced in lung CAFs, because they share the TME with high glycolytic lung cancer cells. The autophagy resulting compounds are released to support surrounding cancer cells (Chaudhri et al. 2013). Furthermore, interactions with bone marrow-derived non-hematopoietic stem cells or skin fibroblasts rescued lung cancer cells, which had mitochondrial defects, leading to reactivation of their mitochondrial function, due to the transfer of mitochondria or mitochondrial DNA from stem/progenitor cells or fibroblasts to lung cancer cells (Spees et al. 2006). All in all, these findings suggest an important association between metabolic reprogramming and the TME interaction. However, improved understanding of details regarding mechanisms of action, the lung TME specific consequences of these interactions and their clinical impacts is essential and needs to be explored in further studies.

16.3.1.1 Role of Oncogenic Mutations (*EGFR, ALK, KRAS*) in Metabolic Reprogramming

Major research focus in lung cancer has been directed to cancer cell intrinsic properties, which has led to the discovery of important driver mutations in oncogenes and/or tumour suppressor genes. Mutations in *EGFR*, *KRAS* and *ALK* rearrangements are mainly found in lung adenocarcinoma, accounting for 30–40% of NSCLCs (Pikor et al. 2013). Thus, mutations in these oncogenes play a role in metabolic reprogramming of cancer cells to support their high energetic demands (Kerr and Martins 2018; Min and Lee 2018).

16.3.1.1.1 Role of *EGFR* Mutations in Metabolic Reprogramming

Approximately 15–30% of NSCLC patients have mutations in exon 19 or 21 of *EGFR*, which constitutively activate the EGFR protein tyrosine kinase domain (Min and Lee 2018; Zhang et al. 2016). The subsequent aberrant activation of signalling pathways promotes mitogenic, pro-survival and pro-invasive phenotypes in cancer cells (Zhang et al. 2010). Additionally, mutant EGFR mediates metabolic reprogramming in NSCLC as enhanced aerobic glycolysis and PPP (pentose phosphate pathway), altered pyrimidine biosynthesis and redox metabolism (Min and Lee 2018; Makinoshima et al. 2014). Combined treatment with erlotinib (EGFR inhibitor) and a glutaminase inhibitor (CB-839) leads to a metabolic crisis in *EGFR* mutant NSCLC cells, resulting in cell death and in rapid tumour regression in mouse NSCLC xenografts (Momcilovic et al. 2017), indicating the need of glutamine as a source for bioenergetics and biosynthesis in *EGFR*-mutated NSCLCs. Another study, shows direct role of EGFR in the stabilization by phosphorylation of stearoyl-CoA desaturase-1 (SCD1), increasing monounsaturated fatty acid synthesis and sustaining cell proliferation (Zhang et al. 2017b). Phosphorylated SCD1 levels were found to be an independent prognostic factor for poor survival in NSCLC (Zhang et al. 2017b). Taken together, these findings indicate that targeting alterations in glucose, glutamine or lipid metabolism could be an alternative therapeutic approach, reinforcing the treatment of *EGFR*-mutated lung adenocarcinomas.

16.3.1.1.2 Role of *ALK* Rearrangements in Metabolic Reprogramming

ALK rearrangement is detected in 3–7% of patients with NSCLC, being *EML4-ALK* the most prevalent *ALK* fusion (Katayama et al. 2015; Hofman 2017). Various ALK inhibitors such as crizotinib and ceritinib have been clinically used for the treatment of patients with lung adenocarcinoma with alterations in *ALK* (Katayama et al. 2015). The impact of *ALK* rearrangements on metabolism in lung adenocarcinoma has not been well characterized, however, a recent study has observed the presence of upregulated glucose metabolism in highly metastatic phenotypes in this subset of lung cancer (Choi et al. 2013).

16.3.1.1.3 Role of *KRAS* Mutations in Metabolic Reprogramming

Mutations in *KRAS* affect ~30% of lung adenocarcinomas but unlike the commonly altered EGFR or ALK proteins, mutant KRAS remains untargetable (Kerr and Martins 2018). *KRAS* is the most frequently mutated oncogene in lung adenocarcinoma and codifies a protein that belongs to the RAS family of GTPases (Hobbs et al. 2016). KRAS is activated through GTP binding, the GTP-bound RAS binds to downstream effectors and triggers activation of multiple signalling pathways such as the RAF-MEK-ERK pathway and the PI3K/Akt pathway, responsible for cell proliferation and survival and consequent tumour growth (Pylayeva-Gupta et al. 2011). Although *KRAS* mutations were directly connected to patient survival in other types of cancer such as colorectal cancer (Phipps et al. 2013), the prognostic relevance of *KRAS* mutations in NSCLC is unclear (Kerr and Martins 2018). Several studies demonstrate the involvement of mutant KRAS in the metabolic remodelling of different types of cancer (Kerr and Martins 2018; Kimmelman 2015; Kawada et al. 2017) with an upregulation of glucose uptake and aerobic glycolysis together with increased glutamine utilization (Ying et al. 2012; Son et al. 2013; Onetti et al. 1997). Proteomic profiles, related to metabolism of NSCLC cell lines carrying intrinsic mutant KRAS, were investigated and compared with those of normal bronchial epithelial cells (Martín-Bernabé et al. 2014). NSCLC cells expressed high levels of enzymes involved in glycolysis (GAPDH, PKM2, LDHA and LDHB; Fig. 16.2) and PPP (G6PD, TKT and 6PGD) compared with non-malignant cells, indicating alterations in glucose metabolism in *KRAS*-mutated NSCLC cells (Martín-Bernabé et al. 2014). In another study, NSCLC cells carrying *KRAS* mutations showed metabolic remodelling with alterations in redox buffering systems and glutamine dependency (Brunelli et al. 2014). The metabolic changes observed *in vitro* were reproduced in a tumour xenograft model bearing the same NSCLC cell line; and again, glutamine and cyst(e)ine metabolism goes together. Moreover, in a study using a mutant *KRAS* lung tumour mouse model, an upregulation of lactate production was observed both *in vivo* and *in vitro*, using those tumors-derived cells (Davidson et al. 2016), showing that this metabolic profile must be fundamental in lung cancer biology. However, lung tumours from these *in vivo* mouse models minimally use glutamine as a carbon source for TCA cycle entry, while *in vitro* there was a dependence on glutamine. In addition, oxidative glucose metabolic enzymes, including PC and PDH (pyruvate dehydrogenase) are necessary for tumour formation and growth in these mouse models (Davidson et al. 2016). Thus, the environmental context needs to be taken into consideration in the study of relevant metabolic alterations, especially in the case of glucose metabolism.

16.4 Targeting Metabolic Reprogramming as a Therapeutic Approach in Lung Cancer

Considering the importance of metabolic alterations in the development and progression of cancer, various agents targeting cancer metabolism have been developed and evaluated under preclinical and clinical studies. Notably, some metabolism-targeting agents including mTOR inhibitors (rapamycin, everolimus and temsirolius) and metformin (AMPK activator and mitochondrial complex I) are now approved for clinical use (Min and Lee 2018) in various types of cancer including glioblastomas, renal cell carcinoma and pancreatic neuroendocrine tumours. Targeting effectors of the signalling pathways downstream of proteins often mutated in lung cancer (eg PI3K), such as AMPK and mTOR can be a good approach to fight these tumours. Accordingly, metformin suppressed the proliferation and increased the radiosensitivity of lung cancer cells (Storozhuk et al. 2013). The inhibition of particular metabolic pathways using recombinant enzymes in order to reduce a specific metabolite have also been developed (Nagarajan et al. 2016; Ott et al. 2013; Karol et al. 2019; Yau et al. 2013) with an impressive clinical success, which is the case of asparaginase to treat certain haematological diseases in children (Karol et al. 2019; Agrawal et al. 2003). In about 50% of SCLC the expression of argininosuccinate synthetase (ASS), an enzyme required for the synthesis of arginine, is abrogated, being those tumours dependent on the uptake of arginine to survive (Kelly et al. 2012). Thus, as it happens with asparagine in some subsets of leukaemia, the systemic treatment with arginine deiminase (degrades arginine) to reduce the availability of arginine to supply SCLC cells is a suitable therapeutic strategy. Hence, recombinant arginine deiminase has been evaluated in phase I and II clinical trials for the treatment of lung cancer (Tran et al. 2012). Despite the various antitumoral approaches described above, most metabolism-targeting agents for lung cancer are still under pre-clinical evaluation (Nagarajan et al. 2016). Although much progress has been made in unravelling the metabolic networks in cancer cells, there is still much to be discovered about how different genetic drivers determine metabolic dependencies in the TME context and how these dependencies can be translated into viable therapeutic approaches.

16.5 Taking Advantage of Lung Cancer Metabolism to Improve and Specify Therapy

The aforementioned metabolic alterations contain numerous intervenients and metabolic pathways that can be used as therapeutic targets (Fig. 16.3), in order to improve and make more specific the treatment of lung cancer. In the next section we will sum up the metabolic alterations in cancer and explore the metabolic targeted approaches, in the cancer cell and in the TME. Whenever it is possible lung cancer will be focused.

16.5.1 Disrupting the Warburg Effect – A Metabolic Approach

Summing up, metabolism is the process whereby biochemicals are turned over to generate energy or are used in the synthesis of macromolecules. In a non-tumour resting cell and in normoxia conditions, energy demands are met as glucose that enters glycolysis is converted to pyruvate producing an energy yield of two molecules of pyruvate. Pyruvate is decarboxylated into acetyl-CoA, in cytoplasm, to be further transported into mitochondria to enter the TCA cycle. After the electron transport chain (ETC), the total energy gain under these conditions is roughly 36 ATP *per* each molecule of glucose (Palsson-

McDermott and O'Neill 2013). However, a markedly increased consumption of glucose by tumours in comparison to the non-proliferating normal tissues was first described more than 90 years ago by Otto Warburg – the Warburg effect (Otto Warburg et al. 1927) – indicating that in a situation of hypoxia or anoxia, a cell has the ability to divert pyruvate away from OXPHOS and converting it into lactate, allowing not only the generation of two molecules of ATP, but also to regenerate NAD^+, which is required as an electron acceptor for the glycolysis to proceed. This switch in the "metabolic preferences" can be explained by the change in the demands for biosynthetic precursors in malignant cells, given its high rate of proliferation. In order to meet these new requirements, tumour cells change their metabolic profile from a comparatively low rate of glycolysis followed by oxidation of glucose derived pyruvate by the TCA cycle, to a high rate of glycolysis followed by lactic acid production (Biswas 2015; Pavlova and Thompson 2016; Palsson-McDermott and O'Neill 2013; Bhattacharya et al. 2016; De Alteriis et al. 2018). However, it does not always mean that cancer cells abrogate OXPHOS, it is commonly supplied by non-glucose derived compounds (Vander Linden and Corbet 2019; Zhang et al. 2019; Mazat and Ransac 2019; Huang et al. 2019). Given that, the Warburg effect is accepted as a common feature of tumour cells and a possible target against cancer (Palsson-McDermott and O'Neill 2013; Bhattacharya et al. 2016; Song et al. 2016).

The GLUT (*SLC2A*) family of membrane transport proteins mediates the import of glucose by a process of facilitative diffusion and it is known to be deregulated in cancer (Thorens and Mueckler 2010; Adekola et al. 2012). It has been observed that cancer cells are more susceptible to glucose deprivation compared with normal cells and studies have demonstrated that inhibition of glucose transport results in apoptosis (Adekola et al. 2012) and can also decrease cancer cell proliferation *in vitro* and tumour growth *in vivo*, in lung cancer models (Liu et al. 2012) and other cancer models (Xu et al. 2014; Jiang et al. 2018); with or without synergistic effects through the combination with existing chemotherapeutic agents. Furthermore, GLUT-1 is responsible for basal glucose transport in all cell types, and it has been shown that its level of expression correlates with the degree of invasion and metastatic potential of tumours. Moreover, other studies have shown how profitable it can be to take advantage of the overexpression of GLUT-1 in cancer cells. Zhou et al. showed that delivery systems designed to be taken up by cancer cells via GLUT-1 protein-mediated endocytosis, inhibited the proliferation of drug-resistant lung cancer cells *in vitro* and *in vivo* models (Zhou et al. 2017).

Furthermore, the main enzymes supporting the Warburg effect are also recognised as promoters of resistance to therapy and are, in consequence, a possible target to overcome chemoresistance in lung cancer. HK-2 is the first rate-limiting enzyme in the glycolytic pathway, it is known to be up-regulated in many cancers and to induce drug resistance (Bhattacharya et al. 2016; Min et al. 2013; Lis et al. 2016). Moreover, Liu et al. showed that the inhibition of HK-2 by shRNA and/or metformin leads to cell apoptosis and decreases tumour growth in a cervical cancer cell model (Liu et al. 2017). Several, studies showed that the inhibition of HK-2 through its silencing or using inhibitors impairs tumour growth and induces cancer cell death *in vitro* and *in vivo* in lung (Wang et al. 2016; Li et al. 2017; Patra et al. 2013) and in other types of cancer (Liu et al. 2017; DeWaal et al. 2018).

Likewise, pyruvate kinase M2 (PKM2) plays a role in the regulation of glycolysis and it has been reported as promoting cell survival and preventing apoptosis by increasing Bcl-xL expression (Kwon et al. 2012) and activating transcription factors such as β-catenin, STAT3 and HIF1 (Gatenby and Gillies 2007; Jang et al. 2013; Vander Heiden et al. 2009). All of these events will contribute for the Warburg effect, tumour growth, angiogenesis, metastasis and evasion to apoptosis (He et al. 2017). Reports about the role of PKM2 expression in resistance to therapy are controversial. Zhu *et al* reported that PKM2 expression was downregulated in cisplatin-resistant cervical cancer cells, suggesting a rewiring of cancer cells energy metabolism.

This results also suggested that PKM2 enhanced sensitivity to cisplatin through interaction with the mTOR signalling pathway in cervical cancer (Zhu et al. 2016). Wang and colleagues concluded that PKM2 overexpression was associated with the resistance of bladder cancer cells to cisplatin, since cancer cells with lower PKM2 activity, by shRNA downregulation or shikonin exposure, became more sensitive to cisplatin (Wang et al. 2017b, 2018). In NSCLC, Yuan et al., suggested that PKM2 knockdown could serve as a chemosensitizer to docetaxel, leading to the inhibition of cell viability, cell cycle arrest at G2/M phase and apoptosis (Yuan et al. 2016). Despite theoretically PKM2 seeming to be a suit able therapeutic target against chemoresistance, contradictory reports have raised concerns on its validity as cancer drug target.

16.5.2 Directing Therapy Towards "Tumour Fellows" – A TME Approach

Tumour resistance to therapy is often driven by cancer stem cells (CSCs). CSCs are not tumour-initiating cells, though they have self-renewal capacity and make up a small proportion of the heterogenous tumour (Rycaj and Tang 2015). However, metastatic relapse after chemotherapy is suggested to be due to therapeutic resistance occurring specifically in CSCs, as their evasion from apoptosis allows the tumour to re-develop after therapy (Yeldag et al. 2018; Dzobo et al. 2016; Zhao 2016). CSCs are thought to remain quiescent most of the time, being protected against drugs toxicity. Several other mechanisms contribute for CSCs resistance to therapy, among them, the ability to actively transporting drugs out of the cell through ATP binding cassette (ABC) transporters on the plasma membrane of cells and on the membranes of exocytic cellular vesicles. CSCs from different cancer types show an increased expression of ABC transporters. Furthermore, drug transporters can co-operate with drug inactivation systems. For example, glutathione can bind to platinum-based drugs such as cisplatin, and this complex is a substrate for ABC transporters (Yeldag et al. 2018; Michael and Doherty 2005).

Furthermore, the EMT program is pointed as a pivotal regulator of CSC phenotype, underlying a putative mechanism chemoresistance (Shibue and Weinberg 2017).

Alterations in the DNA damage response of cancer cells can have both positive and negative effects on chemoresistance. The upregulation of some proteins involved in DNA repair can also promote resistance to chemotherapeutics, as any DNA damage caused by drugs that would otherwise promote apoptosis, is repaired. Conversely, downregulation of DNA damage response proteins can promote cell cycle progression, even in the presence of DNA errors, leading to genomic instability. The p53, the main cell cycle and DNA damage responsive protein, when mutated it promotes resistance to drugs as cisplatin, doxorubicin, gemcitabine, and tamoxifen (Hientz et al. 2017), and it is known to be overexpressed in CSCs (Yeldag et al. 2018).

The movement of drugs from the bloodstream throughout the TME is affected by hypoxia. The glycolytic shift that occurs in response to low oxygen, leads to the production of lactate, and therefore a low extracellular pH. This acidic environment can lead to the electrostatic charge of drugs, limiting their ability to cross the hydrophobic plasma membrane (Yeldag et al. 2018; Sriraman et al. 2014). Paclitaxel, for example, was shown to have weakened cytotoxic effects in a pH 6.5 (Vukovic and Tannock 1997). Additionally, the reduced levels of oxygen slow down, but do not fully arrest the proliferation of cancer cells. Since many chemotherapeutics target highly proliferating cells, drug efficacy is reduced in these conditions. Additionally, in its anti-apoptotic role, HIF-1α upregulates the expression of the anti-apoptotic protein survivin and downregulates the expression of the pro-apoptotic proteins Bcl-2 like protein 4 (BAX) and Bax-like BH3 protein (BID), as well as the activity of caspases (Yeldag et al. 2018). Pro-survival pathways are also induced through HIF-1α-induced modulation of the expression of VEGF (Yeldag et al. 2018). A tumour presents an increased and abnormal vasculature given the deregulated angiogenesis. Since drugs must diffuse from the blood vessels to cancer cells, this aberrant leaky vasculature is by itself, a mechanism of resistance to therapy. Mathematical modelling of the effect of blood vessel architecture on drug delivery to tumours has suggested that the excess of vessel connectivity decreases the ability of drugs to exit the vasculature (Yeldag et al. 2018).

Numerous studies have shown that CAFs promote drug resistance in several types of tumour, such as breast (Amornsupak et al. 2014), ovarian (Yan et al. 2016), pancreatic (Queiroz et al. 2014) and colorectal cancer (Gonçalves-Ribeiro et al. 2016). The molecular interaction between cancer cells and CAFs may be a key in the regulation of resistance to cancer cell-targeted chemotherapy. CAFs contribute to cancer progression by secreting CAF-specific proteins, cytokines, growth factors and ECM components. In lung cancer, the tumour-stroma cross talk was implicated in mediating resistance to EGFR-TKIs. For example, sharing fibroblast-derived hepatocyte growth factor (HGF) to cancer cells, induces gefitinib resistance in NSCLC with EGFR-activating mutations (Mittal et al. 2016a; Ishii 2017; Ying et al. 2015). Wang and colleagues reported that CAFs significantly enhanced cisplatin resistance in lung cancer cells through the activation of ANXA3/JNK signalling pathway (Wang et al. 2019a).

Moreover, studies have demonstrated that exosomes promote tumorigenesis and chemoresistance in a variety of cancers. Previously, oncogenesis promoting proteins have been demonstrated to be transferred between cancer cells through exosomes. Besides proteins, exosomes have been previously demonstrated to shuttle nucleic acids from a donor to recipient cells, which suggests that exosomes are important mediators in the exchange of genetic information and compounds within the TME (Valadi et al. 2007). Lobb et al., reported that mesenchymal-derived exosomes can transfer chemoresistant traits of donor cells to recipient cells, resulting in a CSC-like phenotype that may be related to the transfer of the EMT-associated transcription factor ZEB1. Concluding, CSC-like lung cells can modify and promote dedifferentiation of epithelial cells via exosome communication (Lobb et al. 2017), which can be a possible therapeutic target to overcome cancer therapy resistance.

16.6 Highlights

Lung cancer is the leading cause of cancer-related mortality in the world, mainly due to late diagnosis and lack of specific and effective therapy. As a highly mortal and heterogenous disease, lung cancer presents a complex evolving TME, which is determinant for cancer cells survival and dissemination, and several entities contribute for the optimal conditions that lead to tumour progression, such as CAFs, endothelial cells, immune cells and the tumour associated hypoxia and inflammation state, recognized as features of TME.

Malignant cells reprogram their metabolism to support tumour growth and proliferation, by developing new capabilities to benefit from metabolites of the TME, either by their uptake through metabolite transporters or by a crosstalk with the neighbouring non-malignant cells. Lung cancer progression further benefits from acknowledged metabolic (mal)adaptations, and metabolic reprogramming is one of the emerging hallmarks of cancer. The *EGFR* and *KRAS* mutations, together with *ALK* rearrangements are mainly found in lung adenocarcinoma and they are associated with metabolic changes.

Aerobic glycolysis, alterations in the PPP, glutamine dependency, accumulation of intermediates of glycolysis and upregulation of lipid and amino acids synthesis were reported in several studies using lung cancer as a model. All these alterations contribute for a makeover from a profile of comparatively low rate of glycolysis followed by the TCA cycle, to a profile of a high rate of glycolysis followed by lactic acid production – the Warburg effect.

Resistance to therapeutic agents is currently a major problem in the treatment of lung cancer. From the studies discussed in this chapter, it is evident that anticancer drug resistance to chemotherapy is often linked to metabolic alterations.

In this chapter, we aimed to review the possibility of targeting metabolic remodelling as a putative effective approach to overcome therapy resistance and specify treatment in lung cancer.

Acknowledgments The authors acknowledge iNOVA-4Health – UID/Multi/04462/2013, a program financially supported by Fundação para a Ciência e Tecnologia/Ministério da Educação e Ciência, through national funds and co-funded by FEDER under the PT2020 Partnership Agreement.

References

Abrams JA, Lee PC, Port JL et al (2008) Cigarette smoking and risk of lung metastasis from esophageal cancer. Cancer Epidemiol Biomark Prev 17:2707–2713. https://doi.org/10.1158/1055-9965.EPI-08-0232

Achek A, Kwon H, Lee B, Yoo TH (2017) TLR4/MD2 specific peptides stalled in vivo LPS-induced immune exacerbation. Biomaterials 126:49–60. https://doi.org/10.1016/j.biomaterials.2017.02.023

Adekola K, Rosen ST, Shanmugam M (2012) Glucose transporters in cancer metabolism. Curr Opin Oncol 24:650–654. https://doi.org/10.1097/CCO.0b013e328356da72

Agrawal NR, Bukowski RM, Rybicki LA et al (2003) A phase I-II trial of polyethylene glycol-conjugated L-asparaginase in patients with multiple myeloma. Cancer 98:94–99. https://doi.org/10.1002/cncr.11480

Altorki NK, Markowitz GJ, Gao D et al (2019) The lung microenvironment: an important regulator of tumour growth and metastasis. Nat Rev Cancer 19:9–31

Amornsupak K, Insawang T, Thuwajit P et al (2014) Cancer-associated fibroblasts induce high mobility group box 1 and contribute to resistance to doxorubicin in breast cancer cells. BMC Cancer 14:955. https://doi.org/10.1186/1471-2407-14-955

Bak SP, Alonso A, Turk MJ, Berwin B (2008) Murine ovarian cancer vascular leukocytes require arginase-1 activity for T cell suppression. Mol Immunol 46:258–268. https://doi.org/10.1016/j.molimm.2008.08.266

Banat G-A, Tretyn A, Pullamsetti SS et al (2015) Immune and inflammatory cell composition of human lung cancer stroma. PLoS One 10:e0139073. https://doi.org/10.1371/journal.pone.0139073

Bhattacharya B, Mohd Omar MF, Soong R (2016) The Warburg effect and drug resistance. Br J Pharmacol 173:970–979. https://doi.org/10.1111/bph.13422

Bindea G, Mlecnik B, Tosolini M et al (2013) Spatiotemporal dynamics of intratumoral immune cells reveal the immune landscape in human cancer. Immunity 39:782–795. https://doi.org/10.1016/J.IMMUNI.2013.10.003

Biswas SK (2015) Metabolic reprogramming of immune cells in cancer progression. Immunity 43:435–449. https://doi.org/10.1016/j.immuni.2015.09.001

Bremnes RM, Dønnem T, Al-Saad S et al (2011) The role of tumor stroma in cancer progression and prognosis: emphasis on carcinoma-associated fibroblasts and non-small cell lung cancer. J Thorac Oncol 6:209–217. https://doi.org/10.1097/JTO.0b013e3181f8a1bd

Brunelli L, Caiola E, Marabese M et al (2014) Capturing the metabolomic diversity of KRAS mutants in non-small-cell lung cancer cells. Oncotarget 5:4722–4731. https://doi.org/10.18632/oncotarget.1958

Cammann C, Rath A, Reichl U et al (2016) Early changes in the metabolic profile of activated CD8 + T cells. BMC Cell Biol 17(1):28. https://doi.org/10.1186/s12860-016-0104-x

Candido J, Hagemann T (2013) Cancer-related inflammation. J Clin Immunol 33:79–84. https://doi.org/10.1007/s10875-012-9847-0

Chang CH, Qiu J, O'Sullivan D et al (2015) Metabolic competition in the tumor microenvironment is a driver of cancer progression. Cell 162:1229–1241. https://doi.org/10.1016/j.cell.2015.08.016

Chapman NM, Shrestha S, Chi H (2017) Metabolism in immune cell differentiation and function. Adv Exp Med Biol 1011:1–85

Chaudhri VK, Salzler GG, Dick SA et al (2013) Metabolic alterations in lung cancer–associated fibroblasts correlated with increased glycolytic metabolism of the tumor. Mol Cancer Res 11:579–592. https://doi.org/10.1158/1541-7786.mcr-12-0437-t

Chen Z, Fillmore CM, Hammerman PS et al (2014a) Non-small-cell lung cancers: a heterogeneous set of diseases. Nat Rev Cancer 14:535–546

Chen P-H, Cai L, Kim HS et al (2014b) Metabolic diversity in human non-small cell lung cancer. Cancer Metab 2. https://doi.org/10.1186/2049-3002-2-s1-p13

Chen F, Zhuang X, Lin L et al (2015) New horizons in tumor microenvironment biology: challenges and opportunities. BMC Med 13:45

Choi H, Paeng JC, Kim DW et al (2013) Metabolic and metastatic characteristics of ALK-rearranged lung adenocarcinoma on FDG PET/CT. Lung Cancer 79:242–247. https://doi.org/10.1016/j.lungcan.2012.11.021

Ciamporcero E, Daga M, Pizzimenti S et al (2018) Crosstalk between Nrf2 and YAP contributes to maintaining the antioxidant potential and chemoresistance in bladder cancer. Free Radic Biol Med 115:447–457. https://doi.org/10.1016/j.freeradbiomed.2017.12.005

Colla R, Izzotti A, De Ciucis C et al (2016) Glutathione-mediated antioxidant response and aerobic metabolism: two crucial factors involved in determining the multi-drug resistance of high-risk neuroblastoma. Oncotarget 7:70715–70737

Cruz-Bermúdez A, Laza-Briviesca R, Vicente-Blanco RJ et al (2019) Cancer-associated fibroblasts modify lung cancer metabolism involving ROS and TGF-β signaling. Free Radic Biol Med 130:163–173. https://doi.org/10.1016/j.freeradbiomed.2018.10.450

Dagher Z, Ruderman N, Tornheim K, Ido Y (2001) Acute regulation of fatty acid oxidation and AMP-activated protein kinase in human umbilical vein endothelial cells. Circ Res 88:1276–1282. https://doi.org/10.1161/hh1201.092998

Dang EV, Barbi J, Yang H-Y et al (2011) Control of TH17/Treg balance by hypoxia-inducible factor 1. Cell 146:772–784. https://doi.org/10.1016/j.cell.2011.07.033

Das Roy L, Pathangey LB, Tinder TL et al (2009) Breast cancer-associated metastasis is significantly increased in a model of autoimmune arthritis. Breast Cancer Res 11:R56. https://doi.org/10.1186/bcr2345

Davidson SM, Papagiannakopoulos T, Olenchock BA et al (2016) Environment impacts the metabolic dependencies of ras-driven non-small cell lung cancer. Cell Metab 23:517–528. https://doi.org/10.1016/j.cmet.2016.01.007

Daynes RA, Jones DC (2002) Emerging roles of PPARs in inflammation and immunity. Nat Rev Immunol 2:748–759

De Alteriis E, Cartenì F, Parascandola P et al (2018) Revisiting the Crabtree/Warburg effect in a dynamic perspective: a fitness advantage against sugar-induced cell death. Cell Cycle 17(6):688–701. https://doi.org/10.1080/15384101.2018.1442622

De Bock K, Georgiadou M, Carmeliet P (2013) Role of endothelial cell metabolism in vessel sprouting. Cell Metab 18:634–647. https://doi.org/10.1016/j.cmet.2013.08.001

DeWaal D, Nogueira V, Terry AR et al (2018) Hexokinase-2 depletion inhibits glycolysis and induces oxidative phosphorylation in hepatocellular carcinoma and sensitizes to metformin. Nat Commun 9:446. https://doi.org/10.1038/s41467-017-02733-4

Dittmann K, Mayer C, Paasch A et al (2015) Nuclear EGFR renders cells radio-resistant by binding mRNA species and triggering a metabolic switch to increase lactate production. Radiother Oncol 116:431–437

Dranka BP, Hill BG, Darley-Usmar VM (2010) Mitochondrial reserve capacity in endothelial cells: the impact of nitric oxide and reactive oxygen species. Free Radic Biol Med 48:905–914. https://doi.org/10.1016/j.freeradbiomed.2010.01.015

Dugnani E, Pasquale V, Bordignon C et al (2017) Integrating T cell metabolism in cancer immunotherapy. Cancer Lett 411:12–18. https://doi.org/10.1016/j.canlet.2017.09.039

Dundar E, Oner U, Peker BC et al (2008) The significance and relationship between mast cells and tumour angiogenesis in non-small cell lung carcinoma. J Int Med Res 36:88–95. https://doi.org/10.1177/147323000803600112

Dzobo K, Senthebane DA, Rowe A et al (2016) Cancer stem cell hypothesis for therapeutic innovation in clinical oncology? Taking the root out, not chopping the leaf. Omi A J Integr Biol 20:681–691. https://doi.org/10.1089/omi.2016.0152

Ebos JML, Kerbel RS (2011) Antiangiogenic therapy: impact on invasion, disease progression, and metastasis. Nat Rev Clin Oncol 8(4):210–221

Fan TWM, Lane AN, Higashi RM et al (2009) Altered regulation of metabolic pathways in human lung cancer discerned by 13C stable isotope-resolved metabolomics (SIRM). Mol Cancer 8:41. https://doi.org/10.1186/1476-4598-8-41

Faubert B, Li KY, Cai L et al (2017) Lactate metabolism in human lung tumors. Cell 171:358–371.e9. https://doi.org/10.1016/j.cell.2017.09.019

Franses JW, Baker AB, Chitalia VC, Edelman ER (2011) Stromal endothelial cells directly influence cancer progression. Sci Transl Med 3:66ra5. https://doi.org/10.1126/scitranslmed.3001542

Frauwirth KA, Riley JL, Harris MH et al (2002) The CD28 signaling pathway regulates glucose metabolism. Immunity 16:769–777. https://doi.org/10.1016/S1074-7613(02)00323-0

Gatenby RA, Gillies RJ (2007) Glycolysis in cancer: a potential target for therapy. Int J Biochem Cell Biol 39:1358–1366. https://doi.org/10.1016/j.biocel.2007.03.021

Giatromanolaki A, Liousia M, Arelaki S et al (2017) Differential effect of hypoxia and acidity on lung cancer cell and fibroblast metabolism. Biochem Cell Biol 95:428–436. https://doi.org/10.1139/bcb-2016-0197

Gnanaprakasam JNR, Sherman JW, Wang R (2017) MYC and HIF in shaping immune response and immune metabolism. Cytokine Growth Factor Rev 35:63–70. https://doi.org/10.1016/j.cytogfr.2017.03.004

Gonçalves-Ribeiro S, Díaz-Maroto NG, Berdiel-Acer M et al (2016) Carcinoma-associated fibroblasts affect sensitivity to oxaliplatin and 5FU in colorectal cancer cells. Oncotarget 7:59766–59780. https://doi.org/10.18632/oncotarget.11121

Graves EE, Maity A, Le QT (2010) The tumor microenvironment in non-small-cell lung cancer. Semin Radiat Oncol 20:156–163

Haemmerle G, Moustafa T, Woelkart G et al (2011) ATGL-mediated fat catabolism regulates cardiac mitochondrial function via PPAR-alpha and PGC-1. Nat Med 17:1076–1085. https://doi.org/10.1038/nm.2439

Hanahan D, Coussens LM (2012) Accessories to the crime: functions of cells recruited to the tumor microenvironment. Cancer Cell 21:309–322

Hanahan D, Weinberg RA (2011) Leading edge review hallmarks of Cancer: the next generation. Cell 144:646–674. https://doi.org/10.1016/j.cell.2011.02.013

Harris AL (2002) Hypoxia — a key regulatory factor in tumour growth. Nat Rev Cancer 2:38–47. https://doi.org/10.1038/nrc704

He X, Du S, Lei T et al (2017) PKM2 in carcinogenesis and oncotherapy. Oncotarget 8:110656–110670. https://doi.org/10.18632/oncotarget.22529

Hensley CT, Faubert B, Yuan Q et al (2016) Metabolic heterogeneity in human lung tumors. Cell 164:681–694. https://doi.org/10.1016/j.cell.2015.12.034

Herbst RS, Onn A, Sandler A (2005) Angiogenesis and lung cancer: prognostic and therapeutic implications.

J Clin Oncol 23:3243–3256. https://doi.org/10.1200/JCO.2005.18.853

Hientz K, Mohr A, Bhakta-Guha D, Efferth T (2017) The role of p53 in cancer drug resistance and targeted chemotherapy. Oncotarget 8:8921–8946. https://doi.org/10.18632/oncotarget.13475

Hirsch FR, Scagliotti GV, Mulshine JL et al (2017) Lung cancer: current therapies and new targeted treatments. Lancet 389:299–311. https://doi.org/10.1016/S0140-6736(16)30958-8

Hobbs GA, Der CJ, Rossman KL (2016) RAS isoforms and mutations in cancer at a glance. J Cell Sci 129:1287–1292. https://doi.org/10.1242/jcs.182873

Hofman P (2017) ALK in non-small cell lung cancer (NSCLC) pathobiology, epidemiology, detection from tumor tissue and algorithm diagnosis in a daily practice. Cancers (Basel) 9:E107

Houghton AM (2013) Mechanistic links between COPD and lung cancer. Nat Rev Cancer 13:233–245. https://doi.org/10.1038/nrc3477

Huang Q, Chen Z, Cheng P et al (2019) LYRM2 directly regulates complex I activity to support tumor growth in colorectal cancer by oxidative phosphorylation. Cancer Lett 455:36–47. https://doi.org/10.1016/j.canlet.2019.04.021

I H, Cho J-Y (2015) Lung. Cancer Biomark:107–170

Inamura K (2017) Lung cancer: understanding its molecular pathology and the 2015 WHO classification. Front Oncol 7. https://doi.org/10.3389/fonc.2017.00193

Ishii G (2017) Crosstalk between cancer associated fibroblasts and cancer cells in the tumor microenvironment after radiotherapy. EBioMedicine 17:7–8. https://doi.org/10.1016/j.ebiom.2017.03.004

Jacobs SR, Herman CE, Maciver NJ et al (2008) Glucose uptake is limiting in T cell activation and requires CD28-mediated Akt-dependent and independent pathways. J Immunol 180:4476–4486. https://doi.org/10.4049/jimmunol.180.7.4476

Jang M, Kim SS, Lee J (2013) Cancer cell metabolism: implications for therapeutic targets. Exp Mol Med 45:e45. https://doi.org/10.1038/emm.2013.85

Ji X, Qian J, Rahman SMJ et al (2018) xCT (SLC7A11)-mediated metabolic reprogramming promotes non-small cell lung cancer progression. Oncogene 37:5007–5019. https://doi.org/10.1038/s41388-018-0307-z

Jiang T, Zhou M-L, Fan J (2018) Inhibition of GLUT-1 expression and the PI3K/Akt pathway to enhance the chemosensitivity of laryngeal carcinoma cells in vitro. Onco Targets Ther 11:7865–7872. https://doi.org/10.2147/OTT.S176818

Justus CR, Sanderlin EJ, Yang LV (2015) Molecular connections between cancer cell metabolism and the tumor microenvironment. Int J Mol Sci 16:11055–11086. https://doi.org/10.3390/ijms160511055

Kargl J, Busch SE, Yang GHY et al (2017) Neutrophils dominate the immune cell composition in non-small cell lung cancer. Nat Commun 9:E468–E469. https://doi.org/10.1038/ncomms14381

Karol SE, Janke LJ, Panetta JC et al (2019) Asparaginase combined with discontinuous dexamethasone improves antileukemic efficacy without increasing osteonecrosis in preclinical models. PLoS One 14:e0216328. https://doi.org/10.1371/journal.pone.0216328

Katayama R, Lovly CM, Shaw AT (2015) Therapeutic targeting of anaplastic lymphoma kinase in lung cancer: a paradigm for precision cancer medicine. Clin Cancer Res 21:2227–2235. https://doi.org/10.1158/1078-0432.CCR-14-2791

Kawada K, Toda K, Sakai Y (2017) Targeting metabolic reprogramming in KRAS-driven cancers. Int J Clin Oncol 22:651–659. https://doi.org/10.1007/s10147-017-1156-4

Ke Q, Costa M (2006) Hypoxia-inducible factor-1 (HIF-1). Mol Pharmacol 70:1469–1480. https://doi.org/10.1124/mol.106.027029

Kelly MP, Jungbluth AA, Wu BW et al (2012) Arginine deiminase PEG20 inhibits growth of small cell lung cancers lacking expression of argininosuccinate synthetase. Br J Cancer 106:324–332. https://doi.org/10.1038/bjc.2011.524

Keohavong P, Kahkonen B, Kinchington E et al (2011) K-ras mutations in lung tumors from NNK-treated mice with lipopolysaccharide-elicited lung inflammation. Anticancer Res 31:2877–2882

Kerr EM, Martins CP (2018) Metabolic rewiring in mutant Kras lung cancer. FEBS J 285:28–41

Kim SH, Song Y, Seo HR (2019) GSK-3β regulates the endothelial-to-mesenchymal transition via reciprocal crosstalk between NSCLC cells and HUVECs in multicellular tumor spheroid models. J Exp Clin Cancer Res 38:46. https://doi.org/10.1186/s13046-019-1050-1

Kimmelman AC (2015) Metabolic dependencies in RAS-driven cancers. Clin Cancer Res 21:1828–1834. https://doi.org/10.1158/1078-0432.CCR-14-2425

Koukourakis MI, Kalamida D, Mitrakas AG et al (2017) Metabolic cooperation between co-cultured lung cancer cells and lung fibroblasts. Lab Investig 97:1321–1331. https://doi.org/10.1038/labinvest.2017.79

Kwon O-H, Kang T-W, Kim J-H et al (2012) Pyruvate kinase M2 promotes the growth of gastric cancer cells via regulation of Bcl-xL expression at transcriptional level. Biochem Biophys Res Commun 423:38–44. https://doi.org/10.1016/J.BBRC.2012.05.063

Lee JH, Bhang DH, Beede A et al (2014) Lung stem cell differentiation in mice directed by endothelial cells via a BMP4-NFATc1-thrombospondin-1 axis. Cell 156:440–455. https://doi.org/10.1016/j.cell.2013.12.039

Lee HN, Ahn SM, Jang HH (2016) Cold-inducible RNA-binding protein promotes epithelial-mesenchymal transition by activating ERK and p38 pathways. Biochem Biophys Res Commun 477:1038–1044. https://doi.org/10.1016/j.bbrc.2016.07.028

Lemjabbar-Alaoui H, Hassan OU, Yang Y-W, Buchanan P (2015) Lung cancer: biology and treatment options. Biochim Biophys Acta – Rev Cancer 1856:189–210. https://doi.org/10.1016/j.bbcan.2015.08.002

Li W, Gao F, Ma X et al (2017) Deguelin inhibits non-small cell lung cancer via down-regulating hexokinases II-mediated glycolysis. Oncotarget 8:32586–32599. https://doi.org/10.18632/oncotarget.15937

Lis P, Dyląg M, Niedźwiecka K et al (2016) The HK2 dependent "Warburg Effect" and mitochondrial oxidative phosphorylation in cancer: targets for effective therapy with 3-bromopyruvate. Molecules 21:E1730. https://doi.org/10.3390/molecules21121730

Liu Y, Cao Y, Zhang W et al (2012) A small-molecule inhibitor of glucose transporter 1 downregulates glycolysis, induces cell-cycle arrest, and inhibits cancer cell growth in vitro and in vivo. Mol Cancer Ther 11:1672–1682. https://doi.org/10.1158/1535-7163.MCT-12-0131

Liu Y, Murray-Stewart T, Casero RA et al (2017) Targeting hexokinase 2 inhibition promotes radiosensitization in HPV16 E7-induced cervical cancer and suppresses tumor growth. Int J Oncol 50:2011–2023. https://doi.org/10.3892/ijo.2017.3979

Lobb RJ, van Amerongen R, Wiegmans A et al (2017) Exosomes derived from mesenchymal non-small cell lung cancer cells promote chemoresistance. Int J Cancer 141:614–620. https://doi.org/10.1002/ijc.30752

Lopes-Coelho F, Gouveia-Fernandes S, Nunes SC, Serpa J (2017) Metabolic dynamics in breast cancer: cooperation between cancer and stromal breast cancer cells. J Clin Breast Cancer Res 1:1–7

Lopes-Coelho F, Gouveia-Fernandes S, Serpa J (2018) Metabolic cooperation between cancer and non-cancerous stromal cells is pivotal in cancer progression. Tumor Biol 40:1–15. https://doi.org/10.1177/1010428318756203

Lujan DA, Ochoa JL, Hartley RS (2018) Cold-inducible RNA binding protein in cancer and inflammation. Wiley Interdiscip Rev RNA 9:e1462. https://doi.org/10.1002/wrna.1462

Lyssiotis CA, Kimmelman AC (2017) Metabolic interactions in the tumor microenvironment. Trends Cell Biol 27:863–875. https://doi.org/10.1016/j.tcb.2017.06.003

Makinoshima H, Takita M, Matsumoto S et al (2014) Epidermal growth factor receptor (EGFR) signaling regulates global metabolic pathways in EGFR-mutated lung adenocarcinoma. J Biol Chem 289:20813–20823. https://doi.org/10.1074/jbc.M114.575464

Mallappa S, Neeli PK, Karnewar S, Kotamraju S (2019) Doxorubicin induces prostate cancer drug resistance by upregulation of ABCG4 through GSH depletion and CREB activation: relevance of statins in chemosensitization. Mol Carcinog 58:1118–1133. https://doi.org/10.1002/mc.22996

Mariathasan S, Hewton K, Monack DM et al (2004) Differential activation of the inflammasome by caspase-1 adaptors ASC and Ipaf. Nature 430:213–218. https://doi.org/10.1038/nature02664

Martín-Bernabé A, Cortés R, Lehmann SG et al (2014) Quantitative proteomic approach to understand metabolic adaptation in non-small cell lung cancer. J Proteome Res 13:4695–4704. https://doi.org/10.1021/pr500327v

Martinez-Outschoorn UE, Lisanti MP, Sotgia F (2014) Catabolic cancer-associated fibroblasts (CAFs) transfer energy and biomass to anabolic cancer cells, Fueling Tumor Growth. Semin Cancer Biol 25:1–13. https://doi.org/10.1016/j.semcancer.2014.01.005

Mazat J-P, Ransac S (2019) The fate of glutamine in human metabolism. The interplay with glucose in proliferating cells. Meta 9:81. https://doi.org/10.3390/metabo9050081

Melosky B, Juergens R, Hirsh V et al (2019) Amplifying outcomes: checkpoint inhibitor combinations in first-line non-small cell lung cancer. Oncologist theoncologist.2019-0027. https://doi.org/10.1634/theoncologist.2019-0027

Michael M, Doherty MM (2005) Tumoral drug metabolism: overview and its implications for cancer therapy. J Clin Oncol 23:205–229. https://doi.org/10.1200/JCO.2005.02.120

Migita T, Narita T, Nomura K et al (2008) ATP citrate lyase: activation and therapeutic implications in non-small cell lung cancer. Cancer Res 68:8547–8554. https://doi.org/10.1158/0008-5472.CAN-08-1235

Min HY, Lee HY (2018) Oncogene-driven metabolic alterations in cancer. Biomol Ther 26:45–56. https://doi.org/10.4062/biomolther.2017.211

Min JW, Il KK, Kim H-A et al (2013) INPP4B-mediated tumor resistance is associated with modulation of glucose metabolism via hexokinase 2 regulation in laryngeal cancer cells. Biochem Biophys Res Commun 440:137–142. https://doi.org/10.1016/j.bbrc.2013.09.041

Mittal V, El Rayes T, Narula N et al (2016a) The microenvironment of lung cancer and therapeutic implications. Adv Exp Med Biol 890:75–110

Mittal V, El Rayes T, Navneet N (2016b) The microenvironment of lung cancer and therapeutic implications. Adv Exp Med Biol 890:75–110

Momcilovic M, Bailey ST, Lee JT et al (2017) Targeted inhibition of EGFR and glutaminase induces metabolic crisis in EGFR mutant lung cancer. Cell Rep 18:601–610. https://doi.org/10.1016/j.celrep.2016.12.061

Morales-Nebreda L, Misharin AV, Perlman H, Scott Budinger GR (2015) The heterogeneity of lung macrophages in the susceptibility to disease. Eur Respir Rev 24:505–509. https://doi.org/10.1183/16000617.0031-2015

Murin S, Inciardi J (2003) Cigarette smoking and the risk of pulmonary metastasis from breast cancer. Chest 119:1635–1640. https://doi.org/10.1378/chest.119.6.1635

Nagarajan A, Malvi P, Wajapeyee N (2016) Oncogene-directed alterations in cancer cell metabolism. Trends Cancer 2:365–377. https://doi.org/10.1016/j.trecan.2016.06.002

Nunes SC, Serpa J (2018) Glutathione in ovarian cancer: a double-edged sword. Int J Mol Sci 19:E1882

Nunes SC, Ramos C, Lopes-Coelho F et al (2018) Cysteine allows ovarian cancer cells to adapt to hypoxia and to escape from carboplatin cytotoxicity. Sci Rep 8:9513. https://doi.org/10.1038/s41598-018-27753-y

O'Reilly S, van Laar JM (2018) Targeting the TLR4-MD2 axis in systemic sclerosis. Nat Rev Rheumatol 14:564–566. https://doi.org/10.1038/s41584-018-0077-6

Onetti R, Baulida J, Bassols A (1997) Increased glucose transport in ras-transformed fibroblasts: a possible role for N-glycosylation of GLUT1. FEBS Lett 407:267–270. https://doi.org/10.1016/S0014-5793(97)00340-2

Ott PA, Carvajal RD, Pandit-Taskar N et al (2013) Phase I/II study of pegylated arginine deiminase (ADI-PEG 20) in patients with advanced melanoma. Investig New Drugs 31:425–434. https://doi.org/10.1007/s10637-012-9862-2

Otto Warburg B, Wind F, Negelein N (1927) The metabolism of tumors in the body. J Gen Physiol 8(6):519–530. https://doi.org/10.1085/jgp.8.6.519

Palsson-McDermott EM, O'Neill LAJ (2013) The Warburg effect then and now: from cancer to inflammatory diseases. BioEssays 35:965–973. https://doi.org/10.1002/bies.201300084

Papandreou I, Cairns RA, Fontana L et al (2006) HIF-1 mediates adaptation to hypoxia by actively downregulating mitochondrial oxygen consumption. Cell Metab 3:187–197. https://doi.org/10.1016/j.cmet.2006.01.012

Patra KC, Wang Q, Bhaskar PT et al (2013) Hexokinase 2 is required for tumor initiation and maintenance and its systemic deletion is therapeutic in mouse models of cancer. Cancer Cell 24:213–228. https://doi.org/10.1016/j.ccr.2013.06.014

Pavlides S, Whitaker-Menezes D, Castello-Cros R et al (2009) The reverse Warburg effect: aerobic glycolysis in cancer associated fibroblasts and the tumor stroma. Cell Cycle 8:3984–4001. https://doi.org/10.4161/cc.8.23.10238

Pavlova NN, Thompson CB (2016) The emerging hallmarks of cancer metabolism. Cell Metab 23:27–47. https://doi.org/10.1016/j.cmet.2015.12.006

Phipps AI, Buchanan DD, Makar KW et al (2013) KRAS-mutation status in relation to colorectal cancer survival: the joint impact of correlated tumour markers. Br J Cancer 108:1757–1764. https://doi.org/10.1038/bjc.2013.118

Pikor LA, Ramnarine VR, Lam S, Lam WL (2013) Genetic alterations defining NSCLC subtypes and their therapeutic implications. Lung Cancer 82:179–189

Pober JS, Sessa WC (2007) Evolving functions of endothelial cells in inflammation. Nat Rev Immunol 7:803–815. https://doi.org/10.1038/nri2171

Pylayeva-Gupta Y, Grabocka E, Bar-Sagi D (2011) RAS oncogenes: weaving a tumorigenic web. Nat Rev Cancer 19:1423–1437. https://doi.org/10.1038/nrc3106

Qin A, Coffey DG, Warren EH, Ramnath N (2016) Mechanisms of immune evasion and current status of checkpoint inhibitors in non-small cell lung cancer. Cancer Med 5:2567–2578. https://doi.org/10.1002/cam4.819

Quail DF, Joyce JA (2013) Microenvironmental regulation of tumor progression and metastasis. Nat Med 19:1423–1437. https://doi.org/10.1038/nm.3394

Queiroz KCS, Shi K, Duitman J et al (2014) Protease-activated receptor-1 drives pancreatic cancer progression and chemoresistance. Int J Cancer 135:2294–2304. https://doi.org/10.1002/ijc.28726

Rabold K, Netea MG, Adema GJ, Netea-Maier RT (2017) Cellular metabolism of tumor-associated macrophages – functional impact and consequences. FEBS Lett 591:3022–3041. https://doi.org/10.1002/1873-3468.12771

Rodriguez PC, Zea AH, DeSalvo J et al (2003) L-arginine consumption by macrophages modulates the expression of CD3 zeta chain in T lymphocytes. J Immunol 171:1232–1239. https://doi.org/10.4049/jimmunol.171.3.1232

Rodriguez PC, Quiceno DG, Ochoa AC (2007) L-arginine availability regulates T-lymphocyte cell-cycle progression. Blood 109:1568–1573. https://doi.org/10.1182/blood-2006-06-031856

Rycaj K, Tang DG (2015) Cell-of-origin of cancer versus cancer stem cells: assays and interpretations. Cancer Res 75:4003–4011. https://doi.org/10.1158/0008-5472.CAN-15-0798

Salem A, Asselin M-C, Reymen B et al (2018) Targeting hypoxia to improve non–small cell lung cancer outcome. JNCI J Natl Cancer Inst 110:14–30. https://doi.org/10.1093/jnci/djx160

Schoors S, Bruning U, Missiaen R et al (2015) Fatty acid carbon is essential for dNTP synthesis in endothelial cells. Nature 520:192–197. https://doi.org/10.1038/nature14362

Sellers K, Fox MP, Ii MB et al (2015) Pyruvate carboxylase is critical for non-small-cell lung cancer proliferation. J Clin Invest 125:687–698. https://doi.org/10.1172/JCI72873

Sen DB, Nolan DJ, Guo P et al (2011) Endothelial-derived angiocrine signals induce and sustain regenerative lung alveolarization. Cell 147:539–553. https://doi.org/10.1016/j.cell.2011.10.003

Serpa J, Dias S (2011) Metabolic cues from the microenvironment act as a major selective factor for cancer progression and metastases formation. Cell Cycle 10:180–181. https://doi.org/10.4161/cc.10.2.14476

Shi LZ, Wang R, Huang G et al (2011) HIF1alpha-dependent glycolytic pathway orchestrates a metabolic checkpoint for the differentiation of TH17 and Treg cells. J Exp Med 208:1367–1376. https://doi.org/10.1084/jem.20110278

Shibue T, Weinberg RA (2017) EMT, CSCs, and drug resistance: the mechanistic link and clinical implications. Nat Rev Clin Oncol 14:611–629. https://doi.org/10.1038/nrclinonc.2017.44

Shigematsu H, Lin L, Takahashi T et al (2005) Clinical and biological features associated with epidermal growth factor receptor gene mutations in lung cancers. JNCI J Natl Cancer Inst 97:339–346. https://doi.org/10.1093/jnci/dji055

Son J, Lyssiotis CA, Ying H et al (2013) Glutamine supports pancreatic cancer growth through a KRAS-regulated metabolic pathway. Nature 496:101–105. https://doi.org/10.1038/nature12040

Song H, Yao E, Lin C et al (2012) Functional characterization of pulmonary neuroendocrine cells in lung development, injury, and tumorigenesis. Proc Natl Acad Sci 109:17531–17536. https://doi.org/10.1073/pnas.1207238109

Song H, Yao E, Lin C et al (2014) Repair and regeneration of the respiratory system: complexity, plasticity, and mechanisms of lung stem cell function. Cell Stem Cell 15:123–138. https://doi.org/10.1016/j.stem.2014.07.012

Song K, Li M, Xu X et al (2016) Resistance to chemotherapy is associated with altered glucose metabolism in acute myeloid leukemia. Oncol Lett 12:334–342. https://doi.org/10.3892/ol.2016.4600

Spees JL, Olson SD, Whitney MJ, Prockop DJ (2006) Mitochondrial transfer between cells can rescue aerobic respiration. Proc Natl Acad Sci 103:1283–1288. https://doi.org/10.1073/pnas.0510511103

Sriraman SK, Aryasomayajula B, Torchilin VP (2014) Barriers to drug delivery in solid tumors. Tissue Barriers 2:e29528. https://doi.org/10.4161/tisb.29528

Stathopoulos GT, Sherrill TP, Han W et al (2008) Host nuclear factor- B activation potentiates lung cancer metastasis. Mol Cancer Res 6:364–371. https://doi.org/10.1158/1541-7786.mcr-07-0309

Stoll G, Kremer M, Bloy N et al (2019) Metabolic enzymes expressed by cancer cells impact the immune infiltrate. Oncoimmunology 8:e1571389. https://doi.org/10.1080/2162402X.2019.1571389

Storozhuk Y, Hopmans SN, Sanli T et al (2013) Metformin inhibits growth and enhances radiation response of non-small cell lung cancer (NSCLC) through ATM and AMPK. Br J Cancer 108:2021–2032. https://doi.org/10.1038/bjc.2013.187

Subramanian J, Govindan R (2008) Molecular genetics of lung cancer in people who have never smoked. Lancet Oncol 9:676–682. https://doi.org/10.1016/S1470-2045(08)70174-8

Teuwen LA, Draoui N, Dubois C, Carmeliet P (2017) Endothelial cell metabolism: an update anno 2017. Curr Opin Hematol 24:240–247. https://doi.org/10.1097/MOH.0000000000000335

Thorens B, Mueckler M (2010) Glucose transporters in the 21st century. Am J Physiol Endocrinol Metab 298:E141–E145. https://doi.org/10.1152/ajpendo.00712.2009

Torok S, Hegedus B, Laszlo V et al (2011) Lung cancer in never smokers. Future Oncol 7:1195–1211. https://doi.org/10.2217/fon.11.100

Tran S, Ready N, Krug LM, Pietanza MC, Jungbluth AA, Pan LS, Venhaus RR, Hoffman EW, Peters A-M, Dukelow K, Bomalaski JS, Wu B-W, LJO (2012) Phase II study of ADI-PEG 20 in patients with relapsed sensitive or refractory small cell lung cancer. J Clin Oncol 30:e17558–e17558

Valadi H, Ekström K, Bossios A et al (2007) Exosome-mediated transfer of mRNAs and microRNAs is a novel mechanism of genetic exchange between cells. Nat Cell Biol 9:654–659. https://doi.org/10.1038/ncb1596

Vander Heiden MG, Cantley LC, Thompson CB (2009) Understanding the Warburg effect: the metabolic requirements of cell proliferation. Science 324:1029–1033. https://doi.org/10.1126/science.1160809

Vander Linden C, Corbet C (2019) Reconciling environment-mediated metabolic heterogeneity with the oncogene-driven cancer paradigm in precision oncology. Semin Cell Dev Biol. https://doi.org/10.1016/j.semcdb.2019.05.016

Vansteenkiste J, Wauters E, Reymen B et al (2019) Current status of immune checkpoint inhibition in early stage NSCLC. Ann Oncol 30:1244–1253. https://doi.org/10.1093/annonc/mdz175

Vegliante R, Di Leo L, Ciccarone F, Ciriolo MR (2018) Hints on ATGL implications in cancer: beyond bioenergetic clues. Cell Death Dis 9:316. https://doi.org/10.1038/s41419-018-0345-z

Vukovic V, Tannock IF (1997) Influence of low pH on cytotoxicity of paclitaxel, mitoxantrone and topotecan. Br J Cancer 75:1167–1172

Wang R, Dillon CP, Shi LZ et al (2011) The transcription factor Myc controls metabolic reprogramming upon T lymphocyte activation. Immunity 35:871–882. https://doi.org/10.1016/J.IMMUNI.2011.09.021

Wang H, Wang L, Zhang Y et al (2016) Inhibition of glycolytic enzyme hexokinase II (HK2) suppresses lung tumor growth. Cancer Cell Int 16(9):9. https://doi.org/10.1186/s12935-016-0280-y

Wang L, Cao L, Wang H et al (2017a) Cancer-associated fibroblasts enhance metastatic potential of lung cancer cells through IL-6/STAT3 signaling pathway. Oncotarget 8:76116–76128. https://doi.org/10.18632/oncotarget.18814

Wang X, Zhang F, Wu X-R (2017b) Inhibition of pyruvate kinase M2 markedly reduces chemoresistance of advanced bladder cancer to cisplatin. Sci Rep 7:45983. https://doi.org/10.1038/srep45983

Wang Y, Hao F, Nan Y et al (2018) PKM2 inhibitor Shikonin overcomes the cisplatin resistance in bladder cancer by inducing necroptosis. Int J Biol Sci 14:1883–1891. https://doi.org/10.7150/ijbs.27854

Wang L, Li X, Ren Y et al (2019a) Cancer-associated fibroblasts contribute to cisplatin resistance by modulating ANXA 3 in lung cancer cells. Cancer Sci 110:1609–1620. https://doi.org/10.1111/cas.13998

Wang R, Lou X, Feng G et al (2019b) IL-17A-stimulated endothelial fatty acid β-oxidation promotes tumor angiogenesis. Life Sci 229:46–56. https://doi.org/10.1016/J.LFS.2019.05.030

Warburg O (1956) On the origin of cancer cells. Science 123:309–314. https://doi.org/10.1126/science.123.3191.309

Warburg O, Minami S (1923) Versuche an Überlebendem Carcinomgewebe. Klin Wochenschr 2:776–777

West AP (2017) Mitochondrial dysfunction as a trigger of innate immune responses and inflammation. Toxicology 391:54–63. https://doi.org/10.1016/j.tox.2017.07.016

Woo CG, Seo S, Kim SW et al (2016) Differential protein stability and clinical responses of *EML4-ALK* fusion variants to various ALK inhibitors in advanced *ALK* -rearranged non–small cell lung cancer. Ann Oncol:mdw693. https://doi.org/10.1093/annonc/mdw693

Xu Y-Y, Wu T-T, Zhou S-H et al (2014) Apigenin suppresses GLUT-1 and p-AKT expression to enhance the chemosensitivity to cisplatin of laryngeal carcinoma Hep-2 cells: an in vitro study. Int J Clin Exp Pathol 7:3938–3947

Yan H, Guo B-Y, Zhang S (2016) Cancer-associated fibroblasts attenuate Cisplatin-induced apoptosis in ovarian cancer cells by promoting STAT3 signaling. Biochem Biophys Res Commun 470:947–954. https://doi.org/10.1016/J.BBRC.2016.01.131

Yau T, Cheng PN, Chan P et al (2013) A phase 1 dose-escalating study of pegylated recombinant human arginase 1 (Peg-rhArg1) in patients with advanced hepatocellular carcinoma. Investig New Drugs 31:99–107. https://doi.org/10.1007/s10637-012-9807-9

Yeldag G, Rice A, Del Río Hernández A (2018) Chemoresistance and the self-maintaining tumor microenvironment. Cancers (Basel) 10. https://doi.org/10.3390/cancers10120471

Ying H, Kimmelman AC, Lyssiotis CA et al (2012) Oncogenic kras maintains pancreatic tumors through regulation of anabolic glucose metabolism. Cell 149:656–670. https://doi.org/10.1016/j.cell.2012.01.058

Ying L, Zhu Z, Xu Z et al (2015) Cancer associated fibroblast-derived hepatocyte growth factor inhibits the paclitaxel-induced apoptosis of lung cancer A549 cells by up-regulating the PI3K/Akt and GRP78 signaling on a microfluidic platform. PLoS One 10:e0129593. https://doi.org/10.1371/journal.pone.0129593

Yuan S, Qiao T, Zhuang X et al (2016) Knockdown of the M2 isoform of pyruvate kinase (PKM2) with shRNA enhances the effect of docetaxel in human NSCLC cell lines in vitro. Yonsei Med J 57:1312–1323. https://doi.org/10.3349/ymj.2016.57.6.1312

Zengin M (2019) Prognostic role of tumour-infiltrating T lymphocytes in stage IIA (T3N0) colon cancer: a broad methodological study in a fairly homogeneous population. Ann Diagn Pathol 41:69–78. https://doi.org/10.1016/j.anndiagpath.2019.05.007

Zhang Z, Stiegler AL, Boggon TJ, Kobayashi S (2010) EGFR-mutated lung cancer: a paradigm of molecular oncology abstract: abbreviations used. Oncotarget 1:497–514

Zhang WC, Ng SC, Yang H et al (2012) Glycine decarboxylase activity drives non-small cell lung cancer tumor-initiating cells and tumorigenesis. Cell 148:259–272. https://doi.org/10.1016/j.cell.2011.11.050

Zhang Y-L, Yuan J-Q, Wang K-F et al (2016) The prevalence of *EGFR* mutation in patients with non-small cell lung cancer: a systematic review and meta-analysis. Oncotarget 7:78985–78993. https://doi.org/10.18632/oncotarget.12587

Zhang Y, Wang DC, Shi L et al (2017a) Genome analyses identify the genetic modification of lung cancer subtypes. Semin Cancer Biol 42:20–30. https://doi.org/10.1016/j.semcancer.2016.11.005

Zhang J, Song F, Zhao X et al (2017b) EGFR modulates monounsaturated fatty acid synthesis through phosphorylation of SCD1 in lung cancer. Mol Cancer 16:127. https://doi.org/10.1186/s12943-017-0704-x

Zhang L, Yao Y, Zhang S et al (2019) Metabolic reprogramming toward oxidative phosphorylation identifies a therapeutic target for mantle cell lymphoma. Sci Transl Med 11:eaau1167. https://doi.org/10.1126/scitranslmed.aau1167

Zhao J (2016) Cancer stem cells and chemoresistance: the smartest survives the raid. Pharmacol Ther 160:145–158. https://doi.org/10.1016/J.PHARMTHERA.2016.02.008

Zhou Y, Wen H, Gu L et al (2017) Aminoglucose-functionalized, redox-responsive polymer nanomicelles for overcoming chemoresistance in lung cancer cells. J Nanobiotechnol 15:87. https://doi.org/10.1186/s12951-017-0316-z

Zhu H, Wu J, Zhang W et al (2016) PKM2 enhances chemosensitivity to cisplatin through interaction with the mTOR pathway in cervical cancer. Sci Rep 6:30788. https://doi.org/10.1038/srep30788

Zimmer AD, Walbrecq G, Kozar I et al (2016) Phosphorylation of the pyruvate dehydrogenase complex precedes HIF-1-mediated effects and pyruvate dehydrogenase kinase 1 upregulation during the first hours of hypoxic treatment in hepatocellular carcinoma cells. Hypoxia 4:135–145. https://doi.org/10.2147/HP.S99044

Hydrogen Sulfide Metabolism and Signaling in the Tumor Microenvironment

17

Alessandro Giuffrè, Catarina S. Tomé, Dalila G. F. Fernandes, Karim Zuhra, and João B. Vicente

Abstract

Hydrogen sulfide (H_2S), while historically perceived merely as a toxicant, has progressively emerged as a key regulator of numerous processes in mammalian physiology, exerting its signaling function essentially through interaction with and/or modification of proteins, targeting mainly cysteine residues and metal centers. As a gaseous signaling molecule that freely diffuses across aqueous and hydrophobic biological milieu, it has been designated the third 'gasotransmitter' in mammalian physiology. H_2S is synthesized and detoxified by specialized endogenous enzymes that operate under a tight regulation, ensuring homeostatic levels of this otherwise toxic molecule. Indeed, imbalances in H_2S levels associated with dysfunctional H_2S metabolism have been growingly correlated with various human pathologies, from cardiovascular and neurodegenerative diseases to cancer. Several cancer cell lines and specimens have been shown to naturally overexpress one or more of the H_2S-synthesizing enzymes. The resulting increased H_2S levels have been proposed to promote cancer development through the regulation of various cancer-related processes, which led to the interest in pharmacological targeting of H_2S metabolism. Herein are summarized some of the key observations that place H_2S metabolism and signaling pathways at the forefront of the cellular mechanisms that support the establishment and development of a tumor within its complex and challenging microenvironment. Special emphasis is given to the mechanisms whereby H_2S helps shaping cancer cell bioenergetic metabolism and affords resistance and adaptive mechanisms to hypoxia.

Keywords

Hydrogen sulfide · Tumor microenvironment · Hypoxia · Cellular bioenergetics · Persulfidation · Cystathionine β-synthase · Cystathionine γ-lyase · 3-mercaptopyruvate sulfurtransferase · Sulfide oxidizing pathway

A. Giuffrè (✉)
CNR Institute of Molecular Biology and Pathology, Rome, Italy
e-mail: alessandro.giuffre@uniroma1.it

C. S. Tomé · D. G. F. Fernandes · J. B. Vicente (✉)
Instituto de Tecnologia Química e Biológica António Xavier, NOVA University of Lisbon, Oeiras, Portugal
e-mail: jvicente@itqb.unl.pt

K. Zuhra
CNR Institute of Molecular Biology and Pathology, Rome, Italy

Department of Biochemical Sciences, Sapienza University of Rome, Rome, Italy

J. Serpa (ed.), *Tumor Microenvironment*, Advances in Experimental Medicine and Biology 1219, https://doi.org/10.1007/978-3-030-34025-4_17

17.1 Hydrogen Sulfide and Cancer

Nothing says more about the excitement and interest of a research field than the lack of a profound knowledge despite a massive accumulation of data in the literature. The link between hydrogen sulfide (H_2S) metabolism in human physiology and cancer is certainly characterized by inferred proposals based essentially on scattered yet sometimes converging phenomenological data. Still, different trends can be observed and summarized. Indeed, several excellent reviews have been already published on the topic (Cao et al. 2019; Hellmich et al. 2015; Szabo 2016). Herein, we attempted to provide an overview of the role that H_2S metabolism and signaling pathways may play in shaping the tumor microenvironment in its numerous defining aspects. Besides summarizing the phenomenological data linking H_2S metabolism with cancer and highlighting key elements of H_2S metabolism and signaling pathways, emphasis is given on the role of the latter in relation to two major aspects of the tumor microenvironment: cellular bioenergetics (Sect. 17.4) and hypoxia (Sect. 17.5).

H_2S was historically merely considered a toxic gas until its recognition in the early twenty-first century as an endogenously generated relevant signaling molecule in mammalian physiology regulating numerous processes within the cardiovascular, respiratory, digestive and central nervous systems (reviewed e.g. in (Wang 2012)). Its particular physicochemical properties allow H_2S to freely diffuse across biological milieu and exert its regulatory and signaling functions mostly through modification of target proteins (detailed below). The reactivity and potential toxicity of H_2S demand a fine balance between its biosynthesis and breakdown, the respective metabolic pathways functioning under a tight regulation. The balance between deleterious and beneficial effects of H_2S obeys to a conceptual bell-shaped model where homeostatic H_2S levels operate in a narrow range of optimal concentrations, whereas too much or too little H_2S may lead to dysfunction and toxicity at a cellular and/or systemic level. Besides H_2S, the related reactive sulfur species (RSS) persulfides (RSSH) and polysulfides ($RS_{(n)}SH$), partly generated by H_2S metabolism enzymes, have been growingly demonstrated to have equally relevant roles in signaling that extend to the etiology of pathological conditions (Mishanina et al. 2015; Cuevasanta et al. 2017; Filipovic et al. 2018; Ida et al. 2014).

Multiple cause-effect links have been increasingly established between altered H_2S metabolism and/or signaling and human diseases, particularly cancer. Pivotal studies on ovarian and colorectal cancer have reported a clear overexpression of H_2S-synthesizing cystathionine β-synthase (CBS) in cancer cell lines and tumor samples with respect to non-tumorigenic cells or normal tumor-adjacent tissue (Bhattacharyya et al. 2013; Szabo et al. 2013). Soon followed a similar association between cancer and increased expression of the three main enzymatic H_2S sources: CBS, cystathionine γ-lyase (CSE) and 3-mercaptopyruvate sulfurtransferase (MST). To date, increased expression of any of or all H_2S-synthesizing enzymes has been demonstrated for colorectal, ovarian, breast, prostate, gastric cancer, as well as lung adenocarcinoma, melanoma, hepatocellular carcinoma, urothelial cell carcinoma of bladder, astrocytoma, neuroblastoma, and glioma (reviewed e.g. in (Cao et al. 2019; Hellmich et al. 2015)). In line with the increased H_2S production resulting from up-regulation of H_2S-synthesizing enzymes in cancer cells, Libiad and co-workers have recently reported higher expression and differences in the cellular localization of enzymes involved in H_2S catabolism (Libiad et al. 2019).

Thus far, the full extent of the implications of altered H_2S metabolism to cancer progression is still a matter of significant dedicated research efforts. While many observations derived from cell biology studies on tumor samples and cellular and animal models bring together common threads concerning the effect of altered H_2S metabolism on cancer, strong molecular studies are lagging behind. Moreover, as pointed out by Cao and co-workers, several studies employ non-physiological concentrations of H_2S donors, unspecific inhibitors that may affect other enzymes than those synthesizing H_2S and cause changes in metabolites other than H_2S, derived from or consumed by the H_2S metabolism pathways (Cao et al. 2019). Nevertheless, mounting evidence posits key roles for H_2S and related RSS in the modulation of several recognized characteristics of cancer, such as dysregulation of cell growth and signaling pathways towards uncontrolled proliferation, evasion of apoptosis, stimulation of angiogenesis, subversion of cell energy limitations, genome instability, and tumor enhanced inflammation. In line with the scope of this book, it becomes logic that H_2S metabolic and signaling pathways help shaping the adaptive changes within the tumor microenvironment that favor its progression (Fig. 17.1). Indeed, increased H_2S production by cancer cells naturally overexpressing H_2S-synthesizing enzymes has been shown to stimulate cellular bioenergetics, enhancing ATP production by oxidative phosphorylation and glycolysis (detailed in Sect. 17.4) (reviewed e.g. in (Giuffrè and Vicente 2018; Szabo et al. 2014)). Promotion of neoangiogenesis by H_2S allows replenishing the tumor microenvironment with the nutrients and oxygen that become scarce with the dysregulated proliferation and growth. Both CBS and CSE have been implicated in controlling angiogenesis in colorectal, ovarian and breast cancer, likely involving the H_2S-mediated persulfidation of ATP-sensitive potassium K_{ATP} channels (Mustafa et al. 2011; Tang et al. 2005), as well as the phosphoinositide-3-kinase/protein kinase B (PI3K/Akt) and the mitogen activated protein kinase (MAPK) signaling pathways (Cai et al. 2007; Papapetropoulos et al. 2009; Zhao et al. 2014). The link between the mutual regulation of H_2S metabolism and hypoxia, and angiogenesis

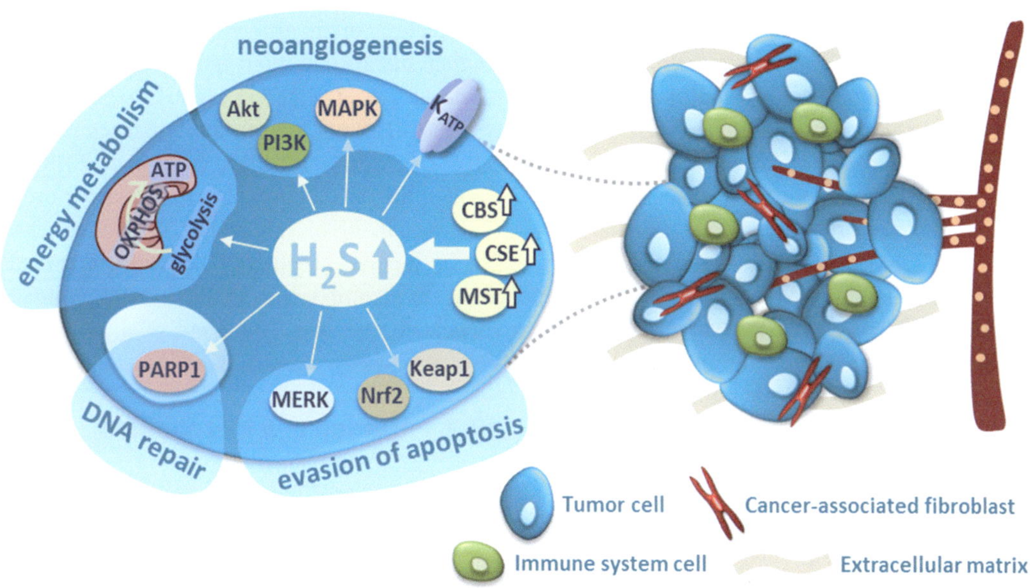

Fig. 17.1 Effects of hydrogen sulfide within the tumor microenvironment. Hydrogen sulfide (H_2S)-synthesizing enzymes (cystathionine β-synthase, CBS, cystathionine γ-lyase, CSE, and 3-mercaptopyruvate sulfurtransferase, MST) have been reported to be upregulated in different cancer cell lines and specimens. The resulting increased H_2S production has been proposed to contribute to modulate the cancer cells adaptation within the complex tumor microenvironment, mainly through persulfidation of key protein targets in signaling pathways that regulate numerous processes, e.g.: neoangiogenesis via the PI3K/Akt and MAPK pathways and by modulation of K_{ATP} channels; evasion of apoptosis via the MERK and Keap1/Nrf2 pathways; DNA repair via the MERK/PARP1 pathway. H_2S also stimulates the tumor cell bioenergetics by directly injecting electrons into the mitochondrial electron transfer chain through sulfide:quinone oxidoreductase, by persulfidation of ATP synthase keeping it in the active state, and by stimulation of glycolysis, particularly by persulfidation of lactate dehydrogenase A. Increased H_2S metabolism in tumors, associated with a higher cysteine flux, has been suggested to contribute to cancer cell chemoresistance

in the context of the tumor microenvironment is detailed in Sect. 17.5.

The evasion of apoptosis and cell cycle acceleration have also been associated with the H_2S-mediated persulfidation of key players in the corresponding signaling cascades, typically activating the respective protein targets. The role of H_2S metabolism in evading apoptosis has been demonstrated for gastric and colorectal cancer, hepatoma and neuroblastoma (Rose et al. 2005; Sekiguchi et al. 2016; Tiong et al. 2010; Zhen et al. 2015), chiefly involving the CSE-H_2S axis in the persulfidation and consequent modulation of protein targets of key pathways (associated also with inflammation) such as: the Keap1-transcription factor nuclear factor erythroid 2-related factor (Nrf2) (Yang et al. 2013), the nuclear factor kappa-light-chain-enhancer of activated B cells (NF-κB) (Zhen et al. 2015), and the extracellular signal-regulated kinase (ERK)-activating protein kinase 1 (MEK1). CBS has been proposed to take part in ferroptosis resistance mechanisms, possibly contributing to evasion of this alternative non-apoptotic cell death mechanism (Wang et al. 2018).

The genome instability of cancer cells triggers the activation of several mechanisms, including DNA repair pathways, possibly by activation of e.g. the MEK1-ERK-Poly [ADP-ribose] polymerase 1 (PARP-1) pathway, whose activity was found to be diminished in the liver and kidney of a CSE knock-out mouse model (Zhao et al. 2014). Szczesny and co-workers have also demonstrated that MST silencing in A549 lung adenocarcinoma cells attenuates the mitochondrial DNA repair rate upon damage (Szczesny et al. 2016).

Another key element for the tumor proliferation concerns the acquisition of resistance to common chemotherapeutic agents. H_2S-synthesizing enzymes have been implicated in the development of chemoresistance phenotypes in ovarian, liver and colorectal cancer cell lines (Bhattacharyya et al. 2013; Stokes et al. 2018; Untereiner et al. 2018). Moreover, the proposal of cysteine-dependent chemoresistance mechanisms in ovarian cancer is based on an enhanced cysteine flux in chemoresistant cancer cells that relies on increased cystine import and enhanced intracellular cysteine catabolism likely via H_2S-synthesizing enzymes (Nunes et al. 2018).

Whereas numerous reports point to a pro-cancer effect of increased expression and activity of H_2S-synthesizing enzymes, it should be noted that exogenous addition of H_2S either with sulfide salts or slow releasers may have anti-cancer effects, depending on the dose and exposure time (Ianaro et al. 2016; Reis et al. 2019). This is not surprising taking into account the proposed bell-shape model that dictates how H_2S, depending on its levels, acts as a signaling molecule or a toxin in human physiology. Indeed, the concentration dependence of H_2S effects simply appears to be shifted in several cancer cell models and patient samples to hint for a higher H_2S metabolic flux in cancer that overall contributes to shape the tumor microenvironment and promote tumor growth and proliferation. This dual nature of H_2S in cancer, nevertheless, offers different options for pharmaceutical therapeutic interventions, either through the development of inhibitors for H_2S-synthesising enzymes (Hellmich et al. 2015; Szczesny et al. 2016; Druzhyna et al. 2016; Zuhra et al. 2019), or through the development of H_2S-releasing drugs (Ianaro et al. 2016; Reis et al. 2019; Wallace et al. 2018), including naturally derived compounds such as components from garlic extracts (Puccinelli and Stan 2017; Yagdi et al. 2016).

17.2 Human H_2S Metabolism

The dual role of H_2S in physiology implies a fine tuning of sulfide metabolism to maintain its homeostatic levels. In mammals, besides the contribution of gut microbiota metabolism and dietary per-/poly-sulfides breakdown, a group of enzymes specialized in the synthesis and catabolism of H_2S regulates the sulfide pool (Fig. 17.2).

17.2.1 H_2S Synthesis

Endogenous H_2S is produced mainly by three enzymes: cystathionine β-synthase (CBS), cystathionine γ-lyase (CSE) and 3-mercatopyru-

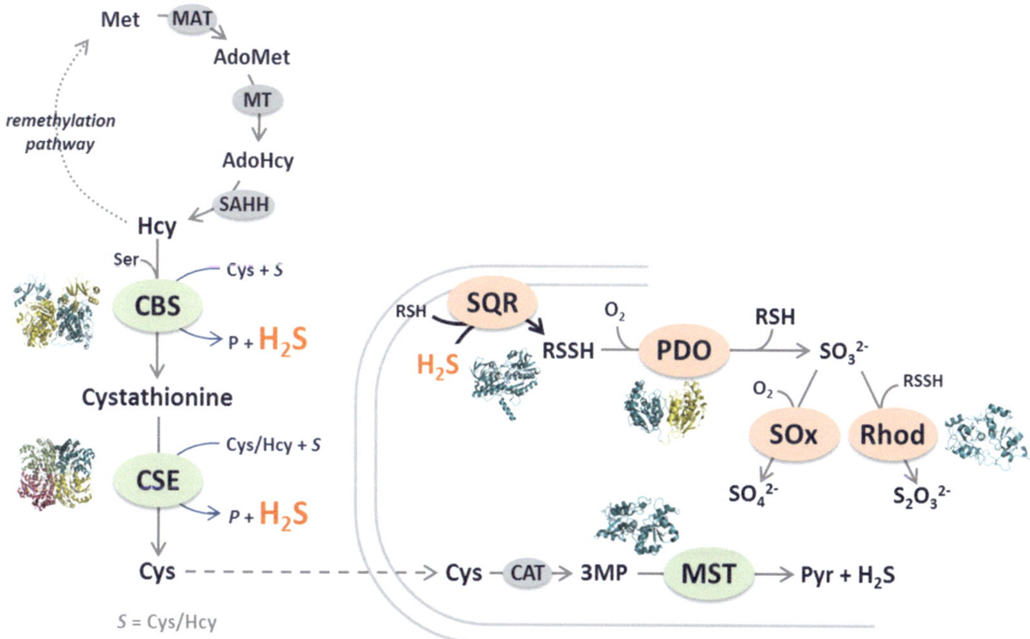

Fig. 17.2 Metabolic pathways of H_2S synthesis and catabolism. Enzymatic production of H_2S is accomplished by cystathionine β-synthase (CBS), cystathionine γ-lyase (CSE) and mercaptopyruvate sulfurtransferase (MST). CBS and CSE participate in the transsulfuration pathway that converts homocysteine into cysteine. Both enzymes catalyze alternative reactions that use cysteine and/or homocysteine as substrates and yield H_2S. MST participates in the cysteine catabolic pathway, where it converts 3-mercaptopyruvate (derived from cysteine via cysteine aminotransferase, CAT) into pyruvate and releases H_2S. Enzymatic breakdown of H_2S is accomplished by the sulfide oxidizing pathway, comprising four mitochondrial enzymes. The first irreversible and rate-limiting step of H_2S catabolism is catalyzed by sulfide:quinone oxidoreductase (SQR) that transfers the sulfur atom to an acceptor (RSH); persulfide dioxygenase (PDO) oxidizes the resulting persulfide (RSSH) and oxygenates the sulfane sulfur to yield sulfite (SO_3^{2-}). Sulfite can be further converted by thiosulfate sulfurtransferase (Rhod) to thiosulfate ($S_2O_3^{2-}$) or by sulfite oxidase (SOx) to sulfate (SO_4^{2-}). MAT: methionine-adenosyl transferase; MT: methyltransferase; SAHH: S-adenosyl homocysteine hydrolase. Protein three-dimensional structures were generated from PDB entries 4COO (CBS), 2NMP (CSE), 4JGT (MST), 6MP5 (SQR), 4CHL (PDO) and 2ORA (Rhod)

vate sulfurtransferase (MST). The tissue distribution of the three enzymes (and thus their contribution to H_2S production) differs: CBS is mainly found in liver, pancreas, kidney and nervous system (Bao et al. 1998; Kabil et al. 2011; Mudd et al. 1965; Saha et al. 2016); CSE is mainly found in liver, kidney and smooth muscle (Kabil et al. 2011; Ogasawara et al. 1994; Yang et al. 2008); MST has a broader distribution, being present in most tissues (Tomita et al. 2016). While the main cellular localization is considered cytosolic for CBS and CSE and mitochondrial for MST, stress conditions can dictate translocation of these enzymes into different cell compartments or even the extracellular milieu. While MST can additionally be detected in the cytosol (Frasdorf et al. 2014), CBS and CSE can be found extracellularly or in the mitochondria upon, for instance, oxidative stress (Fu et al. 2012; Teng et al. 2013).

MST participates in the cysteine catabolism pathway, producing H_2S as a product of its reaction with 3-mercaptopyruvate (3MP). It is a monomeric protein (297 amino acids; 33 kDa) composed of two structurally nearly identical domains between which lie the 3MP binding site and a catalytic cysteine residue (Cys_{248}) (Yadav et al. 2013). Upon reaction with 3MP, Cys_{248} (Cys-SH) becomes persulfidated (Cys-SSH) and prone to react with reductant molecules (such as glutathione, L-homocysteine or thioredoxin) that release H_2S after accepting the sulfane sulfur atom.

CBS and CSE are both PLP-dependent homotetrameric proteins. CBS monomers (551 amino acids; 61 kDa) consist of three domains: an N-terminal heme-binding domain, a central pyridoxal 5′-phosphate (PLP)-binding domain, and a C-terminal S-adenosyl-L-methionine (AdoMet)-binding domain (Ereno-Orbea et al. 2013). CSE monomers (405 amino acids; 44 kDa) consist of an N-terminal PLP-binding domain and a C-terminal domain (Sun et al. 2009). The canonical reactions catalyzed by CBS and CSE constitute the transsulfuration pathway (Fig. 17.2): CBS catalyzes the condensation of L-homocysteine and L-serine to yield cystathionine; CSE then converts cystathionine into L-cysteine, α-ketobutyrate and ammonia. Production of H_2S by CBS and CSE results from alternative reactions that use L-cysteine and/or L-homocysteine as substrates.

Similarly to its canonical reaction, CBS can catalyze the condensation of L-homocysteine and L-cysteine to yield cystathionine, releasing H_2S instead of H_2O. Also, in the presence of L-cysteine, CBS and CSE produce H_2S *via* β-replacement/β-elimination or α,β-elimination, respectively. CSE catalyzes the same alternative reactions as CBS, and can additionally use L-homocysteine in β-replacement/β-elimination reactions that yield H_2S. The extent of H_2S generation by CBS and CSE thus depends on the balance between canonical *versus* alternative reactions, which in turn is affected by tissue expression levels, substrate availability and regulation at protein level. Indeed, CBS activity can be modulated by post-translational modifications and allosteric regulators. CBS activity has been shown to be increased by glutathionylation at Cys_{346}, proposed to serve as a redox sensor to boost the transsulfuration pathway under oxidative conditions towards formation of cysteine and, ultimately, glutathione (Niu et al. 2015). An opposite effect was observed for another redox sensor reported for CBS consisting of a CXXC motif in the catalytic domain of CBS ($C_{272}PGC_{275}$), which allosterically induces ~2-3-fold higher CBS activity upon reduction of this cysteine disulfide with DTT (Niu et al. 2018). The two flanking domains of CBS also have a regulatory function. The C-terminal domain adopts an autoinhibitory conformation in the resting enzyme, blocking the entrance of the catalytic site (McCorvie et al. 2014). Binding of the allosteric activator AdoMet to each C-terminal domain of adjacent monomers triggers their dimerization, facilitating the access of substrates to the active site and activating the protein (McCorvie et al. 2014; Ereno-Orbea et al. 2014). The N-terminal domain binds a heme moiety that mediates CBS regulation through changes in its redox and ligand state. Reduction of the heme promotes enzyme inactivation. This ferrous form is also able to bind NO and CO that displace the iron endogenous ligands (Cys_{52} and His_{65}), inhibiting CBS (Vicente et al. 2014, 2016a, b; Banerjee and Zou 2005). The forty N-terminal residues in CBS constitute an intrinsically disordered peptide suggested to represent a second heme binding site (*via* Cys_{15} and His_{22}), although its physiological relevance is not clear (Kumar et al. 2018). Interestingly, despite the distance between the two regulatory domains (>30 Å), intercommunication between heme- and AdoMet-modulation is observed, with AdoMet binding enhancing CO- and NO-mediated inhibition (Vicente et al. 2016b). The sequential reactions of CBS and CSE in the transsulfuration pathway and the fact that they share the same substrates in H_2S-generating reactions implies that regulation of one enzyme will affect the other through substrate/product accumulation/depletion, which will favor one biochemical pathway over the other. Besides, it has been shown that inhibition of CBS may result in overall higher H_2S production through CSE which presents a higher catalytic efficiency using the same H_2S-originating substrates (Banerjee 2017).

17.2.2 H_2S Catabolism

In mammals, the sulfide-oxidizing pathway – comprising four mitochondrial enzymes – is responsible for H_2S catabolism (Fig. 17.2). The first irreversible and limiting step in H_2S degradation is catalyzed by sulfide:quinone oxidoreductase (SQR). SQR oxidizes H_2S, transferring the sulfur atom to an acceptor molecule (such as glutathione) that becomes persulfidated, and using coenzyme Q (CoQ) as an electron acceptor. SQR is an integral membrane flavoprotein (450 amino acids; 50 kDa), located at the inner mitochondrial membrane. The active site is accessible through the matrix-facing surface, where H_2S reacts with a catalytic disulfide (Cys_{201}/Cys_{379}) and the extracted electrons are transferred via a flavin adenine dinucleotide (FAD) cofactor to a CoQ molecule bound to hydrophobic pocket accessible from the membrane-facing surface (Jackson et al. 2019). In the second step of H_2S catabolism, the glutathione persulfide released by SQR is taken up by persulfide dioxygenase (ETHE1 or PDO) that catalyzes the oxidation of the sulfane sulfur and yields sulfite using a non-heme iron cofactor. In the only crystal structure of human ETHE1, a cysteinyl sulfinic acid (C_{247}-SO_2H) is present 15 Å away from the active site, although the catalytic relevance of this observation is still unknown (Pettinati et al. 2015). SQR-derived persulfidated glutathione and ETHE1-derived sulfite can be further converted to thiosulfate by thiosulfate sulfurtransferase (Rhod). In the last step of H_2S catabolism, sulfite oxidase (SOx) converts sulfite to sulfate, using molybdenum and heme-iron as cofactors, a H_2O molecule as oxygen atom donor and cytochrome *c* as electron acceptor (Libiad et al. 2014). The coupling of the mitochondrial H_2S oxidation pathway with the respiratory chain through SQR-mediated reduction of CoQ at low sulfide concentrations, and the H_2S-mediated inhibition of complex IV (Petersen 1977) at high sulfide concentrations suggest implications on cellular bioenergetics upon different pathophysiological conditions (detailed in Sect. 17.4).

17.3 H_2S-Mediated Signaling

As mentioned above, H_2S exerts its numerous signaling and regulatory functions mainly by interacting with and/or directly modifying target proteins by two distinct mechanisms: (i) interaction with metal centers, mostly heme moieties, and (ii) protein persulfidation.

17.3.1 H_2S and Heme Proteins

As a small signaling molecule, H_2S can interact with heme proteins. These reactions are very complex and determined by many factors, such as the H_2S concentration, the redox and ligation state of the heme iron, the heme pocket environment, the protonation state of the bound sulfide and the presence/absence of O_2 or reducing agents in solution, as reviewed e.g. in (Giuffrè and Vicente 2018; Nagy 2015). H_2S can bind to heme-Fe(III) as such or as HS^- (deprotonated state of H_2S), generating heme-Fe(III)-H_2S or heme-Fe(III)-HS^-, respectively. The stability of each adduct depends on the protein residues in the heme surroundings: nonpolar residues can stabilize heme-Fe(III)-H_2S by limiting the deprotonation of the ligand; in contrast, basic residues can mediate the deprotonation of the H_2S, yielding heme-Fe(III)-HS^-. Moreover, bound sulfide can reduce heme-Fe(III), which can result in the heme-Fe(II)-HS· radical adduct. The reduced heme-Fe(II)-HS· species can further react with excess HS^-, resulting in a Fe(II)-S-S·$^-$ species, which may react with HS· to originate polysulfides, or with O_2 and H_2O in several steps to yield thiosulfate (Vitvitsky et al. 2015).

Alternatively, H_2S can react with a heme-Fe(II)-O_2 complex likely *via* a heme-Fe(IV) = O ferryl intermediate, yielding a sulfheme derivative, with the sulfur atom being incorporated into one of the porphyrin pyrrole rings. The reaction is favored in the presence of H_2O_2, thus implicating the formation of higher valent heme iron intermediates (Nagy 2015; Pietri et al. 2011; Rios-Gonzalez et al. 2014). Although the mechanistic details of sulfheme formation are still to be fully clarified, these reactions have been described for different heme proteins such as globins (particularly hemoglobin and myoglobin), heme-based sensors, peroxidases and catalase (reviewed e.g. in (Rios-Gonzalez et al. 2014)). Formation of the sulfheme derivative of hemoglobin (Hb), designated as sulfhemoglobin, results from the insertion of a sulfur atom into the heme B pyrrole. This derivative is irreversibly formed and has lower O_2 affinity, thus being considered to contribute to sulfide-derived toxicity (Rios-Gonzalez et al. 2014). Alternatively, the accumulation of ferric hemoglobin (metHb) in the blood, designated as methemoglobinemia, has been suggested to protect against sulfide toxicity in mice, by promoting sulfide disposal (Smith and Gosselin 1966). Furthermore, Hb can also be a source of physiologically relevant sulfane sulfur products resulting from sulfide oxidation, namely thiosulfate and glutathione persulfide (Vitvitsky et al. 2015, 2017).

The historical hallmark of H_2S interacting with heme proteins concerns the inhibition of mitochondrial cytochrome *c* oxidase (CcOX). Indeed, CcOX is considered the main target of the three gasotransmitters (H_2S, NO, and CO), being inhibited through different mechanisms and with different kinetics (Vicente et al. 2016a; Cooper and Brown 2008). CO only binds to fully-reduced heme a_3-Cu_B active site, whereas NO-mediated inhibition can proceed *via* binding to the single-electron reduced active site (Giuffrè et al. 2002) through two reaction pathways, depending on the oxygen tension and electron flux (Mastronicola et al. 2003; Sarti et al. 2012). The mechanism whereby H_2S inhibits CcOX does not involve binding to ferrous heme a_3. Rather, it is hypothesized that CcOX in turnover with O_2 is primarily targeted by H_2S at the oxidized or reduced Cu_B, followed by intramolecular sulfide transferred to ferric a_3 (Vicente et al. 2016a; Nicholls et al. 2013). Irrespectively of the mechanism, H_2S mediated CcOX inhibition is fast, potent and reversible.

17.3.2 Persulfidation of Protein Cysteine Residues

As mentioned in Sect. 17.1, many of the signaling effects commonly attributed to H_2S have been growingly assigned to per- and poly-sulfides. Formation of protein-bound persulfides occurs through the modification of cysteine side chains of target proteins, often upon reaction with free reactive low-molecular weight (LMW) persulfides (RSSH), such as glutathione persulfide (GSSH) and cysteine persulfide (CysSSH) (Fig. 17.3) (Kasamatsu et al. 2016; Millikin et al. 2016; Bianco et al. 2016).

Such sulfane sulfur-containing metabolites are actually generated both by H_2S-synthesizing and -catabolyzing enzymes, namely through: production of CysSSH (or homocysteine persulfide) by CBS and CSE using cystine (or homocystine) and by MST using cysteine (or homocysteine) as sulfur-accepting co-substrate, and production of GSSH by SQR or by MST (Kimura et al. 2017). The CBS- and CSE-catalyzed CysSSH-producing reactions result also in longer polythiolated products that can react with glutathione to yield GSSH (Yadav et al. 2016). Other possibilities to generate protein-bound cysteine persulfides include the reaction of oxidized cysteine residues directly with H_2S, or the reaction of protein-bound cysteine thiols with sulfhydryl radical (generated by reaction of H_2S with metal centers) and subsequently with O_2. Protein persulfidation requires the appropriate environment surrounding the target cysteine, a favorable cellular redox status and availability of free glutathione/cysteine/H_2S (Giuffrè and Vicente 2018). A recent report (Akaike et al. 2017) posits that mammalian cysteinyl-tRNA synthetases (CARSs), in particular the mitochondrial isoform CARS2, are the main source of free and protein-bound CysSSH (and $CysSS_{(n)}H$), the latter resulting from co-translational insertion of previously per- or polysulfidated cysteine (Fig. 17.3). It is also suggested that CySSH is formed in the mitochondria, prior to being released into the cytosol to exert its effects. Whereas CARS2 is possibly the major CysSSH source under physiological conditions, CBS and CSE still play a major role in CysSSH synthesis in pathophysiological conditions. In the latter case, cystine levels are increased in line with oxidative and electrophilic stress, such as in cancer, where the glutamate/cystine xCT antiporter is often upregulated (Ida et al. 2014; Akaike et al. 2017). Besides its signaling and regulatory function, persulfidation also protects thiol-containing res-

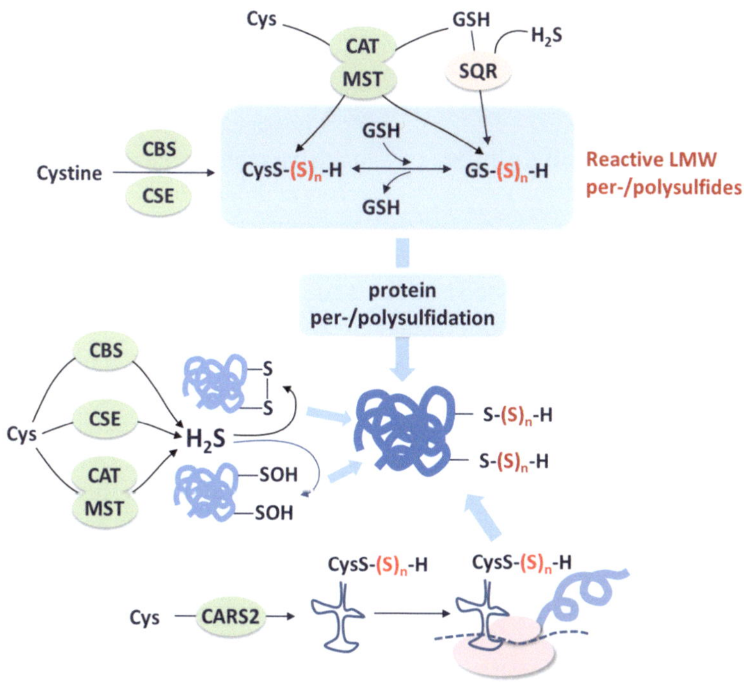

Fig. 17.3 **H_2S-mediated signaling via protein persulfidation**. Persulfidation of protein cysteine residues can occur through two plausible mechanisms: posttranslational modification of cysteine residues or co-translational incorporation of cysteine persulfide (CysSSH) through CysSSH-bound tRNA derived from the mitochondrial protein cysteinyl-tRNA synthetase (CARS2). Cystathionine β-synthase (CBS), cystathionine γ-lyase (CSE) and 3-mercatopyruvate sulfurtransferase (MST) generate CysSSH from either cystine (CBS and CSE) or cysteine (CAT-MST). CysSSH can then react with glutathione (GSH), yielding glutathione persulfide (GSSH). Sulfane sulfur can next be reversibly transferred to other thiols (such as GSH or protein-SH) to form the corresponding persulfide/polysulfide species. GSSH is also generated via SQR-mediated H_2S oxidation. H_2S produced by CBS, CSE or MST can also react with oxidized cysteine residues in target proteins, either in the disulfide or sulfenic form, to generate the corresponding persulfides

idues against irreversible chemical modification by oxidants and electrophiles (Kasamatsu et al. 2016). For cellular signaling, a key advantage of a post-translational modification such as persulfidation is that this modification is reversed by reaction of the resulting derivatives with reducing agents (*e.g.* glutathione), proteins (*e.g.* thioredoxin or glutaredoxin) or *via* re-formation of a disulfide bond initiated by nucleophilic attack (Mishanina et al. 2015). Deficiency in persulfidated proteins has been associated with various pathologies, like cancer and cardiovascular disease (Giuffrè and Vicente 2018; Paul and Snyder 2015). On the other hand, cancer cells have higher expression of H_2S-synthesizing enzymes and likely overproduce LMW persulfides owing to the oxidative environment. The latter could endow cancer cells with chemoresistance mechanisms, since it has been proposed that free LMW persulfides have a higher affinity than their thiol counterparts to form *s*-conjugates with exogenous electrophilic molecules (Cuevasanta et al. 2017; Goncalves-Dias et al. 2019), such as common alkylating/oxidative chemotherapeutic drugs.

17.4 Effect of H_2S on Cellular Bioenergetics

Cellular bioenergetics is a trademark of the dual role of H_2S in human physiology. Indeed, H_2S has a bell-shaped effect on bioenergetics, contributing to ATP synthesis at lower concentrations while being cytotoxic and potentially lethal at higher levels (Fig. 17.4).

17.4.1 Dual Effect of H_2S on Cellular Respiration

While the detailed molecular mechanisms behind H_2S toxicity are not yet fully understood, it is well recognized that it reversibly binds a heme moiety of CcOX, resulting in mitochondrial respiration impairment (Vicente et al. 2016a). Although CcOX inhibition occurs with a relatively low K_i value (K_i = 0.2 μM at pH 7.4), as reported working on isolated enzyme (Petersen 1977), inhibition of the electron transport chain using isolated mitochondria or cultured cells is usually observed at much higher concentrations (up to tens of micromolar) (Leschelle et al. 2005).

Consistently, mammalian cells are equipped with the sulfide oxidizing pathway, able to couple energy production with sulfide detoxification (Lagoutte et al. 2010). As mentioned in Sect. 17.2, the rate-limiting reaction of this pathway is catalyzed by SQR, which transfers a sulfur atom to an acceptor such as GSH and concomitantly the sulfide-derived electrons to CoQ. H_2S oxidation by SQR thus supplies electron equivalents to the mitochondrial electron transport chain and consequently stimulates ATP production *via* oxidative phosphorylation (Libiad et al. 2014). From a methodological point of view, the sulfide oxi-

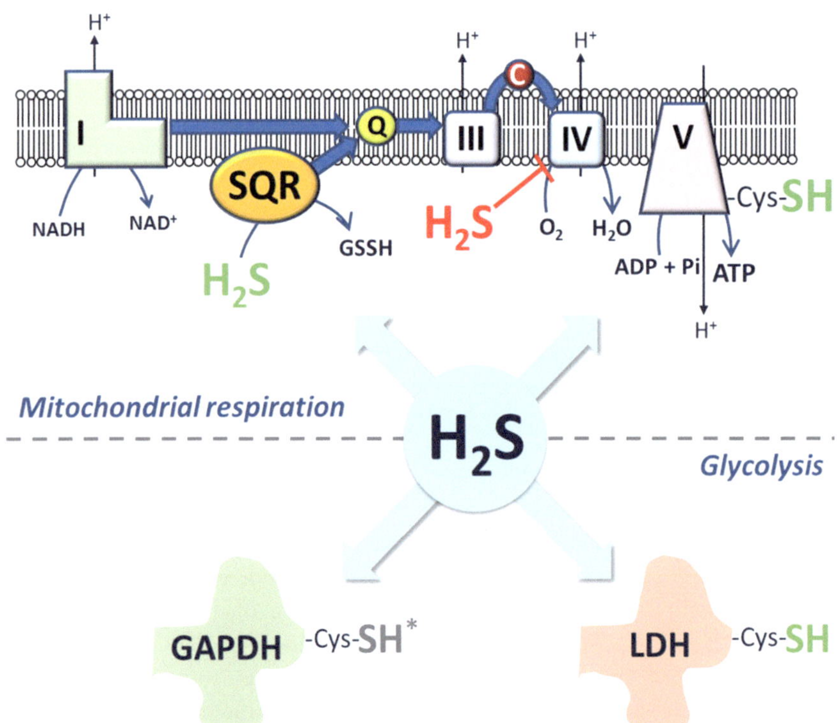

Fig. 17.4 **H_2S effect on cellular bioenergetics.** At low concentrations H_2S stimulates the mitochondrial electron transport chain by acting as a metabolic fuel. H_2S-derived electrons are transferred to coenzyme Q (CoQ) by sulfide:quinone oxidoreductase (SQR). CoQ is then re-oxidized by complex III and electrons are shuttled via cytochrome *c* to complex IV, where O_2 is eventually reduced to H_2O. Complexes I, III and IV contribute to generating a proton electrochemical gradient which is the driving force for ATP synthesis catalyzed by ATPase (complex V). H_2S stimulates directly complex V activity by persulfidation, which contributes to maintain the enzyme in its catalytically active conformation. At higher concentrations H_2S binds cytochrome *c* oxidase (complex IV), thus leading to impairment of electron transfer. In the cytosol, H_2S modulates the activity of glycolytic enzymes by mediating persulfidation of glyceraldehyde-3-phosphate dehydrogenase (GAPDH; with opposite effects reported) and lactate dehydrogenase (LDH)

dizing activity of living cells or isolated mitochondria can be studied by following O_2 consumption at increasing concentration of H_2S administered as e.g. sodium sulfide (Na_2S). Because of its ability to inhibit CcOX at high concentrations, sulfide is usually supplied with a continuous infusion at a selected rate rather than a single bolus. Sulfide-dependent O_2 consumption can be arrested by inhibiting complex III with antimycin or complex IV with cyanide, as sulfide-derived electrons enter the electron chain transport at CoQ level (Szabo et al. 2014; Malagrino et al. 2019).

Notably, the sulfide oxidizing activity varies among different cell types, ranging from undetectable in nervous system cells to high in colon cells (Linden et al. 2012; Vitvitsky et al. 2012; Fagerberg et al. 2014). Indeed, colonocytes are physiologically exposed to high sulfide concentrations (in the range of millimolar) produced by the gut microbiota. Therefore, it is not surprising that the sulfide oxidizing pathway enzymes are significantly expressed and precisely localized in colonic tissue (Libiad et al. 2019) and that the corresponding sulfide disposal activity is possibly the maximal in the organism (Goubern et al. 2007). Overall, this process couples the oxidation of H_2S with ATP production, consuming ~ 0.75 O_2 molecules (0.25 by CcOX and 0.5 by ETHE1) per molecule of H_2S (Lagoutte et al. 2010). In terms of energy yield, this process displays a relatively low efficiency, particularly because of the additional consumption of O_2 by ETHE1. On the other hand, considering that H_2S is highly diffusible through biological membranes and its bioavailability is tightly controlled by the interplay between H_2S-synthesizing and the H_2S-consuming enzymes, it has been suggested that this molecule may act mainly as an 'emergency' substrate of the electron transport chain when Krebs cycle-derived electron donors are insufficient to fulfil the energy demand (Szabo et al. 2014). Enhanced sulfide metabolism has been extensively associated to cancer survival by stimulation of energy supply (Szabo 2016). For instance, both the cytosolic enzymes CBS and CSE were reported to accumulate in mitochondria when cells are exposed to hypoxic conditions, a common factor within the tumor microenvironment. The resulting increase in mitochondrial sulfide availability was suggested to support ATP production and provide protection against oxidative stress (Teng et al. 2013; Wang et al. 2014). Isolated mitochondria from HCT 116 colorectal cancer cells treated with cysteine displayed enhanced O_2 consumption, which was suppressed by shRNA mediated CBS silencing or its pharmacological inhibition with aminooxyacetic acid (AOAA). This observation suggested that H_2S derived from CBS, but not from CSE, significantly contributes to cancer cell respiration (Szabo et al. 2013). Consistently, in ovarian cancer cells CBS inhibition resulted in mitochondrial impairment with concomitant overproduction of reactive oxygen species (ROS) (Bhattacharyya et al. 2013).

Probably, among the three H_2S-synthetizing enzymes, the most relevant for mitochondrial bioenergetics at physiological conditions is MST (Abdollahi Govar et al. 2019; Augsburger and Szabo 2018). Currently, two MST isoforms are known, with comparable enzymatic activity, MST-Iso1 localized in the cytosol and MST-Iso2, which exists both in the cytosol and in mitochondria (Frasdorf et al. 2014). The interest on this enzyme arose from its privileged localization, from where the produced H_2S is readily available as a substrate for the electron transport chain via SQR. Indeed, it has been shown in isolated liver mitochondria and cultured murine hepatoma cells that mitochondrial bioenergetics is stimulated upon treatment with low concentrations of the MST substrate 3MP. Basal cellular bioenergetics was reduced upon silencing of either MST or SQR, supporting the hypothesis that mitochondrial respiration is in part sustained by the 3MP-derived sulfide and its oxidation catalyzed by SQR (Modis et al. 2013). Beyond the ability of H_2S to act as a metabolic fuel, H_2S-mediated protein persulfidation has been shown to be involved, at different levels, in the regulation of cellular bioenergetics. Indeed, it was discovered that the mitochondrial inner membrane protein ATP synthase (complex V) is physiologically regulated by persulfidation and this post-translational modification may maintain this enzyme in its activated state (Modis et al. 2016) (Fig. 17.4).

17.4.2 H_2S-Mediated Energy Metabolism Reprogramming in Tumor Cells

Tumors undergo a well-known energy metabolism reprogramming (Warburg effect), promoting glycolysis instead of the more efficient oxidative phosphorylation pathway, even under aerobic conditions (Warburg 1956a, b). Indeed, cancer cells seem to express the isoform A of lactate dehydrogenase (LDH-A), that catalyzes the conversion of pyruvate to lactate (thus resulting in a higher NAD^+/NADH ratio), more than they express LDH-B, that catalyzes the opposite reaction (Valvona et al. 2016). Recently, it was observed that colon cancer cells treated with a sulfide releaser display increased lactate levels and, consistently, higher LDH-A catalytic activity, which was shown to be positively modulated via persulfidation. Accordingly, H_2S-mediated stimulation of glycolysis, and also of oxidative phosphorylation, seems to be LDH-A-dependent, thus making H_2S a pivotal regulator of cancer bioenergetics (Untereiner et al. 2018) (Fig. 17.4).

Modulation of glycolysis through persulfidation of glyceraldehyde-3-phosphate dehydrogenase (GAPDH) mediated by H_2S or related sulfane sulfur species has been proposed, although opposite effects regarding enzyme activation/inactivation have been reported (Jarosz et al. 2015; Mustafa et al. 2009) (Fig. 17.4). H_2S has been shown to stimulate mitochondrial ATP production in human endothelial cells (ECs), resulting in enhanced proliferation and migration. Conversely, MST pharmacological inhibition or its silencing suppresses microvessel growth (Abdollahi Govar et al. 2019). Interestingly, ECs stimulation induced by treatment with vascular endothelial growth factor (VEGF) is abrogated in the presence of inhibitors of CSE or K_{ATP}, suggesting that this process is at least in part mediated by K_{ATP} channels opening by persulfidation as mentioned in Sect. 17.1 (Papapetropoulos et al. 2009). The pro-angiogenic action of H_2S thus indirectly promotes solid tumor bioenergetics by increasing nutrients supply, including glucose and O_2.

In summary, H_2S seems to be a polyhedral actor being involved at different levels on cellular bioenergetics, stimulating mitochondrial ATP production, glycolysis and angiogenesis. On the other hand, in line with its Janus-faced character, at higher concentration it impairs the electron transport chain resulting in mitochondrial respiration failure.

17.5 H_2S and Hypoxia in the Tumor Microenvironment

The microenvironment of solid tumors is typically characterized by low O_2 tension. Hypoxia, while representing a challenge for cancer cells in some respects, in many others is beneficial, promoting cancer progression. O_2 deprivation has been recognized not only to promote cancer development and spreading by stimulating neoangiogenesis and metastasization, but also to increase resistance of cancer cells to treatments with chemotherapeutic agents or irradiation (see (Muz et al. 2015) for a review). Though based on a rather limited number of studies, there is growing evidence that H_2S plays a role in cancer cells under hypoxic conditions, as reviewed below.

Hypoxia has drastic effects on gene expression and metabolism in cancer cells (reviewed in (Masson and Ratcliffe 2014; Xie and Simon 2017; Samanta and Semenza 2018)) (Fig. 17.5a). Hypoxia-inducible factor-1 (HIF-1) is a master regulator of gene expression in response to changes in O_2 levels. This transcription factor regulates the expression of numerous genes, many of which are involved in fundamental processes occurring in cancer cells (Semenza 2006). Under normoxic conditions, HIF-1α is targeted by prolyl-hydroxylases (PHD) and committed to degradation. The rates of hydroxylation by PHD are, however, low under hypoxic conditions which thereby promote accumulation of HIF-1α (Yee Koh et al. 2008). HIF-1α was found to be

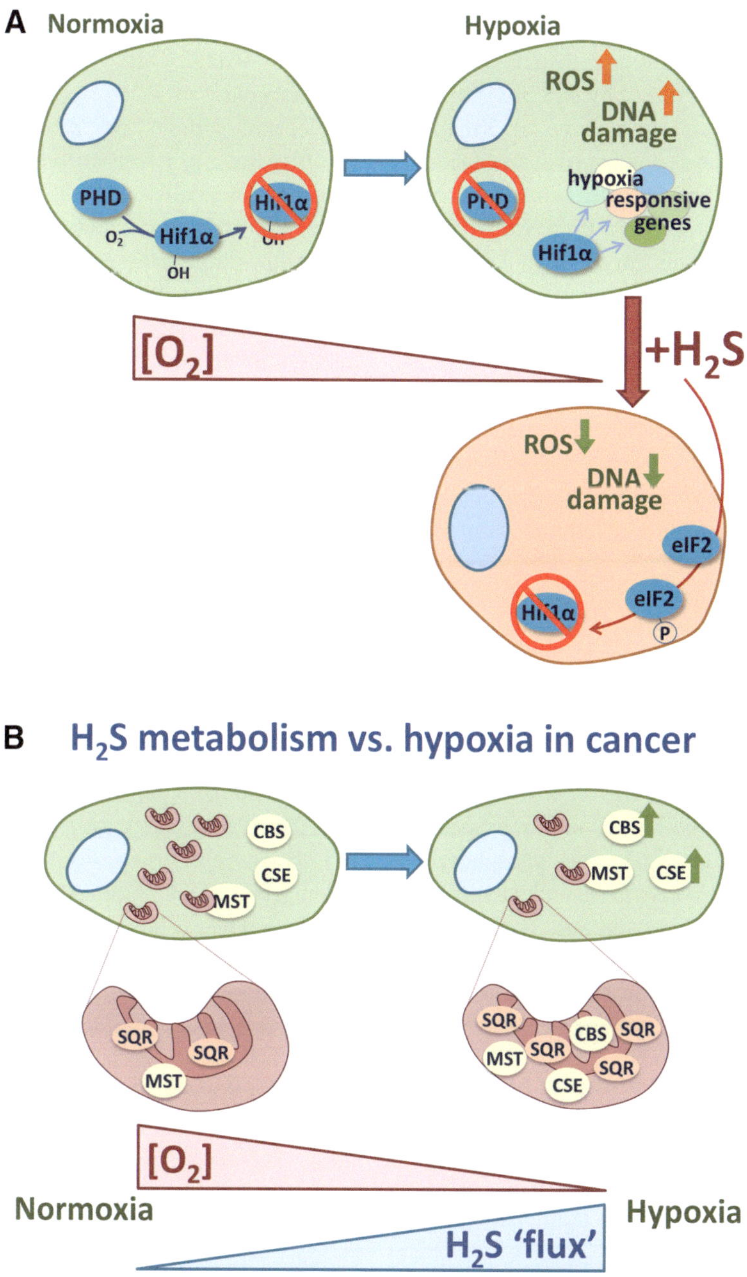

Fig. 17.5 **H_2S and hypoxia.** (**a**) The master gene regulator hypoxia-inducible factor-1α (HIF-1α), while being committed by prolyl-hydroxylases (PHD) to ubiquitin-mediated proteasomal degradation under normoxic conditions, under hypoxic conditions tends to accumulate in the cell in response to PHD activity impairment. In hypoxia, however, in addition to affording protection from oxidative stress and DNA damage, H_2S downregulates HIF-1α synthesis via phosphorylation of the eukaryotic translation initiation factor 2α (eIF2α). (**b**) Hypoxia up-regulates the expression of the H_2S-synthesizing enzymes cystathionine β-synthase (CBS) and cystathionine γ-lyase (CSE) and promotes their partial translocation into mitochondria, where 3-mercaptopyruvate sulfurtransferase (MST) is partly located. Concomitantly, mitochondria become less abundant, but enriched in the H_2S-consuming enzyme sulfide:quinone oxidoreductase (SQR) to protect the electron transport chain from H_2S poisoning

up-regulated by H_2S in *Caenorhabditis elegans* under normoxic conditions (Budde and Roth 2010), but down-regulated by H_2S under hypoxia in several human cell lines (Kai et al. 2012). The latter observation was confirmed and expanded in another study (Wu et al. 2012) where H_2S, exogenously administered as NaHS, proved to reduce HIF-1α protein levels in different human cell lines (HEK293T, Hep3B and EA.hy926) under both hypoxia and hypoxia-mimetic conditions in a dose- and time-dependent manner (Fig. 17.5a). The same authors investigated the molecular mechanism of HIF-1α down-regulation by H_2S and found that H_2S does not work at the transcriptional level, as shown by real-time PCR analysis, or stimulating the ubiquitin-proteasomal degradation pathway, but rather inhibits the translation of HIF-1α by promoting phosphorylation of the eukaryotic translation initiation factor 2α (eIF2α). eIF2α is a component of the eIF-2 complex that is required to start protein synthesis. When eIF2α is phosphorylated at Ser51, formation of the eIF-2 complex is impaired and HIF-1α translation inhibited (Yee Koh et al. 2008) (Fig. 17.5a). To be noted that repression of HIF-1α translation by H_2S via phosphorylated eIF2α was shown to take place under hypoxic but not hypoxia-mimetic conditions. To add further complexity, a mutual control between HIF-1α and H_2S seems to occur at low O_2 tension as HIF-1α was shown to stimulate CBS expression in hypoxia (Takano et al. 2014).

The hypoxic microenvironment is known to increase resistance of cancer cells not only to chemotherapeutic agents, but also to radiation-induced killing (Rockwell et al. 2009). Working on human hepatoma HepG2 cells, both endogenously produced and exogenous H_2S was reported to play a role in enhancing cancer cell radioresistance in hypoxia (Zhang et al. 2011). Compared to control cells not treated with H_2S-donors or inhibitors of H_2S synthesis, irradiated hypoxic cells were found to display reduced damages in the presence of NaHS and elevated damages in the presence of the inhibitors of H_2S-synthesis AOAA and propargylglycine (PPG). The radioprotective effect of NaHS was shown to be concentration dependent up to 100 μM and suggested to involve activation of the K_{ATP} channels, as shown using the selective inhibitor glibenclamide.

Neoangiogenesis, the development of new blood vessels, is a well-known strategy through which tumoral cells support their proliferation. A drawback of such a strategy, however, is that blood flow through these newly formed vessels is less constant than in the normal vasculature. Due to blood flow intermittence, tumoral cells are exposed to alternating conditions of hypoxia and re-oxygenation which are known to boost the formation of ROS, eventually leading to oxidative damage. Under these conditions, H_2S is likely to play a protective role. H_2S is indeed known to protect cells against ischemia/reperfusion damages (Bos et al. 2015; Jensen et al. 2017) through a variety of mechanisms to date only partly understood. Apart from exerting its well-known vasodilative and antioxidant effects, H_2S was reported to protect from hypoxia-induced proteostasis disruption, as shown working on *Caenorhabditis elegans* (Fawcett et al. 2015), and from mitochondrial DNA damages (Szczesny et al. 2016) which are known to occur in response to ischemia/reperfusion. A functional H_2S catabolism seems to be required to assure protection from hypoxia, as shown by knocking SQR down in Hepa1–6 cells (Hine et al. 2015). Accordingly, some studies highlighted a protective role of thiosulfate, one of the products of H_2S catabolism, against ischemia/reperfusion damage (Hine et al. 2015; Leskova et al. 2017; Marutani et al. 2015). Under hypoxic (but not under normoxic) conditions, H_2S either endogenously produced or exogenously administered as NaHS was also found to sustain energy metabolism in vascular smooth-muscle cells via stimulation of mitochondrial ATP production (Fu et al. 2012). In this context, it is noteworthy that the H_2S-precursor cysteine was recently reported to favor in several ovarian cancer cell lines adaptation to hypoxic conditions and resistance to the widely used chemotherapeutic agent carboplatin (Nunes et al. 2018). In the same study, cysteine was found to be the prevalent thiol compound in the ascitic fluid from patients with advanced ovarian cancer, and higher levels of cysteine and homocysteine (another precursor of H_2S) were intriguingly found in the serum of patients with ovarian cancer compared to healthy individuals.

Based on the evidence presented above, under hypoxic conditions, cancer cells seem to benefit from endogenously produced or exogenous H_2S in many regards, and H_2S can be viewed as contributing to adaptation of cancer cells to their harsh microenvironment. However, the occurrence of H_2S under hypoxia does not always represent a benefit for cancer cells in that hypoxia, under some circumstances, can exacerbate some of the detrimental effects of H_2S. For instance, in the ovarian cancer cell line A2780, it was found that the H_2S released from the donor GYY4137 induces disruption of calcium homeostasis causing endoplasmic reticulum stress and, ultimately, apoptosis, the effect being more pronounced in hypoxic than in normoxic conditions (Lencesova et al. 2016).

For cancer cells, particularly under hypoxic conditions, it is imperative to finely control H_2S bioavailability so that H_2S exerts its protective effects without inducing cytotoxicity. Control of intracellular H_2S levels in hypoxia relies on multiple processes and is based on a rather intricate interplay between H_2S and O_2 which led to recognition of H_2S as an O_2 sensor (Olson et al. 2006). As reviewed in (Olson 2015), O_2 can affect H_2S bioavailability essentially by regulating the protein levels and cellular localization of the H_2S-synthesizing enzymes, the efficacy of H_2S breakdown through the mitochondrial sulfide-oxidizing pathway and even the availability of inhibitors of H_2S synthesis, such as CO or NO. On the other hand, H_2S can regulate O_2 levels either enhancing or inhibiting mitochondrial O_2 consumption (see Sect. 17.4). Hypoxia is therefore expected to have a strong influence on H_2S bioavailability. O_2 deprivation not only is expected to result into a higher chemical stability of H_2S, but it was also shown to stimulate H_2S synthesis by enhancing the expression of H_2S-synthesizing enzymes (Wang et al. 2014; Takano et al. 2014), releasing inhibition of CBS and CSE by CO (Morikawa et al. 2012; Yuan et al. 2015) and accumulating CBS in mitochondria (Teng et al. 2013) (Fig. 17.5b). The effect of hypoxia on mitochondrial H_2S breakdown was also investigated in a few studies. Working initially on immortalized cells derived from alveolar macrophages (Matallo et al. 2014) and, then, on CHO cells (Abou-Hamdan et al. 2016), the mitochondrial sulfide-oxidizing activity was measured at different O_2 tensions and found to be lower at lower O_2 concentration, as expected. A more recent investigation assessed the effects of chronic (24 h) exposure to hypoxia (1% O_2) on the ability of cells to dispose sulfide at mitochondrial level, using SW480 colorectal cancer cells as a model (Malagrino et al. 2019). In line with previous studies (Solaini et al. 2010; Wu and Chen 2015; Zhang et al. 2008), hypoxia was found to reduce the mitochondrial mass and the overall ability of cells to dispose sulfide. However, considering their lower mitochondrial content, hypoxia-treated cells were found to contain mitochondria with higher maximal sulfide-detoxifying capacity and higher SQR levels, compared to untreated cells. The hypoxia-induced enrichment in SQR of mitochondria was suggested to have a protective role, preventing poisoning of mitochondria by enhanced production of sulfide under hypoxic conditions (Malagrino et al. 2019) (Fig. 17.5b). All in all, in hypoxic cancer cells H_2S is expected to occur at higher levels, as a result of its higher chemical stability, enhanced synthesis and diminished breakdown through the mitochondrial pathway. The higher bioavailability of H_2S could be beneficial for cancer cells, as long as they undergo adaptive mechanisms preventing H_2S toxicity.

Summing up, hypoxia as a common feature of tumor microenvironment has a strong impact on the bioavailability of H_2S, which in turn appears to play a role in adaptation of cancer cells to hypoxic conditions, further pointing to the H_2S-producing and -consuming enzymes as possible drug targets. Experimental research in this field is still in its infancy and complicated by the fact that not only hypoxia and H_2S can cause different effects between normal and tumoral cells, and even between different kinds of tumoral cells (Bianco et al. 2017), but sometimes different observations are made depending on the experimental design used to set-up hypoxic conditions. More efforts are therefore needed to shed light on the intricate interplay between H_2S and hypoxia in the framework of cancer biology, a promising research field with a great potential in terms of development of innovative anti-cancer strategies.

Acknowledgments The Authors are grateful for funding from Ministero dell'Istruzione, dell'Università e della Ricerca of Italy (PNR-CNR Aging Program 2012–2014 and PRIN 20158EB2CM_003). iNOVA4Health Research Unit (LISBOA-01-0145-FEDER-007344), which is cofunded by Fundação para a Ciência e Tecnologia/ Ministério da Ciência e do Ensino Superior, through national funds, and by FEDER under the PT2020 Partnership Agreement, is acknowledged by CST and JBV.

References

Abdollahi Govar A et al (2019) 3-Mercaptopyruvate sulfurtransferase supports endothelial cell angiogenesis and bioenergetics. Br J Pharmacol (in press)

Abou-Hamdan A et al (2016) Positive feedback during sulfide oxidation fine-tunes cellular affinity for oxygen. Biochim Biophys Acta 1857(9):1464–1472

Akaike T et al (2017) Cysteinyl-tRNA synthetase governs cysteine polysulfidation and mitochondrial bioenergetics. Nat Commun 8(1):1177

Augsburger F, Szabo C (2018) Potential role of the 3-mercaptopyruvate sulfurtransferase (3-MST)-hydrogen sulfide (H2S) pathway in cancer cells. Pharmacol Res:104083

Banerjee R (2017) Catalytic promiscuity and heme-dependent redox regulation of H_2S synthesis. Curr Opin Chem Biol 37:115–121

Banerjee R, Zou CG (2005) Redox regulation and reaction mechanism of human cystathionine-beta-synthase: a PLP-dependent hemesensor protein. Arch Biochem Biophys 433(1):144–156

Bao L et al (1998) Identification and tissue distribution of human cystathionine beta-synthase mRNA isoforms. Arch Biochem Biophys 350(1):95–103

Bhattacharyya S et al (2013) Cystathionine beta-synthase (CBS) contributes to advanced ovarian cancer progression and drug resistance. PLoS One 8(11):e79167

Bianco CL et al (2016) The chemical biology of the persulfide (RSSH)/perthiyl (RSS·) redox couple and possible role in biological redox signaling. Free Radic Biol Med 101:20–31

Bianco S et al (2017) Hypoxia and hydrogen sulfide differentially affect normal and tumor-derived vascular endothelium. Redox Biol 12:499–504

Bos EM et al (2015) Hydrogen sulfide: physiological properties and therapeutic potential in ischaemia. Br J Pharmacol 172(6):1479–1493

Budde MW, Roth MB (2010) Hydrogen sulfide increases hypoxia-inducible factor-1 activity independently of von Hippel-Lindau tumor suppressor-1 in C. elegans. Mol Biol Cell 21(1):212–217

Cai WJ et al (2007) The novel proangiogenic effect of hydrogen sulfide is dependent on Akt phosphorylation. Cardiovasc Res 76(1):29–40

Cao X et al (2019) A review of hydrogen sulfide synthesis, metabolism, and measurement: is modulation of hydrogen sulfide a novel therapeutic for cancer? Antioxid Redox Signal 31(1):1–38

Cooper CE, Brown GC (2008) The inhibition of mitochondrial cytochrome oxidase by the gases carbon monoxide, nitric oxide, hydrogen cyanide and hydrogen sulfide: chemical mechanism and physiological significance. J Bioenerg Biomembr 40(5):533–539

Cuevasanta E, Moller MN, Alvarez B (2017) Biological chemistry of hydrogen sulfide and persulfides. Arch Biochem Biophys 617:9–25

Druzhyna N et al (2016) Screening of a composite library of clinically used drugs and well-characterized pharmacological compounds for cystathionine beta-synthase inhibition identifies benserazide as a drug potentially suitable for repurposing for the experimental therapy of colon cancer. Pharmacol Res 113(Pt A):18–37

Ereno-Orbea J et al (2013) Structural basis of regulation and oligomerization of human cystathionine beta-synthase, the central enzyme of transsulfuration. Proc Natl Acad Sci U S A 110(40):E3790–E3799

Ereno-Orbea J et al (2014) Structural insight into the molecular mechanism of allosteric activation of human cystathionine beta-synthase by S-adenosylmethionine. Proc Natl Acad Sci U S A 111(37):E3845–E3852

Fagerberg L et al (2014) Analysis of the human tissue-specific expression by genome-wide integration of transcriptomics and antibody-based proteomics. Mol Cell Proteomics 13(2):397–406

Fawcett EM et al (2015) Hypoxia disrupts proteostasis in Caenorhabditis elegans. Aging Cell 14(1):92–101

Filipovic MR et al (2018) Chemical biology of H_2S signaling through persulfidation. Chem Rev 118(3):1253–1337

Frasdorf B, Radon C, Leimkuhler S (2014) Characterization and interaction studies of two isoforms of the dual localized 3-mercaptopyruvate sulfurtransferase TUM1 from humans. J Biol Chem 289(50):34543–34556

Fu M et al (2012) Hydrogen sulfide (H_2S) metabolism in mitochondria and its regulatory role in energy production. Proc Natl Acad Sci U S A 109(8):2943–2948

Giuffrè A, Vicente JB (2018) Hydrogen sulfide biochemistry and interplay with other gaseous mediators in mammalian physiology. Oxidative Med Cell Longev 2018:6290931

Giuffrè A et al (2002) Nitric oxide reacts with the single-electron reduced active site of cytochrome c oxidase. J Biol Chem 277(25):22402–22406

Goncalves-Dias C et al (2019) Mercapturate pathway in the tubulocentric perspective of diabetic kidney disease. Nephron:1–7

Goubern M et al (2007) Sulfide, the first inorganic substrate for human cells. FASEB J 21(8):1699–1706

Hellmich MR et al (2015) The therapeutic potential of cystathionine beta-synthetase/hydrogen sulfide inhibition in cancer. Antioxid Redox Signal 22(5):424–448

Hine C et al (2015) Endogenous hydrogen sulfide production is essential for dietary restriction benefits. Cell 160(1–2):132–144

Ianaro A, Cirino G, Wallace JL (2016) Hydrogen sulfide-releasing anti-inflammatory drugs for chemoprevention and treatment of cancer. Pharmacol Res 111:652–658

Ida T et al (2014) Reactive cysteine persulfides and S-polythiolation regulate oxidative stress and redox signaling. Proc Natl Acad Sci U S A 111(21):7606–7611

Jackson MR, Loll PJ, Jorns MS (2019) X-ray structure of human sulfide: quinone oxidoreductase: insights into the mechanism of mitochondrial hydrogen sulfide oxidation. Structure 27(5):794–805. e4

Jarosz AP et al (2015) Glyceraldehyde 3-phosphate dehydrogenase (GAPDH) is inactivated by S-sulfuration in vitro. Free Radic Biol Med 89:512–521

Jensen AR et al (2017) Hydrogen sulfide: a potential novel therapy for the treatment of ischemia. Shock 48(5):511–524

Kabil O et al (2011) The quantitative significance of the transsulfuration enzymes for H_2S production in murine tissues. Antioxid Redox Signal 15(2):363–372

Kai S et al (2012) Hydrogen sulfide inhibits hypoxia-but not anoxia-induced hypoxia-inducible factor 1 activation in a von hippel-lindau- and mitochondria-dependent manner. Antioxid Redox Signal 16(3):203–216

Kasamatsu S et al (2016) Redox signaling regulated by cysteine persulfide and protein polysulfidation. Molecules 21(12):E1721

Kimura Y et al (2017) 3-Mercaptopyruvate sulfurtransferase produces potential redox regulators cysteine- and glutathione-persulfide (Cys-SSH and GSSH) together with signaling molecules H_2S_2, H_2S_3 and H_2S. Sci Rep 7(1):10459

Kumar A et al (2018) Heme interaction of the intrinsically disordered N-terminal peptide segment of human cystathionine-beta-synthase. Sci Rep 8(1):2474

Lagoutte E et al (2010) Oxidation of hydrogen sulfide remains a priority in mammalian cells and causes reverse electron transfer in colonocytes. Biochim Biophys Acta 1797(8):1500–1511

Lencesova L et al (2016) Hypoxic conditions increases H(2)S-induced ER stress in A2870 cells. Mol Cell Biochem 414(1–2):67–76

Leschelle X et al (2005) Adaptative metabolic response of human colonic epithelial cells to the adverse effects of the luminal compound sulfide. Biochim Biophys Acta 1725(2):201–212

Leskova A et al (2017) Role of thiosulfate in hydrogen sulfide-dependent redox signaling in endothelial cells. Am J Physiol Heart Circ Physiol 313(2):H256–H264

Libiad M et al (2014) Organization of the human mitochondrial hydrogen sulfide oxidation pathway. J Biol Chem 289(45):30901–30910

Libiad M et al (2019) Hydrogen sulfide perturbs mitochondrial bioenergetics and triggers metabolic reprogramming in colon cells. J Biol Chem 294(32):12077–12090

Linden DR et al (2012) Sulphide quinone reductase contributes to hydrogen sulphide metabolism in murine peripheral tissues but not in the CNS. Br J Pharmacol 165(7):2178–2190

Malagrino F et al (2019) Hydrogen sulfide oxidation: adaptive changes in mitochondria of SW480 colorectal cancer cells upon exposure to hypoxia. Oxidative Med Cell Longev 2019:8102936

Marutani E et al (2015) Thiosulfate mediates cytoprotective effects of hydrogen sulfide against neuronal ischemia. J Am Heart Assoc 4(11):e002125

Masson N, Ratcliffe PJ (2014) Hypoxia signaling pathways in cancer metabolism: the importance of co-selecting interconnected physiological pathways. Cancer Metab 2(1):3

Mastronicola D et al (2003) Control of respiration by nitric oxide in Keilin-Hartree particles, mitochondria and SH-SY5Y neuroblastoma cells. Cell Mol Life Sci 60(8):1752–1759

Matallo J et al (2014) Sulfide-inhibition of mitochondrial respiration at very low oxygen concentrations. Nitric Oxide 41:79–84

McCorvie TJ et al (2014) Inter-domain communication of human cystathionine beta-synthase: structural basis of S-adenosyl-L-methionine activation. J Biol Chem 289(52):36018–36030

Millikin R et al (2016) The chemical biology of protein hydropersulfides: studies of a possible protective function of biological hydropersulfide generation. Free Radic Biol Med 97:136–147

Mishanina TV, Libiad M, Banerjee R (2015) Biogenesis of reactive sulfur species for signaling by hydrogen sulfide oxidation pathways. Nat Chem Biol 11(7):457–464

Modis K et al (2013) Intramitochondrial hydrogen sulfide production by 3-mercaptopyruvate sulfurtransferase maintains mitochondrial electron flow and supports cellular bioenergetics. FASEB J 27(2):601–611

Modis K et al (2016) S-Sulfhydration of ATP synthase by hydrogen sulfide stimulates mitochondrial bioenergetics. Pharmacol Res 113(Pt A):116–124

Morikawa T et al (2012) Hypoxic regulation of the cerebral microcirculation is mediated by a carbon monoxide-sensitive hydrogen sulfide pathway. Proc Natl Acad Sci U S A 109(4):1293–1298

Mudd SH et al (1965) Transsulfuration in mammals. Microassays and tissue distributions of three enzymes of the pathway. J Biol Chem 240(11):4382–4392

Mustafa AK et al (2009) H2S signals through protein S-sulfhydration. Sci Signal 2(96):ra72

Mustafa AK et al (2011) Hydrogen sulfide as endothelium-derived hyperpolarizing factor sulfhydrates potassium channels. Circ Res 109(11):1259–1268

Muz B et al (2015) The role of hypoxia in cancer progression, angiogenesis, metastasis, and resistance to therapy. Hypoxia (Auckl) 3:83–92

Nagy P (2015) Mechanistic chemical perspective of hydrogen sulfide signaling. Methods Enzymol 554:3–29
Nicholls P et al (2013) Sulfide inhibition of and metabolism by cytochrome c oxidase. Biochem Soc Trans 41(5):1312–1316
Niu WN et al (2015) S-glutathionylation enhances human cystathionine beta-synthase activity under oxidative stress conditions. Antioxid Redox Signal 22(5):350–361
Niu W et al (2018) Allosteric control of human cystathionine beta-synthase activity by a redox active disulfide bond. J Biol Chem 293(7):2523–2533
Nunes SC et al (2018) Cysteine allows ovarian cancer cells to adapt to hypoxia and to escape from carboplatin cytotoxicity. Sci Rep 8(1):9513
Ogasawara Y, Isoda S, Tanabe S (1994) Tissue and subcellular distribution of bound and acid-labile sulfur, and the enzymic capacity for sulfide production in the rat. Biol Pharm Bull 17(12):1535–1542
Olson KR (2015) Hydrogen sulfide as an oxygen sensor. Antioxid Redox Signal 22(5):377–397
Olson KR et al (2006) Hydrogen sulfide as an oxygen sensor/transducer in vertebrate hypoxic vasoconstriction and hypoxic vasodilation. J Exp Biol 209(Pt 20):4011–4023
Papapetropoulos A et al (2009) Hydrogen sulfide is an endogenous stimulator of angiogenesis. Proc Natl Acad Sci U S A 106(51):21972–21977
Paul BD, Snyder SH (2015) H_2S: a novel gasotransmitter that signals by sulfhydration. Trends Biochem Sci 40(11):687–700
Petersen LC (1977) The effect of inhibitors on the oxygen kinetics of cytochrome c oxidase. Biochim Biophys Acta 460(2):299–307
Pettinati I et al (2015) Crystal structure of human persulfide dioxygenase: structural basis of ethylmalonic encephalopathy. Hum Mol Genet 24(9):2458–2469
Pietri R, Roman-Morales E, Lopez-Garriga J (2011) Hydrogen sulfide and hemeproteins: knowledge and mysteries. Antioxid Redox Signal 15(2):393–404
Puccinelli MT, Stan SD (2017) Dietary bioactive diallyl trisulfide in cancer prevention and treatment. Int J Mol Sci 18(8):E1645
Reis A, Stern A, Monteiro HP (2019) S-nitrosothiols and H_2S donors: potential chemo-therapeutic agents in cancer. Redox Biol 27:101190
Rios-Gonzalez BB et al (2014) Hydrogen sulfide activation in hemeproteins: the sulfheme scenario. J Inorg Biochem 133:78–86
Rockwell S et al (2009) Hypoxia and radiation therapy: past history, ongoing research, and future promise. Curr Mol Med 9(4):442–458
Rose P et al (2005) Hydrogen sulfide protects colon cancer cells from chemopreventative agent beta-phenylethyl isothiocyanate induced apoptosis. World J Gastroenterol 11(26):3990–3997
Saha S et al (2016) Cystathionine beta-synthase regulates endothelial function via protein S-sulfhydration. FASEB J 30(1):441–456
Samanta D, Semenza GL (2018) Metabolic adaptation of cancer and immune cells mediated by hypoxia-inducible factors. Biochim Biophys Acta Rev Cancer 1870(1):15–22
Sarti P et al (2012) The chemical interplay between nitric oxide and mitochondrial cytochrome c oxidase: reactions, effectors and pathophysiology. Int J Cell Biol 2012:571067
Sekiguchi F et al (2016) Endogenous hydrogen sulfide enhances cell proliferation of human gastric cancer AGS cells. Biol Pharm Bull 39(5):887–890
Semenza GL (2006) Regulation of physiological responses to continuous and intermittent hypoxia by hypoxia-inducible factor 1. Exp Physiol 91(5):803–806
Smith RP, Gosselin RE (1966) On the mechanism of sulfide inactivation by methemoglobin. Toxicol Appl Pharmacol 8(1):159–172
Solaini G et al (2010) Hypoxia and mitochondrial oxidative metabolism. Biochim Biophys Acta 1797(6–7):1171–1177
Stokes E et al (2018) Efflux inhibition by H_2S confers sensitivity to doxorubicin-induced cell death in liver cancer cells. Life Sci 213:116–125
Sun Q et al (2009) Structural basis for the inhibition mechanism of human cystathionine gamma-lyase, an enzyme responsible for the production of H(2)S. J Biol Chem 284(5):3076–3085
Szabo C (2016) Gasotransmitters in cancer: from pathophysiology to experimental therapy. Nat Rev Drug Discov 15(3):185–203
Szabo C et al (2013) Tumor-derived hydrogen sulfide, produced by cystathionine-beta-synthase, stimulates bioenergetics, cell proliferation, and angiogenesis in colon cancer. Proc Natl Acad Sci U S A 110(30):12474–12479
Szabo C et al (2014) Regulation of mitochondrial bioenergetic function by hydrogen sulfide. Part I. Biochemical and physiological mechanisms. Br J Pharmacol 171(8):2099–2122
Szczesny B et al (2016) Inhibition of hydrogen sulfide biosynthesis sensitizes lung adenocarcinoma to chemotherapeutic drugs by inhibiting mitochondrial DNA repair and suppressing cellular bioenergetics. Sci Rep 6:36125
Takano N et al (2014) Hypoxia-inducible factors regulate human and rat cystathionine beta-synthase gene expression. Biochem J 458(2):203–211
Tang G et al (2005) Direct stimulation of K(ATP) channels by exogenous and endogenous hydrogen sulfide in vascular smooth muscle cells. Mol Pharmacol 68(6):1757–1764
Teng H et al (2013) Oxygen-sensitive mitochondrial accumulation of cystathionine beta-synthase mediated by Lon protease. Proc Natl Acad Sci U S A 110(31):12679–12684
Tiong CX, Lu M, Bian JS (2010) Protective effect of hydrogen sulphide against 6-OHDA-induced cell injury in SH-SY5Y cells involves PKC/PI3K/Akt pathway. Br J Pharmacol 161(2):467–480

Tomita M, Nagahara N, Ito T (2016) Expression of 3-mercaptopyruvate sulfurtransferase in the mouse. Molecules 21(12):E1707
Untereiner AA et al (2018) Drug resistance induces the upregulation of H_2S-producing enzymes in HCT116 colon cancer cells. Biochem Pharmacol 149:174–185
Valvona CJ et al (2016) The regulation and function of lactate dehydrogenase a: therapeutic potential in brain tumor. Brain Pathol 26(1):3–17
Vicente JB et al (2014) NO∗ binds human cystathionine beta-synthase quickly and tightly. J Biol Chem 289(12):8579–8587
Vicente JB et al (2016a) Bioenergetic relevance of hydrogen sulfide and the interplay between gasotransmitters at human cystathionine beta-synthase. Biochim Biophys Acta 1857(8):1127–1138
Vicente JB et al (2016b) S-Adenosyl-l-methionine modulates CO and NO∗ binding to the human H_2S-generating enzyme cystathionine beta-synthase. J Biol Chem 291(2):572–581
Vitvitsky V, Kabil O, Banerjee R (2012) High turnover rates for hydrogen sulfide allow for rapid regulation of its tissue concentrations. Antioxid Redox Signal 17(1):22–31
Vitvitsky V et al (2015) Sulfide oxidation by a noncanonical pathway in red blood cells generates thiosulfate and polysulfides. J Biol Chem 290(13):8310–8320
Vitvitsky V et al (2017) Structural and mechanistic insights into hemoglobin-catalyzed hydrogen sulfide oxidation and the fate of polysulfide products. J Biol Chem 292(13):5584–5592
Wallace JL et al (2018) Hydrogen sulfide-releasing therapeutics: translation to the clinic. Antioxid Redox Signal 28(16):1533–1540
Wang R (2012) Physiological implications of hydrogen sulfide: a whiff exploration that blossomed. Physiol Rev 92(2):791–896
Wang M, Guo Z, Wang S (2014) Regulation of cystathionine gamma-lyase in mammalian cells by hypoxia. Biochem Genet 52(1–2):29–37
Wang L et al (2018) A pharmacological probe identifies cystathionine beta-synthase as a new negative regulator for ferroptosis. Cell Death Dis 9(10):1005
Warburg O (1956a) On respiratory impairment in cancer cells. Science 124(3215):269–270
Warburg O (1956b) On the origin of cancer cells. Science 123(3191):309–314
Wu H, Chen Q (2015) Hypoxia activation of mitophagy and its role in disease pathogenesis. Antioxid Redox Signal 22(12):1032–1046
Wu B et al (2012) Hydrogen sulfide inhibits the translational expression of hypoxia-inducible factor-1alpha. Br J Pharmacol 167(7):1492–1505
Xie H, Simon MC (2017) Oxygen availability and metabolic reprogramming in cancer. J Biol Chem 292(41):16825–16832
Yadav PK et al (2013) Structure and kinetic analysis of H_2S production by human mercaptopyruvate sulfurtransferase. J Biol Chem 288(27):20002–20013
Yadav PK et al (2016) Biosynthesis and reactivity of cysteine persulfides in signaling. J Am Chem Soc 138(1):289–299
Yagdi E et al (2016) Garlic-derived natural polysulfanes as hydrogen sulfide donors: friend or foe? Food Chem Toxicol 95:219–233
Yang G et al (2008) H_2S as a physiologic vasorelaxant: hypertension in mice with deletion of cystathionine gamma-lyase. Science 322(5901):587–590
Yang G et al (2013) Hydrogen sulfide protects against cellular senescence via S-sulfhydration of Keap1 and activation of Nrf2. Antioxid Redox Signal 18(15):1906–1919
Yee Koh M, Spivak-Kroizman TR, Powis G (2008) HIF-1 regulation: not so easy come, easy go. Trends Biochem Sci 33(11):526–534
Yuan G et al (2015) Protein kinase G-regulated production of H2S governs oxygen sensing. Sci Signal 8(373):ra37
Zhang H et al (2008) Mitochondrial autophagy is an HIF-1-dependent adaptive metabolic response to hypoxia. J Biol Chem 283(16):10892–10903
Zhang J et al (2011) Hydrogen sulfide contributes to hypoxia-induced radioresistance on hepatoma cells. J Radiat Res 52(5):622–628
Zhao K et al (2014) S-sulfhydration of MEK1 leads to PARP-1 activation and DNA damage repair. EMBO Rep 15(7):792–800
Zhen Y et al (2015) Exogenous hydrogen sulfide exerts proliferation/anti-apoptosis/angiogenesis/migration effects via amplifying the activation of NF-kappaB pathway in PLC/PRF/5 hepatoma cells. Int J Oncol 46(5):2194–2204
Zuhra K et al (2019) Screening pyridine derivatives against human hydrogen sulfide-synthesizing enzymes by orthogonal methods. Sci Rep 9(1):684

18 Ovarian Cancer Biomarkers: Moving Forward in Early Detection

Vasco D. B. Bonifácio

Abstract

Ovarian cancer is a silent cancer which rate survival mainly relays in early stage detection. The discovery of reliable ovarian cancer biomarkers plays a crucial role in the disease management and strongly impact in patient's prognosis and survival. Although having many limitations CA125 is a classical ovarian cancer biomarker, but current research using proteomic or metabolomic methodologies struggles to find alternative biomarkers, using non-invasive our relatively non-invasive sources such as urine, serum, plasma, tissue, ascites or exosomes. Metabolism and metabolites are key players in cancer biology and its importance in biomarkers discovery cannot be neglected. In this chapter we overview the state of art and the challenges facing the use and discovery of biomarkers and focus on ovarian cancer early detection.

Keywords

Cancer biomarkers · Ovarian cancer · Early detection · Urine biomarkers · Proteomics · Metabolomics

V. D. B. Bonifácio (✉)
IBB – Institute for Bioengineering and Biosciences, Instituto Superior Técnico, Universidade de Lisboa, Lisbon, Portugal
e-mail: vasco.bonifacio@tecnico.ulisboa.pt

Abbreviation

ApoA-1	Apolipoprotein A-1
AS	Advanced Stage
CA125	Cancer Antigen 125
COL3A1	Collagen alpha-1 (III)
EMILIN2	Elastin microfibril interfacer 2
EphA8	Ephrin receptor A8
ESD	Early Stage Disease
FDA	Food and Drug Administration
FGA	Fibrinogen alpha
FGB NT	Fibrinogen beta NT
FSH	Follicle-Stimulating Hormone
FTL	Ferritin light chain
HE4	Human Epididymis protein 4
Hp	Haptoglobin
HSP27	Heat Shock Protein 27
IGF-2	Insulin-like Growth Factor 2
IR-2α	Interleukin-2 Receptor α
ITIH4	Inter-α-Trypsin Inhibitor heavy chain H4
NAGK	*N*-acetyl-D-glucosamine kinase
NM23-H1	Nucleoside diphosphate kinase A
PA28	Reg-alpha fragment
PEBP1	Phosphatidylethanolamine Binding Protein 1
PP1	Protein phosphatase-1
Prx-II	Peroxiredoxin II
PSMA6	Proteasome alpha-6
SCEH	Mitochondrial short-chain enoyl-CoA hydratase

J. Serpa (ed.), *Tumor Microenvironment*, Advances in Experimental Medicine and Biology 1219,
https://doi.org/10.1007/978-3-030-34025-4_18

TTR	Transthyretin
TVUS	Transvaginal Ultrasonography
VWF	von Willebrand factor

18.1 Introduction

By definition, a cancer biomarker is any molecular or biochemical alteration that can be measured and effectively used in a clinical scenario for cancer detection, diagnosis, prognosis, and prediction of therapeutic response (Patriotis et al. 2017). Tied to this definition, as a rule of thumb, in the occurrence of false positive and false negative results it is paramount that any diagnostic method leaves little room for error. Cancer early detection is absolutely crucial in patient's treatment and survival, and the discovery of cancer biomarkers operated a revolution in cancer screening. Moreover, if sensitive (probability of correctly identifying presence) and specific (probability of correctly identifying absence), they hold a great promise in the therapeutic roadmap, reducing costs and cancer-related mortality and morbidity. However, despite the huge efforts made in this field, only a few cancer biomarkers are currently FDA approved (Anderson and Anderson 2002). A scenario that needs urgent attention from regulators. Considering the fact that cancer biomarkers can be explored in relatively noninvasive body fluids or excretions, such as blood, saliva, sputum, upper digestive tract effusion, urine and stool, its investigation should be of utmost priority. Aside a lower concentration in these fluids, if compared with cancer tissues, the remarkable developments in proteomics in the last few years will certainly surpass this issue (Fig. 18.1).

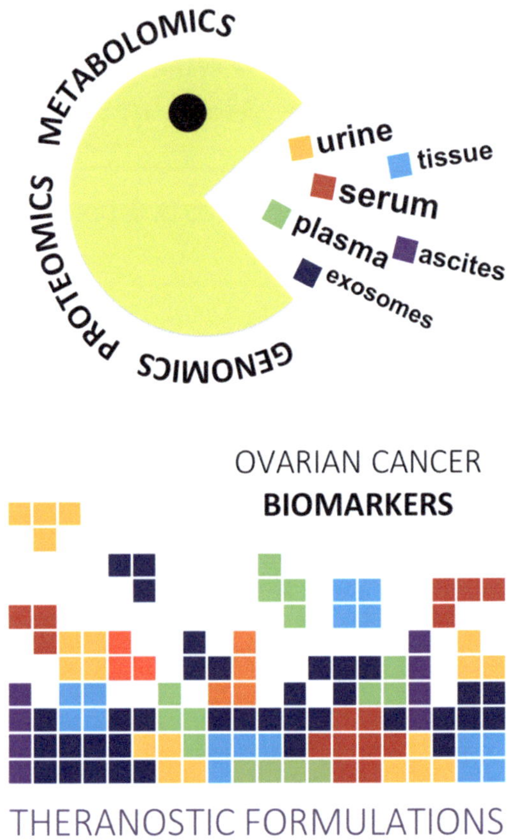

Fig. 18.1 Scheme showing the main players in ovarian cancer biomarkers discovery towards theranostics

Regarding cancer early detection, a special focus in silent cancers is urgently needed. This is the case of ovarian carcinoma, a very aggressive and highly lethal cancer. Currently, CA125 and Human Epididymis protein 4 (HE4) are the only two markers that have been approved by the FDA for monitoring treatment and detecting disease recurrence. Very recently, OVERA® was approved as referral or triage test for patients presenting with ovarian mass. The test is as second generation of the previous version OVA1® and is a combination of CA125, HE4, apolipoprotein A-1, follicle-stimulating hormone (FSH), and transferrin (sensitivity 91% and specificity 69%) (Coleman et al. 2016). In this new formulation transthyretin and β-2-microglolin were substituted by transferrin and FSH.

18.2 Ovarian Carcinoma: A Silent Killer

Ovarian cancer is the most lethal gynaecological cancer and is the fifth cause of cancer related death among western women (Ferlay et al. 2013). Notwithstanding a better awareness of the disease, aggressive cytoreductive surgery, and new chemotherapeutics, mortality rates haven't changed in the last 30 years. This is partly due to the difficulty in detecting early stage disease (ESD) as well as the lack of effective therapeutic options for advanced stages (AS – stage III/IV). ESD 5-year overall survival is 90% *vs* 20–25% in AS. However, 70% of patients are diagnosed in AS. Therefore, subtle symptoms, leading to later stages with high dissemination of the disease, lack of adequate screening and chemoresistance, all contribute to poor diagnostics and treatment failure.

Since the events leading to AS are poorly understood and often contradictory (Rei et al. 2014; Zhao et al. 2015), early detection is undoubtedly a crucial issue. Despite many efforts on finding ovarian cancer biomarkers, early diagnosis methods still rely on CA125 serum measurement (Scholler and Urban 2007), but many others have been disclosed as alternatives (see Table 18.1), both protein- (Zhang et al. 2010; Elzek and Rodland 2015) and gene-based biomarkers (Zhang et al. 2011).

CA125 is a mucin-type glycoprotein (MUC16) that is elevated in 83% of patients with ovarian cancer, but only in 50–60% of patients in stage I. Overexpression in other cancers, and benign diseases of the ovaries and reproductive tract (*e.g.* endometriosis), menstruation and pregnancy, is a problem (Bast et al. 1998). CA125 is also overexpressed in hematologic malignancies such as Hodgkin's and non-Hodgkin's lymphomas and is used as a biomarker in these cases (Russo et al. 2007; Wang et al. 2009). Thus, due to low reliability (70% sensitivity and 87% specificity) is not recommended for population screening (Moss et al. 2005). In alternative, transvaginal ultrasonography (TVUS) screening has been also extensively investigated. TVUS produces ovarian images by applying ultrasound waves (5–7.5 MHz) across the vaginal wall and allow identification of abnormal, malignant or benign, morphological changes that are non-identifiable upon physical examination. Yet, current data suggest that this approach alone not only lacks specificity, but may be of little value to diagnose ovarian carcinomas before they metastasize (Horiuchi et al. 2003). However, the combination of TVUS with the measurement of multiple serum markers was shown to be advantageous in ESD (Woolas et al. 1993). Proteomic approaches based on a panel of three biomarkers, combined

Table 18.1 Examples of protein-based ovarian cancer biomarkers

Biomarker	References
Serum and Plasma	
CA125	Bast et al. (1998)
CA125 and soluble IR-2α	Hurteau et al. (1995)
CA125 and Prostasin	Mok et al. (2001)
CA125 and ApoA-1, TTR, ITIH4	Zhang et al. (2004)
Apolipoprotein A1	Kozak et al. (2005)
TTR	Kozak et al. (2005)
Leptin, Prolactin, IGF-2	Mor et al. (2005)
Osteopontin	Mor et al. (2005); Ye et al. (2006)[a]
Amyloid A1	Helleman et al. (2008); Moshkovskii et al. (2005)[b]
Catabolic fragments of complement factors, EMILIN2, VWF, PEBP1	Scholler et al. (2008)[c]
Afamin	Jackson et al. (2007)
β-Hemoglobin	Kozak et al. (2005)
Transferrin	Kozak et al. (2005); Ahmed et al. (2005)
Hp, Hp-1 precursor, Hp-α subunit	Ahmed et al. (2005); Ahmed et al. (2004); Ye et al. (2003)
Fibrinopeptide-A	Bergen et al. (2003)
HE4	Lin et al. (2012); Wang et al. (2014); Diavatis and Papanikolaou (2016)
Claudine-4	Li et al. (2009b)[b]
Urine	
Mesothelin	Badgwell et al. (2007)
Eosinophil protein X	Ye et al. (2006)
FGA, FGB NT and COL3A1 fragments	Petri et al. (2009)
Angiostatin	Drenberg et al. (2010)
Tissue	
PA28	Lemaire et al. (2007)
NM23-H1, NAGK, Annexin-1, PP1, FTL, PSMA6	An et al. (2006)
EphA8	Liu et al. (2016)
Prx-II, Prx-III, HSP27, HSP60, SCEH, Prohibitin	Li et al. (2009a)
Exosomes	
HSP70	Gobbo et al. (2016)

[a]Urine; [b]Plasma; [c]Ascites

with CA125 measurements, distinguished patients with stage I/II ovarian cancer from healthy controls with a specificity of 94% (Zhang et al. 2004). More recently, using a multiplex, bead-based, immunoassay system, by concentration analysis of CA125, leptin, prolactin, osteopontin, insulin-like growth factor II, macrophage inhibitory factor and HE4, a high overall sensitivity of 94.3% and specificity of 92.3% were found (Gschwantler-Kaulich et al. 2017). This is a good example how cancer biomarkers, analyzed together, may give a reliable response to such a complex cancer. The screening of non-protein biomarkers was been also performed. Remarkably, lysophosphatidic acid (LPA) and other lysophospholipids (LPL) appear to be useful diagnostic and prognostic ovarian cancer biomarkers (Sutphen et al. 2004).

18.3 Metabolism and Cancer Biomarkers

Cancer development, progression, and therapeutic response is highly dependent on the tumor microenvironment, and is known that in complex process the metabolism of cancer cells is reprogrammed from that of normal cells (Yang 2017). As a consequence of these abnormal events, metabolic dysregulation leads to genes overexpression or silencing, which ultimately impact on alien proteins or metabolites. In this scenario, cancer biomarkers emerge as metabolic patterns, thus allowing disease monitoring. The quest for metabolic biomarkers of cancer has been intense over the last decade, but translation to clinics has been precluded mainly due to lack of a clear and proven mechanistic link to cancer metabolism (Muthu and Nordström 2019). Nevertheless, ongoing investigation clearly points toward an intimate relation between biomarkers, metabolic dysregulation and cancer mortality (Akinyemiju et al. 2018). This is a transversal evidence to many cancer types (Bosco et al. 2015), but for ovarian cancer in particular is worth noting the impact of impaired glucose metabolism (Lambe et al. 2011; Melvin et al. 2012). Nevertheless, the shift between glycolysis and glutaminolysis has been shown as fundamental for increased ovarian cancer aggressiveness (Yang et al. 2014). Several studies, addressing the blockade of glucose and glutamine degradation, disclosed this route as a new therapeutic strategy to disturb cancer cells viability (Guo et al. 2016). Besides being an important energy and biomass source, glutamine is crucial in the regulation of the redox state, in normal and cancer cells (Lopes-Coelho et al. 2016). Glutamine-derived glutamate together with cysteine and glycine constitute the glutathione (GSH) tripeptide, which is the main small thiol working as a free radical scavenger in mammal cells. The increased levels of GSH was already proven as a mechanism of chemoresistance in ovarian cancer; and the inhibition of its synthesis is a way of re-sensitizing cancer cells to alkylating drugs (Lopes-Coelho et al. 2016; Bruntz et al. 2019). Hence in some ovarian cancer subtypes GSH metabolism is considered a valuable target (Ogiwara et al. 2019). In addition, the levels of cysteine and homo-cysteine, in peripheral blood, were pointed out as putative markers for early detection of ovarian cancer (Nunes et al. 2018). In this onset, metabolites from tumor origin are a new paradigm in cancer aggressiveness and tumor repopulation and represent a great potential for theranostics (Dando et al. 2019; Collins et al. 2017).

18.4 Urine: A Universal Source for Cancer Biomarkers?

Urine is one of the most popular biofluids used in cancer biomarker discovery. Apart from its highly attractive noninvasive nature, urine has also the advantage of having a much less complex protein profile than blood, being those more stable. The potential of urine analysis has a source of biomarkers has been investigated in different cancers, namely prostate (Goo and Goodlett 2010), renal cell carcinoma (Kim et al. 2009), bladder (Shirodkar and Lokeshwar 2009), breast (Slupsky et al. 2010) and ovarian (Ye et al. 2006). The potential of urine as a source of ovarian cancer biomarkers (Chambers and Vanderhyden 2006) is remarkable. Following a two-step proteomic approach two non-cancer-specific biomarkers were found, eosinophil protein X, a glycosylated form of an eosinophil-derived neurotoxin (2 times elevated), and carboxylic acid-terminated fragments of osteopontin. Interestingly, eosinophil protein X has been shown to have antitumor and antiangiogenic activity via induction of apoptosis of endothelial cells and osteopontin is correlated with systemic inflammation. Although poorly understood, the relevance of osteopontin in ovarian cancer biology may lead to important advances in ovarian cancer early detection. Potential urine biomarkers include also HE4 (Li et al. 2009c) and Bcl-2 (B-cell lymphoma 2) (Anderson et al. 2009). HE4 has only a specificity of 87% and a sensitivity of 74% but is still under consideration for use in larger trials (Grayson et al. 2019). Urine metabolomics is another emergent topic in ovarian cancer biomarker that deserve a more detailed investigation (Jiang et al. 2015), since urinary metabolic profiling shows changes in metabolite concentrations that can be specifically correlated not only with ovarian cancer (98% sensitivity, 99% specificity) but also with breast cancer (100% sensitivity, 93% specificity) (Slupsky et al. 2010). Currently, known urine biomarkers require further validation and still lack efficiency to detect ovarian cancer in stages I and II, but may be useful in combination with other non-urinary biomarkers and TVUS.

18.5 Future Perspectives

Albeit the enormous work developed on ovarian cancer biomarkers discovery, we are still far from an optimal (noninvasive, with low false positives) early detection test. Nevertheless, in the last years we have witnessed many advances in this field, either exploring imaging, protein profiles, specific symptoms, or combinations of these (Cohen et al. 2014). Despite controversy, CA125 is still the most used biomarker in ovarian cancer detection. Nevertheless, its recommendation must be carefully weighted in the case of remission after first-line therapy since these patients do not benefit from routine measurements during follow-up, being this a cause of distress (Krell et al. 2017). It is well known that heterogeneity in cancer cells negatively influences treatment efficacy and survival of patients. In this sense, single-cell analysis is foreseen as novel platform to progress the investigation of more specific biomarkers to identify and target cancer stem cells. In particular, ovarian cancer single-cell analysis revealed two major subsets of cells characterized by stromal gene expression patterns and epithelial gene expression signature (Radpour and Forouharkhou 2018). Tumor acidity is a well-known and primary regulator of cancer immunity (Huber et al. 2017). Regarding theranostic approaches, many nanoformulations (*e.g.* dendrimers, liposomes) explore the tumor microenvironment acidity (Feng et al. 2018; Fernandes et al. 2018). In ovarian cancer treatment, the co-delivery of carboplatin and paclitaxel was already investigated using cross-linked multilamellar liposomes (Zhang et al. 2017). Therefore, theranostics is a strategy of special importance in cancer eradication, and we cannot neglect the truly fundamental role of cancer metabolism investigation as these biomarkers may be explored in targeted drug delivery (Tanasova et al. 2018; Kutova et al. 2019). Another interesting emergent approach is the use of ovarian cancer derived exosomes (Yang et al. 2019), which may operate a revolution in theranostics. In perspective, metabolomics is certainly the task of force in the discovery of new ovarian cancer biomarkers (Plewa et al. 2019). Also, in this field ascitic fluid seems to be an underestimated fluid that may hold a rich source for ovarian cancer biomarkers (Mills et al. 1988; Bharti et al. 2017; Nunes et al. 2018). Substantial advances have been made in order to understand the molecular bases of ovarian cancer, but the current status of the disease remains a challenge for researchers and clinicians. Overall, the information gathered so far will be crucial in future biomarker discovery and validation studies or promising candidates, which we hope soon translate to clinic.

References

Ahmed N, Barker G, Oliva KT, Hoffmann P, Riley C, Reeve S, Smith AI, Kemp BE, Quinn MA, Rice GE (2004) Proteomic-based identification of haptoglobin-1 precursor as a novel circulating biomarker of ovarian cancer. Br J Cancer 91:129–140

Ahmed N, Oliva KT, Barker G, Hoffmann P, Reeve S, Smith IA, Quinn MA, Rice GE (2005) Proteomic tracking of serum protein isoforms as screening biomarkers of ovarian cancer. Proteomics 5:4625–4636

Akinyemiju T, Moore JX, Judd SE, Pisu M, Goodman M, Howard VJ, Long L, Safford M, Gilchrist SC, Cushman M (2018) Pre-diagnostic biomarkers of metabolic dysregulation and cancer mortality. Oncotarget 9:16099–16109

An HJ, Kim DS, Park YK, Kim SK, Choi YP, Kang S, Ding B, Cho NH (2006) Comparative proteomics of ovarian epithelial tumors. J Proteome Res 5:1082–1090

Anderson NL, Anderson NG (2002) The human plasma proteome: history, character, and diagnostic prospects. Mol Cell Proteomics 1:845–867

Anderson NS, Bermudez Y, Badgwell D, Chen R, Nicosia SV, Bast RC Jr, Kruk PA (2009) Urinary levels of Bcl-2 are elevated in ovarian cancer patients. Gynecol Oncol 112(1):60–67

Badgwell D, Lu Z, Cole L, Fritsche H, Atkinson EN, Somers E, Allard J, Moore RG, Lu KH, Bast RC Jr (2007) Urinary mesothelin provides greater sensitivity for early stage ovarian cancer than serum mesothelin, urinary hCG free beta subunit and urinary hCG beta core fragment. Gynecol Oncol 106(3):490–497

Bast RC Jr, Xu FJ, Yu YH, Barnhill S, Zhang Z, Mills GB (1998) CA125: the past and the future. Int J Biol Markers 13:179–187

Bergen HR 3rd, Vasmatzis G, Cliby WA, Johnson KL, Oberg AL, Muddiman DC (2003) Discovery of ovarian cancer biomarkers in serum using NanoLC electrospray ionization TOF and FT-ICR mass spectrometry. Dis Markers 19:239–249

Bharti SK, Wildes F, Hung CF, Wu TC, Bhujwalla ZM, Penet MF (2017) Metabolomic characterization of experimental ovarian cancer ascitic fluid. Metabolomics 13:113

Bosco C, Wulaningsih W, Melvin J, Santaolalla A, De Piano M, Arthur R, Van Hemelrijck M (2015) Metabolic serum biomarkers for the prediction of cancer: a follow-up of the studies conducted in the Swedish AMORIS study. ecancer 9:555

Bruntz RC, Belshoff AC, Zhang Y, Macedo JKA, Higashi RM, Lane AN, Fan TW (2019) Inhibition of anaplerotic glutaminolysis underlies selenite toxicity in human lung cancer. Proteomics 12:e1800486

Chambers AF, Vanderhyden BC (2006) Ovarian cancer biomarkers in urine. Clin Cancer Res 12(2):323–327

Cohen JG, White M, Cruz A, Farias-Eisner R (2014) In 2014, can we do better than CA125 in the early detection of ovarian cancer? World J Biol Chem 5(3):286–300

Coleman R, Herzog T, Chan D, Munroe D, Pappas T, Smith A, Zhang Z, Wolf J (2016) Validation of a second-generation multivariate index assay for malignancy risk of adnexal masses. Am J Obstet Gynecol 215:82.e1–82.e11

Collins RRJ, Patel K, Putnam WC, Kapur P, Rakheja D (2017) Oncometabolites: a new paradigm for oncology, metabolism, and the clinical laboratory. Clin Chem 63(12):1812–1820

Dando I, Pozza ID, Ambrosini G, Torrens-Mas M, Butera G, Mullappilly N, Pacchiana R, Palmieri M, Donadelli M (2019) Oncometabolites in cancer aggressiveness and tumour repopulation. Biol Rev 94(4). https://doi.org/10.1111/brv.12513

Diavatis S, Papanikolaou A (2016) Level of HE4 is correlated with diagnosis of struma ovarii: a case report. Am J Case Rep 17:459–461

Drenberg CD, Saunders BO, Wilbanks GD, Chen R, Nicosia RF, Kruk PA, Nicosia SV (2010) Urinary angiostatin levels are elevated in patients with epithelial ovarian cancer. Gynecol Oncol 117(1):117–124

Elzek MA, Rodland KD (2015) Proteomics of ovarian cancer: functional insights and clinical applications. Cancer Metastasis Rev 34:83–96

Feng L, Dong Z, Tao D, Zhang Y, Liu Z (2018) The acidic tumor microenvironment: a target for smart cancer nano-theranostics. Natl Sci Rev 5(2):269–286

Ferlay J, Steliarova-Foucher E, Lortet-Tieulent J, Rosso S, Coebergh JW, Comber H, Forman D, Bray F (2013) Cancer incidence and mortality patterns in Europe: estimates for 40 countries in 2012. Eur J Cancer 49:1374–1403

Fernandes C, Suares D, Yergeri MC (2018) Tumor microenvironment targeted nanotherapy. Front Pharmacol 9:1230

Gobbo J, Marcion G, Cordonnier M, Dias AMM, Pernet N, Hammann A, Richaud S, Mjahed H, Isambert N, Clausse V, Rébé C, Bertaut A, Goussot V, Lirussi F, Ghiringhelli F, de Thonel A, Fumoleau P, Seigneuric R, Garrido C (2016) Restoring anticancer immune response by targeting tumor-derived exosomes with a HSP70 peptide aptamer. J Natl Cancer Inst 108(3).:djv330

Goo YA, Goodlett DR (2010) Advances in proteomic prostate cancer biomarker discovery. J Proteome 73:1839–1850

Grayson K, Gregory E, Khan G, Guinn B-A (2019) Urine biomarkers for the early detection of ovarian cancer – are we there yet? Biomark Cancer 11:1179299X19830977

Gschwantler-Kaulich D, Weingartshofer S, Rappaport-Fürhauser C, Zeillinger R, Pils D, Muhr D, Braicu EI, Kastner MT, Tan YY, Semmler L, Sehouli J, Singer CF (2017) Diagnostic markers for the detection of ovarian cancer in BRCA1 mutation carriers. PLoS One 12(12):e0189641

Guo L, Zhou B, Liu Z, Xu Y, Lu H, Xia M, Guo E, Shan W, Chen G, Wang C (2016) Blockage of glutaminolysis enhances the sensitivity of ovarian cancer cells to PI3K/mTOR inhibition involvement of STAT3 signaling. Tumour Biol 37(8):11007–11015

Helleman J, van der Vlies D, Jansen MP, Luider TM, van der Burg ME, Stoter G, Berns EM (2008) Serum proteomic patterns for ovarian cancer monitoring. Int J Gynecol Cancer 18:985–995

Horiuchi A, Itoh K, Shimizu M, Nakai I, Yamazaki T, Kimura K, Suzuki A, Shiozawa I, Ueda N, Konishi I (2003) Towards understanding the natural history of ovarian carcinoma development: a clinicopathologic approach. Gynecol Oncol 88:309–317

Huber V, Camisaschi C, Berzi A, Ferro S, Lugini L, Triulzi T, Tuccitto A, Tagliabue E, Castelli C, Rivoltini L (2017) Cancer acidity: an ultimate frontier of tumor immune escape and a novel target of immunomodulation. Semin Cancer Biol 43:74–89

Hurteau JA, Woolas RP, Jacobs IJ, Oram DC, Kurman CC, Rubin LA, Nelson DL, Berchuck A, Bast RC Jr, Mills GB (1995) Soluble interleukin-2 receptor α is elevated in sera of patients with benign ovarian neoplasms and epithelial ovarian cancer. Cancer 76:1615–1620

Jackson D, Craven RA, Hutson RC, Graze I, Lueth P, Tonge RP, Hartley JL, Nickson JA, Rayner SJ, Johnston C, Dieplinger B, Hubalek M, Wilkinson N, Perren TJ, Kehoe S, Hall GD, Daxenbichler G, Dieplinger H, Selby PJ, Banks RE (2007) Proteomic profiling identifies afamin as a potential biomarker for ovarian cancer. Clin Cancer Res 13:7370–7379

Jiang T, Lin Y, Yin H, Wang S, Sun Q, Zhang P, Bi W (2015) Correlation analysis of urine metabolites and clinical staging in patients with ovarian cancer. Int J Clin Exp Med 8(10):18165–18171

Kim K, Aronov P, Zakharkin SO, Anderson D, Perroud B, Thompson IM, Weiss RH (2009) Urine metabolomics analysis for kidney cancer detection and biomarker discovery. Mol Cell Proteomics 8:558–570

Kozak KR, Su F, Whitelegge JP, Faull K, Reddy S, Farias-Eisner R (2005) Characterization of serum biomarkers for detection of early stage ovarian cancer. Proteomics 5:4589–4596

Krell D, Battistino FS, Benafif S, Ganegoda L, Hall M, Rustin GJS (2017) Audit of CA125 follow-up after first-line therapy for ovarian cancer. Int J Gynecol Cancer 27(6):1118–1122

Kutova OM, Guryev EL, Sokolova EA, Alzeibak R, Balalaeva IV (2019) Cancers (Basel) 11(1):68

Lambe M, Wigertz A, Garmo H, Walldius G, Jungner I, Hammar N (2011) Impaired glucose metabolism and diabetes and the risk of breast, endometrial, and ovarian cancer. Cancer Causes Control 22(8):1163–1171

Lemaire R, Menguellet SA, Stauber J, Marchaudon V, Lucot JP, Collinet P, Farine MO, Vinatier D, Day R, Ducoroy P, Salzet M, Fournier I (2007) Specific MALDI imaging and profiling for biomarker hunting and validation: fragment of the 11S proteasome activator complex, Reg alpha fragment, is a new potential ovary cancer biomarker. J Proteome Res 6:4127–4134

Li XQ, Zhang SL, Cai Z, Zhou Y, Ye TM, Chiu JF (2009a) Proteomic identification of tumor-associated protein in ovarian serous cystadenocarcinoma. Cancer Lett 275:109–116

Li J, Sherman-Baust CA, Tsai-Turton M, Bristow RE, Roden RB, Morin PJ (2009b) Claudin-containing exosomes in the peripheral circulation of women with ovarian cancer. BMC Cancer 9:244

Li J, Dowdy S, Tipton T, Podratz K, Lu WG, Xie X, Jiang SW (2009c) HE4 as a biomarker for ovarian and endometrial cancer management. Expert Rev Mol Diagn 9(6):555–566

Lin J, Qin J, Li X, Dong P, Yin B (2012) Diagnostic value of human epididymis protein 4 compared with mesothelin for ovarian cancer: a systematic review and meta-analysis. Asian Pac. J Cancer Prev 13(11):5427–5432

Liu X, Xu Y, Jin Q, Wang W, Zhang S, Wang X, Zhang Y, Xu X, Huang J (2016) EphA8 is a prognostic marker for epithelial ovarian cancer. Oncotarget 7(15):20801–20809

Lopes-Coelho F, Gouveia-Fernandes S, Gonçalves LG, Nunes C, Faustino I, Silva F, Félix A, Pereira SA, Serpa J (2016) HNF1β drives glutathione (GSH) synthesis underlying intrinsic carboplatin resistance of ovarian clear cell carcinoma (OCCC). Tumour Biol 37(4):4813–4829

Melvin JC, Seth D, Holmberg L, Garmo H, Hammar N, Jungner I, Walldius G, Lambe M, Wigertz A, Van Hemelrijck M (2012) Lipid profiles and risk of breast and ovarian cancer in the Swedish AMORIS study. Cancer Epidemiol Biomark Prev 21(8):1381–1384

Mills GB, May C, McGill M, Roifman CM, Mellors A (1988) A putative new growth factor in ascitic fluid from ovarian cancer patients: identification, characterization, and mechanism of action. Cancer Res 48(5):1066–1071

Mok SC, Chao J, Skates S, Wong K, Yiu GK, Muto MG, Berkowitz RS, Cramer DW (2001) Prostasin, a potential serum marker for ovarian cancer: identification through microarray technology. J Natl Cancer Inst 93:1458–1464

Mor G, Visintin I, Lai Y, Zhao H, Schwartz P, Rutherford T, Yue L, Bray-Ward P, Ward DC (2005) Serum protein markers for early detection of ovarian cancer. Proc Natl Acad Sci U S A 102:7677–7682

Moshkovskii SA, Serebryakova MV, Kuteykin-Teplyakov KB, Tikhonova OV, Goufman EI, Zgoda VG, Taranets IN, Makarov OV, Archakov AI (2005) Ovarian cancer marker of 11.7 kDa detected by proteomics is a serum amyloid A1. Proteomics 5:3790–3797

Moss EL, Hollingworth J, Reynolds TM (2005) The role of CA125 in clinical practice. J Clin Pathol 58:308–312
Muthu M, Nordström A (2019) Current status and future prospects of clinically exploiting cancer-specific metabolism – why is tumor metabolism not more extensively translated into clinical targets and biomarkers? Int J Mol Sci 20(6):1385
Nunes SC, Ramos C, Lopes-Coelho F, Sequeira CO, Silva F, Gouveia-Fernandes S, Rodrigues A, Guimarães A, Silveira M, Abreu S, Santo VE, Brito C, Félix A, Pereira SA, Serpa J (2018) Cysteine allows ovarian cancer cells to adapt to hypoxia and to escape from carboplatin cytotoxicity. Sci Rep 8(1):9513
Ogiwara H, Takahashi K, Sasaki M, Kuroda T, Yoshida H, Watanabe R, Maruyama A, Makinoshima H, Chiwaki F, Sasaki H, Kato T, Okamoto A, Kohno T (2019) Targeting the vulnerability of glutathione metabolism in ARID1A-deficient cancers. Cancer Cell 35(2):177–190.e8
Patriotis C, Maruvada P, Srivastav S (2017) Molecular detection and diagnosis of Cancer. In: Colemann WB, Tsongalis GJ (eds) The molecular basis of human cancer, 2nd edn. Springer, New York
Petri AL, Simonsen AH, Yip TT, Hogdall E, Fung ET, Lundvall L, Hogdall C (2009) Three new potential ovarian cancer biomarkers detected in human urine with equalizer bead technology. Acta Obstet Gynecol Scand 88:18–26
Plewa S, Horała A, Dereziński P, Nowak-Markwitz E, Matysiak J, Kokot ZJ (2019) Wide spectrum targeted metabolomics identifies potential ovarian cancer biomarkers. Life Sci 222:235–244
Radpour R, Forouharkhou F (2018) Single-cell analysis of tumors: creating new value for molecular biomarker discovery of cancer stem cells and tumor-infiltrating immune cells. World J Stem Cells 10(11):160–171
Rei M, Gonçalves-Sousa N, Lança T, Thompson RG, Mensurado S, Balkwill FR, Kulbe H, Pennington DJ, Silva-Santos B (2014) Murine $CD27^{(-)}$ $V\gamma6^{(+)}$ $\gamma\delta$ T cells producing IL-17A promote ovarian cancer growth via mobilization of protumor small peritoneal macrophages. Proc Natl Acad Sci U S A 111(34):E3562–E3570
Russo F, Lastoria S, Svanera G, Capobianco G, de Chiara A, di Francia R, Squame E, de Martinis F, Pinto A (2007) Long-term follow-up study on the role of serum CA-125 as a prognostic factor in 221 newly diagnosed patients with Hodgkin's lymphoma. Leukemia Lymphoma 48(4):723–730
Scholler N, Urban (2007) CA125 in ovarian cancer. Biomark Med 1(4):513–523
Scholler N, Gross JA, Garvik B, Wells L, Liu Y, Loch CM, Ramirez AB, McIntosh MW, Lampe PD, Urban N (2008) Use of cancer-specific yeast-secreted in vivo biotinylated recombinant antibodies for serum biomarker discovery. J Transl Med 6:41
Shirodkar SP, Lokeshwar VB (2009) Potential new urinary markers in the early detection of bladder cancer. Curr Opin Urol 19:488–493
Slupsky CM, Steed H, Wells TH, Dabbs K, Schepansky A, Capstick V, Faught W, Sawyer MB (2010) Urine metabolite analysis offers potential early diagnosis of ovarian and breast cancers. Clin Cancer Res 16(23):5835–5841
Sutphen R, Xu Y, Wilbanks GD, Fiorica J, Grendys EC Jr, LaPolla JP, Arango H, Hoffman MS, Martino M, Wakeley K, Griffin D, Blanco RW, Cantor AB, Xiao YJ, Krischer JP (2004) Lysophospholipids are potential biomarkers of ovarian cancer. Cancer Epidemiol Biomark Prev 13(7):1185–1191
Tanasova M, Begoyan VV, Weselinski LJ (2018) Targeting sugar uptake and metabolism for cancer identification and therapy: an overview. Curr Top Med Chem 18:467–483
Wang M-L, Huang O, Yang T-X (2009) IgE myeloma with elevated level of serum CA125. J Zhejiang Univ Sci B 10(7):559–562
Wang J, Gao J, Yao H, Wu Z, Wang M, Qi J (2014) Diagnostic accuracy of serum HE4, CA125 and ROMA in patients with ovarian cancer: a meta-analysis. Tumour Biol 35(6):6127–6138
Woolas RP, Xu FJ, Jacobs IJ, Yu YH, Daly L, Berchuck A, Soper JT, Clarke-Pearson DL, Oram DH, Bast RC Jr (1993) Elevation of multiple serum markers in patients with stage I ovarian cancer. J Natl Cancer Inst 85(21):1748–1751
Yang LV (2017) Tumor microenvironment and metabolism. Int J Mol Sci 18(12):2729
Yang L, Moss T, Mangala LS, Marini J, Zhao H, Wahlig S, Armaiz-Pena G, Jiang D, Achreja A, Win J, Roopaimoole R, Rodriguez-Aguayo C, Mercado-Uribe I, Lopez-Berestein G, Liu J, Tsukamoto T, Sood AK, Ram PT, Nagrath D (2014) Metabolic shifts toward glutamine regulate tumor growth, invasion and bioenergetics in ovarian cancer. Mol Syst Biol 10:728
Yang C, Kim HS, Song G, Lim W (2019) The potential role of exosomes derived from ovarian cancer cells for diagnostic and therapeutic approaches. J Cell Physiol:1–11
Ye B, Cramer DW, Skates SJ, Gygi SP, Pratomo V, Fu L, Horick NK, Licklider LJ, Schorge JO, Berkowitz RS, Mok SC (2003) Haptoglobin-alpha subunit as potential serum biomarker in ovarian cancer: identification and characterization using proteomic profiling and mass spectrometry. Clin Cancer Res 9:2904–2911
Ye B, Skates S, Mok SC, Horick NK, Rosenberg HF, Vitonis A, Edwards D, Sluss P, Han WK, Berkowitz RS, Cramer DW (2006) Proteomic-based discovery and characterization of glycosylated eosinophil-derived-neurotoxin and C-terminal osteopontin fragments for ovarian cancer in urine. Clin Cancer Res 12:432–441

Zhang Z, Bast RC Jr, Yu Y, Li J, Sokoll LJ, Rai AJ, Rosenzweig JM, Cameron B, Wang YY, Meng XY, Berchuck A, Van Haaften-Day C, Hacker NF, de Bruijn HW, van der Zee AG, Jacobs IJ, Fung ET, Chan DW (2004) Three biomarkers identified from serum proteomic analysis for the detection of early stage ovarian cancer. Cancer Res 64:5882–5890

Zhang B, Barekati Z, Kohler C, Radpour R, Asadollahi R, Holzgreve W, Zhong XY (2010) Proteomics and biomarkers for ovarian cancer diagnosis. Ann Clin Lab Sci 40(3):218–225

Zhang B, Cai FF, Zhong XY (2011) An overview of biomarkers for the ovarian cancer diagnosis. Eur J Obstet Gyn R B 158(2):119–123

Zhang X, Liu Y, Kim YJ, Mac J, Zhuang R, Wang P (2017) Co-delivery of carboplatin and paclitaxel via cross-linked multilamellar liposomes for ovarian cancer treatment. RSC Adv 7:19685–19693

Zhao S, Yumei M, Xianghua H (2015) Trefoil factor 1 elevates the malignant phenotype of mucinous ovarian cancer cell through Wnt/β-catenin signaling. Int J Clin Exp Pathol 8(9):10412–10419

Part IV

Metabolomics: A New Way of Screening Cancer

19 Exploring Cancer Metabolism: Applications of Metabolomics and Metabolic Phenotyping in Cancer Research and Diagnostics

Gonçalo Graça, Chung-Ho E. Lau, and Luís G. Gonçalves

Abstract

Altered metabolism is one of the key hallmarks of cancer. The development of sensitive, reproducible and robust bioanalytical tools such as Nuclear Magnetic Resonance Spectroscopy and Mass Spectrometry techniques offers numerous opportunities for cancer metabolism research, and provides additional and exciting avenues in cancer diagnosis, prognosis and for the development of more effective and personalized treatments. In this chapter, we introduce the current state of the art of metabolomics and metabolic phenotyping approaches in cancer research and clinical diagnostics.

Keywords

Metabolomics · Cancer metabolism · NMR spectroscopy · Mass spectrometry · Diagnostics · Metabolic imaging · Biofluids · Tissues

G. Graça (✉) · C.-H. E. Lau
Department of Metabolism, Digestion and Reproduction, Faculty of Medicine, Imperial College London, London, UK
e-mail: g.gomes-da-graca@imperial.ac.uk

L. G. Gonçalves (✉)
Proteomics of Non-Model Organisms Lab, ITQB Nova-Instituto de Tecnologia Química e Biológica António Xavier, Universidade Nova de Lisboa, Oeiras, Portugal
e-mail: lgafeira@itqb.unl.pt

19.1 Metabolic Alterations in Cancer

It is well known that cancer cells undergo metabolic reprogramming in order to sustain the anabolic requirements of tumorigenesis and cellular proliferation. This can be achieved by mutations in genes regulating oncogenic signaling pathways ultimately interfering with the expression of key metabolic enzymes (Pavlova and Thompson 2016). Generally speaking, cancer cells tend to display enhanced uptake of glucose, amino acids such as glutamine and other nutrients, increased reliance on glycolysis for ATP production, TCA cycle intermediates for biosynthesis and NADPH production (Pavlova and Thompson 2016). These metabolic changes also bring about alterations in metabolite-driven gene regulation and metabolic interactions with the tumor microenvironment (Pavlova and Thompson 2016), which in turn will have implications in tumor progression and invasiveness.

The most well known metabolic change in cancer occurs in central metabolism with the increased use of aerobic glucose metabolism in which cellular glucose import is increased to generate ATP and lactic acid, known as the Warburg effect (Sanderson and Locasale 2018). The enhanced glucose consumption was the basis of the positron-emission tomography (PET) imaging in which a glucose analogue, ^{18}F-fluoro-

J. Serpa (ed.), *Tumor Microenvironment*, Advances in Experimental Medicine and Biology 1219,
https://doi.org/10.1007/978-3-030-34025-4_19

2-deoxyglucose (FDG), is used to detect tumor activity, enabling cancer diagnosis, staging and treatment follow-up (Zhu et al. 2011). Another important metabolic alteration is the increased glutamine uptake, which was shown to be implicated in important pathways such as the synthesis of NADPH and as a source of nitrogen in the biosynthesis of non-essential amino acids and nucleotides (Pavlova and Thompson 2016). Glutamine can also play an important role in the cellular import of essential amino acids such as leucine, isoleucine, valine, methionine, tyrosine, tryptophan, and phenylalanine by acting as an antiporter through the LAT1 membrane transporter (Pavlova and Thompson 2016). Similarly to PET-FDG, ^{18}F-fluoroglutamine, a compound analogous to glutamine has been tested in diagnostic PET imaging (Dunphy et al. 2018), and is particularly useful as an alternative to FDG in tissues where glucose utilization is physiologically high, such as in brain tissue. The described applications are some examples of techniques aimed at a few metabolic alterations that motivated the development of diagnostic imaging techniques which are nowadays in current clinical practice. By looking simultaneously to all possible metabolites in a tissue or body fluid, metabolomics, metabolic profiling and phenotyping techniques aim at exploring other tissue- or tumor-specific metabolic alterations, ultimately contributing to the knowledge of disease mechanisms and to the development of diagnostic tools. The current chapter aims at introducing the reader to the field of metabolomics and metabolic phenotyping and illustrate some of the most important applications in the study of cancer metabolism and diagnostic.

19.2 Metabolomics and Metabolic Phenotyping

Metabolic phenotyping/metabolomics aims to take a holistic view of a biological sample and is broadly defined as the comprehensive measurement and fingerprinting of low molecular weight compounds in biological samples to understand their roles in cellular functions and diseases. Whilst the terminology of metabolomics/metabonomics were introduced in the late 1990s, the concept of utilizing the distinctive color, odor or taste of human urine for clinical applications were documented as early as the sixteenth century. However, it was not until the twentieth century when molecular entities were effectively elucidated from biological samples. Separation of metabolites in urine samples achieved by 2D paper chromatography was reported in 1956 (Dalgliesh 1956) and the technique was successfully applied to identify metabolites associated with cystinuria, argininosuccinic aciduria and Hartnup disease. Development of gas chromatography meant that by the 1970s up to around 250 volatile components could be detected in urine and breath samples. NMR spectroscopy was first applied in 1967 to identify a urinary metabolite associated with an inborn error of metabolism (Tanaka and Isselbacher 1967), and the profiling of multiple chemicals in urine or blood samples by NMR were first reported by Nicholson et al. in 1984. The advent of information technology and the explosion of computational infrastructures in the early 1990s meant statistical techniques were now being developed in earnest to address the data analytics challenges – helping to analyze and visualize the multivariate datasets. Wishart et al. have made great progress in defining the composition of the human serum metabolome, and later the human urine metabolome, and in setting up the Human Metabolome Database (HMDB) containing background information and spectral data of a large collection of compounds (Wishart et al. 2013). More recently, we see a number of well-resourced specialist metabolic phenotyping centers being setup to support large-scale, high-throughput metabolomics serving researchers across the biomedical research communities.

The first studies that look for a large set of metabolites in the context of cancer can be dated to the 1980s, but the technical developments and recognizing the importance of metabolic alterations in cancer, lead to an increased interest in metabolomics applied to cancer research since 2004 (Fig. 19.1). In the first years the number of publications using MS or NMR approaches were

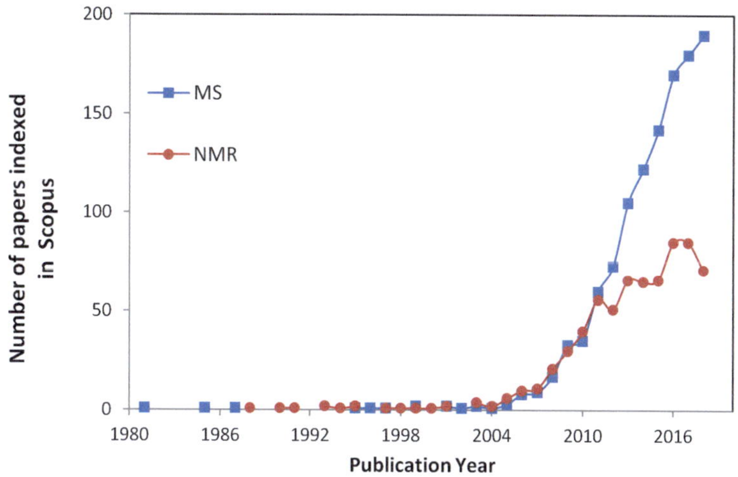

Fig. 19.1 Number of papers using metabolomic approaches in cancer research by year (search was performed on March 2019 in Scopus limited to original papers that mention (("nuclear magnetic resonance" OR NMR) AND Cancer AND (metabolomics OR metabonomics OR "metabolic profile")) or (("Mass spectrometry" OR MS) AND Cancer AND (metabolomics OR metabonomics OR "metabolic profile")) in the title, abstract or keyword)

very similar. However since 2012 the number of papers with a MS approach outperforms the studies using NMR (Fig. 19.1). The preference for MS in metabolomics studies results from its higher sensitivity, smaller sample size demands and relatively lower operational costs. The technology developments in MS instrumentation, software databases and tools introduced in the last decade, has also permitted an increase in resolution and ease of analysis (Amberg et al. 2017; Emwas 2015; Bingol 2018). Nevertheless, NMR spectroscopy continues to have an important role in cancer metabolomics and more importantly, the combination of NMR and MS approaches provides additional metabolite coverage (Psychogios et al. 2011; Wishart et al. 2013).

Important information about metabolic pathways and fluxes can be drawn from metabolic studies using stable isotope tracers, which can be considered as one of the next-generation applications of metabolomics. In this field, NMR methodologies have unique advantages, since it allows the determination of the position of the isotopomers from isotopically enriched metabolites, the identification and structure elucidation of unknown metabolites as well as the analysis of metabolic pathways dynamics *in vivo* and *in situ* in cell culture, tissues and whole organisms (Fan et al. 2012; Fan and Lane 2016).

Metabolomics studies are usually divided into two categories: targeted, where only selected metabolites are analyzed, e.g. from one single metabolic pathway; and untargeted studies which do not focus on any particular set of metabolites and all signals from either NMR or MS are analyzed.

19.2.1 Nuclear Magnetic Resonance (NMR) Spectroscopy

Nuclear Magnetic Resonance Spectroscopy (NMR) is a powerful and versatile analytical technique. It is used in diverse fields from the structural elucidation of macromolecules and small molecules to the quantification of metabolites present in a sample. NMR was discovered in the mid 1940s, by two different groups (Purcell et al. 1946; Bloch et al. 1946) and from the beginning it was used to characterize molecules, with the first commercial spectrometer developed in 1952 (Marion 2013). The basic principle of NMR involves the atomic nucleus. A nucleus with a non-zero nuclear spin (an odd atomic number),

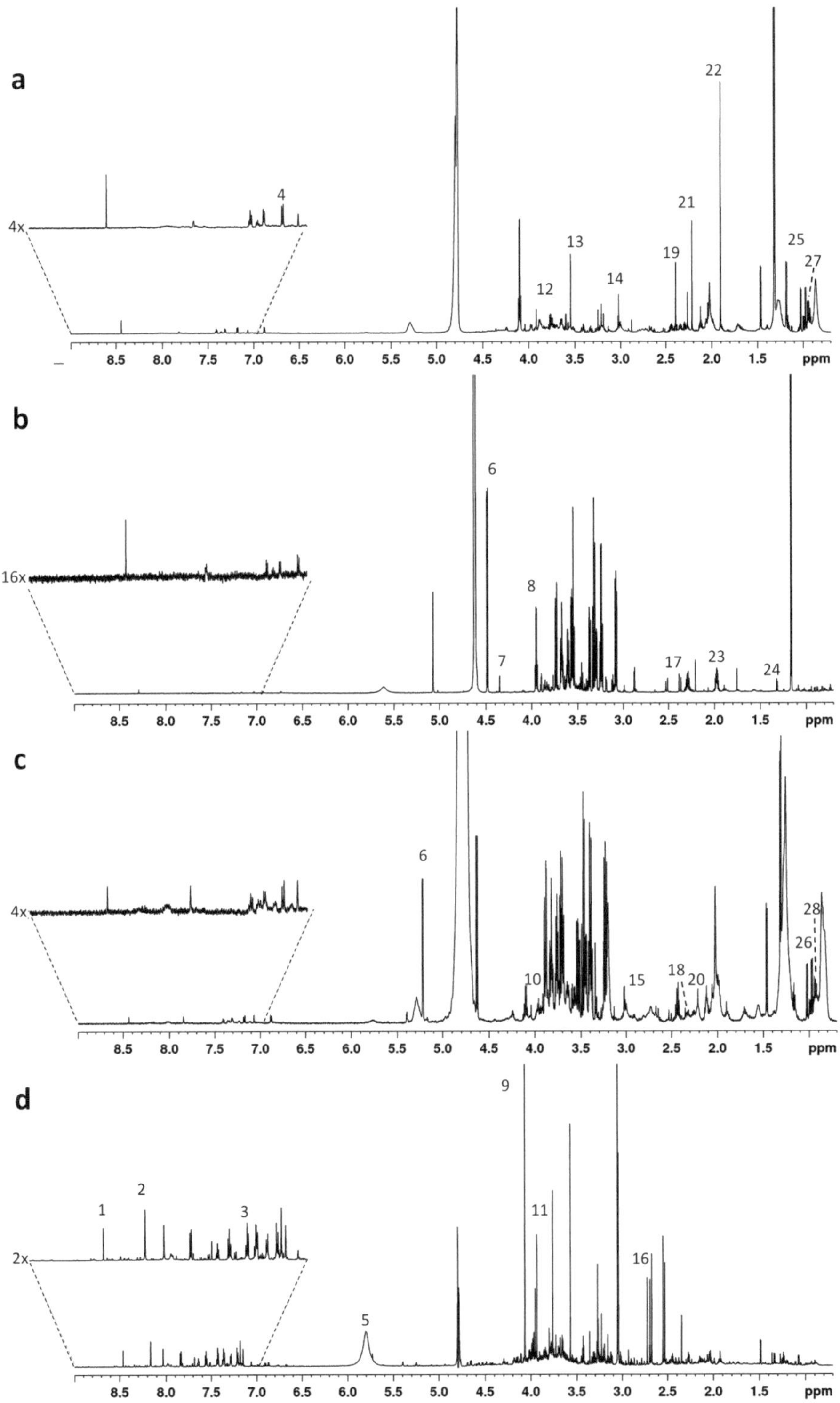

Fig. 19.2 Example of NMR spectra of different biofluids used in cancer NMR metabolomics: (**a**) Ascitic fluid from an ovarian tumor patient; (**b**) Cerebrospinal fluid (CSF) from a non-Hodgkin lymphoma patient; (**c**) Serum from an ovarian tumor patient; and (**d**) Urine from a paraganglioma patient. All spectra were acquired in an 800 MHz spectrometer at 298 K, except the serum which was acquired in a 600 MHz spectrometer at 310 K.

when placed in an external magnetic field can absorb and re-emit radiofrequency with a frequency characteristic of the magnetic field acting on the nucleus. The magnetic field of each nucleus in the sample depends on the action of the external magnetic field and the weak magnetic fields each nucleus in its vicinity (Marion 2013). As a result, each nucleus in a molecule has a resonance at a characteristic frequency, which allows the identification and structural characterization of a molecule.

NMR spectroscopy is extremely useful for studying biological systems, since one of the most sensitive nuclei, the hydrogen isotope (^{1}H or proton), has a natural abundance of almost 100%. The protons with similar molecular environment are called equivalent and produce signals in the ^{1}H-NMR spectrum at specific frequencies (Fig. 19.2). The signal intensity is directly proportional to the number of protons that originate the signal, and also to the concentration of the molecule in the sample. The proximity of other nuclei inside the molecule also produces signal splitting (multiplicity) which varies according to the number of nuclei in the vicinity. These characteristics make NMR spectroscopy a very popular and powerful technique for metabolomics. Moreover, NMR is highly reproducible even between different spectrometers (at the same magnetic field strength and similar hardware configurations) and/or operators (Dona et al. 2014). It is also a very versatile technique, making it possible to analyze intact tissues or biofluids, in most of the cases, with minimal sample preparation (Fig. 19.2). The sample is not consumed in the analysis, thus it can be reanalyzed for as long as it remains stable. The application of different NMR techniques enables the identification of unknown compounds and its structure elucidation (Graça et al. 2019). NMR versatility to study biological systems is extended beyond proton to other magnetic nuclei present in organic molecules (e.g. ^{31}P, ^{13}C or ^{15}N) (Gowda and Raftery 2017). An example is the use of ^{31}P-NMR in prostate cancer, which can be used to measure the changes in phospholipid contents in the prostate tissues induced by the carcinoma (Cornel et al. 1993; Komoroski et al. 2011). The major drawback of NMR is the low sensitivity when compared to MS techniques, as it detects compounds with concentrations of >50 μM while MS compounds with concentrations >10–100 nM (Emwas et al. 2019). Recent advances to improve NMR sensitivity included developments of novel pulse sequences, new probes, spectrometers with higher magnetic fields strengths and by applying enhanced signal polarization techniques such as dynamic nuclear polarization (DNP) (Ardenkjaer-Larsen et al. 2015).

Another challenge in NMR metabolomics is spectral resolution. Biological samples contain hundreds to thousands of metabolites which produce hundreds of NMR signals leading to significant signal overlap (Fig. 19.2), which makes metabolite identification and concentration determination difficult tasks. These challenges have been tackled through the development of spectral deconvolution software such as Chenomx (Chenomx Inc., Edmonton, Canada) and BATMAN (Hao et al. 2012) and comprehensive spectral databases, such as HMDB (Wishart et al. 2013). Two-dimensional NMR experiments (2D NMR), where signals are dispersed into more than one frequency dimension, constitute also an essential tool for metabolite identification in such complex samples (Emwas 2015; Graça et al. 2019). Despite their advantages, the use of 2D NMR as a profiling platform in metabolomics has been hindered by the long experimental times of 1 to several hours. For this reason, 1D NMR experiments are still routinely used and reported in vast majority of the studies. Nevertheless, new

Fig. 19.2 (continued) A noesygpprld pulse program was used for the CSF and urine, while in serum and ascitic fluid a cpmgprld pulse program was used to suppress the signals from macromolecules (proteins and lipoproteins). Some of the metabolites detected are indicated: 1-formate, 2-histidine, 3-phenylalanine, 4-tyrosine, 5-urea, 6-glucose, 7-ascorbate, 8-lactate, 9-creatinine 10-myo-inositol, 11-creatine, 12-phosphocreatine, 13-glycine, 14-choline, 15-phosphocholine, 16-dimethylamine, 17-citrate, 18-glutamate, 19-pyruvate, 20-acetoacetate, 21-acetone, 22-acetate, 23-glutamine, 24-alanine, 25-3-hydroxybutyrate, 26-valine, 27-leucine, 28-isoleucine

fast 2D techniques which reduce significantly the 2D spectral acquisition time, have been introduced in biofluids NMR metabolomics with very promising results. These include ultrafast (UF) NMR and non-uniform sampling (NUS) (Guennec et al. 2014; Marchand et al. 2017).

19.2.2 Mass Spectrometry Methods

Mass spectrometry (MS) is a very popular and powerful analytical technique which has found particular use for the elucidation of molecular structures and quantitative analysis of small molecules such as metabolites. MS was introduced in the beginning of the twentieth century and gained widespread popularity in the late 1950s. The technique is based on the detection of charged molecules in the gas phase (Glish and Vachet 2003). Because, not all compounds are easily ionizable and volatile, the technique was limited to gaseous samples for several years since its introduction. It was not until the late 1980s that new ionization techniques such as electrospray ionization (ESI) and matrix-assisted desorption-ionization (MALDI) enabled the direct ionization of molecules from the liquid and solid samples, respectively, into the gas phase (Glish and Vachet 2003). While ESI enables the generation of ions from liquid samples by generation of charged microparticles after passing the sample through a charged needle, in MALDI, the sample is mixed with a light absorbing compound (matrix) which, when excited with laser light promotes de ionization and displacement of ions from the sample (Glish and Vachet 2003; Bodzon-Kulakowska and Suder 2016). Other important group of ionization techniques are the ambient-pressure ionization techniques which enable, for instance, the withdrawal of ions directly from solid samples into the mass spectrometer (Glish and Vachet 2003; Hänel et al. 2019). Among these, desorption-electrospray ionization (DESI) technique has special importance in mass spectrometry imaging, as it will be described in more detail in Sect. 19.3.7.3. In DESI, ions are withdrawn from the sample by a jet of gas and charged micro droplets usually oriented at an angle close to 45° to sample surface, then injected into the ESI MS source (Hänel et al. 2019). Another ionization technique with important applications in imaging is secondary ion MS (SIMS), in which ions are extracted from the sample surface (secondary ions) after collision with primary ions from an inorganic ion beam (Bodzon-Kulakowska and Suder 2016).

Ionization occurs at the inlet of the mass spectrometer, known as the source. After ionization, the ions are transmitted to the mass analyzer, which is composed of a series of charged metal plates under vacuum, where the ions are separated according to their mass-to-charge ratio (m/z), before hitting the detector (Glish and Vachet 2003). A mass spectrum, a representation of the ion abundance as a function of each ion m/z is then produced (Fig. 19.3a).

In comparison to NMR spectroscopy, MS is more sensitive, also requiring lower amounts of sample. On the other hand, the sample is consumed during analysis because the ions are lost after reaching the detector.

The MS spectrum of a pure compound can be very simple as the one shown in Fig. 19.3a. However, in applications such as metabolomics analysis, where complex mixtures are analyzed, the spectra can become quite convoluted and difficult to interpret. As an example of such complexity an MS spectrum of negatively charged molecules from a human blood serum lipid extract is shown in Fig. 19.3b. Apart from the complexity stemming from the peak overlap, in practice the most abundant ions can suppress the ionization of other ions. To resolve such problems, MS is usually coupled with online compound physical separation such as gaseous- or liquid-chromatography, the techniques being termed gas- or liquid- chromatography – MS (GC-MS and LC-MS respectively) or capillary electrophoresis (CE-MS). GC-MS is commonly applied to volatile samples, whereas LC-MS and CE-MS are usually employed to analyze liquid samples and solid sample extracts (Emwas 2015). An example of a LC-MS chromatogram from a human blood serum lipid extract sample is shown in Fig. 19.3c. It is clear that the resolution and number of observed peaks (ions) increased quite dramatically in comparison to the direct MS analysis (Fig. 19.3c).

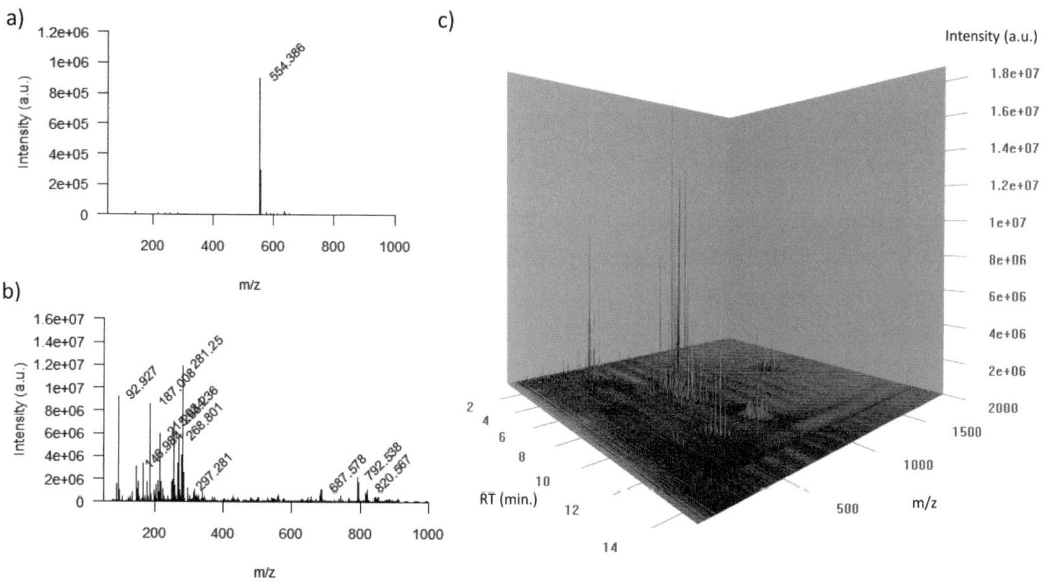

Fig. 19.3 Mass spectra: (**a**) Leucine enkephalin peptide acquired by electrospray ionization in negative ion mode (ESI⁻); (**b**) human blood serum lipid extract acquired in ESI⁻, (**c**) LC-MS chromatogram of human blood serum lipid extract acquired in positive ion mode (ESI⁺), where mass spectra are acquired continuously during chromatographic separation; *RT* retention time

Another important feature to MS-based techniques, is that multiple ions can be generated from a single compound. As an example, in ESI one organic compound can ionize by capturing or releasing a proton or by forming adducts with other ions already present in the sample (e.g. $[M + H]^+$, $[M + Na]^+$, $[M\text{-}H]^-$, $[M + \text{Formate}]^-$, where M represents the organic compound). After ionization, some molecules also break into charged fragments. Moreover, due to the sensitivity of the technique, several forms of each compound containing one or more naturally occurring isotopes such as ^{13}C (1% abundance), isotopologues, can also be detected increasing the complexity of the spectrum. These factors will lead to higher number of peaks than detected molecules in MS datasets, which is something that needs to be accounted for when interpreting the data. Metabolite identification from MS spectra is therefore a non-trivial task. Often, the analyst will need to perform searches with the measured m/z values on publicly available databases and, eventually, run additional ion fragmentation experiments to get more insight into the molecule identity (Emwas 2015).

19.2.3 Statistical Data Analysis

Metabolomics experiments generate large quantities of data composed of thousands of variables if simultaneous measurements are collected as in untargeted metabolomics experiments. In most cases, metabolomics datasets need additional processing, such as spectral baseline correction, peak alignment, normalization and variable scaling before statistical analysis can be performed. These operations are required to correct for sample dilution, sample preparation and/or analytical bias and to scale the relevant contributions of each variable (Emwas 2015). Adequate statistical analysis tools are then employed to extract meaningful information from the data. Both univariate and multivariate statistical approaches can be utilized for these purposes. However, care should be taken when using univariate analysis tests for untargeted metabolomics data. In those cases, multiple tests are usually performed simultaneously and the risk of false-discovery results is high. In those cases suitable multiple correction strategies should be used (Broadhurst and Kell 2006).

Multivariate analysis (MVA) are the most commonly used methods in untargeted metabolomics. They have the double advantage of generating interpretable statistical summaries of the data, which are necessary to pursue biological and physiological interpretations, and also enabling the development of predictive models of the disease under investigation. MVA methods can be divided into unsupervised methods, where no *a priori* sample classification or patient information is taken into account in the analysis; and supervised methods, where the information regarding patient diagnostic is included in the analysis (Trygg et al. 2007). Examples of unsupervised analysis methods are Principal Component Analysis (PCA) and Hierarchical Clustering, which are used to investigate similarities between samples and trends in the data. Supervised methods such as Partial Least Squares – Discriminant Analysis (PLS-DA) and related variants (e.g. Orthogonal PLS-DA), Random Forests, Support Vector Machines and other machine learning approaches are commonly used to develop classification models and look for metabolites that correlate with the disease studied (Trygg et al. 2007; Gromski et al. 2015). Supervised methods are obtained in two-stages: (1) model training, where samples of known class are used to generate classification models; and (2) model validation, where subsets of training data or an external sample set (cross-validation and test-set validation, respectively) are used to test the model classification performance (Trygg et al. 2007; Gromski et al. 2015).

19.3 Applications to Cancer Diagnostics

Metabolomics studies in cancer diagnosis usually involve the comparison of matched groups of patients (at one or more stages of cancer) *versus* healthy control or benign cases. One or several types of biological material are obtained and analyzed, whose selection is based on the affected organ(s). A remarkable collection of studies on the application of metabolomics to study cancer is available in the literature. It is beyond the scope of this chapter to provide a systematic review of all studies performed to date for all cancer types. Instead applications of metabolomics and metabolic phenotyping to diagnosis, prognosis and treatment monitoring of major cancer types are illustrated with studies from 2008 to 2019 period, organized by sample type.

19.3.1 Blood Serum and Plasma

Blood serum and plasma are the most studied biological fluids in cancer metabolomics, as they reflect metabolite levels entering systemic circulation and directly provide an accessible snapshot of the physiological condition of an individual without the need of tissue biopsies. Blood metabolites could be valuable biomarkers for early disease detection. For example, the median survival interval of patients with pancreatic cancer is currently less than 12 months and one study has identified elevated blood branched-chain amino acids as an early risk factor in human pancreatic adenocarcinoma development (Mayers et al. 2014). Using data from targeted LC-MS methods comparing plasma samples from 450 patients to their matched controls collected before the onset of the disease, the same study found that elevated blood branched-chain amino acids were associated with a twofold increase in future risk in developing pancreatic cancer (Mayers et al. 2014). Multiple reports indicated that alteration of circulating amino acids including tryptophan, glutamine, glutamate, phenylalanine and branched chain amino acids and lysophosphatidylcholine (C18:0, C18:2) could potentially serve as useful diagnostic biomarkers for pancreatic tumors (Sakai et al. 2016; Fukutake et al. 2015; Akita et al. 2016) which may be linked to pancreatic adenocarcinoma-associated cachexia, insulin resistance or hyperglycemia.

Tumors located in different organ sites could have distinctive footprint on the blood metabolome. For example elevated levels of circulating ketone bodies (including 3-hydroxybutyric acid), sugars and free fatty acids, and lower levels of glycolytic and TCA metabolites have previously been reported in ovarian cancer patients (N = 158) which may be the consequence of increased activity of fatty acid oxidation in specific tumor organ sites (Hilvo et al. 2016). Furthermore, it has been shown that metabolomics performed on plasma and serum samples could be applied to monitor treatment response in patients. A number of studies have demonstrated that pharmacodynamic response to inhibitors targeting oncogenic signaling could be successfully monitored in patient's plasma samples. One such study was able to show that changes in phosphatidylcholines and sphingomyelins levels were observed in responder patients with advanced melanoma treated with a mitogen-activated protein kinase (MEK) inhibitor, and that pre treatment levels of a panel of lipids were predictive of inhibitor treatment response (Ang et al. 2017). Furthermore, it has been shown in a separate study that time and dose-dependent response to Phosphoinositide 3-kinases (PI3K) inhibitor could be observed in patients enrolled in a phase I dose-escalation trial (Ang et al. 2016), demonstrating plasma metabolomics could be a valuable resource for translating and validating preclinical findings in patients. Also, plasma metabolomics have been applied to predict future cancer risk. A Danish study analyzed plasma samples from 838 women by ^{1}H-NMR, where half of the women had developed breast cancer between the time of enrolment in the study and the follow-up date. The inclusion of the NMR data in the risk predictive model increased its sensitivity and specificity to above 80%, and glycerol, ethanol and formate, were amongst the metabolites contributing to the prediction model (Bro et al. 2015).

19.3.2 Urine

Urine is a noninvasive, accessible and concentration and volume-rich biofluid for clinicians to collect, and many urinary metabolomics studies have focused on tumors located in the urinary tract. For example, it has been reported that the levels of metabolites involved in glycolysis and fatty acid oxidation are altered in patients with bladder tumors (N = 138) compared to control subjects (N = 121) and this may be related to changes to carnitine transferase and pyruvate dehydrogenase complex expression in the patient group (Jin et al. 2014). One study has identified dopamine 4-sulfate, aspartyl-histidine, and tyrosyl-methionine to be discriminatory between non-muscle invasive bladder cancer patients (N = 167) and healthy controls (N = 117), with higher levels of tryptophan metabolites in urine found patients with higher grade tumor (Cheng et al. 2018). Kidney cancer has also been investigated, and in particular urinary levels of acylcarnitines have been found to discriminate patients with low- and high-grade tumors (Ganti et al. 2012).

In addition, urine has been applied to study tumors that are remote from the urinary tract, and has been successful in differentiating patients with malignancies ranging from prostate, lung and gastrointestinal cancers such as gastric cancer, from their healthy controls (Dinges et al. 2019). For example, with NMR metabolomics urinary 2-hydroxyisobutyrate, 3-indoxylsulfate, and alanine, were identified as discriminatory between patients with gastric cancer (N = 43), healthy individuals (N = 40) and nonmalignant gastric conditions (N = 40) with a classification accuracy of 95% as indicated through the area under the receiver operating characteristic curve (Chan et al. 2016). In addition, some reports have indicated that tumors of distinct organ systems could have unique urine metabolic signatures (Woo et al. 2009; Slupsky et al. 2010), which would be an important consideration if urine metabolomics were to be utilized for cancer diagnostics in clinics.

Urine metabolomics has also been applied to examine the treatment effects of chemotherapy. For example, one study used 2D ^{1}H-^{1}H J-resolved NMR data to follow the effects of cisplatin in patients with non-small-cell lung cancer (N = 5) and show that cisplatin alters urinary amino acids levels (Doskocz et al. 2015).

19.3.3 Cerebrospinal Fluid

Cerebrospinal fluid is traditionally the fluid of choice to study neurological conditions. However, it has been shown to be an important source of biomarkers of malignant cell invasion to the leptomeninges, which is a relatively rare condition of late stage solid and hematologic cancers. In this context, two separate studies inspected CSF metabolic composition by ^{1}H-NMR spectroscopy in leptomeningeal invasion from lung cancer and B-cell non-Hodgkin lymphoma and found metabolite alterations related to the presence of malignant cells in CSF (An et al. 2015; Graça et al. 2017). An et al. compared CSF samples from controls affected by neurologic conditions (N = 41) with samples from patients diagnosed with leptomeningeal carcinomatosis from lung adenocarcinoma (N = 26). Changes in the levels of myo-inositol, creatine, lactate, alanine and citrate were the most discriminatory CSF metabolites between the two groups of patients (An et al. 2015). These authors also reported a good correlation between the metabolic profile and the grading of radiological leptomeningeal enhancement accessed by magnetic resonance imaging (MRI), suggesting the potential utility of CSF metabolic profile in grading of leptomeningeal carcinomatosis (An et al. 2015). Graça et al. compared the CSF metabolic profiles of B-cell non-Hodgkin lymphoma patients with positive (N = 5) and negative (N = 13) diagnosis of leptomeningeal invasion. Among the most significant metabolite alterations glycine, alanine, pyruvate, acetylcarnitine, carnitine, phenylalanine as well as protein signals seemed to be increased in the positively diagnosed patients (Graça et al. 2017). The authors also found that leptomeningeal invasion chemotherapy treatment produced sharp decreases in the levels of those metabolites in a group of follow-up positively diagnosed patients (Graça et al. 2017).

19.3.4 Ascitic Fluid

Malignant ascites is the abnormal buildup of tumor-cell containing fluid in the abdomen, the ascitic fluid (Sangisetty and Miner 2012). The presence of malignant ascites is generally signal of an advanced stage of the disease and poor prognostic in ovarian, uterine, colorectal and pancreatic cancers (Garrison et al. 1986). Because ascitic fluid can also accumulate in other diseases, such as cirrhosis, it is important to devise a quick method to determine the causes of ascite origin in cases. Some studies investigated the origin of ascitic fluid using metabolomics, by comparing ovarian carcinoma patients with cirrhotic patients showing promising results (Bala et al. 2008; Shender et al. 2014). Important differences were observed in the levels of fatty acids, cholesterol, ceramide, glycerol-3-phosphate, glucose, and glucose-3-phosphate between ovarian cancer patients (N = 10) and cirrhotic patient (N = 5) in a study using GC-MS (Shender et al. 2014). In a study using ^{1}H-NMR 3-hydroxybutyric acid, lactate, citrate, and tyrosine were the metabolites that discriminated between ovarian cancer (N = 15) and cirrhotic patients (N = 47) (Bala et al. 2008). ^{1}H-NMR metabolomics was also applied to identify the metabolic differences induced by chemotherapy in ovarian serous carcinoma effusions, indicating that the ascitic fluid levels of glucose and lipids increase while the levels of lactate and β-hydroxybutyrate decrease after chemotherapy (N = 35) when compared with ascitic fluid before chemotherapy (N = 44) (Vettukattil et al. 2013).

Animal models have also been used to investigate the development of ascites and ascitic fluid in cancer. A metabolomics study of two murine xenograft ovarian carcinoma models, one with a mouse ID8-vascular endothelial growth factor (VEGF)-Defb29 cell line (N = 8) and the human OVCAR3 cell line (N = 5), was carried out to characterize the malignant ascites metabolic features (Bharti et al. 2017). Despite the two cell lines lead to different metabolic profiles, some metabolites were common to both xenograft models: β-hydroxybutyric acid, maleic acid and citrate (Bharti et al. 2017).

19.3.5 Exhaled Breath Analysis

The analysis of exhaled breath is an established non-invasive technique for specific applications such as alcoholemia and *Helicobacter Pylori* testing (measurement of ^{13}C urea). Its application to cancer diagnostic focuses on the measurement of endogenous volatile organic compounds (VOCs), which can be defined as carbon-containing volatile compounds at room temperature (Hanna et al. 2019). Due to their physico-chemical properties, VOCs are well detected and measured using MS-related techniques, notably GC-MS but also direct-MS measurements (Hanna et al. 2019).

Exhaled breath VOC analysis can provide means of early diagnosis and patient stratification, particularly in population groups at higher risk for cancer development, e.g. smokers or individuals exposed to volatile and particulate contaminants; but also for patients presenting non-specific symptoms associated with cancer. Therefore, it can help clinicians decide on more invasive diagnostic or imaging procedures. The non-invasiveness of exhaled breath analysis can also lead to more patient enrolment (Hanna et al. 2019).

Applications of exhaled breath analysis in cancer seem particularly suitable in early diagnosis of cancer from the respiratory and digestive systems such as lung (Fu et al. 2014; Li et al. 2015), gastroesophageal (Kumar et al. 2015), oral cavity (Bouza et al. 2017) and laryngeal cancers (Garcia et al. 2014) since the affected organs have direct contact with breath. Nevertheless, some authors have also explored the application to cancers from distant organs such as liver, breast, prostate or ovarian (Qin et al. 2010; Barash et al. 2015; Peng et al. 2010; Amal et al. 2015).

The most popular application is by far the discrimination of groups of lung cancer patients from control subjects. Two representative studies have reported that VOCs analyses provided sensitivity values close or above 90% and specificity values above 80% for discrimination between controls and lung cancer patients (Fu et al. 2014; Li et al. 2015). Levels of carbonyl compounds levels were found elevated in patients with lung tumors (N = 85) (Li et al. 2015), whereas the concentrations of 2-butanone, 2-hydroxyacetaldehyde, 3-hydroxy-2-butanone, and 4-hydroxyhexenal in the exhaled breath of lung cancer patients (N = 97) were found significantly higher than in the exhaled breath of healthy smoker and non-smoker controls (N = 88) (Fu et al. 2014). Another interesting application is related to esophagogastric cancer. In a representative study, Kumar et al. identified 12 VOCs (pentanoic acid, hexanoic acid, phenol, methyl phenol, ethyl phenol, butanal, pentanal, hexanal, heptanal, octanal, nonanal, and decanal) increased in exhaled breath from esophageal (N = 48) and gastric adenocarcinoma (N = 33) when compared to non-cancer controls (N = 129), which provides specificity and sensitivity values for patient discrimination above 80% (Kumar et al. 2015).

Nanoarray-based sensor technology developments are also making it possible to measure breath VOCs. This technology has several advantages over GC-MS, particularly regarding operational costs and portability. It has been tested in the analysis of breath analysis from patients with several types of cancers, such as ovarian cancer (Amal et al. 2015), gastric cancers (Amal et al. 2016) as well as lung, breast, colorectal and prostate cancers (Peng et al. 2010), with discrimination performances similar to those of GC-MS.

Regardless of the application of exhaled breath analysis in cancer, additional analytical bias assessment and the introduction of standardized sampling procedures are key elements in the development and transitioning of the technique and its results to clinical applications (Hanna et al. 2019).

19.3.6 Other Noninvasive Biological Matrices: Saliva, Sputum, and Feces

In addition to the biological fluids/ matrices described above, metabolomics investigations have also been performed in numerous other matrix types. For example, saliva obtained from oral, breast and pancreatic cancer patients has successfully been analyzed (Sugimoto et al. 2010). Oncogenic MYC has been reported to regulate polyamine biosynthesis leading to accumulation in cancer cells, and Asai et al. have used CE-MS for detecting polyamines, and found spermine, *N*-acetylspermidine, and

N-acetylspermine levels in saliva successfully discriminate patients with pancreatic cancer (N = 39) from controls (N = 26) (Asai et al. 2018). Similarly, the levels of several polyamines in saliva have also been found elevated in relapsed breast cancer patients (N = 22) in another study using targeted LC-MS (Tsutsui et al. 2013).

Sputum consists of mucus produced in the respiratory tract and is potentially relevant for the diagnosis of lung cancer. There are currently very limited cancer metabolomics literature available on sputum, however, one study has successfully utilized flow infusion MS and GC–MS for distinguishing their 34 lung cancer patients and 33 healthy controls (Cameron et al. 2016).

Feces, rather like urine is readily available and information rich, as it contains undigested food passed from the gastrointestinal tract (GI) and metabolite compositions reflect dietary habits, mammalian– gut microbial interactions, as well as health status of the GI tract. Compared to health controls, colorectal cancer patients may have altered levels of acetate, butyrate, propionate, isovalerate, isobutyrate, valerate, and bile acids in their feces (Lin et al. 2016; Le Gall et al. 2018).

19.3.7 Biopsy and Cytology Material

Tissue biopsies and cytology aspirates are obtained from tumors and their metastases to confirm the diagnosis, molecular typing and staging which are performed at cyto- and histopathological analysis. The metabolomic analysis of such materials offers complementary metabolic information for further disease characterization and phenotyping. Moreover, it can be used as a diagnostic tool on its own. As mentioned in previous sections, both NMR and MS techniques are suitable for analysis of tissues and cells, either by analysis of extracts or intact material.

19.3.7.1 Analysis of Cells and Tissue Extracts

Cell and tissue extractions break up cellular structures and releases metabolites for in-depth or targeted biochemical analysis, for instance with focus on lipids or in hydrophilic metabolites. However, extractions may have reproducibility issues. For this reason, extraction procedures must ensure an effective arrest of cellular metabolism and minimize metabolite loss. Nevertheless, the analysis of cell and tissue extracts have been a valuable resource in *in vitro* tumor metabolism studies. One such studies is the study of isocitrate hydrogenase mutation in specific types of tumors. Isocitrate dehydrogenase 1 and 2 (IDH1/2) are enzymes important for energy metabolism, redox control and DNA methylation. Mutations in the genes encoding for these enzymes are frequent, including in majority of gliomas (Yan et al. 2009) and cartilage tumours (Pansuriya et al. 2011), and can be found in a significant portion of acute myeloid leukaemia (Molenaar et al. 2015). In a landmark paper, Dang et al. has shown that tumors harboring IDH1/2 mutations gain the ability to convert α-ketoglutarate to 2-hydroxyglutarate, leading to accumulation in 2-hydroxyglutarate in tumor cells. Comparing to the wild type gliomas, 2-hydroxyglutarate level in IDH mutant human tumors increased by 100-fold (Dang et al. 2009). The gain-of-function mutations are phenotypically specific and, in fact, 2-hydroxyglutarate could be detected directly *in vivo* in patients with glioma using magnetic resonance spectroscopy (MRS) acquired in MRI instruments (Choi et al. 2012). The conversion of α-ketoglutarate to 2-hydroxyglutarate could be measured in real time *in vivo* by using the same methodology with increased sensitivity through substrate dynamic nuclear polarization (DNP-MRS) (Chaumeil et al. 2013).

19.3.7.2 Analysis of Intact Tissues

A specific NMR technique, high-resolution magic-angle spinning (HRMAS), allows the analysis of micro-grams of tissue biopsies with similar resolution of liquid NMR (Emwas 2015). HRMAS can be used for the analysis of human and animal tumor tissues *ex vivo*, however, freezing delay time should be minimized as it could adversely bias analysis. Significant metabolite changes have been observed in samples frozen after 30 min of resection, and some metabolites are affected by prolonged experiment time due to sample spinning and degradation (Haukaas et al. 2016). Nevertheless, HRMAS can be useful in identifying diagnostic markers if experiments were designed and samples were handled with care. This has been illustrated for some types of cancer such as prostate or colorectal. Indeed, using tissue samples, spermine, spermidine, choline, kynurenine, sarcosine, citrate have been proposed as potential candidates as markers of diagnosis or staging in prostate tumors (de Vogel et al. 2014; Sreekumar et al. 2009; McDunn et al. 2013; Liu et al. 2015; Giskeodegard et al. 2013). Increased levels of lactate, taurine, and isoglutamine and decreased levels of lipids/triglycerides have been found in colorectal cancer (N = 88) relative to healthy mucosa (N = 83) (Mirnezami et al. 2014).

Surgical evaluation of tumor margins is routinely performed during tumor-extracting surgery, in which the surgeon decides on the extent of malignant tissue to extract while trying to maintain healthy tissue intact. This delicate process is usually assisted by a trained histopathologist who analyses the frozen surgically extracted tissues by light microscopy. The whole process needs to be done quickly while the patient is under anesthesia (Ifa and Eberlin 2016; Hänel et al. 2019).

Although HRMAS NMR could be applied in the analysis of tumor margins (Bathen et al. 2013; Paul et al. 2018), there is a great advantage in using MS techniques due to their higher sensitivity and smaller sample amounts requirement. Ambient-ionization MS techniques seem the most useful as they allow the acquisition of MS spectra in real-time and are easily operated by non-specialists, which gives the technique great advantage in surgical tumor diagnostics (Ifa and Eberlin 2016; Hänel et al. 2019).

Several ambient-ionization MS techniques have been introduced in cancer tissue analysis such as DESI, rapid evaporative MS (REIMS), “MasSpec pen” and picosecond infrared laser (PIRL) with some promising results towards intact *ex vivo* sample analysis (Hänel et al. 2019). While all of them focus on the analysis of lipid content, each one has specific characteristics regarding the amount of sample consumed, cross-contamination, preanalytical issues, surface scanning and transferability towards clinical diagnostic application (Hänel et al. 2019). REIMS is the most popular of ambient-ionization MS techniques because it is also applicable *in vivo* (Balog et al. 2013).

The most well known setup of REIMS, known as “iKnife” or “intelligent scalpel”, has been used intra-surgically. It consists of a hand-held device connected to an electrosurgical instrument which transfers the aerosols produced by cutting through the tissue directly into the MS spectrometer. The MS spectrum produced contains a signature of the lipidome profile of the tissue being cut (Fig. 19.4). MS spectra, collected in real-time, are immediately tested in a multivariate discriminant model (trained on real benign and malignant tissue spectra from samples classified via histopathology) giving a classification of the tissue cut by the surgeon (Balog et al. 2013). The iKnife has been tested on hundreds of patients with several types of tumor such as liver, lung, colorectal, breast, gynecologic, glioma, glioblastoma as well as in metastasis from lung and colon cancer to the brain enabling classification of sampled tissues with high sensitivity (90–98%) and specificity (94–100%) values (Balog et al. 2013; St John et al. 2017; Phelps et al. 2018). A version of the iKnife procedure was also introduced in the endoscopic analysis of colon polyps (Balog et al. 2015). The major disadvantages of REIMS compared to the above mentioned methods are sample consumption, possible cross-contamination and analyte degradation during tissue cutting due to the high temperatures generated (Hänel et al. 2019).

The PIRL method is a promising method for *in vivo* applications and it has some advantages over REIMS. PIRL uses infrared laser to cut through tissue which enable MS spectra to be obtained from smaller areas of tissue and even single cells and avoid damaging adjacent tissue (Hänel et al. 2019). It also has the potential to achieve better spatial resolution *in vivo* compared to REIMS (Hänel et al. 2019).

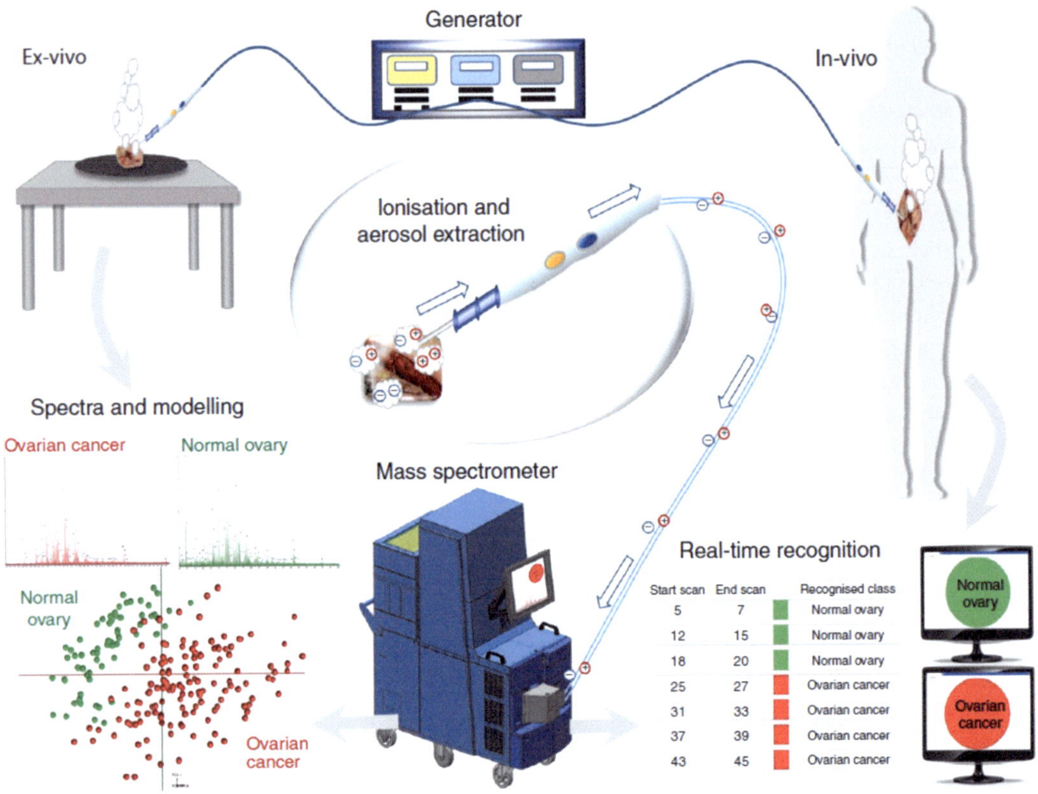

Fig. 19.4 Rapid evaporative mass spectrometry "iKnife" analysis of intact tissue applied to ovarian cancer. Electrical current, produced from the generator, is applied to the tissue and the resultant charged particles are extracted through the custom-designed hand-piece and drawn into the REIMS atmospheric inlet and analyzed in a Xevo G2-XS mass spectrometer to produce tissue-specific mass spectra, which are then subjected to multivariate statistical analysis using Principal Component – Linear Discriminant Analysis (PC-LDA). Within 1–2 s, real-time tissue diagnosis is displayed on a screen for the surgeon to see. Adapted with permission from Phelps et al. (2018) under Creative Commons Attribution 4.0 License

19.3.7.3 MS Imaging of Intact Tissue

Perhaps one of the most interesting applications of MS is imaging (MSI). In MSI, samples are prepared into fine slices or smears, much like in histologic preparations. Then the sample surface is scanned in small areas (10–200 μm) corresponding to image pixels, and ions are withdrawn and analyzed in the mass spectrometer (Bodzon-Kulakowska and Suder 2016). One MS spectrum is acquired from every small area (pixel) of the sample. An image can be then generated by mapping the intensity of any selected ion into the optical image of the tissue (Fig. 19.5).

In order to get into the fine molecular imaging detail, MSI spectrometers are equipped with ionization techniques such as MALDI, DESI or SIMS and very high resolution detectors such as time-of-flight (TOF), orbitrap or ion-cyclotron resonance, to ensure both high image and MS resolutions. Although SIMS provides higher sensitivity and resolution than MALDI and DESI, the latter two being "soft" ionization methods find more wide-spread application in tumor tissue MSI (Bodzon-Kulakowska and Suder 2016).

Due to the fine molecular detail provided, MSI has an enormous potential both as a diagnostic and research tool in cancer and can be viewed as a form of augmented histology. Indeed changes in metabolites such as lipids as those illustrated

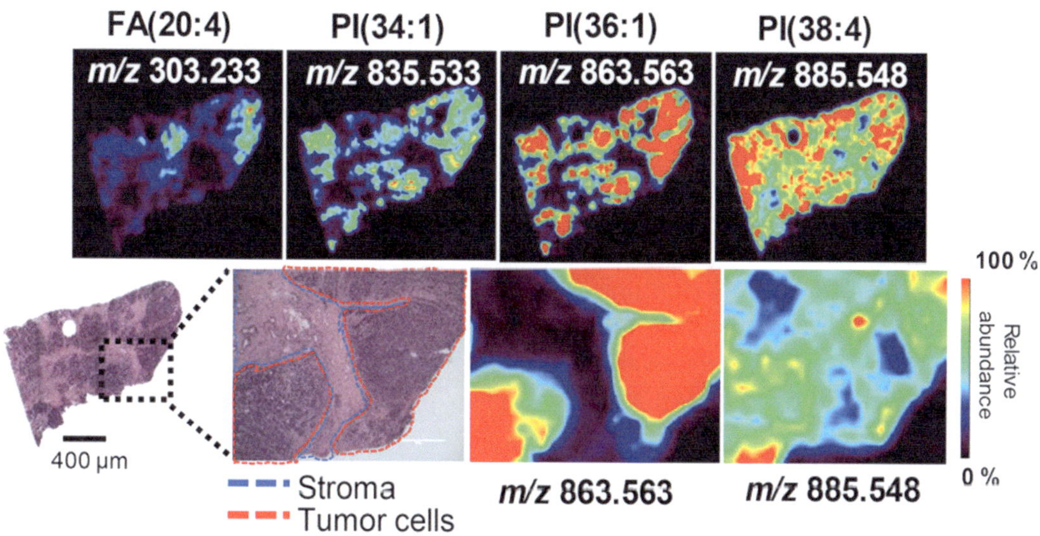

Fig. 19.5 Negative-ion-mode DESI-MS images of a breast tissue sample from an invasive ductal carcinoma patient. Upper panel shows images of specific fatty acid (FA) and phosphatidylinositols (PI) ions highlighting their distribution in the tissue slice. Bottom panel shows the Hematoxylin and Eosin staining optical imaging; expansions of the sectioned tissue shows the delimited stromal and tumoral cells areas and abundance of PI(36:1) and PI(38:4) ions. Lipid species are described by the numbers of fatty acid chain carbons and double bonds. Adapted with permission from Porcari, et al. (2018). Copyright 2018 American Chemical Society

in Fig. 19.5 can be mapped into tumor tissue sections, providing finer detail about tumor heterogeneity and help in the diagnosis of invasive ductal carcinoma (Porcari et al. 2018). MSI also permits the *in situ* study of metabolic pathways that may be altered due to reprogramming. For instance, Sun et al. effectively mapped several metabolites from tumor-associated metabolic pathways, including proline biosynthesis, glutamine metabolism, uridine metabolism, histidine metabolism, fatty acid biosynthesis, and polyamine biosynthesis in tissues from 256 esophageal cancer patients, thus helping to uncover abnormal expression of enzymes pyrroline-5-carboxylate reductase 2 (PYCR2) and uridine phosphorylase 1 (UPase1) in esophageal squamous cell carcinoma (Sun et al. 2019).

Finally, MSI has found an increasing applicability in pharmaceutical research and drug development in oncology, particularly in drug biodistribution, pharmacodynamic biomarker research and in toxicology assessment studies (Goodwin and Webborn 2015).

19.4 Final Remarks and Future Prospects

Metabolomics and metabolic phenotyping are established tools in the study of cancer metabolism. They have benefited from technological developments in both NMR and MS analytical instrumentation coupled with state-of-the art data analysis, particularly in the last decade. Both analytical platforms seem well suited for the development of diagnostic methods in cancer. However, the higher investment and operational costs of NMR hinder its wide-spread adoption. One exception is *in vivo* NMR spectroscopy (MRS), which can be performed in diagnostic MRI instruments and, in fact, it is an approved diagnostic tool to investigate certain types of brain tumors (Horská and Barker 2010). However, in comparison with *ex vivo* NMR, *in vivo* MRS has limited resolution and sensitivity which are factors that may have limited the translation of *ex vivo* discoveries to *in vivo* diagnostic MRS. The introduction of hyperpolarized substrates using

DNP techniques, showed very promising results in preclinical studies and may form the basis for future metabolic imaging applications using NMR (Julià-Sapé et al. 2019). On the other hand, MS-based techniques have been present in clinical chemistry laboratories as diagnostic tool for several decades, especially are used in drug monitoring, newborn screening and in the diagnosis of metabolic diseases (Hänel et al. 2019), which make them ideally suited to metabolomics/metabolic profiling based diagnostic applications. Major advances in MS-based approaches such as intra-operative MS and MS imaging are opening the door for real clinical applications. Nevertheless, there is still a long road ahead until the development of truly diagnostic metabolomics approaches in cancer comes to fruition in the clinics, particularly if less-invasive and early diagnosis applications are to be considered.

The number of published studies in metabolomics/ metabolic phenotyping applications in oncology is already vast, covering a wide range of malignancies at different stages of the disease, across numerous types of biological samples and diverse patient/subject background and of varying sample size. As the amount of scientific literature grows, putting all the information into context in order to draw meaningful conclusions useful for diagnostic application becomes a challenge. This is in part due to the varying study designs, different reporting details of patient data, diverging sample preparation and acquisition protocols as well as insufficient reporting of analytical bias, which makes knowledge integration (for instance through meta-analysis) a difficult task. Therefore, standardized reporting of study design, sampling, experimental protocols, metadata and rigorous metabolite identification and analytical bias reporting would facilitate knowledge integration and would also help promote replication studies which are needed for biomarker validation.

References

Akita H, Ritchie SA, Takemasa I et al (2016) Serum metabolite profiling for the detection of pancreatic cancer: results of a large independent validation study. Pancreas 45:1418–1423

Amal H, Shi DY, Ionescu R et al (2015) Assessment of ovarian cancer conditions from exhaled breath. Int J Cancer 136:E614–E622. https://doi.org/10.1002/ijc.29166

Amal H, Leja M, Funka K et al (2016) Detection of precancerous gastric lesions and gastric cancer through exhaled breath. Gut 65:400–407

Amberg A, Riefke B, Schlotterbeck G et al (2017) NMR and MS methods for metabolomics. Methods Mol Biol 1641:229–258. https://doi.org/10.1007/978-1-4939-7172-5_13

An YJ, Cho HR, Kim TM et al (2015) An NMR metabolomics approach for the diagnosis of leptomeningeal carcinomatosis in lung adenocarcinoma cancer patients. Int J Cancer 136:162–171

Ang JE, Pandher R, Ang JC et al (2016) Plasma metabolomic changes following PI3K inhibition as pharmacodynamic biomarkers: preclinical discovery to phase I trial evaluation. Mol Cancer Ther 15:1412–1424

Ang JE, Pal A, Asad YJ et al (2017) Modulation of plasma metabolite biomarkers of the MAPK pathway with MEK inhibitor RO4987655: pharmacodynamic and predictive potential in metastatic melanoma. Mol Cancer Ther 16:2315–2323

Ardenkjaer-Larsen J-H, Boebinger GS, Comment A et al (2015) Facing and overcoming sensitivity challenges in biomolecular NMR spectroscopy. Angew Chem Int Ed Engl 54:9162–9185

Asai Y, Itoi Y, Sugimoto M et al (2018) Elevated polyamines in saliva of pancreatic cancer. Cancers 10:E43. https://doi.org/10.3390/cancers10020043

Bala L, Sharma A, Yellapa RK et al (2008) ^{1}H NMR spectroscopy of ascitic fluid: discrimination between malignant and benign ascites and comparison of the results with conventional methods. NMR Biomed 21:606–614

Balog J, Sasi-Szabó L, Kinross J et al (2013) Intraoperative tissue identification using rapid evaporative ionization mass spectrometry. Sci Transl Med 5:194ra93. https://doi.org/10.1126/scitranslmed.3005623

Balog J, Kumar S, Alexander J et al (2015) In vivo endoscopic tissue identification by rapid evaporative ionization mass spectrometry (REIMS). Angew Chem 54:11059–11062

Barash O, Zhang W, Halpern JM et al (2015) Differentiation between genetic mutations of breast cancer by breath volatolomics. Oncotarget 6:44864–44876. https://doi.org/10.18632/oncotarget.6269

Bathen TF, Geurts B, Sitter B et al (2013) Feasibility of MR metabolomics for immediate analysis of resection margins during breast cancer surgery. PLoS One 8:e61578. https://doi.org/10.1371/journal.pone.0061578

Bharti SK, Wildes F, Hung C-F et al (2017) Metabolomic characterization of experimental ovarian cancer ascitic fluid. Metabolomics 113. https://doi.org/10.1007/s11306-017-1254-3

Bingol K (2018) Recent advances in targeted and untargeted metabolomics by NMR and MS/NMR methods. High-Throughput 7:E9. https://doi.org/10.3390/ht7020009

Bloch F, Hansen WW, Packard M (1946) The nuclear induction experiment. Phys Rev 70:474–485. https://doi.org/10.1103/PhysRev.70.474

Bodzon-Kulakowska A, Suder P (2016) Imaging mass spectrometry: instrumentation, applications, and combination with other visualization techniques. Mass Spectrom Rev 35:147–169

Bouza M, Gonzalez-Soto J, Pereiro R et al (2017) Exhaled breath and oral cavity VOCs as potential biomarkers in oral cancer patients. J Breath Res 11:016015. https://doi.org/10.1088/1752-7163/aa5e76

Bro R, Kamstrup-Nielsen MH, Engelsen SB et al (2015) Forecasting individual breast cancer risk using plasma metabolomics and biocontours. Metabolomics 11:1376–1380

Broadhurst DI, Kell DB (2006) Statistical strategies for avoiding false discoveries in metabolomics and related experiments. Metabolomics 2:171–196

Cameron SJS, Lewis KE, Beckmann M et al (2016) The metabolomic detection of lung cancer biomarkers in sputum. Lung Cancer 94:88–95

Chan AW, Mercier P, Schiller D et al (2016) ^{1}H-NMR urinary metabolomic profiling for diagnosis of gastric cancer. Br J Cancer 114:59–62

Chaumeil MM, Larson PEZ, Yoshihara HAI et al (2013) Non-invasive in vivo assessment of IDH1 mutational status in glioma. Nat Commun 4:2429

Cheng XM, Liu XY, Liu X et al (2018) Metabolomics of non-muscle invasive bladder cancer: biomarkers for early detection of bladder cancer. Front Oncol 8:494. https://doi.org/10.3389/fonc.2018.00494

Choi C, Ganji SK, DeBerardinis RJ et al (2012) 2-hydroxyglutarate detection by magnetic resonance spectroscopy in subjects with IDH-mutated gliomas. Nat Med 18:624–629

Cornel EB, Smits GA, Oosterhof GO et al (1993) Characterization of human prostate cancer, benign prostatic hyperplasia and normal prostate by in vitro ^{1}H and ^{31}P magnetic resonance spectroscopy. J Urol 150:2019–2024

Dalgliesh CE (1956) Two-dimensional paper chromatography of urinary indoles and related substances. Biochem J 64:481–485

Dang L, White DW, Gross S et al (2009) Cancer-associated IDH1 mutations produce 2-hydroxyglutarate. Nature 462:739–744

de Vogel S, Ulvik A, Meyer K et al (2014) Sarcosine and other metabolites along the choline oxidation pathway in relation to prostate cancer-a large nested case-control study within the JANUS cohort in Norway. Int J Cancer 134:197–206

Dinges SS, Hohm A, Vandergrift LA et al (2019) Cancer metabolomic markers in urine: evidence, techniques and recommendations. Nat Rev Urol 16:339–362

Dona AC, Jiménez B, Schäfer H et al (2014) Precision high-throughput proton NMR spectroscopy of human urine, serum, and plasma for large-scale metabolic phenotyping. Anal Chem 86:9887–9894

Doskocz M, Marchewka Z, Jeż M et al (2015) Preliminary study on J-resolved NMR method usability for toxic kidney's injury assessment. Adv Clin Exp Med 24:629–635

Dunphy MPS, Harding JJ, Venneti S (2018) In vivo PET assay of tumor glutamine flux and metabolism: in-human trial of ^{18}F-(2S,4R)-4-fluoroglutamine. Radiology 287:667–675

Emwas AH (2015) The strengths and weaknesses of NMR spectroscopy and mass spectrometry with particular focus on metabolomics research. Methods Mol Biol 1277:161–193

Emwas AH, Roy R, McKay RT et al (2019) NMR spectroscopy for metabolomics research. Meta 9. https://doi.org/10.3390/metabo9070123

Fan TW-M, Lane AN (2016) Applications of NMR spectroscopy to systems biochemistry. Prog Nucl Magn Reson Spectrosc 92-93:18–53

Fan TW-M, Lorkiewicz PK, Sellers K et al (2012) Stable isotope-resolved metabolomics and applications for drug development. Pharmacol Ther 133:366–391

Fu XA, Li MX, Knipp RJ et al (2014) Noninvasive detection of lung cancer using exhaled breath. Cancer Med 3:174–181

Fukutake N, Ueno M, Hiraoka N et al (2015) A novel multivariate index for pancreatic cancer detection based on the plasma free amino acid profile. PLoS One 10:e0132223. https://doi.org/10.1371/journal.pone.0132223

Ganti S, Taylor SL, Kim K et al (2012) Urinary acylcarnitines are altered in human kidney cancer. Int J Cancer 130:2791–2800

Garcia RA, Morales V, Martin S et al (2014) Volatile organic compounds analysis in breath air in healthy volunteers and patients suffering epidermoid laryngeal carcinomas. Chromatographia 77:501–509

Garrison RN, Kaelin LD, Galloway RH et al (1986) Malignant ascites. Clinical and experimental observations. Ann Surg 203:644–651

Giskeodegard GF, Bertilsson H, Selnaes KM et al (2013) Spermine and citrate as metabolic biomarkers for assessing prostate cancer aggressiveness. PLoS One 8:e62375. https://doi.org/10.1371/journal.pone.0062375

Glish GL, Vachet RW (2003) The basics of mass spectrometry in the twenty-first century. Nat Rev Drug Dis 2:140–150

Goodwin RJA, Webborn PJH (2015) Future directions of imaging MS in pharmaceutical R&D. Bioanalysis 7(20):2667–2673

Graça G, Desterro J, Sousa J et al (2017) Identification of putative biomarkers for leptomeningeal invasion in B-cell non-Hodgkin lymphoma by NMR metabolomics. Metabolomics 13:136

Graça G, Serrano-Contreras JI, Chekmeneva E (2019) Nuclear magnetic resonance spectroscopy: pulse sequences for chemical analysis. In: Worsfold P, Poole C, Townshend A, Miró M (eds) Encyclopedia of analytical science, vol 7, 3rd edn. Elsevier, Amsterdam, pp 354–365

Gromski PS, Muhamadali H, Ellis DI et al (2015) A tutorial review: metabolomics and partial least squares-discriminant analysis - a marriage of convenience or a shotgun wedding. Anal Chim Acta 879:10–23

Guennec AL, Giraudeau P, Caldarelli S (2014) Evaluation of fast 2D NMR for metabolomics. Anal Chem 86:5946–5954

Hänel L, Kwiatkowski M, Heikaus L et al (2019) Mass spectrometry-based intraoperative tumor diagnostics. Future Sci OA 5:FSO373. https://doi.org/10.4155/fsoa-2018-0087

Hanna GB, Boshier PR, Markar SR et al (2019) Accuracy and methodological challenges of volatile organic compound-based exhaled breath tests for cancer diagnosis: a systematic review and meta-analysis. JAMA Oncol 5:e182815. https://doi.org/10.1001/jamaoncol.2018.2815

Hao J, Astle W, De Iorio M et al (2012) BATMAN – an R package for the automated quantification of metabolites from nuclear magnetic resonance spectra using a Bayesian model. Bioinformatics 28:2088–2090

Haukaas TH, Moestue SA, Vettukattil R et al (2016) Impact of freezing delay time on tissue samples for metabolomic studies. Front Oncol 6:17. https://doi.org/10.3389/fonc.2016.00017

Hilvo M, de Santiago I, Gopalacharyulu P et al (2016) Accumulated metabolites of hydroxybutyric acid serve as diagnostic and prognostic biomarkers of ovarian high-grade serous carcinomas. Cancer Res 76:796–804

Horská A, Barker PB (2010) Imaging of brain tumors: MR spectroscopy and metabolic imaging. Neuroimaging Clin N Am 20(3):293–310

Ifa DR, Eberlin LS (2016) Ambient ionization mass spectrometry for cancer diagnosis and surgical margin evaluation. Clin Chem 62:111–123

Jin X, Yun SJ, Jeong P et al (2014) Diagnosis of bladder cancer and prediction of survival by urinary metabolomics. Oncotarget 5:1635–1645

Julià-Sapé M, Candiota AP, Arús C (2019) Cancer metabolism in a snapshot: MRS(I). NMR Biomed 11:e4054. https://doi.org/10.1002/nbm.4054

Komoroski RA, Holder JC, Pappas AA et al (2011) ^{31}P NMR of phospholipid metabolites in prostate cancer and benign prostatic hyperplasia. Magn Reson Med 65:911–913

Kumar S, Huang J, Abbassi-Ghadi N et al (2015) Mass spectrometric analysis of exhaled breath for the identification of volatile organic compound biomarkers in esophageal and gastric adenocarcinoma. Ann Surg 262:981–990

Le Gall G, Guttula K, Kellingray L et al (2018) Metabolite quantification of faecal extracts from colorectal cancer patients and healthy controls. Oncotarget 9:33278–33289

Li MX, Yang DK, Brock G et al (2015) Breath carbonyl compounds as biomarkers of lung cancer. Lung Cancer 90:92–97

Lin Y, Ma CC, Liu CK et al (2016) NMR-based fecal metabolomics fingerprinting as predictors of earlier diagnosis in patients with colorectal cancer. Oncotarget 7:29454–29464

Liu W, Bai XF, Liu YJ et al (2015) Topologically inferring pathway activity toward precise cancer classification via integrating genomic and metabolomic data: prostate cancer as a case. Sci Rep 5:13192

Marchand J, Martineau E, Guitton Y et al (2017) Multidimensional NMR approaches towards highly resolved, sensitive and high-throughput quantitative metabolomics. Curr Opin Biotechnol 43:49–55

Marion D (2013) An introduction to biological NMR spectroscopy. Mol Cell Proteomics 12:3006–3025

Mayers JR, Wu C, Clish CB et al (2014) Elevation of circulating branched-chain amino acids is an early event in human pancreatic adenocarcinoma development. Nat Med 20:1193–1198

McDunn JE, Li Z, Adam KP et al (2013) Metabolomic signatures of aggressive prostate cancer. Prostate 73:1547–1560

Mirnezami R, Jiménez B, Li JV et al (2014) Rapid diagnosis and staging of colorectal cancer via high-resolution magic angle spinning nuclear magnetic resonance (HR-MAS NMR) spectroscopy of intact tissue biopsies. Ann Surg 259:1138–1149

Molenaar RJ, Thota S, Nagata Y et al (2015) Clinical and biological implications of ancestral and non-ancestral IDH1 and IDH2 mutations in myeloid neoplasms. Leukemia 29:2134–2142

Nagana Gowda GA, Raftery D (2017) Recent advances in NMR-based metabolomics. Anal Chem 89:490–510

Nicholson JK, O'Flynn MP, Sadler PJ et al (1984) Proton-nuclear-magnetic-resonance studies of serum, plasma and urine from fasting normal and diabetic subjects. Biochem J 217:365–375

Pansuriya TC, van Eijk R, d'Adamo P et al (2011) Somatic mosaic IDH1 and IDH2 mutations are associated with enchondroma and spindle cell hemangioma in Ollier disease and Maffucci syndrome. Nat Genet 43:1256–1261

Paul A, Kumar S, Raj A et al (2018) Alteration in lipid composition differentiates breast cancer tissues: a ^{1}H HRMAS NMR metabolomic study. Metabolomics 14:119

Pavlova NN, Thompson CB (2016) The emerging hallmarks of cancer metabolism. Cell Metab 23:27–47. https://doi.org/10.1016/j.cmet.2015.12.006

Peng G, Hakim M, Broza YY et al (2010) Detection of lung, breast, colorectal, and prostate cancers from exhaled breath using a single array of nanosensors. Br J Cancer 103:542–551

Phelps DL, Balog J, Gildea LF et al (2018) The surgical intelligent knife distinguishes normal, borderline and malignant gynaecological tissues using rapid evaporative ionisation mass spectrometry (REIMS). Br J Cancer 118:1349–1358

Porcari AM, Zhang J, Garza KY et al (2018) Multi-center study using desorption-electrospray-ionization-mass-spectrometry imaging for breast cancer diagnosis. Anal Chem 90:11324–11332

Psychogios N, Hau DD, Peng J et al (2011) The human serum metabolome. PLoS One 6:e16957. https://doi.org/10.1371/journal.pone.0016957

Purcell EM, Torrey HC, Pound RV (1946) Resonance absorption by nuclear magnetic moments in a solid. Phys Rev 69.37–38

Qin T, Liu H, Song Q et al (2010) The screening of volatile markers for hepatocellular carcinoma. Cancer Epidemiol Biomark Prev 19:2247–2253

Sakai A, Suzuki M, Kobayashi T et al (2016) Pancreatic cancer screening using a multiplatform human serum metabolomics system. Biomark Med 10:577–586

Sanderson SM, Locasale JW (2018) Revisiting the Warburg effect: some tumors hold their breath. Cell Metab 28:669–670

Sangisetty SL, Miner TJ (2012) Malignant ascites: a review of prognostic factors, pathophysiology and therapeutic measures. World J Gastrointest Surg 4:87–95

Shender VO, Pavlyukov MS, Ziganshin RH et al (2014) Proteome-metabolome profiling of ovarian cancer ascites reveals novel components involved in intercellular communication. Mol Cell Proteomics 13:3558–3571

Slupsky CM, Steed H, Wells TH et al (2010) Urine metabolite analysis offers potential early diagnosis of ovarian and breast cancers. Clin Cancer Res 16:5835–5841

Sreekumar A, Poisson LM, Rajendiran TM et al (2009) Metabolomic profiles delineate potential role for sarcosine in prostate cancer progression. Nature 457:910–914

St John ER, Balog J, Mckenzie JS et al (2017) Rapid evaporative ionisation mass spectrometry of electrosurgical vapours for the identification of breast pathology: towards an intelligent knife for breast cancer surgery. Breast Cancer Res 19:59. https://doi.org/10.1186/s13058-017-0845-2

Sugimoto M, Wong DT, Hirayama A et al (2010) Capillary electrophoresis mass spectrometry-based saliva metabolomics identified oral, breast and pancreatic cancer-specific profiles. Metabolomics 6:78–95

Sun C, Lia T, Song X et al (2019) Spatially resolved metabolomics to discover tumor associated metabolic alterations. PNAS 116:52–57

Tanaka K, Isselbacher KJ (1967) The isolation and identification of N-isovalerylglycine from urine of patients with isovaleric acidemia. J Biol Chem 242:2966–2972

Trygg J, Holmes E, Lundstedt T (2007) Chemometrics in metabonomics. J Proteome Res 6:469–479

Tsutsui H, Mochizuki T, Inoue K et al (2013) High-throughput LC-MS/MS based simultaneous determination of polyamines including N-acetylated forms in human saliva and the diagnostic approach to breast cancer patients. Anal Chem 85:11835–11842

Vettukattil R, Hetland TE, Flørenes VA et al (2013) Proton magnetic resonance metabolomic characterization of ovarian serous carcinoma effusions: chemotherapy-related effects and comparison with malignant mesothelioma and breast carcinoma. Hum Pathol 44:1859–1866

Wishart DS, Jewison T, Guo AC et al (2013) HMDB 3.0 — the human metabolome database in 2013. Nucleic Acids Res 41:D801–D807. https://doi.org/10.1093/nar/gks1065

Woo HM, Kim KM, Choi MH et al (2009) Mass spectrometry based metabolomic approaches in urinary biomarker study of women's cancers. Clin Chim Acta 400:63–69

Yan H, Parsons DW, Jin GL et al (2009) IDH1 and IDH2 mutations in gliomas. N Engl J Med 360:765–773

Zhu A, Lee D, Shim H (2011) Metabolic PET imaging in cancer detection and therapy response. Semin Oncol 38:55–69

Part V

Animal Models: Addressing Cancer Microenvironment

Animal Models to Study Cancer and Its Microenvironment

20

N. Mendes, P. Dias Carvalho, F. Martins, S. Mendonça, A. R. Malheiro, A. Ribeiro, J. Carvalho, and S. Velho

Abstract

Cancers are complex tissues composed by genetically altered cancer cells and stromal elements such as inflammatory/immune cells, fibroblasts, endothelial cells and pericytes, neuronal cells, and a non-cellular component, the extracellular matrix. The complex network of interactions and crosstalk established between cancer cells and the supportig cellular and non-cellular components of the microenvironment are of extreme importance for tumor initiation and progression, strongly impacting the course and the outcome of the disease. Therefore, a better understanding of the tumorigenic processes implies the combined study of the cancer cell and the biologic, chemical and mechanic constituents of the tumor microenvironment, as their concerted action plays a major role in the carcinogenic pathway and is a key determinant of the efficacy of anti-cancer treatments. The use of animal models (e.g. Mouse, Zebrafish and Drosophila) to study cancer has greatly impacted our understanding of the processes governing initiation, progression and metastasis and allowed the discovery and pre-clinical validation of novel cancer treatments as it allows to recreate tumor development in a more pathophysiologic environment.

Keywords

Tumor microenvironment (TME) · Animal models · Mouse · Zebrafish · Drosophila · Cancer progression · Metastasis

20.1 Background

Cancer is a composite and heterogeneous disease in which multiple factors act together to promote a malignant phenotype. In fact, cancers are complex tissues composed by genetically altered cancer cells and stromal elements such as inflammatory/immune cells, fibroblasts, endothelial cells and pericytes, neuronal cells, and a non-cellular component, the extracellular matrix (Fig. 20.1) (Hanahan and Coussens 2012). The complex network of interactions and crosstalk established between cancer cells

The original version of the chapter has been revised. A correction to this chapter can be found at https://doi.org/10.1007/978-3-030-34025-4_24

N. Mendes (✉) · P. Dias Carvalho · F. Martins
S. Mendonça · A. Ribeiro · J. Carvalho · S. Velho (✉)
i3S, Instituto de Investigação e Inovação em Saúde, Porto, Portugal

IPATIMUP, Instituto de Patologia Molecular e Imunologia da Universidade do Porto, Porto, Portugal
e-mail: nmendes@ipatimup.pt; svelho@ipatimup.pt

A. R. Malheiro
i3S, Instituto de Investigação e Inovação em Saúde, Porto, Portugal

IBMC, Instituto de Biologia Molecular e Celular da Universidade do Porto, Porto, Portugal

J. Serpa (ed.), *Tumor Microenvironment*, Advances in Experimental Medicine and Biology 1219,
https://doi.org/10.1007/978-3-030-34025-4_20

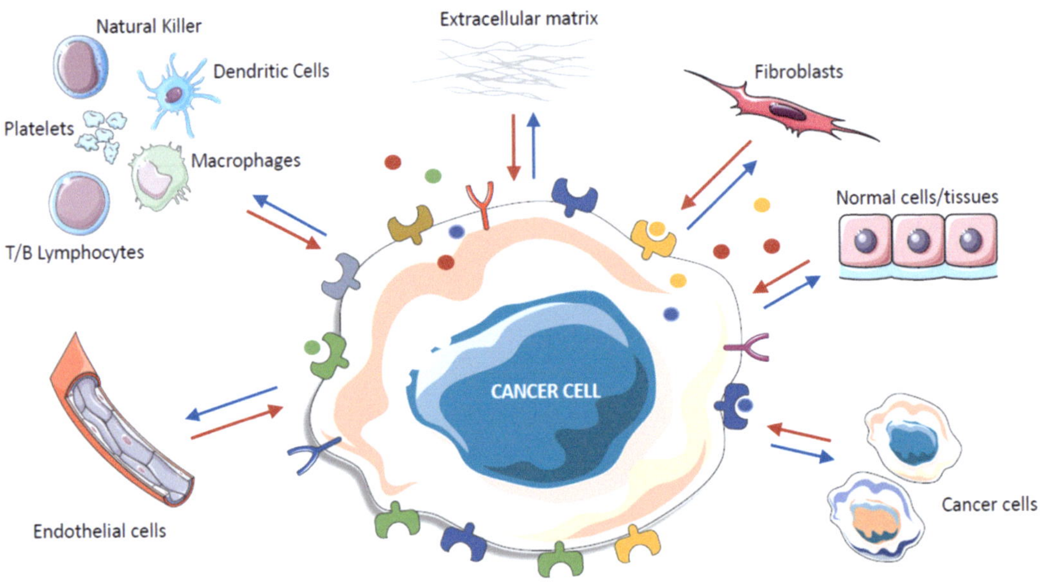

Fig. 20.1 Interactions at the tumor microenvironment. Cancer cells interact with the tumor microenvironment components (e.g. other cancer and normal cells, or stromal cells such as fibroblasts, immune cells, blood vessels) to gain survival advantages and to build up an aggressive phenotype. Through direct or indirect (via soluble factors) interactions, cancer cells actively modulate the properties of the surrounding elements, inducing a pro-tumorigenic microenvironment. The pro-tumorigenic microenvironment feeds back to the cancer cell, enhancing its malignant behavior. Understanding the complexity of cancer can only be achieved by applied experimental approaches that combine the study of the cancer cell and the biologic, chemical and mechanic constituents of the tumor microenvironment

and the supporting cellular and non-cellular components are of extreme importance for tumor initiation and progression (Fiori et al. 2019). They dictate the biologic, chemical and mechanical properties of the cancer tissue, strongly impacting the course and the outcome of the disease. For example, increased interstitial pressure, perturbations in structure and function of the extracellular matrix, hypoxia, host and tumor cells immune response interaction and angiogenesis are all phenomena that promote cancer and metastases (Lindner 2014). Unraveling the molecular mediators of this crosstalk is fundamental to identify new therapeutic targets to abrogate cancer progression (Dias Carvalho et al. 2018).

Therefore, a better understanding of the tumorigenic processes implies the combined study of the cancer cell and the biologic, chemical and mechanic constituents of the tumor microenvironment, as their concerted action plays a major role in the carcinogenic pathway and is a key determinant of the efficacy of anti-cancer treatments (Hanahan and Weinberg 2000).

20.2 Animal Models in Tumor Microenvironment Cancer Research

Prior to the development of animal models, *in vitro* cell culture systems using cell lines derived from human tumors were the primary model to study cancer. This system provided, and still provides, valuable information on the molecular mechanisms underlying cancer. However, the inability to examine pathophysiological interactions among tumor cells and between tumor cells and their microenvironment, currently known as key aspects of tumor development and progression, represent a major limitation of the model. The use of animal models to study cancer overcomes this limitation as it allows to recreate tumor development in a more pathophysiologic environment. Consequently, their use has greatly impacted our understanding of the processes governing initiation, progression and metastasis and allowed the discovery and pre-clinical validation of novel cancer treatments (Day et al. 2015).

Distinct animal models can be used in cancer research. Over the last decades the most common species used is the *Mus musculus,* vulgarly known as the laboratory mouse. However, other models have also been applied, including the Zebrafish – *Danio rerio*, and the common fruit fly – *Drosophila melanogaster*. A variety of approaches can be applied depending on the aim of the study. In this chapter the most commonly used animal models for the study of tumorigenesis and metastasis will be described.

20.2.1 Mouse Models

Understanding the complexities of cancer demands a versatile experimental approach within the context of the whole animal (Van Dyke and Jacks 2002). For that purpose, mice are excellent biologic mimetics of the human physiology as both species follow similar embryonic developmental stages and their bodies have the same kind of organs and complex regulatory mechanisms. Importantly, mice and human genomes share a high degree of homology, and there is an overlap in the function of their genes. Moreover, as mice have a shorter lifespan, it makes possible to study development and progression of diseases such as cancer in a feasible period of time. Additionally, mice are small, relatively economical to maintain, and easy to manipulate, making them the ideal laboratory animal model (www.jax.org/why-the-mouse/excellent-models).

The most common approaches for modelling cancer in mouse can be divided in four major categories: (1) Cell line-Derived Xenografts (CDX), (2) Patient-Derived Xenograft (PDX), (3) chemically-induced, and (4) Genetically Engineered Mouse Models (GEMMs) (Fig. 20.2).

20.2.1.1 Cell Line-Derived Xenografts

The CDX cancer model was introduced during the 1980s by Fidler and colleagues (Fidler and Hart 1982) marking the start of a new era of research. The use of murine and human tumor cell lines to induce tumors in immunocompetent or, more often, immunocompromised mice, respectively, allowed not only the understanding of the tumor growth kinetics but also the comprehension of metastatic progression and the expand of therapeutic approaches. A xenograft model generated by the injection of cancer cell lines subcutaneously into immunodeficient mice is the most commonly used model as it offers the advantages of being easy to generate and shows consistent tumor growth among animals (Jung et al. 2018) (Fig. 20.2a). Moreover, the advances in site specific cell inoculations allowed the induction of tumors into the specific organs (e.g. gastric cancer cell lines inoculated in the gastric wall) – orthotopic tumors (Fig. 20.2b). Transplantation of cancer cell lines in the organ of origin captures the original characteristics of the organ-specific tumor microenvironment, allowing to replicate the invasion properties of cancer cells and original metastasis pathway. However, these models are laborious and require an expensive monitoring scheme with access to high performance imaging equipment (Day et al. 2015).

Although CDX models are of extreme relevance in understanding cancer mechanisms, in preclinical drug development (Teicher 2006), and for analyzing resistance mechanisms (Garraway and Jänne 2012), they also present caveats that limit their successful translation into patients' tumor reality. Cancer cells exhibit variable degrees of genomic instability, resulting in heterogeneous subsets of cells that increase the complexity of a given cancer. By using cancer cell lines grown in a Petri dish for several generations, CDX models do not recapitulate this complex tumor heterogeneity. As a consequence of the inability to recreate cellular and genetic heterogeneity, CDXs have failed to predict human efficacy for most therapeutic targets, a fact supported by the low FDA approval rates for cancer drugs (approximately 5–7%), especially for solid tumors (Sharpless and Depinho 2006). Moreover, xenograft tumor often cannot recapitulate the tumor microenvironment. In particular, the requirement of immunodeficient mouse strains impairs the study of the anti-tumor host immune response (Jung et al. 2018), constituting a problem when the aim of a particular study is focused in immunomodulatory mechanisms, one of the major players of the tumor microenvironment.

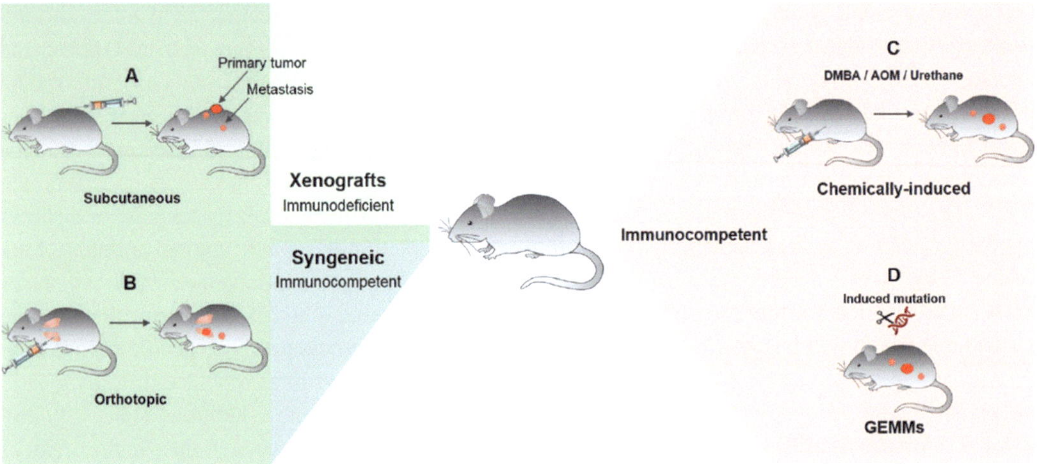

Fig. 20.2 Mouse as models in cancer research: Cell line-derived mouse models can be divided in xenograft and syngeneic models. In xenograft models (or heterologous models) the tumors induced are mostly of Human origin (or not from the same animal species from the host. For this reason immunodeficient mouse strains are used to avoid implant rejection. In contrast, the tumors induced in syngeneic models are derived from mouse cell lines origin in immunocompetent hosts. A syngeneic mouse model provides an effective approach for studying how cancer therapies perform in the presence of a functional immune system (**a**) ***Subcutaneous model***: tumor cells are subcutaneously implanted, frequently, on the flank of animals where they grow and form palpable and measurable tumors. (**b**) ***Orthotopic model***: tumor cells are surgically implanted into their organ of origin. The organ-specific microenvironment induces tumor growth similar to that of the original tumor. As an example, the figure illustrates inoculation of lung cancers cells directly in the lung. If transplanted sample is from a solid patient tumor the technique performed is a Patient-Derived Xenograft. Depending on the experimental design this models can be metastatic or non-metastatic. (**c**) ***Chemically-induced cancer mouse models***: specific strains of mice have different susceptibility to develop cancer when treated with tissue-specific chemical carcinogens. For example, azoxymethane (AOM) is used to induce colon cancer; 7,12-dimethylbenz[a]anthracene (DMBA) is used to induce mammary tumors and urethane to induce lung cancer. (**d**) ***Genetically Engineered Mouse Cancer Models (GEMMs)***: tumors develop in immunocompetent mouse in the presence of an induced mutation in a constitutive or inducible form

20.2.1.2 Syngeneic Mouse Models to Study the Tumor-Host Immune System Interaction

To overcome the weakness of CDX in recreating the immune system, the use of syngeneic tumors is of extreme relevance. In this case, tumors are usually generated by subcutaneous implantation of major histocompatibility complex (MHC)-matched tumor cell lines, derived from mouse spontaneous or induce tumors, in the original, immune competent mouse strain (Fig. 20.2a). Although lacking tumor heterogeneity and, if subcutaneous, the organ specific tumor microenvironment, they offer the possibility to study the interplay between cancer cells and the host immune system and are commonly used to characterize efficacy and mechanisms of cancer immunotherapies (Yu et al. 2018). There are several mouse cell lines from different cancer models currently used to form syngeneic tumors in distinct mouse strains (Table 20.1) Interestingly, it has been shown that, depending on the cell line, syngeneic tumors possess a unique tumor-immune infiltrate profile (Table 20.1) that can be probed with immunotherapies to inform on anti-tumor mechanisms and treatment strategies in human tumors with similar profiles (Lechner et al. 2013; Yu et al. 2018). An important aspect to take into consideration when using syngeneic tumor models is that the tumor immune phenotype is highly modulated by the organ specific microenvironment surrounding the tumor. Organ specific elements such as resident and recruited immune cells, endothelial cells, fibroblasts, extracellular matrix proteins, and others can uniquely shape tumor proliferation, vascularization, metastasis,

Table 20.1 Mouse-derived cancer cell lines commonly used in syngeneic mouse models.

Cell line	Cancer model	Mouse strain	Tumor-immune infiltrate profile	References
RENCA	Renal adenocarcinoma	BALB/c	Highly immunogenic; Highly immune-infiltrated	Yu et al. (2018)
CT26	Colon carcinoma	BALB/c	Highly immunogenic; immune infiltrated	Lechner et al. (2013), Yu et al. (2018)
MC38	Colon carcinoma	C57BL/6	Immune-excluded	Ganesh and Massagué (2018), Mariathasan et al. (2018)
EMT6	Breast carcinoma (triple negative subtype)	BALB/c	Immune-excluded	Mariathasan et al. (2018), Yu et al. (2018)
E0771	Breast carcinoma (triple negative subtype)	C57BL/6	Poorly immunogenic; Poorly infiltrated	Hoover et al. (2012)
4T1	Breast carcinoma (basal subtype)	BALB/c	Highly immunogenic	Szatmári et al. (2006)
B16F10	Melanoma	C57BL/6	Poorly immunogenic; Immune-excluded	Yu et al. (2018)
Pan02	Pancreatic carcinoma	C57BL/6	Poorly immunogenic	Gnerlich et al. (2010)
MAD109	Lung carcinoma	BALB/c	Poorly immunogenic	Szatmári et al. (2006)
LLC	Lung carcinoma	C57BL/6	Poorly immunogenic	Szatmári et al. (2006)
GL261	Glioblastoma	C57BL/6	Moderately immunogenic	Szatmári et al. (2006), Yi et al. (2013)
ONC26M4	Glioblastoma	FVBN	Not defined	Szatmári et al. (2006)

as well as tumor-immune infiltration and consequently can shape the response to immunotherapy (Yu et al. 2018). Accordingly, previous studies have shown that orthotopic implantation of RENCA or CT26 as well as spontaneous lung or pancreatic tumors from genetically engineered mouse models yield a more immunosuppressed tumor-immune infiltrate profile that is not as responsive to immunotherapy when compared to immunotherapy responsive subcutaneously implanted tumors (Devaud et al. 2014).

20.2.1.3 Patient-Derived Xenograft Models

In order to overcome some of the issues related to the CDX models, patient-derived xenograft (PDX) models were established by Fiebig et al. in 1984 (Fiebig et al. 1984). The success of PDX models was first demonstrated in 1988 (Mattern et al. 1988) when certain chemotherapeutic agents, such as alkaloids and anti-metabolites used in both, mice and human patients, shown similar responses. In contrast, the results obtained with the NCI60- based CDX models treated with numerous cytotoxic agents were not so impressive (J. I. Johnson et al. 2001), highlighting the importance and better correlation in PDX models. PDX models offer several advantages over CDX models as early passages can retain the stromal composition and the histological and molecular heterogeneity of the original tumor (Hidalgo et al. 2014; Siolas and Hannon 2013; Tentler et al. 2012), replicating in more detail the human disease complexity and overcoming the clonal selection that CDX tumors imply.

PDX are obtained by immediate transplantation of small pieces of tumors (2–3 mm^3), obtained by surgery or biopsy, heterotopically (subcutaneous or in the renal capsule) or orthotopically (in the same organ as the original tumors) in a recipient immunocompromised mouse (Yada et al. 2018) (Fig. 20.2a). Depending on the organ of origin, orthotopic models may be difficult to generate, but they display a more similar microenvironment to that of the original tumor which may impact the course and outcome of the disease (Jung et al. 2018). For example, it was reported that orthotopic PDX models generated from pancreatic cancers showed increased incidence of metastases, compared with heterotopic subcutaneous models (Fu et al. 1992; Yada et al. 2018).

Over the last years, one of the main constrictions of using these models has been the access to clinical samples. However, the collection and freezing of PDX expanded tumors, at early passages, allows the creation of biobanks. This approach can overcome the need of a continuous collection of tissues directly from patients as PDXs can be successfully engrafted in new mouse after thawing (Calles et al. 2013). In this case, a continuous monitoring and correlation between the histopathological and molecular patterns of both, engrafted and primary human tumor, are of major relevance to detect deviations. Other limitations include the time to establish a PDX model from a patient which can take as long as 6 months (or longer), and the fact that there are still some tumor types, such as prostate cancers, that are difficult to establish as PDX models (Choi et al. 2018).

Regarding the study of the tumor microenvironment using PDX models there are unavoidable limitations. A major point of concern regarding PDX models is that almost all stromal cells derived from the human tumor cannot proliferate continuously and, gradually over time, the human stromal compartment is completely replaced by the mouse stroma (Cassidy et al. 2015). This drawback limits the capacity to use PDX models to study all cancer cell-stroma interactions due to the species-species differences regarding the recognition of murine ligands by human receptors and vice versa. This is, for example, the case of the human Met receptor that does not recognize the mouse met ligand so paracrine met signaling is not recapitulated (Williams 2018). Also, mouse prolactin (PRL) antagonizes the human PRL receptor, thereby impairing the ability of PRL positive human tumors to grow in mice (Utama et al. 2009; Williams 2018). Additional PDX models are highly vascularized which does not always reflect the vascularization state in humans raising concerns on the use of these models to study the efficacy of anti-angiogenic drugs (Dong et al. 2013; Williams 2018). Despite all the limitations, it has been reported that human cancer cells still interact and modulate the murine stroma. In a colorectal cancer PDX model, cancer cells were able to re-organize the normal, quiescent murine stromal cells into a pro-tumorigenic phenotype, supporting human CRC growth (Chao et al. 2017). In a different report, the murine stromal transcripts derived from CRC PDXs recapitulated the prognostic mesenchymal gene signature of human CRC tumors (Isella et al. 2015). Moreover, it was also demonstrated that, despite the early replacement of the human CRC stroma by the murine cells (at the second passage), the metabolic profiles of both stromal and cancer cells remained stable for at least four generations in comparison to the original patient material (Blomme et al. 2018). Taken together, these data indicate that, at least for CRC PDX models, appropriate reciprocal paracrine signaling between the cancer cell and the murine stromal cells still occurs (Chao et al. 2017), enabling to some extent the use of these models to study cancer cell-stromal interactions.

Another obvious limitation of PDX models is that, in order to circumvent implant rejection, tumors develop in mice with a compromised immune system, such as nude mice (T cell-deficient), severe combined immunodeficient mice (SCID; T- and B cell-deficient) and extremely immunodeficient mice [non-obese diabetic (NOD)SCID, NOG mice (NOD.Cg-$Prkdc^{scid}Il2rg^{tm1Sug}$/ShiJic), and NSG mice (NOD.Cg-$Prkdc^{scid}Il2rg^{tm1Wjl}$/SzJ); T-, B-, and NK cell-deficient;]. Therefore, studies addressing cancer cell-host immune response interactions and the effect of immunotherapy using PDX models raises serious concerns (Hidalgo et al. 2011). Establishing PDX models in hosts with some functioning immune cells enables a model that retains at least some of the stroma-tumor interactions found in a natural setting (Williams 2018; Yada et al. 2018). The iPDX, which differs from the traditional PDX primarily in that the experiments are conducted in the first passages to avoid replacement of the human stroma by murine stroma, may help to overcome this limitation. In this model, human tumor infiltrating lymphocytes are still present in the TME, and retain the same immunosuppressive features as in the original context allowing to study the human species-specific interaction among tumor and immune cells. Although good to study the effect of immunotherapies, the iPDX models still show some limitations: (i) long-term experiments cannot be performed (after 3–4 weeks postengraftment, murine innate cells partially replace their human counterparts); (ii) because tumors cannot be passaged, the amount of starting material limits the number of animals to enter the study (generally enough for 6–12 mice); (iii) iPDX models do not allow to study immune cell recruitment (Sanmamed et al. 2016).

PDX establishment in a humanized mouse (displaying a human fully competent immune system) represents another powerful tool to overcome the lack of a functional immune system (Zhao et al. 2018). Humanized PDX models are generated by implanting fresh human tumor fragments in NOD-*scid Il2rg*$^{-/-}$ (NSG) mice with a type I human leucocyte antigen matched human immune system. Briefly, NSG mice are irradiated with whole body gamma irradiation, and human CD34^{+} hematopoietic stem cells are intravascularly injected into NSG mice at 5 weeks. Engraftment of human HSCs is monitored by flow cytometry and, if successful, patient tumor tissue is then transplanted. Humanized PDX models allow the study of human anti-tumor immune responses and the response to a variety of therapies, including immunotherapies (Choi et al. 2018; Zhao et al. 2018). However, these models are expensive and time consuming, and it is very difficult to obtain CD34^{+} cells from the cancer patient. Therefore, an allogenic immune approach is usually used, limiting the ideal situation of using the same immune system from which the PDX was derived (Choi et al. 2018).

20.2.1.4 Chemical Carcinogenesis Mouse Models

Chemically-induced mouse models are generated by exposing the mouse to certain chemical compounds with carcinogenic potential through several administration routes, mimicking environmentally-induced human tumors in sensitive organs (Liu et al. 2015) (Fig. 20.2c). Tumors are formed in immune competent strains allowing to study the effect of the immune system, and because tumors develop *in loco*, the influence of the natural environment of the organ is kept along disease development and progression. Moreover, and similarly to environmentally-induced human tumors, chemical carcinogenesis mouse models carry a high mutation burden and are therefore uniquely able to recreate the heterogeneity of genetic and epigenetic events known as critical determinants of individual patient prognosis, responses to therapy, or development of drug resistance (McCreery et al. 2015). These models are particularly interesting to study cancer of naturally exposed organs such as skin, lung, the esophagus, head and neck and gastrointestinal tract (McCreery et al. 2015). These models offer additional advantages including non- or less invasiveness, reproducibility and abundant tumor burden, and the capacity to model several types of cancer (Liu et al. 2015) (a list of chemicals used to model carcinogenesis in different organs can be found elsewhere (Kemp 2015; Liu et al. 2015).

20.2.1.5 Genetically Engineered Mouse Cancer Models (GEMMs): Germline GEMMs and Non-Germline GEMM Models

Genetically engineered mouse models (GEMMs), created by engineering the mouse germline to partially mimic the molecular events found in human tumors, have contributed significantly to the field of cancer research as they provide a very complete picture of cancer development (Day et al. 2015). GEMMs develop *de novo* tumors in a pathophysiologic microenvironment, keeping the natural conditions of the organ. Importantly, tumors arise in an immune-proficient context, being the gold standard model for evaluation and optimization of immunomodulatory therapies (Kersten et al. 2017). Given that GEMMs capture both tumor cell-intrinsic and cell-extrinsic factors that drive *de novo* tumor initiation and progression, they more faithful recapitulate the histopathological and molecular features of the corresponding human disease, displaying cellular and genetic heterogeneity and commonly developing spontaneous metastatic disease (Kersten et al. 2017). These features award GEMMs a relevant role to study candidate cancer genes and drug targets, to assess therapy efficacy and mechanisms of drug resistance, and to discover predictive biomarkers. Moreover, GEMMS are attractive tools to dissect the impact of the tumor microenvironment (Singh et al. 2012) in cancer progression, and are particularly appealing to study the effect that specific cancer-associated genetic alterations play in the interaction of the cancer cell with the tumor microenvironment.

20.2.1.5.1 Germline GEMMs

In the last decade, several techniques have been developed to engineer a germline GEMM with high precision. By editing the genome of embryonic stem cells on zygotes, mice are programmed to develop diseases such cancer by *knocking-down* or *knocking-in* specific tumor suppressor genes or oncogenes, respectively (Kersten et al. 2017; Walrath et al. 2010) (Fig. 20.2d). Recent advances in the clustered regularly interspaced short palindromic repeats (CRISPR)/cas9 technology came to accelerate germline and somatic engineering providing a powerful, versatile and efficient tool to expand the variety of available GEMs (Sánchez-Rivera and Jacks 2015).

GEMs can be divided in constitutive or inducible which depends on the strategy they are obtained, and the genes of interest can be expressed or ablated in defined tissues or cellular subtypes (Katigbak et al. 2018). Inducible models can be obtained by combining cell specific expression of transcription factors (eg. doxycycline-modulated tet-transactivators) or recombinases (cre-lox, per example) with related cis elements linked to a target gene, or by expressing proteins fused with hormone-responsive domain (eg. tamoxifen inducible estrogen receptor domain) (Day et al. 2015). In many cases the best cancer model is achieved by combination of multiple distinct inducible models by crossing and backcrossing of different mouse GEMMs permitting, for example, the comparison with the human counterpart (Young et al. 2011).

Still, GEMMS are difficult to obtain for large scale studies with, for example, cancer drugs candidates due to its high cost, ethical issues, long breeding and genotyping protocols and synchronous tumorigenesis among individuals but also require high imaging technology in order to select mice bearing similarly sized tumors to enroll in a well-designed experiment (Varticovski et al. 2007).

20.2.1.5.2 Non-Germline GEMMs

Non-Germline GEM Models or GEM-Derived Allograft Models were developed to overcome the difficulties that researchers found in experimental design with conventional GEMMs. The development of GEM-Derived Allograft Models (GDAs) are based in the PDX transplantation technology. Fragments derived from GEM tumors are expanded by transplantation, subcutaneously or orthotopically, into immunocompetent syngeneic hosts. In the same way as PDXs, these expanded tumors can be banked to produce large cohorts for large scale studies (Heyer et al. 2010) (Fig. 20.3).

In GDAs immune system functionality is maintained as well as the interaction of tumor cells and their intrinsic microenvironment. GDA's are also one of the best models to study metastases as they occur from a single tumor in an immunocompetent mouse. These tumors, but not exclusive in GDA model, can be surgically resected to extend life in mice and promote the implantation of tumor cells niches in distant organs, mimicking the human condition (Kersten et al. 2017; Mendes et al. 2017).

20.2.2 Other Animal Models

Modeling cancer in mice has provided valuable information on the diverse aspects regulating tumor development and revealed as invaluable tools for therapeutic tests. However, due to the complexity of cancer development in mammals, simpler model organisms, such as the Zebrafish, *Danio rerio* and the fruit fly, *Drosophila melanogaster*, are being utilized to provide insights into the molecular mechanisms involved (Richardson and Portela 2018).

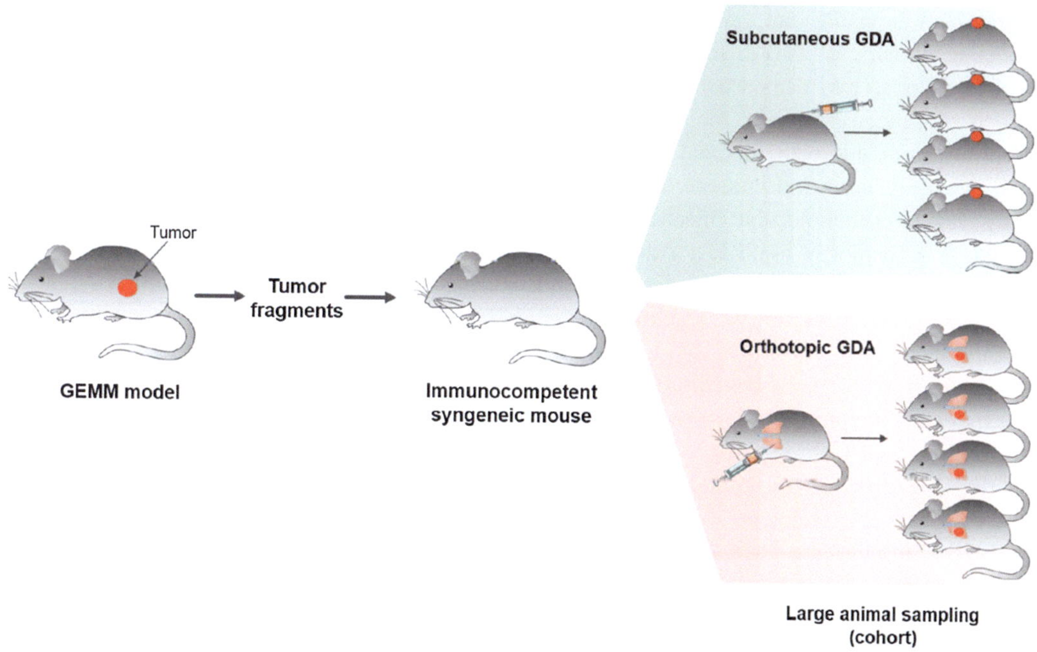

Fig. 20.3 GEMM-Derived Allograft Model (GDA) establishment: tumor fragments are obtained from cancers that developed in GEMMs. Tumors are then induced in immunocompetent syngeneic mice either by subcutaneous or orthotopic transplantation. This technique will allow a higher animal sampling (cohort), normalization of tumor growth kinetics among individuals in the experimental protocol and the refinement of the study

20.2.2.1 Zebrafish

Zebrafish is emerging as a versatile *in vivo* model to understand the mechanisms of cancer development, and to promote drug discovery as it offers a number of features, such as its rapid development, its small size, high number of offspring, and tractable genetics, complementing the classical studies done in mice. Moreover, embryos and larvae are optical transparent thus offering unique conditions for *in vivo* imaging (Kirchberger et al. 2017; Mione and Trede 2010), and the development of the "casper" fish, a transparent adult zebrafish model, greatly facilitate the analysis of transplanted and endogenous tumors in the adult animals (White et al. 2008).

Xenotransplantation of human or mouse cancer cells or even patient-derived tumor tissue into zebrafish embryos and larvae is also possible as zebrafish embryos lack a fully developed immune system. In these models, a detailed *in vivo* examination of cell-cell and cell-stromal interactions within the context of neoplastic cell survival, angiogenesis, migration, invasion, and metastasis is potentiated by the optical transparency of the embryos and the availability of multiple zebrafish lines that express fluorescent proteins in normal tissues (Shive 2013). Xenotransplanted zebrafish embryos also facilitate a high-throughput system for evaluating novel drug therapies *in vivo*, as they are less time-consuming than mouse transplantation studies, embryos easily absorb compounds from the water, and only small volumes of small molecules are needed to test effective compounds (Amatruda and Patton 2008; Fior et al. 2017).

There are a large array of transgenic zebrafish lines expressing oncogenes, and other genetic mutants of tumor suppressor genes, associated with the development of several tumor models including melanoma, leukemia, pancreatic cancer, sarcomas, intestinal hyperplasia among other solid tumors (Mione and Trede 2010). These tumor models resemble their human counterparts, both histologically and genetically, positioning zebrafish as a viable and valuable system for modeling human cancers (Liu and Leach 2011).

Several studies have demonstrated that zebrafish models of cancer are useful tools to study the effect of tumor microenvironment components, such as angiogenesis and immune responses, as many regulatory molecular mechanisms are shared (Kirchberger et al. 2017). Zebrafish has been used to study the mechanisms behind implantation and rejection of xenotransplanted human cancer cell lines. It has also been shown that tumor formed in transgenic zebrafish are able to recruit immune cells such as neutrophils, lymphocytes and macrophages, which infiltrate the tumor and act as pro-tumorigenic factors. Moreover, specific immune lineages can be inhibited through the use of morpholinos (Antonio et al. 2015). A major drawback of these models is that in the embryo stage, when imaging is facilitated, the immune system is not yet fully functional (Kirchberger et al. 2017).

20.2.2.2 Drosophila

Over the last decade, the fruit fly *Drosophila melanogaster* has become an important model system for cancer studies. The reduced redundancy in the *Drosophila* genome compared with that of humans, the conservation in the processes driving cancer development between the two species, and the ability to conduct large-scale genetic screens by performing genetic changes in specific cells and tissues are features that award *Drosophila* a relevant role as a model organism to study not only tumor cell-autonomous (intrinsic) but also non-tumor autonomous (extrinsic) molecular mechanisms mediating carcinogenesis (Miles et al. 2011; Parvy et al. 2018). Several studies have demonstrated that, in fact, *Drosophila* is a relevant model for studying cancer and its interactions with the TME. In the *Drosophila*, tumors are generated in the imaginal discs of the larvae which mostly comprise immune cells, the fat body (functionally similar to the mammalian liver and adipose tissues) and the trachea (analogous to the vertebrate vasculature) (Parvy et al. 2018).

The *Drosophila* has been a useful model to study the molecular mechanisms mediating both pro- and anti-tumoral immunity. As in humans, chronic inflammation in the *Drosophila* is associated with tumor initiation and progression towards a metastatic stage. However, the *Drosophila* immune system differs from the human counterpart as it only possesses innate immunity which includes three main cells types—plasmatocytes, lamellocytes, and crystal cells—commonly called hemocytes (Parvy et al. 2018). Still, the genes involved in the *Drosophila* innate immunity are homologous or very similar to genes implicated in mammalian innate immune responses (Hoffmann JA 2003 Nature). Resembling the human tumor context, high numbers of circulating hemocytes were found in tumor bearing animals which also showed enlarged lymph glands as a result of increased hemocyte proliferation, and hemocytes were found to infiltrate epithelial tumors (Bangi 2013; Pastor-Pareja et al. 2008).

Using *Drosophila* as a model, it is also possible to model and study neovascularization of tumors as malignant cells also recruit vessels to oxygenate the tumor mass (Mirzoyan et al. 2019). In the fly, the tracheal system, an interconnected tubular network, is currently considered to be the functional analogue of the mammalian vascular system, promoting oxygen spread throughout the body by the tracheal system, an interconnected tubular network whose regulation is significantly analogue to that of mammalian vascular tree (Grifoni et al. 2015).

As in human cancers, *Drosophila* tumor cells are sensitive to oxygen deprivation, releasing factors that promote an angiogenesis-like process, tracheogenesis, to promote oxygenation. Nevertheless, it has been shown that the tracheal system may also be involved in the production and/or transportation of growth factors acting locally or systemically, and on cancer cell spread thus supporting cancer growth and the transport of cancer cells during metastatic disease (Grifoni et al. 2015).

Using *Drosophila* as a model system, it is also possible to investigate the links between tumor development and an altered metabolism associated to diet and obesity factors. By providing a high sugar diet to the fly it is possible to generate a phenotype resembling the insulin resistance condition found in humans. In this context, it was observed that small clones of noninvasive tumor cells evade diet-induced systemic insulin resistance, becoming highly proliferative and metastatic (Parvy et al. 2018).

Acknowledgements AR, SM, FM were hired through FEDER funds through the Operational Programme for Competitiveness Factors (COMPETE 2020), Programa Operacional de Competitividade e Internacionalização (POCI), Programa Operacional Regional do Norte (Norte 2020), European Regional Development Fund (ERDF) and by National Funds through the Portuguese Foundation for Science and Technology (FCT), under the projects POCI-01-0145-FEDER-031354, POCI-01-0145-FEDER-016390, and NORTE-01-0145-FEDER-000029, respectively. SV and JC were hired by IPATIMUP under norma transitória do DL n.° 57/2016 alterada pela Law n.° 57/2017.

References

Amatruda JF, Patton EE (2008) Genetic models of cancer in zebrafish. Int Rev Cell Mol Biol 271:1–34. https://doi.org/10.1016/S1937-6448(08)01201-X

Antonio N, Bønnelykke-Behrndtz ML, Ward LC, Collin J, Christensen IJ, Steiniche T et al (2015) The wound inflammatory response exacerbates growth of pre-neoplastic cells and progression to cancer. EMBO J 34(17):2219–2236. https://doi.org/10.15252/embj.201490147

Bangi E (2013) Drosophila at the intersection of infection, inflammation, and cancer. Front Cell Infect Microbiol 3:103. https://doi.org/10.3389/fcimb.2013.00103

Blomme A, Van Simaeys G, Doumont G, Costanza B, Bellier J, Otaka Y et al (2018) Murine stroma adopts a human-like metabolic phenotype in the PDX model of colorectal cancer and liver metastases. Oncogene 37(9):1237–1250. https://doi.org/10.1038/s41388-017-0018-x

Calles A, Rubio-Viqueira B, Hidalgo M (2013) Primary human non-small cell lung and pancreatic tumorgraft models–utility and applications in drug discovery and tumor biology. Curr Protoc Pharmacol Chapter 14(1), Unitas 14.26–14.26.21. https://doi.org/10.1002/0471141755.ph1426s61

Cassidy JW, Caldas C, Bruna A (2015) Maintaining tumor heterogeneity in patient-derived tumor Xenografts. Cancer Res 75(15):2963–2968. https://doi.org/10.1158/0008-5472.CAN-15-0727

Chao C, Widen SG, Wood TG, Zatarain JR, Johnson P, Gajjar A et al (2017) Patient-derived Xenografts from colorectal carcinoma: a temporal and hierarchical study of murine stromal cell replacement. Anticancer Res 37(7):3405–3412. https://doi.org/10.21873/anticanres.11707

Choi Y, Lee S, Kim K, Kim S-H, Chung Y-J, Lee C (2018) Studying cancer immunotherapy using patient-derived Xenografts (PDXs) in humanized mice. Exp Mol Med 50(8):99. https://doi.org/10.1038/s12276-018-0115-0

Day C-P, Merlino G, Van Dyke T (2015) Preclinical mouse cancer models: a maze of opportunities and challenges. Cell 163(1):39–53. https://doi.org/10.1016/j.cell.2015.08.068

Devaud C, Westwood JA, John LB, Flynn JK, Paquet-Fifield S, Duong CPM et al (2014) Tissues in different anatomical sites can sculpt and vary the tumor microenvironment to affect responses to therapy. Mol Ther 22(1):18–27. https://doi.org/10.1038/mt.2013.219

Dias Carvalho P, Guimarães CF, Cardoso AP, Mendonça S, Costa ÂM, Oliveira MJ, Velho S (2018) KRAS oncogenic signaling extends beyond cancer cells to orchestrate the microenvironment. Cancer Res 78(1):7–14. https://doi.org/10.1158/0008-5472.CAN-17-2084

Dong Z, Imai A, Krishnamurthy S, Zhang Z, Zeitlin BD, Nör JE (2013) Xenograft tumors vascularized with murine blood vessels may overestimate the effect of anti-tumor drugs: a pilot study. PLoS One 8(12):e84236. https://doi.org/10.1371/journal.pone.0084236

Fidler IJ, Hart IR (1982) Biological diversity in metastatic neoplasms: origins and implications. Science 217(4564):998–1003

Fiebig HH, Schuchhardt C, Henss H, Fiedler L, Löhr GW (1984) Comparison of tumor response in nude mice and in the patients. Behring Inst Mitt 74:343–352

Fior R, Póvoa V, Mendes RV, Carvalho T, Gomes A, Figueiredo N, Ferreira MG (2017) Single-cell functional and chemosensitive profiling of combinatorial colorectal therapy in zebrafish xenografts. Proc Natl Acad Sci USA 114(39):E8234–E8243. https://doi.org/10.1073/pnas.1618389114

Fiori ME, Di Franco S, Villanova L, Bianca P, Stassi G, De Maria R (2019) Cancer-associated fibroblasts as abettors of tumor progression at the crossroads of EMT and therapy resistance. Mol Cancer 18(1):70. https://doi.org/10.1186/s12943-019-0994-2

Fu X, Guadagni F, Hoffman RM (1992) A metastatic nude-mouse model of human pancreatic cancer constructed orthotopically with histologically intact patient specimens. Proc Natl Acad Sci USA 89(12):5645–5649

Ganesh K, Massagué J (2018) TGF-β inhibition and immunotherapy: checkmate. Immunity 48(4):626–628. https://doi.org/10.1016/j.immuni.2018.03.037

Garraway LA, Jänne PA (2012) Circumventing cancer drug resistance in the era of personalized medicine. Cancer Discov 2(3):214–226. https://doi.org/10.1158/2159-8290.CD-12-0012

Gnerlich JL, Mitchem JB, Weir JS, Sankpal NV, Kashiwagi H, Belt BA et al (2010) Induction of Th17 cells in the tumor microenvironment improves survival in a murine model of pancreatic cancer. J Immunol 185(7):4063–4071. https://doi.org/10.4049/jimmunol.0902609

Grifoni D, Sollazzo M, Fontana E, Froldi F, Pession A (2015) Multiple strategies of oxygen supply in Drosophila malignancies identify tracheogenesis as a novel cancer hallmark. Sci Rep 5(1):9061. https://doi.org/10.1038/srep09061

Hanahan D, Coussens LM (2012) Accessories to the crime: functions of cells recruited to the tumor microenvironment. Cancer Cell 21(3):309–322. https://doi.org/10.1016/j.ccr.2012.02.022

Hanahan D, Weinberg RA (2000) The hallmarks of cancer. Cell 100(1):57–70

Heyer J, Kwong LN, Lowe SW, Chin L (2010) Non-germline genetically engineered mouse models for translational cancer research. Nat Rev Cancer 10(7):470–480. https://doi.org/10.1038/nrc2877

Hidalgo M, Bruckheimer E, Rajeshkumar NV, Garrido-Laguna I, De Oliveira E, Rubio-Viqueira B et al (2011) A pilot clinical study of treatment guided by personalized tumorgrafts in patients with advanced cancer. Mol Cancer Ther 10(8):1311–1316. https://doi.org/10.1158/1535-7163.MCT-11-0233

Hidalgo M, Amant F, Biankin AV, Budinská E, Byrne AT, Caldas C et al (2014) Patient-derived Xenograft models: an emerging platform for translational cancer research. Cancer Discov 4(9):998–1013. https://doi.org/10.1158/2159-8290.CD-14-0001

Hoover RG, Gullickson G, Kornbluth J (2012) Natural killer lytic-associated molecule plays a role in controlling tumor dissemination and metastasis. Front Immunol 3(393). https://doi.org/10.3389/fimmu.2012.00393

Isella C, Terrasi A, Bellomo SE, Petti C, Galatola G, Muratore A et al (2015) Stromal contribution to the colorectal cancer transcriptome. Nat Genet 47(4):312–319. https://doi.org/10.1038/ng.3224

Johnson JI, Decker S, Zaharevitz D, Rubinstein LV, Venditti JM, Schepartz S et al (2001) Relationships between drug activity in NCI preclinical in vitro and in vivo models and early clinical trials. Br J Cancer 84(10):1424–1431. https://doi.org/10.1054/bjoc.2001.1796

Jung J, Seol HS, Chang S (2018) The generation and application of patient-derived Xenograft model for Cancer research. Cancer Res Treat 50(1):1–10. https://doi.org/10.4143/crt.2017.307

Katigbak A, Robert F, Paquet M, Pelletier J (2018) Inducible genome editing with conditional CRISPR/Cas9 mice. G3 (Bethesda) 8(5):1627–1635. https://doi.org/10.1534/g3.117.300327

Kemp CJ (2015) Animal models of chemical carcinogenesis: driving breakthroughs in Cancer research for 100 years. Cold Spring Harb Protoc 2015(10):865–874. https://doi.org/10.1101/pdb.top069906

Kersten K, de Visser KE, van Miltenburg MH, Jonkers J (2017) Genetically engineered mouse models in oncology research and cancer medicine. EMBO Mol Med 9(2):137–153. https://doi.org/10.15252/emmm.201606857

Kirchberger S, Sturtzel C, Pascoal S, Distel M (2017) Quo natas, Danio?-recent progress in modeling cancer in zebrafish. Front Oncol 7(186). https://doi.org/10.3389/fonc.2017.00186

Lechner MG, Karimi SS, Barry-Holson K, Angell TE, Murphy KA, Church CH et al (2013) Immunogenicity of murine solid tumor models as a defining feature of in vivo behavior and response to immunotherapy. J Immunother 36(9):477–489. https://doi.org/10.1097/01.cji.0000436722.46675.4a

Lindner D (2014) Animal models and the tumor microenvironment: studies of tumor-host symbiosis. Semin Oncol 41(2):146–155. https://doi.org/10.1053/j.seminoncol.2014.02.004

Liu S, Leach SD (2011) Zebrafish models for cancer. Annu Rev Pathol 6(1):71–93. https://doi.org/10.1146/annurev-pathol-011110-130330

Liu Y, Yin T, Feng Y, Cona MM, Huang G, Liu J et al (2015) Mammalian models of chemically induced primary malignancies exploitable for imaging-based preclinical theragnostic research. Quant Imaging Med Surg 5(5):708–729. https://doi.org/10.3978/j.issn.2223-4292.2015.06.01

Mariathasan S, Turley SJ, Nickles D, Castiglioni A, Yuen K, Wang Y et al (2018) TGFβ attenuates tumour response to PD-L1 blockade by contributing to exclusion of T cells. Nature 554(7693):544–548. https://doi.org/10.1038/nature25501

Mattern J, Bak M, Hahn EW, Volm M (1988) Human tumor xenografts as model for drug testing. Cancer Metastasis Rev 7(3):263–284

McCreery MQ, Halliwill KD, Chin D, Delrosario R, Hirst G, Vuong P et al (2015) Evolution of metastasis revealed by mutational landscapes of chemically induced skin cancers. Nat Med 21(12):1514–1520. https://doi.org/10.1038/nm.3979

Mendes N, Tortosa F, Valente A, Marques F, Matos A, Morais TS et al (2017) Vivo performance of a ruthenium-cyclopentadienyl compound in an Orthotopic triple negative breast Cancer model. Anti Cancer Agents Med Chem 17(1):126–136

Miles WO, Dyson NJ, Walker JA (2011) Modeling tumor invasion and metastasis in Drosophila. Dis Model Mech 4(6):753–761. https://doi.org/10.1242/dmm.006908

Mione MC, Trede NS (2010) The zebrafish as a model for cancer. Dis Model Mech 3(9–10):517–523. https://doi.org/10.1242/dmm.004747

Mirzoyan Z, Sollazzo M, Allocca M, Valenza AM, Grifoni D, Bellosta P (2019) Drosophila melanogaster: a model organism to study cancer. Front Genet 10(51). https://doi.org/10.3389/fgene.2019.00051

Parvy J-P, Hodgson JA, Cordero JB (2018) Drosophila as a model system to study nonautonomous mechanisms affecting tumour growth and cell death. Biomed Res Int 2018(9):7152962–7152913. https://doi.org/10.1155/2018/7152962

Pastor-Pareja JC, Wu M, Xu T (2008) An innate immune response of blood cells to tumors and tissue damage in Drosophila. Dis Model Mech 1(2–3):144–154. discussion 153. https://doi.org/10.1242/dmm.000950

Richardson HE, Portela M (2018) Modelling cooperative tumorigenesis in Drosophila. Biomed Res Int 2018(4):4258387–4258329. https://doi.org/10.1155/2018/4258387

Sánchez-Rivera FJ, Jacks T (2015) Applications of the CRISPR-Cas9 system in cancer biology. Nat Rev Cancer 15(7):387–395. https://doi.org/10.1038/nrc3950

Sanmamed MF, Chester C, Melero I, Kohrt H (2016) Defining the optimal murine models to investigate immune checkpoint blockers and their combination with other immunotherapies. Ann Oncol 27(7):1190–1198. https://doi.org/10.1093/annonc/mdw041

Sharpless NE, Depinho RA (2006) The mighty mouse: genetically engineered mouse models in cancer drug development. Nat Rev Drug Discov 5(9):741–754. https://doi.org/10.1038/nrd2110

Shive HR (2013) Zebrafish models for human cancer. Vet Pathol 50(3):468–482. https://doi.org/10.1177/0300985812467471

Singh M, Murriel CL, Johnson L (2012) Genetically engineered mouse models: closing the gap between preclinical data and trial outcomes. Cancer Res 72(11):2695–2700. https://doi.org/10.1158/0008-5472.CAN-11-2786

Siolas D, Hannon GJ (2013) Patient-derived tumor xenografts: transforming clinical samples into mouse models. Cancer Res 73(17):5315–5319. https://doi.org/10.1158/0008-5472.CAN-13-1069

Szatmári T, Lumniczky K, Désaknai S, Trajcevski S, Hídvégi EJ, Hamada H, Sáfrány G (2006) Detailed characterization of the mouse glioma 261 tumor model for experimental glioblastoma therapy. Cancer Sci 97(6):546–553. https://doi.org/10.1111/j.1349-7006.2006.00208.x

Teicher BA (2006) Tumor models for efficacy determination. Mol Cancer Ther 5(10):2435–2443. https://doi.org/10.1158/1535-7163.MCT-06-0391

Tentler JJ, Tan AC, Weekes CD, Jimeno A, Leong S, Pitts TM et al (2012) Patient-derived tumour xenografts as models for oncology drug development. Nat Rev Clin Oncol 9(6):338–350. https://doi.org/10.1038/nrclinonc.2012.61

Utama FE, Tran TH, Ryder A, LeBaron MJ, Parlow AF, Rui H (2009) Insensitivity of human prolactin receptors to nonhuman prolactins: relevance for experimental modeling of prolactin receptor-expressing human cells. Endocrinology 150(4):1782–1790. https://doi.org/10.1210/en.2008-1057

Van Dyke T, Jacks T (2002) Cancer modeling in the modern era: progress and challenges. Cell 108(2):135–144

Varticovski L, Hollingshead MG, Robles AI, Wu X, Cherry J, Munroe DJ et al (2007) Accelerated preclinical testing using transplanted tumors from genetically engineered mouse breast cancer models. Clin Cancer Res 13(7):2168–2177. https://doi.org/10.1158/1078-0432.CCR-06-0918

Walrath JC, Hawes JJ, Van Dyke T, Reilly KM (2010) Genetically engineered mouse models in cancer research. Adv Cancer Res 106:113–164. https://doi.org/10.1016/S0065-230X(10)06004-5

White RM, Sessa A, Burke C, Bowman T, LeBlanc J, Ceol C et al (2008) Transparent adult zebrafish as a tool for in vivo transplantation analysis. Cell Stem Cell 2(2):183–189. https://doi.org/10.1016/j.stem.2007.11.002

Williams JA (2018) Using PDX for preclinical Cancer drug discovery: the evolving field. J Clin Med 7(3):41. https://doi.org/10.3390/jcm7030041

Yada E, Wada S, Yoshida S, Sasada T (2018) Use of patient-derived xenograft mouse models in cancer research and treatment. Future Sci OA 4(3):FSO271. https://doi.org/10.4155/fsoa-2017-0136

Yi L, Zhou C, Wang B, Chen T, Xu M, Xu L, Feng H (2013) Implantation of GL261 neurospheres into C57/BL6 mice: a more reliable syngeneic graft model for research on glioma-initiating cells. Int J Oncol 43(2):477–484. https://doi.org/10.3892/ijo.2013.1962

Young NP, Crowley D, Jacks T (2011) Uncoupling cancer mutations reveals critical timing of p53 loss in sarcomagenesis. Cancer Res 71(11):4040–4047. https://doi.org/10.1158/0008-5472.CAN-10-4563

Yu JW, Bhattacharya S, Yanamandra N, Kilian D, Shi H, Yadavilli S et al (2018) Tumor-immune profiling of murine syngeneic tumor models as a framework to guide mechanistic studies and predict therapy response in distinct tumor microenvironments. PLoS One 13(11):e0206223. https://doi.org/10.1371/journal.pone.0206223

Zhao Y, Shuen TWH, Toh TB, Chan XY, Liu M, Tan SY et al (2018) Development of a new patient-derived xenograft humanised mouse model to study human-specific tumour microenvironment and immunotherapy. Gut 67(10):1845–1854. https://doi.org/10.1136/gutjnl-2017-315201

Modulating the Metabolic Phenotype of Cancer Microenvironment

21

Inês Matias, Sérgio Dias, and Tânia Carvalho

Abstract

This chapter provides a brief overview of the methods to study and modulate the metabolic phenotype of the tumor microenvironment, including own research work to demonstrate the impact that metabolic shifts in the host have on cancer. Firstly, we briefly discuss the relevance of using animal models to address this topic, and also the importance of acknowledging that animals have diverse metabolic phenotypes according to species, and even with strain, age or sex. We also present original data to highlight the impact that changes in metabolic phenotype of the microenvironment have on tumor progression. Using an acute leukemia mouse xenograft model and high-fat diet we show that a shift in the host metabolic phenotype, induced by high-fat feeding, significantly impacts on tumor progression. The mechanism through which this occurs involves a direct effect of the increased levels of circulating lipoproteins in both tumor and non-neoplastic cells.

Keywords

Murine models · Cancer microenvironment · Metabolic remodeling · Cholesterol impact in cancer progression

I. Matias · S. Dias · T. Carvalho (✉)
Instituto de Medicina Molecular João Lobo Antunes, Universidade de Lisboa, Lisbon, Portugal
e-mail: taniacarvalho@medicina.ulisboa.pt

21.1 Modulating the Metabolic Status of the Cancer Host

For selecting the fit-for-purpose assay(s) to study adaptation of cancer as a response to metabolic changes in the microenvironment, we need to acknowledge that not every technique provides useful information in every model. Concerning complexity versus tractableness of a given model, there are trade-offs that need to consider in metabolic experiments. *In vitro* culture systems are experimentally tractable, but a rather simple model compared to the biological context of human tumors. Animal models, on the other hand, are inherently more complex but can recapitulate cancer onset and/or progression and their metabolic phenotype can be easily manipulated. So these can be used to dissect how multiple cell types interact in the tumor microenvironment. Although alternative methods should always be considered, currently, these are still supplementary to the use of animals in biomedical research. The availability of numerous models with distinct and well characterized metabolic phenotypes surely makes investigations on the impact that a particular microenvironmental shift has in cancer relatively straightforward.

When *in vivo* assays are chosen, differences between the model and the modelled organisms should be acknowledged. Various animal species have been thoroughly used to exploit the relevance that shifts in tissue metabolism have

J. Serpa (ed.), *Tumor Microenvironment*, Advances in Experimental Medicine and Biology 1219,
https://doi.org/10.1007/978-3-030-34025-4_21

on tumor progression, always with the perspective of translating findings to humans, but the metabolic phenotype of the host will diverge according to species, and even with strain, age or sex (Barthold 2004), and this will in turn impact on the cross-talk host-tumor. When comparing metabolic rate of mice *versus* man, for example, that of mice is sevenfold higher (Demetrius 2005). Additionally, several metabolic features are already known to distinguish rodents from humans, including bile acid synthesis pathways (Russell 2003), substrate selectivity for cytochrome P450 (Martignoni et al. 2006; Muruganandan and Sinal 2008) or even mitochondrial fatty acid oxidation (Bergen and Mersmann 2018). Glycogen content in the muscle of mice corresponds to only 10% of the total glycogen in human muscle (Nandi 2004), while glucose plays a key role as an energy source in most mammals, but its importance in fish appears to be limited (Zhang et al. 2018). Zhang et al. found that most metabolic genes are conserved in vertebrates, and variances in carbohydrate utilization between mammals and fish are attributable to insulin association regulators and transport proteins. Also, different inbred mouse strains also differ in their metabolic phenotypes even at the substrain level. A striking example is that of C57BL/6J (The Jackson Laboratory) being more predisposed to obesity and diabetes than C57BL/6N (NIH) due to a single mutation in the *Nnt* gene (nicotinamide nucleotide transhydrogenase) of C57BL/6J (Rossmeisl et al. 2003; Berglund et al. 2008). With all of this in mind, deep knowledge on the idiosyncrasies of different animal species and strains of laboratory mice is imperative, but this can in fact be used in our favor when studying tumor adaptation to metabolic shifts of the microenvironment.

Another important challenge is that often the metabolic phenotype that researchers aim at modeling in the host is not one, but a constellation of metabolic abnormalities. That is the case of metabolic syndrome, a clinical phenotype in humans that combines central obesity to elevated plasma triglyceride levels, reduced high-density lipoproteins, increased blood pressure, and/or increased fasting plasma glucose (Alberti et al. 2006). Concerning mouse models of metabolic syndrome, there is not a single one that mimics exactly all aspects of the human disease. There are many naturally occurring and gene-targeted mutations in mice associated with obesity and other metabolic defects, and selection of the fit-for-purpose model must take into account their specific attributes (Grubb et al. 2014). Besides genetic models, manipulation of the metabolic phenotype of a host can also be chemically or diet-induced (Nandi 2004; Savage 2009; Kennedy et al. 2010; Lee et al. 2014; Rozman et al. 2014; Ruzzenente et al. 2016) (Fig. 21.1).

Next, we present original research data showing how dietary changes may impact on the progression of malignant tumors in a murine model of acute lymphoblastic leukemia (ALL). We used a high-fat/high-cholesterol and cholate 3-week feeding to alter the metabolic phenotype of the mouse, mimicking a 'Western-type' diet. Then we induced leukemia by xenotransplantation and assessed disease progression.

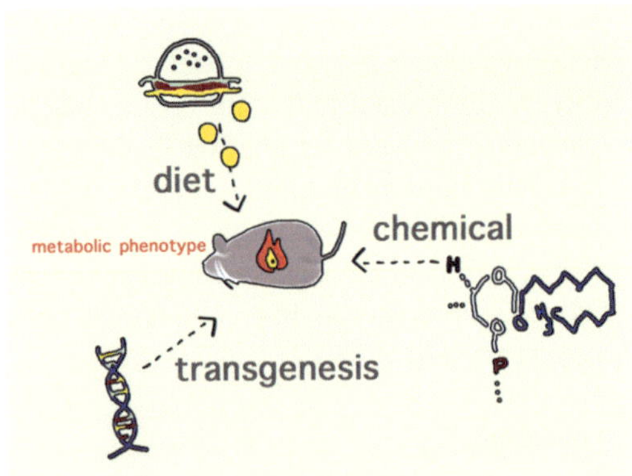

Fig. 21.1 An illustration of common strategies used to alter the metabolic phenotype of the tumor the microenvironment, i.e. of the experimental animals.
Variations in metabolic status of the tumor microenvironment can be achieve with diet, chemically or through transgenesis

21.2 Outcome of Acute Lymphoblastic Leukemia in a High-Fat Microenvironment

Acute lymphoblastic leukemia (ALL) is a cancer of white blood cells, generally classified in 3 subtypes depending on the exact cell type that it originates from: B-cell ALL, T-cell ALL or mixed lineage (with lymphocytic and myeloid features) (Swerdlow et al. 2017). It affects all of the bone marrow in the body and, with progression, spreads to other organs, such as the liver, spleen, lymph nodes and also central nervous system (CNS). CNS metastasis in ALL is a major obstacle to cure, accounting for 30–40% of initial relapse (Pui and Howard 2008), and it displays as leptomeningeal disease or, more rarely, parenchymal infiltration. It can be seen at diagnosis, in 3–5% of adult ALL patients, and at relapse, in 5–7% (Surapaneni et al. 2002; Cortes 2001). The presence of leptomeningeal infiltration therefore predicts for systemic disease recurrence and it is associated with poor outcome (Kaplan et al. 1990; Nayar et al. 2017).

Disorders of the lipid metabolism are very common in man and lipid profile in the blood is extensively used to infer risk for certain diseases, namely cardiovascular. Parameters include plasma total cholesterol, total triglycerides, high-density lipoprotein-associated cholesterol (HDL), and low-density lipoprotein-associated cholesterol (LDL), the latter being the primary lipid parameter whose elevation is associated with disease. As discussed in previous sections and chapters, cancer is associated with multiple metabolic abnormalities of the host, and lipid metabolism is not an exception.

Epidemiological studies alluded to the possible connection between lipids/cholesterol levels and disease burden and relapse in acute leukemia patients (Scribano et al. 1996; Butturini et al. 2007). Concerning ALL, altered plasma lipid profile has been observed in pediatric leukemia, at diagnosis (Scribano et al. 1996; Moschovi et al. 2004; Kuliszkiewicz-Janus et al. 2008). Hypertriglyceridemia, reduction in high density lipoprotein cholesterol (HDL-C), and/or an increase in low density lipoprotein cholesterol (LDL) have also been reported in adult ALL patients (Moschovi et al. 2004; Grau et al. 2016). The return of serum lipids and lipoproteins towards normal limits during remission supports correlation of these lipid abnormalities with primary disease activity (Moschovi et al. 2004).

As to test the hypothesis that, in ALL, lipid metabolism may be directly associated with certain disease features and/or with altered disease progression, we conducted an experiment using inbred mice xenotransplanted with a human cell line of B-cell ALL, with high-fat diet as the metabolic cue. First, we confirmed that our stimulus (high-fat diet) was associated with an altered metabolic phenotype, in naïve mice. Second, we characterized the leukemia xenograft model in terms of pattern of dissemination. Third, we finally combined the two variables, with ALL mice being allocated either to the standard diet control group, or to the experimental group of high-fat. Disease progression was then monitored and results are presented below.

21.2.1 High-Fat Diet Results in Elevated Circulating Levels of LDL Cholesterol

After a 30-day high cholesterol feeding trial, conducted in BALB/c SCID mice (female, 8–10 weeks), total cholesterol, LDL, and HDL levels were assessed in the blood. High-fat diet resulted in a significant increase in total cholesterol levels in plasma and, most importantly, LDL that is one of the major 5 groups of lipoproteins and the major cell source of cholesterol to cells, was elevated up to 5.5-fold in animals on high-cholesterol diet (Fig. 21.2a). Our results corroborate that of others also describing murine models of hypercholesterolemia (Paigen 1995; Gomes et al. 2010).

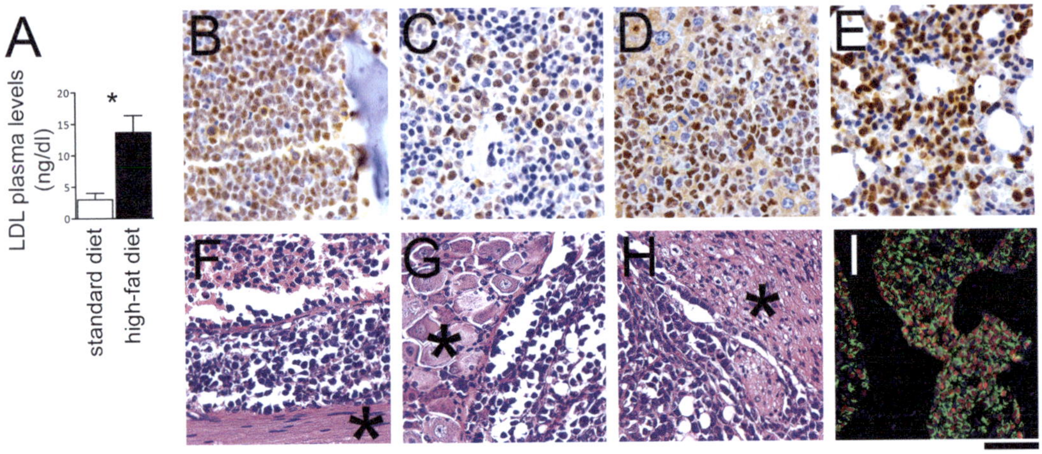

Fig. 21.2 Phenotypes of high-fat diet and of xenograft murine model of acute lymphoblastic leukemia
(**a**) Values of serum LDL-cholesterol at baseline (standard diet) and on high-fat diet, expressed as milligram per deciliter, show a 5.5-fold increase in LDL in the blood of BALB/c SCID mice (female, 8–10 weeks) after 30 days on high-fat diet. *Results are expressed as mean ± s.e.m. Statistical analysis corresponds to two tailed unpaired student t test;*: $p < 0.05$.* (**b–e**) Leukemia cells immunostained for TdT are seen to infiltrate bone marrow (**b**), spleen (**c**), liver (**d**) and lung (**e**). *DAB counterstained with Harris Hematoxylin. Original magnification 40×, bar = 100 μm.* (**f–i**) Leukemia cells are seen to invade and expand in the leptomeninges of brain and spinal cord (*, **f**), cuffing ganglions (*, **g**), and also cranial and peripheral nerves (*, **h**). Immunostaining of leukemia cells for TdT (red) and vimentin (green) (**i**). *Hematoxylin and Eosin (f–h) and immunofluorescence (i). Original magnification 40×, bar = 100 μm*

21.2.2 The Xenograft Mouse Model of Acute Leukemia Is Characterized by Metastasis to the Central Nervous System (CNS)

BALB/c SCID mice (female, 8–10 weeks) were injected in the tail vein with a suspension of leukemia cells. Twelve days after injection mice were euthanized, necropsy was performed and organs were harvested for routine histological analysis. There was marked expansion of the hematopoietic organs (bone marrow and spleen) by leukemia cells (Fig. 21.2b, c), which were also seen to invade liver and lung (Fig. 21.2d, e). Nature of these cells was confirmed though immune-positivity for TdT (Terminal deoxynucleotidyl Transferase), a specialized DNA polymerase expressed in immature, pre-B, pre-T lymphoid cells, and acute lymphoblastic leukemia/lymphoma cells.

Analysis of the peripheral and CNS showed marked and diffuse infiltration of the leptomeninges of brain and spinal cord (Fig. 21.2f), similar to what is described for the leptomeningeal disease in humans; as well as marked tumor cell adhesion and cuffing of peripheral and cranial nerves by tumor cells (Fig. 21.3g, h). To confirm that these cells corresponded to the xenografted cell line, we performed immunofluorescence for TdT and also for Vimentin, using an antibody that is cross-reactive only with the human protein, and not with murine vimentin. Cells infiltrating nervous system were double positive for these markers (Fig. 21.2i).

21.2.3 High-Fat Diet Is Associated with Enhanced Tumor Progression and Metastasis to the Central Nervous System (CNS)

BALB/c SCID mice (female, 8–10 weeks) were allocated either to the control group, fed standard diet, or to the experimental group of high-fat diet. After 30 days on these, feeding regimens mice

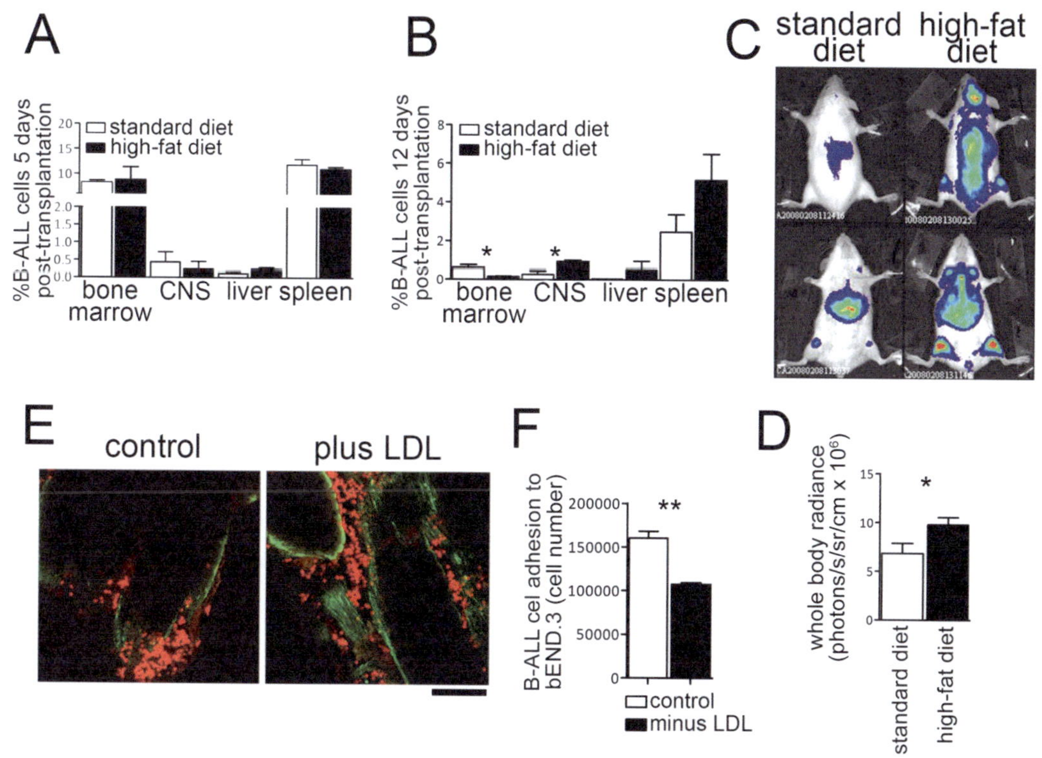

Fig. 21.3 High-fat diet is associated with enhanced tumor progression and metastasis to the central nervous system, and LDL favors leukemia cell adhesion to nerves and endothelium
(**a** and **b**) Flow cytometry analysis of the percentage of B-ALL cells in the bone marrow, liver, spleen and nervous system (including brain, leptomeninges and intracranial segments of cranial nerves) at days 5 and 12 post-xenotransplantation, shows a significant increase in the percentage of tumor cells infiltrating the nervous system of mice on high-fat diet. (**c** and **d**) Bioluminescence imaging also shows stronger and more disseminated signal and increased luciferase activity with high-fat diet (**c**). The graphic corresponds to light emission values (photons per steradian per square centimeter) for each group (**d**). *Results are expressed as mean ± s.e.m. Statistical analysis corresponds to two tailed unpaired student t test;*: $p < 0.05$.* (**e**) Confocal microscopy after *ex vivo* co-culture of mouse cranial nerves (green) and leukemia cells (red) with/without LDL enrichment shows increased adhesion of leukemic cells upon LDL exposure. *Immunofluorescence for S100 (green) and vimentin (red); original magnification 20×, bar = 200 μm.* (**f**) Adhesion of leukemia cells to mouse brain endothelial cells is diminished upon exposure to LDL free media. *Results are expressed as mean ± s.e.m. Statistical analysis corresponds to two tailed unpaired student t test;**: $p < 0.01$*

were injected in the tail vein with a suspension of leukemia cells expressing luciferase and GFP, and further ascribed to 2 sub-groups, to monitor disease progression. One group was sacrificed at day 5 and another at day 12 post-injection, and quantification of percentage of cells infiltrating the different organs and tissues was performed through flow cytometry analysis of GFP-positive cells. At day 5 there was marked expansion of leukemia cells in the bone marrow and spleen of mice from both groups (standard and high-fat diet), with no difference in tumor load between them. Infiltration of liver and central nervous system at this time-point was minimal (Fig. 21.3a). At day 12 however high-fat diet mice showed significant infiltration of the CNS, quantified by flow cytometry (Fig. 21.3b) but also evident in the IVIS LUMINA system. Mice on high-fat were seen to display an intense signal overlying head, spinal cord, liver and hind legs (Fig. 21.3c). This stronger and more disseminated luciferase activity, when quantified, confirmed presence of a significantly higher tumor burden in high-fat diet, compared to standard diet mice (Fig. 21.3d).

Interestingly, less tumor cells were seen to infiltrate the bone marrow of high-fat diet mice at day 12, compared with control, either because at late stages of the disease, leukemic cells in the bone marrow undergo necrosis (which is frequently seen in these models, data not shown), or because the host microenvironment is favoring the exit of these cells from the hematopoietic organs and metastasis in secondary organs, namely CNS. Further experiments will be necessary to properly study tumor cell dynamics in this model.

21.2.4 LDL-Cholesterol Confers Peripheral Nerves and Endothelial Cells a More Adhesive Phenotype

Severe leptomeningeal disease seen in high-fat diet mice could be due to an effect of high cholesterol on the microenvironment, on the tumor cells, or in both. To try to uncouple these effects we conducted *ex vivo* and *in vitro* assays using co-cultures of cranial nerves, endothelial cells and tumor cells.

Co-culture adhesion assays were performed by adding leukemia cells (same cell line used for *in vivo* assays) to mouse cranial nerves or a cell line of mouse brain endothelial cells (bEND.3). Cranial nerves were collected at necropsy from naïve BALB/c SCID mice and co-cultured with leukemia cells with/without LDL enrichment. After 24 h, cells were vigorously washed and the remaining elements (nerves and adherent leukemia cells) were stained for S100 (a neuronal marker) and vimentin (mesenchymal marker also expressed in leukemia cells). Confocal microscopy showed vimentin-positive leukemia cells adherent to S100-positive nerves and it was clear that co-cultures enriched with LDL showed increased tumor cell adhesion to the nerves (Fig. 21.3e).

Similar rationale was used for a co-culture experiment with leukemia cells and brain-derived endothelial cells (bEND.3), where cells were cultured in control or LDL-free media for 24 h. The number of leukemia cells adherent to the endothelial cells were counted and LDL depletion was associated with a significant decrease in tumor cell adhesion to endothelium (Fig. 21.3f).

21.3 Discussion

Here we discussed the various strategies available to study and modulate the metabolic phenotype of the tumor microenvironment, in order to address the impact that metabolic shifts in the host have on the tumor. Animals models are invaluable to address this topic, as they provide means to study interaction between cancer cells and the multiple non-neoplastic cells that make up for the tumor microenvironment. There are numerous readily available genetic and non-genetic models that can be used for this purpose, including those induced by simple dietary changes.

To demonstrate the significant impact that subtle changes in the host metabolic phenotype may have on tumor progression, we combined an acute leukemia mouse xenograft model with high-fat feeding. We objectively showed that a high-fat diet regimen leads to persistently altered plasma lipid profiles, mostly at the cost of high levels of LDL cholesterol, and this change is associated with increased tumor progression and metastasis to the CNS (Fig. 21.2). This phenotype seems to be, at least in part, associated with direct effects of LDL cholesterol in the microenvironment, i.e. in nerve cells and endothelial cells, and/or tumor cells (Fig. 21.3), making tumor cells more adhesive to the neuronal and vascular compartments, and favoring invasion and disease spread (Fig. 21.4).

We have further investigated into the molecular mechanisms of this observation and found that LDL confers survival, adhesion and migration advantages to tumor cells; and that, concomitantly, this perturbed lipidemia also modulates the microenvironment, resulting in the upregulation of specific chemo attractive factor(s) and receptors in endothelium and nerves, namely fractalkine. Fractalkine is a transmembrane chemokine (CX3CL1/Neurotactin expressed in endothelial cells and neurons, which mediates

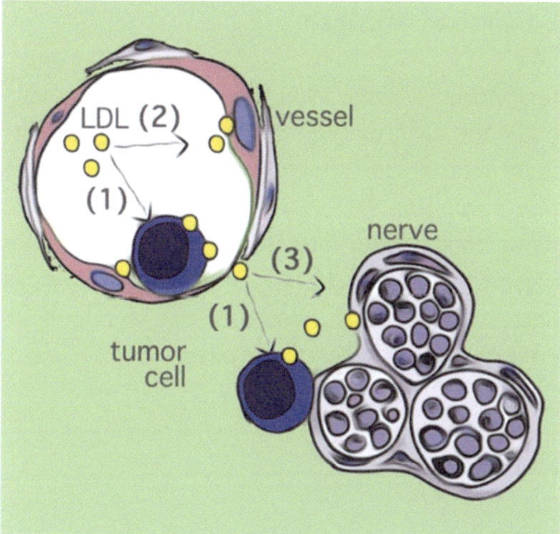

Fig. 21.4 An illustration describing how increased circulating LDL alters the microenvironment and tumor cells, favoring leukemia cell adhesion to endothelium and metastasis to the central nervous system
Low density lipoprotein (LDL) triggers an adhesive phenotype in tumor cells (1) and tumor microenvironment – vascular (2) and neuronal (3) compartments, favoring metastasis to central nervous system

adhesion by leukocytes and also leukemia and other tumor cells, through its receptor CX3CR1 (data not shown).

A topic that was not addressed in this work relates to the impact that these changes in the metabolic status of the host and of tumor microenvironment also have on the different sub-types of immune, endothelial and stromal cells attracted to the tumor site; which in turn modulate cancer cell growth and invasion (Picard et al. 1986; Harjes et al. 2012; Kouidhi et al. 2018).

In sum, environmental cues and their impact on different host cell types surely codetermine whether a single cancer cell progresses to macro metastasis or remains dormant. Unraveling this interplay may help develop strategies for prevention and treatment of cancer metastasis through modulating the metabolic phenotype of the host.

Acknowledgements Project funded by Fundação para a Ciência e a Tecnologia (FCT)/Ministério da Ciência, Tecnologia e Ensino Superior (MCTES) through Fundos do Orçamento de Estado (UID/BIM/50005/2019).

Materials and Methods

Animal Experiments

All animal experiments were performed in BALB/c SCID mice (female, 8–10 weeks old), with the approval of the Instituto Gulbenkian de Ciencia Animal Ethics Committee and Review Board. High-cholesterol feeding consisted on a 30-day regimen on a special high-fat/high-cholesterol/cholate diet (HFC/0.5% cholate, Ssniff Spezialdiäten GmbH) with food and water ad libitum) as described previously (Gomes et al. 2010); mice fed the standard diet were used as control. Acute lymphoblastic leukemia xenograft model consisted of a sub-lethal irradiation (250rad) 24 h prior to Luciferase-GFP+ ALL cells (human 697 pre-B cell line) xenotransplant (1 × 107 cells, tail vein injection). For luciferase imaging, mice were anaesthetized with ketamine/xylazine, injected IV with 150 μg luciferin g^{-1} and routinely scanned after 5 min in IVIS Lumina (Caliper Life Sciences); quantification was performed with Living Image software (Caliper Life Sciences), to obtain the radiance (photons per s per cm^2 per steradian, i.e. photons s^{-1} cm^{-2} sr^{-1}) over each region of interest, in all animals from each condition tested.

For tumor cell quantification mice were sacrificed with CO_2 narcosis at days 5 and 12 post-xenotransplantation and flow cytometric analysis was performed for GFP-positive cells in bone marrow, spleen, liver and central nervous system (brain) by using FACS Calibur (BD Biosciences). Analysis was carried out using Cell Quest software.

Cell Lines and In Vitro/Ex Vivo Studies

ALL cells (human 697 pre-B cell line) were stably transduced with lentiviral vectors driving the expression of Luciferase and GFP, kindly provided by Dr. Luigi Naldini. After transduction cells were sorted in a FacsAria Multicolor cell sorter (BD Biosciences).

Trigeminal nerves harvested from naïve mice and incubated for 24 h in serum free media with LDL (100 μg/ml, Calbiochem); after which they were washed and co-cultures with B-ALL cells (1×10^6) for 24 h in LDL-free serum. Cells were then immunostained for Vimentin and S100 (M0725 and Z0311, Dako Cytomation), and imaged in a fluorescence microscope (Zeiss Axiovert 200M). Adhesion to b.END3 cells' monolayers (70% confluence) was conducted for 24 h in RPMI with lipoprotein-deficient serum (S5394, Sigma), with/without LDL (100 μg/ml). Cells were counted in 5 high-power fields after gently washing off non-adherent cells with PBS.

Histopathology

Mice were sacrificed with CO_2 narcosis day 12 post-xenotrasplant. PB was collected from the heart in EDTA–coated tubes (Multivette 600; Sarstedt). Plasma was obtained by centrifugation at 4 °C and 1500 g for 20 min and used for the determination of total cholesterol, LDL-C and HDL-C, all measured in the Architect ci8200 analyzer (Abbott Diagnostics). Bone marrow was flushed with 500 μL of phosphate-buffered saline in the form of a fine cell suspension, and centrifuged at 180 g for 5 min. PB and BM cells were used for determination of the percentage of circulating GFP+ B-ALL cells. Necropsy was performed and femur, tibia, lung, liver, brain, cranial nerves and spinal cord were collected, fixed in 10% formalin, decalcified in rapid bone decalcifier (Perudo00-008; Eurobio, Les Ulis, France), paraffin embedded and stained with hematoxylin and eosin. Immunofluorescence and immunohistochemistry were performed after antigen retrieval (Dako PT link, pH 6, 95 °C, 20 min), the later using routine protocols with the ChemMate Dako EnVision detection kit (Dako Cytomation) employing Peroxidase/Diaminobenzidine. Antibodies used were TdT, Vimentin, S100 (A3524, M0725 and Z0311, Dako Cytomation).

Statistical Analysis

GraphPad software (San Diego CA, www.graphpad.com) was used to analyze the data. Linear regression analysis was performed to determine the correlation between bichemical and clinical parameters. Intergroup statistical analysis was performed using one or two-way ANOVA or unpaired t-test, both two-tailed. All were considered significant when the p values $\leq .05$.

References

Alberti KGMM, Zimmet P, Shaw J (2006) Metabolic syndrome – a new world-wide definition. A consensus statement from the international diabetes federation. Diabet Med 23(5):469–480

Barthold SW (2004) Genetically altered mice: phenotypes, no phenotypes, and faux phenotypes. Genetica 122(1):75–88

Bergen WG, Mersmann HJ (2018) Comparative aspects of lipid metabolism: impact on contemporary research and use of animal models. J Nutr. https://doi.org/10.1093/jn/135.11.2499

Berglund ED, Li CY, Poffenberger G, Ayala JE, Fueger PT, Willis SE, Jewell MM, Powers AC, Wasserman DH (2008) Glucose metabolism in vivo in four commonly used inbred mouse strains. Diabetes. https://doi.org/10.2337/db07-1615

Butturini AM, Dorey FJ, Lange BJ, Henry DW, Gaynon PS, Fu C, Franklin J, Siegel SE, Seibel NL, Rogers PC, Sather H, Trigg M, Bleyer WA, Carroll WL (2007) Obesity and outcome in pediatric acute lymphoblastic leukemia. J Clin Oncol. https://doi.org/10.1200/JCO.2006.07.7792

Cortes J (2001) Central nervous system involvement in adult acute lymphocytic leukemia. Hematol Oncol Clin North Am. https://doi.org/10.1016/S0889-8588(05)70203-3

Demetrius L (2005) Of mice and men. When it comes to studying ageing and the means to slow it down, mice are not just small humans. EMBO Rep 6 Spec No:S39–S44. https://doi.org/10.1038/sj.embor.7400422

Gomes AL, Carvalho T, Serpa J, Torre C, Dias S (2010) Hypercholesterolemia promotes bone marrow cell mobilization by perturbing the SDF-1:CXCR4 axis. Blood. https://doi.org/10.1182/blood-2009-08-240580

Grau G, Portillo N, Almaraz RL, Echebarria A, Adán R, Rodriguez A, Vela A, Astigarraga I, Rica I (2016) Severe hypertriglyceridemia in pediatric oncology patient. Horm Res Paediatr. https://doi.org/10.1159/000449142

Grubb SC, Bult CJ, Bogue MA (2014) Mouse phenome database. Nucleic Acids Res. https://doi.org/10.1093/nar/gkt1159

Harjes U, Bensaad K, Harris AL (2012) Endothelial cell metabolism and implications for cancer therapy. Br J Cancer 107(8):1207–1212

Kaplan JG, DeSouza TG, Farkash A, Shafran B, Pack D, Rehman F, Fuks J, Portenoy R (1990) Leptomeningeal metastases: comparison of clinical features and laboratory data of solid tumors, lymphomas and leukemias. J Neurooncol. https://doi.org/10.1007/BF02341153

Kennedy AJ, Ellacott KLJ, King VL, Hasty AH (2010) Mouse models of the metabolic syndrome. Dis Model Mech. https://doi.org/10.1242/dmm.003467

Kouidhi S, Ayed FB, Elgaaied AB (2018) Targeting tumor metabolism: a new challenge to improve immunotherapy. Front Immunol 9:353

Kuliszkiewicz-Janus M, Małecki R, Mohamed AS (2008) Lipid changes occuring in the course of hematological cancers. Cell Mol Biol Lett. https://doi.org/10.2478/s11658-008-0014-9

Lee HY, Jeong KH, Choi CS (2014) In-depth metabolic phenotyping of genetically engineered mouse models in obesity and diabetes. Mamm Genome

Martignoni M, Groothuis GMM, de Kanter R (2006) Species differences between mouse, rat, dog, monkey and human CYP-mediated drug metabolism, inhibition and induction. Expert Opin Drug Metab Toxicol. https://doi.org/10.1517/17425255.2.6.875

Moschovi M, Trimis G, Apostolakou F, Papassotiriou I, Tzortzatou-Stathopoulou F (2004) Serum lipid alterations in acute lymphoblastic leukemia of childhood. J Pediatr Hematol Oncol. https://doi.org/10.1097/00043426-200405000-00006

Muruganandan S, Sinal CJ (2008) Mice as clinically relevant models for the study of cytochrome P450-dependent metabolism. Clin Pharmacol Ther. https://doi.org/10.1038/clpt.2008.50

Nandi A (2004) Mouse models of insulin resistance. Physiol Rev. https://doi.org/10.1152/physrev.00032.2003

Nayar G, Ejikeme T, Chongsathidkiet P, Elsamadicy AA, Blackwell KL, Clarke JM, Lad SP, Fecci PE, Nayar G, Ejikeme T, Chongsathidkiet P, Elsamadicy AA, Blackwell KL, Clarke JM, Lad SP, Fecci PE, Nayar G, Ejikeme T, Chongsathidkiet P, Elsamadicy AA, Blackwell KL, Clarke JM, Lad SP, Fecci PE (2017) Leptomeningeal disease: current diagnostic and therapeutic strategies. Oncotarget. https://doi.org/10.18632/oncotarget.20272

Paigen B (1995) Genetics of responsiveness to high-fat and high-cholesterol diets in the mouse. Am J Clin Nutr 62(2):458S–462S

Picard O, Rolland Y, Poupon MF (1986) Fibroblast-dependent tumorigenicity of cells in nude mice: implication for implantation of metastases. Cancer Res 46(7):3290–3294

Pui CH, Howard SC (2008) Current management and challenges of malignant disease in the CNS in paediatric leukaemia. Lancet Oncol 9(3):257–268

Rossmeisl M, Rim JS, Koza RA, Kozak LP (2003) Variation in type 2 diabetes – related traits in mouse strains susceptible to diet-induced obesity. Diabetes. https://doi.org/10.2337/diabetes.52.8.1958

Rozman J, Klingenspor M, Hrabě de Angelis M (2014) A review of standardized metabolic phenotyping of animal models. Mamm Genome 25(9–10):497–507

Russell DW (2003) The enzymes, regulation, and genetics of bile acid synthesis. Annu Rev Biochem. https://doi.org/10.1146/annurev.biochem.72.121801.161712

Ruzzenente B, Rötig A, Metodiev MD (2016) Mouse models for mitochondrial diseases. Hum Mol Genet 25(R2):R115–R122

Savage DB (2009) Mouse models of inherited lipodystrophy. Dis Model Mech. https://doi.org/10.1242/dmm.002907

Scribano D, Baroni S, Pagano L, Zuppi C, Leone G, Giardina B (1996) Return to normal values of lipid pattern after effective chemotherapy in acute lymphoblastic leukemia. Haematologica 81(4):343–345

Surapaneni UR, Cortes JE, Thomas D, O'Brien S, Giles FJ, Koller C, Faderl S, Kantarjian H (2002) Central nervous system relapse in adults with acute lymphoblastic leukemia. Cancer. https://doi.org/10.1002/cncr.10265

Swerdlow SH, Campo E, Harris NL, Jaffe ES, Pileri SA, Stein H, Thiele J, Vardiman J (2017) WHO classification of tumours of haematopoietic and lymphoid tissues. International Agency for Research on Cancer, Lyon

Zhang Y, Qin C, Yang L, Lu R, Zhao X, Nie G (2018) A comparative genomics study of carbohydrate/glucose metabolic genes: from fish to mammals. BMC Genomics. https://doi.org/10.1186/s12864-018-4647-4

22 Modeling of Solid-Tumor Microenvironment in Zebrafish (Danio Rerio) Larvae

Yuxiao Yao, Lei Wang, and Xu Wang

Abstract

The zebrafish larvae have emerged as a powerful model for studying tumorigenesis *in vivo*, with remarkable conservation with mammals in genetics, molecular and cell biology. Zebrafish tumor models bear the significant advantages of optical clarity in comparison to that in the mammalian models, allowing non-invasive investigation of the tumor cell and its microenvironment at single-cell resolution. Here we review recent progressions in the field of zebrafish models of solid tumor diseases in two main categories: the genetically engineered tumor models in which all cells in the tumor microenvironment are zebrafish cells, and xenograft tumor models in which the tumor microenvironment is composed of zebrafish cells and cells from other species. Notably, the zebrafish patient-derived xenograft (zPDX) models can be used for personalized drug assessment on primary tumor biopsies, including the pancreatic cancer. For the future studies, a series of high throughput drug screenings on the library of transgenic zebrafish models of solid tumor are expected to provide systematic database of oncogenic mutation, cell-of-origin, and leading compounds; and the humanization of zebrafish in genetics and cellular composition will make it more practical hosts for zPDX modeling. Together, zebrafish tumor model systems are unique and convenient *in vivo* platforms, with great potential to serve as valuable tools for cancer researches.

Keywords

Tumor microenvironment · Animal model · Zebrafish · Transgenesis · Xenograft · Chimeric antigen receptor (CAR) T-cells

Y. Yao · L. Wang · X. Wang (✉)
Cancer Metabolism Laboratory, Cancer Institute, Fudan University Shanghai Cancer Center, Shanghai, China

Key Laboratory of Metabolism and Molecular Medicine, Ministry of Education, Department of Biochemistry and Molecular Biology, School of Basic Medical Sciences, Fudan University, Shanghai, China
e-mail: 12111010009@fudan.edu.cn; 17211010023@fudan.edu.cn; wangxu2013@fudan.edu.cn

22.1 Introduction

Due to the advantages of transparency, ectogenesis, and inexpensive costs, zebrafish have been extensively used as model organisms for genetics, developmental biology, and human disease modeling (Van Slyke et al. 2018). In comparison to the mammals and rodents, the genetic modifications on zebrafish are usually easier and faster, and researchers can perform *in vivo* time-elapse recording and realtime cell tracing under micros-

J. Serpa (ed.), *Tumor Microenvironment*, Advances in Experimental Medicine and Biology 1219, https://doi.org/10.1007/978-3-030-34025-4_22

copies in a noninvasive way (MacRae and Peterson 2015; Yee et al. 2015). Besides, the small size and high fertility save significant human resource, time and funds, and allow small research teams to perform high throughput screening for target genes or small molecular compounds in an affordable way. During the past few decades, the zebrafish research community has been steadily developed globally, and NIH has listed zebrafish as the third most popular vertebrate model animals, after the mouse and rat (Van Slyke et al. 2018).

Among all the diseases, cancer is a leading cause of death in both developing and developed countries (Bray et al. 2018). Animal models are indispensable for cancer research because it is very difficult to modeling tumor microenvironment *in vitro*. However, only a small proportion of cancer researchers employed zebrafish for their *in vivo* studies. Concerning the potential differences between zebrafish and mammals in genetics and physiology, most *in vivo* cancer models are established in mammals, including mice, rats, rabbits, dogs, and primates (Gardner et al. 2016; DE Fatima et al. 2018; Cekanova and Rathore 2014; White et al. 2013; Sonoshita and Cagan 2017). Nevertheless, zebrafish is the only mini model animal in the vertebrate kingdom, and the intrinsic features of zebrafish may facilitate the investigation of scientific questions in cancer research that cannot be easily answered in rodents. In the past few years, zebrafish have been increasingly used for modeling tumorigenesis via two main strategies, genetic modification and xenografting (Kirchberger et al. 2017) (Fig. 22.1). The genetic models are classified into transgenic models and mutagenic models, in which the oncogenic or cancer repressor pathways were genetically activated or disrupted respectively, mimicking the human cancer patients in genetics (Fig. 22.1). On the other sides, the xenograft models may be generated by transplanting stabilized mammalian cell lines or primary tumor tissues from mammals into embryonic, larval or adult zebrafish (Fig. 22.1). From the aspects of the tumor microenvironment, the two modeling approaches also reflect the differences in cellular compositions in the tumor, since the genetic models only carry zebrafish cells, while the xenograft models harbor cells from different species.

In fact, hematopoietic malignancy (leukemia) is one of the mostly-studied and well-studied cancer diseases in zebrafish, and the progression in zebrafish leukemia models has been frequently reviewed (Kwan and North 2017; He et al. 2017). In this article, we will describe the recent studies in the zebrafish models of solid tumors. Solid tumors are defined by an abnormal mass of tumorous tissue and can occur to a variety of organs (Allen-Rhoades et al. 2018). Solid tumors in organs that are evolutionally-conserved between zebrafish and human can be modeled via genetic approaches, including liver cancer, pancreatic cancer, colon cancer, and melanoma. Other solid tumors like lung cancer and breast cancer can only be modeled in zebrafish by xenografting.

22.2 Review on the Genetic Zebrafish Models of Solid Tumor Diseases: Mutants and Transgenic Lines

22.2.1 Oncogenic Mutation Models in Zebrafish

It is well established that genetic mutations induced by environmental or endogenous stimuli cause cancer (Lord and Ashworth 2012; Tubbs and Nussenzweig 2017). Given the tremendous progress in our understanding of the genomic alterations in human cancer patients, the genetic models of solid tumor in zebrafish directly simulate those oncogenic features. In the old time, those oncogenic mutations may be randomly induced by chemical compounds like *N*-ethyl-*N*-nitrosourea (ENU) and random insertions via retrovirus or transponson, followed by phenotype screening, mapping, and sequencing. Nowadays, with the discovery and prevalence of genome editing tools, researchers can easily generate zebrafish mutants of cancer repressor genes. So far, many zebrafish mutant lines have been identified to be oncogenic, including *tp53* (Berghmans

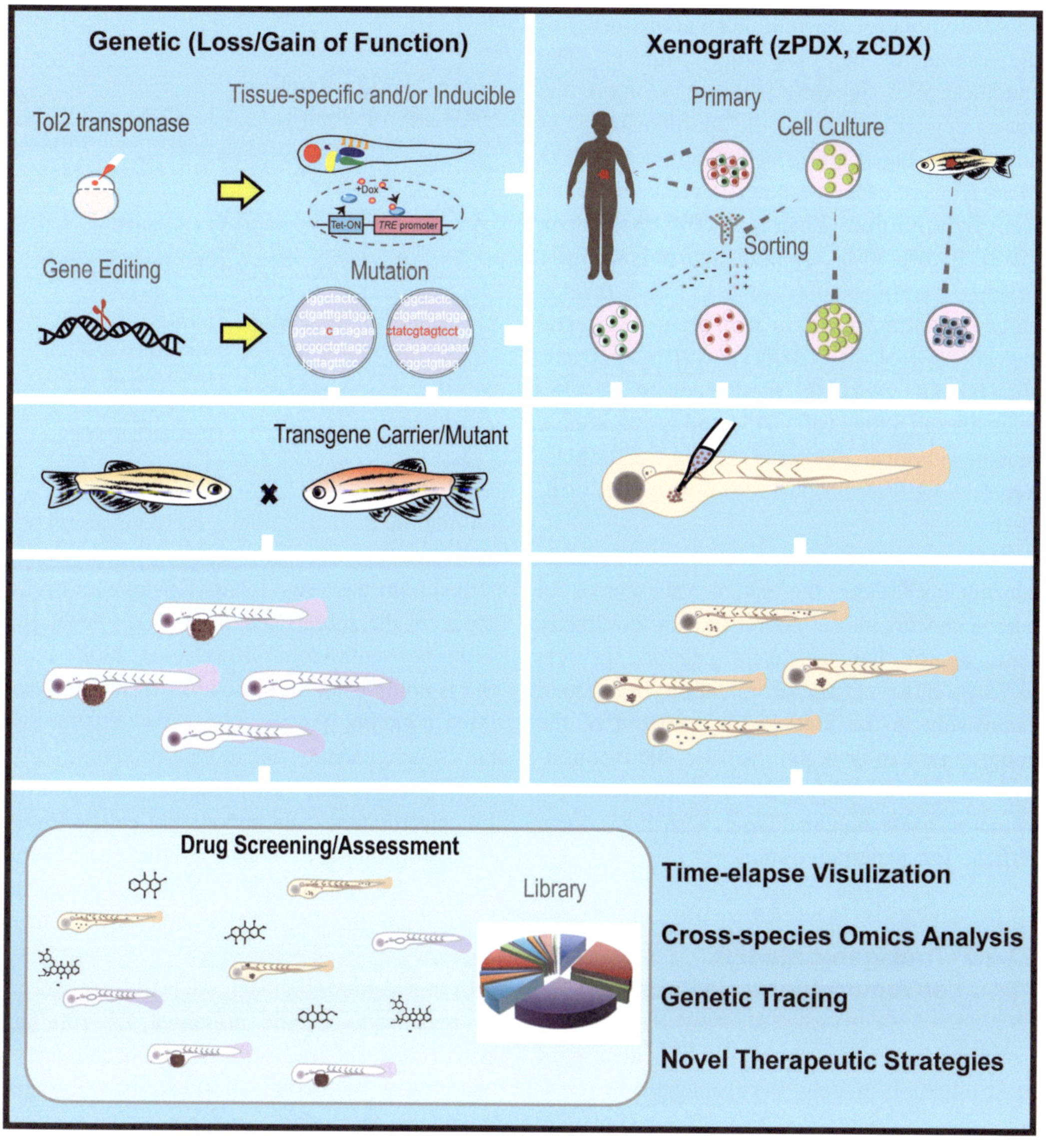

Fig. 22.1 Zebrafish models of tumor diseases

et al. 2005), *apc* (Rai et al. 2010), *rb1* (Solin et al. 2015), *nf1a/b* (He et al. 2016), *ptena/b* (den Hertog 2016), and *brca2* (Shive et al. 2014).

TP53 is the most frequently mutated genes in almost all human cancer, and the loss-of-function *tp53* mutant zebrafish spontaneously develop malignant peripheral nerve sheath tumors (Berghmans et al. 2005), nephroblastoma (Shive et al. 2014), melanomas (Kim et al. 2017), eye tumors (den Hertog 2016), and ovarian tumors (Shive et al. 2010). The successful induction of tumors in *tp53* mutants may rely on the accumulation of additional mutations during the hierarchic evolution of tumor initiating clones (Hanahan and Weinberg 2011). The incidences of tumorigenesis in *tp53* mutant zebrafish are significantly increased by directly introducing additional oncogenic mutations in genes like *brca2* or *ptena/b* (den Hertog 2016; Shive et al. 2014), confirming the "multiple-hits" theory of tumorigenesis (Knudson 2001; Belikov 2017).

APC is one of the inhibitory components of the canonical Wnt signaling pathway. The canonical Wnt signaling pathway is one of the master regulators during the morphogenesis at embryonic stages, and is ectopically activated in many types of cancer, especially those originated from digestive tract (Zhan et al. 2017; Wang et al. 2012). In zebrafish, *apc* mutants spontaneously develop gastrointestinal tumors, which display epigenetic alternations in relevant to colorectal cancer (Rai et al. 2010; Mir et al. 2016). Similarly, *rb1* mutant zebrafish spontaneously develop brain tumor (Solin et al. 2015), and *nf1a/b* double mutant zebrafish randomly develop neuroblastoma at 4 wpf (weeks post fertilization) (He et al. 2016).

However, although the mutations in the zebrafish models mimic the genetic aberrances in human cancers, those models are not the exactly same as the human cancer patients. In real patients, most oncogenic mutations are found exclusively in the focal tissues, instead of the entire organism or organ. Besides, the inconsistent occurrence and frequency of the cancerous events in those mutant models also made them difficult for quantitative study.

22.2.2 Transgenic Models of Tumorigenesis in Zebrafish

In comparison to the spontaneous incidences of most mutation models, the incidences in transgenic models can reach up to 100%, making them ideal for quantitative investigations. The combination of tissue-specific promoter and compound-inducible genetic switches allowed the dosage-dependent activation of oncogenic pathways in certain cell types, and has been well administrated for modeling certain cancer types. Here we will summarize the representative transgenic models generated for liver cancer and pancreatic cancer as examples (Table 22.1).

22.2.2.1 Transgenic Zebrafish Models of Liver Cancer

Liver is the major organ of metabolism and its functions are highly conserved between zebrafish and mammals. Interestingly, early larval zebrafish liver is very similar to human liver in shape, as they both have two lobes (left lobe and right lobe), but the zebrafish at later stages develop a third lobe (ventral lobe) (Korzh et al. 2008). Each lobe is composited of basic units of liver lobule, which contains parenchymal cells (hepatocytes and cholangiocytes) and nonparenchymal cells (fibroblasts, stellate cells, Kupffer cells, neutrophil, macrophage, and endothelial cells). Based on histopathology, hepatocellular carcinoma (HCC) and intrahepatic cholangiocarcinoma (ICCA) are the two major subtypes of liver cancer.

Table 22.1 Transgenic zebrafish models of liver cancer and pancreatic cancer

Tumor types	Transgenic/ Mutation	Reference
Liver cancer	*kras*^{G12V}	Nguyen et al. (2012), Yan et al. (2017a, b)
	XMRK	Li et al. (2012)
	Myc	Li et al. (2013)
	tgfb1a	Yan et al. (2017a, b)
	edn1	Lu et al. (2014)
	HBx scr/tp53	Lu et al. (2013)
	CTNNB^{mut}	Yao et al. (2018), Evason et al. (2015)
	nras^{61K}	Wang et al. (2017)
Pancreatic cancer	*KRAS*^{G12V}	Park et al. (2008)
	MYCN	Yang et al. (2004)

In transgenic zebrafish models of liver cancer, the hepatocyte-specific promoter *fabp10a* has been extensively used as the tissue-specific driver (Nguyen et al. 2011). Besides, two dosage-dependent transgenic switches, doxycycline-inducible TetOn system and mifepristone-inducible LexPR system, were employed to conditionally activate the transcription of downstream oncogenes (Nguyen et al. 2012; Yao et al. 2018). In addition, a variety of oncogenes, including *kras*^{V12}, *nras*^{K61}, *tgfb1a*, *edn1*, *xmrk* (a xiphophorus version of mutated *egfrb*), mouse *Myc*, human *CTNNB1, and HBx/src*, have been used as the effector genes to induce carcinogenesis (Yao et al. 2018; Yan et al. 2017a, b; Li et al. 2012, 2013; Lu et al. 2013, 2014; Evason et al. 2015; Wang et al. 2017). In most cases, the oncogenic insults to the hepatocytes induced HCC. However, in cases that *tgfb1a* and *nras*^{K61} was overexpressed, ICCA was also detected as the outcome, indicating the existence of transdifferentiation from hepatic lineage into biliary lineage cells.

Those transgenic models have a tumor microenvironment composed of zebrafish cells, and can be used to investigate the interaction between tumor cells and non-tumor cells/tissues. In the Ras-induced HCC model *Tg(fabp10a:TetOn; TRE:Egfp-kras*^{V12}*)*, researchers found that the tumors were more heavily infiltrated with neutrophils and macrophages in male versus female, which are caused by the increased cortisol production, demonstrating the feasibility to use zebrafish cancer model for investigating the immune and endocrine microenvironment (Yan et al. 2017a, b). In our recent collaborating work with Gong's group, we also showed that the leptin secreted by the zebrafish tumorous liver can directly induce wasting of non-tumor tissues including the skeletal muscle (Yang et al. 2019).

22.2.2.2 Transgenic Zebrafish Models of Pancreatic Cancer

Pancreatic cancer can be divided into two major groups. 95% pancreatic cancer occurs in the pancreatic tissue that produces digestive enzymes, known as pancreatic ductal adenocarcinoma (PDAC), and the rest are mainly pancreatic neuroendocrine tumors (PanNETs) (Klimstra et al. 2010).

Pancreatic ductal adenocarcinoma (PDAC) is the fourth leading cause of cancer-related death, and the 5-year survival rates for PDAC remain around 5% all over the years. Genome sequencing revealed the major oncogenic drivers are mutations in genes like *KRAS*, *TP53*, *CDKN2A* and components in the TGFβ pathway (Mueller et al. 2018). Two transgenic zebrafish models of pancreatic cancer, *Tg(ptf1a:eGFP-KRAS*^{G12V}*)* (Park et al. 2008) and *Tg(ptf1a:Gal4-VP16; UAS:eGFP-KRAS*^{G12V}*)* (Liu and Leach 2011), were generated by Leach's group. Although those transgenic zebrafish did not use the transgenic switching strategies, they were able to be stabilized and induced pancreatic tumors with high penetration. Two thirds of *Tg(ptf1a:eGFP-KRAS*^{G12V}*)* develop tumor by 9 mpf (months post fertilization), and some of the tumors displayed liver & gut invasion and ovarian metastasis (Park et al. 2008). Around 50% *Tg(ptf1a:Gal4-VP16; UAS:eGFP-KRAS*^{G12V}*)* developed pancreatic tumors by 5 mpf, with the earliest tumorigenesis detectable at 2 mpf (Liu and Leach 2011). The application of Gal4/UAS transgenic allowed the generation of new pancreatic cancer models by combining the *ptf1a:Gal4-VP16* driver with lineages expressing new oncogenes by *UAS* promoter (Liu and Leach 2011). Our recent study also indicated that the random silencing of *UAS* promoter would produce a chimeric/mosaic expression pattern, which may be a good simulation of the cancer heterogeneity (Yao et al. 2018).

Besides, targeted expression of MYCN in pancreatic islet via *z-myod* promoter induced neuroendocrine carcinoma (Yang et al. 2004). Those tumors express insulin mRNA, and pancreatic exocrine cells and ducts can be observed within the tumor tissues (Yang et al. 2004).

22.3 Review on the Xenograft Zebrafish Models of Solid Tumor Diseases: zCDX and zPDX

Xenograft quickly generates several tumor models carrying tumor cells/tissues from identical donors, and has significantly improved our understanding of the tumorigenesis, heterogeneity, and metastasis. Traditionally, xenograft was performed by injecting human or rodent's malignant cells into immune-compromised mice. However, maintenance of immune-compromised mouse models in SPF environment can be costly and time-consuming, and the xenografts in the mouse models cannot be directly observed *in vivo*. In the past few years, many groups have investigated the strategy to transplant fluorescent mammalian cells into zebrafish larvae or adults with optically clear *casper* background or PTU treatment for direct visualization.

Table 22.2 Human cancer cell lines for zCDX models

Tumor types	Cell lines	Reference
Melanoma	C8161	Lee et al. (2005)
	A375	Smith et al. (2013)
	Mel270	van der et al. (2015)
	OMM2.3	van der et al. (2015)
	OMM2.5	van der et al. (2015)
	92.1	van der et al. (2015)
	OMM1	van der et al. (2015)
	MM96L	Ikonomopoulou et al. (2018)
Liver cancer	Bel-7402	Hou et al. (2013)
	HepG2	Yang et al. (2017)
	Hep3B	Avci et al. (2018)
	SKHep1	Avci et al. (2018)
	Huh7	Avci et al. (2018)
Pancreatic cancer	Mia PaCa-2	Guo et al. (2015)
	BxPC-3	Guo et al. (2015)
Prostate cancer	C4-2B	Wagner et al. (2010)
	DU-145	Chiavacci et al. (2015)
	PC3-CTR	Xu et al. (2018)
Lung cancer	H1299	Moshal et al. (2011)
	A549	Leung et al. (2017)
	NCI-H2009	Tan et al. (2014)
Colon cancer	SW480	Fior et al. (2017)
	SW620	Fior et al. (2017)
	HT29	Fior et al. (2017)
	HCT116	Fior et al. (2017)
	Hke3	Fior et al. (2017)

22.3.1 zCDX (Zebrafish Cell-Line-Derived Xenograft) Models

Most xenograft zebrafish models are cell-line-derived xenograft (zCDX), and dozens of mammalian cell lines have been tested for generating zCDX models with various outcomes (Table 22.2). Zebrafish at different stages were chosen, including blastula stage, 24 hpf (hours post fertilization), and 48 hpf up to adulthood. The locations for microinjection included yolk sac, duct of cuvier, caudal vein, pericardial cavity, perivitelline space, and brain ventricles (Barriuso et al. 2015; Veinotte et al. 2014; Nicoli and Presta 2007). Melanoma C8161 cells was one of the first human cell lines that were injected into zebrafish embryos at blastula stage and they were found to be able to survive, proliferate, and migrate in zebrafish hosts (Lee et al. 2005). Other melanoma cell lines, A375 (Smith et al. 2013), Mel270 (van der et al. 2015), OMM2.3 (van der et al. 2015), OMM2.5 (van der et al. 2015), 92.1 (van der et al. 2015), OMM1 (van der et al. 2015), and MM96L (Ikonomopoulou et al. 2018) were also injected into yolk sac and/or circulation of zebrafish, and the xenografts displayed different behaviors.

Several human liver cancer cell lines were used to establish zCDX models, including Bel-7402 (Hou et al. 2013), HepG2 (Yang et al. 2017), Hep3B (Avci et al. 2018), SKHep1 (Avci et al. 2018), and Huh7 (Avci et al. 2018). Interestingly, the later three human liver cancer cell lines were injected into the yolk sac of *ache*$^{-/-}$ zebrafish, and found that acetylcholine accumulation supports the cell growth (Avci et al. 2018). Such experiments indicated that zebrafish host can be genetically modified to provide favorable endocrine microenvironment for tumor xenograft.

To model pancreatic cancer in zCDX models, we previous injected two human pancreatic cancer cell lines, Mia PaCa-2 and BxPC-3, into the zebrafish larvae, and use the models to assess a new candidate drug U0126 targeting *Kras* mutation (Guo et al. 2015). In the pancreatic zCDX models, the transgenic background *Tg(flk1:Egfp)* was introduced to label the zebrafish vascular endothelial cells, and we observed a tumor microenvironment composited of human cancer cells and zebrafish blood vessels (Guo et al. 2015). Besides, many other human cancer cell lines including prostate cancer cell lines C4-2B (Wagner et al. 2010), DU-145 (Chiavacci et al. 2015), and PC3-CTR (Xu et al. 2018) were also employed in zCDX models for studying miRNA functions or for high throughput drug screening. The zebrafish do not have lung, but the human non-small cell lung carcinoma cells H1299 (Moshal et al. 2011), NCI-H2009 (Tan et al. 2014), and A549 (Leung et al. 2017) lung cancer cells can be injected as xenograft, and their proliferation and migration upon different genetic manipulation or drug treatment were assessed. The long list of solid tumor cell lines that have been adopted for zCDX also contains colon cancer cell lines (Fior et al. 2017), ovarian carcinomas (Latifi et al. 2011), gliomas (Yang et al. 2013), breast cancer cells (Drabsch et al. 2013; Wu et al. 2018), retinoblastomas (Jo et al. 2013), and ewing sarcomas (Ban et al. 2014; van der et al. 2014; Franzetti et al. 2017).

22.3.2 zPDX (Zebrafish Patient-Derived Xenograft) Models

Recently, the primary cells derived from human cancer patients were also used for producing xenograft models in zebrafish (zPDX). One of the first attempts was to inject primary culture of bone metastasis derived a breast cancer patient into zebrafish embryos (Mercatali et al. 2016). The cells were observed to extravasate from the vessels, and were engrafted into the caudal hematopoietic tissues, indicating the feasibility of using zebrafish as hosts for PDX modeling (Mercatali et al. 2016). A few other primary human cells used for zPDX models include melanoma cells (four patients) (Waster et al. 2017), gastric cancer cells (nine patients) (Wu et al. 2017), neuroendocrine tumor cells (eight patients) (Gaudenzi et al. 2017), and colorectal cancer cells (eleven patients) (Fior et al. 2017). The drug responses in those zPDX were also analyzed together with the clinic outcomes, indicating certain level of expected correlations (Fior et al. 2017; Wu et al. 2017).

The drug assessment on primary cells for precision medicine may also be performed in *in vitro* cell/organoid culture and mouse PDX models. However, the *in vitro* culture never forms a microenvironment closer to the *in vivo* condition. The mouse PDX models, on the other side, usually took months to obtain enough number of models for multiple drug assessment, and patients carrying fast-developing tumor like pancreatic cancer may not benefit from the assay. In zebrafish zPDX models, CM-Dil or other Dil dyes were generally used to quickly label the primary cells to allow the direct observation of cellular behavior upon drug treatment, and the whole procedure may only take several days, displaying significant advantage for clinical purpose. We previously tested the application of such strategy in assessing drug responses of primary pancreatic cancer (Fig. 22.2). We harvested the fresh pancreatic cancer tissues, digested them into single cells or small cell groups via both collagenase and steel mesh, stained them with CM-Dil, and injected the cells into 48 hpf *Tg(flk1:eGFP)* zebrafish yolk sac. For each larva, 50–80 cells were microinjected, and about 300 viable zebrafish patient-derived xenografted (zPDX) models can be obtained per patient within 30 min. Different drugs or drug combinations [Gemcitabine (7.5 μg/mL), Gemcitabine (3.75 μg/mL) & DDP (Cisplatin, 0.5 μg/mL), Gemcitabine (3.75 μg/mL) & PTX (Paclitaxel, 0.5 μg/mL)] were administrated to the zPDX models, followed by imaging 3 days later for assessing the drug responses. The whole process took only 4 days and the rest primary cells were cryopreserved and may be recovered for generating more zPDXs for further assessment (Fig. 22.2).

However, the fluorescent dyes CM-Dil diffuse in the yolk sac after the death of the tumor cells, significantly affecting the quantitative analysis. Besides, the conditions of the original solid tumor samples are complex, and are usually composited of several different cell types, including cancer cells, cancer stem cells, cancer-associated fibroblasts, endothelial cells, pericytes, and infiltrative lymphocytes ((Hanahan and Weinberg 2011)). Therefore, the fluorescence alterations did not necessary represent the drug responses in cancer cells, and it is difficult to extract significant information. To optimize the zPDX for clinical pancreatic cancer, in our recent studies, we proposed a novel heterogeneous zPDX modeling strategy for better standardized quantitative analyses (Wang et al. 2019). Lentivirus were used to label the isolated cancer cells and tumor-associated fibroblasts

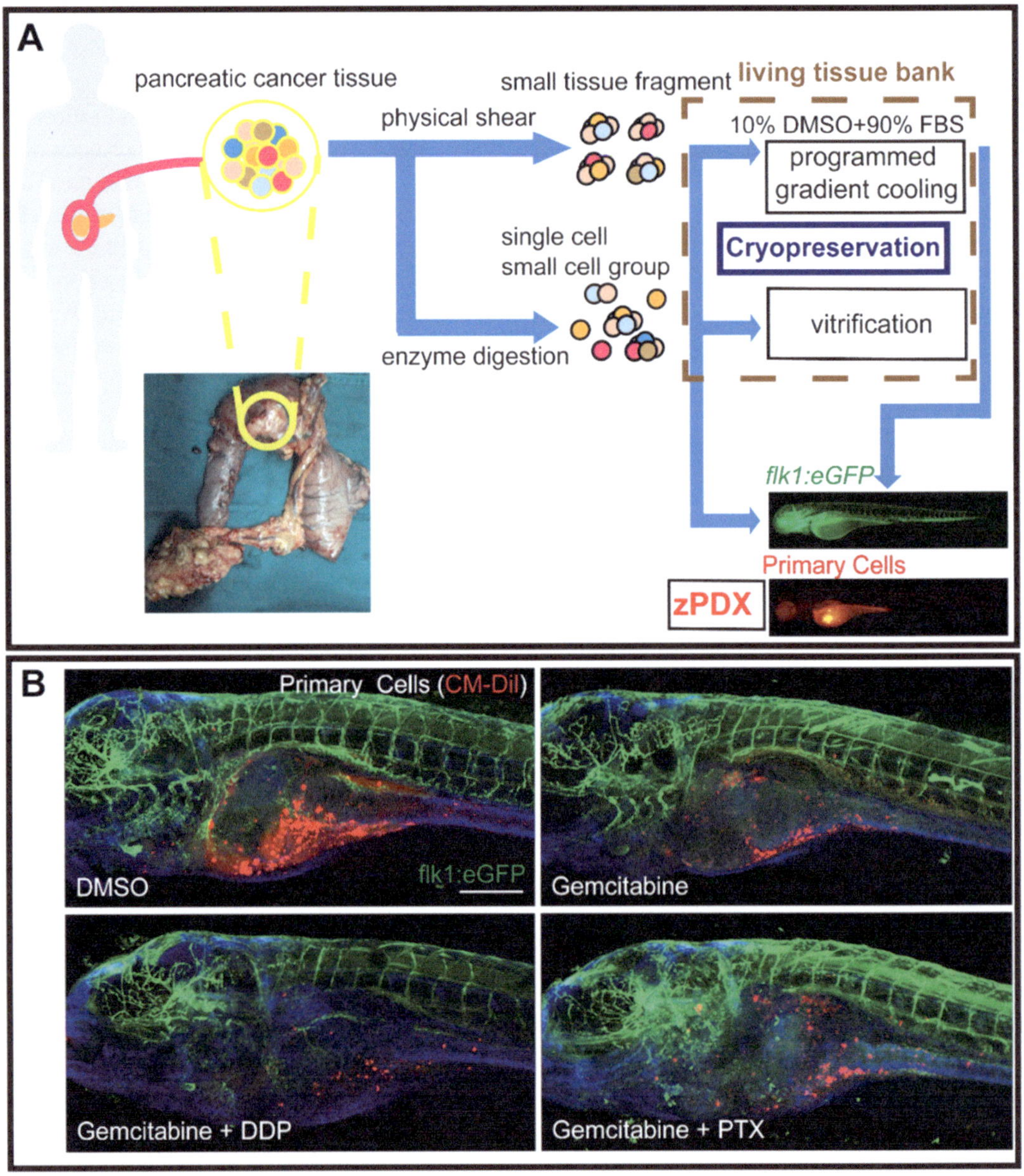

Fig. 22.2 The pancreatic cancer zPDX models for drug assessment. (**a**) Procedures of pancreatic cancer zPDX modeling; (**b**) Representative images of the pancreatic cancer zPDX models upon different drug treatment conditions, Scale bar: 400 μm

in different fluorescence separately. Both human cells were pre-sorted from the primary tissues, and were co-injected into the zebrafish after labelling. The new model better mimics the cellular composition of the tumor microenvironment, and significantly improves the consistency of the results via comparative analyses (Wang et al. 2019). Besides, unlike the Dil labeling dyes, the fluorescent proteins are degraded after the cell death, serving as indirect indicators of the cell viability (Wang et al. 2019).

22.4 Perspective of Future Directions

22.4.1 Systematic High Throughput Drug Screening on the Library of Transgenic Zebrafish Models

Both the genetic events and where those events take place (cell-of-origin) determine the pathological and molecular features of tumors that correlate with drug response in patients (Visvader 2011; Rycaj and Tang 2015). Transgenic zebrafish models mimic different subtypes of human solid tumors in genetics and cellular origin, and the tumorous organs can be labeled by fluorescent protein expression, allowing high throughput compound screening via direct imaging (Yao et al. 2017).

Different promoter-specific Gal4 lines and oncogene-specific UAS lines can be outcrossed to form a library of matrix combination. Such a library may cover all major subtypes of tumor diseases, and facilitates the investigation of the crosstalking among different oncogenic pathways. Morphology-based high throughput screening may be performed in the library of models, and a database of cell-of-origin/oncogenes/compound responses can be documented for new drug research and development.

22.4.2 CRISPR/Cas9 Based Mutagenesis for Tumor Modeling in Zebrafish

The applications of the CRISPR/Cas9 system provide an alternative and fast means of inducing tumor formation *in vivo*, as exemplified in mouse cancer modeling studies in multiple organs (Maddalo et al. 2014; Maresch et al. 2016; Platt et al. 2014; Romero et al. 2017; Sanchez-Rivera et al. 2014; Zuckermann et al. 2015). In CRISPR/Cas9-mediated tumorigenesis, we cannot control the exact mutations the sgRNAs will induce and the cancer cells can be highly heterogeneous. However, the randomness makes the CRISPR/Cas9-induced models similar to the real molecular features in human cancer patients.

Recently, we integrated the RNA endonuclease Csy4 into the Tet-On-regulated CRISPR/Cas9 system to facilitate the precise expression of the entire genome editing components in specific tissues upon doxycycline induction. By injecting the vector system (*TRE:Csy4-2A-Cas9* & *TRE:mKate2-Triplex-sgRNA*) into the zygotes of *Tg(fabp10a:TetOn;TRE:Egfp-kras*^{v12}*)*, we were able to disrupt the oncogenes exclusively in the hepatocytes (Wang et al. 2018). The same strategy can be used to achieve tissue-specific knockout of cancer repressor genes, and generate zebrafish cancer models with focal mutagenesis.

22.4.3 Humanization of Zebrafish as a Better Host for Human Xenograft

Most xenograft zebrafish models are short-term models with very limited experimental windows. The human cells xenografted in zebrafish cannot be used for lineage passage like those in mouse models. To improve the viability of the human cells in zebrafish, we need to perform humanization of zebrafish. The most common humanization is immune-comprised modification. Recently, several immune-comprised zebrafish lines have been generated, including *rag2*, *prkdc*, and *jak3* mutants. *Rag2* gene is responsible for the normal development of T cells, and the $rag2^{E450fs}$ immune-comprised zebrafish can be used for long-term allogeneic transplantation of melanoma cells from $Tg(mitfa{:}BRAF^{V600E}; mitfa^{-/-}; tp53^{-/-})$ (Tang et al. 2014, 2016). Similarly, deficiency in zebrafish *prkdc* results in loss of mature T and B cells, and knockout of *jak3* causes loss of T cells and putative NK cells. Both $prkdc^{D3612fs}$ and $jak3^{P369fs}$ significantly improved the successful engraft or xenograft of fluorescently labeled malignant cells (Moore et al. 2016). Notably, among all three immune-comprised mutants, $prkdc^{D3612fs}$ mutant seems to be the best host, since it is the only line that was maintained as homozygotes and survived well post-injection (Moore et al. 2016). However, it remains unclear whether we can cross those mutants and obtain a better combined immune-compromised zebrafish with T, B, and NK cells depleted, and how to perform zPDX experiments using immune-compromised zebrafish in a SPF-like environment is unexplored.

In addition to the humanization in genetics, we may also humanize zebrafish in cellular composition. The lymphocytes play critical roles in tumor microenvironments, and modulations of the immune/inflammation activities have been extensively investigated in the past few years, and demonstrate remarkable therapeutic values (Balar and Weber 2017; Pettitt et al. 2018). Recently, an interesting study demonstrated the possibility to generate the mammal-zebrafish hematopoietic tissue chimeras (Parada-Kusz et al. 2018), providing the hope to rebuild a humanized immune microenvironment or to introduce human monocytes into the immune-comprised zebrafish. However, such a strategy has not been tested yet. On the other side, to observe the potential interaction between the human immune cells and tumor cells under zebrafish physiological environment, we performed an experiment by injecting CAR-jurkat T cells into a zCDX model bearing tumor cells. The zCDX models were generated by xenografting human liver cancer MHCC-97H cells, which express high level of MET (Fig. 22.3a). The jurkat T cells were transfected by a CAR structure specifically recognizing MET on the cell surface (Fig. 22.3b, c). In this experiment, we first injected a hundred MHCC-97H cells into the yolk sac of a 2.5 dpf PTU-treated zebrafish. After 3 h, a single CAR-T cell was injected to the tumor mass. After another 12 h, we performed a confocal scan and found that the cell membrane of the CAR-T cell attached closely to the membranes of two tumor cells (Fig. 22.3d). However, we did not observe the proliferation of the CAR-jurkat T cells during our 5-days observation. To better study the behaviors of human monocytes in zebrafish, further genetic modification on zebrafish hosts may be required. One potential solution is to introduce transgenic expression cascades of human cell growth factors and IL2 into the zebrafish genome, to support the proliferation and activation of human monocytes.

To conclude, zebrafish is a unique but highly-relevant modeling platform for cancer research. The genetic models serve as useful tools for studying mechanisms and for performing high-throughput drug screening, and the xenograft models bear the capacity to mimic a human-like tumor microenvironment in zebrafish. Zebrafish host is a natural 3D medium and "multiple tissue chip", and the humanization in both genetics and cellular composition (immune system) will further improve the survival and growth of human cells, allowing long-term observations. In the future, the zPDX/zCDX models based on the humanized zebrafish larvae may be used not only to test compound-based therapeutic regimens, but also to assess different strategies of immune therapy *in vivo*.

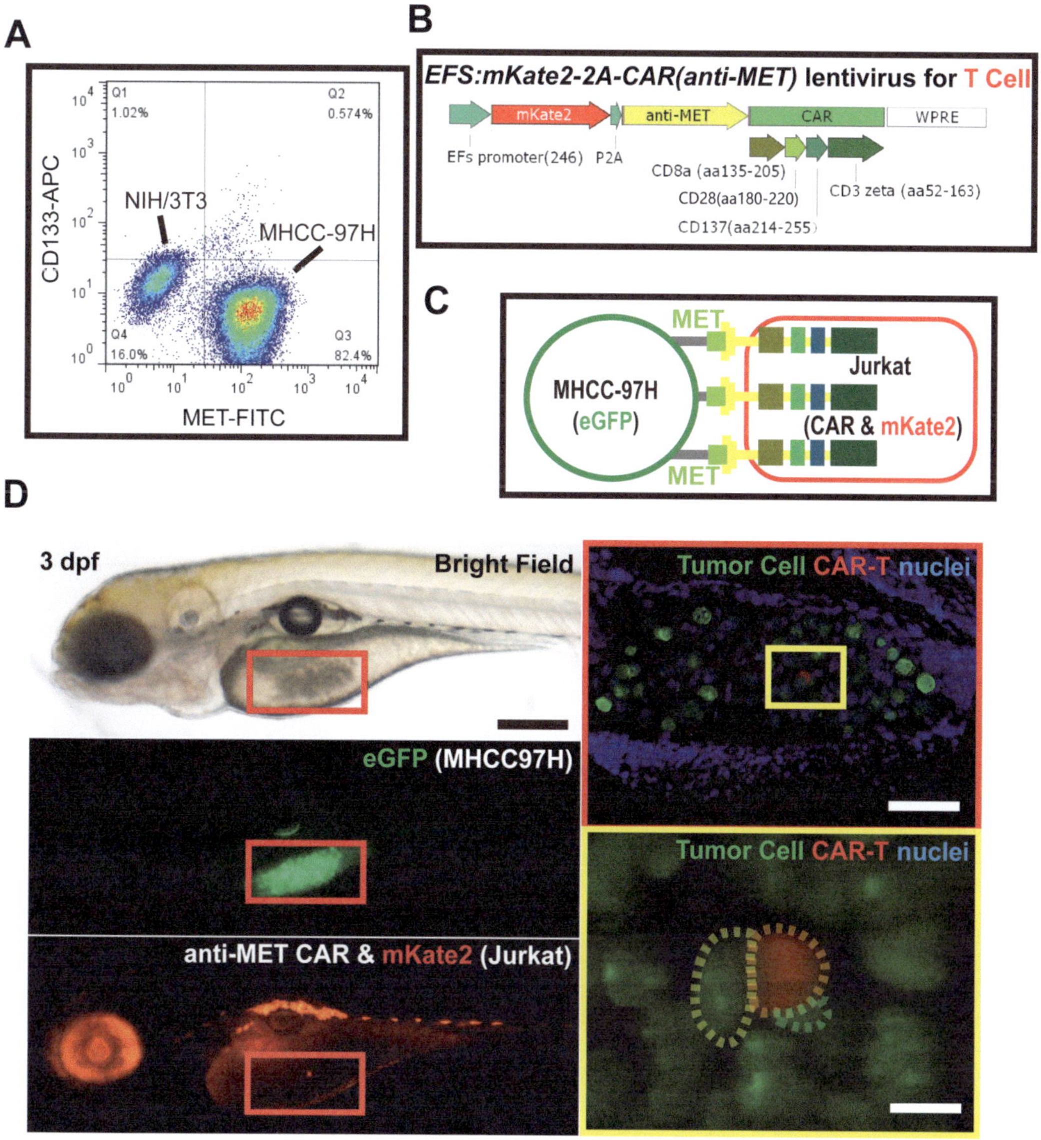

Fig. 22.3 The CAR-T binds to the liver cancer cells in double zCDX model. (**a**) Flow cytometry of MHCC-97H and NIH/3 T3 using CD133 and MET antibodies; (**b**) Lentivirus structure for transfecting jurkat T cells; (**c**) Illustration of the reorganization and binding of liver cancer cells by CAR-T cells; (**d**) zCDX carrying 100 MHCC-97H liver cancer cells and a single CAR-jurkat T cells. Scale bars: 400 μm (Left), 60 μm (Right Top), 12 μm (Right Bottom)

Financial Support National Natural Science Foundation of China 81402582 and 81802333.
Natural Science Foundation of Shanghai 14YF1400600 and 18ZR1404500.
Natural Science Foundation of Guangdong Province 2018A030310053.

References

Allen-Rhoades W, Whittle SB, Rainusso N (2018) Pediatric solid tumors of infancy: an overview. Pediatr Rev 39(2):57–67. https://doi.org/10.1542/pir.2017-0057

Avci ME, Keskus AG, Targen S, Isilak ME, Ozturk M, Atalay RC, Adams MM, Konu O (2018) Development of a novel zebrafish xenograft model in ache mutants using liver cancer cell lines. Sci Rep 8(1):1570. https://doi.org/10.1038/s41598-018-19817-w

Balar AV, Weber JS (2017) PD-1 and PD-L1 antibodies in cancer: current status and future directions. Cancer Immunol Immunother 66(5):551–564. https://doi.org/10.1007/s00262-017-1954-6

Ban J, Aryee DN, Fourtouna A, van der Ent W, Kauer M, Niedan S, Machado I, Rodriguez-Galindo C, Tirado OM, Schwentner R, Picci P, Flanagan AM, Berg V, Strauss SJ, Scotlandi K, Lawlor ER, Snaar-Jagalska E, Llombart-Bosch A, Kovar H (2014) Suppression of deacetylase SIRT1 mediates tumor-suppressive NOTCH response and offers a novel treatment option in metastatic Ewing sarcoma. Cancer Res 74(22):6578–6588. https://doi.org/10.1158/0008-5472.CAN-14-1736

Barriuso J, Nagaraju R, Hurlstone A (2015) Zebrafish: a new companion for translational research in oncology. Clin Cancer Res 21(5):969–975. https://doi.org/10.1158/1078-0432.CCR-14-2921

Belikov AV (2017) The number of key carcinogenic events can be predicted from cancer incidence. Sci Rep 7(1):12170. https://doi.org/10.1038/s41598-017-12448-7

Berghmans S, Murphey RD, Wienholds E, Neuberg D, Kutok JL, Fletcher CD, Morris JP, Liu TX, Schulte-Merker S, Kanki JP, Plasterk R, Zon LI, Look AT (2005) tp53 mutant zebrafish develop malignant peripheral nerve sheath tumors. Proc Natl Acad Sci U S A 102(2):407–412. https://doi.org/10.1073/pnas.0406252102

Bray F, Ferlay J, Soerjomataram I, Siegel RL, Torre LA, Jemal A (2018) Global cancer statistics 2018: GLOBOCAN estimates of incidence and mortality worldwide for 36 cancers in 185 countries. CA Cancer J Clin 68(6):394–424. https://doi.org/10.3322/caac.21492

Cekanova M, Rathore K (2014) Animal models and therapeutic molecular targets of cancer: utility and limitations. Drug Des Devel Ther 8:1911–1921. https://doi.org/10.2147/DDDT.S49584

Chiavacci E, Rizzo M, Pitto L, Patella F, Evangelista M, Mariani L, Rainaldi G (2015) The zebrafish/tumor xenograft angiogenesis assay as a tool for screening anti-angiogenic miRNAs. Cytotechnology 67(6):969–975. https://doi.org/10.1007/s10616-014-9735-y

DE Fatima FBDC, DE Castro SAC, Muniz J, RC DA, Lamarao LM, DE Fatima AMNC, DE Assumpcao PP, Burbano RR (2018) Deregulation of the SRC family tyrosine kinases in gastric carcinogenesis in non-human primates. Anticancer Res 38(11):6317–6320. https://doi.org/10.21873/anticanres.12988

den Hertog J (2016) Tumor suppressors in zebrafish: from TP53 to PTEN and beyond. Adv Exp Med Biol 916:87–101. https://doi.org/10.1007/978-3-319-30654-4_4

Drabsch Y, He S, Zhang L, Snaar-Jagalska BE, Ten DP (2013) Transforming growth factor-beta signalling controls human breast cancer metastasis in a zebrafish xenograft model. Breast Cancer Res 15(6):R106. https://doi.org/10.1186/bcr3573

Evason KJ, Francisco MT, Juric V, Balakrishnan S, Lopez PMP, Gordan JD, Kakar S, Spitsbergen J, Goga A, Stainier DY (2015) Identification of chemical inhibitors of beta-catenin-driven liver tumorigenesis in zebrafish. PLoS Genet 11(7):e1005305. https://doi.org/10.1371/journal.pgen.1005305

Fior R, Povoa V, Mendes RV, Carvalho T, Gomes A, Figueiredo N, Ferreira MG (2017) Single-cell functional and chemosensitive profiling of combinatorial colorectal therapy in zebrafish xenografts. Proc Natl Acad Sci U S A 114(39):E8234–E8243. https://doi.org/10.1073/pnas.1618389114

Franzetti GA, Laud-Duval K, van der Ent W, Brisac A, Irondelle M, Aubert S, Dirksen U, Bouvier C, de Pinieux G, Snaar-Jagalska E, Chavrier P, Delattre O (2017) Cell-to-cell heterogeneity of EWSR1-FLI1 activity determines proliferation/migration choices in Ewing sarcoma cells. Oncogene 36(25):3505–3514. https://doi.org/10.1038/onc.2016.498

Gardner HL, Fenger JM, London CA (2016) Dogs as a model for cancer. Annu Rev Anim Biosci 4:199–222. https://doi.org/10.1146/annurev-animal-022114-110911

Gaudenzi G, Albertelli M, Dicitore A, Wurth R, Gatto F, Barbieri F, Cotelli F, Florio T, Ferone D, Persani L, Vitale G (2017) Patient-derived xenograft in zebrafish embryos: a new platform for translational research in neuroendocrine tumors. Endocrine 57(2):214–219. https://doi.org/10.1007/s12020-016-1048-9

Guo M, Wei H, Hu J, Sun S, Long J, Wang X (2015) U0126 inhibits pancreatic cancer progression via the KRAS signaling pathway in a zebrafish xenotransplantation model. Oncol Rep 34(2):699–706. https://doi.org/10.3892/or.2015.4019

Hanahan D, Weinberg RA (2011) Hallmarks of cancer: the next generation. Cell 144(5):646–674. https://doi.org/10.1016/j.cell.2011.02.013

He S, Mansour MR, Zimmerman MW, Ki DH, Layden HM, Akahane K, Gjini E, de Groh ED, Perez-Atayde AR, Zhu S, Epstein JA, Look AT (2016) Synergy

between loss of NF1 and overexpression of MYCN in neuroblastoma is mediated by the GAP-related domain. Elife 5:5. https://doi.org/10.7554/eLife.14713
He S, Jing CB, Look AT (2017) Zebrafish models of leukemia. Methods Cell Biol 138:563–592. https://doi.org/10.1016/bs.mcb.2016.11.013
Hou Y, Chu M, Du FF, Lei JY, Chen Y, Zhu RY, Gong XH, Ma X, Jin J (2013) Recombinant disintegrin domain of ADAM15 inhibits the proliferation and migration of Bel-7402 cells. Biochem Biophys Res Commun 435(4):640–645. https://doi.org/10.1016/j.bbrc.2013.05.037
Ikonomopoulou MP, Fernandez-Rojo MA, Pineda SS, Cabezas-Sainz P, Winnen B, Morales R, Brust A, Sanchez L, Alewood PF, Ramm GA, Miles JJ, King GF (2018) Gomesin inhibits melanoma growth by manipulating key signaling cascades that control cell death and proliferation. Sci Rep 8(1):11519. https://doi.org/10.1038/s41598-018-29826-4
Jo DH, Son D, Na Y, Jang M, Choi JH, Kim JH, Yu YS, Seok SH, Kim JH (2013) Orthotopic transplantation of retinoblastoma cells into vitreous cavity of zebrafish for screening of anticancer drugs. Mol Cancer 12:71. https://doi.org/10.1186/1476-4598-12-71
Kim IS, Heilmann S, Kansler ER, Zhang Y, Zimmer M, Ratnakumar K, Bowman RL, Simon-Vermot T, Fennell M, Garippa R, Lu L, Lee W, Hollmann T, Xavier JB, White RM (2017) Microenvironment-derived factors driving metastatic plasticity in melanoma. Nat Commun 8:14343. https://doi.org/10.1038/ncomms14343
Kirchberger S, Sturtzel C, Pascoal S, Distel M (2017) Quo natas, Danio?-Recent progress in modeling cancer in zebrafish. Front Oncol 7:186. https://doi.org/10.3389/fonc.2017.00186
Klimstra DS, Modlin IR, Coppola D, Lloyd RV, Suster S (2010) The pathologic classification of neuroendocrine tumors: a review of nomenclature, grading, and staging systems. Pancreas 39(6):707–712. https://doi.org/10.1097/MPA.0b013e3181ec124e
Knudson AG (2001) Two genetic hits (more or less) to cancer. Nat Rev Cancer 1(2):157–162. https://doi.org/10.1038/35101031
Korzh S, Pan X, Garcia-Lecea M, Winata CL, Pan X, Wohland T, Korzh V, Gong Z (2008) Requirement of vasculogenesis and blood circulation in late stages of liver growth in zebrafish. BMC Dev Biol 8:84. https://doi.org/10.1186/1471-213X-8-84
Kwan W, North TE (2017) Netting novel regulators of hematopoiesis and hematologic malignancies in zebrafish. Curr Top Dev Biol 124:125–160. https://doi.org/10.1016/bs.ctdb.2016.11.005
Latifi A, Abubaker K, Castrechini N, Ward AC, Liongue C, Dobill F, Kumar J, Thompson EW, Quinn MA, Findlay JK, Ahmed N (2011) Cisplatin treatment of primary and metastatic epithelial ovarian carcinomas generates residual cells with mesenchymal stem cell-like profile. J Cell Biochem 112(10):2850–2864. https://doi.org/10.1002/jcb.23199
Lee LM, Seftor EA, Bonde G, Cornell RA, Hendrix MJ (2005) The fate of human malignant melanoma cells transplanted into zebrafish embryos: assessment of migration and cell division in the absence of tumor formation. Dev Dyn 233(4):1560–1570. https://doi.org/10.1002/dvdy.20471
Leung A, Veinotte CJ, Melong N, Oh MH, Chen K, Enfield K, Backstrom I, Warburton C, Yapp D, Berman JN, Bally MB, Lockwood WW (2017) In vivo validation of PAPSS1 (3′-phosphoadenosine 5′-phosphosulfate synthase 1) as a cisplatin-sensitizing therapeutic target. Clin Cancer Res 23(21):6555–6566. https://doi.org/10.1158/1078-0432.CCR-17-0700
Li Z, Huang X, Zhan H, Zeng Z, Li C, Spitsbergen JM, Meierjohann S, Schartl M, Gong Z (2012) Inducible and repressable oncogene-addicted hepatocellular carcinoma in Tet-on xmrk transgenic zebrafish. J Hepatol 56(2):419–425. https://doi.org/10.1016/j.jhep.2011.07.025
Li Z, Zheng W, Wang Z, Zeng Z, Zhan H, Li C, Zhou L, Yan C, Spitsbergen JM, Gong Z (2013) A transgenic zebrafish liver tumor model with inducible Myc expression reveals conserved Myc signatures with mammalian liver tumors. Dis Model Mech 6(2):414–423. https://doi.org/10.1242/dmm.010462
Liu S, Leach SD (2011) Screening pancreatic oncogenes in zebrafish using the Gal4/UAS system. Methods Cell Biol 105:367–381. https://doi.org/10.1016/B978-0-12-381320-6.00015-1
Lord CJ, Ashworth A (2012) The DNA damage response and cancer therapy. Nature 481(7381):287–294. https://doi.org/10.1038/nature10760
Lu JW, Yang WY, Tsai SM, Lin YM, Chang PH, Chen JR, Wang HD, Wu JL, Jin SL, Yuh CH (2013) Liver-specific expressions of HBx and src in the p53 mutant trigger hepatocarcinogenesis in zebrafish. PLoS One 8(10):e76951. https://doi.org/10.1371/journal.pone.0076951
Lu JW, Liao CY, Yang WY, Lin YM, Jin SL, Wang HD, Yuh CH (2014) Overexpression of endothelin 1 triggers hepatocarcinogenesis in zebrafish and promotes cell proliferation and migration through the AKT pathway. PLoS One 9(1):e85318. https://doi.org/10.1371/journal.pone.0085318
MacRae CA, Peterson RT (2015) Zebrafish as tools for drug discovery. Nat Rev Drug Discov 14(10):721–731. https://doi.org/10.1038/nrd4627
Maddalo D, Manchado E, Concepcion CP, Bonetti C, Vidigal JA, Han YC, Ogrodowski P, Crippa A, Rekhtman N, de Stanchina E, Lowe SW, Ventura A (2014) In vivo engineering of oncogenic chromosomal rearrangements with the CRISPR/Cas9 system. Nature 516(7531):423–427. https://doi.org/10.1038/nature13902
Maresch R, Mueller S, Veltkamp C, Ollinger R, Friedrich M, Heid I, Steiger K, Weber J, Engleitner T, Barenboim M, Klein S, Louzada S, Banerjee R, Strong A, Stauber T, Gross N, Geumann U, Lange S, Ringelhan M, Varela I, Unger K, Yang F, Schmid RM, Vassiliou GS, Braren R, Schneider G, Heikenwalder M, Bradley A,

Saur D, Rad R (2016) Multiplexed pancreatic genome engineering and cancer induction by transfection-based CRISPR/Cas9 delivery in mice. Nat Commun 7:10770. https://doi.org/10.1038/ncomms10770

Mercatali L, La Manna F, Groenewoud A, Casadei R, Recine F, Miserocchi G, Pieri F, Liverani C, Bongiovanni A, Spadazzi C, de Vita A, van der Pluijm G, Giorgini A, Biagini R, Amadori D, Ibrahim T, Snaar-Jagalska E (2016) Development of a patient-derived xenograft (PDX) of breast cancer bone metastasis in a zebrafish model. Int J Mol Sci 17(8). https://doi.org/10.3390/ijms17081375

Mir R, Pradhan SJ, Patil P, Mulherkar R, Galande S (2016) Wnt/beta-catenin signaling regulated SATB1 promotes colorectal cancer tumorigenesis and progression. Oncogene 35(13):1679–1691. https://doi.org/10.1038/onc.2015.232

Moore JC, Tang Q, Yordan NT, Moore FE, Garcia EG, Lobbardi R, Ramakrishnan A, Marvin DL, Anselmo A, Sadreyev RI, Langenau DM (2016) Single-cell imaging of normal and malignant cell engraftment into optically clear prkdc-null SCID zebrafish. J Exp Med 213(12):2575–2589. https://doi.org/10.1084/jem.20160378

Moshal KS, Ferri-Lagneau KF, Haider J, Pardhanani P, Leung T (2011) Discriminating different cancer cells using a zebrafish in vivo assay. Cancers (Basel) 3(4):4102–4113. https://doi.org/10.3390/cancers3044102

Mueller S, Engleitner T, Maresch R, Zukowska M, Lange S, Kaltenbacher T, Konukiewitz B, Ollinger R, Zwiebel M, Strong A, Yen HY, Banerjee R, Louzada S, Fu B, Seidler B, Gotzfried J, Schuck K, Hassan Z, Arbeiter A, Schonhuber N, Klein S, Veltkamp C, Friedrich M, Rad L, Barenboim M, Ziegenhain C, Hess J, Dovey OM, Eser S, Parekh S, Constantino-Casas F, de la Rosa J, Sierra MI, Fraga M, Mayerle J, Kloppel G, Cadinanos J, Liu P, Vassiliou G, Weichert W, Steiger K, Enard W, Schmid RM, Yang F, Unger K, Schneider G, Varela I, Bradley A, Saur D, Rad R (2018) Evolutionary routes and KRAS dosage define pancreatic cancer phenotypes. Nature 554(7690):62–68. https://doi.org/10.1038/nature25459

Nguyen AT, Emelyanov A, Koh CH, Spitsbergen JM, Lam SH, Mathavan S, Parinov S, Gong Z (2011) A high level of liver-specific expression of oncogenic Kras(V12) drives robust liver tumorigenesis in transgenic zebrafish. Dis Model Mech 4(6):801–813. https://doi.org/10.1242/dmm.007831

Nguyen AT, Emelyanov A, Koh CH, Spitsbergen JM, Parinov S, Gong Z (2012) An inducible kras(V12) transgenic zebrafish model for liver tumorigenesis and chemical drug screening. Dis Model Mech 5(1):63–72. https://doi.org/10.1242/dmm.008367

Nicoli S, Presta M (2007) The zebrafish/tumor xenograft angiogenesis assay. Nat Protoc 2(11):2918–2923. https://doi.org/10.1038/nprot.2007.412

Parada-Kusz M, Penaranda C, Hagedorn EJ, Clatworthy A, Nair AV, Henninger JE, Ernst C, Li B, Riquelme R, Jijon H, Villablanca EJ, Zon LI, Hung D, Allende ML (2018) Generation of mouse-zebrafish hematopoietic tissue chimeric embryos for hematopoiesis and host-pathogen interaction studies. Dis Model Mech 11(11):dmm034876. https://doi.org/10.1242/dmm.034876

Park SW, Davison JM, Rhee J, Hruban RH, Maitra A, Leach SD (2008) Oncogenic KRAS induces progenitor cell expansion and malignant transformation in zebrafish exocrine pancreas. Gastroenterology 134(7):2080–2090. https://doi.org/10.1053/j.gastro.2008.02.084

Pettitt D, Arshad Z, Smith J, Stanic T, Hollander G, Brindley D (2018) CAR-T cells: a systematic review and mixed methods analysis of the clinical trial landscape. Mol Ther 26(2):342–353. https://doi.org/10.1016/j.ymthe.2017.10.019

Platt RJ, Chen S, Zhou Y, Yim MJ, Swiech L, Kempton HR, Dahlman JE, Parnas O, Eisenhaure TM, Jovanovic M, Graham DB, Jhunjhunwala S, Heidenreich M, Xavier RJ, Langer R, Anderson DG, Hacohen N, Regev A, Feng G, Sharp PA, Zhang F (2014) CRISPR-Cas9 knockin mice for genome editing and cancer modeling. Cell 159(2):440–455. https://doi.org/10.1016/j.cell.2014.09.014

Rai K, Sarkar S, Broadbent TJ, Voas M, Grossmann KF, Nadauld LD, Dehghanizadeh S, Hagos FT, Li Y, Toth RK, Chidester S, Bahr TM, Johnson WE, Sklow B, Burt R, Cairns BR, Jones DA (2010) DNA demethylase activity maintains intestinal cells in an undifferentiated state following loss of APC. Cell 142(6):930–942. https://doi.org/10.1016/j.cell.2010.08.030

Romero R, Sayin VI, Davidson SM, Bauer MR, Singh SX, LeBoeuf SE, Karakousi TR, Ellis DC, Bhutkar A, Sanchez-Rivera FJ, Subbaraj L, Martinez B, Bronson RT, Prigge JR, Schmidt EE, Thomas CJ, Goparaju C, Davies A, Dolgalev I, Heguy A, Allaj V, Poirier JT, Moreira AL, Rudin CM, Pass HI, Vander HM, Jacks T, Papagiannakopoulos T (2017) Keap1 loss promotes Kras-driven lung cancer and results in dependence on glutaminolysis. Nat Med 23(11):1362–1368. https://doi.org/10.1038/nm.4407

Rycaj K, Tang DG (2015) Cell-of-origin of cancer versus cancer stem cells: assays and interpretations. Cancer Res 75(19):4003–4011. https://doi.org/10.1158/0008-5472.CAN-15-0798

Sanchez-Rivera FJ, Papagiannakopoulos T, Romero R, Tammela T, Bauer MR, Bhutkar A, Joshi NS, Subbaraj L, Bronson RT, Xue W, Jacks T (2014) Rapid modelling of cooperating genetic events in cancer through somatic genome editing. Nature 516(7531):428–431. https://doi.org/10.1038/nature13906

Shive HR, West RR, Embree LJ, Azuma M, Sood R, Liu P, Hickstein DD (2010) brca2 in zebrafish ovarian development, spermatogenesis, and tumorigenesis. Proc Natl Acad Sci U S A 107(45):19350–19355. https://doi.org/10.1073/pnas.1011630107

Shive HR, West RR, Embree LJ, Golden CD, Hickstein DD (2014) BRCA2 and TP53 collaborate in tumorigenesis in zebrafish. PLoS One 9(1):e87177. https://doi.org/10.1371/journal.pone.0087177

Smith MP, Ferguson J, Arozarena I, Hayward R, Marais R, Chapman A, Hurlstone A, Wellbrock C (2013) Effect of SMURF2 targeting on susceptibility to MEK inhibitors in melanoma. J Natl Cancer Inst 105(1):33–46. https://doi.org/10.1093/jnci/djs471

Solin SL, Shive HR, Woolard KD, Essner JJ, McGrail M (2015) Rapid tumor induction in zebrafish by TALEN-mediated somatic inactivation of the retinoblastoma1 tumor suppressor rb1. Sci Rep 5:13745. https://doi.org/10.1038/srep13745

Sonoshita M, Cagan RL (2017) Modeling human cancers in drosophila. Curr Top Dev Biol 121:287–309. https://doi.org/10.1016/bs.ctdb.2016.07.008

Tan DS, Haaland B, Gan JM, Tham SC, Sinha I, Tan EH, Lim KH, Takano A, Krisna SS, Thu MM, Liew HP, Ullrich A, Lim WT, Chua BT (2014) Bosutinib inhibits migration and invasion via ACK1 in KRAS mutant non-small cell lung cancer. Mol Cancer 13:13. https://doi.org/10.1186/1476-4598-13-13

Tang Q, Abdelfattah NS, Blackburn JS, Moore JC, Martinez SA, Moore FE, Lobbardi R, Tenente IM, Ignatius MS, Berman JN, Liwski RS, Houvras Y, Langenau DM (2014) Optimized cell transplantation using adult rag2 mutant zebrafish. Nat Methods 11(8):821–824. https://doi.org/10.1038/nmeth.3031

Tang Q, Moore JC, Ignatius MS, Tenente IM, Hayes MN, Garcia EG, Torres YN, Bourque C, He S, Blackburn JS, Look AT, Houvras Y, Langenau DM (2016) Imaging tumour cell heterogeneity following cell transplantation into optically clear immune-deficient zebrafish. Nat Commun 7:10358. https://doi.org/10.1038/ncomms10358

Tubbs A, Nussenzweig A (2017) Endogenous DNA damage as a source of genomic instability in Cancer. Cell 168(4):644–656. https://doi.org/10.1016/j.cell.2017.01.002

van der Ent W, Jochemsen AG, Teunisse AF, Krens SF, Szuhai K, Spaink HP, Hogendoorn PC, Snaar-Jagalska BE (2014) Ewing sarcoma inhibition by disruption of EWSR1-FLI1 transcriptional activity and reactivation of p53. J Pathol 233(4):415–424. https://doi.org/10.1002/path.4378

van der Ent W, Burrello C, de Lange MJ, van der Velden PA, Jochemsen AG, Jager MJ, Snaar-Jagalska BE (2015) Embryonic zebrafish: different phenotypes after injection of human uveal melanoma cells. Ocul Oncol Pathol 1(3):170–181. https://doi.org/10.1159/000370159

Van Slyke CE, Bradford YM, Howe DG, Fashena DS, Ramachandran S, Ruzicka L (2018) Using ZFIN: data types, organization, and retrieval. Methods Mol Biol 1757:307–347. https://doi.org/10.1007/978-1-4939-7737-6_11

Veinotte CJ, Dellaire G, Berman JN (2014) Hooking the big one: the potential of zebrafish xenotransplantation to reform cancer drug screening in the genomic era. Dis Model Mech 7(7):745–754. https://doi.org/10.1242/dmm.015784

Visvader JE (2011) Cells of origin in cancer. Nature 469(7330):314–322. https://doi.org/10.1038/nature09781

Wagner DS, Delk NA, Lukianova-Hleb EY, Hafner JH, Farach-Carson MC, Lapotko DO (2010) The in vivo performance of plasmonic nanobubbles as cell theranostic agents in zebrafish hosting prostate cancer xenografts. Biomaterials 31(29):7567–7574. https://doi.org/10.1016/j.biomaterials.2010.06.031

Wang X, Kopinke D, Lin J, McPherson AD, Duncan RN, Otsuna H, Moro E, Hoshijima K, Grunwald DJ, Argenton F, Chien CB, Murtaugh LC, Dorsky RI (2012) Wnt signaling regulates postembryonic hypothalamic progenitor differentiation. Dev Cell 23(3):624–636. https://doi.org/10.1016/j.devcel.2012.07.012

Wang J, Leng X, Wang G, Wan X, Cao H (2017) The construction of intrahepatic cholangiocarcinoma model in zebrafish. Sci Rep 7(1):13419. https://doi.org/10.1038/s41598-017-13815-0

Wang J, Fei F, Berberoglu MA, Sun S, Wang L, Dong Z, Wang X (2018) Csy4-based vector system enables conditional chimeric gene editing in zebrafish without interrupting embryogenesis. J Mol Cell Biol 10(6):586–588. https://doi.org/10.1093/jmcb/mjy017

Wang L, Chen H, Fei F, He X, Sun S, Lv K, Yu B, Long J, Wang X (2019) Patient-derived heterogeneous xenograft model of pancreatic cancer using zebrafish larvae as hosts for comparative drug assessment. J Vis Exp 146. https://doi.org/10.3791/59507

Waster P, Orfanidis K, Eriksson I, Rosdahl I, Seifert O, Ollinger K (2017) UV radiation promotes melanoma dissemination mediated by the sequential reaction axis of cathepsins-TGF-beta1-FAP-alpha. Br J Cancer 117(4):535–544. https://doi.org/10.1038/bjc.2017.182

White R, Rose K, Zon L (2013) Zebrafish cancer: the state of the art and the path forward. Nat Rev Cancer 13(9):624–636. https://doi.org/10.1038/nrc3589

Wu JQ, Zhai J, Li CY, Tan AM, Wei P, Shen LZ, He MF (2017) Patient-derived xenograft in zebrafish embryos: a new platform for translational research in gastric cancer. J Exp Clin Cancer Res 36(1):160. https://doi.org/10.1186/s13046-017-0631-0

Wu Q, Zheng K, Huang X, Li L, Mei W (2018) Tanshinone-IIA-based analogues of imidazole alkaloid act as potent inhibitors to block breast cancer invasion and metastasis in vivo. J Med Chem 61:10488–10501. https://doi.org/10.1021/acs.jmedchem.8b01018

Xu W, Foster BA, Richards M, Bondioli KR, Shah G, Green CC (2018) Characterization of prostate cancer cell progression in zebrafish xenograft model. Int J Oncol 52(1):252–260. https://doi.org/10.3892/ijo.2017.4189

Yan C, Yang Q, Shen HM, Spitsbergen JM, Gong Z (2017a) Chronically high level of tgfb1a induction causes both hepatocellular carcinoma and cholangiocarcinoma via a dominant Erk pathway in zebrafish. Oncotarget 8(44):77096–77109. https://doi.org/10.18632/oncotarget.20357

Yan C, Yang Q, Gong Z (2017b) Tumor-associated neutrophils and macrophages promote gender disparity in hepatocellular carcinoma in zebrafish. Cancer Res 77(6):1395–1407. https://doi.org/10.1158/0008-5472.CAN-16-2200

Yang HW, Kutok JL, Lee NH, Piao HY, Fletcher CD, Kanki JP, Look AT (2004) Targeted expression of human MYCN selectively causes pancreatic neuroendocrine tumors in transgenic zebrafish. Cancer Res 64(20):7256–7262. https://doi.org/10.1158/0008-5472.CAN-04-0931

Yang XJ, Cui W, Gu A, Xu C, Yu SC, Li TT, Cui YH, Zhang X, Bian XW (2013) A novel zebrafish xenotransplantation model for study of glioma stem cell invasion. PLoS One 8(4):e61801. https://doi.org/10.1371/journal.pone.0061801

Yang J, Pei H, Luo H, Fu A, Yang H, Hu J, Zhao C, Chai L, Chen X, Shao X, Wang C, Wu W, Wan L, Ye H, Qiu Q, Peng A, Wei Y, Yang L, Chen L (2017) Non-toxic dose of liposomal honokiol suppresses metastasis of hepatocellular carcinoma through destabilizing EGFR and inhibiting the downstream pathways. Oncotarget 8(1):915–932. https://doi.org/10.18632/oncotarget.13687

Yang Q, Yan C, Wang X, Gong Z (2019) Leptin induces muscle wasting in a zebrafish kras-driven hepatocellular carcinoma (HCC) model. Dis Model Mech 12(2):dmm038240. https://doi.org/10.1242/dmm.038240

Yao Y, Sun S, Fei F, Wang J, Wang Y, Zhang R, Wu J, Liu L, Liu X, Cui Z, Li Q, Yu M, Dang Y, Wang X (2017) Screening in larval zebrafish reveals tissue-specific distribution of fifteen fluorescent compounds. Dis Model Mech 10(9):1155–1164. https://doi.org/10.1242/dmm.028811

Yao Y, Sun S, Wang J, Fei F, Dong Z, Ke AW, He R, Wang L, Zhang L, Ji MB, Li Q, Yu M, Shi GM, Fan J, Gong Z, Wang X (2018) Canonical Wnt signaling remodels lipid metabolism in zebrafish hepatocytes following Ras oncogenic insult. Cancer Res 78(19):5548–5560. https://doi.org/10.1158/0008-5472.CAN-17-3964

Yee NS, Ignatenko N, Finnberg N, Lee N, Stairs D (2015) Animal models of cancer biology. Cancer Growth Metastasis 8(Suppl 1):115–118. https://doi.org/10.4137/CGM.S37907

Zhan T, Rindtorff N, Boutros M (2017) Wnt signaling in cancer. Oncogene 36(11):1461–1473. https://doi.org/10.1038/onc.2016.304

Zuckermann M, Hovestadt V, Knobbe-Thomsen CB, Zapatka M, Northcott PA, Schramm K, Belic J, Jones DT, Tschida B, Moriarity B, Largaespada D, Roussel MF, Korshunov A, Reifenberger G, Pfister SM, Lichter P, Kawauchi D, Gronych J (2015) Somatic CRISPR/Cas9-mediated tumour suppressor disruption enables versatile brain tumour modelling. Nat Commun 6:7391. https://doi.org/10.1038/ncomms8391

Part VI

In Vitro and Ex Vivo Cancer Models

In Vitro and Ex Vivo Models – The Tumor Microenvironment in a Flask

23

Catarina Pinto, Marta F. Estrada, and Catarina Brito

Abstract

Experimental tumor modeling has long supported the discovery of fundamental mechanisms of tumorigenesis and tumor progression, as well as provided platforms for the development of novel therapies. Still, the attrition rates observed today in clinical translation could be, in part, mitigated by more accurate recapitulation of environmental cues in research and preclinical models. The increasing understanding of the decisive role that tumor microenvironmental cues play in the outcome of drug response urges its integration in preclinical tumor models. In this chapter we review recent developments concerning *in vitro* and *ex vivo* approaches.

Keywords

Cancer models · Tumor explants · 3D cell cultures · Experimental tumor modeling · Tumor microenvironment

Authors Catarina Pinto and Marta F. Estrada have equally contributed to this chapter.

C. Pinto · M. F. Estrada · C. Brito (✉)
iBET, Instituto de Biologia Experimental e Tecnológica, Oeiras, Portugal

Instituto de Tecnologia Química e Biológica António Xavier, Universidade Nova de Lisboa, Oeiras, Portugal
e-mail: anabrito@ibet.pt

23.1 Tumor Microenvironment

Cancer progression has been compared to the development of multicellular organisms where mechanisms controlling cell division, cell-fate determination and tissue organization are deregulated (Huch and Rawlins 2017). Understanding the physiological processes that are being co-opted by tumors to thrive in harsh physicochemical conditions and avoid cell death, to acquire unrestricted growth capacity and the propensity to invade adjacent tissues and disseminate throughout the host, is crucial for rational design of therapies. These properties emerge in different tumor types through distinct mechanisms that result from tumor cell intrinsic and extrinsic properties (Hanahan and Weinberg 2011).

Over the years, the dissection of the oncogenic pathways mediating malignancy has revealed tissue or cancer type-specific genetic vulnerabilities that constitute prime candidates for targeted therapies. The exploitation of such targets has drastically improved cancer treatment. Yet, these remain in many cases poorly efficient due to the acquisition of drug resistance mechanisms (Hanahan and Coussens 2012), which are the main obstacle in cancer treatment (Sun 2015). The identification of the underlying mechanisms is crucial to overcome current shortcomings and improve clinical outcomes (Sun 2015).

Cancer cell intrinsic mechanisms supporting acquired resistance include upregulation of drug

J. Serpa (ed.), *Tumor Microenvironment*, Advances in Experimental Medicine and Biology 1219,
https://doi.org/10.1007/978-3-030-34025-4_23

efflux pumps and increased drug metabolism, as well as compensatory loss of specific oncogenes and emergence of apoptotic defects (Sun 2015). Also, intra and inter-tumor heterogeneity across cancer types and tissue of origin promotes tumor evolution and therapeutic resistance (McGranahan and Swanton 2017). Nevertheless, tumor heterogeneity is also a result of the cells present in the surrounding microenvironment, both in terms of composition and activation states (Quail and Joyce 2013). Extrinsic determinants such as cytokines, growth factors and extracellular matrix (ECM) secreted by the cells in the tumor microenvironment (TME), can enhance or dampen the effects of genetic and epigenetic alterations in the epithelial compartment (Costa et al. 2018), which directly impact tumor evolution and disease recurrence (McGranahan and Swanton 2017).

The TME is a complex network of different cell types and soluble factors, embedded in an ECM which provides physical support, allows cell migration and modulates cell signaling (Sun 2015). Under normal physiological conditions, the microenvironment maintains tissue architecture and restricts cell growth, thus inhibiting tumor initiation and progression (Bussard et al. 2010). The concept of "seed and soil", first proposed by Stephen Paget in 1889, states that tumor cells may only lead to tumor outgrowth when a supportive microenvironment emerges (Quail and Joyce 2013). Thus, while tumor initiation appears mostly inevitable, its progression into a malignant state could potentially be managed with full knowledge of the intervening partners (Bissell and Hines 2011). The cellular players include endothelial and perivascular cells, adipocytes and fibroblasts, and engaged immune cells including macrophages, dendritic cells, NK cells, myeloid derived suppressor cells (MSDC) and T and B cells. These acquire phenotypic and functional characteristics that are distinct from the tissue resident counterparts in the healthy tissue, and are reminiscent of a tissue recovering from wound healing (Ronca et al. 2018). Understanding the phenotypic and functional diversity of stromal and immune cells within the TME, as well as the different axis mediating this crosstalk, is essential for cancer prevention, detection and treatment (Bissell and Hines 2011). Deciphering the intricate networks established between the different players will provide knowledge into the biological mechanisms behind acquired anticancer therapy resistance, along with a rational for combinatorial therapies (Sun 2015). These have also been instrumental in uncovering important new targets for therapeutic intervention, namely immunotherapies (Sun 2015).

23.2 Experimental Tumor Modeling

Experimental tumor modeling has long supported the discovery of fundamental mechanisms of tumorigenesis and tumor progression and provided platforms for the development of novel therapies. Still, the attrition rates observed today in clinical translation could be, in part, mitigated by more accurate recapitulation of environmental cues in research and preclinical models (Gu and Mooney 2015). Since no model can fully mimic the real system, these must be chosen while considering the balance between their limitations and the necessary complexity to support the objectives of the study (Thomas et al. 2016).

The models used in cancer research can be divided in three categories: *in vitro*, *ex vivo* and *in vivo* model systems.

23.2.1 *In Vitro* Models

In vitro model systems are developed using cell lines or dissociated primary cells that can be cultured either in two-dimensional (2D) or in three-dimensional (3D) culture systems (Fig. 23.1).

23.2.1.1 2D Tumor Cell Models

Two-dimensional (2D) cell culture systems provide the necessary throughput for fast screening of multiple compounds and for the identification of cell-cell interactions (Nyga et al. 2016; McMillin et al. 2010; Straussman et al. 2012). These cultures are easy to implement and provide cheap and robust models, amenable for high throughput screening (HTS) (Weigelt et al.

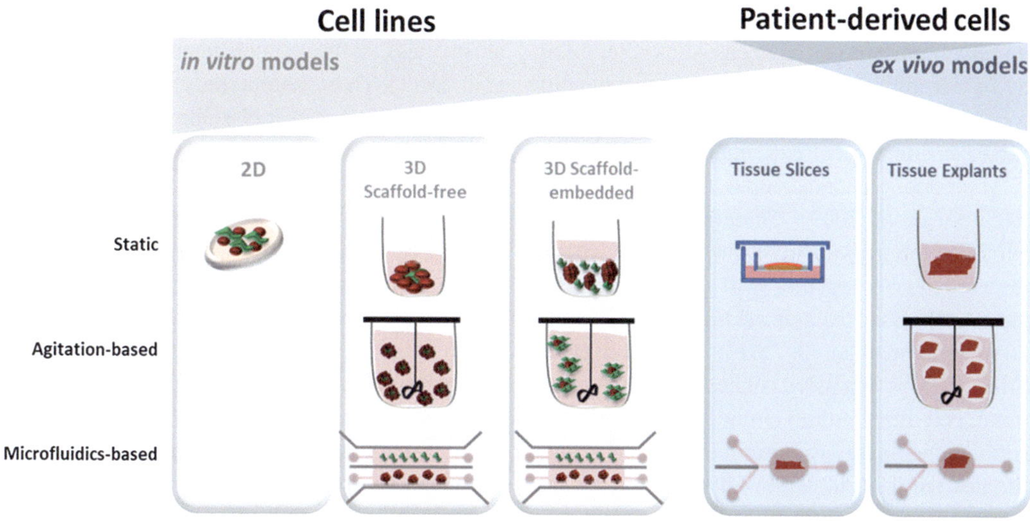

Fig. 23.1 *In vitro* and *ex vivo* models of the Tumor Microenvironment

2014). Many ground-breaking discoveries have been made on 2D cultures, such as the responsiveness of ER+ breast cancer cells to anti-estrogen therapies (Unger et al. 2014). Nonetheless, in many cases, the translation of the obtained results to patients has proven challenging. Despite showing considerable efficacy in 2D culture assays, these models poorly predict drug response and lead to an overwhelming number of possible targets that, are proven ineffective in clinical trials, prolonging patients' survival by a few months or a couple of years (Bissell and Hines 2011; Gu and Mooney 2015). Additionally, although cell lines cultured in 2D recapitulate many molecular pathways and genetic events described *in vivo*, cell growth rates and cell-cell and cell-ECM interactions differ from those observed *in vivo* (Mürdter et al. 2006). It is now evident that, in 2D systems, cell-plastic connections prevail over cell-cell and cell-ECM interactions (Pampaloni et al. 2007). Thus, the need for more predictive drug discovery assays has prompted advances in cell culture techniques that permitted a fast evolution of complex *in vitro* cell models attempting to recapitulate tumor architecture (Lovitt et al. 2016). Pioneering work by Mina Bissel evidenced that complex systems can be exploited to uncover molecular mechanisms of tumorigenesis and invasion. Blockade of integrin-beta 1 (ITGB1) reverted the malignant phenotype of breast cancer (BC) cells, in a three-dimensional (3D) culture setting, forming reverted acini and re-establishing E-cadherin (CDH1)/catenin beta 1 (CTNNB1) complexes (Weaver et al. 1997). This phenomena had not been observed in 2D cultures, and showed that epithelial tumor cells can change polarity in a microenvironment-dependent manner (Inman and Bissell 2010). In contrast, 3D cultures have the potential to recapitulate the cell-cell and cell-ECM interactions and the diffusion gradients (oxygen, nutrients, metabolites, soluble factors and drugs) described in human tumors, and are also amenable for HTS (Weigelt et al. 2014; Nath and Devi 2016).

23.2.1.2 3D Tumor Cell Models

Most used 3D tumor models rely on tumor spheroids (Benien and Swami 2014), which have been employed in drug screening and to study tumor cell function, angiogenesis and tumor-immune interactions (Katt et al. 2016). These constitute high-throughput tools to select drug candidates and decrease animal experimentation, as they present lower cell proliferation rates and higher resistance to treatment than 2D cultures; thus better resembling the drug response observed in solid tumors (Friedrich et al. 2009). Tumor spheroids are formed by self-aggregation of cells in culture and present characteristics resembling the *in vivo* tumors, namely three-dimensionality in cell structure and polarity (Inman and Bissell 2010), tensile forces (Levental et al. 2009) and ECM production and accumulation (Bissell et al. 2005; Xu et al. 2009). Therefore, cell-cell and cell-ECM interactions in 3D are closer to what is observed *in vivo*, together with gradients of nutrients, metabolites, oxygen and also ECM accumulation (Hickman et al. 2014; Weiswald et al. 2015). Finally, extensive studies have shown that tumor cell lines cultured under 3D conditions exhibit gene expression profiles closer to patient samples (Katt et al. 2016; Hirschhaeuser et al. 2010).

Three dimensional cultures can be generated by many different methods, such as magnetic levitation, bioprinting or single-cell aggregation into spheroids, by either static or agitation-based systems (Rijal and Li 2016; Shafiee et al. 2015). The static systems include hanging drop or free-floating aggregation in ultra-low attachment plates. Agitation-based systems comprise orbital shaking, stirred-tank and rotating wall systems (Hickman et al. 2014; Breslin and O'Driscoll 2013). After the aggregation phase, spheroids can be cultured as mono or co-cultures in scaffold-free conditions or embedded in scaffolds (Fig. 23.1; Shafiee et al. 2015). In scaffold-free conditions it becomes highly challenging to recapitulate the spatial distribution of the different cell types, as observed in the tissue. One example is the co-culture of breast cancer cells with fibroblasts in a 3D rotary bioreactor, that resulted in the formation of a core of fibroblasts surrounded by epithelial tumor cells (Kaur et al. 2011). Although this model was used to investigate how tumor cells invaded the inner stromal compartment, it does not resemble the *in vivo* tumor-stromal organization of *in situ* breast carcinomas (Thottassery et al. 2004). Alternatively, scaffolds provide physical support to cells, allow cell migration and aim to reflect aspects of the ECM. In these systems it is possible to recapitulate some aspects of the tumor-stromal architecture and mimic the heterotypic cell-cell crosstalk through autocrine and paracrine signaling mechanisms and direct cell-cell interactions (Dolznig et al. 2011). Thus, heterotypic cultures are developed mostly in scaffold-embedded conditions. Progress in bioengineering of cells and culture systems, mostly stirred-tank and microfluidic based systems (Fig. 23.1) has enabled the development of cellular models that support the heterotypic and spatial interactions between cells, which was only possible in *in vivo* settings (Chen 2016).

Scaffolds can be derived from natural materials or synthetic polymers. The synthetic polymers include polycaprolactone (PCL), poly(lactic-glycolic acid) (PLGA) or poly(ethylene glycol) (PEG). The naturally-derived scaffolds can be non-inert: decellularized ECM, matrigel, collagen, gelatin; or inert: alginate, chitosan or silk fibroin (Rijal and Li 2016). Models based on scaffolds have been employed in the dissection of complex molecular cues. One example was the co-culture of prostate epithelial tumor cells and fibroblasts to address the effect of paracrine signaling on tumor progression. In this model, tumor and stromal cells were subjected to a dual step microencapsulation in alginate hydrogel. Tumor cells were confined in the inner compartment whilst stromal cells were in the outer compartment. Although direct epithelial-stromal interactions were not allowed, both cell types communicated through paracrine signaling, which resulted in dysregulated levels of E-cadherin in the epithelial compartment of co-cultures but not in mono-cultures (Fang et al. 2013). Another example was the triple co-culture of hepatocarcinoma spheroids with fibroblasts and endothelial cells to investigate induction of angiogenesis that conjugated a starPEG-heparin hydrogel with adhesion ligands (fibronectin and laminin), growth factors and MMP-sensitive sequences (Chwalek et al. 2014). This system supported vessel formation observed by the growth and polarization of endothelial cells into blood vessels. Amongst the natural-derived scaffolds the most used is collagen, which is the most abundant protein in the human connective tissue and highly produced in cancerous states (Rijal and Li 2016). Cell behavior is highly influenced by the type of matrix, which poses a main challenge in the choice of the correct scaffold. For instance, when mammary carcinomas were cultured in parallel either in collagen I or in matrigel, cell dissemination occurred in the collagen I gels only. Healthy mammary epithelial fragments showed similar migratory behavior, but cell migration was transient and blocked by integrin binding to newly synthesized laminin-111. In contrast, no cell dissemination was observed, when the same carcinomas were cultured in matrigel (Shamir et al. 2012).

Numerous 3D *in vitro* tumor models have been developed in the past decade, aiming at depicting features of the tumor microenvironment. Most heterotypic culture systems use only two cell types due to the complexity of identifying the interactions between the different cells and the origin of the secreted factors (Weigelt et al. 2014). Typically, these cultures combine tumor cells with cells present in the tumor microenvironment, such as, fibroblasts, adipocytes, macrophages, or endothelial cells (Weigelt et al. 2014). One example is the *in vitro* modulation of colon carcinoma by fibroblasts, when cultured in collagen I gels (Dolznig et al. 2011). In this co-culture system, tumor cells were embedded as spheroids and stromal cells as single cells. Fibroblasts migrated towards the tumor spheroids and, through the concerted action of paracrine signaling and direct cell-cell interactions, modulated tumor progression. Fibroblasts induced tumor cell invasive spreading, both collective and as single cells, and led to the activation of several signaling pathways, such as Ras and NFkB signaling, which are associated with more aggressive stages of the disease.

Most available models were developed by making use of bioactive scaffolds that modulate tumor and stromal cell phenotype, namely through the presence of cytokines, growth factors, and ECM content and cross-linking (Sung and Beebe 2014; Mafi et al. 2012; Velez et al. 2017). Alternatively, inert biocompatible scaffolds provide physical support to cells, allow cell-to-cell communication, accumulation of secreted factors and cell migration (Andersen et al. 2015). Nonetheless, the stiffness of the matrix also influences cell behavior through tensional forces that promote activation of specific signaling pathways (Chaudhuri et al. 2014). One example is the activation of fibroblasts into myofibroblasts upon culture in a PEG hydrogel with higher stiffness (Smithmyer et al. 2014).

At our lab, we have combined alginate microencapsulation with agitation-based culture systems to develop 3D cellular models that could overcome some of these constraints and contribute to uncover key aspects of the heterotypic crosstalk within the tumor microenvironment (Estrada et al. 2016). MCF-7 breast cancer cells were co-cultured as spheroids with human fibroblasts within alginate microcapsules. Fibroblasts were surrounding the tumor spheroids, creating an architecture resembling ductal carcinoma *in situ*, with distinct epithelial and stromal compartments. Another limitation of most models is that these are typically generated in non-scalable culture systems with poor robustness, no control of the physicochemical parameters and allowing only for end-point analysis (Weigelt et al. 2014; Kimlin et al. 2013; Haycock 2011). The stirred culture systems used in this work (Estrada et al. 2016) allowed for long term culture which proved essential for the development of tumor phenotypes associated with disease progression, in the co-culture setting. At the initial stages of culture, small lumina surrounded by polarized cells could be observed in the tumor spheroids, which was previously described for human breast cancer tumors of the luminal subtype. Along culture time, a reduced expression of estrogen receptor and membranous E-cadherin was observed, concomitant with a loss in cell polarity and increased cell migration. The accumulation of secreted pro-inflammatory cytokines and collagen was observed in the stromal compartment, which likely contributed to the remodeling of the cellular compartments. Furthermore, increased collective cell migration and enhanced angiogenic potential measured in co-cultures, using the chicken embryo chorioallantoic membrane (CAM) angiogenesis assay, further suggested phenotypic alterations typical of advanced stages of cancer. The fact that these phenomena were not observed in monoculture of MCF-7 spheroids highlights the importance of the epithelial stromal interaction for tumor development.

In the past years, organoid cultures have been proposed as an alternative method for expansion of primary human adult stem cells in 3D, followed by differentiation in the several cell lineages present in the adult tissue (Dutta et al. 2017). Organoids can be derived from almost all organs, from both healthy and diseased patients. With this method, it became possible to modulate both organogenesis and disease, and to unravel new cellular and molecular mechanisms that have applications both in basic biology and in translational medicine (Dutta et al. 2017). One example was the development of organoids from the four molecular subtypes of BC, with maintenance of expression of the nuclear hormone receptors (ER and PR) and the Human Epidermal Growth Factor Receptor 2 (Her2). Although ER expression was diminished, tumors remained sensitive to anti-estrogen therapies, as Her2+ tumors were also sensitive to Herceptin (anti-Her2) treatments. Analysis of mutational signature, driver gene and copy number alterations, revealed that organoid cultures were comparable to the original tumor (Korving et al. 2017). The authors demonstrated that this model is predictive of the patient response to a given therapy suggesting it as a suitable tool for co-clinical assays. However, most organoid-based cellular models lack stromal and immune cells, and do not maintain the original tissue architecture, which are key players in modulating drug response and resistance mechanisms. Another drawback is the highly variable cell expansion times and the use of undefined animal-derived matrices, which are known to influence drug response profiles (Stock et al. 2016). As an alternative, a method for culturing mature mammary tissues as human breast organoids by using a defined matrix has been proposed (Sokol et al. 2016). Hydrogels were constituted by collagen I, hyaluronan, laminin and fibronectin, components of the breast ECM (Sokol et al. 2016). In another study, a fully defined matrix was designed to support the growth of intestinal organoids (Gjorevski et al. 2016). This matrix was based on PEG hydrogels functionalized with adhesion molecules (fibronectin, laminin, type IV collagen, hyaluronic acid and perlecan) and substrate peptides. The binding to fibronectin motifs was enough to promote survival and proliferation of intestinal stem cells (ISC), which was further enhanced by higher matrix stiffness (1.3 kPa) in a Yesa ssociated protein (YAP)-dependent mechanism. Matrix softening (300 Pa) and enrichment with laminin-based adhesion molecules was needed to induce ISC differentiation and organoid formation.

In vitro models can be generated from cell lines or patient-derived cells. *Ex vivo* models are generated from patient-derived tissue that is partially processed to generate slices or explants. Commonly employed 3D culture strategies include scaffold-free and scaffold embedding approaches, in static, agitation-based or microfluidics-based culture systems.

23.2.1.3 Heterotypic Crosstalk – Introducing the Immune Component

Most heterotypic cell cultures incorporate cancer-associated fibroblasts (CAF), which promote an inflammatory microenvironment and are associated to drug resistance, though other cell types have been used, namely mesenchymal stem cells (Chen et al. 2009). Upon contact with transformed cells, fibroblasts present an activated phenotype, with altered contractile and secretory profiles when compared with fibroblasts from normal tissue (Hirt et al. 2014), and a gene expression profile similar to fibroblasts involved in wound healing (Chang et al. 2005; Chang et al. 2004). Cancer-mediated fibroblast "education" into CAF is achieved through the secretion of growth factors, such as TGFB1, PDGF and fibroblast growth factor 2 (FGF2), whose production is also induced on fibroblasts at later stages (Ronca et al. 2018). CAF accumulation at the TME correlates with increased risk of relapse and reduced anti-tumor immunity in several cancer types (Costa et al. 2018; Finak et al. 2008; Toullec et al. 2010). Besides prognosis, these can also instruct therapeutic response. A gene expression signature characteristic of reactive stroma could predict resistance to neoadjuvant chemotherapy (Farmer et al. 2009), indicating that combination therapies with drugs targeting CAF could contribute to overcome therapeutic resistance.

In addition to fibroblasts, immune cells have been shown to be key players in modulating tumor progression, namely, through the complex interplay between tumor cells, fibroblasts and immune cells. CAF were also shown to induce suppression of antitumor immunity (Kraman et al. 2010). CAFs are the main source of ECM at the tumor site, namely collagen, fibronectin and laminin, and induce desmoplasia in advanced carcinomas (Dumont et al. 2013). It has been proposed that the dense matrix formed, limits lymphocyte access to tumor sites (Salmon et al. 2012). In another study, depletion of CAF correlated with increased differentiation of T helper (Th) 1 cells, reduced recruitment of M2-like macrophages and was also correlated with increased CD8 + T effector cell infiltration and activity (Turley et al. 2015). Stromal expression of C-C motif chemokine ligand (CCL) 2, CCL3, CCL4, CCL5 has been shown to influence macrophage distribution and composition within the TME by recruitment of blood monocytes and immature myeloid cells (Murdoch 2004). CAF also contribute to the maintenance of an immunosuppressive TME through the production of CXCL8, IL4 and IL6, which polarize macrophages towards an M2-like phenotype (Kim et al. 2012a). Adaptive immunity is also dampened by the stromal compartment. Tumor infiltrating CD8+ T cells are often located in the adjacent tumor areas and in direct contact with the fibroblast population. These CAF can directly decrease tumor antigen-specific CD8+ T cells through cross-presentation of antigens in concomitance with immune checkpoint expression (programmed cell death 1 ligand 2 – *PDCD1LG2* – and Fas ligand – *FASLG*) (Lakins et al. 2018). TGFB1 secretion also inhibits CD8 + T and effector memory cells by inhibiting T cell receptor (TCR)-CD28 signaling (Broderick and Bankert 2006; Ahmadzadeh and Rosenberg 2005), and thymic stromal lymphopoietin (TSLP) directs T cell into a Th2 phenotype (De Monte et al. 2011).

Immune cell infiltration at the tumor site is largely dependent on the tissue of origin and on the cancer subtype. These cells have complex interactions to modulate immune surveillance, through different states of activation or polarization that can either be anti- or pro-tumorigenic. In breast cancer, the infiltrated immune cells are mostly macrophages and lymphocytes, but their presence depends largely on the tumor sub-type. Typically, basal-like breast cancers have high macrophage and lymphocyte infiltrate, whilst ER^+ tumors, have a reduced immune infiltrate due to the low mutagenicity of these tumors (Stanton and Disis 2016). In colorectal cancer, different macrophage populations have been identified when comparing the invasive front with other regions of the tumor (Oliveira et al. 2017). More specifically, the invasive front was rich in anti-inflammatory macrophages (M2-like) that were not detected in other regions of the tumor. Macrophages are able to stimulate fibroblast proliferation, activation and survival, through secretion of high amounts of TGFβ and IGF-1, resulting in increased collagen synthesis and matrix cross-linking (Lech and Anders 2013).

To study some of these complex interactions, 3D heterotypic cellular models may provide the necessary complementary solution in terms of functionality, complexity and throughput, between *in vivo* experimental models or standard *in vitro* approaches and clinical oncology (Hirt et al. 2014). Spheroids appear uniquely qualified to decipher tumor-immune interactions. Transcriptional analysis of mesothelioma spheroids vs monolayers showed that the majority of upregulated genes are related with immune response, wound healing, lymphocyte stimulation and response to cytokine stimulation, while downregulated genes mainly include promotion of apoptosis (Nyga et al. 2016; Kim et al. 2012b). Early reports show that cancer cells in spheroids show increased migratory capacity when co-cultured with M2-like macrophages, resembling to some extent the effects described for TAM in human tumors (Hauptmann et al. 1993).

Moreover, the reciprocal crosstalk between breast cancer spheroids from different subtypes and monocytes was shown to be dependent on the aggressiveness of each subtype (Chimal-Ramírez et al. 2013). The co-culture with breast cancer spheroids from more aggressive phenotypes, namely basal-like breast cancer, induced a pro-tumorigenic phenotype in the macrophage population, with consequent increased invasive potential for the tumor cells (Chimal-Ramírez et al. 2013), features already described for these tumors *in vivo* (Jiang and Shapiro 2014). Tumor spheroids also present diminished antigen presenting capacity and decreased proliferation (Hirt et al. 2014). This leads to lower immunogenicity when compared to monolayers, decreasing tumor cell sensitivity to lymphocyte effector functions, which is also a prominent feature in different solid tumors *in vivo* (Nyga et al. 2016; Hirt et al. 2014; Turley et al. 2015). Additionally, high lactate production is a key mechanism of action leading to an immunosuppressive TME in models encompassing tumor spheroids (Hirt et al. 2014).

Innovative strategies for research and preclinical studies of tumor-stroma-immune cell interactions pose a difficult challenge for the tumor modeling field (Hirt et al. 2014). Traditional co-culture techniques have brought significant insights into the crosstalk between cells from the TME. Conditioned media experiments have been extensively employed to study interactions between different cells; however, these preclude direct cell-cell interactions and dynamic crosstalk between the different cell types. Indirect co-cultures, on the other hand, allow for reciprocal and dynamic crosstalk between the different components but still, only through soluble factors (Regier et al. 2016). Finally, direct co-culture assays add an extra layer of complexity on the readouts since either gene expression or protein analysis on individual cell subsets requires previous separation of the cell types; thus most analysis is ultimately based on imaging and secreted factors (Regier et al. 2016). Nevertheless, the contribution of cell-cell and cell-ECM components is crucial for the tumorigenic phenotype and thus such interaction should be included in preclinical models. Several *in vivo* models have also been used to study tumor-TME interactions.

The fast development of immunotherapies, is driving the need for preclinical models that incorporate the immunological state of the TME, while maintaining compatibility with drug screening platforms and allowing straightforward functionality assessment (Hirt et al. 2014). At our lab, we further explored the previously developed culture platform, based on alginate microencapsulation combined with long term culture in stirred systems, to develop a triple co-culture system comprising cancer spheroids, CAF and myeloid cells (Rebelo et al. 2018). We demonstrated that this culture recapitulates aspects of the invasive and immunosuppressive environment present at the tumor site, namely the accumulation of cytokines/chemokines (IL4, IL10, IL13, CCL22, CCL24, CXCL1), ECM elements (collagen type I, IV and fibronectin) and matrix metalloproteinases (MMP1/9). This cellular model supported cell migration and promoted cell-cell interactions within the alginate capsules. Using this system, we were showed that both THP1 and human peripheral-blood derived monocytes could infiltrate the tumor mass and polarize into an anti-inflammatory M2-like macrophage phenotype expressing CD68, CD163 and CD206. This resembled the tumor-associated macrophage phenotype described for human non-small cell lung cancer. Furthermore, this culture system was amenable to drug challenge, and the therapeutic response could be assessed separately within each cellular component. Importantly, immunomodulatory agents could be studied within this setting. The CSF1R inhibitor BLZ945 decreased the percentage of M2-like macrophages in cultures, while not affecting viability, as previously described. Thus, this constitutes one of the first available tools *in vitro* to study tumor-stroma-immune interactions and myeloid cell plasticity, a feature that is still lacking in most current preclinical models. Recently, a cellular model based on the co-culture of CRC organoids and peripheral blood lymphocytes from the same patient was developed, which could be used to isolate tumor-reactive T cells and study sensitivity to various immunotherapies

(Dijkstra et al. 2018). The lack of TME cells in current organoid models has also been tackle by a different group, which co-cultured patient-derived organoids with tumor infiltrating lymphocytes using an air-liquid interface (ALI) method (Neal et al. 2018). Using this method, they were able to model the response to immune checkpoint inhibitors (Neal et al. 2018).

23.2.2 Ex Vivo Models

Ex vivo cultures of freshly isolated tumor samples aim to preserve the original tissue architecture, the heterogeneity, and the surrounding microenvironment. The use of tumor biopsies or resected tumor sections embedded in a matrix can maintain heterogeneity of tumor cell populations and be a potential tool for assessment of patient-specific therapies (Katt et al. 2016). *Ex vivo* cultures include tissue slices and explant cultures. Although these have been exploited, their application for drug discovery faces technical problems and limitations in sample material (Lovitt et al. 2016). In both culture settings, cells maintain cell-cell communication and the interactions with the microenvironment for a short period (Mürdter et al. 2006), not allowing long-term monitoring of disease progression nor interrogation of the long-term effects of drug treatments. On the other hand, primary material is difficult to obtain, and the result interpretation is very complex, due the inter- and intra-patient tumor heterogeneity (Nath and Devi 2016). Overall, improvement of these methods towards extension of culture time would allow evaluation of the long-term effects of drug treatments and the mechanisms of drug resistance.

Explant cultures have been successfully implemented for both healthy and cancerous tissues. In order to maintain the cell-matrix and the cell-cell interactions between the various cell types, the isolation and culture methods had to be optimized for each tissue type. Breast explants obtained from reduction mammoplasties could be successfully cultured for 7 days in low adhesion plates (Tanos et al. 2013). In this system, cells maintained viability, the original tissue architecture, the cell-to-cell communication and hormone responsiveness. Specifically, luminal epithelial cells maintained apico-basal polarity and were surrounded by a thin layer of myoepithelial cells. Although the expression of the nuclear hormone receptors, ER and PR, decreased along the 7 days of culture, epithelial cell proliferation was detected in response to estradiol stimulation. With this system, RANKL was identified as a paracrine mediator of PR in inducing cell proliferation. The maintenance of the breast architecture was key to maintain the paracrine signaling, as these mechanisms were not observed for breast cancer cell lines nor for dissociated breast tissue. This work reinforces the need to maintain the original tissue architecture to address hormone action in the breast (Tanos et al. 2013). A recent report describes the extension of the culture time of breast cancer explants up to 15 days, by employing a perfusion-based bioreactor (Spagnoli et al. 2017). In this system, tissue integrity and cell viability was maintained, although cellularity was decreased along time. Moreover, hormone responsiveness was only evaluated by quantification of explant cellularity (H&E staining) upon prolonged Fulvestrant treatment and expression of ER downstream targets was not assessed. In another study, colorectal and head and neck cancer explant cultures were proposed as a co-clinical tool for prediction of patient-specific drug-response (Majumder et al. 2015). In this model system, explants were cultured in a matrix-specific scaffold, which was adapted for each tumor type and grade. The authors report culture the explants for 72 h, with maintenance of the initial tumor phenotype. Tumor explants were cultured with the patient autologous serum, to guarantee the supply of the original cytokines and growth factors. The results obtained were incorporated in a machine learning algorithm, which allowed the prediction of the tumor response to a specific drug treatment.

Typical tissue slice cultures are 200 μm thick and can be patient or PDX-derived. Several approaches have been proposed for culturing of tissue slices: immersion in culture medium, maintenance in an air-liquid interface (Mürdter et al. 2006; Davies et al. 2015). Previous studies have demonstrated that tissue slices can be cultured for at least 4 days, with good cellularity and maintenance of the original tissue architecture (Mürdter et al. 2006). These authors demonstrated that culturing slices in a stirred system (approx. 150 rpm) prevents oxygen zonation, which was a limitation observed in previous studies. Another study tried to improve the maintenance of cell viability and original tissue architecture by culturing PDX-derived slices on an air-liquid interface, on top of a filter (Davies et al. 2015). The physical support provided by this scaffold reduced apoptotic cell death and vacuolation. However, the authors employed a static culture system, which led to oxygen zonation and differential hormone receptor expression within the slice: cells in contact with the filter became hypoxic, as shown by the expression of HIF1α, and lost ER expression. In contrast, the normoxic region presented ER^+ cells and mouse macrophages (Davies et al. 2015). To solve the issue, slices were cultured with intermittent immersion in the culture media. The use of this dynamic system promoted an increase of oxygen distribution homogeneity (Davies et al. 2015).

In combination with *in vitro* models, the use of *ex vivo* models might help uncover variables inherent to tumor heterogeneity and to corroborate hypothesis formulated from studies employing less complex *in vitro* models. The concerted action of the multiple cell types, growth factors and ECM present in the native tumor environment, might modulate tumor cell behavior in a different manner to what was identified in the *in vitro* model. However, due to the complexity of result interpretation in explant cultures, the analytical tools used should also be improved. Routine histopathological analysis of patient samples, used for comparison with the results obtained from explant cultures, is performed by immunohistochemistry staining against specific molecular markers, such as, ER, PR, Her2 and KI67 (Russnes et al. 2017). This technique allows the visual evaluation of the sample and a semi-quantification of the number and intensity of the stained cells. However, inter-observer variation is very high, and the number of sections analyzed is low. To tackle this issue, slide scanners coupled with image analysis software are being developed to speed up the analysis process and to generate quantitative data. This will allow clinicians to increase the number of analyzed slides and to improve robustness of the conclusions taken. Machine learning tools for automatic image analysis and accurate quantification of stained sections, still need to be improved to enable software to distinguish between different cell types and the intracellular location of each stained protein. Additionally, in recent years single cell analysis tools with spatial resolution have been proposed.

23.3 Conclusions

The advances in pharmacogenomics over the recent years aim at providing a rational stratification of patients for a given therapy (Nieto et al. 2016). This could revolutionize precision medicine by uncovering the most relevant genetic drivers and cellular pathways mediating disease progression and therapeutic response in a genetically defined subset of patients (Nieto et al. 2016). However, the development of therapies for a given tumor has often proven an inaccurate and complex process, as well as has therapeutic biomarker identification, which is often slow and inefficient (Sun 2015). Following *in vitro* observation of a given mechanism or pathway, standard animal testing follows, using mostly mouse models (Hoarau-Véchot et al. 2018). Since the majority of these models presents compromised immune systems and offers non-human tumor-stromal interactions, the concurrence rate of clinical translation can be as low as 8%. Moreover, the pain and discomfort these animals undergo are growing concerns (Voskoglou-Nomikos et al. 2003; Mak et al. 2014).

Thus, the development of accurate *in vitro* cellular models aims at bridging the gap between traditional *in vitro* and *in* vivo models. By providing tumor phenotypes that better resemble the *in vivo* setting, and by eliminating the interspecies discrepancies of cellular interactions in *in vivo* models, these could allow more accurate disease modeling and drug testing. The development and application of strategies that could target specifically the cell of interest, within such heterogeneous environments, would facilitate the process and generate safer therapies, with fewer side-effects.

Altogether, these models should help predict disease progression and drug response mechanisms and contribute to a deeper understanding on the effects of stromal and immune cells on tumor cell behavior. Routine analysis should be performed by using a set of analytic tools for result validation, instead of focusing on one type of assay only, which is typically done in research labs, hospitals and pharma industry. The era of translational medicine has promoted collaborations between researchers and physicians, which is contributing to a deeper understanding of the disease and to the implementation of new analytical tools in hospitals. Although proof-of-concept is still ongoing, there is great expectation on the use of these and other models for co-clinical trials both for evaluation of first line treatments (short-time assays), and for prediction of disease relapse mechanisms (long-term assays). Finally, pharma industry is already implementing some of these models as pre-clinical tools, instead of using the simplistic mono-cultures performed in 2D. On the other hand, the potential to provide cues to explain patients' differential response to treatment and to be the basis for therapies that prolong disease free survival and patient's quality of life should not be underestimated (Tang et al. 2016).

Acknowledgments The authors acknowledge iNOVA-4Health – UID/Multi/04462/2013, a program financially supported by Fundação para a Ciência e Tecnologia/ Ministério da Educação e Ciência, through national funds and co-funded by FEDER under the PT2020 Partnership Agreement.

References

Ahmadzadeh M, Rosenberg SA (2005) TGF-beta1 attenuates the acquisition and expression of effector function by tumor antigen-specific human memory CD8 T cells. J Immunol 174:5215–5223

Andersen T et al (2015) 3D Cell Culture in Alginate Hydrogels. Microarrays 4:133–161

Benien P, Swami A (2014) 3D tumor models: history, advances and future perspectives. Future Oncol 10:1311–1327

Bissell MJ, Hines WC (2011) Why don't we get more cancer? A proposed role of the microenvironment in restraining cancer progression. Nat Med 17:320–329

Bissell MJ et al (2005) Microenvironmental regulators of tissue structure and function also regulate tumor induction and progression: the role of extracellular matrix and its degrading enzymes. Cold Spring Harb Symp Quant Biol 70:343–356

Breslin S, O'Driscoll L (2013) Three-dimensional cell culture: the missing link in drug discovery. Drug Discov Today 18:240–249

Broderick L, Bankert RB (2006) Membrane-associated TGF-beta1 inhibits human memory T cell Signaling in malignant and nonmalignant inflammatory microenvironments. J Immunol 177:3082–3088

Bussard KM et al (2010) Reprogramming human cancer cells in the mouse mammary gland. Cancer Res 70:6336–6343

Chang HY et al (2004) Gene expression signature of fibroblast serum response predicts human cancer progression: similarities between tumors and wounds. PLoS Biol 2:e7

Chang HY et al (2005) From the cover: robustness, scalability, and integration of a wound-response gene expression signature in predicting breast cancer survival. Proc Natl Acad Sci 102:3738–3743

Chaudhuri O et al (2014) Extracellular matrix stiffness and composition jointly regulate the induction of malignant phenotypes in mammary epithelium. Nat Mater 13:970–978

Chen CS (2016) 3D biomimetic cultures: the next platform for cell biology. Trends Cell Biol 26:798–800

Chen A et al (2009) Endothelial cell migration and vascular endothelial growth factor expression are the result of loss of breast tissue polarity. Cancer Res 69:6721–6729

Chimal-Ramírez GK et al (2013) MMP1, MMP9, and COX2 expressions in Promonocytes are induced by breast cancer cells and correlate with collagen degradation, transformation-like morphological changes in MCF-10A acini, and tumor aggressiveness. Biomed Res Int 2013:1–15

Chwalek K et al (2014) Glycosaminoglycan-based hydrogels to modulate heterocellular communication in in vitro angiogenesis models. Sci Rep 4:4–11

Costa A et al (2018) Fibroblast heterogeneity and immunosuppressive environment in human breast Cancer. Cancer Cell 33:463–479.e10

Davies EJ et al (2015) Capturing complex tumour biology in vitro: histological and molecular characterisation of precision cut slices. Sci Rep 5:17187

De Monte L et al (2011) Intratumor T helper type 2 cell infiltrate correlates with cancer-associated fibroblast thymic stromal lymphopoietin production and reduced survival in pancreatic cancer. J Exp Med 208:469–478

Dijkstra KK et al (2018) Generation of tumor-reactive T cells by co-culture of peripheral blood lymphocytes and tumor organoids. Cell 174:1586–1598.e12

Dolznig H et al (2011) Modeling colon adenocarcinomas in vitro. Am J Pathol 179:487–501

Dumont N et al (2013) Breast fibroblasts modulate early dissemination, tumorigenesis, and metastasis through alteration of extracellular matrix characteristics. Neoplasia 15:249–IN7
Dutta D et al (2017) Disease modeling in stem cell-derived 3D organoid systems. Trends Mol Med 23:393–410
Estrada MF et al (2016) Modelling the tumour microenvironment in long-term microencapsulated 3D co-cultures recapitulates phenotypic features of disease progression. Biomaterials 78:50–61
Fang X et al (2013) Novel 3D co-culture model for epithelial-stromal cells interaction in prostate cancer. PLoS One 8:1–10
Farmer P et al (2009) A stroma-related gene signature predicts resistance to neoadjuvant chemotherapy in breast cancer. Nat Med 15:68–74
Finak G et al (2008) Stromal gene expression predicts clinical outcome in breast cancer. Nat Med 14:518–527
Friedrich J et al (2009) Spheroid-based drug screen: considerations and practical approach. Nat Protoc 4:309–324
Gjorevski N et al (2016) Designer matrices for intestinal stem cell and organoid culture. Nature 539:560–564
Gu L, Mooney DJ (2015) Biomaterials and emerging anticancer therapeutics: engineering the microenvironment. Nat Rev Cancer 16:56–66
Hanahan D, Coussens LM (2012) Accessories to the crime: functions of cells recruited to the tumor microenvironment. Cancer Cell 21:309–322
Hanahan D, Weinberg RA (2011) Hallmarks of cancer: the next generation. Cell 144:646–674
Hauptmann S et al (1993) Macrophages and multicellular tumor spheroids in co-culture: a three-dimensional model to study tumor-host interactions. Evidence for macrophage-mediated tumor cell proliferation and migration. Am J Pathol 143:1406–1415
Haycock JW (2011) 3D cell culture: a review of current approaches and techniques. Methods Mol Biol 695:1–15
Hickman JA et al (2014) Three-dimensional models of cancer for pharmacology and cancer cell biology: capturing tumor complexity in vitro/ex vivo. Biotechnol J 9:1115–1128
Hirschhaeuser F et al (2010) Multicellular tumor spheroids: an underestimated tool is catching up again. J Biotechnol 148:3–15
Hirt C et al (2014) "In vitro" 3D models of tumor-immune system interaction. Adv Drug Deliv Rev 79–80:145–154
Hoarau-Véchot J et al (2018) Halfway between 2D and animal models: are 3D cultures the ideal tool to study cancer-microenvironment interactions? Int J Mol Sci 19:181
Huch M, Rawlins EL (2017) Cancer: tumours build their niche. Nature 545:292–293
Inman JL, Bissell MJ (2010) Apical polarity in three-dimensional culture systems: where to now? J Biol 9:2
Jiang X, Shapiro DJ (2014) The immune system and inflammation in breast cancer. Mol Cell Endocrinol 382:673–682
Katt ME et al (2016) In vitro tumor models: advantages, disadvantages, variables, and selecting the right platform. Front Bioeng Biotechnol 4:12
Kaur P et al (2011) Human breast cancer histoid: an in vitro 3-dimensional co-culture model that mimics breast cancer tissue. J Histochem Cytochem 59:1087–1100
Kim JH et al (2012a) The role of myofibroblasts in upregulation of S100A8 and S100A9 and the differentiation of myeloid cells in the colorectal cancer microenvironment. Biochem Biophys Res Commun 423:60–66
Kim H et al (2012b) Changes in global gene expression associated with 3D structure of Tumors: an ex vivo matrix-free mesothelioma spheroid model. PLoS One 7:e39556
Kimlin LC et al (2013) In vitro three-dimensional (3D) models in cancer research: an update. Mol Carcinog 52:167–182
Korving J et al (2017) A living biobank of breast cancer organoids captures disease heterogeneity. Cell 172:373–386.e10
Kraman M et al (2010) Suppression of antitumor immunity by stromal cells expressing fibroblast activation protein-alpha. Science 330:827–830
Lakins MA et al (2018) Cancer-associated fibroblasts induce antigen-specific deletion of CD8+ T cells to protect tumour cells. Nat Commun 9:948
Lech M, Anders HJ (2013) Macrophages and fibrosis: how resident and infiltrating mononuclear phagocytes orchestrate all phases of tissue injury and repair. Biochim Biophys Acta Mol basis Dis 1832:989–997
Levental KR et al (2009) Matrix crosslinking forces tumor progression by enhancing integrin signaling. Cell 139:891–906
Lovitt CJ et al (2016) Cancer drug discovery: recent innovative approaches to tumor modeling. Expert Opin Drug Discovery 11:885–894
Mafi P et al (2012) Evaluation of biological protein-based collagen scaffolds in cartilage and musculoskeletal tissue engineering – a systematic review of the literature. Curr Stem Cell Res Ther 7:302–309
Majumder B et al (2015) Predicting clinical response to anticancer drugs using an ex vivo platform that captures tumour heterogeneity. Nat Commun 6:1–14
Mak IWY et al (2014) Lost in translation: animal models and clinical trials in cancer treatment. Am J Transl Res 6:114–118
McGranahan N, Swanton C (2017) Clonal heterogeneity and tumor evolution: past, present, and the future. Cell 168:613–628
McMillin DW et al (2010) Tumor cell-specific bioluminescence platform to identify stroma-induced changes to anticancer drug activity. Nat Med 16:483–489
Murdoch C (2004) Mechanisms regulating the recruitment of macrophages into hypoxic areas of tumors and other ischemic tissues. Blood 104:2224–2234
Mürdter TE et al (2006) Short term culture of breast cancer tissues to study the activity of the anticancer drug taxol in an intact tumor environment. BMC Cancer 6:1–11

Nath S, Devi GR (2016) Three-dimensional culture systems in cancer research: focus on tumor spheroid model. Pharmacol Ther 163:94–108
Neal JT et al (2018) Organoid modeling of the tumor immune microenvironment. Cell 175:1972–1988.e16
Nieto MA et al (2016) EMT: 2016. Cell 166:21–45
Nyga A et al (2016) The next level of 3D tumour models: immunocompetence. Drug Discov Today 21:1421–1428
Oliveira MJ et al (2017) Decellularized human colorectal cancer matrices polarize macrophages towards an anti-inflammatory phenotype promoting cancer cell invasion via CCL18. Biomaterials 124:211–224
Pampaloni F et al (2007) The third dimension bridges the gap between cell culture and live tissue. Nat Rev Mol Cell Biol 8:839–845
Quail DF, Joyce JA (2013) Microenvironmental regulation of tumor progression and metastasis. Nat Med 19:1423–1437
Rebelo SP et al (2018) 3D-3-culture: a tool to unveil macrophage plasticity in the tumour microenvironment. Biomaterials 163:185–197
Regier MC et al (2016) Progress towards understanding heterotypic interactions in multi-culture models of breast cancer. Integr Biol 8:684–692
Rijal G, Li W (2016) 3D scaffolds in breast cancer research. Biomaterials 81:135–156
Ronca R et al (2018) Paracrine interactions of cancer-associated fibroblasts, macrophages and endothelial cells. Curr Opin Oncol 30:45–53
Russnes HG et al (2017) Breast cancer molecular stratification: from intrinsic subtypes to integrative clusters. Am J Pathol 187:2152–2162
Salmon H et al (2012) Matrix architecture defines the preferential localization and migration of T cells into the stroma of human lung tumors. J Clin Invest 122:899–910
Shafiee H et al (2015) Engineering cancer microenvironments for in vitro 3-D tumor models. Mater Today 18:539–553
Shamir ER et al (2012) ECM microenvironment regulates collective migration and local dissemination in normal and malignant mammary epithelium. Proc Natl Acad Sci 109:E2595–E2604
Smithmyer ME et al (2014) Hydrogel scaffolds as in vitro models to study fibroblast activation in wound healing and disease. Biomater Sci 2:634–650
Sokol ES et al (2016) Growth of human breast tissues from patient cells in 3D hydrogel scaffolds. Breast Cancer Res 18:1–13
Spagnoli GC et al (2017) Ex-vivo assessment of drug response on breast cancer primary tissue with preserved microenvironments. Oncoimmunology. https://doi.org/10.1080/2162402x.2017.1331798
Stanton SE, Disis ML (2016) Clinical significance of tumor-infiltrating lymphocytes in breast cancer. J Immunother Cancer 4:1–7
Stock K et al (2016) Capturing tumor complexity in vitro: comparative analysis of 2D and 3D tumor models for drug discovery. Sci Rep 6:28951
Straussman R et al (2012) Tumour micro-environment elicits innate resistance to RAF inhibitors through HGF secretion. Nature 487:500–504
Sun Y (2015) Translational horizons in the tumor microenvironment: harnessing breakthroughs and targeting cures. Med Res Rev 35(2):408–436
Sung KE, Beebe DJ (2014) Microfluidic 3D models of cancer. Adv Drug Deliv Rev 79:68–78
Tang H et al (2016) Immunotherapy and tumor microenvironment. Cancer Lett 370:85–90
Tanos T et al (2013) Progesterone/RANKL is a major regulatory axis in the human breast. Sci Transl Med 5:182ra55
Thomas RM et al (2016) Concepts in cancer modeling: a brief history. Cancer Res 76:5921–5925
Thottassery JV et al (2004) Breast fibroblasts modulate epithelial cell proliferation in three-dimensional in vitro co-culture. Breast Cancer Res 7:R46–R59
Toullec A et al (2010) Oxidative stress promotes myofibroblast differentiation and tumour spreading. EMBO Mol Med 2:211–230
Turley SJ et al (2015) Immunological hallmarks of stromal cells in the tumour microenvironment. Nat Rev Immunol 15:669–682
Unger C et al (2014) Modeling human carcinomas: physiologically relevant 3D models to improve anti-cancer drug development. Adv Drug Deliv Rev 79:50–67
Velez DO et al (2017) 3D collagen architecture induces a conserved migratory and transcriptional response linked to vasculogenic mimicry. Nat Commun 8(1):1651
Voskoglou-Nomikos T et al (2003) Clinical predictive value of the in vitro cell line, human xenograft, and mouse allograft preclinical cancer models. Clin Cancer Res 9:4227–4239
Weaver VM et al (1997) Reversion of the malignant phenotype of human breast cells in three-dimensional culture and in vivo by integrin blocking antibodies. J Cell Biol 137:231–245
Weigelt B et al (2014) The need for complex 3D culture models to unravel novel pathways and identify accurate biomarkers in breast cancer. Adv Drug Deliv Rev 69–70:42–51
Weiswald LB et al (2015) Spherical cancer models in tumor biology. Neoplasia 17:1–15
Xu R et al (2009) Sustained activation of STAT5 is essential for chromatin remodeling and maintenance of mammary-specific function. J Cell Biol 184:57–66

Correction to: Animal Models to Study Cancer and Its Microenvironment

N. Mendes, P. Dias Carvalho, F. Martins, S. Mendonça, A. R. Malheiro, A. Ribeiro, J. Carvalho, and S. Velho

Correction to:
Chapter 20 in: J. Serpa (ed.), *Tumor Microenvironment*, Advances in Experimental Medicine and Biology 1219, https://doi.org/10.1007/978-3-030-34025-4_20

The original version of this chapter has been revised. In the Acknowledgments section the reference PTDC/MED-ONC/31354/2017 has been replaced with POCI-01-0145-FEDER-031354.

The updated version of this chapter can be found at
https://doi.org/10.1007/978-3-030-34025-4_20

J. Serpa (ed.), *Tumor Microenvironment*, Advances in Experimental Medicine and Biology 1219,
https://doi.org/10.1007/978-3-030-34025-4_24